VOLUME FIVE HUNDRED AND NINETY SIX

METHODS IN ENZYMOLOGY

Measurement and Analysis of Kinetic Isotope Effects

METHODS IN ENZYMOLOGY

Editors-in-Chief

ANNA MARIE PYLE
Departments of Molecular, Cellular and Developmental Biology and Department of Chemistry
Investigator, Howard Hughes Medical Institute
Yale University

DAVID W. CHRISTIANSON
Roy and Diana Vagelos Laboratories
Department of Chemistry
University of Pennsylvania
Philadelphia, PA

Founding Editors

SIDNEY P. COLOWICK and NATHAN O. KAPLAN

VOLUME FIVE HUNDRED AND NINETY SIX

METHODS IN ENZYMOLOGY

Measurement and Analysis of Kinetic Isotope Effects

Edited by

MICHAEL E. HARRIS
University of Florida,
Gainesville, FL, United States

VERNON E. ANDERSON
Colesville, MD, United States

ELSEVIER

ACADEMIC PRESS
An imprint of Elsevier

Academic Press is an imprint of Elsevier
50 Hampshire Street, 5th Floor, Cambridge, MA 02139, United States
525 B Street, Suite 1800, San Diego, CA 92101–4495, United States
The Boulevard, Langford Lane, Kidlington, Oxford OX5 1GB, United Kingdom
125 London Wall, London, EC2Y 5AS, United Kingdom

First edition 2017

ISBN: 978-0-12-812273-0
ISSN: 0076-6879

For information on all Academic Press publications
visit our website at https://www.elsevier.com/books-and-journals

Publisher: Zoe Kruze
Acquisition Editor: Zoe Kruze
Editorial Project Manager: Helene Kabes and Alina Cleju
Production Project Manager: Magesh Kumar Mahalingam and Divya Krishna Kumar
Cover Designer: Greg Harris

Typeset by SPi Global, India

CONTENTS

Contributors *xi*
Preface *xv*

1. Binding Isotope Effects for Interrogating Enzyme–Substrate Interactions **1**
Christopher F. Stratton, Myles B. Poulin, and Vern L. Schramm
1. Introduction 2
2. Equipment 6
3. Materials 6
4. Measurement of Equilibrium BIEs Using the RED Device 9
5. BIE Interpretation and Examples 17
6. Conclusions 20
Acknowledgments 20
References 20

2. Chemical Ligation and Isotope Labeling to Locate Dynamic Effects **23**
Alan F. Scott, Louis Y.P. Luk, and Rudolf K. Allemann
1. Introduction 24
2. A General Overview of the Approach 26
3. Hints and Variations on the Strategy for Other Proteins 35
4. Conclusion 38
References 39

3. Measurement of Enzyme Isotope Effects **43**
Svetlana A. Kholodar, Ananda K. Ghosh, and Amnon Kohen
1. Introduction 44
2. Methods 45
3. Summary 76
Acknowledgments 77
References 77

4. Extracting Kinetic Isotope Effects From a Global Analysis of Reaction Progress Curves **85**

Sam Hay

1. Introduction 86
2. (Global) Nonlinear Fitting 87
3. Global Fitting of a Time Course Described by Integrated Rate Equations 91
4. Modeling and Fitting Time Courses Using Numerical Integration of Rate Equations 94
5. Model Selection and Error Analysis 98
6. Specific Examples 102
7. Summary 109

Acknowledgments 109
References 109

5. Natural Isotope Abundance in Metabolites: Techniques and Kinetic Isotope Effect Measurement in Plant, Animal, and Human Tissues **113**

Illa Tea and Guillaume Tcherkez

1. Introduction 114
2. Compound-Specific ^{13}C and ^{15}N Analysis: History and Overview 120
3. Compound-Specific ^{15}N Analysis 124
4. Compound-Specific ^{13}C Analysis 129
5. KIE Measurement From LC-co-IRMS or GC-C-IRMS Data 133
6. Conclusions and Perspectives 142

Acknowledgments 142
References 142

6. Measurement of Kinetic Isotope Effects in an Enzyme-Catalyzed Reaction by Continuous-Flow Mass Spectrometry **149**

Kenneth M. Roberts and Paul F. Fitzpatrick

1. Introduction 149
2. Development of a Continuous-Flow Mass Spectrometry System 150
3. Measurement of KIEs 158
4. Conclusions 160

Acknowledgments 161
References 161

7. Primary Deuterium Kinetic Isotope Effects From Product Yields: Rationale, Implementation, and Interpretation **163**

Tina L. Amyes and John P. Richard

1. Introduction 164
2. Enzyme-Catalyzed Decarboxylation of OMP 166

3. Experimental Protocol 168
4. Interpretation of Results 173
5. Other Enzymes 174
Acknowledgment 175
References 175
Further Reading 177

8. Measurement and Prediction of Chlorine Kinetic Isotope Effects in Enzymatic Systems **179**

Lukasz Szatkowski, Rabindra Nath Manna, Anna Grzybkowska, Rafał Kamiński, Agnieszka Dybala-Defratyka, and Piotr Paneth

1. Introduction 180
2. Experimental Determination of Cl-KIEs 181
3. Theoretical Predictions of Cl-KIEs 186
4. Summary 209
References 210

9. Kinetic Deuterium Isotope Effects in Cytochrome P450 Reactions **217**

Frederick P. Guengerich

1. Introduction and History of Cytochrome P450 Kinetic Isotope Effects 218
2. Types of Experiments 219
3. Choice of Enzymatic Parameters 225
4. KIEs for Various P450 Reactions 226
5. Examples of P450 KIE Studies 228
6. In Vivo KIEs With P450s 231
7. Consideration of P450 KIEs in Drug Development 232
8. Summary and Conclusions 233
Acknowledgment 234
References 234

10. Use of Isotopes and Isotope Effects for Investigations of Diiron Oxygenase Mechanisms **239**

Rahul Banerjee, Anna J. Komor, and John D. Lipscomb

1. Introduction 240
2. General Methods 240
3. Methane Monooxygenase 249
4. CmlI Oxygenase 273
5. Conclusions 286
References 286

11. Characterization of Substrate, Cosubstrate, and Product Isotope Effects Associated With Enzymatic Oxygenations of Organic Compounds Based on Compound-Specific Isotope Analysis **291**

Sarah G. Pati, Hans-Peter E. Kohler, and Thomas B. Hofstetter

1. Introduction 292
2. Deriving Apparent Kinetic Isotope Effects From Measurements of Stable Isotope Ratios 294
3. Experimental Approaches for Determining Isotope Fractionation During Oxygenation Reactions 307
4. Instrumentation for Stable Isotope Analysis by Isotope Ratio Mass Spectrometry 313
5. Summary and Conclusion 323

Acknowledgments 323

References 323

12. pH-Free Measurement of Relative Acidities, Including Isotope Effects **331**

Charles L. Perrin

1. Introduction to NMR Titration 332
2. Multicomponent NMR Titration 333
3. Motivation: Applications to the Reverse Anomeric Effect 337
4. Applications of Other NMR Titrations to Relative Acidities 341
5. Measurement of Binding Constants 344
6. Heavy-Atom Isotope Effects on Acidity 350
7. Secondary Deuterium Isotope Effects on Acidity 351
8. Secondary Deuterium Isotope Effects on Binding 355
9. Further Studies of Secondary Deuterium Isotope Effects, Toward Elucidating Their Origin 356
10. Advantages of NMR Titration 359
11. Some Practical Considerations 360
12. Summary 364

Acknowledgment 364

References 364

13. Isotope Ratio Monitoring ^{13}C Nuclear Magnetic Resonance Spectrometry for the Analysis of Position-Specific Isotope Ratios **369**

Richard J. Robins, Gérald S. Remaud, and Serge Akoka

1. Introduction 370
2. Generic Aspects of the Methodology 371
3. Examples of Methods 381
4. Summary and Conclusions 397

References 398

14. Applications of Isotope Ratio Mass Spectrometry in Sports Drug Testing Accounting for Isotope Fractionation in Analysis of Biological Samples **403**

Thomas Piper and Mario Thevis

1. Introduction 404
2. Sample Preparation for Urinary Steroids 406
3. Quality Control Measures in Sports Drug Testing 414
4. Combination of Hydrogen and Carbon Isotope Ratios 422
5. Conclusion 428

References 429

15. Kinetic Isotope Effect Analysis of RNA 2′-*O*-Transphosphorylation **433**

Michael E. Harris, Darrin M. York, Joseph A. Piccirilli, and Vernon E. Anderson

1. Introduction 434
2. Synthesis of Isotopomer RNA Substrates Enriched in ^{18}O at the Reacting Phosphoryl Oxygens 439
3. Use of Electrospray Ionization Quadrupole/Time-of-Flight Mass Spectrometry to Measure ^{18}O KIEs for RNA 2′-*O*-transphosphorylation 441
4. Future Directions 453

References 454

16. Theory and Application of the Relationship Between Steady-State Isotope Effects on Enzyme Intermediate Concentrations and Net Rate Constants **459**

Mark W. Ruszczycky and Hung-wen Liu

1. Introduction 460
2. Modeling Steady-State Isotope Effects 461
3. An Example: DesII 472
4. Theory 479

Acknowledgments 496

References 496

17. Determining Carbon Kinetic Isotope Effects Using Headspace Analysis of Evolved CO_2 **501**

Scott O.C. Mundle, Barbara Sherwood Lollar, and Ronald Kluger

1. Basic Experimental Procedures 504
2. Continuous Pressure-Monitored Headspace Analysis 510
3. Control Reactions 514
4. Improving Data Quality 515
5. Conclusions 519

Acknowledgments 519

References 519

18. Kinetic Isotope Effect Studies to Elucidate the Reaction Mechanism of RNA-Modifying Enzymes 523

Govardhan R. Veerareddygari and Eugene G. Mueller

1. Introduction 524
2. KIE Studies on TruB and RluA 526
3. Equipment 526
4. Chemicals 527
5. Protocol 528
6. Concluding Thoughts 544

Acknowledgments 545
References 545
Further Reading 546

19. Measurement of Kinetic Isotope Effects by Continuously Monitoring Isotopologue Ratios Using NMR Spectroscopy 547

Natalia Sannikova, Andrew R. Lewis, and Andrew J. Bennet

1. Introduction 548
2. Procedure 549
3. Data Processing 560
4. Data Analysis 567
5. Summary and Conclusion 568

References 568

Author Index *573*
Subject Index *599*

CONTRIBUTORS

Serge Akoka
University of Nantes—CNRS UMR6230, Nantes, France

Rudolf K. Allemann
School of Chemistry, Cardiff University, Cardiff, United Kingdom

Tina L. Amyes
University at Buffalo, SUNY, Buffalo, NY, United States

Vernon E. Anderson
Case Western Reserve University School of Medicine, Cleveland, OH, United States

Rahul Banerjee
University of Minnesota, Minneapolis, MN, United States

Andrew J. Bennet
Simon Fraser University, Burnaby, BC, Canada

Agnieszka Dybala-Defratyka
Lodz University of Technology, Faculty of Chemistry, Institute of Applied Radiation Chemistry, Lodz, Poland

Paul F. Fitzpatrick
University of Texas Health Science Center, San Antonio, TX, United States

Ananda K. Ghosh
University of Iowa, Iowa City, IA, United States

Anna Grzybkowska
Elucidation of Biosynthesis by Isotopic Spectrometry Group, CEISAM, University of Nantes—CNRS UMR6230, Nantes, France

Frederick P. Guengerich
Vanderbilt University School of Medicine, Nashville, TN, United States

Michael E. Harris
University of Florida, Gainesville, FL, United States

Sam Hay
BBSRC/EPSRC Centre for Synthetic Biology of Fine and Speciality Chemicals (SYNBIOCHEM), Manchester Institute of Biotechnology and School of Chemistry, The University of Manchester, Manchester, United Kingdom

Thomas B. Hofstetter
Eawag, Swiss Federal Institute of Aquatic Science and Technology, Dübendorf; Institute of Biogeochemistry and Pollutant Dynamics (IBP), ETH Zürich, Zürich, Switzerland

Rafał Kamiński
Lodz University of Technology, Faculty of Chemistry, Institute of Applied Radiation Chemistry, Lodz, Poland

Svetlana A. Kholodar
University of Iowa, Iowa City, IA, United States

Ronald Kluger
University of Toronto, Toronto, ON, Canada

Amnon Kohen
University of Iowa, Iowa City, IA, United States

Hans-Peter E. Kohler
Eawag, Swiss Federal Institute of Aquatic Science and Technology, Dübendorf, Switzerland

Anna J. Komor
University of Minnesota, Minneapolis, MN, United States

Andrew R. Lewis
Simon Fraser University, Burnaby, BC, Canada

John D. Lipscomb
University of Minnesota, Minneapolis, MN, United States

Hung-wen Liu
University of Texas at Austin, Austin, TX, United States

Barbara Sherwood Lollar
University of Toronto, Toronto, ON, Canada

Louis Y.P. Luk
School of Chemistry, Cardiff University, Cardiff, United Kingdom

Rabindra Nath Manna
Indian Association for the Cultivation of Science, Physical Chemistry, Kolkata, West Bengal, India

Eugene G. Mueller
University of Louisville, Louisville, KY, United States

Scott O.C. Mundle
Great Lakes Institute for Environmental Research, University of Windsor, Windsor, ON, Canada

Piotr Paneth
Lodz University of Technology, Faculty of Chemistry, Institute of Applied Radiation Chemistry, Lodz, Poland

Sarah G. Pati
Eawag, Swiss Federal Institute of Aquatic Science and Technology, Dübendorf; Institute of Biogeochemistry and Pollutant Dynamics (IBP), ETH Zürich, Zürich, Switzerland

Charles L. Perrin
University of California, San Diego, La Jolla, CA, United States

Joseph A. Piccirilli
The University of Chicago, Chicago, IL, United States

Thomas Piper
Center for Preventive Doping Research—Institute of Biochemistry, German Sport University Cologne, Cologne, Germany

Myles B. Poulin
Albert Einstein College of Medicine, Bronx, NY, United States

Gérald S. Remaud
University of Nantes—CNRS UMR6230, Nantes, France

John P. Richard
University at Buffalo, SUNY, Buffalo, NY, United States

Kenneth M. Roberts
University of South Carolina Aiken, Aiken, SC, United States

Richard J. Robins
University of Nantes—CNRS UMR6230, Nantes, France

Mark W. Ruszczycky
University of Texas at Austin, Austin, TX, United States

Natalia Sannikova
Simon Fraser University, Burnaby, BC, Canada

Vern L. Schramm
Albert Einstein College of Medicine, Bronx, NY, United States

Alan F. Scott
School of Chemistry, Cardiff University, Cardiff, United Kingdom

Christopher F. Stratton
Albert Einstein College of Medicine, Bronx, NY, United States

Lukasz Szatkowski
University of Arizona, Department of Chemistry and Biochemistry, Tucson, AZ, United States

Guillaume Tcherkez
Research School of Biology, Australian National University, Canberra, ACT, Australia

Illa Tea
Research School of Biology; Cancer Metabolism and Genetics Group, John Curtin School of Medical Research, Australian National University, Canberra, ACT, Australia; EBSI Team, CEISAM, University of Nantes-CNRS UMR 6230, Nantes, France

Mario Thevis
Center for Preventive Doping Research—Institute of Biochemistry, German Sport University Cologne, Cologne, Germany

Govardhan R. Veerareddygari
University of Louisville, Louisville, KY, United States

Darrin M. York
Rutgers University, Piscataway, NJ, United States

PREFACE

Over 100 years ago, two lines of corroborating evidence led to the discovery of isotopes: identification of the decomposition products of radioactive elements and the detection by primordial mass spectrometry of two types of neon (Aston, 1920; Rutherford & Soddy, 1903; Thomson, 1912). In the 1920s, isotopes of many elements were identified ending with the isolation of deuterium in 1932 (Urey, Brickwedde, & Murphy, 1932). Almost immediately these isotopes were conscripted into the service of biology as tracers of metabolism. Not until the 1950s were the effects of isotopes on the rates of reaction given a theoretical basis of interpretation by Bigeleisen (Bigeleisen & Mayer, 1947).

The application of kinetic isotope effects to the study of enzyme reactions adopted this theoretical approach. Volumes devoted to the study of kinetic isotope effects have appeared with some regularity starting with Melander's text in 1960 (Melander, 1960). In the middle of the 1970s, two volumes inspired renewed interest in the application of kinetic isotope effects to the study of enzyme reactions: The *Sixth Steenbock Symposium on Isotope Effects on Enzyme Reactions* (Cleland, O'Leary, & Northrop, 1977) and Gandour and Schowen's *Transition States of Biochemical Processes* (Gandour & Schowen, 1978). Methodological and theoretical advances focusing on using isotope effects to understand biological catalysis were highlighted in Cook's *Enzyme Mechanism from Isotope Effects* (Cook, 1991). Isotopes and isotope effects continue to see broad application in chemical and geochemical research as reviewed in *Isotope Effects in the Chemical, Geological, and Bio Sciences* edited by Wolfsberg and colleagues (Wolfsberg, Van Hook, Paneth, & Rebelo, 2009). The most recent review volume published in 2006 was edited by Limbach and Kohen, *Isotope Effects in Chemistry and Biology* (Kohen & Limbach, 2005).

The historic background of isotope effect studies continues to inform the research advances described in the contributions to this volume. The application of both radioactive isotopes as described in the chapters by Stratton et al., and of mass spectrometry remains the major analytical approaches used to achieve the necessary level of precision necessary for quantitative measurement of kinetic isotope effects. Originally only applicable to gases, the development of ionization methodologies applicable to both highly charged and macromolecular substrates opens up new horizons in isotope studies. The new higher resolution spectrometers and innovative methods of introducing analytes into isotope ratio mass spectrometers have dramatically expanded the scope of isotope effect studies. Some of these methods are reflected in contributions of chapters by Szatkowski et al., Harris et al., Pati et al., Tea & Tcherkez, and Kluger. Scintillation counting methods have been refined and developed for application to multiplexed studies as described by Schramm. In the middle of the 20th century the different nuclear spin properties of isotopes became a feature of NMR analyses. The demands of high quantitative precision required of kinetic isotope effect studies initially limited the application of NMR methods. In this volume, several new applications that have been developed relying on the higher resolution and stability of modern NMR spectrometers are highlighted in the chapters by Bennet, Perrin, and Robins. In the case of large deuterium effects, direct determination of rate constants is possible and the chapter by Hay illustrates progress curve analysis useful which can be very powerful in many cases and applicable to a broad range of enzyme assays.

The inclusion of computational methods to model kinetic isotope effects (KIEs) would require a similar sized volume to survey the advances that have been made in quantum chemistry methods of modeling transition states (or modifications to transition state theory to accommodate tunneling, etc.). Nonetheless, theoretical models are so integral to isotope effect studies they do appear in some of the contributions, notably in Ruszczycky and Liu's contribution on measuring kinetic isotope effects on intermediate distributions and their interpretation as effects on net rate constants. The coordination between theory and experiment is also important to description of RNA 2′-O-transphosphorylation transition states by Harris and colleagues. The contribution by Stratton et al. on binding isotope effects (BIEs) summarizes the range of enzyme BIEs and interpretations based on computation. Indeed, the importance of combined KIE analysis with computation to test mechanistic predictions is illustrated very clearly by the overview of theory and experiment by Kohen and colleagues aimed at understanding heavy enzyme isotope effects, and by Allemann describing chemical ligation methods to experimentally locate these effects.

Similarly, it would be intractable to comprehensively review the impact that isotope effect measurement and analysis has had on revealing the chemical details of enzyme catalysis. The long history of deriving enzyme mechanism from isotope effects would itself provide material for an independent volume. The broad applicability of isotope effects is nonetheless illustrated by the range of systems examined in this volume including orotidine 5′-monophosphate decarboxylase (Richard), *N*-acetylpolyamine oxidase (Fitzpatrick), ribonuclease (Harris), dehalogenase (Dybala-Defratyka), fatty acid hydroxylation (Robins), enzymes of steroid biosynthesis (Piper), dioxygenase (Hofstetter), and the radical SAM enzyme DesII (Ruszczycky). Application of isotope effect analyses to probe new reactions is illustrated by the contribution by Richard aimed at reactions in which hydron is incorporated into the reaction product from solvent water. The kind of insights possible for specific systems are expertly surveyed by Guengerich for cytochrome P450 catalysis, Lipscomb and colleagues for diiron oxygenases, and Mueller for pseudouridine synthase. For each of the enzymes used in the methods described herein unique challenges had to be overcome that together can provide a collection of useful guidelines for exploration of many other systems. The field continues to be advanced by clever adaptation of technologies to measure isotope effects as shown in the chapter by Fitzpatrick showing how continuous flow mass spectrometry can resolve effects inaccessible by any other means.

We hope that this volume will serve as a guide to new developments that take advantage of the increased instrumental capabilities that have become available in recent years. This volume also reflects the expansion of kinetic isotope effects into the environmental and drug metabolism fields where we hope they will find greater interest and application. While we can already anticipate newer methods from higher resolution spectrometers, we believe this volume can help expand the interest and application of isotope effect studies to a greater understanding of the biological world.

Michael E. Harris and
Vernon E. Anderson

REFERENCES

Aston, F. W. (1920). XLIV. The constitution of atmospheric neon. *Philosophical Magazine*, *39*(232), 449–455.

Bigeleisen, J., & Mayer, M. G. (1947). Calculation of equilibrium constants for isotopic exchange reactions. *The Journal of Chemical Physics*, *15*(5), 261–267.

Cleland, W. W., O'Leary, M. H., & Northrop, D. B. (1977). *Isotope effects on enzyme-catalyzed reactions: Proceedings of the sixth annual Harry Steenbock symposium*. Baltimore: University Park Press.

Cook, P. F. (1991). *Enzyme mechanism from isotope effects*. Boca Raton, FL: Taylor & Francis.

Gandour, R. D., & Schowen, R. L. (1978). *Transition states of biochemical processes*. New York: Springer.

Kohen, A., & Limbach, H. H. (2005). *Isotope effects in chemistry and biology*. Boca Raton, FL: Taylor & Francis.
Melander, L. (1960). *Isotope effects on reaction rates*. New York, NY: Ronald Press.
Rutherford, E., & Soddy, F. (1903). LX. Radioactive change. *Philosophical Magazine Series 6*, 5(29), 576–591.
Thomson, J. J. (1912). XIX. Further experiments on positive rays. *Philosophical Magazine*, *24*(140), 209–253.
Urey, H. C., Brickwedde, F. G., & Murphy, G. M. (1932). A hydrogen isotope of mass 2. *Physical Review*, *39*(1), 164–165.
Wolfsberg, M., Van Hook, W. A., Paneth, P., & Rebelo, L. P. N. (2009). *Isotope effects in the chemical, geological, and bio sciences*. Dordrecht: Springer.

CHAPTER ONE

Binding Isotope Effects for Interrogating Enzyme–Substrate Interactions

Christopher F. Stratton[2,3], Myles B. Poulin[2,4], Vern L. Schramm[1]
Albert Einstein College of Medicine, Bronx, NY, United States
[1]Corresponding author: e-mail address: vern.schramm@einstein.yu.edu

Contents

1. Introduction 2
2. Equipment 6
3. Materials 6
   3.1 Buffers 7
   3.2 Radiolabeled Substrates 8
4. Measurement of Equilibrium BIEs Using the RED Device 9
   4.1 Determining Substrate Equilibration Time 12
   4.2 Determination of Substrate Stability 13
   4.3 Measurement of BIEs on Binary Complex Association 14
   4.4 Measurement of BIEs on Ternary Complex Association 15
   4.5 Scintillation Counting and Spectral Deconvolution 16
5. BIE Interpretation and Examples 17
6. Conclusions 20
Acknowledgments 20
References 20

Abstract

Equilibrium binding isotope effects (BIEs) report on the bond vibrational status of enzyme substrates in the Michaelis complex prior to the transition state and how they differ from the solution state. Accordingly, BIEs provide an experimental means of interrogating enzyme–substrate interactions and inform on the influence of enzyme-mediated atomic distortions in modulating substrate reactivity. In this chapter, we outline a rapid equilibrium dialysis method that our lab has used to measure BIEs for several enzyme systems. Implementation of the rapid equilibrium dialysis approach is described in the context of our recent studies on the substrate bonding environment for the human protein lysine *N*-methyltransferase NSD2. A summary of BIE effects provides context for the range of experimental values.

[2] These authors contributed equally to this work.
[3] Present address: Pfizer Inc., Groton, CT 06340, United States.
[4] Present address: University of Maryland, College Park, MD 20742, United States.

Methods in Enzymology, Volume 596
ISSN 0076-6879
http://dx.doi.org/10.1016/bs.mie.2017.07.019

1. INTRODUCTION

Equilibrium binding isotope effects (BIEs) report on differences in the bond vibrational environment of enzyme substrates in the free and bound states preceding catalysis. As such, BIEs inform on substrate structure in the Michaelis complex and provide an experimental approach to probe substrate reactivity prior to the transition state (TS). BIEs have been measured for several enzyme systems, and these data have been leveraged in conjunction with quantum chemical calculations to elucidate the effects of protein binding on substrate conformation and the role of enzyme–substrate (ES) interactions in promoting catalysis (Anderson, 2005; Schramm, 2007; Swiderek & Paneth, 2013).

The expression of BIEs on enzyme binding results from perturbing the vibrational state of the substrate at the position of isotopic substitution (Gaylord & Eliel, 1960). For systems in which the free and bound substrates occupy distinct bonding environments, an observable BIE occurs when differential equilibria are established for substrates bearing either "light" or "heavy" isotopes at a fixed position (Swiderek & Paneth, 2013). A normal BIE (>1.0) results from preferential binding of the substrate bearing the light isotope and is generally reported as either a ratio of the equilibrium constants for the isotopically substituted substrates (e.g., $K_{eq}^{Light}/K_{eq}^{Heavy}=1.05$) or as a positive percentage (e.g., 5%). Normal BIEs occur when the bonding environment of the isotopically substituted position is less constrained in the bound state than in the free state, resulting in a decreased force constant for this bond upon enzyme binding (Fig. 1A). By contrast, inverse BIEs (<1.0) occur when binding of the substrate bearing the heavy isotope is favored and are commonly expressed as a ratio of equilibrium constants (e.g., $K_{eq}^{Light}/K_{eq}^{Heavy}=0.95$) or as a negative percentage value (e.g., −5%). Inverse BIEs indicate an increased force constant for the isotopically substituted bond in the bound state and illustrate that the enzyme binding site presents a more constrained bonding environment relative to the solution state (Fig. 1B).

Kinetic isotope effects (KIEs) inform on bond vibrational changes between the ground state (GS) and TS of a chemical reaction and remain one of the most powerful experimental techniques for interrogating TS structure. KIEs on enzyme reactions measured by the internal competition of labeled substrates report on k_{cat}/K_M (V/K) and include contributions from all steps of the reaction up to and including the first irreversible chemical step (Northrop, 1981). As such, V/K KIEs reflect not only the intrinsic

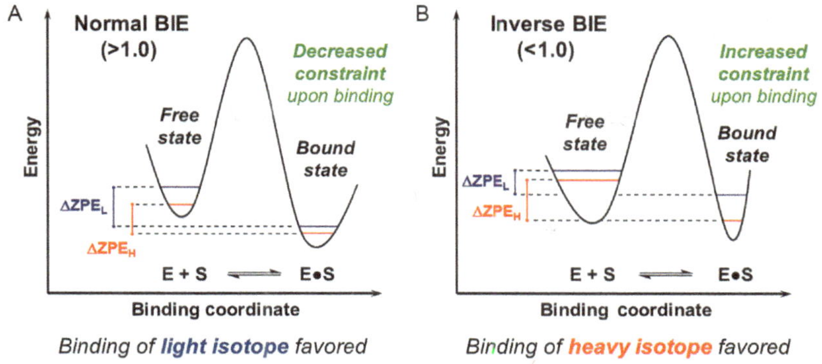

Fig. 1 Energetic considerations for equilibrium BIEs. Equilibrium isotope effects on enzyme binding occur due to changes in the bonding environment surrounding the position of isotopic substitution in the substrate. As isotopic substitution perturbs the vibrational modes of a chemical bond without altering the potential energy landscape, differential equilibria for isotopically labeled substrates can be understood in the context of changes in vibrational zero-point energies. (A) Normal BIEs occur when the bonding environment of the isotopically substituted position is less constrained in the bound state than in the free state, which results in a decreased force constant for this bond in the ES complex. Binding of the substrate bearing the light isotope is preferred due to a greater change in vibrational free energy as compared to that of the substrate bearing the heavy isotope. (B) Inverse BIEs occur when the bonding environment of the isotopically substituted position is more constrained in the bound state than in the free state, which results in an increased force constant for this bond in the ES complex. Binding of the substrate bearing the heavy isotope is preferred due to a greater change in vibrational free energy as compared to that of the substrate bearing the light isotope. *BIE*, binding isotope effect; *E*, enzyme; *S*, substrate; ZPE_H, zero-point energy for substrate bearing the heavy isotope; ZPE_L, zero-point energy for substrate bearing the light isotope.

value (the KIE on chemistry alone), but may also contain contributions from isotopically sensitive steps associated with substrate binding, including forward commitment and BIEs (Fig. 2; Ruszczycky & Anderson, 2006). Accordingly, experimental determination of both *V*/*K* KIEs and equilibrium BIEs provides information on the relative bonding environments for the substrates of an enzymatic reaction in the free state, the ES complex, and at the TS. Such data inform on the role played by enzymes in modulating the GS and TS structures of the reactants to promote catalysis.

Equilibrium BIEs are generally measured under competitive conditions wherein substrates bearing either a light or heavy isotope at the position of interest are mixed and compete for the enzyme binding site. BIE values are then determined by comparing the isotopic ratios of the substrates in the free state to that of the bound state as measured by either liquid scintillation counting (for radioisotopes, e.g., ^{3}H and ^{14}C) or mass spectrometry (for stable isotopes, e.g., ^{2}H, ^{13}C, or ^{18}O). Several experimental approaches have been

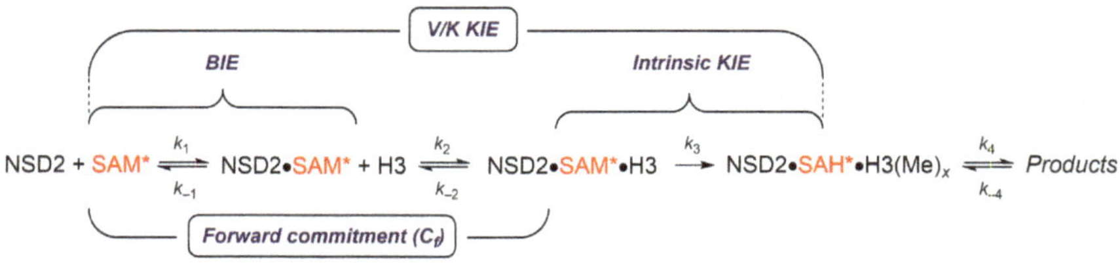

Fig. 2 The relationship between KIEs, forward commitment, and BIEs using the NSD2 reaction as an example. NSD2 catalyzes the SAM-dependent monomethylation and dimethylation of histone H3 (lysine 36) to produce SAH and the methylated histone products. The generalized scheme illustrates that the reaction mechanism is an ordered process, with the sequential equilibrium binding of SAM (k_1/k_{-1}) and histone H3 (k_2/k_{-2}), followed by an irreversible methyl transfer step (k_3) and product dissociation (k_4/k_{-4}). When KIEs for SAM are measured using the internal competition of isotopically labeled substrates (*), *V/K* values report on changes in bond vibrational status between SAM free in solution and at the TS. Both C_f and BIEs have the potential to convolute KIE measurements such that the observed *V/K* KIEs may not reflect the intrinsic values (the KIE on chemistry alone). Forward commitment for SAM represents the fraction of ES complex (i.e., NSD2•SAM*•H3) that partitions over the barrier to product vs the fraction that dissociates back to free enzyme and substrate. BIEs relate to any discrimination in binding between the isotopically labeled substrates. Theoretically, as the limit of C_f approaches infinity (i.e., if every molecule of SAM that binds to NSD2 proceeds over the catalytic barrier without the opportunity to first equilibrate back into solution), any discrimination between isotopes is not kinetic in nature, and thus, the *V/K* KIE approaches the BIE. Experimental determination of *V/K* KIEs, C_f, and BIEs for the NSD2 reaction allows for derivation of intrinsic KIE values (Poulin, Schneck, Matico, Hou, et al., 2016; Poulin, Schneck, Matico, McDevitt, et al., 2016). *BIE*, binding isotope effect; *ES*, enzyme–substrate; *Me*, methyl; *SAH*, *S*-adenosyl-L-homocysteine; *SAM*, *S*-adenosyl-L-methionine; *TS*, transition state; *V/K KIE*, kinetic isotope effect on k_{cat}/K_M.

developed to partition the free and bound substrate states for BIE measurements (vide infra). Common to many of these methods is the passage of an equilibrated mixture of the enzyme and isotopically labeled substrates across a semipermeable membrane with a molecular weight cutoff sufficient to prevent diffusion of the ES complex (i.e., bound fraction) while allowing free passage of the unbound substrate. General considerations for measuring BIEs on enzyme–ligand association include commercial or synthetic access to the requisite isotopically labeled substrates, which are commonly used in stoichiometric excess of the enzyme to promote formation of the ES complex. In addition, high enzyme concentrations (i.e., μ*M* to m*M*) are often required to provide measurable concentrations of the enzyme-bound substrate fraction.

The first isotope effects on binding were reported by Anderson and coworkers (LaReau, Wan, & Anderson, 1989) for the association of [4-^{2}H]NAD$^+$ and [4-^{3}H]NAD$^+$ with lactate dehydrogenase (LDH). In this work, the authors partitioned free and bound substrates by adding a solution of LDH and the [4-^{x}H]NAD^{+} to one side of a dual-chamber dialysis cell and allowing the free substrates to equilibrate across a membrane that was impermeable to the ES complex. The equilibrium dialysis method has been used to measure BIEs for a number of other enzyme systems, including phosphoenolpyruvate carboxylase (Gawlita, Caldwell, O'Leary, Paneth, & Anderson, 1995) and pyruvate kinase (Gawlita et al., 1995). The primary consideration for this approach is the time required for the system to equilibrate across the dialysis membrane, which demands that both the protein and substrate remain stable at room temperature for an extended period (i.e., 5–8 h).

Our lab developed an ultrafiltration method for measuring BIEs using a custom-made Plexiglas multiwell apparatus, which was constructed such that a layer of dialysis membrane is sandwiched between two halves of the device to form a division between the superior and inferior aspects of each well (Schramm, 1976). In the ultrafiltration approach, an equilibrated solution of ES complex and free substrate is added to the superior aspect of a well and the pressure of inert gas is used to push approximately half of the solution volume through the dialysis membrane. The ES complex is retained in the superior aspect of the well, while the free substrate passes through the membrane and can be isolated from the inferior aspect of the well. Relative to equilibrium dialysis methods, the ultrafiltration approach generally allows for more rapid partitioning of the free substrate and ES complex (~1 h). However, filtration through membranes with low-molecular-weight cutoffs and the use of high protein concentrations can often increase filtration times. In addition, as the force for filtration is supplied by pressurized gas to the superior aspect of the wells, some evaporation of the solution is possible which could require correction of reaction concentrations.

The ultrafiltration approach has been used to measure equilibrium BIEs for a number of enzyme systems, including brain hexokinase (Lewis & Schramm, 2003a, 2003b), thymidine phosphorylase (Birck & Schramm, 2004), purine nucleoside phosphorylase (Murkin, Tyler, & Schramm, 2008), orotate phosphoribosyltransferase (Zhang & Schramm, 2011), and uridine phosphorylase (Silva, Kipp, & Schramm, 2012). Recently, the ultrafiltration method for BIE measurements was employed by Klinman and coworkers in the study of ES interactions for catechol *O*-methyltransferase (Zhang, Kulik, Martinez, & Klinman, 2015) and glycine *N*-methyltransferase (Zhang & Klinman, 2016). Technical application of the ultrafiltration approach to BIEs is detailed in these publications and has also been detailed for direct ligand binding studies (Schramm, 1976).

Other methods that have been implemented to partition free and bound substrate fractions for BIE measurements include the use of protein concentrator (Griswold, Castro, Fisher, & Toney, 2012) and G-25 size-exclusion spin tubes (Linscott et al., 2016). In the approach using concentrator tubes, partitioning of the ES complex and free substrate occurs based on passage through a semipermeable membrane. However, in contrast to both the equilibrium dialysis and ultrafiltration methods, the solution of free substrate that passes through the membrane is removed from the inferior aspect of the system by centrifugation, which prevents the equilibration of free substrate back across the membrane after initial diffusion. Partitioning of the unbound and bound substrate fractions by G-25 spin columns occurs due to retention of the free small-molecule substrate on the resin and early elution of the ES complex. This approach leads to complete separation of the ES complex and free substrate fraction, which could disrupt the initial equilibrium state of the system, particularly for weakly binding substrates.

Recently, our lab reported BIEs on substrate binding for dihydropteroate synthase (Stratton, Namanja-Magliano, Cameron, & Schramm, 2015) and the human protein lysine *N*-methyltransferase NSD2 (Poulin, Schneck, Matico, Hou, et al., 2016) using a commercially available rapid equilibrium dialysis (RED) device (Thermo Fisher Scientific). In this chapter, we describe an equilibrium dialysis approach for measuring BIEs using the RED device as implemented in our study on the effects of nucleosome binding on the substrate bonding environment of NSD2.

2. EQUIPMENT

RED device with 8-kDa molecular weight cutoff (Thermo Fisher Scientific)
Micropipettes with tips
Temperature-controlled orbital shaker with attachment for holding 96-well plates
Scintillation vials (20 mL)
Vortex agitator
Tri-Carb 2910TR scintillation counter (Perkin-Elmer)
HPLC with UV detector and reverse-phase C-18 column (4.6 mm × 250 mm)
Benchtop microcentrifuge
SpeedVac centrifugal concentrator

3. MATERIALS

The buffer mixtures and substrates used to measure BIEs on the association of *S*-adenosyl-L-methionine (SAM) and NSD2 in binary and ternary complexes are summarized in Tables 1–7.

3.1 Buffers

Table 1 10× Buffer

Component	**Final Concentration (m*M*)**	**Stock**	**Amount**
Tris–HCl (pH 9.0)	500	1 *M*	5 mL
$MgCl_2$	50	1 *M*	500 μL
Dithiothreitol (DTT)	20	—	31 mg

Note: Add Milli-Q water to a final volume of 10 mL. Store as 500 μL aliquots at −20°C and avoid repeated freeze/thaw cycles to prevent oxidation of DTT.

Table 2 Equilibration Buffer

Component	**Final Concentration**	**Stock**	**Amount**
10× Buffer	50 m*M* Tris–HCl, 5 m*M* $MgCl_2$, 2 m*M* DTT	10×	30 μL
Glycerol	8% (v/v)	50% (v/v)	48 μL

Note: Add Milli-Q water to a final volume of 300 μL.

Table 3 Sample Buffer

Component	**Final Concentration**	**Stock**	**Amount**
10× Buffer	1×	10×	10 μL
Glycerol	8% (v/v)	50% (v/v)	16 μL
[5′-$^3H_2$]SAM or [Me-$^3H_3$]SAM	400,000 cpm	—	—
[8-^{14}C]SAM	400,000 cpm	—	—
Cold SAM	20 μ*M*	—	—

Note: Add Milli-Q water to a final volume of 100 μL. Glycerol is used to stabilize H3K36M nucleosomes in the ternary complex BIE measurements and is included in all buffers to keep the composition consistent between experiments. "Cold SAM" refers to a stock of substrate that does not bear a radioisotope label.

Table 4 Binary Complex Sample Buffer

Component	**Final Concentration**	**Stock**	**Amount**
10× Buffer	1×	10×	9.5 μL
Glycerol	8% (v/v)	50% (v/v)	16 μL
[5′-$^3H_2$]SAM or [Me-$^3H_3$]SAM	≥100,000 cpm	—	—
[8-^{14}C]SAM	≥100,000 cpm	—	—
Cold SAM	20 μ*M*	—	—
NSD2 enzyme	4 μ*M*	80 μ*M*	5 μL

Note: Add Milli-Q water to a final volume of 100 μL. Glycerol is used to stabilize H3K36M nucleosomes in the ternary complex BIE measurements and is included in all buffers to keep the composition consistent between experiments. "Cold SAM" refers to a stock of substrate that does not bear a radioisotope label. The NSD2 enzyme is stored in 1× buffer.

Table 5 Ternary Complex Sample Buffer

Component	Final Concentration	Stock	Amount
10 × Buffer	1 ×	10 ×	9.5 μL
[5′-$^3H_2$]SAM or [Me-$^3H_3$]SAM	≥100,000 cpm	—	—
[8-^{14}C]SAM	≥100,000 cpm	—	—
Cold SAM	20 μ*M*	—	—
NSD2 enzyme	4 μ*M*	80 μ*M*	5 μL
H3K36M polynucleosome	5 μ*M*	12.5 μ*M*	40 μL

Note: Add Milli-Q water to a final volume of 100 μL. Glycerol is used to stabilize the H3K36M nucleosomes. The NSD2 enzyme is stored in 1 × buffer.

Table 6 HPLC Buffer A

Component	Final Concentration	Amount
Ammonium formate	50 m*M*	3.15 g
Milli-Q water	—	800 mL

Note: Adjust the pH to 4.0 with formic acid. Add Milli-Q water to a final volume of 1000 mL. Pass through a 0.22-μm filter before use.

Table 7 HPLC Buffer B

Component	Final Concentration (v/v)	Amount
Acetonitrile	50%	500 mL
Milli-Q water	49.9%	499 mL
Formic acid	0.1%	1 mL

Note: Pass through a 0.22-μm filter before use.

3.2 Radiolabeled Substrates

Measuring equilibrium BIEs by the competitive radiolabel approach requires a pair of substrates labeled with heavy or light isotopes. BIEs on the binding of SAM and NSD2 were measured using either [Me-$^3H_3$] SAM or [5′-$^3H_2$]SAM as the heavy substrates, and [8-^{14}C]SAM as the light substrate. The remote 8-^{14}C label in the light substrate is used to report on ^{1}H at the position of interest, and ^{14}C can be shown to not exhibit an isotope effect on binding (Lewis & Schramm, 2006). When the labeled SAMs are mixed, changes in the ratio of light/heavy substrate can be quantified by monitoring $^{14}C/^3H$ levels via dual channel scintillation counting. Detailed procedures for the chemoenzymatic synthesis of isotopically labeled SAM substrates have been described elsewhere (Du, Wang, & Schramm, 2016; Poulin, Du, & Schramm, 2015).

Notes: The sulfonium center of biosynthetically derived SAM is generated in the (*S*) configuration. Prolonged incubation of SAM at room temperature and neutral pH can lead to epimerization of the sulfonium center to produce a mixture of the (*S*)- and (*R*)-isomers. Commercial SAM preparations contain appreciable quantities of the (*R*)-isomer, as well as additional breakdown products. As such, use of commercial SAM for BIE measurements should be avoided. Ideally, SAM substrates should be prepared enzymatically and stored at −80°C to retain the (*S*)-isomer.

4. MEASUREMENT OF EQUILIBRIUM BIES USING THE RED DEVICE

The RED device (Thermo Fisher Scientific) consists of a polypropylene base with a standard 96-well plate footprint that holds 48 dialysis inserts. Each dialysis insert is constructed as two wells separated by a vertical cylinder of cellulose membrane. The volume contained within the cylinder of dialysis membrane will be referred to as the "sample well," and the outer volume surrounding the dialysis membrane will be referred to as the "buffer well" (Fig. 3).

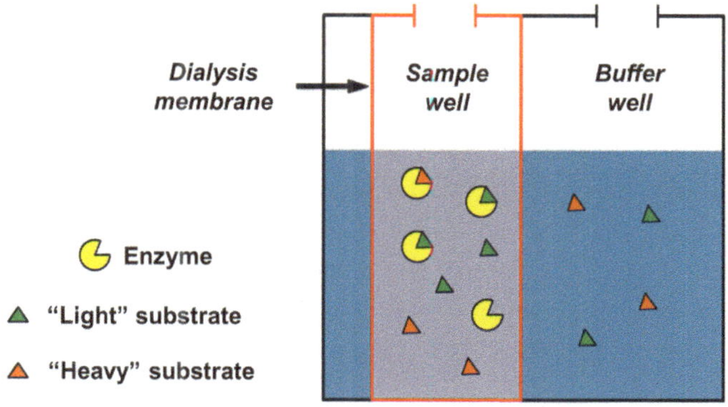

Fig. 3 Schematic of a RED device dialysis insert at equilibrium. The graphic represents a cross section of the dialysis inserts used to measure BIEs on the association of SAM and NSD2. In the context of measurements with NSD2, the "light" substrate was [8-^{14}C]SAM (where the remote 8-^{14}C label was used to report on ^{1}H at the position of interest) and the "heavy" substrate was either [Me-$^{3}H_3$]SAM or [5′-$^{3}H_2$]SAM.

In the BIE experiment with SAM and NSD2, unbound substrate (i.e., [^{3}H]SAM and [^{14}C]SAM) can freely diffuse across the 8-kDa cutoff membrane, whereas the enzyme and ES complex (i.e., bound substrate) cannot. Following equilibration of the substrate across the dialysis membrane, the sample well contains counts per minute (cpm) corresponding to both the bound ($^{3}\mathrm{H}_{\mathrm{sw}}^{\mathrm{bound}}$; $^{14}\mathrm{C}_{\mathrm{sw}}^{\mathrm{bound}}$) and unbound ($^{3}\mathrm{H}_{\mathrm{sw}}^{\mathrm{unbound}}$; $^{14}\mathrm{C}_{\mathrm{sw}}^{\mathrm{unbound}}$) substrate in a total volume V_{sw}. By contrast, the buffer well contains cpm for only the unbound substrate ($^{3}\mathrm{H}_{\mathrm{bw}}$; $^{14}\mathrm{C}_{\mathrm{bw}}$) in a total volume V_{bw}. Accordingly, at equilibrium, the cpm *per unit volume* for the unbound substrate in the sample well is equivalent to that of the buffer well:

$$\frac{^{3}\mathrm{H}_{\mathrm{sw}}^{\mathrm{unbound}}}{V_{\mathrm{sw}}}=\frac{^{3}\mathrm{H}_{\mathrm{bw}}}{V_{\mathrm{bw}}}\quad\text{and}\quad\frac{^{14}\mathrm{C}_{\mathrm{sw}}^{\mathrm{unbound}}}{V_{\mathrm{sw}}}=\frac{^{14}\mathrm{C}_{\mathrm{bw}}}{V_{\mathrm{bw}}}\tag{1}$$

The cpm corresponding to bound substrate in the sample well can be expressed as the difference between the total cpm in the sample well ($^{3}\mathrm{H}_{\mathrm{sw}}^{\mathrm{total}}$; $^{14}\mathrm{C}_{\mathrm{sw}}^{\mathrm{total}}$) and the cpm for the unbound substrate in the sample well, as illustrated by Eq. (2):

$$^{3}\mathrm{H}_{\mathrm{sw}}^{\mathrm{bound}}={}^{3}\mathrm{H}_{\mathrm{sw}}^{\mathrm{total}}-{}^{3}\mathrm{H}_{\mathrm{sw}}^{\mathrm{unbound}}\quad\text{and}\quad{}^{14}\mathrm{C}_{\mathrm{sw}}^{\mathrm{bound}}={}^{14}\mathrm{C}_{\mathrm{sw}}^{\mathrm{total}}-{}^{14}\mathrm{C}_{\mathrm{sw}}^{\mathrm{unbound}}\tag{2}$$

By combining Eqs. (1) and (2), cpm for the bound substrate in the sample well can be expressed as a relation of the cpm for substrate in the buffer well:

$$^{3}\mathrm{H}_{\mathrm{sw}}^{\mathrm{bound}}={}^{3}\mathrm{H}_{\mathrm{sw}}^{\mathrm{total}}-{}^{3}\mathrm{H}_{\mathrm{bw}}\left(\frac{V_{\mathrm{sw}}}{V_{\mathrm{bw}}}\right)\quad\text{and}\quad{}^{14}\mathrm{C}_{\mathrm{sw}}^{\mathrm{bound}}={}^{14}\mathrm{C}_{\mathrm{sw}}^{\mathrm{total}}-{}^{14}\mathrm{C}_{\mathrm{bw}}\left(\frac{V_{\mathrm{sw}}}{V_{\mathrm{bw}}}\right)\tag{3}$$

BIEs can be expressed as a relation of equilibrium constants for the "light" substrate (e.g., [8-^{14}C]SAM, where the remote ^{14}C reports on ^{1}H at the position on interest) and the "heavy" substrate (e.g., [Me-^{3}H$_{3}$]SAM or [5′-^{3}H$_{2}$]SAM). Alternatively, BIEs can be defined as a relation of the ^{14}C/^{3}H ratio for the total bound substrate and the ^{14}C/^{3}H ratio for the total unbound substrate, as illustrated in Eq. (4).

$$\mathrm{BIE}=\frac{^{14}K_{\mathrm{eq}}}{^{3}K_{\mathrm{eq}}}=\frac{\left(\frac{^{14}\mathrm{C}}{^{3}\mathrm{H}}\right)_{\mathrm{bound}}}{\left(\frac{^{14}\mathrm{C}}{^{3}\mathrm{H}}\right)_{\mathrm{unbound}}}\tag{4}$$

Substituting Eqs. (1) and (3) into Eq. (4) allows for the expression of Eq. (5):

$$\mathrm{BIE}=\frac{\dfrac{^{14}\mathrm{C}_{\mathrm{sw}}^{\mathrm{bound}}}{^{3}\mathrm{H}_{\mathrm{sw}}^{\mathrm{bound}}}}{\dfrac{^{14}\mathrm{C}_{\mathrm{sw}}^{\mathrm{unbound}}+{}^{14}\mathrm{C}_{\mathrm{bw}}}{^{3}\mathrm{H}_{\mathrm{sw}}^{\mathrm{unbound}}+{}^{3}\mathrm{H}_{\mathrm{bw}}}}=\frac{\dfrac{^{14}\mathrm{C}_{\mathrm{sw}}^{\mathrm{total}}-{}^{14}\mathrm{C}_{\mathrm{bw}}\left(\dfrac{V_{\mathrm{sw}}}{V_{\mathrm{bw}}}\right)}{^{3}\mathrm{H}_{\mathrm{sw}}^{\mathrm{total}}-{}^{3}\mathrm{H}_{\mathrm{bw}}\left(\dfrac{V_{\mathrm{sw}}}{V_{\mathrm{bw}}}\right)}}{\dfrac{^{14}\mathrm{C}_{\mathrm{bw}}\left(\dfrac{V_{\mathrm{sw}}}{V_{\mathrm{bw}}}\right)+{}^{14}\mathrm{C}_{\mathrm{bw}}}{^{3}\mathrm{H}_{\mathrm{bw}}\left(\dfrac{V_{\mathrm{sw}}}{V_{\mathrm{bw}}}\right)+{}^{3}\mathrm{H}_{\mathrm{bw}}}}=\frac{\dfrac{^{14}\mathrm{C}_{\mathrm{sw}}^{\mathrm{total}}-{}^{14}\mathrm{C}_{\mathrm{bw}}\left(\dfrac{V_{\mathrm{sw}}}{V_{\mathrm{bw}}}\right)}{^{3}\mathrm{H}_{\mathrm{sw}}^{\mathrm{total}}-{}^{3}\mathrm{H}_{\mathrm{bw}}\left(\dfrac{V_{\mathrm{sw}}}{V_{\mathrm{bw}}}\right)}}{\dfrac{^{14}\mathrm{C}_{\mathrm{bw}}\left(\dfrac{V_{\mathrm{sw}}}{V_{\mathrm{bw}}}+1\right)}{^{3}\mathrm{H}_{\mathrm{bw}}\left(\dfrac{V_{\mathrm{sw}}}{V_{\mathrm{bw}}}+1\right)}}$$

$$=\frac{\dfrac{^{14}\mathrm{C}_{\mathrm{sw}}^{\mathrm{total}}}{^{14}\mathrm{C}_{\mathrm{bw}}}-\dfrac{V_{\mathrm{sw}}}{V_{\mathrm{bw}}}}{\dfrac{^{3}\mathrm{H}_{\mathrm{sw}}^{\mathrm{total}}}{^{3}\mathrm{H}_{\mathrm{bw}}}-\dfrac{V_{\mathrm{sw}}}{V_{\mathrm{bw}}}} \quad (5)$$

If volumes v_{sw} and v_{bw} are removed from the sample well and buffer well (respectively), then Eq. (5) can be modified to give Eq. (6)

$$\mathrm{BIE}=\frac{\dfrac{^{14}\mathrm{C}_{\mathrm{sw}}^{\mathrm{total}}\left(\dfrac{V_{\mathrm{sw}}}{v_{\mathrm{sw}}}\right)}{^{14}\mathrm{C}_{\mathrm{bw}}\left(\dfrac{V_{\mathrm{bw}}}{v_{\mathrm{bw}}}\right)}-\left(\dfrac{V_{\mathrm{sw}}}{V_{\mathrm{bw}}}\right)}{\dfrac{^{3}\mathrm{H}_{\mathrm{sw}}^{\mathrm{total}}\left(\dfrac{V_{\mathrm{sw}}}{v_{\mathrm{sw}}}\right)}{^{3}\mathrm{H}_{\mathrm{bw}}\left(\dfrac{V_{\mathrm{bw}}}{v_{\mathrm{bw}}}\right)}-\left(\dfrac{V_{\mathrm{sw}}}{V_{\mathrm{bw}}}\right)}=\frac{\left(\dfrac{^{14}\mathrm{C}_{\mathrm{sw}}^{\mathrm{total}}\times v_{\mathrm{bw}}}{^{14}\mathrm{C}_{\mathrm{bw}}\times v_{\mathrm{sw}}}-1\right)\dfrac{V_{\mathrm{sw}}}{V_{\mathrm{bw}}}}{\left(\dfrac{^{3}\mathrm{H}_{\mathrm{sw}}^{\mathrm{total}}\times v_{\mathrm{bw}}}{^{3}\mathrm{H}_{\mathrm{bw}}\times v_{\mathrm{sw}}}-1\right)\dfrac{V_{\mathrm{sw}}}{V_{\mathrm{bw}}}}=\frac{\dfrac{^{14}\mathrm{C}_{\mathrm{sw}}^{\mathrm{total}}}{^{14}\mathrm{C}_{\mathrm{bw}}}-\dfrac{v_{\mathrm{sw}}}{v_{\mathrm{bw}}}}{\dfrac{^{3}\mathrm{H}_{\mathrm{sw}}^{\mathrm{total}}}{^{3}\mathrm{H}_{\mathrm{bw}}}-\dfrac{v_{\mathrm{sw}}}{v_{\mathrm{bw}}}} \quad (6)$$

Scintillation counting for BIEs measurements is carried out using equivolume aliquots from the sample and buffer wells (vide infra). As such, $v_{\mathrm{sw}}=v_{\mathrm{bw}}$ and Eq. (6) can be reduced to give Eq. (7):

$$\mathrm{BIE}=\frac{\dfrac{^{14}\mathrm{C}_{\mathrm{sw}}^{\mathrm{total}}}{^{14}\mathrm{C}_{\mathrm{bw}}}-1}{\dfrac{^{3}\mathrm{H}_{\mathrm{sw}}^{\mathrm{total}}}{^{3}\mathrm{H}_{\mathrm{bw}}}-1} \quad (7)$$

4.1 Determining Substrate Equilibration Time

The rapid equilibrium dialysis method for measuring BIEs requires an accurate determination of the time required for equilibration of the unbound substrate across the dialysis membrane. Equilibration times will vary depending on substrate size, buffer viscosity, temperature, shaking velocity, and the molecular weight cutoff of the dialysis membrane. As a result, it is important to measure the equilibration time for each substrate system being evaluated. The methods below describe the procedure for measuring the equilibration time for SAM, which should be carried out in duplicate to ensure the results are reproducible.

Methods

1. Prepare sample buffer (Table 3) with at least 400,000 cpm for both the [^{3}H] and [^{14}C]SAM substrates and the final chemical concentration of SAM at 20 μ*M*.
2. Transfer 100 μL of sample buffer to the sample well of a RED device dialysis insert.

   The sample well is identified by a red ring on top of the dialysis insert.
3. Add 300 μL of equilibration buffer (Table 2) to the buffer well in the same insert of the RED device and cover the well openings with a sealing film.

   The buffer well is identified by a white ring on top of the dialysis insert. The manufacturer recommends using 300 μL of equilibration buffer to ensure optimal surface area coverage of the dialysis membrane.
4. Transfer the RED device to an orbital shaker and agitate at 250 rpm. Ensure the RED cassette is securely held in the orbital shaker to prevent tipping.

   For measurements with SAM and NSD2, the RED device is shaken at 25° C to minimize isomerization of the SAM sulfonium center. However, shaking the RED device at 37° C is recommended to reduce the time required for equilibration, but requires that both the protein and substrates are stable at elevated temperatures for the incubation period.
5. Remove 10 μL aliquots from both the sample well and buffer well of the dialysis insert every 30–60 min, and transfer the aliquots to separate 20 mL scintillation vials containing 490 μL of Milli-Q water.
6. Add 10 mL of UltimaGold™ scintillation cocktail (Perkin-Elmer) to each scintillation vial and vortex the vials for 30 s to ensure the sample and scintillation cocktail are thoroughly mixed.

   Allow the samples to settle in the dark for at least 1 h to eliminate bubbles that could interfere with scintillation counting.
7. Read each sample in a dual channel liquid scintillation counter as described in Section 4.5.

The equilibration time is reached when the cpm values for ^3H and ^{14}C in the sample well and buffer well of the dialysis insert are equal. The $^3H/^{14}C$ ratio of the substrate in the sample well should be equivalent to that of the substrate in the buffer well, indicating that dialysis does not result in isotopic fractionation.

Notes: Whenever possible, use at least 400,000 cpm of each [^{3}H]-labeled and [^{14}C]-labeled substrates. This ensures that once the system reaches equilibrium, a 10-μL aliquot contains at least 10,000 cpm of each labeled substrate, which will give optimal statistics in the scintillation counting.

4.2 Determination of Substrate Stability

Substrate stability is an important consideration when measuring BIEs by equilibrium dialysis. To ensure that the substrate does not react or degrade during the equilibration, an independent technique must be used to verify the composition of the ligand under the conditions used for the BIE measurement. In the study of BIEs with NSD2 an HPLC-based method was used to investigate the stability of SAM in both binary and ternary complex incubations, as described later.

Methods

1. Prepare 100 μL of either the binary complex buffer (Table 4) or the ternary complex buffer (Table 5) in a microcentrifuge tube and incubate at 25°C.
2. Remove 10 μL aliquots every 30–60 min (until the previously determined equilibration time is reached) and inactivate the enzyme by adding 1 μL of 100 m*M* H_2SO_4.
3. Freeze the aliquots on dry ice to prevent any decomposition before HPLC analysis.
4. Thaw each aliquot immediately prior to analysis and pellet the protein by centrifuging at 16,000 × *g* for 2 min in a benchtop microcentrifuge.
5. Dilute the supernatant to 100 μL with filtered Milli-Q water and transfer the solution to a low-retention HPLC auto sampler vial.
6. Analyze each sample for degradation and/or isomerization of SAM via HPLC equipped with a reverse-phase C-18 column (4.6 mm × 250 mm) using the method outlined in Table 8. Monitor UV absorbance at 262 nm and collect peaks corresponding to the (*S*)-isomer of SAM (4.8–5.2 min), (*R*)-isomer of SAM (5.2–6.0 min), and *S*-adenosyl-L-homocysteine (SAH, 15–16 min) in 20 mL scintillation vials.
7. Dry the HPLC samples in a SpeedVac centrifugal concentrator for approximately 2–4 h until no liquid remains in the scintillation vials.

Table 8 HPLC Method for Stability Analysis of SAM Substrates

Time (min)	Flow Rate (mL min^{-1})	% Buffer A	% Buffer B
0	1	100	0
4	1	100	0
12	1	40	60
20	1	40	60
22	1	100	0

Note: The compositions of Buffer A and Buffer B are given in Tables 6 and 7, respectively.

8. Dissolve each sample in 500 μL of Milli-Q water and add 10 mL of UltimaGold™ scintillation fluid (Perkin-Elmer). Vortex each vial for 30 s each to ensure the sample and scintillation cocktail are thoroughly mixed.

   Allow the samples to settle in the dark for at least 1 h to eliminate bubbles that could interfere with scintillation counting.
9. Read each sample in a dual channel liquid scintillation counter as described in Section 4.5.

Notes: Following the full incubation period, ≥95% of the counts should remain in the peak corresponding to the (*S*)-isomer of SAM. Any isomerization to the (*R*)-isomer of SAM or decomposition to SAH can convolute the interpretation of the BIE measurements by introducing multiple species that compete for binding to the enzyme.

4.3 Measurement of BIEs on Binary Complex Association

Binary complex BIEs report directly on differences in the bonding environment of enzyme substrates while free in solution and when bound to the enzyme. Importantly, BIEs on binary complex formation can only be measured for multisubstrate enzyme systems wherein the substrate binding is unordered and for systems with ordered mechanisms in which the substrate of interest binds first. The methods below describe the measurement of BIEs on association of the SAM•NSD2 binary complex.

Methods

1. Add 100 μL of binary complex sample buffer (Table 4) to the sample well of a dialysis insert in the RED device.

   The concentration of substrate and enzyme used should be selected so that between 15% and 40% of the total light substrate is bound in the ES complex.

2. Add 300 μL of equilibration buffer (Table 2) to the buffer well in the same insert of the RED device and cover the well openings with a sealing film.
3. Place the RED device in a temperature-controlled orbital shaker and agitate at 250 rpm for 4.5 h at 25°C.

   An incubation temperature of 25°C is used to minimize the isomerization of SAM (<5%) over the course of the equilibration. Using an incubation time that is 30 min longer than the equilibration time will ensure that equilibrium has been reached before samples are removed and analyzed.
4. Remove 50 μL aliquots from both the sample well and buffer well of the dialysis insert and transfer the aliquots to separate 20 mL liquid scintillation vials containing 450 μL of Milli-Q water.
5. Add 10 mL of UltimaGold™ scintillation cocktail (Perkin-Elmer) to each vial and vortex the vials for 30 s to ensure the sample and scintillation cocktail are well mixed.

   Allow the samples to settle for at least 1 h in the dark until there are no bubbles remaining in the scintillation cocktail that could interfere with scintillation counting.
6. Read each sample in a dual channel liquid scintillation counter as described in Section 4.5.

4.4 Measurement of BIEs on Ternary Complex Association

BIEs on ternary complex association can be measured for multisubstrate enzyme systems by using cosubstrate analogs that mimic the structure and binding of the natural cosubstrate, but do not allow for catalytic turnover. BIEs on association of the SAM•NSD2•nucleosome ternary complex were carried out using a recombinant nucleosome substrate analog in which the histone H3 lysine 36 residues were mutated to methionine (H3K36M). The H3K36M polynucleosome binds to NSD2, but does not undergo the methyl transfer reaction with SAM. The procedure for preparation of the H3K36M nucleosome has been described elsewhere (Poulin, Schneck, Matico, Hou, et al., 2016).

Methods

The procedure for measuring BIEs on association of the SAM•NSD2•nucleosome ternary complex is identical to the methods described for the SAM•NSD2 binary complex except that the ternary complex buffer (Table 5) is used in place of the binary complex buffer.

Notes: When measuring ternary complex BIEs, it is important to verify that the cosubstrate analog does not interfere with scintillation counting or react with the substrate bearing the isotopic labels under the experimental conditions.

4.5 Scintillation Counting and Spectral Deconvolution

Scintillation counting as described in this chapter was carried out using a Tri-Carb 2910TR scintillation counter (Perkin-Elmer), which is a dual channel instrument that registers signal for ^{3}H in Channel A and the signal for ^{14}C in both Channel A and Channel B. Accordingly, the raw scintillation counting data must be deconvoluted to accurately determine isotopic ratios.

Methods

1. The channels on the scintillation counter should be set such that >99% of all counts for the [^{3}H]-labeled substrate register in Channel A, and 100% of the counts for the [^{14}C]-labeled substrate are accounted for between Channel A and Channel B.

   For measurements with [^{3}H]- and [^{14}C]-labeled SAM substrates, Channel A was set to 0–25 meV and Channel B set to 25–400 meV.

2. For optimal scintillation counting statistics, samples should contain ~10,000 cpm of each isotope. Counting should be carried out using 10 cycles at 10 min per cycle, and the average cpm value for each isotope over these 10 cycles should be used in the BIE calculations.

   Measuring a total of 1,000,000 counts per isotope was found to give optimal counting efficiency, and collecting counts beyond this value did not significantly improve the sample measurement accuracy.

3. Use a sample of [8-^{14}C]SAM to determine the proportion of ^{14}C counts in Channel A and Channel B as defined by Eq. (8), where CPM_A and CPM_B are the counts registered in Channel A and Channel B, respectively.

$$r = CPM_A / CPM_B \tag{8}$$

4. The raw scintillation count data for mixtures of [8-^{14}C]SAM with either [Me-$^{3}H_3$]SAM or [5′-$^{3}H_2$]SAM can be deconvoluted to provide total cpm values for ^{14}C and ^{3}H using Eqs. (9) and (10), respectively.

$$^{14}C = CPM_B + (CPM_B \times r) \tag{9}$$
$$^{3}H = CPM_A - (CPM_B \times r) \tag{10}$$

5. The total ^{14}C and ^{3}H cpm values measured for each sample can then be used to calculate the equilibrium BIEs using Eq. (7).

   BIE measurements should be carried out in triplicate and averaged between at least two independent experiments. To allow for the interpretation of BIE data in the context of experimental error, BIE values should be reported with the standard deviation or standard error for the averaged data set. When standard error is used, the number of replicates in the dataset should also be reported.

5. BIE INTERPRETATION AND EXAMPLES

Equilibrium BIE values can be interpreted using quantum chemical calculations to investigate the molecular-scale bonding environment of a ligand when bound to its protein target (Lewis & Schramm, 2003a, 2003b; Murkin et al., 2007; Poulin, Schneck, Matico, Hou, et al., 2016; Schramm, 2007; Stratton et al., 2015; Zhang & Schramm, 2011). The nature of the calculations required to interpret BIE values depends largely on the protein–ligand system under investigation. In general, density functional theory (DFT) calculations as implemented in Gaussian 09 (Frisch et al., 2009) are used to model conformations of the ligand when free in solution and when bound in the ES complex. Then, predicted equilibrium BIEs are calculated by comparing the scaled vibrational frequencies for the free and bound ligands using the ISOEFF98 program (Anisimov & Paneth, 1999). The structure of the free ligand is generally optimized in water at physiological pH. Conformations of the protein-bound ligand will vary depending on the specific binding interactions under evaluation, but using input coordinates from cocrystal structures of the protein and ligand, if available, is a reasonable starting point.

Though detailed methods for the computational modeling of BIE data is beyond the scope of this chapter, the conceptual aspects of a few investigations are outlined below. For example, the influence of hydrogen bonding interactions on equilibrium BIEs has been modeled by modulating the distance between the hydrogen atom of interest in the substrate and hydrogen bond donor residues in the enzyme active site. This approach has been used to model equilibrium BIEs for glucose binding to human brain hexokinase (Lewis & Schramm, 2003a), as well as NSD2 (Poulin, Schneck, Matico, Hou, et al., 2016). Protein-mediated atomic distortions are another important factor impacting the expression of equilibrium BIEs on protein–ligand association. Geometric distortions can be modeled by freezing specific dihedral angles in the substrate when optimizing the structure of the bound ligand. In addition, steric compaction has been modeled by modulating the distance between the atom of interest in the substrate and specific active site residues (Lewis & Schramm, 2003a), or by constraining specific bond lengths in the substrate (Poulin, Schneck, Matico, Hou, et al., 2016; Stratton et al., 2015). Specific examples of measured BIEs and interpretations based on quantum chemical calculations are presented in Table 9 and the references cited therein.

Table 9 Examples of Experimental BIEs and Interpretations

Enzyme	Substrate/Inhibitor	Measured BIEs			Comments
			Binary	Ternary	
NSD2 (Poulin, Schneck, Matico, Hou, et al., 2016)	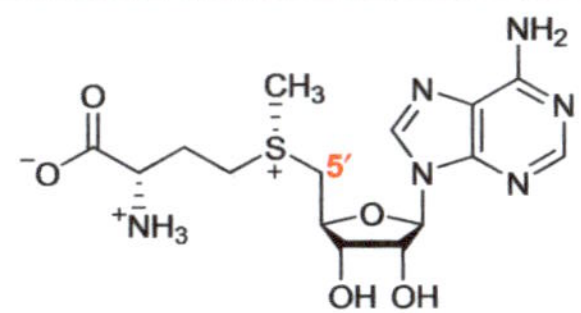 SAM	$[Me\text{–}^3H_3]$ $[5'\text{–}^3H_2]$	0.650 0.973	0.990 0.982	Inverse BIEs for $[Me\text{–}^3H_3]$ and $[5'\text{–}^3H_2]$ SAM in binary complex with NSD2 illustrate a constrained bonding environment. DFT modeling links the observed BIEs with steric compaction. BIEs with a nucleosome substrate analog indicate this constraint lessens in the ternary complex
DHPS (Stratton et al., 2015)	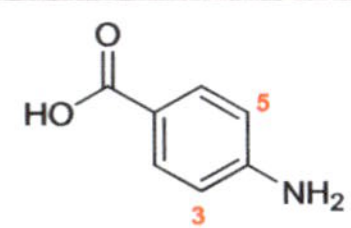 *p*ABA	$[3,5\text{–}^3H]$	—	0.828[a]	An inverse BIE for $[3,5\text{–}^3H]$ *p*ABA with DHPS from *P. falciparum* illustrates a constrained bonding environment in the ternary complex. DFT modeling suggests the observed BIE results from steric compression of the C3–H3/C5–H5 bonds
Hexokinase (Lewis & Schramm, 2003a)	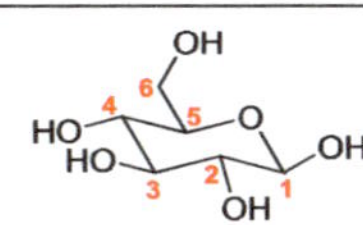 Glucose	$[1\text{–}^3H]$ $[2\text{–}^3H]$ $[3\text{–}^3H]$ $[4\text{–}^3H]$ $[5\text{–}^3H]$ $[6,6\text{–}^3H_2]$	1.027 0.927 1.027 1.051 0.988 1.065	1.013[b] 0.929 1.031 1.052 0.997 1.034	Systematic BIEs inform on the bonding environment of glucose in binary and ternary complexes with hexokinase. Modeling studies using DFT calculations suggest the observed BIEs reflect GS destabilization of the substrate

Thymidine phosphorylase (Birck & Schramm, 2004)	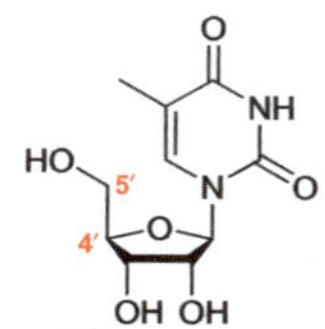 Thymidine	[4′-^{3}H] [5′-^{3}H]	— —	0.997[c] 1.060	The BIE for [5′-^{3}H]thymidine with thymidine phosphorylase is equivalent to the *V*/*K* KIE, indicating the effect is not kinetic in nature. DFT modeling indicates the observed BIE arises from torsional restriction of the substrate upon binding
Uridine phosphorylase (Silva et al., 2012)	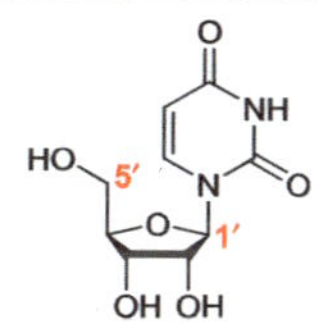 Uridine	[1′-^{3}H] [5′-^{3}H]	0.953 —	0.947[c] 0.979	DFT methods were used to study the influence of uridine ribosyl ring pucker and bond dihedral angles on predicted BIEs. The measured BIEs for [1′-^{3}H] and [5′-^{3}H]uridine with uridine phosphorylase are consistent with GS stabilization in the ES complex
PNP (Murkin et al., 2008)	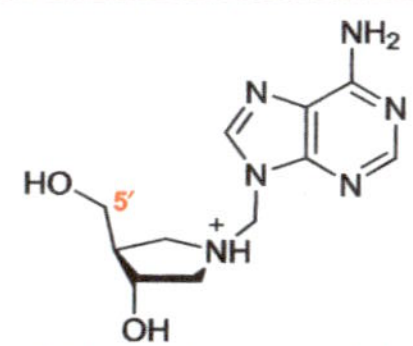 DADMe-ImmH	[5′-^{3}H]	1.292	—	BIEs for the TS analog [5′-^{3}H]DADMe-ImmH with PNP probe bond distortions of the reactants at the TS. DFT calculations suggest the large 5′-^{3}H BIE results from torsional restriction and hyperconjugation effects stemming from H-bonding of the 5′OH with active site residues

[a]Ternary complex BIE was measured using a 6-hydroxymethylpterin pyrophosphate, a nonreactive analog of the natural substrate 6-hydroxymethyl-7,8-dihydropterin pyrophosphate.
[b]Ternary complex BIEs were measured using β-γ, CH_2-ATP, a nonreactive analog of ATP.
[c]Ternary complex BIEs were measured using sulfate as a substrate analog of phosphate.
DADMe-ImmH, 4′-deaza-1′-aza-2′-deoxy-9-methylene immucillin-H; *DFT*, density functional theory; *DHPS*, dihydropteroate synthase; *ES*, enzyme–substrate; *GS*, ground state; *H-bonding*, hydrogen bonding; *KIE*, kinetic isotope effect; *Me*, methyl; *NSD2*, human protein lysine *N*-methyltransferase; *pABA*, *para*-aminobenzoic acid; *PNP*, purine nucleoside phosphorylase; *SAM*, *S*-adenosyl-L-methionine; *TS*, transition state.

6. CONCLUSIONS

Equilibrium BIEs offer an experimental approach for probing the vibrational state and bonding environment of enzyme substrates in the Michaelis complex and inhibitors in enzyme–inhibitor complexes. Despite the utility of this technique within the context of enzymology, BIEs have only been measured for relatively few enzymes. The principles governing BIEs for enzymes are also applicable to receptors and small-molecule ligands, though BIEs on receptor–ligand interactions have not yet, to our knowledge, been reported. We hope the accessible approach outlined in this chapter will support the extension of these methods to new protein systems and provide readers with additional tools to interrogate the role of ES interactions in modulating substrate structure and reactivity.

ACKNOWLEDGMENTS

We thank Dr. D. Randal Kipp (Relay Therapeutics) for helpful discussions regarding the use of the RED device for equilibrium BIE measurements. This work was supported by research Grants GM041916 and CA135405 from the National Institutes of Health.

REFERENCES

Anderson, V. E. (2005). Quantifying energetic contributions to ground state destabilization. *Archives of Biochemistry and Biophysics*, *433*, 27–33.

Anisimov, V., & Paneth, P. (1999). ISOEFF98. A program for studies of isotope effects using Hessian modifications. *Journal of Mathematical Chemistry*, *26*, 75–86.

Birck, M. R., & Schramm, V. L. (2004). Binding causes the remote [5′-^{3}H]thymidine kinetic isotope effect in human thymidine phosphorylase. *Journal of the American Chemical Society*, *126*, 6882–6883.

Du, Q., Wang, Z., & Schramm, V. L. (2016). Human DNMT1 transition state structure. *Proceedings of the National Academy of Sciences of the United States of America*, *113*, 2916–2921.

Frisch, M. J., Trucks, G. W., Schlegel, H. B., Scuseria, G. E., Robb, M. A., Cheeseman, J. R., et al. (2009). *Gaussian 09*. Wallingford, CT: Gaussian, Inc.

Gawlita, E., Caldwell, W. S., O'Leary, M. H., Paneth, P., & Anderson, V. E. (1995). Kinetic isotope effects on substrate association: Reactions of phosphoenolpyruvate with phospho-enolpyruvate carboxylase and pyruvate kinase. *Biochemistry*, *34*, 2577–2583.

Gaylord, N. G., & Eliel, E. L. (1960). *Isotope effects on reaction rates*. New York: Ronald Press.

Griswold, W. R., Castro, J. N., Fisher, A. J., & Toney, M. D. (2012). Ground-state electronic destabilization via hyperconjugation in aspartate aminotransferase. *Journal of the American Chemical Society*, *134*, 8436–8438.

LaReau, R. D., Wan, W., & Anderson, V. E. (1989). Isotope effects on binding of NAD^{+} to lactate dehydrogenase. *Biochemistry*, *28*, 3619–3624.

Lewis, B. E., & Schramm, V. L. (2003a). Binding equilibrium isotope effects for glucose at the catalytic domain of human brain hexokinase. *Journal of the American Chemical Society*, *125*, 4785–4798.

Lewis, B. E., & Schramm, V. L. (2003b). Glucose binding isotope effects in the ternary complex of brain hexokinase demonstrate partial relief of ground-state destabilization. *Journal of the American Chemical Society, 125*, 4672–4673.

Lewis, B. E., & Schramm, V. L. (2006). Enzymatic binding isotope effects and the interaction of glucose with hexokinase. In A. Kohen & H.-H. Limbach (Eds.), *Isotope effects in chemistry and biology* (pp. 1019–1053). Boca Raton: CRC Taylor & Francis.

Linscott, J. A., Kapilashrami, K., Wang, Z., Senevirathne, C., Bothwell, I. R., Blum, G., et al. (2016). Kinetic isotope effects reveal early transition state of protein lysine methyltransferase SET8. *Proceedings of the National Academy of Sciences of the United States of America, 113*, e8369–e8378.

Murkin, A. S., Birck, M. R., Rinaldo-Matthis, A., Shi, W., Taylor, E. A., Almo, S. C., et al. (2007). Neighboring group participation in the transition state of human purine nucleoside phosphorylase. *Biochemistry, 46*, 5038–5049.

Murkin, A. S., Tyler, P. C., & Schramm, V. L. (2008). Transition-state interactions revealed in purine nucleoside phosphorylase by binding isotope effects. *Journal of the American Chemical Society, 130*, 2166–2167.

Northrop, D. B. (1981). The expression of isotope effects on enzyme-catalyzed reactions. *Annual Review of Biochemistry, 50*, 103–131.

Poulin, M. B., Du, Q., & Schramm, V. L. (2015). Chemoenzymatic synthesis of (36)S isotopologues of methionine and *S*-Adenosyl-L-methionine. *The Journal of Organic Chemistry, 80*, 5344–5347.

Poulin, M. B., Schneck, J. L., Matico, R. E., Hou, W., McDevitt, P. J., Holbert, M., et al. (2016). Nucleosome binding alters the substrate bonding environment of histone H3 lysine 36 methyltransferase NSD2. *Journal of the American Chemical Society, 138*, 6699–6702.

Poulin, M. B., Schneck, J. L., Matico, R. E., McDevitt, P. J., Huddleston, M. J., Hou, W., et al. (2016). Transition state for the NSD2-catalyzed methylation of histone H3 lysine 36. *Proceedings of the National Academy of Sciences of the United States of America, 113*, 1197–1201.

Ruszczycky, M. W., & Anderson, V. E. (2006). Interpretation of *V/K* isotope effects for enzymatic reactions exhibiting multiple isotopically sensitive steps. *Journal of Theoretical Biology, 243*, 328–342.

Schramm, V. L. (1976). Comparison of initial velocity and binding data for allosteric adenosine monophosphate nucleosidase. *The Journal of Biological Chemistry, 251*, 3417–3424.

Schramm, V. L. (2007). Binding isotope effects: Boon and bane. *Current Opinion in Chemical Biology, 11*, 529–536.

Silva, R. G., Kipp, D. R., & Schramm, V. L. (2012). Constrained bonding environment in the Michaelis complex of *Trypanosoma cruzi* uridine phosphorylase. *Biochemistry, 51*, 6715–6717.

Stratton, C. F., Namanja-Magliano, H. A., Cameron, S. A., & Schramm, V. L. (2015). Binding isotope effects for *para*-aminobenzoic acid with dihydropteroate synthase from *Staphylococcus aureus* and *Plasmodium falciparum*. *ACS Chemical Biology, 10*, 2182–2186.

Swiderek, K., & Paneth, P. (2013). Binding isotope effects. *Chemical Reviews, 113*, 7851–7879.

Zhang, J., & Klinman, J. P. (2016). Convergent mechanistic features between the structurally diverse *N*- and *O*-methyltransferases: Glycine *N*-methyltransferase and catechol *O*-methyl-transferase. *Journal of the American Chemical Society, 138*, 9158–9165.

Zhang, J., Kulik, H. J., Martinez, T. J., & Klinman, J. P. (2015). Mediation of donor-acceptor distance in an enzymatic methyl transfer reaction. *Proceedings of the National Academy of Sciences of the United States of America, 112*, 7954–7959.

Zhang, Y., & Schramm, V. L. (2011). Ground-state destabilization in orotate phosphoribosyl-transferases by binding isotope effects. *Biochemistry, 50*, 4813–4818.

CHAPTER TWO

Chemical Ligation and Isotope Labeling to Locate Dynamic Effects

Alan F. Scott, Louis Y.P. Luk, Rudolf K. Allemann[1]
School of Chemistry, Cardiff University, Cardiff, United Kingdom
[1]Corresponding author: e-mail address: allemannrk@cardiff.ac.uk

Contents

1. Introduction 24
2. A General Overview of the Approach 26
 2.1 Production of N-Terminal Thioester Peptide EcDHFR (1–28) 28
 2.2 Production of C-Terminal Cysteine Peptide EcDHFR-A29C (29–159) 31
 2.3 Chemical Ligation to Form Full-Length EcDHFR 33
 2.4 Peptide Mass Spectrometry 34
 2.5 Circular Dichroism Spectroscopy 34
 2.6 Steady-State Kinetic Measurements 34
 2.7 Presteady-State Kinetic Measurements 35
3. Hints and Variations on the Strategy for Other Proteins 35
 3.1 Production of Peptides With a C-Terminal α-Thioester 35
 3.2 Production of Cysteine Peptides With an N-Terminal Cysteine 36
 3.3 What to Do When There Is No Native Cysteine at the Ligation Site? 37
 3.4 Multiple Segments 38
4. Conclusion 38
References 39

Abstract

Heavy isotope labeling of enzymes slows protein motions without disturbing the electrostatics and can therefore be used to probe the role of dynamics in enzyme catalysis. To identify the structural elements responsible for dynamic effects, individual segments of an enzyme can be labeled and the resulting effect on the kinetics of the reaction can be measured. Such hybrid isotopomers can be constructed by expressed protein ligation, in which complementary labeled and unlabeled peptide segments are prepared by recombinant gene expression and linked by means of chemical ligation. The construction of such hybrid isotopomers is exemplified here with the paradigmatic enzyme dihydrofolate reductase (DHFR) from *Escherichia coli*.

Methods in Enzymology, Volume 596
ISSN 0076-6879
http://dx.doi.org/10.1016/bs.mie.2017.06.040

1. INTRODUCTION

Kinetic analysis of isotopically labeled enzymes is a powerful approach to investigate the role of protein dynamics in enzyme catalysis. Non-exchangeable isotopes of entire enzymes, typically ^{1}H, ^{12}C, and ^{14}N, can be replaced with heavy counterparts such as ^{2}H, ^{13}C, and ^{15}N (Kipp, Silva, & Schramm, 2011; Silva, Murkin, & Schramm, 2011). Heavy isotopic substitution slows protein motions from femtosecond (fs) bond vibrations to millisecond (ms) structural changes without perturbing the electrostatic properties of the enzyme (Kipp et al., 2011; Silva et al., 2011). Accordingly, the impact of protein motions on enzyme catalysis can be assessed through the measurement of the "enzyme kinetic isotope effect" (enzyme KIE), which is defined as the ratio of the chemical transformation rate constant for the natural abundance (light) enzyme to that of the isotopically enriched, (heavy) enzyme. Several enzymes including dihydrofolate reductase (DHFR) (Francis, Sapienza, Lee, & Kohen, 2016; Luk et al., 2013; Luk, Ruiz-Pernía, et al., 2015; Ruiz-Pernia et al., 2013; Wang, Antoniou, Schwartz, & Schramm, 2016; Wang, Singh, Czekster, Kohen, & Schramm, 2014), HIV protease (Kipp et al., 2011), purine nucleoside phosphorylase (Silva et al., 2011), alanine racemase (Toney, Castro, & Addington, 2013), and pentaerythritol tetranitrate reductase (Longbotham, Hardman, Görlich, Scrutton, & Hay, 2016; Pudney et al., 2013) have been studied using this approach. For a number of these enzymes isotopic substitution altered the reaction rates, and this change was suggested as evidence for the existence of "promoting vibrations," where fast dynamic motions promote the actual chemical reaction (Antoniou, Ge, Schramm, & Schwartz, 2012; Kipp et al., 2011; Pudney et al., 2013; Silva et al., 2011). However, in DHFR from *Escherichia coli* (EcDHFR) the measurable effect on the chemical reaction is small and is not related to quantum tunneling (Luk et al., 2013). While experiments involving complete isotopic substitution can provide insight into the properties of a whole enzyme, they cannot locate the mechanistic origin of dynamic coupling to enzyme catalysis. Hence, using EcDHFR as a prototype the technique of "heavy" enzyme labeling was further developed to label defined segment(s) of an enzyme (Luk, Loveridge, & Allemann, 2015; Luk, Ruiz-Pernía, et al., 2015).

EcDHFR has long been used as a model to investigate the role of protein motions in enzyme catalysis (Kohen, 2015; Luk, Loveridge, et al., 2015). EcDHFR catalyzes the transfer of a hydride from NADPH and a proton from the solvent to the C-6 and N-5 positions of dihydrofolate (DHF), respectively, yielding tetrahydrofolate (THF) and NADP+ as products (Fig. 1A).

EcDHFR possesses three mobile loops (Fig. 1B): the M20 loop (residues 9–24), the FG loop (residues 116–132), and the GH loop (residues 142–149) (Sawaya & Kraut, 1997). Conformational changes coordinated by these mobile segments are shown to be important for the physical steps of EcDHFR catalysis, including substrate binding and product release (Sawaya & Kraut, 1997). Protein isotope labeling has been used to investigate the role of protein dynamics in EcDHFR (Luk et al., 2013). Under steady-state conditions, where the physical step of product release is rate limiting (Fierke, Johnson, & Benkovic, 1987), a slight reduction in the catalytic

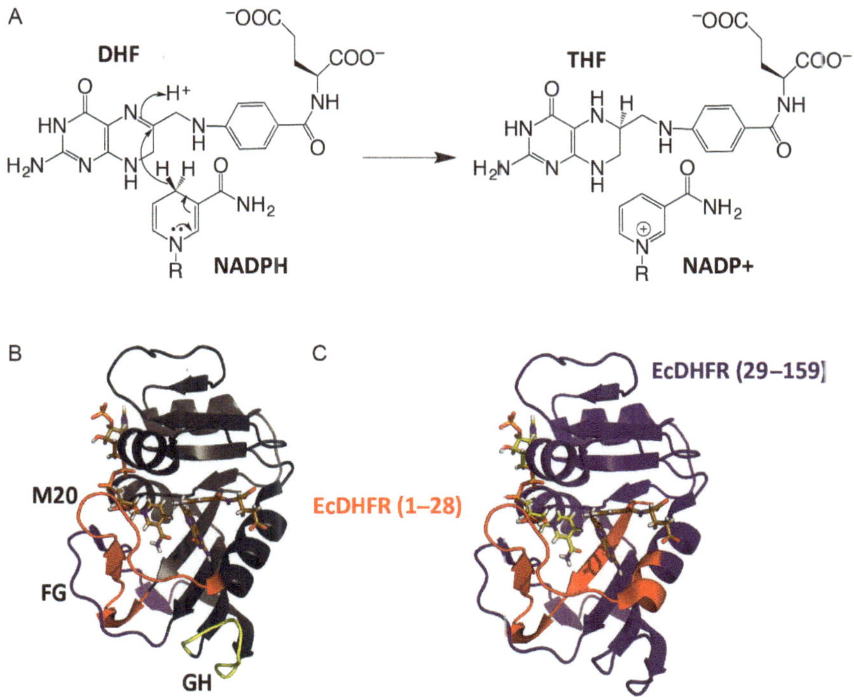

Fig. 1 Reaction catalyzed by DHFR. (A) The *pro-R* hydride from the C4 position of NADPH is transferred to the C6 position of dihydrofolate (DHF) with concomitant protonation of N-5 to yield tetrahydrofolate (THF). (B) EcDHFR crystal structure (PDB ID: 4pdj) with the three mobile loops: M20 loop (residues 9–24), FG loop (residues 116–132), and GH loop (residues 142–149). (C) EcDHFR crystal structure (PDB ID: 4pdj) with residues 1–28 labeled *red* and residues 29–159 labeled *blue*. These two regions are the two parts of the two hybrid isotopomers constructed in this case study.

rate was observed for the heavy enzyme, thus confirming that isotopic substitution slowed protein motions. Presteady-state kinetic measurements revealed a slight reduction in the rate of hydride transfer for the heavy enzyme, implying that protein motions are important for the chemical step too. Interestingly, computational analysis showed that neither the reaction barrier nor the contribution of quantum tunneling to the chemical step was affected by isotopic substitution. This result rules out the possibility that promoting motions play a significant role in driving tunneling or modulating the barrier in EcDHFR catalysis. Instead, computational analysis indicates that dynamic motions affect the frequency of recrossing events on the transition state dividing surface of the free energy landscape. Protein isotope labeling of DHFRs from psychro-, thermo-, and hyperthermophilic organisms also suggested that dynamic coupling is detrimental to catalysis and may indeed have been minimized by natural selection (Luk et al., 2013; Luk, Loveridge, et al., 2015; Luk, Ruiz-Pernía, et al., 2015; Ruiz-Pernia et al., 2016, 2013). Interestingly, combined experimental and computational studies revealed a higher frequency of recrossing events when EcDHFR becomes catalytically compromised by site-directed mutations (Ruiz-Pernia et al., 2016, 2013).

To locate the origin of dynamic effects, an experimental approach based on chemical ligation and segmental protein isotope labeling was developed (Luk, Loveridge, et al., 2015; Luk, Ruiz-Pernía, et al., 2015). Two hybrid isotopomers of EcDHFR were prepared: one with the N-terminal segment (residues 1–28) isotopically labeled, with the remainder of the enzyme retaining isotopes of natural abundance, and the other where only the C-terminal segment (residues 29–159) of EcDHFR was isotopically labeled (Fig. 1C). Kinetic investigations of these enzymes provided strong experimental evidence that ruled out contributions of dynamic coupling from the M20 loop to DHFR chemistry.

2. A GENERAL OVERVIEW OF THE APPROACH

The EcDHFR hybrid is initially produced as two separate linear peptides that are subsequently connected via chemical ligation, a peptide bond-forming reaction that proceeds via *trans*-thioesterification (Fig. 2). One peptide is referred to as the N-terminal thioester peptide and consists of residues 1–28 that include the mobile M20 loop (residues 9–23), and the other is termed the C-terminal cysteine peptide (residues 29–159) which carries an A29C mutation to facilitate the chemical ligation.

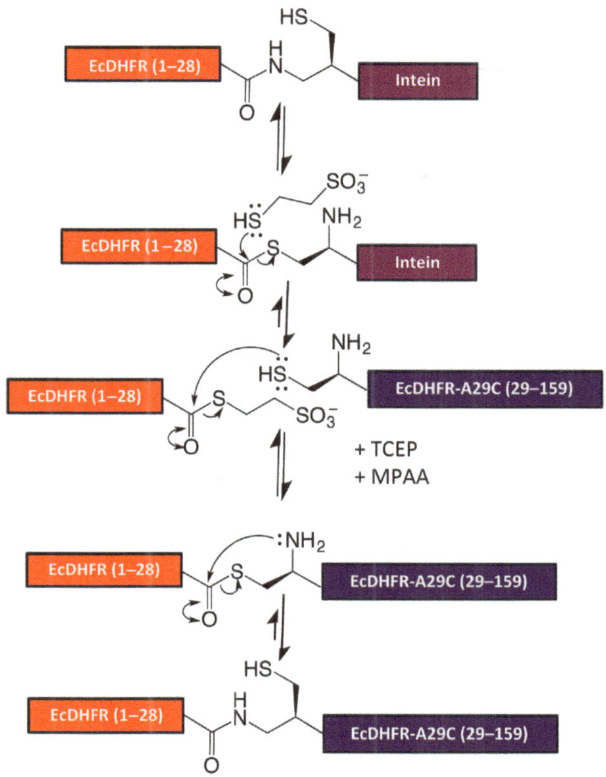

Fig. 2 Schematic of the intein-mediated chemical ligation of two EcDHFR fragments: the thioester peptide EcDHFR (1–28) and the cysteine peptide EcDHFR-A29C (29–159).

In principle, isotopically labeled peptide segments can be prepared by solid-phase synthesis using isotopically labeled amino acids as building blocks. However, considering the relatively low reaction yield and the high cost of the building blocks (e.g., ^{15}N,^{13}C, labeled Fmoc glycine $2540/1 g; Cambridge Isotope Laboratories Inc.), we adapted an alternative recombinant "expressed protein ligation" approach to prepare the isotopically labeled segments. The N-terminal thioester peptide is expressed in fusion with the GyrA intein, a protein segment from *Mycobacterium xenopi* that excises itself after chemical induction to create a C-terminal thioester. The C-terminal cysteine peptide is linked to the thrombin recognition sequence Leu-Val-Pro-Arg-Gly-Ser and a hexa-histidine tag for easy purification. To facilitate the thioesterification, an N-terminal cysteine is generated by replacing alanine 29 with cysteine. Following ligation of the two

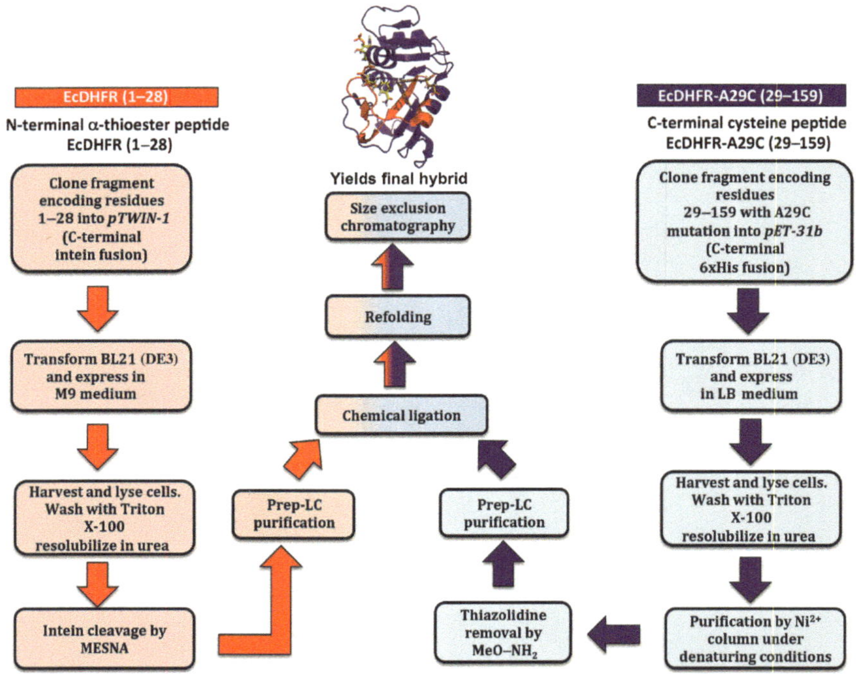

Fig. 3 Reaction overview of the steps required to construct EcDHFR from intein-linked EcDHFR (1–28) and EcDHFR-A29C (29–159).

peptides, in vitro refolding is used to produce a functional enzyme. The general scheme for expressed peptide ligation is shown in Fig. 3.

2.1 Production of N-Terminal Thioester Peptide EcDHFR (1–28)

EcDHFR (1–28) is produced in fusion with gyrA and then converted into the corresponding thioester peptide. The gyrA gene is included in the commercially available *pTWIN-1* expression vector (NEB™), which enables intein fusion at either the N- or C-terminal end of the target peptide/protein (Fig. 4). The cDNA for EcDHFR (1–28) is inserted between the *Nde*I and *Sap*I sites so the resulting peptide is produced in fusion with gyrA at the C-terminal end. Good expression is obtained in BL21(DE3), but the resulting protein is found in inclusion bodies. The chimeric protein is resolubilized in 8.0 *M* urea and then refolded by dilution into thioester exchange buffer. The intein is cleaved with sodium-2-mercaptoethanesulfonate (MESNA) to yield an α-thioester at the C-terminal end and is further purified by reverse-phase HPLC.

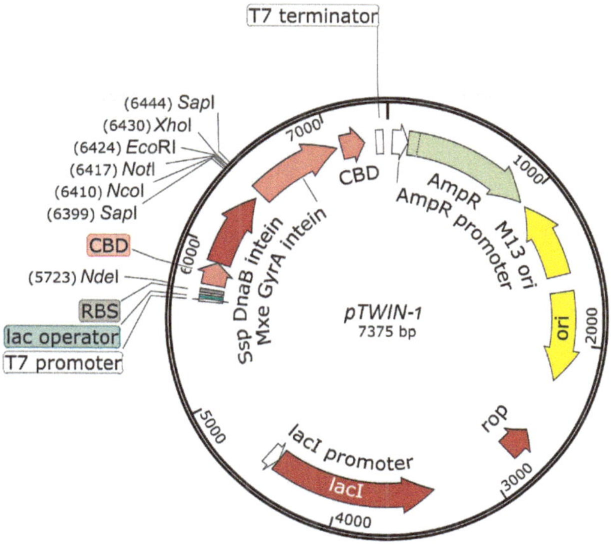

Fig. 4 *pTWIN* vector (NEB™).

Equipment Needed

- Temperature-controlled rotary shaker with clamp for 2 L Erlenmeyer flask
- Autoclave
- High-speed centrifuge and rotors for 1 L and 50 mL samples
- Spectrophotometer
- HPLC with a reverse-phase C18 column (e.g., Phenomenex C-18, 250 × 10 mm, 10 μm, 110 Å)
- Electrospray mass spectrometer (e.g., ion trap)
- SDS-PAGE equipment
- pH meter

Materials Required

- M9 media
- Isotopically labeled feedstocks including:
  - [$^2H_2$, 99.9%] H_2O
  - [^{15}N, 98%] NH_4Cl
  - [$^{13}C_6$, 99%; 1,2,3,4,5,6,6-$^2H_7$, 97%] glucose
- 1.0 *M* IPTG stock filter sterilized (0.5 m*M* working concentration)
- Ampicillin 100 mg/mL stock filter sterilized (100 μg/mL working concentration)
- Resuspension buffer: 25 m*M* HEPES pH 7.5, 150 m*M* NaCl

- 1% Triton X-100
- 8.0 *M* Urea
- Hydrochloric acid
- Thioester exchange buffer: 25 m*M* HEPES pH 8.5, 500 m*M* NaCl, 1 m*M* EDTA, 200 m*M* MESNA
- HPLC Buffer A: 0.1% TFA in H_2O
- HPLC Buffer B: 0.08% TFA in acetonitrile

1. An overnight culture at 37°C with shaking is grown in M9 media + 100 μg/mL ampicillin.
2. The cells are used to inoculate 1.0 L M9 media + 100 μg/mL ampicillin. The culture is grown at 37°C with shaking until OD_{600} reaches 0.6. Gene expression is induced with 0.5 m*M* IPTG and growth continues at 37°C until $OD_{600} = 2.0$.
3. Cells are harvested by centrifugation at 3000 × *g* at 4°C for 15 min and resuspended in "resuspension buffer" (25 m*M* HEPES pH 7.5, 150 m*M* NaCl). Typically 10 mL of resuspension buffer is used per 1 L culture.

Tip: Cells may be stored at −20°C until required.

4. Lysis is achieved by sonication for 5 min with bursts of 5 s on and 10 s off at 40% amplitude with the tube held in an ice–water slurry. The lysate is clarified by centrifugation at 30,000 × *g* for 15 min at 4°C.
5. The pellet is washed with 1% Triton X-100 and resolubilized in 8 *M* urea.
6. The intein is cleaved and the thioester generated by diluting the sample to 2 *M* urea using "thioester exchange buffer" (25 m*M* HEPES pH 8.5, 500 m*M* NaCl, 1 m*M* EDTA, 200 m*M* MESNA) and incubation overnight at 4°C.

Tip: It is important not to substitute MESNA for other reducing agents such as dithiothreitol or β-mercaptoethanol as the thioester is hydrolyzed via an intramolecular rearrangement (Gates, Stephan, Lee, & Kent, 2013).

7. The thioester peptide is precipitated by acidification to pH 3.5 at 4°C and collected by centrifugation at 30,000 × *g* for 15 min.
8. The peptide is further purified by HPLC using a reverse-phase C18 column (Phenomenex C-18, 250 × 10 mm, 10 μm, 110 Å) using a linear gradient 30%–70% B over 120 min (0.33% increase of B per min) at 1 mL/min. Buffer A: 0.1% TFA in H_2O; Buffer B: 0.08% TFA in MeCN.

Mass spectrometry is required to assess the purity and confirm the molecular mass of the peptide. See Section 2.5. Typical yield is 3–7 mg peptide per L of cell culture.

Tip: As the thioester is unstable, it is not recommended for storage.

Production of isotopically labeled peptide is performed in the same manner as above except that [$^{2}H_2$, 99.9%] H_2O, [^{15}N, 98%] NH_4Cl, and [$^{13}C_6$, 99%; 1,2,3,4,5,6,6-$^{2}H_7$, 97%] glucose are used to the replace the ingredients of natural isotope abundance during the preparation of the M9 media. MESNA is not labeled.

2.2 Production of C-Terminal Cysteine Peptide EcDHFR-A29C (29–159)

The C-terminal cysteine peptide consists of residues 29–159 where alanine at position 29 is replaced with cysteine to allow the *trans*-thioesterification reaction during chemical ligation. The cDNA for this peptide is subcloned into a *pET-31b* plasmid (Novagen) in frame with the 5′-end of a thrombin cleavage site and a 6xHis tag-coding region. The fusion gene is then expressed in BL21(DE3) and the resultant peptide purified from the cell lysate by Ni^{2+} affinity chromatography under denaturing conditions. An endogenously generated thiazolidine adduct at the N-terminal cysteine is removed and the peptide is further purified by HPLC.

Equipment Needed

- Temperature-controlled rotary shaker with clamp for 2 L Erlenmeyer flask
- Autoclave
- High-speed centrifuge and rotors for 1 L and 50 mL samples
- Spectrophotometer
- HPLC with a reverse-phase C18 column (e.g., Phenomenex C-18, 250 × 10 mm, 10 μm, 110 Å)
- Electrospray mass spectrometer
- SDS-PAGE equipment
- pH meter

Materials Required

- M9 media
- Isotopically labeled feedstocks including:
  - [$^{2}H_2$, 99.9%] H_2O
  - [^{15}N, 98%] NH_4Cl
  - [$^{13}C_6$, 99%; 1,2,3,4,5,6,6-$^{2}H_7$, 97%] glucose
- 1.0 *M* IPTG stock filter sterilized (0.5 m*M* working concentration)
- Ampicillin 100 mg/mL stock filter sterilized (100 μg/mL working concentration)
- Resuspension buffer: 25 m*M* Tris (pH 8.0), 150 m*M* NaCl
- 1% Triton X-100
- 8.0 *M* Urea

- Elution buffer: 25 m*M* Tris–HCl pH 8.0, 150 m*M* NaCl, 500 m*M* imidazole
- Deprotection buffer: 250 m*M* MeO–NH_2, 50 m*M* TCEP pH 8.0
- HPLC Buffer A: 0.1% TFA in H_2O
- HPLC Buffer B: 0.08% TFA in MeCN

1. An overnight culture of BL21(DE3) transformed with *pET-31b* harboring the peptide gene is grown at 37°C with shaking in M9 media with 100 μg/mL ampicillin.
2. The starter culture is used to inoculate 1 L M9 media + 100 μg/mL ampicillin. The culture is grown at 37°C with shaking until the OD_{600} reaches 0.6. Gene expression is induced with 1.0 m*M* IPTG and growth is allowed to continue at reduced temperature 30°C until $OD_{600}=2.0$.
3. Cells are harvested by centrifugation at 3000 × *g* for 15 min at 4°C and resuspended in resuspension buffer (25 m*M* Tris (pH 8.0) with 150 m*M* NaCl) at a density of 10 mL per 1 L culture volume.

Tip: Cells may be stored at −20°C until required.

4. Lysis is achieved by sonication for 5 min with bursts of 5 s on and 10 s off at 40% amplitude with tubes placed in an ice–water slurry. The lysate is clarified by centrifugation at 30,000 × *g* for 15 min at 4°C.
5. The pellet is washed with 10 mL 1% Triton X-100 and resolubilized in 10 mL "resuspension buffer" containing 8.0 *M* urea.
6. The resolubilized protein is clarified by centrifugation at 30,000 × *g* for 15 min and loaded onto a 5 mL Ni^{2+} sepharose column preequilibrated in 5 c.v of "resuspension buffer" (25 m*M* Tris–HCl pH 8.0, 150 m*M* NaCl). The column is washed with 10 c.v. of "resuspension buffer" to remove unbound proteins. The cysteine peptide is eluted in a minimal volume of "elution buffer" (25 m*M* Tris–HCl pH 8.0, 150 m*M* NaCl, 500 m*M* imidazole).
7. As N-terminal cysteine can form thiazolidine adducts with pyruvate, as evidenced by mass spectrometry, the eluted peptide is incubated with stirring overnight at 37°C in "deprotection buffer" (250 m*M* MeO–NH_2, 50 m*M* TCEP pH 8.0), in a modification of the protocol previously reported (Ottesen, Bar-Dagan, Giovani, & Muir, 2008). Methoxylamine deprotects the N-terminal cysteine and the TCEP is a reducing agent to protect the peptide.
8. The peptide is purified by reverse-phase HPLC on a C18 column as already described for the N-terminal EcDHFR fragment.

Mass spectrometry is required to assess the purity and confirm the molecular mass of the peptide. See Section 2.5. Typical yield is 12–18 mg peptide per L of cell culture.

Production of isotopically labeled peptide is performed in the same manner as above except that [$^2H_2$, 99.9%] H_2O, [^{15}N, 98%] NH_4Cl, and [$^{13}C_6$, 99%; 1,2,3,4,5,6,6-$^2H_7$, 97%] glucose are used to replace the corresponding ingredients with isotopes of natural abundance during the preparation of M9 media.

2.3 Chemical Ligation to Form Full-Length EcDHFR

Equipment Required

- pH meter
- Dialysis tubing 10 kDa cutoff
- High-speed centrifuge, rotor, and tubes for 50 mL sample
- ÄKTA™ FPLC
- Superdex S-75 column (G.E. Healthcare)

Materials Required

- Ligation buffer: 6.0 *M* guanidine HCl, 0.2 *M* Na_2PO_4, 10 m*M* tris (2-carboxyethyl)phosphine (TCEP) hydrochloride, 10 m*M* 4-mercaptophenylacetic acid (MPAA)
- Refolding buffer: 50 m*M* K_iPO_4 pH 7.0, 100 m*M* NaCl, 5 m*M* DTT
- SE buffer: 50 m*M* K_iPO_4 pH 7.0, 5 m*M* DTT
- Nitrogen cylinder

1. N-terminal thioester peptide (1.31 μmol) and C-terminal cysteine peptide (1.60 μmol) are mixed in ligation buffer (6.0 *M* guanidine HCl, 0.2 *M* Na_2PO_4, 10 m*M* TCEP hydrochloride, 10 m*M* MPAA). TCEP is required as a reducing agent to prevent unwanted cysteine oxidation, and MPAA is included as an accelerating reagent for chemical ligation.
2. The pH is adjusted to 7.0 and the reaction is incubated at room temperature for 24 h. Typical yield of ligated peptide is 10% of starting material. The yield can be improved by optimizing incubation time, temperature, pH, and concentration of MPAA.
3. The ligated peptide is refolded using overnight dialysis against "refolding buffer" (50 m*M* K_iPO_4 pH 7.0, 100 m*M* NaCl, 5 m*M* DTT) purged under nitrogen. Precipitated peptides are removed by centrifugation at 30,000 × *g* for 30 min at 4°C.

Tip: Refolding can also be achieved by rapid dilution into the refolding buffer.

4. The refolded protein is purified by size exclusion chromatography using a Superdex-75 (G.E. Healthcare) equilibrated in "SE buffer" (50 m*M* K_iPO_4 pH 7.0, 5 m*M* DTT) at a flow rate of 1 mL/min at 18°C.

2.4 Peptide Mass Spectrometry

Peptide mass spectrometry was achieved using an ACE® C18 (100 × 2.1 mm, 10 μm, 300 Å) column on an Ultimate 3000 UPLC with a linear gradient of 5%–95% B over 30 min (A × 0.1% formic acid in H_2O, B × 0.1% formic acid in MeCN). Mass was recorded on a Bruker amaZonSL MS ion trap. Typically, 10 μL of a 10 μ*M* solution is injected onto the column.

2.5 Circular Dichroism Spectroscopy

Equipment Required

- Vacuum degassing system for buffers
- Applied Photophysics Chirascan spectrophotometer
- Quartz cuvettes 10 mm

Materials Required

- 10 m*M* K_iPO_4 pH 7.0 (deoxygenated)

It is important to confirm whether the hybrid isotopomer is correctly folded and structurally identical to the native enzyme. Hence, the circular dichroism (CD) spectra of both the wild type and the hybrid must be compared.

1. Protein (16 μ*M*) was diluted into deoxygenated buffer containing 10 m*M* K_iPO_4 pH 7.0.
2. Spectra were recorded on an Applied Photophysics Chirascan spectrometer in 10 mm quartz cuvettes at 20°C under N_2. Spectra were measured between 109 and 400 nm with a scan speed of 50 nm/min, 0.5 nm data pitch, 1 nm bandwidth, and 0.5 s response time.

2.6 Steady-State Kinetic Measurements

Equipment Required

- Vacuum degassing system for buffers
- Shimadzu UV-2401 PC spectrophotometer
- Quartz cuvettes 1 mL

Materials Required

- Assay buffer: 50 m*M* K_iPO_4 pH 7.0, 100 m*M* NaCl, 5 m*M* DTT
- NADPH (Fisher)
- Dihydrofolate (H_2F) (Sigma)

The kinetic rates are determined by monitoring the consumption of NADPH during the reaction from the decrease in NADPH absorbance at 340 nm. Data are recorded on a Shimadzu UV-2401 PC spectrophotometer and the absorbance units are converted into concentration using the

Beer–Lambert law with the extinction coefficient ε_{340} (NADPH + H_2F) of 11,800 M^{-1} cm^{-1}. Assay buffer contains 50 m*M* K_iPO_4 pH 7.0, 100 m*M* NaCl, 5 m*M* DTT.

Tip: Chemically ligated enzymes are liable to oxidation; therefore, all buffers should be thoroughly degassed and 5 m*M* DTT or other reducing agent added before use.

1. Enzyme (0.3–60 n*M*) is incubated with 250 μ*M* NADPH on ice for 2 min.
2. Enzyme, NADPH, and 100 μ*M* H_2F are rapidly mixed in a 1 mL quartz cuvette and the absorbance at 340 nm monitored for 1 min.
3. For measurement of Michaelis constants, concentrations of NADPH were varied between 3 and 100 μ*M* and H_2F between 0.5 and 100 μ*M*, while keeping the concentration of the other reactant fixed at saturating concentrations.

2.7 Presteady-State Kinetic Measurements

Equipment Required

- Vacuum degassing system for buffers
- Shimadzu UV-2401 PC spectrophotometer
- Quartz cuvettes 1 mL

Materials Required

- Assay buffer: 50 m*M* K_iPO_4 pH 7.0, 100 m*M* NaCl, 5 m*M* DTT
- NADPH (Fisher)
- Dihydrofolate (H_2F) (Sigma)

1. Enzyme (7.5 μ*M*) and NADPH (1.5 μ*M*) are incubated for 5 min in 100 m*M* K_iPO_4 pH 7.0, 100 m*M* NaCl, 5 m*M* DTT.
2. The enzyme and NADPH solutions are rapidly mixed in a Hi-Tech Scientific stopped-flow spectrophotometer, and the rate of depletion of NADPH monitored by fluorescence decreases at 340 nm. Measurements should be repeated 12 times. Rate constants are extracted by fitting the kinetic data to the equation for a double exponential decay.

3. HINTS AND VARIATIONS ON THE STRATEGY FOR OTHER PROTEINS

3.1 Production of Peptides With a C-Terminal α-Thioester

The most direct method for generating a peptide with an α-thioester is by MESNA-induced thiolysis of a peptide–intein fusion. However, it is important to note that the amino acid directly adjacent to the intein fusion can impact on the efficiency of the thiolysis reaction. In the case of the *gyrA* intein, proline and serine significantly hinder the reaction, while aspartic acid causes premature cleavage in vivo (Southworth, Amaya, Evans, Xu, & Perler, 1999). A full list of amino acids and their effects on *gyrA* cleavage can be found in an article by Southworth et al. (1999). Cleavage efficiency can be optimized by variation of temperature, pH, MESNA concentration, reaction time, and peptide concentration. Buffer additives such as glycerol, kosmotropes, detergents, and limited concentrations of chaotropes may be required to keep the protein in solution. The cleavage reaction is compatible with up to 4 *M* urea (Sydor, Mariano, Sideris, & Nock, 2002).

If recombinant production of the peptide is hampered by poor gene expression or protein insolubility, these limitations may be overcome by protein refolding as described above. If difficulties are encountered in either the thiolysis reaction or the expression step, synthetic peptides can be produced using solid-phase peptide synthesis. The main disadvantages of synthetic peptides are the expense of isotope labeling and the size limitation for efficient synthesis (~50 amino acids). Although Fmoc methodology is usually preferred for peptide synthesis, as it bypasses the use of hydrogen fluoride during the peptide cleavage step, the basic conditions used for deprotection of the amino group cause thioester cleavage. There are strategies to overcome this limitation, but they have not been thoroughly explored with respect to the optimization of yield and sequence applicability (Blanco-Canosa & Dawson, 2008; Ingenito, Bianchi, Fattori, & Pessi, 1999; Noisier & Albericio, 2015; Raz & Rademann, 2011; Zheng, Tang, Qi, Wang, & Liu, 2013). Therefore, synthetic strategies that rely on the use of Boc chemistry are usually preferred for the synthesis of thioester peptides (Noisier & Albericio, 2015).

3.2 Production of Cysteine Peptides With an N-Terminal Cysteine

In the current example, the cysteine peptide was produced in *E. coli*, relying on in vivo cleavage of the N-terminal methionine by methionine aminopeptidase (MAP). This enzyme cleaves the N-terminal methionine in peptides with small side chains at the second position, including cysteine, but the results are sequence dependent (Ben-Bassat et al., 1987; Frottin et al., 2006; Hirel, Schmitter, Dessen, Fayat, & Blanquet, 1989; Xiao, Zhang, Nacev, Liu, & Pei, 2010). Inefficient cleavage of methionine in certain peptides and potential alkylation of the cysteine by endogenous pyruvate are potential issues for the method described. Optimization of culture conditions can improve MAP activity (Gentle, De Souza, & Baca, 2004). Low yields of the peptide may also result if the peptide is small (<10 kDa) due to inefficient expression of small genes. Both issues can be overcome by the addition of a leader sequence, followed by a protease site that allows for cleavage of the leader sequence to reveal an N-terminal cysteine. Leader sequences that encode glutathione-*S*-transferase (GST) or maltose binding protein (di Guana, Lib, Riggsa, & Inouyeb, 1988; Smith & Johnson, 1988) will also enhance gene expression levels and protein solubility. Suitable cleavage sites include those recognized by thrombin (Liu, Xu, Dutta, & Cowburn, 2008), as well as the proteases from foot and mouth disease virus (Busch et al., 2008), and tobacco etch virus (Tolbert, Franke, & Wong, 2005). Alternatively, the peptide can be produced in fusion with an N-terminal intein using a vector such as *pTWIN-1* (NEB™) (Fig. 4). Cleavage of the intein with dithiothreitol or β-mercaptoethanol reveals an N-terminal cysteine (Southworth et al., 1999). The cysteine peptide can of course be made chemically but, as mentioned above, the high costs of isotopically labeled protected amino acid building blocks and the often-low yields limit the usefulness of this approach.

3.3 What to Do When There Is No Native Cysteine at the Ligation Site?

Chemical ligation requires an N-terminal cysteine on C-terminal peptide. If the required peptide lacks a native cysteine, mutagenesis can be used to replace the wild-type residue with a cysteine but such a replacement can affect the physical and catalytic properties of the enzyme. Substitutions of solvent-exposed serines, methionines, or alanines are usually well tolerated. If necessary, the cysteine can be converted to alanine after ligation by desulfurization (Yan & Dawson, 2001). However, this chemistry will affect all cysteine residues unless the peptide has been synthetically made to give the cysteines, a protecting group that can be removed after desulfurization (Deng, Zhang, Wang, Schneewind, & Kent, 2014). For synthetic peptides there are also amino acids with thiolated side chains (e.g., β-mercapto-phenylalanine) that can be inserted into the sequence as substitutes for cysteine and desulfurized after native chemical ligation to give the corresponding natural amino acid (Crich & Banerjee, 2007). In a similar fashion cysteine auxiliaries such as a 1-phenyl-2-mercaptoethyl group may be tethered to the N-terminus of an amino acid to facilitate thioester-mediated ligation. Afterward, they are cleaved by acid hydrolysis or photolysis (Botti, Carrasco, & Kent, 2001; Nadler, Nadler, Hansen, & Diederichsen, 2015; Offer & Dawson, 2000).

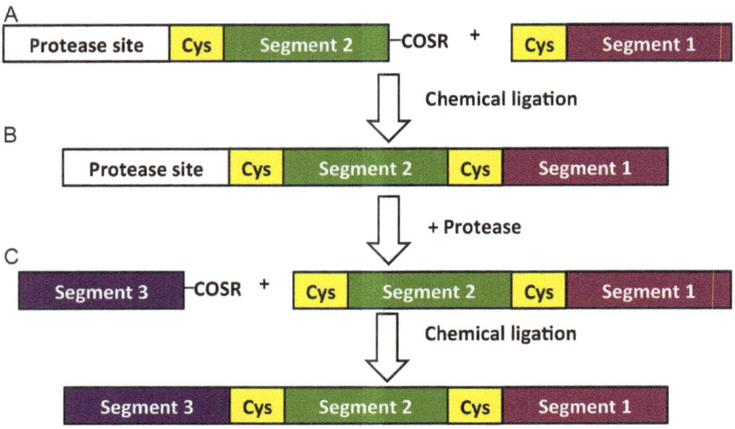

Fig. 5 Chemical ligation of multiple segments. (A) The internal segment has a C-terminal thioester that ligates to a segment with an N-terminal cysteine. (B) Addition of a protease deprotects the N-terminal cysteine on the internal segment. (C) The third segment with a C-terminal thioester is ligated to the newly exposed N-terminal cysteine on the internal segment.

3.4 Multiple Segments

Chemical ligation can also be used to join more than two peptide segments. This operation is necessary when an internal protein segment or multiple discontinuous segments need to be labeled. In such cases, the peptides for the internal segments require both a C-terminal thioester group and an N-terminal cysteine residue and so need to be protected to avoid self-ligation. This can be achieved by a cleavable leader sequence that is removed by a protease after the first ligation (Fig. 5) (Cotton, Ayers, Xu, & Muir, 1999).

4. CONCLUSION

A method has been described for the construction of EcDHFR hybrid isotopomers where either the N-terminal residues, incorporating the M20 loop, are labeled with heavy isotopes or the remaining C-terminal residues are labeled. In recent work (Luk, Loveridge, et al., 2015; Luk, Ruiz-Pernía, et al., 2015), these isotopomers were used to isolate the origin of the dynamic effects observed previously when the entire enzyme was labeled (Luk et al., 2013). Under steady-state conditions, where product release is rate limiting (Fierke et al., 1987), a small, temperature-dependent heavy enzyme KIE was reported (Luk et al., 2013). This KIE was reproduced in the hybrid isotopomer in which the N-terminal 1–28 segment was exclusively labeled, but was not observed when only the C-terminal region (residues 29–159) was labeled (Luk, Loveridge, et al., 2015; Luk, Ruiz-Pernía, et al., 2015). These observations confirm the involvement of the highly mobile M20 loop (residues 9–24) in product release. Presteady-state kinetics were used to study the effect of isotopic substitution on the chemical step. When the whole enzyme was labeled, a small but temperature-dependent heavy enzyme KIE was observed that increased from 0.92 to 1.07 over 5–35°C (Luk et al., 2013). Interestingly, the hybrid isotopomer with only the N-terminal 28 residues labeled did not show any measurable heavy enzyme KIE under the same conditions, despite containing the highly mobile M20 loop (Luk, Loveridge, et al., 2015; Luk, Ruiz-Pernía, et al., 2015). Selective labeling of the C-terminal segment (residues 29–159) restored the temperature-dependent heavy enzyme KIE to the level observed when the entire enzyme was labeled. This result showed that the dynamic effects on hydride transfer are most likely to be a consequence of global motions rather than localized motions in the mobile M20 loop. This is in stark contrast to the dynamic effects on the physical step of product release, which were shown to result from motions in the M20 loop. Chemical ligation has the potential to locate other regions of EcDHFR that affect the catalytic efficiency of the enzyme. In conclusion, segmental isotope labeling of enzymes is a powerful approach for studying protein dynamics, particularly for locating the origin of coupled motions. This method is highly adaptable and widely applicable to enzymes other than EcDHFR.

REFERENCES

Antoniou, D., Ge, X., Schramm, V. L., & Schwartz, S. D. (2012). Mass modulation of protein dynamics associated with barrier crossing in purine nucleoside phosphorylase. *Journal of Physical Chemistry Letters*, *3*, 3538–3544.

Ben-Bassat, A., Bauer, K., Chang, S. Y., Myambo, K., Boosman, A., & Chang, S. (1987). Processing of the initiation methionine from proteins: Properties of the *Escherichia coli* methionine aminopeptidase and its gene structure. *Journal of Bacteriology*, *169*, 751–757.

Blanco-Canosa, J. B., & Dawson, P. E. (2008). An efficient Fmoc-SPPS approach for the generation of thioester peptide precursors for use in native chemical ligation. *Angewandte Chemie*, *120*, 6957–6961.

Botti, P., Carrasco, M. R., & Kent, S. B. H. (2001). Native chemical ligation using removable Nα-(1-phenyl-2-mercaptoethyl) auxiliaries. *Tetrahedron Letters*, *42*, 1831–1833.

Busch, G. K., Tate, E. W., Gaffney, P. R. J., Rosivatz, E., Woscholski, R., & Leatherbarrow, R. J. (2008). Specific N-terminal protein labelling: Use of FMDV 3C pro protease and native chemical ligation. *Chemical Communications*, *266*, 3369–3371.

Cotton, G. J., Ayers, B., Xu, R., & Muir, T. W. (1999). Insertion of a synthetic peptide into a recombinant protein framework: A protein biosensor. *Journal of the American Chemical Society*, *121*, 1100–1101.

Crich, D., & Banerjee, A. (2007). Native chemical ligation at phenylalanine. *Journal of the American Chemical Society*, *129*, 10064–10065.
Deng, F.-K., Zhang, L., Wang, Y.-T., Schneewind, O., & Kent, S. B. H. (2014). Total chemical synthesis of the enzyme sortase AΔN59 with full catalytic activity. *Angewandte Chemie International Edition*, *53*, 4662–4666.
di Guana, C., Lib, P., Riggsa, P. D., & Inouyeb, H. (1988). Vectors that facilitate the expression and purification of foreign peptides in *Escherichia coli* by fusion to maltose-binding protein. *Gene*, *67*, 21–30.
Fierke, C. A., Johnson, K. A., & Benkovic, S. J. (1987). Construction and evaluation of the kinetic scheme associated with dihydrofolate reductase from *Escherichia coli*. *Biochemistry*, *26*, 4085–4092.
Francis, K., Sapienza, P. J., Lee, A. L., & Kohen, A. (2016). The effect of protein mass modulation on human dihydrofolate reductase. *Biochemistry*, *55*, 1100–1106.
Frottin, F., Martinez, A., Peynot, P., Mitra, S., Holz, R. C., Giglione, C., et al. (2006). The proteomics of N-terminal methionine cleavage. *Molecular & Cellular Proteomics*, *5*, 2336–2349.
Gates, Z. P., Stephan, J. R., Lee, D. J., & Kent, S. B. H. (2013). Rapid formal hydrolysis of peptide-[small alpha]thioesters. *Chemical Communications*, *49*, 786–788.
Gentle, I. E., De Souza, D. P., & Baca, M. (2004). Direct production of proteins with N-terminal cysteine for site-specific conjugation. *Bioconjugate Chemistry*, *15*, 658–663.
Hirel, P. H., Schmitter, M. J., Dessen, P., Fayat, G., & Blanquet, S. (1989). Extent of N-terminal methionine excision from *Escherichia coli* proteins is governed by the side-chain length of the penultimate amino acid. *Proceedings of the National Academy of Sciences of the United States of America*, *86*, 8247–8251.
Ingenito, R., Bianchi, E., Fattori, D., & Pessi, A. (1999). Solid phase synthesis of peptide C-terminal thioesters by Fmoc/t-Bu chemistry. *Journal of the American Chemical Society*, *121*, 11369–11374.
Kipp, D. R., Silva, R. G., & Schramm, V. L. (2011). Mass-dependent bond vibrational dynamics influence catalysis by HIV-1 protease. *Journal of the American Chemical Society*, *133*, 19358–19361.
Kohen, A. (2015). Dihydrofolate reductase as a model for studies of enzyme dynamics and catalysis. *F1000Research*, *4*. F1000 Faculty Rev-1464.
Liu, D., Xu, R., Dutta, K., & Cowburn, D. (2008). N-terminal cysteinyl proteins can be prepared using thrombin cleavage. *FEBS Letters*, *582*, 1163–1167.
Longbotham, J. E., Hardman, S. J. O., Görlich, S., Scrutton, N. S., & Hay, S. (2016). Untangling heavy protein and cofactor isotope effects on enzyme-catalyzed hydride transfer. *Journal of the American Chemical Society*, *138*, 13693–13699.
Luk, L. Y. P., Javier Ruiz-Pernía, J., Dawson, W. M., Roca, M., Loveridge, E. J., Glowacki, D. R., et al. (2013). Unraveling the role of protein dynamics in dihydrofolate reductase catalysis. *Proceedings of the National Academy of Sciences of the United States of America*, *110*, 16344–16349.
Luk, L. Y. P., Loveridge, E. J., & Allemann, R. K. (2015). Protein motions and dynamic effects in enzyme catalysis. *Physical Chemistry Chemical Physics*, *17*, 30817–30827.
Luk, L. Y. P., Ruiz-Pernía, J. J., Adesina, A. S., Loveridge, E. J., Tuñõn, I., Moliner, V., et al. (2015). Chemical ligation and isotope labeling to locate dynamic effects during catalysis by dihydrofolate reductase. *Angewandte Chemie International Edition*, *54*, 9016–9020.
Nadler, C., Nadler, A., Hansen, C., & Diederichsen, U. (2015). A photocleavable auxiliary for extended native chemical ligation. *European Journal of Organic Chemistry*, *2015*, 3095–3102.
Noisier, A. F. M., & Albericio, F. (2015). Advance in ligation techniques for peptide and protein synthesis. In M. Ryadnov & E. Farkas (Eds.), *A specialist periodical report: Vol. 39. Amino acids, peptides and proteins* (pp. 1–20). Cambridge: The Royal Society of Chemistry.

Offer, J., & Dawson, P. E. (2000). Nα-2-Mercaptobenzylamine-assisted chemical ligation. *Organic Letters*, *2*, 23–26.

Ottesen, J. J., Bar-Dagan, M., Giovani, B., & Muir, T. W. (2008). An amalgamation of solid phase peptide synthesis and ribosomal peptide synthesis. *Peptide Science*, *90*, 406–414.

Pudney, C. R., Guerriero, A., Baxter, N. J., Johannissen, L. O., Waltho, J. P., Hay, S., et al. (2013). Fast protein motions are coupled to enzyme H-transfer reactions. *Journal of the American Chemical Society*, *135*, 2512–2517.

Raz, R., & Rademann, J. (2011). Fmoc-based synthesis of peptide thioesters for native chemical ligation employing a tert-butyl thiol linker. *Organic Letters*, *13*, 1606–1609.

Ruiz-Pernia, J. J., Behiry, E., Luk, L. Y. P., Loveridge, E. J., Tunon, I., Moliner, V., et al. (2016). Minimization of dynamic effects in the evolution of dihydrofolate reductase. *Chemical Science*, *7*, 3248–3255.

Ruiz-Pernia, J. J., Luk, L. Y. P., García-Meseguer, R., Martí, S., Loveridge, E. J., Tuñón, I., et al. (2013). Increased dynamic effects in a catalytically compromised variant of *Escherichia coli* dihydrofolate reductase. *Journal of the American Chemical Society*, *135*, 18689–18696.

Sawaya, M. R., & Kraut, J. (1997). Loop and subdomain movements in the mechanism of *Escherichia coli* dihydrofolate reductase: Crystallographic evidence. *Biochemistry*, *36*, 586–603.

Silva, R. G., Murkin, A. S., & Schramm, V. L. (2011). Femtosecond dynamics coupled to chemical barrier crossing in a Born-Oppenheimer enzyme. *Proceedings of the National Academy of Sciences of the United States of America*, *108*, 18661–18665.

Smith, D. B., & Johnson, K. S. (1988). Single-step purification of polypeptides expressed in *Escherichia coli* as fusions with glutathione S-transferase. *Gene*, *67*, 31–40.

Southworth, M. W., Amaya, K., Evans, T. C., Xu, M. Q., & Perler, F. B. (1999). Purification of proteins fused to either the amino or carboxy terminus of the Mycobacterium xenopi gyrase A intein. *Biotechniques*, *27*, 110–114. 116, 118–20.

Sydor, J. R., Mariano, M., Sideris, S., & Nock, S. (2002). Establishment of intein-mediated protein ligation under denaturing conditions: C-terminal labeling of a single-chain antibody for biochip screening. *Bioconjugate Chemistry*, *13*, 707–712.

Tolbert, T. J., Franke, D., & Wong, C.-H. (2005). A new strategy for glycoprotein synthesis: Ligation of synthetic glycopeptides with truncated proteins expressed in *E. coli* as TEV protease cleavable fusion protein. *Bioorganic & Medicinal Chemistry*, *13*, 909–915.

Toney, M. D., Castro, J. N., & Addington, T. A. (2013). Heavy-enzyme kinetic isotope effects on proton transfer in alanine racemase. *Journal of the American Chemical Society*, *135*, 2509–2511.

Wang, Z., Antoniou, D., Schwartz, S. D., & Schramm, V. L. (2016). Hydride transfer in DHFR by transition path sampling, kinetic isotope effects, and heavy enzyme studies. *Biochemistry*, *55*, 157–166.

Wang, Z., Singh, P., Czekster, C. M., Kohen, A., & Schramm, V. L. (2014). Protein mass-modulated effects in the catalytic mechanism of dihydrofolate reductase: Beyond promoting vibrations. *Journal of the American Chemical Society*, *136*, 8333–8341.

Xiao, Q., Zhang, F., Nacev, B. A., Liu, J. O., & Pei, D. (2010). Protein N-terminal processing: Substrate specificity of *Escherichia coli* and human methionine aminopeptidases. *Biochemistry*, *49*, 5588–5599.

Yan, L. Z., & Dawson, P. E. (2001). Synthesis of peptides and proteins without cysteine residues by native chemical ligation combined with desulfurization. *Journal of the American Chemical Society*, *123*, 526–533.

Zheng, J.-S., Tang, S., Qi, Y.-K., Wang, Z.-P., & Liu, L. (2013). Chemical synthesis of proteins using peptide hydrazides as thioester surrogates. *Nature Protocols*, *8*, 2483–2495.

CHAPTER THREE

Measurement of Enzyme Isotope Effects

Svetlana A. Kholodar, Ananda K. Ghosh, Amnon Kohen[1]
University of Iowa, Iowa City, IA, United States
[1]Corresponding author: e-mail address: amnon-kohen@uiowa.edu

Contents

1. Introduction 44
2. Methods 45
   2.1 Isotopically Labeled Enzymes and Their Physiochemical Properties 45
   2.2 Structural Characterization of Isotopically Labeled Enzymes 53
   2.3 Effect of Enzyme Isotopic Labeling on Catalysis 55
3. Summary 76
Acknowledgments 77
References 77

Abstract

Enzyme isotope effects, or the kinetic effects of "heavy" enzymes, refer to the effect of isotopically labeled protein residues on the enzyme's activity or physical properties. These effects are increasingly employed in the examination of the possible contributions of protein dynamics to enzyme catalysis. One hypothesis assumed that isotopic substitution of all ^{12}C, ^{14}N, and nonexchangeable ^{1}H by ^{13}C, ^{15}N, and ^{2}H, would slow down protein picosecond to femtosecond dynamics without any effect on the system's electrostatics following the Born–Oppenheimer approximation. It was suggested that reduced reaction rates reported for several "heavy" enzymes accords with that hypothesis. However, numerous deviations from the predictions of that hypothesis were also reported. Current studies also attempt to test the role of individual residues by site-specific labeling or by labeling a pattern of residues on activity. It appears that in several systems the protein's fast dynamics are indeed reduced in "heavy" enzymes in a way that reduces the probability of barrier crossing of its chemical step. Other observations, however, indicated that slower protein dynamics are electrostatically altered in isotopically labeled enzymes. Interestingly, these effects appear to be system dependent, thus it might be premature to suggest a general role of "heavy" enzymes' effect on catalysis.

Methods in Enzymology, Volume 596
ISSN 0076-6879
http://dx.doi.org/10.1016/bs.mie.2017.06.033

1. INTRODUCTION

Effects of isotopic substitution of enzyme's substrates or solvents are a powerful and well-established tool for elucidation of various aspects of enzyme mechanisms (Cleland, 2005; Cook, 1991; Quinn, 2005). This approach provides information on the chemical and kinetic mechanism of the enzymatic reactions under study, as well as geometric and electronic characteristics of various transition states along the catalytic cascade. Effects of isotopic substitutions in the catalyst itself (denoted "heavy enzymes"), however, are still an emerging topic in modern enzymology. Enzyme isotope effects were postulated to reflect a link between enzyme's femtosecond-picosecond dynamics and the catalyzed chemical transformation, but may also affect other steps along the catalytic cycle.

The role and implications of protein dynamics in catalysis is one of the old, yet hot topics in contemporary enzymology, as deliberated in a several recent reviews (Callender & Dyer, 2015; Klinman, 2015; Kohen, 2015; Schramm, 2015). The hypothesis that nonstatistical dynamics of the enzyme at the femtosecond-picosecond timescale are coupled to transition-state formation originated from the theoretical analyses which predicted that motions of protein residues facilitate passage over the transition state barrier in a several enzymatic reactions (e.g., Caratzoulas, Mincer, & Schwartz, 2002; Mincer & Schwartz, 2003; Saen-Oon, Quaytman-Machleder, Schramm, & Schwartz, 2008; Schwartz & Schramm, 2009). Additionally, ample studies that assumed transient statistical (thermally equilibrated) dynamics on the same fast timescale also predicted coupling to the chemical transformation (e.g., Agarwal, Billeter, Rajagopalan, Benkovic, & Hammes-Schiffer, 2002; Dybala-Defratyka, Paneth, Banerjee, & Truhlar, 2007; Renger & Marcus, 2002; Roca, Moliner, Tuñón, & Hynes, 2006; Warshel, 1984; Warshel & Bora, 2016). In fact, decades before these computational studies, a number of theoretical models involving dynamical correlation between the enzyme and its chemical reaction emerged to rationalize the enormous catalytic power of the enzymes (Careri & Gratton, 1986; Ikegami, 1986; Lumry & Gregory, 1986; Somogyi & Damjanovich, 1986). Since then, numerous experimental studies attempted to shed light on the involvement of fast protein dynamics in enzymatic catalysis (e.g., Hay, Johannissen, Hothi, Sutcliffe, & Scrutton, 2012; Knapp, Rickert, & Klinman, 2002; Kohen, Cannio, Bartolucci, & Klinman, 1999; Luk et al., 2013; Pudney, Johannissen, Sutcliffe, Hay, & Scrutton, 2010; Singh,

Francis, & Kohen, 2015). Those studies did not carefully address the differences between catalyzed and uncatalyzed reactions, thus the term "catalysis" here actually refer solely to enzyme-catalyzed reactions.

A recent approach used to test the effect of dynamics on chemical catalysis was pioneered by Schramm in 2011 (Silva, Murkin, & Schramm, 2011), who introduced a concept of Born–Oppenheimer (BO) or "heavy" enzyme. This approach assumes that replacement of all ^{12}C, ^{14}N, and non-exchangeable protons by ^{13}C, ^{15}N, and ^{2}H reduces femtosecond bond vibration frequencies, while leaving the electrostatic properties of the system unaltered. Using purine nucleoside phosphorylase (PNP) as a model system, Schramm and coworkers demonstrated that, in agreement with computational predictions, "heavy" PNP exhibited decreased probability of the transition-state barrier crossing manifested in the diminished rate of the chemical step, while maintaining the rest of its characteristics (e.g., transition-state structure and steady-state parameters related to substrate binding and product release) equivalent to its "light" version. Studies by others expanded the application of this approach to other enzymatic systems (Table 1) as recently reviewed by Świderek, Ruiz-Pernia, Moliner, and Tunon (2014). In some of these cases, the increase in the protein mass was accompanied by slower on-enzyme chemistry, which accords with "heavy" enzyme distortion of coupling of the vibrational modes of protein to the transition state in the "light," natural, enzyme. However, many of the examined enzymatic systems also exhibited significant isotope effects on kinetic parameters other than the rate of the chemical step, therefore casting doubt on validity of the BO approximation in isotopically substituted enzymes, and keeping the interpretation of findings with "heavy" enzymes open to ongoing examination. In this chapter, a system-dependent approach to analysis of enzyme isotope effects is summarized, and the potential applications of "heavy" enzymes as a complimentary tool to study the contribution of protein dynamics to its catalyzed chemical step are discussed.

2. METHODS

2.1 Isotopically Labeled Enzymes and Their Physiochemical Properties

Long before, the introduction of the BO enzyme concept, labeling of catalytic and noncatalytic proteins with stable isotopes had been widely utilized to advance structural analysis by nuclear magnetic resonance (NMR) spectroscopy (Zhang & van Ingen, 2016), neutron reflectometry

Table 1 Summary of the Enzymatic Reactions Examined by Enzyme Isotope Effect Measurement

Key:
1. Enzyme
2. Type of chemical step examined by enzyme isotope effect
3. Brief mechanism of the reaction (marked in *red* is a chemical step examined by enzyme isotope effect; a step marked with "RLS" when it is a rate-limiting step in the reaction)
4. Reference(s)

---

1. Purine nucleoside phosphorylase (PNP)
2. Phosphorolysis
3. 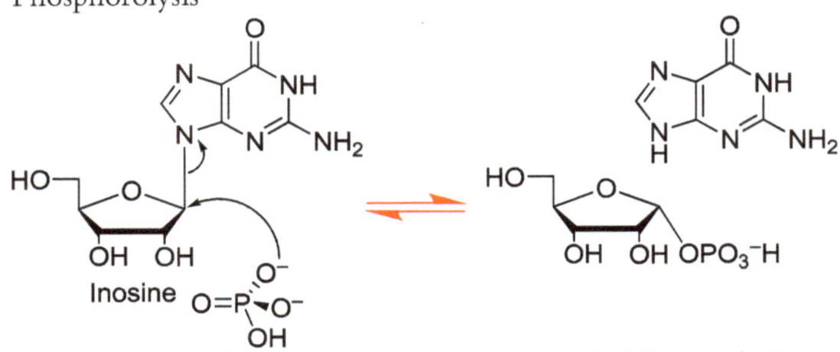

4. Antoniou, Ge, Schramm, and Schwartz (2012), Silva et al. (2011), Suarez and Schramm (2015), and Zoi et al. (2016)

---

1. HIV-1 protease
2. Hydrolysis
3. 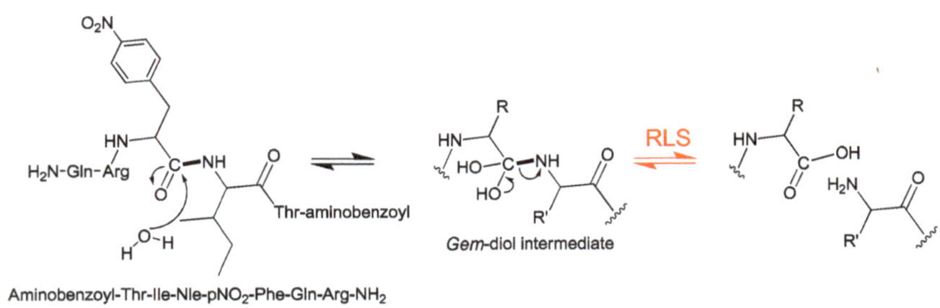

4. Kipp, Silva, and Schramm (2011) and Krzemińska, Moliner, and Świderek (2016)

---

1. Alanine racemase (AR)
2. Proton transfers
3. 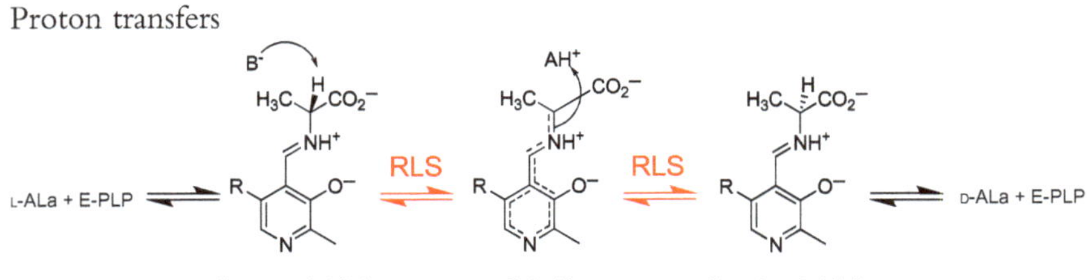

4. Toney, Castro, and Addington (2013)

Table 1 Summary of the Enzymatic Reactions Examined by Enzyme Isotope Effect Measurement—cont'd

1. Pentaerythritol Tetranitrate Reductase (PETNR)
2. Hydride transfer (reductive half reaction)
3. 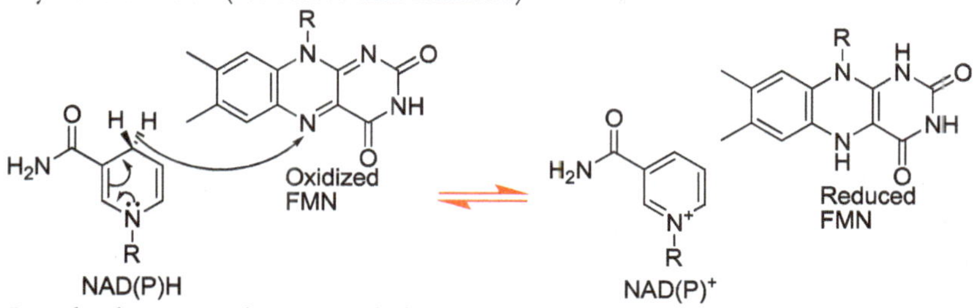

4. Longbotham, Hardman, Gorlich, Scrutton, and Hay (2016) and Pudney et al. (2013)

---

1. Lactate dehydrogenase (LDH)
2. Hydride and proton transfers
3. 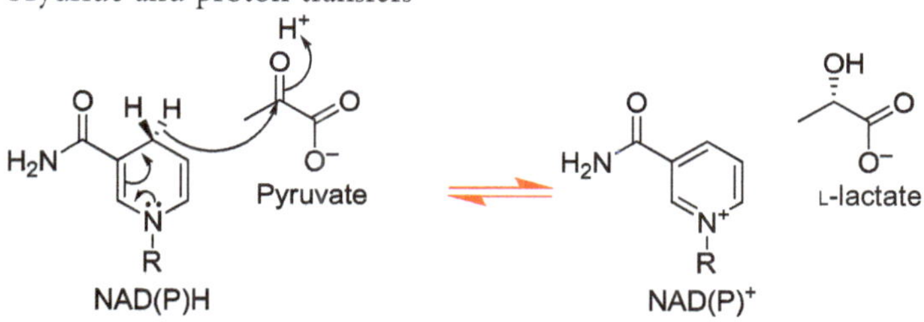

4. Masterson and Schwartz (2013), Wang, Antoniou, Schwartz, and Schramm (2016), and Wang, Chang, and Schramm (2016)

---

1. Dihydrofolate reductase (DHFR)
2. Hydride transfer
3. 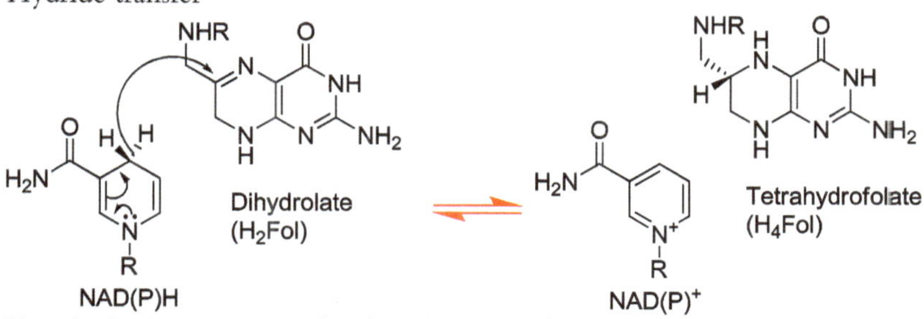

4. Francis, Sapienza, Lee, and Kohen (2016), Luk et al. (2013), Luk, Loveridge, and Allemann (2014) Luk, Ruiz-Pernia, et al. (2014), Luk et al. (2015), Ruiz-Pernia et al. (2013), Wang, Antoniou, et al. (2016), Wang, Chang, et al. (2016), and Wang, Singh, Czekster, Kohen, and Schramm (2014)

(Heinrich, 2016), small-angle neutron scattering (Haertlein et al., 2016), proteomic mass spectrometry (Chahrour, Cobice, & Malone, 2015), and site-directed vibrational spectroscopy (Peuker et al., 2016). A comprehensive collection of the methods involved in production of the eukaryotic and prokaryotic proteins with various labeling patterns including uniform isotopic labeling (Duff, Wilde, Rekas, Lake, & Holden, 2015; Hoopes, Elberson, Preston, Reddy, & Kelman, 2015; Le, Brown, Simon, & Dayie, 2015), amino acid-specific labeling (Lin et al., 2015), site-specific labeling in which a single labeled amino acid residue is introduced (Peuker et al., 2016), and segmental labeling (Michel & Allain, 2015), has been recently published in *Methods in Enzymology* and elsewhere. In fact, almost every known labeling pattern has been utilized in enzyme isotope effect studies except for the site-specific labeling. General methods for the corresponding labeling patterns are briefly outlined in Table 2.

The goal of nonuniform labeling (Luk et al., 2015; Suarez & Schramm, 2015) and current attempts to achieve site-specific amino acid labeling is to examine whether particular amino acids contribute specifically to the catalyzed chemistry. These experiments could test, for example, if only the mass of residues that constitute protein promoting vibrations (according to calculations by Schwartz and coworkers; e.g., Basner & Schwartz, 2004; Mincer & Schwartz, 2003; Núñez, Wing, Antoniou, Schramm, & Schwartz, 2006) is responsible for the effect of "heavy" enzymes, rather than the mass of the whole protein (see more details in Section 2.3.3 of this chapter). The reason behind examination of the different uniform labeling patterns, on the other hand, is primarily to demonstrate that the "heavy" enzyme effect on catalysis is a pure mass effect, and is thus independent of other effects of isotopic substitution (e.g., alteration of electronic potential surface due to anharmonic effects, secondary effects of protein–ligand or protein–protein interactions, etc.). It appears that the local effect of isotopic labeling closely follows the BO approximation. For example, the C—H bond has an IR stretch around $3000\,cm^{-1}$, while C—D bonds present a red-shift to $\sim 2100\,cm^{-1}$ which is close to $3000 \cdot 2^{-1/2}$ as predicted by the BO approximation. While these effects reflect phenomena at a 10-fs timescale, special concern has to be associated with potential alteration of electrostatic and steric properties of the protein at the ps timescale (i.e., $10–200\,cm^{-1}$), which are also predicted to be coupled to the catalyzed chemistry (Basner & Schwartz, 2004; Hatcher, Soudackov, & Hammes-Schiffer, 2004). Upon perdeuteration of the protein, several effects on

Table 2 Summary of Isotope Labeling Techniques Used in Enzyme Isotope Effect Studies

Overview of the Labeling Method	Enzyme and Mass Increase (%)	References
Uniform labeling		
• *E. coli* BL21(DE3) cells conditioned to grow in $^{2}H_2O$ (D_2O) are transformed with a vector encoding sequence for an enzyme of interest. • Resulting cells are grown and the protein is expressed in the minimal media (M9 or M63) prepared in D_2O in case of ^{2}H- and ^{13}C,^{15}N,^{2}H-labeling, or in H_2O in case of ^{15}N and ^{13}C,^{15}N-labeling, and supplemented with [1,2,3,4,5,6,6-$^{2}H_7$]glucose for ^{2}H-labeling,[a] $^{15}NH_4Cl$ for ^{15}N-labeling, [U-$^{13}C_6$]glucose and $^{15}NH_4Cl$ for ^{13}C,^{15}N-labeling, and [U-$^{13}C_6$, 1,2,3,4,5,6,6-$^{2}H_7$]glucose and $^{15}NH_4Cl$ for ^{13}C,^{15}N,^{2}H-labeling. • Protein is purified and stored using protocols established for its unlabeled variant in buffered H_2O. • Corresponding "light" version of the enzyme is prepared using the same gene construct in BL21 (DE3) cells grown on a minimal media supplemented with "light" nutrients.	^{2}H-*Bs*DHFR (5.0) *Hs*PNP (5.3) *Bs*AR (5.5)	Luk, Loveridge, et al. (2014) and Luk, Ruiz-Pernia, et al. (2014), Suarez and Schramm (2015), and Toney et al. (2013)
	^{15}N-HIV-1 protease (1.2), PETNR (1.3)	Kipp et al. (2011) and Pudney et al. (2013)
	^{13}C,^{15}N-*Bs*DHFR (5.5), *Hs*PNP (5.6)	Luk, Loveridge, et al. (2014), Luk, Ruiz-Pernia, et al. (2014), and Suarez and Schramm (2015)
	^{13}C,^{15}N,^{2}H-*Hs*PNP (9.9), HIV-1 protease (11.6), PETNR (9.8), *Hs*LDH (11.2), *Ec*DHFR (11.0), *Tm*DHFR (10.6), *Bs*DHFR (11.1), *Hs*DHFR (11.3)	Silva et al. (2011) Zoi et al. (2016) Kipp et al. (2011) Krzemińska et al. (2016) Pudney et al. (2013) Longbotham et al. (2016) Wang, Antoniou, et al. (2016) and Wang, Chang, et al. (2016) Wang et al. (2014) Luk et al. (2013) Ruiz-Pernia et al. (2013) Luk, Loveridge, et al. (2014) and Luk, Ruiz-Pernia, et al. (2014) Luk, Loveridge, et al. (2014) and Luk, Ruiz-Pernia, et al. (2014) Francis et al. (2016)

Continued

Table 2 Summary of Isotope Labeling Techniques Used in Enzyme Isotope Effect Studies—cont'd

Overview of the Labeling Method	Enzyme and Mass Increase (%)	References
Amino acid-specific labeling		
• *E. coli* strain auxotrophic for a specific amino acid and containing vector encoding sequence for an enzyme of interest is grown in a minimal media supplemented with appropriate antibiotics, glucose, salts, vitamins, and a *labeled amino acid*. • Expressed protein is purified and stored according to the known protocols for its unlabeled analog. • Corresponding "light" version of the enzyme is prepared using the same gene construct in BL21(DE3) cells grown in a minimal media supplemented with appropriate unlabeled nutrients.	[^{15}N, ^{2}H]His$_8$-PNP (0.19)	Suarez et al. (2015)
Segmental labeling		
• Isotopic hybrids as well an unlabeled version of the enzyme are prepared using native chemical ligation as one of the several available methods (Michel & Allain, 2015). • Uniformly ^{13}C,^{15}N,^{2}H-labeled and natural abundance precursors (N- and/or C-terminal fragments) are produced by recombinant expression in *E. coli* in the presence of appropriately labeled nutrients. • Due to necessity of having N-terminal cysteine on the C-terminal fragment, introduction of a point mutation to the final hybrid and unlabeled protein structure is a likely situation. Therefore, special care has to be taken to reassure absence of structural alterations caused by this mutation.	^{13}C,^{15}N,^{2}H-[1–28]*Ec*DHFR (1.8), ^{13}C,^{15}N,^{2}H-[29–159]*Ec*DHFR (8.9)	Luk et al. (2015)

[a]Perdeuterated glycerol was used in case of *Bacilus stearothermophilius* alanine racemase (*Bs*AR) as the carbon source (Toney et al., 2013).

protein electrostatics and conformational landscape (i.e., non-BO) should be considered, including:

(1) Changes in hydrophilicity due to deuteration: Increased hydrophobicity of C—D bonds in comparison to C—H bonds arises due to an approximately 0.005 Å (Kuchitsu & Bartell, 1962) shorter distance and the associated decrease in the electrostatic dipole of the C—D bond. This effect originates from the lower vibrational energy of the heavier isotope and the anharmonicity of the C—H stretch. Well-known manifestations of this phenomenon include isotope effects on nuclear shielding in NMR spectroscopy (Hansen, 1988) and reverse-phase high-pressure liquid chromatography separation of protiated and deuterated pairs of compounds (Turowski et al., 2003). In contrast, deuterium substitutions in aromatic compounds was reported to decrease their observed lipophilicity (increased hydrophilicity) based on results of chromatographic separations and presumably due to the weakening of aromatic edge-to-face and alkyl–π interactions between deuterated solutes and the stationary phase (Turowski et al., 2003). However, some studies demonstrated nonexistent D/H isotope effect for CH–π interactions (Zhao et al., 2012), so this effect might be system specific. Reported effects of hydrophobicity change in proteins upon perdeuteration include altered tendency to subunit association (e.g., lowered for phycocyanines; Chen, Liu, MacColl, & Berns, 1983), but increased for glutathione *S*-transferase (GST) (Brockwell et al., 2001), and decrease in thermal stability or stability to chemical denaturation (Brockwell et al., 2001; Chen et al., 1983; Hattori, Crespi, & Katz, 1965; Liu, Hanson, Langan, & Viola, 2007; Meilleur, Contzen, Myles, & Jung, 2004).

(2) Secondary equilibrium isotope effects on acidity: NMR titration studies demonstrated that aliphatic carboxylic acids with deuterated β- and/or γ-carbon are less acidic than their corresponding "light" versions (e.g., for CH_3CD_2COOH and CD_3CH_2COOH $\Delta pK_a = pK_a^D = -pK_a^H =$ 0.034 and 0.007 correspondingly). Likewise, β-deuteration of aliphatic amines was found to increase their basicity (up to $\Delta pK_a = 0.206$ for $(CD_3)_3NH^+$) (Perrin, 2010; Perrin & Dong, 2007). While the local effect is consistent with BO approximation, alteration of acidities of individual residues in the context of the entire protein might perturb its isoelectric point, local charges, and charge distribution across the protein, which may in turn alter the protein potential surface, conformational landscape, or cause other non-BO effects. Examples of this effect include shift of the pH optimums of the perdeuterated plant ribonuclease (Rokop, Gajda, Parmerter, Crespi, & Katz, 1969) and haloalkane dehydrogenase (Liu et al., 2007). In the latter case, the hydrogen bond network in the enzyme active site was altered to exclude the crucial nucleophilic water, giving rise to a nonproductive conformation.

In contrast to the effect of perdeuteration, the effect of "heavy" atom isotopic substitution on physiochemical properties of the proteins is less discussed in literature due to fewer examples of rigorous characterization of proteins with such substitutions and more subtle corresponding effects. Only a few examples of secondary ^{13}C and ^{15}N isotope effects on pK_as of carboxylic acids have been reported, and these do not exceed ca. 0.3% (Perrin, 2010). Therefore, significant alteration of the pH optimum is not anticipated upon these substitutions.

(3) Steric isotope effects: As described earlier, C—D bonds have a diminished steric requirement relative to C—H bonds. Within the confined environment of the enzyme active site, steric isotope effects might be observed both at the ground state and the transition state. This effect is well documented in the nonenzymatic host–guest systems where binding of the less bulky deuterated versions of the guests is often preferred, resulting in an inverse isotope effect on the binding equilibrium constant (Haino, Fukuta, Iwamoto, & Iwata, 2009; Rechavi, Scarso, & Rebek, 2004). In the case of enzymatic systems, deuteration of the protein is favored in a sterically crowded transition state resulting in an inverse enzyme isotope effect on the rate of catalysis reported for perdeuterated GST (Brockwell et al., 2001) and $^{13}C,^{15}N,^{2}H$-labeled pentaerythritol tetranitrate reductase (PETNR) (Longbotham et al., 2016; Pudney et al., 2013).

From the practical perspective, upon protein-labeling with deuterium, an incomplete back-exchange of the solvent-exchangeable deuteriums to protiums requires the experimentalists′ attention. Expression of the uniformly deuterated or $^{13}C,^{15}N,^{2}H$-labeled proteins results in deuterium substitution of all protiums including exchangeable ones. Subsequent protein purification in buffered H_2O usually results in exchange of most solvent-exchangeable deuterons to protons, but some would require partial protein denaturation to back exchange. As an example, the resistance of 10 heteroatom-bound deuteriums to exchange was reported in the case of preparation of perdeuterated GST (Brockwell et al., 2001). This exchange is critical, since contribution of the solvent isotope effects would preclude meaningful interpretation of the enzyme isotope effects. Therefore assessment of isotopic composition by mass spectrometry and by $^{1}H,^{15}N$-HSQC is required not only to confirm the incorporation of the labels, but also to verify the absence of the deuteriums in the exchangeable positions.

2.2 Structural Characterization of Isotopically Labeled Enzymes

Examination of the degree of structural alteration upon isotopic substitution is a critical step in every study involving isotopically labeled enzymes. Enzyme isotope effect studies commonly utilize circular dichroism (CD) spectroscopy to test for altered folding. CD is a first choice for rapid examination of the secondary structure and folding characteristics of proteins, and the protocols associated with it are described elsewhere (Greenfield, 2007a, 2007b). The method is advantageous as it requires $\leq$20 μg of protein per measurement, and allows analysis in physiologically relevant buffers at various temperatures within a few hours. Typically, comparison of CD spectra for the natural-isotope abundance and "heavy" (^{13}C,^{15}N,^{2}H-labeled) enzymes confirms structural identity as reported for HIV-1 protease (Kipp et al., 2011), *Escherichia coli* dihydrofolate reductase (*Ec*DHFR) (Luk et al., 2013) and its N23PP/S148A mutant (Ruiz-Pernia et al., 2013), *Thermotoga maritima* DHFR (*Tm*DHFR) (Luk, Loveridge, et al., 2014), *Bacillus stearothermophilus* DHFR (*Bs*DHFR) (Luk, Ruiz-Pernia, et al., 2014), and segmentally labeled *Ec*DHFR (Luk et al., 2015). However, in a different study, it was observed that the isoelliptic point is altered in the CD spectra for "heavy" *Ec*DHFR, suggesting that the increase in the protein mass modulated protein conformational ensembles (Wang et al., 2014). Example of CD spectra of "light" and "heavy" *Ec*DHFRs (denoted as l-DHFR and h-DHFR) recorded at different temperatures is provided in Fig. 1A.

Additionally, CD spectroscopy is utilized to assess and compare thermal stability of the "light" and "heavy" enzyme upon plotting temperature dependencies of the mean residue ellipticity (θ) at a certain wavelength as illustrated in Fig. 1B for *Ec*DHFR (Wang et al., 2014). Alternatively, thermal stabilities are assessed using differential scanning calorimetry (Brockwell et al., 2001) or differential scanning fluorimetry (Golden, Attwood, Duff, Meilleur, & Vrielink, 2015) techniques.

NMR spectroscopy is a more powerful, but a less common technique used to test whether the labeled enzyme is isostructural to the "light" version as was reported in case of ^{13}C,^{15}N,^{2}H-labeled PETNR and its "light" version represented by ^{15}N-labeled PETNR (Pudney et al., 2013). Fig. 2 illustrates NMR evidence for the isostructural nature of ^{15}N-PETNR and h-PETNR.

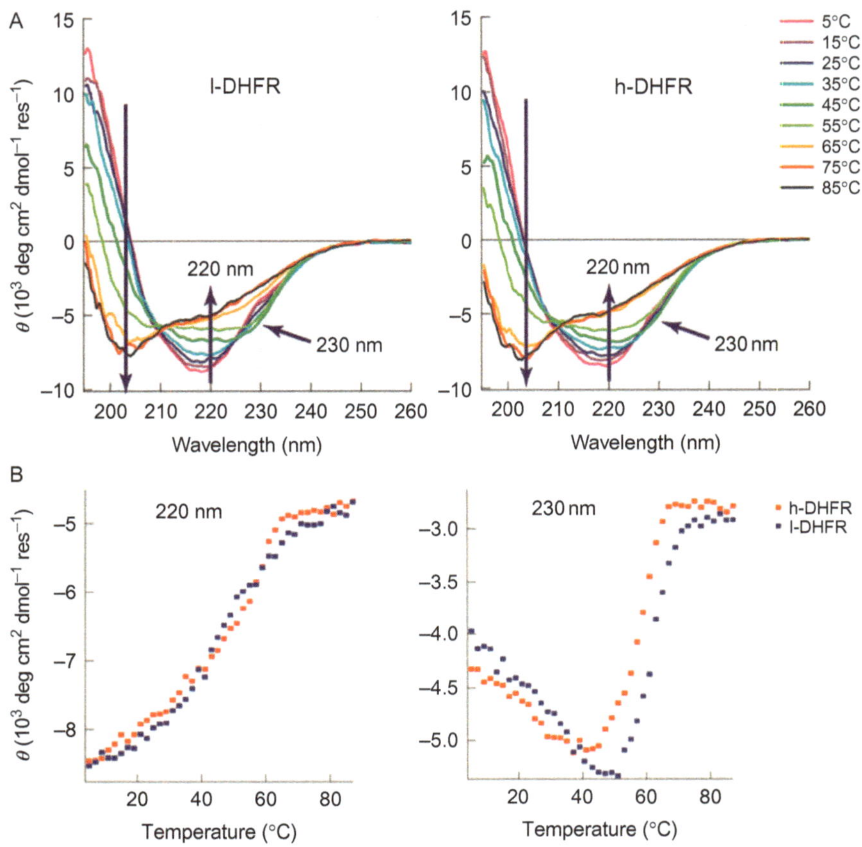

Fig. 1 (A) Temperature-dependent CD spectra for "light" DHFR (l-DHFR, *left*) and ^{13}C,^{15}N,^{2}H-labeled DHFR (h-DHFR, *right*). Altered isoelliptic point at 225 nm for h-DHFR indicates altered protein conformations in the 5–45 C range. (B) Temperature dependencies of the mean residual ellipticities (θ) at 220 nm (*left*) and 230 nm (*right*) for l-DHFR (*blue*) and h-DHFR (*red*). Absorption at 220 nm primarily corresponds to the peptide bond, therefore similar melting curves for l- and h-DHFRs at 220 nm (*left*) indicate the same overall fold and thermal stability for both enzymes. Shift of the melting curve for the h-DHFR to lower temperatures when monitored at 230 nm (*right*) suggests protein mass-modulated alteration of protein conformations (specifically in adenosine-binding domain of DHFR). *Reprinted with permission from Wang, Z., Singh, P., Czekster, C. M., Kohen, A., & Schramm, V. L. (2014). Protein mass-modulated effects in the catalytic mechanism of dihydrofolate reductase: beyond promoting vibrations.* Journal of American Chemical Society, 136, *8333–8341. Copyright 2014 American Chemical Society.*

X-ray crystallography has not yet been used to assess the effect of isotopic labeling on the structure in the enzyme isotope effect studies. However, several studies report X-ray crystal structures of the perdeuterated enzymes including *Staphylococcal* nuclease (Gamble, Clauser, & Kossiakoff, 1994),

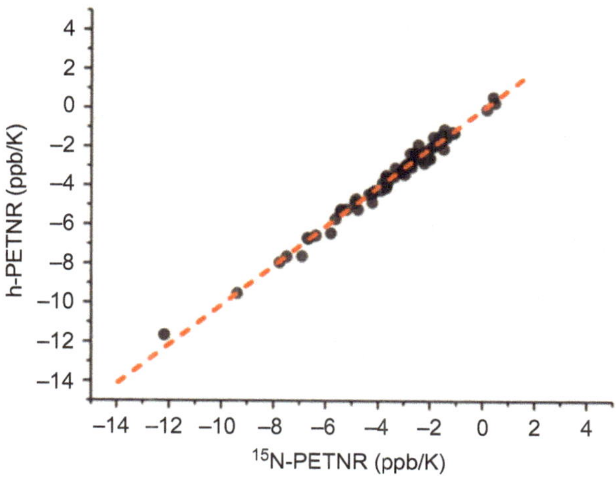

Fig. 2 NMR evidence for structural identity of h-PETNR and ^{15}N-PETNR. Correlation of NMR temperature coefficients (temperature dependencies of the backbone NH chemical shifts) for h-PETNR and ^{15}N-PETNR residues. *Reprinted with permission from Pudney, C. R., Guerriero, A., Baxter, N. J., Johannissen, L. O., Waltho, J. P., Hay, S., et al. (2013). Fast protein motions are coupled to enzyme H-transfer reactions.* Journal of American Chemical Society, 135, *2512–2517. Copyright 2013 American Chemical Society.*

carbonic anhydrase (Budayova-Spano et al., 2006), cytochrome P450cam (Meilleur, Dauvergne, Schlichting, & Myles, 2005), haloalkane dehalogenase (Liu et al., 2007), arginase I (Di Costanzo, Moulin, Haertlein, Meilleur, & Christianson, 2007), aldose reductase (Blakeley et al., 2008), diisopropyl fluorophosphatase (Blum et al., 2010), and cholesterol oxidase (Golden et al., 2015). In all of these cases, with the exception of cholesterol oxidase (Golden et al., 2015), "light" and perdeuterated versions of the corresponding enzymes were found to crystallize under identical conditions, in the same space group, resulting in superimposable structures with virtually identical parameters including B factors.

2.3 Effect of Enzyme Isotopic Labeling on Catalysis

2.3.1 Effect on Steady-State Kinetic Parameters

One of the first observation of enzyme isotope effect on the rate of catalysis dates back to a 1969 kinetic study of perdeuterated bacterial alkaline phosphatase exhibiting a 1.8-fold decrease in V_{max} in comparison to its protiated analog (Rokop et al., 1969). The isotope effect was rationalized by a lack of vibrational energy available at the active site of perdeuterated enzyme. Interestingly, the same study reported absence of analogous isotope effect for the plant ribonuclease. These two different outcomes underscore the importance of a system-specific experimental approach toward evaluation and interpretation of enzyme isotope effects. Since V_{max} of alkaline phosphatase is limited by the rate of the chemical step (Han & Coleman, 1995; Hull, Halford, Gutfreund, & Sykes, 1976) and V_{max} of ribonucleases is commonly limited by the rate of product release (Doucet, Watt, & Loria, 2009; Hsieh, Walker, Fierke, & Engelke, 2009), the latter case likely did not reflect the effect of protein deuteration on the chemical step.

Recent protein isotope effect studies, driven by the hypothesis that protein subpicosecond dynamics promote formation of the transition state, primarily consider kinetic parameter(s) associated with the chemical step per se (e.g., rate constants and/or intrinsic kinetic isotope effects) as a probe of enzyme mass effects. However, examination of the corresponding effects on the steady-state or presteady-state kinetic parameters of the turnover and/or ligand binding is a common way to test BO behavior of the "heavy" enzymes. Table 3 summarizes effects of the uniform ^{13}C,^{15}N,^{2}H-labeling of various enzymes on the steady-state kinetic parameters of the corresponding enzymatic reactions illustrated in Table 1.

Special considerations have to be taken when interpreting enzyme isotope effects on steady-state kinetic parameters, since the chemical step is commonly not the rate-limiting step of the enzymatic reaction due to kinetic complexity associated with the multistep nature of every enzyme's catalytic cycle. Hence, a prerequisite for the interpretation of enzyme isotope effect study is an assessment of the rate-limiting step, or otherwise of the microscopic rate constant for this step embedded in the steady-state parameters. Based on the knowledge of the kinetic mechanism and enzyme isotope effect on steady-state kinetic parameters, several potential situations were reported:

(1) Chemical step is not rate limiting for any steady-state parameter and steady-state kinetic parameters are not altered by protein mass. Examples reported in literature include studies of PNP (Silva et al., 2011; Suarez & Schramm, 2015) and N23PP/S148A mutant of *Ec*DHFR (Ruiz-Pernia et al., 2013). Fig. 3A provides illustration of the superimposable steady-state saturation curves (Michaelis-Menten curves) for the "light" and "heavy" PNPs (Silva et al., 2011). Since for these systems steady-state kinetic parameters are not governed by the rate of chemical step, but rather by the rate of physical steps such as ligands binding and/or release, absence of enzyme isotope effect on steady-state kinetic parameters supports that substrate (or product) binding and/or release, and therefore electrostatics of the system are not perturbed upon

Table 3 Summary of Reported Enzyme Isotope Effects on Steady-State Kinetic Parameters (*l* and *h* Superscripts Designate Natural (Light) Enzyme and ^{13}C,^{15}N,^{2}H-Labeled Enzyme (Heavy) Correspondingly Unless Stated Otherwise; Marked in *Bold* Are Nonunity Isotope Effects on Steady-State Kinetic Parameters Associated With Steps Other Than Chemistry)

		Enzyme Isotope Effect on Steady-State Kinetic Parameter[a]		
Enzyme	**Substrate(s)**	k^l_{cat}/k^h_{cat}	K^l_m/K^h_m	**References**
Chemical step is not rate limiting				
*Hs*PNP	Inosine	1.0 ± 0.2	1.1 ± 0.2	Silva et al. (2011)
	Guanosine	1.0 ± 0.1	0.9 ± 0.2	
E258D/ L261A *Hs*PNP	Guanosine	1.0 ± 0.1	1.1 ± 0.2	Zoi et al. (2016)
*Ec*DHFR	DHF	**1.09 ± 0.02**	**0.74 ± 0.08**	Wang et al. (2014)
	NADPH	**1.12 ± 0.04**	1.1 ± 0.2	
	DHF	1.04 ± 0.03[b]	**1.98 ± 0.38**[b]	Luk et al. (2013)
	NADPH		**1.3 ± 0.2**[b]	
N23PP/ S148A *Ec*DHFR	DHF	1.0 ± 0.1	1.0 ± 0.2	Ruiz-Pernia et al. (2013)
	NADPH		1.1 ± 0.3	
*Tm*DHFR	DHF	**1.36 ± 0.05**[a]	1.5 ± 0.5[b]	Luk, Loveridge, et al. (2014) and Luk, Ruiz-Pernia, et al. (2014)
	NADPH		0.77 ± 0.3[b]	
*Bs*DHFR	DHF	**2.4 ± 0.2**[a]	0.7 ± 0.2[b]	Luk, Loveridge, et al. (2014) and Luk, Ruiz-Pernia, et al. (2014)
	NADPH		1.0 ± 0.2[b]	
*Hs*DHFR	DHF	**1.44 ± 0.02**	1.07 ± 0.4	Francis et al. (2016)
	NADPH	**1.45 ± 0.02**	1.06 ± 0.26	
Chemical step is rate limiting				
HIV-1 protease	Aminobenzoyl-Thr-Ile-Nle-pNO_2-Phe-	1.19 ± 0.05	**0.62 ± 0.07**	Kipp et al. (2011)
	Gln-Arg-NH_2			
AR[c]	D-Alanine	1.32 ± 0.04	1.0 ± 0.1	Toney et al. (2013)
	L-Alanine	1.21 ± 0.03	0.9 ± 0.1	

[a]At 25°C unless stated otherwise.
[b]At 20°C.
[c]Enzyme isotope effects are measured for perdeuterated AR.

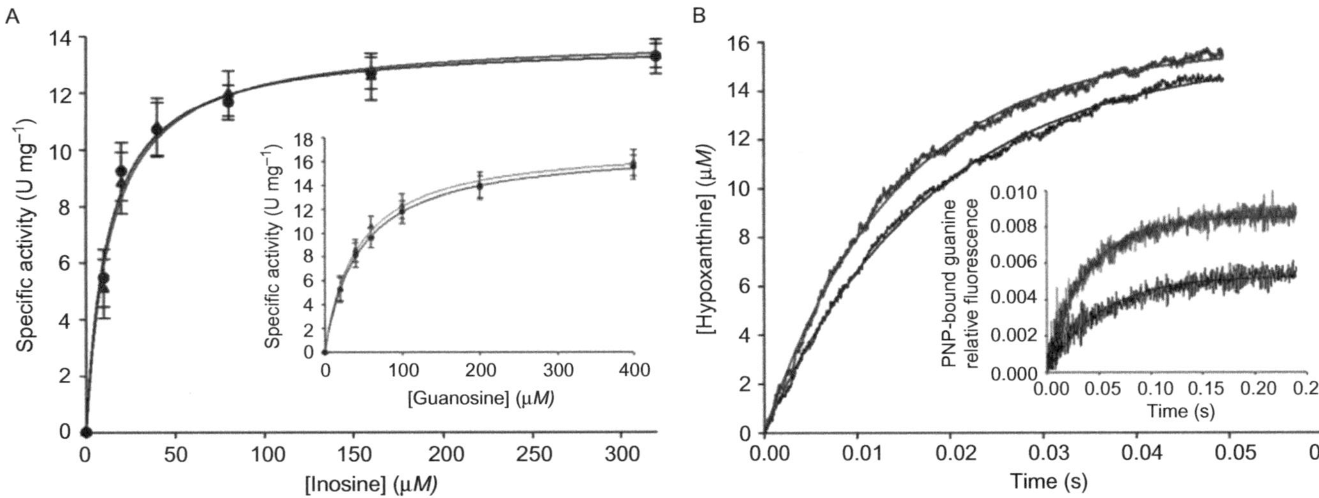

Fig. 3 (A) Michaelis–Menten curves for inosine (or guanosine in the *insert*) phosphorolysis catalyzed by "light" (*blue*) and "heavy" (*black*) PNPs. (B) Stopped-flow traces for single-turnover experiments for inosine (or guanosine in the *insert*) phosphorolysis by "light" (*blue*) and "heavy" (*black*) PNPs. *Reprinted with permission from Silva, R. G., Murkin, A. S., & Schramm, V. L. (2011). Femtosecond dynamics coupled to chemical barrier crossing in a Born-Oppenheimer enzyme.* Proceedings of the National Academy of Sciences of the United States of America, 108, *18661–18665. Copyright (2011) National Academy of Sciences.*

isotopic labeling of the protein, in agreement with the BO approximation. However, alternative techniques could be utilized to unmask enzyme isotope effects on the chemical step, including altered reaction conditions under which the chemical step becomes rate-limiting, presteady-state kinetics measurements (Fig. 3B, see also Section 2.3.2.2 of this chapter), substrate intrinsic kinetic isotope effects (i.e., kinetic isotope effects on an elementary reaction of an isotopically substituted substrate; see details in Section 2.3.2.3 of this chapter), and combinations of these techniques.

(2) Chemical step is not rate limiting for any steady-state parameter, yet steady-state parameters are affected by protein mass. This case was reported for DHFRs from different organisms (Francis et al., 2016; Luk, Loveridge, et al., 2014; Luk, Ruiz-Pernia, et al., 2014; Wang et al., 2014). In these cases, an increase in the mass of the protein resulted in normal observed kinetic isotope effect (i.e., isotope effect >1) for the parameter k_{cat}, consistent with a diminished rate constant for product release, as was demonstrated for *Hs*DHFR (Francis et al., 2016). Most of the DHFR variants did not exhibit a significant enzyme isotope effect on the K_M parameter of either of its substrates; however, an inverse isotope effect of K_M (i.e., isotope effect <1) was observed for *Ec*DHFR due to slower binding and faster release of the substrate for the "heavy" analog of the enzyme (Wang et al., 2014). Fig. 4A and B

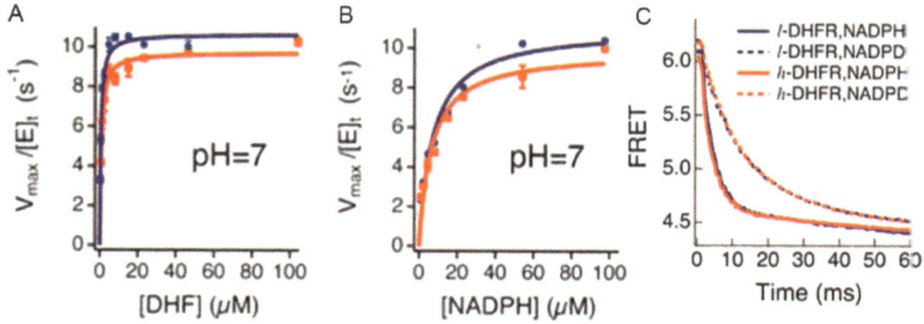

Fig. 4 Michaelis–Menten curves for dihydrofolate (DHF) reduction catalyzed by l-DHFR (*blue*) and h-DHFR (*red*) at varying concentrations of DHF (A) or NADPH (B), in the presence of saturating concentrations of NADPH or DHF correspondingly. Normal enzyme KIE is apparent for the parameter k_{cat}. (C) Presteady-state single-turnover transients for l-DHFR and h-DHFR are equivalent with either NADPH (*solid curve*) or NADPD (*dashed curve*) as the cosubstrate. *Reprinted with permission from Wang, Z., Singh, P., Czekster, C. M., Kohen, A., & Schramm, V. L. (2014). Protein mass-modulated effects in the catalytic mechanism of dihydrofolate reductase: beyond promoting vibrations.* Journal of American Chemical Society, 136, *8333–8341. Copyright 2014 American Chemical Society.*

illustrates enzyme isotope effects on steady-state kinetic parameters for *Ec*DHFR (Wang et al., 2014). Consequently, the term BO enzyme is not fully justified for the DHFR variants, as protein mass modulations alters protein–ligand interactions in the ground state, which could indicate modified conformational ensembles and/or electrostatics of the protein.

(3) Chemical step is rate limiting (or partially rate limiting) for some steady-state parameters, and steady-state kinetic parameters associated with nonchemical steps are not altered by protein mass. The only reported example of this scenario is for alanine racemase (AR) (Toney et al., 2013). In this case no enzyme isotope effect was observed on the parameter K_M, supporting that substrate binding is not sensitive to isotopic substitutions in the protein. However, the enzyme isotope effect on the parameter k_{cat} was assumed to probe the chemical events.

(4) Chemistry is rate limiting (or partially rate limiting) for some steady-state parameters, and steady-state parameters associated with nonchemical steps are altered by protein mass. This case is illustrated by the studies of HIV-1 protease (Kipp et al., 2011). This system exhibited diminished k_{cat} and an inverse enzyme isotope effect on K_M. Whereas the effect on k_{cat} agrees with an overall trend of diminished rate of chemistry in the "heavy" enzymes (in accordance with the BO idea), the effect on K_M suggests that ground-state interactions with the substrates were also perturbed upon isotopic substitution in the protein (a non-BO effect for the most part) (Fig. 5).

In the most common case, enzyme isotope effects are reported on the steady-state kinetic parameters k_{cat}, K_M, and k_{cat}/K_M; however, another steady-state kinetic parameter which can be used as a probe of enzyme isotope effect is a forward commitment to catalysis (C_f). C_f represents a probability of a Michaelis complex (enzyme–substrate complex poised for catalysis) to pass the chemical barrier as opposed to the probability to partition back to unbound enzyme and substrate (Northrop, 1981). For the most simple kinetic model of the enzymatic reaction corresponding to Scheme 1, the forward commitment to catalysis is:

$$C_f = \frac{k_2}{k_{-1}} \tag{1}$$

C_f can be calculated from the values of observed KIEs and intrinsic KIEs (as described in Section 2.3.2.3 of this chapter), or measured directly by the isotope-trapping method (Rose, 1980). Since the value of C_f is governed

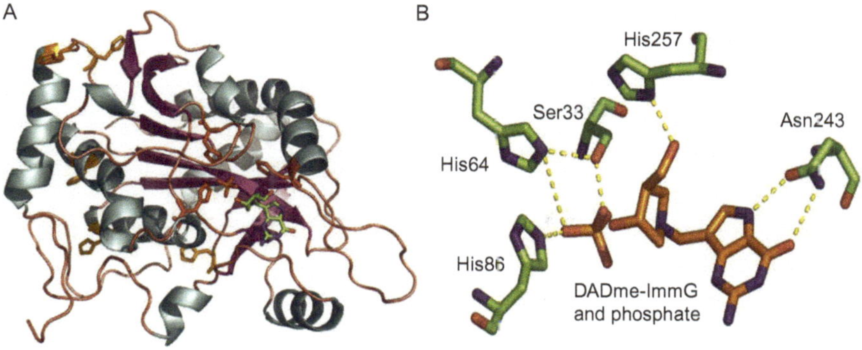

Fig. 5 Location of eight His residues site specifically labeled in [^{15}N, ^{2}H]His$_8$-PNP. (A) A crystal structure of PNP monomer with eight histidine residues highlighted. Five histidines (*orange*) are located in the periphery, away from the active site, and three (*red*) are located in the active site forming hydrogen bonds to the substrates. (B) Active site of PNP featuring three histidines (His65, His86, and His257) involved in direct contact with phosphate and the transition state analog DADMe-ImmG (PDB ID 3PHB). *Reprinted with permission from Suarez, J. & Schramm, V. L. (2015). Isotope-specific and amino acid-specific heavy atom substitutions alter barrier crossing in human purine nucleoside phosphorylase.* Proceedings of the National Academy of Sciences of the United States of America, 112, *11247–11251. Copyright (2015) National Academy of Sciences.*

$$\mathrm{E} \underset{k_{-1}}{\overset{k_1[\mathrm{S}]}{\rightleftharpoons}} \mathrm{E{\cdot}S} \xrightarrow{k_2} \mathrm{E} + \mathrm{P}$$

Scheme 1 Minimal kinetic model for an enzymatic reaction.

by the chemical step's rate constant (e.g., k_2) as well as a nonchemical step rate constant (e.g., k_{-1}), special considerations upon using C_f must be taken into account as illustrated by the case of PNP (Silva et al., 2011). Estimation of the C_f for the "light" and "heavy" versions of PNP revealed 15%- and 30%-reduced barrier crossing efficiency in the "heavy" enzyme for inosine and guanosine substrates, correspondingly. Justification of the minor contribution from substrate release (denominator of C_f) to the C_f was provided by the fact that the substrate K_M was not perturbed in the "heavy" enzyme (Table 3).

General trends for the enzyme isotope effects on steady-state kinetic parameters are summarized in Table 4. The absence of ground-state effects may serve to support the hypothesis that enzyme isotope effects on the chemical step are caused by vibrational alteration of the protein, and not electrostatic perturbation of the system. The presence of ground-state-effects, on the other hand, may emphasize limitations of the hypothesis,

Table 4 Summary of the General Trends for Enzyme Isotope Effects on Steady-State Kinetic Parameters (Marked in *Bold* Are Effects That May Not Agree With the BO Approximation)

	Enzyme Isotope Effect on		
Rate-Limiting Step	k_{cat}	K_m	**Examples (References)**
Nonchemical step (conformational changes associated with ligands binding or release)	Unity	Unity	*Hs*PNP (Silva et al., 2011, Zoi et al., 2016), N23PP/S148A *Ec*DHFR (Ruiz-Pernia et al., 2013)
	Normal	Unity	*Tm*DHFR (Luk, Loveridge, et al., 2014; Luk, Ruiz-Pernia, et al., 2014), *Bs*DHFR (Luk, Loveridge, et al., 2014; Luk, Ruiz-Pernia, et al., 2014), *Hs*DHFR (Francis et al., 2016)
		Inverse	*Ec*DHFR (Wang et al., 2014)
Chemical step	Normal	Unity	*Bs*AR (Toney et al., 2013)
		Inverse	HIV-1 protease (Kipp et al., 2011)

but does not preclude its use if transition-state effects can be isolated experimentally. In fact, examination of enzyme isotope effects on steady-state parameters is a convenient due to the fact that relatively little amount of enzyme needed and the analysis is fast. Thus, the approach of using steady-state kinetics is common strategy; however, other measurements may better isolate effects on the chemical step in the catalytic cascade.

2.3.2 Effect on Chemical Step

2.3.2.1 Computational Studies

Enzyme isotope effects can be measured on any kinetic or thermodynamic parameter of the enzymatic reaction under study; however, effects of isotopic labeling of the enzyme on the chemical step per se are of special interest, since they can be directly correlated with computer simulations examining that step. As it was mentioned in Section 1 of this chapter, computational simulations and specifically application of the transition path sampling (TPS) (Bolhuis, Chandler, Dellago, & Geissler, 2002) to enzymatic reactions (Núñez et al., 2006; Schwartz & Schramm, 2009) predicted a role for nonstatistical dynamics of the protein on the femtosecond-picosecond timescale in its catalysis of the chemical step. Indeed, those and earlier studies inspired the idea that perturbation of these motions upon isotopic substitutions of protein atoms might provide experimental support to the computational predictions. The rational is that bond activation requires a good resonance between the two events (protein motion and barrier crossing) and that this resonance is likely to be abolished for the "heavy" enzyme. That idea is not different than the expectation that thermal dynamics fluctuations (e.g., Umbrella Sampling assuming no role for nonstatistical vibrations) at the picosecond timescale would resonate with transition state sampling in the "light" enzyme, and could be disturbed in the "heavy" protein (Agarwal et al., 2002; Dybala-Defratyka et al., 2007; Hammes-Schiffer, 2002; Radkiewicz & Brooks, 2000; Renger & Marcus, 2002; Roca et al., 2006; Warshel, 1984; Warshel & Bora, 2016). Only recently the TPS approach was actually used to test a "heavy" enzyme (Masterson & Schwartz, 2013; Zoi et al., 2016), but the molecular-mechanical (MM) region was not parametrized for "heavy" isotopes, limiting the practicality of the results (no measurable parameter was predicted). On the other hand, the more common transition-state theory (TST)-based calculations were also used to examine "heavy" enzymes assuming nonstatistical events (Krzemińska et al., 2016; Luk et al., 2013; Ruiz-Pernia et al., 2013). While these calculations predicted and reproduced measurable parameters (rates and protein KIEs), they also used a MM region not parameterized for the heavier isotopes. This limitation, common to both TPS and TST methods, is practically equivalent to assuming a BO effect of the isotopically labeled enzyme at the MM region of QM/MM calculations (where QM stands for quantum mechanical calculations). Examples of systems that were examined with both theoretical and experimental approaches are summarized in Table 5. There are at least two important outcomes that can be provided by computational studies of enzyme isotope effects:

(1) Theoretical values of enzyme isotope effects using TST and Umbrella Sampling.

The major contribution to the enzyme isotope effects computed using variational transition-state theory (VTST) (Truhlar & Garrett, 1984) with tunneling and recrossing corrections was a difference in the recrossing transmission coefficients between "light" and "heavy" proteins. Consequently, it was concluded that variation in the rate constant for the "heavy" and the "light" enzyme is due to the altered vibrational frequency of the protein motions caused by isotopic labeling. Therefore, the agreement between experimental and theoretical values of enzyme isotope effect for HIV-1 protease (Krzemińska et al., 2016), *Ec*DHFR (Luk et al., 2013), its N23PP/S148A mutant (Ruiz-Pernia et al., 2013), and *Bs*DHFR (Luk, Loveridge, et al., 2014; Luk, Ruiz-Pernia, et al., 2014) (Table 6) reinforces the interpretation of enzyme isotope effects as a result of the "heavy" protein altering

Table 5 Summary of the Enzymatic Systems in Which Enzyme Isotope Effects Were Examined by Both Experimental and Theoretical Methods

Enzyme	Experimental Studies	Computational Studies
PNP	Silva et al. (2011), Suarez & Schramm (2015), and Zoi et al. (2016)	Antoniou et al. (2012) and Zoi et al. (2016)
HIV-1 protease	Kipp et al. (2011)	Krzemińska et al. (2016)
LDH	Wang, Antoniou, et al. (2016) and Wang, Chang, et al. (2016)	Masterson & Schwartz (2015)
*Hs*DHFR	Francis et al. (2016)	Masterson & Schwartz (2015)
*Ec*DHFR	Wang et al. (2014)	Wang, Antoniou, et al. (2016) and Wang, Chang, et al. (2016)
	Luk et al. (2013)	
N23PP/S148A *Ec*DHFR	Ruiz-Pernia et al. (2013)	
*Bs*DHFR	Luk, Loveridge, et al. (2014) and Luk, Ruiz-Pernia, et al. (2014)	

the resonance of protein dynamics with the chemical transformation catalyzed by the enzyme.

(2) Insights into molecular basis of enzyme isotope effects on transition state formation

Computational simulations allow for prediction of underlying atomic motions which lead to altered barrier crossing probability in the "heavy" enzyme. For example, it was demonstrated in the case of PNP that mass effects modulate motions of two specific amino acid residues of the active site that polarize the N-ribosidic bond and stabilize the purine leaving group (Antoniou et al., 2012). This prediction agreed with an experimental enzyme isotope effect study of amino acid-specific-labeled PNP (Suarez & Schramm, 2015). While this agreement is of interest, the calculations did not attempt to assess "heavy" enzyme or any measurable parameter (rate or KIE).

2.3.2.2 Experimental Studies: Enzyme Isotope Effects on the Rate of the Chemical Step

In enzymatic reactions, the rate of the chemical step of interest is commonly masked by other steps including binding of the substrate(s), release of

Table 6 Summary of Experimental and Theoretical Enzyme KIEs on the Rate Constant of Chemical Step (k^l/k^h, Where k^l and k^h are Rate Constants of the Chemical Step Catalyzed by "Light" Enzyme and "Heavy" Enzyme Correspondingly Unless Stated Otherwise)

Enzyme	Substrate(s)	Experimental (Theoretical) Enzyme KIEs[a]	References
PNP	Inosine	1.25 ± 0.07[b]	Silva et al. (2011)
	Guanosine	1.37 ± 0.09[b]	
E258D/ L261A PNP	Guanosine	0.997 ± 0.005[b]	Zoi et al. (2016)
HIV-1 protease	Aminobenzoyl-Thr-Ile-Nle-pNO_2-Phe-Gln-Arg-NH_2	1.19 ± 0.05[c] (1.042–1.228)[d]	Kipp et al. (2011) and Krzemińska et al. (2016)
AR[e]	D-Ala	1.32 ± 0.04[c]	Toney et al. (2013)
	[2-^{2}H]-D-Ala	1.40 ± 0.04[c]	
	L-Ala	1.21 ± 0.03[c]	
	[2-^{2}H]-L-Ala	1.67 ± 0.05[c]	
PETNR	FMN + NADH	1.11 ± 0.09[b]	Pudney et al. (2013)
	FMN + NADPH	1.29 ± 0.02[b]	
	FMN + NADP^{2}H	0.89 ± 0.02[b]	Longbotham et al. (2016)
	h-FMN[f] + NADP^{2}H	0.92 ± 0.02[b]	
LDH	Lactate + NADH + H^+	1.11 ± 0.01[b,g]	Wang, Antoniou, et al. (2016) and Wang, Chang, et al. (2016)
	Lactate + NADH + $^2H^+$	1.22 ± 0.01[b,g]	
	Lactate + NAD^{2}H + H^+	1.14 ± 0.01[b,g]	
	Lactate + NAD^{2}H + $^2H^+$	1.24 ± 0.02[b,g]	
*Ec*DHFR	DHF + NADPH	1.02 ± 0.03[h]	Wang et al. (2014)
	DHF + NADP^{2}H	1.00 ± 0.03[h]	
	DHF + NADPH	0.92 ± 0.04 to 1.18 ± 0.09[b,i] 1.10 ± 0.05[a] (1.16[a])	Luk et al. (2013)

Continued

Table 6 Summary of Experimental and Theoretical Enzyme KIEs on the Rate Constant of Chemical Step (k^l/k^h, Where k^l and k^h are Rate Constants of the Chemical Step Catalyzed by "Light" Enzyme and "Heavy" Enzyme Correspondingly Unless Stated Otherwise)—cont'd

Enzyme	Substrate(s)	Experimental (Theoretical) Enzyme KIEs	References
N23PP/ S148A *Ec*DHFR	DHF + NADPH	$1.37 \pm 0.03^{b,j}$ $(1.26 \pm 0.04)^{j}$	Ruiz-Pernia et al. (2013)
*Tm*DHFR		0.992 ± 0.012 to $1.014 \pm 0.012^{b,k}$	Luk, Loveridge, et al. (2014) and Luk, Ruiz-Pernia, et al. (2014)
*Bs*DHFR		1.65 ± 0.09 to $1.09 \pm 0.06^{b,l}$ (1.53 ± 0.09 to 0.98 ± 0.04)	Luk, Loveridge, et al. (2014) and Luk, Ruiz-Pernia, et al. (2014)

[a]At 25°C unless state otherwise.
[b]Calculated from k_{chem} (single turnover).
[c]Calculated from k_{cat}.
[d]Computed from two different transition states.
[e]Enzyme isotope effects are measured for perdeuterated AR.
[f]^{13}C,^{15}N,^{2}H-labeled FMN.
[g]At pH/pD 4.3.
[h]Determined from the pre-steady state burst phase.
[i]In the range of 7–40°C.
[j]At 30°C.
[k]In the range of 5–65°C.
[l]In the range of 5–45°C.

product(s), conformational changes of intermediate complexes, and chemical steps. Therefore, measurement of the enzyme isotope effect on the rate of the chemical step requires special attention. Several situations and corresponding solutions can be considered:

(1) *Reaction is rate limited by the rate of the chemical step.* In this case steady-state kinetic parameter k_{cat} can be used as a direct probe of enzyme isotope effect on the rate constant of the chemical step. This scenario is illustrated by the studies of HIV-1 protease (Kipp et al., 2011). Reaction catalyzed by this enzyme is rate limited by the breakdown of the *gem*-diol intermediate (Table 1). Therefore enzyme isotope effect of 1.19 on the steady-state parameter k_{cat} reflects enzyme isotope effect on the rate constant of chemical step (Table 6). This value was in a good agreement with recently calculated values (1.042–1.228) of this isotope effect using Ensemble-Averaged VTST (Krzemińska et al., 2016).

In the study of AR, the parameter k_{cat} was also used as a probe of perdeuterated protein isotope effect on the chemical step (Toney et al., 2013). However, in this case, two consecutive proton transfer steps are jointly rate limiting the reaction (Table 1); therefore, the possibility that the enzyme isotope effect might be reporting on one, but not the other proton transfer, or a combination of both, cannot be excluded. Qualitatively, it was suggested that the initial proton transfer from the external aldimine is indeed affected by the mass of protein upon isolation of this step in measurements of substrate KIE (i.e., KIE for an isotopically labeled substrate [2-^{2}H]-L-Ala). Since deuterium KIE on this step was increased for the "heavy" AR (Table 6), it was concluded that both the enzyme and the substrate are vibrationally coupled in at least the first partially rate-limiting step (for more details, refer to Section 2.3.2.3 of this chapter). The invoked concept refers to commonly used interpretation for the violation of the rule of geometric mean (i.e., a proposition of isotope effect theory, which implies that isotope effects are independent of each other; Bigeleisen, 1955).

(2) *Reaction is rate limited by the rate of product release, but the presteady-state rate constant solely reflects the rate constant of the chemical step per se.* In this case the rate constant of the chemical step can be assessed by measuring presteady-state kinetics under single-turnover or burst conditions using rapid-mixing techniques. Details of these methods are summarized elsewhere (Fierke & Hammes, 1995; Johnson, 1992). This strategy was used to unmask the rate constant of chemistry in the enzyme isotope effect studies of PNP (Silva et al., 2011; Suarez & Schramm, 2015; Zoi et al., 2016), PETNR (Longbotham et al., 2016; Pudney et al., 2013), LDH (Wang, Chang, et al., 2016), and DHFRs from *E. coli*, *B. stearothermophilus*, and *T. maritima* (Luk et al., 2013; Luk, Loveridge, et al., 2014; Luk, Ruiz-Pernia, et al., 2014; Luk et al., 2015; Ruiz-Pernia et al., 2013; Wang et al., 2014). Fig. 3 illustrates a contrast between absence of enzyme isotope effect on steady-state kinetics (Fig. 3A) and its unmasking under presteady-state conditions (Fig. 3B) where the chemical step is rate limiting for PNP (Silva et al., 2011). The opposite is demonstrated in Fig. 4 where steady-state (Fig. 4A and B), but not presteady-state (Fig. 4C) kinetics of *Ec*DHFR are affected in the "heavy" enzyme (Wang et al., 2014).

In certain cases, additional care should be taken to expose the rate constant of chemistry under single-turnover conditions, such as variation in pH (note that a change in pH will not alter the rate constant of chemical step per se, but it will rather change the fraction of correctly protonated form of the enzyme) which was required in the studies of *Hs*LDH (Wang,

Antoniou, et al., 2016; Wang, Chang, et al., 2016) and *Ec*DHFR (Luk et al., 2013; Wang et al., 2014). In the latter case of *Ec*DHFR (Luk et al., 2013; Wang et al., 2014), the observed presteady-state rate constant k_{chem} at any given pH did not represent the rate constant of the chemical step per se due to kinetic complexity associated with the need to protonate the substrate prior to hydride transfer (Liu, Francis, et al., 2014; Wang et al., 2014). Another limitation of the presteady-state kinetics measurement as a method to assess the rate of chemical step is a requirement for this rate to be slower than the dead time of the rapid-mixing instrument. This limitation is illustrated in the case of *Hs*DHFR for which the reaction was complete within the 2-ms-long mixing time of the stopped-flow spectrophotometer (Francis et al., 2016).

(3) *Chemistry is not rate-limiting even under presteady-state conditions.* In this case complete unmasking of the rate of the chemistry *might not be possible* using presteady-state kinetics measurements and/or by changing the pH. This case is represented to a certain degree by studies of *Ec*DHFR (and possibly by DHFRs from other organisms) discussed earlier (Luk et al., 2013; Wang et al., 2014), where presteady-state kinetics were used to estimate enzyme isotope effect on the chemical step. More rigorous assessment of the enzyme isotope effect on the chemical step in these systems is achieved upon measurement of intrinsic substrate KIEs and their temperature dependencies (Francis et al., 2016; Wang et al., 2014) as discussed in the following Section 2.3.2.3 of this chapter.

Table 6 summarizes enzyme isotope effects on the rate constants of chemical step for those systems where the rate of chemistry could be assessed (at least partially). Typical values of an enzyme isotope effect on the rate constant of the chemical step are found in the range of 1.1–1.7. Two major exceptions are: (i) DHFRs from various organisms exhibiting absence of enzyme isotope effect on the rate of chemistry under physiological conditions (the pH-independent rate was not assessed). Accordingly, computational (Dametto, Antoniou, & Schwartz, 2012) and experimental (Liu, Layfield, et al., 2014) studies suggested that rate acceleration by *Ec*DHFR is assisted by the electrostatic environment of the active and not by the fast protein motions coupled to the reaction coordinate. Additionally, at nonphysiological temperatures *Ec*DHFR and *Bs*HDFR exhibited small enzyme isotope effects on the rate constant of chemical step (Luk et al., 2013; Luk, Loveridge, et al., 2014; Luk, Ruiz-Pernia, et al., 2014) suggesting a shift in their conformational ensembles under those conditions; and (ii) PETNR exhibits an inverse enzyme isotope effect on the rate constant of deuteride transfer, but normal for the rate constant of hydride transfer (Longbotham et al., 2016; Pudney et al., 2013). In this case inverse enzyme isotope effect could result from reduced steric strain in the deuterium-labeled active site and consequent relief of the crowded transition state as discussed in Section 2.1 of this chapter. Alternatively, development of tighter interactions in the transition state for the deuteride transfer would also produce an inverse secondary KIE as suggested by Świderek et al. (2014).

2.3.2.3 Experimental Studies: Enzyme Isotope Effects on Substrate Intrinsic Kinetic Isotope Effects

A KIE is defined as the ratio of the rate constants for two reactions that differ only in the isotopic composition of reactants (isotopologues) and can be expressed with the following equation:

$$\mathrm{KIE} = \frac{k_{\mathrm{light}}}{k_{\mathrm{heavy}}} \tag{2}$$

where k_{light} is the rate constant with "light" isotope and k_{heavy} is the rate constant with "heavy" isotope. The substrate KIE that arises from a single chemical step (e.g., bond cleavage or formation, change in hybridization or protonation state, etc.) is defined as intrinsic KIE ($\mathrm{KIE}_{\mathrm{int}}$). Experimental measurements in enzymatic systems typically report on observed KIE ($\mathrm{KIE}_{\mathrm{obs}}$). Since the intrinsic KIE does not by definition include the kinetic complexities inherent to observed KIEs, it offers an important means of probing activation of the specific bond under investigation. The intrinsic KIE is mathematically extracted from the experimental observed KIE using the following equation:

$$\mathrm{KIE}_{\mathrm{obs}} = \frac{\mathrm{KIE}_{\mathrm{int}} + C}{1 + C} \tag{3}$$

where C is the commitment to catalysis that arises due to kinetic complexity (e.g., Eq. (1) represents the commitment to catalysis for the minimal kinetic model of enzymatic reaction). Three of the methods used to assess the intrinsic KIE from its observed value are: (i) the Northrop method (Ghosh, Islam, Krueger, Abeysinghe, & Kohen, 2015; Northrop, 1991; Singh, Islam, & Kohen, 2016) (useful for reactions involving hydrogen transfer), or (ii) the commitment factors can be assessed using the isotope trapping method (Rose, 1980) and used to calculate the intrinsic KIE, or (iii) the multiple isotope effect method (Cook & Cleland, 2007).

Measurement of substrate intrinsic KIEs is the most direct way to probe the chemical step of the reaction, and therefore the enzyme isotope effects on this step as well. There are two major approaches toward using substrates intrinsic KIEs in the enzyme isotope effect studies:

(1) *Correlation between substrate and enzyme isotope effects.*

Values of intrinsic substrate KIEs (KIEs due to isotopic substitution in the substrate on the rate constant of the chemical step) report on the geometric and electronic features of the transition state of the reaction (Schramm, 1998). In studies of PNP (Silva et al., 2011), it was found that all substrate intrinsic KIEs are essentially identical to those measured for the corresponding "heavy" enzymes. It was suggested that the geometry of the transition state is not perturbed upon protein isotopic labeling. Similarly, for *Ec*DHFR (Wang et al., 2014) and *Hs*DHFR (Francis et al., 2016), no enzyme isotope effect was observed at relevant temperatures for intrinsic KIEs or their temperature dependence. In these cases, it was concluded that the enzyme isotope effect reports on the altered probability of finding transition state geometry rather than on altered geometry itself. In several other systems including PETNR (Longbotham et al., 2016; Pudney et al., 2013), LDH (Wang, Antoniou, et al., 2016; Wang, Chang, et al., 2016), and AR (Toney et al., 2013) substrate intrinsic KIEs were increased in "heavy" enzymes, implying that the nature of the transition state was altered. Absence of such phenomenon in the case of PNP might indicate that it is specific to the systems where the chemical step is a transfer of a "light" particle (proton or hydride) as hypothesized by Toney et al. (2013). Several studies that reported a synergy between substrate and enzyme isotope effects, rationalized it as a consequence of coupling between the enzyme vibrational motions and the reaction coordinate (Toney et al., 2013; Wang, Antoniou, et al., 2016; Wang, Chang, et al., 2016) (Table 6). The data, however, could also reflect a change in the conformational ensemble or other effects associated with altered electrostatics.

(2) *Temperature dependencies of intrinsic substrate KIEs.*

Temperature dependence of intrinsic KIEs probes the involvement of protein dynamics in catalysis for enzymatic hydrogen transfer (H^+, $H^\bullet$, or H^-) between donor and acceptor carbons. If the temperature dependence of intrinsic KIE is close to zero, then the sampling of the donor–acceptor distance (DAD) is fast and has a narrow distribution. Steep temperature dependency, however, indicates slow sampling frequencies (Francis et al., 2016; Kohen, 2015; Singh et al., 2016; Wang et al., 2014).

The temperature dependence of intrinsic KIEs can be quantified from fitting to the Arrhenius equation, Eq. (4):

$$\mathrm{KIE_{int}} = \frac{k_l}{k_h} = \frac{A_l}{A_h} \exp\left(\frac{-\Delta E_{a,l-h}}{RT}\right) \tag{4}$$

where ΔE_a is the difference in energy of activation between the different isotopologues, the subscript l denotes the "light" isotope, and the subscript h denotes the "heavy" isotope; A_l/A_h is the isotope effect on the pre-exponential Arrhenius factors; T is the absolute temperature; and R is the universal gas constant. The slope of the Arrhenius plot (ΔE_a) is close to zero for accurate, narrowly distributed transition states. Indeed, most natural (light) enzymes exhibit temperature-independent intrinsic KIEs, suggesting narrow distribution of DADs at the reactive state for the hydrogen transfer. If the protein vibrations are coupled to the bond activations in the natural-isotope abundance "light" enzyme, isotopic substitution of this enzyme could disrupt those motions and thus modulate the DAD sampling at the reactive state. This modulation in the DADs fluctuations would be reflected in a change in the temperature dependence of intrinsic KIEs for "heavy" enzymes. Since "heavy" proteins have altered vibrational frequency, an alteration in the temperature dependence of intrinsic KIEs could reveal that the fast vibrations, which had a role in hydrogen transfer in a natural enzyme, are distorted in a "heavy" enzyme.

Current literature that uses temperature dependence of intrinsic KIEs as a tool to study "heavy" enzymes includes studies of DHFR (Francis et al., 2016; Luk et al., 2013; Luk, Loveridge, et al., 2014; Luk, Ruiz-Pernia, et al., 2014; Luk et al., 2015; Ruiz-Pernia et al., 2013; Wang et al., 2014) and PETNR (Longbotham et al., 2016; Pudney et al., 2013). Studies of "heavy" *Ec*DHFR revealed that the hydride-transfer intrinsic substrate KIEs are temperature independent and identical to "light" *Ec*DHFR at physiological temperatures (25–45°C). Below 25°C (Wang et al., 2014), the "heavy" *Ec*DHFR had steeper temperature dependence of intrinsic substrate KIEs, which may suggest intrinsic phase transition of the "heavy" enzyme at low temperature (i.e., reactive conformational distribution of the "heavy" enzyme may be altered at lower temperatures resulting in a change in DAD fluctuations, which in turn would alter the temperature dependence of intrinsic KIEs). Furthermore, in the case of *Hs*DHFR, hydride-transfer intrinsic KIEs remain temperature independent for both the "heavy" and "light" *Hs*DHFR throughout the temperature range, suggesting equally narrow DADs distributions for these enzymes (Francis et al., 2016). In contrast, studies of "heavy" PETNR (^{13}C-, ^{15}N-, ^{2}H-, and ^{15}N-labeled) revealed an alteration in the temperature dependence of KIEs due to isotope labeling of the protein (Pudney et al., 2013). The authors rationalized these observations as the vibrational coupling of chemistry to vibrational modes of the protein.

The case of DHFRs demonstrates the problems associated with such studies. TPS calculations (Dametto et al., 2012) found no protein-promoting vibrations (PPVs) for the *Ec*DHFR and significant PPVs for *Hs*DHFR (Masterson & Schwartz, 2015). The last reference suggested that "heavy" enzymes could be used to test that prediction, but studies of intrinsic substrate KIE for *Ec*DHFR (Wang et al., 2014) and *Hs*DHFR (Francis et al., 2016) found no enzyme isotope effect on intrinsic substrate KIEs for either of these enzymes at physiological temperatures. This discrepancy can be rationalized by the weak sensitivity of dynamics at the picosecond timescale to the minor change of mass upon isotopic labeling of the enzyme (~10%). For example, sampling dynamics of 100 cm^{-1} in the "light" enzyme would only be reduced to ~94 cm^{-1}, which might not be detectable experimentally (Francis et al., 2016).

2.3.3 Regio-Specific Effects of Enzyme Isotopic Labeling on Catalysis

Labeling a specific pattern of amino acids in the protein is another approach that is used in enzyme isotope effect studies to examine the mass-dependent effects on catalysis (e.g., BO effects) vs effects that are not proportional to the mass of the protein (e.g., non-BO effects), and the effect of specific protein regions. Major approaches in this direction involve:

(1) Correlation between protein mass and enzyme isotope effects on the rate of chemical step.

The approach was predominantly utilized to isolate effects of deuteration from the global protein mass effect. As discussed in Section 2.1 of this chapter, altered character of C—D bond in comparison to C—H bond might perturb electrostatic and steric properties of the system, which in turn might modulate formation of the transition state. In order to exclude this non-BO effect for PNP, three partially labeled enzymes, such as ^{2}H,^{13}C,^{15}N-, ^{13}C,^{15}N-, and ^{2}H-enriched proteins were produced, followed by examination of enzyme isotope effect on the chemical step or other kinetic properties. The BO approximation for "heavy" enzymes requires that the property under study (e.g., reaction rate) would be correlated to the mass of the protein, as found in the case of *Hs*PNP (Table 7) (Suarez & Schramm, 2015).

Table 7 Effects of Protein Mass Increase (% Addition to Natural Enzyme) for Different Uniform Protein-Labeling Patterns on the Enzyme Isotope and/or Mass Effects on Rates of the Chemical Step

Enzyme (Reference)		Mass Increase % (Enzyme Isotope Effect on Rate)				References
		^{15}N	^{2}H	$^{13}C,^{15}N$	$^{2}H,^{13}C,^{15}N$	
Isolation of the effect of ^{2}H-labeling						
*Hs*PNP		N/A	5.3 (1.18 ± 0.03)	5.6 (1.19 ± 0.08)	9.9 (1.36 ± 0.09)	Silva et al. (2011) and Suarez and Schramm (2015)
*Bs*DHFR	25°C	N/A	5.0 (1.18 ± 0.03)	5.5 (1.21 ± 0.03)	11.1 (1.14 ± 0.04)	Luk, Loveridge, et al. (2014) and Luk, Ruiz-Pernia, et al. (2014)
	5°C	N/A	5.0 (1.42 ± 0.05)	5.5 (1.24 ± 0.04)	11.1 (1.65 ± 0.09)	
Isolation of the effect of ^{15}N-labeling						
HIV-1 protease		1.2 (1.03 ± 0.04)	N/A	N/A	11.6 (1.19 ± 0.05)	Kipp et al. (2011)
PETNR		1.3 (1.01 ± 0.02)	N/A	N/A	9.8 (1.29 ± 0.02)	Pudney et al. (2013)

In contrast, for *Bs*DHFR, where a similar strategy was used, no analogous correlation was found at 25°C (Table 7) (Luk, Loveridge, et al., 2014; Luk, Ruiz-Pernia, et al., 2014). The result was rationalized by minimal dynamic coupling for this enzyme under physiological conditions. However, a mass-modulated trend of enzyme isotope effects was revealed at lower nonphysiological temperatures for this enzyme (Table 7). Two other cases outlined in Table 7 illustrate isolation of the effect of ^{15}N-labeling on the chemical step. For both HIV-1 protease (Kipp et al., 2011) and PETNR (Pudney et al., 2013), a negligible increase in mass associated with uniform ^{15}N-labeling was accompanied by absence of a measurable enzyme isotope effect on the chemical step.

(2) Spatial resolution of dynamic effects on enzyme catalysis.

This approach is utilized to test whether the enzyme isotope effects are caused by the global protein dynamics, local motions of specific residues, or specific domains of the protein. For example, protein promoting vibrations may involve several residues that stretch across the protein, but not all residues. Additionally, site-directed mutagenesis can be used to complement these methods. Two illustrative cases are as follows:

(a) Case of amino acid-specific-labeled *Hs*PNP (Suarez & Schramm, 2015).

Following the demonstration that the mass of *Hs*PNP is correlated to its reaction rate (Suarez & Schramm, 2015), it was interesting to test the effect of an active site histidine (His257) in catalysis. TPS simulations suggested that the dynamics of His257 are coupled to the reaction coordinate (S_N1 substitution on C1 of the substrate) (Antoniou et al., 2012; Núñez et al., 2006). While isotopic labeling of a single residue is challenging, *Hs*PNP in which all eight histidines were uniformly ^{15}N,^{2}H-labeled (only nonexchangeable hydrogens) was generated (Fig. 6). Observation of a significant on-enzyme KIE (1.13 ± 0.02) for [^{2}H,^{15}N]$_8$-His-PNP upon negligible global mass increase (only 0.19%) supported the hypothesis that alteration of these specific residues is responsible for much of the disruption of dynamics of the chemical barrier crossing, which indirectly accords with the protein promoting vibration prediction. This study in turn was followed up by computational analyses confirming that His257 is less mobile in the "heavy" PNP (Zoi et al., 2016). Interestingly, the same study reports experimental results supporting recovery of the dynamics of His257 in "heavy" PNP upon mutation of 2 s layer residues in its proximity. This recovery was manifested in the absence of enzyme KIE for the corresponding mutant (Table 6).

(b) Case of segmentally labeled *Ec*DHFR (Luk et al., 2015).

This case tested if the small but measurable enzyme isotope effect for this enzyme (1.13 ± 0.08 at 35°C) results from altered fast promoting protein

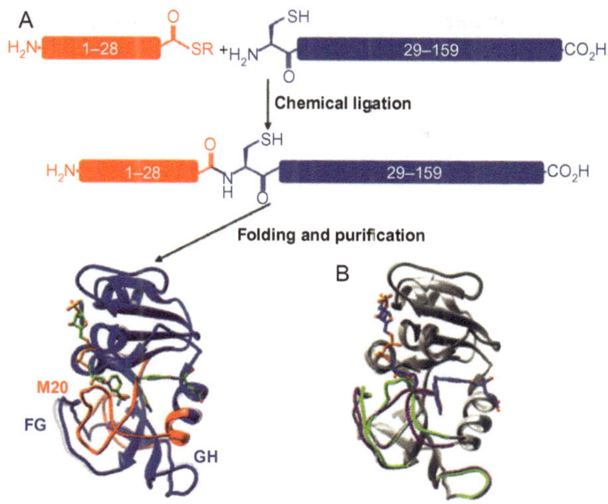

Fig. 6 (A) Synthetic scheme for chemical ligation of the N-terminal thioester peptide (*red*, residues 1–28) and C-terminal cysteine peptide (*blue*, residues 29–159) of *Ec*DHFR. Structural elements (M20 loop, comprising part of the active site, and FG and GH loops, which are involved in electrostatic preorganization of the active site) are indicated. (B) X-ray crystal structures of *Ec*DHFR in the closed (PDB ID 1RX2) and occluded (PDB ID 1RX4) conformations. The M20, FG, and GH loops are in *green* (*closed*) and *purple* (*occluded*) conformations with the ligands ($NADP^+$ and folate) shown only in the closed conformation. *From Luk, L. Y., Ruiz-Pernia, J. J., Adesina, A. S., Loveridge, E. J., Tuñon, I., Moliner, V.,* et al. *(2015). Chemical ligation and isotope labeling to locate dynamic effects during catalysis by dihydrofolate reductase.* Angewandte Chemie International Edition English, 54, *9016–9020. Copyright © 2000 by John Wiley Sons, Inc. Reprinted by permission of John Wiley & Sons, Inc.*

dynamics of only part of the active site, or from perturbed dynamics of the remaining protein. Two "heavy" segments of the protein were prepared: $^{13}C,^{15}N,^{2}H$-[1–28]*Ec*DHFR in which residues 1–28 corresponding to a segment of the active site were modified, and $^{13}C,^{15}N,^{2}H$-[29–159] *Ec*DHFR that includes "heavy" atom substitution at other parts of the active site and the rest of the protein. These two segments were chemically ligated to the complementary "light" polypeptide segments to produce full-length *Ec*DHFR with different isotopic labeling patterns (Fig. 6).

It was found that active site-labeled *Ec*DHFR ("heavy" 1–28; "light" 29–159) exhibited absence of enzyme KIE on the hydride-transfer rate constant (under single turnover conditions), whereas "light" 1–28; "heavy" 29–159 could account for entire observed "heavy" enzyme KIE on the same kinetic parameter. The absence of promoting dynamics in one segment is in accordance with TPS simulation predicting no protein promoting vibrations in *Ec*DHFR (Dametto et al., 2012), but the effect of the other does not. The latter was rationalized in terms of perturbation of efficiency of electrostatic preorganization required to reach the transition state. Another interpretation is that the lack of effect of "heavy" *Ec*DHFR in intrinsic substrate KIEs and their temperature dependence at 35°C (Wang et al., 2014) accords with lack of protein promoting vibrations, although the lack of enzyme isotope effect on intrinsic substrate KIEs in *Hs*DHFR (Francis et al., 2016) does not match predictions from calculations for that enzyme (Masterson & Schwartz, 2015). A simulation that will predict or reproduce experimental data may assist reaching a more solid conclusion.

3. SUMMARY

Enzyme isotope effect studies have recently been gaining popularity as an attempt to provide experimental evidence for computationally predicted roles of proteins' fast dynamics in enzyme catalysis. This account summarizes methods of preparation and structural characterization of "heavy" enzymes, as well as reported effects of isotopic labeling on physiochemical properties of the proteins. This chapter further presents a summary of approaches toward measurement and interpretation of enzyme isotope effects. Specifically, case studies of several mechanistically unrelated enzymes, PNP, HIV-1 protease, AR, LDH, PETNR, and DHFR, are considered in a systematic way rather than on a case-by-case basis. Enzyme isotope effects on the steady-state kinetic parameters are discussed with emphasis on whether these are effects on the chemical step per se or mostly represent effects on other catalytic events. Commonly, only the former reflects effects that may test vibrational coupling of an enzyme and its catalyzed chemical reaction. A critical support for this test is provided by agreement with computational simulations. Several recent experimental methods also examined spatial resolution of the enzyme isotope effects down to protein segments and specific amino acid residues. We conclude that experimental examination of isotopically substituted enzyme, in combination with computational studies, could be a used to disentangle the role of fast protein dynamics in catalysis. Great care is however needed to distinguish between effects on the chemical step vs other kinetic barrier crossings, and between vibrational effects (associated with subnanosecond dynamics) vs electrostatic alteration of the isotopically labeled protein (associated with slower motions of the protein and its interactions with ligands). Consequently, several factors are critical for meaningful interpretation of the enzyme isotope effects requires: (1) comprehensive knowledge of the chemical and kinetic mechanism of the enzyme under investigation, (2) well-established protocols for assaying kinetic and thermodynamic parameters of the reaction catalyzed by this enzyme, (3) reliable means of structural characterization of the enzyme under study, and (4) computational simulations of the system under study, and preferably reproduction or prediction of measurable parameters (reaction rates, KIEs, and/or their temperature dependencies) of the reaction catalyzed by the "light" and the "heavy" enzyme.

ACKNOWLEDGMENTS

This work was supported by NIH GM065368 and NSF CHE-0133117.

REFERENCES

Agarwal, P. K., Billeter, S. R., Rajagopalan, P. R., Benkovic, S. J., & Hammes-Schiffer, S. (2002). Network of coupled promoting motions in enzyme catalysis. *Proceedings of the National Academy of Sciences of the United States of America, 99*, 2794–2799.

Antoniou, D., Ge, X., Schramm, V. L., & Schwartz, S. D. (2012). Mass modulation of protein dynamics associated with barrier crossing in purine nucleoside phosphorylase. *The Journal of Physical Chemistry Letters, 3*, 3538–3544.

Basner, J. E., & Schwartz, S. D. (2004). Donor–acceptor distance and protein promoting vibration coupling to hydride transfer: A possible mechanism for kinetic control in isozymes of human lactate dehydrogenase. *The Journal of Physical Chemistry B, 108*, 444–451.

Bigeleisen, J. J. (1955). Statistical mechanics of isotopic systems with small quantum corrections. I. General considerations and the rule of the geometric mean. *The Journal of Chemical Physics, 23*, 2264–2267.

Blakeley, M. P., Ruiz, F., Cachau, R., Hazemann, I., Meilleur, F., Mitschler, A., et al. (2008). Quantum model of catalysis based on a mobile proton revealed by subatomic x-ray and neutron diffraction studies of h-aldose reductase. *Proceedings of the National Academy of Sciences of the United States of America, 105*, 1844–1848.

Blum, M. M., Tomanicek, S. J., John, H., Hanson, B. L., Ruterjans, H., Schoenborn, B. P., et al. (2010). X-ray structure of perdeuterated diisopropyl fluorophosphatase (DFPase): Perdeuteration of proteins for neutron diffraction. *Acta Crystallographica Section F, Structural Biology and Crystallization Communications, 66*, 379–385.

Bolhuis, P. G., Chandler, D., Dellago, C., & Geissler, P. L. (2002). Transition path sampling: Throwing ropes over rough mountain passes, in the dark. *Annual Review of Physical Chemistry, 53*, 291–318.

Brockwell, D., Yu, L., Cooper, S., McCleland, S., Cooper, A., Attwood, D., et al. (2001). Physicochemical consequences of the perdeuteriation of glutathione S-transferase from S. japonicum. *Protein Science, 10*, 572–580.

Budayova-Spano, M., Fisher, S. Z., Dauvergne, M. T., Agbandje-McKenna, M., Silverman, D. N., Myles, D. A., et al. (2006). Production and X-ray crystallographic analysis of fully deuterated human carbonic anhydrase II. *Acta Crystallographica Section F, Structural Biology and Crystallization Communications, 62*, 6–9.

Callender, R., & Dyer, R. B. (2015). The dynamical nature of enzymatic catalysis. *Accounts of Chemical Research, 48*, 407–413.

Caratzoulas, S., Mincer, J. S., & Schwartz, S. D. (2002). Identification of a protein-promoting vibration in the reaction catalyzed by horse liver alcohol dehydrogenase. *Journal of the American Chemical Society, 124*, 3270–3276.

Careri, G., & Gratton, E. (1986). The statistical time correlation approach to enzyme action: The role of hydration. In G. R. Welch (Series Ed.) & I. Prigogine & G. Nicolis (Vol. Eds.), *The fluctuating enzyme: Vol. 5. Nonequilibrium problems in the physical sciences and biology* (pp. 1–191). Hoboken, NJ: John Wiley & Sons.

Chahrour, O., Cobice, D., & Malone, J. (2015). Stable isotope labelling methods in mass spectrometry-based quantitative proteomics. *Journal of Pharmaceutical and Biomedical Analysis, 113*, 2–20.

Chen, C.-H., Liu, I. W., MacColl, R., & Berns, D. S. (1983). Differences in structure and stability between normal and deuterated proteins (phycocyanin). *Biopolymers, 22*, 1223–1233.

Cleland, W. W. (2005). The use of isotope effects to determine enzyme mechanisms. *Archives of Biochemistry and Biophysics, 433*, 2–12.

Cook, P. F. (1991). *Enzyme mechanism from isotope effects.* Boca Raton: CRC Press, pp. 181–202.

Cook, P. F., & Cleland, W. W. (2007). *Enzyme kinetics and mechanism.* London, New York: Garland Science.

Dametto, M., Antoniou, D., & Schwartz, S. D. (2012). Barrier crossing in dihydrofolate reductasedoes not involve a rate-promoting vibration. *Molecular Physics, 110*, 531–536.

Di Costanzo, L., Moulin, M., Haertlein, M., Meilleur, F., & Christianson, D. W. (2007). Expression, purification, assay, and crystal structure of perdeuterated human arginase I. *Archives of Biochemistry and Biophysics, 465*, 82–89.

Doucet, N., Watt, E. D., & Loria, J. P. (2009). The flexibility of a distant loop modulates active site motion and product release in ribonuclease A. *Biochemistry, 48*, 7160–7168.

Duff, A. P., Wilde, K. L., Rekas, A., Lake, V., & Holden, P. J. (2015). Chapter one—Robust high-yield methodologies for ^{2}H and $^{2}H/^{15}N/^{13}C$ labeling of proteins for structural investigations using neutron scattering and NMR. In K. Zvi (Ed.), *Methods in enzymology: Vol. 565* (pp. 3–25). Cambridge, MA: Academic Press.

Dybala-Defratyka, A., Paneth, P., Banerjee, R., & Truhlar, D. G. (2007). Coupling of hydrogenic tunneling to active-site motion in the hydrogen radical transfer catalyzed by a coenzyme B12-dependent mutase. *Proceedings of the National Academy of Sciences of the United States of America, 104*, 10774–10779.

Fierke, C. A., & Hammes, G. G. (1995). Transient kinetic approaches to enzyme mechanisms. *Methodsin enzymology: Vol. 249* (pp. 3–37). Cambridge, MA: Academic Press.

Francis, K., Sapienza, P. J., Lee, A. L., & Kohen, A. (2016). The effect of protein mass modulation on human dihydrofolate reductase. *Biochemistry, 55*, 1100–1106.

Gamble, T. R., Clauser, K. R., & Kossiakoff, A. A. (1994). The production and X-ray structure determination of perdeuterated staphylococcal nuclease. *Biophysical Chemistry, 53*, 15–25.

Ghosh, A. K., Islam, Z., Krueger, J., Abeysinghe, T., & Kohen, A. (2015). The general base in the thymidylate synthase catalyzed proton abstraction. *Physical Chemistry Chemical Physics, 17*, 30867–30875.

Golden, E., Attwood, P. V., Duff, A. P., Meilleur, F., & Vrielink, A. (2015). Production and characterization of recombinant perdeuterated cholesterol oxidase. *Analytical Biochemistry, 485*, 102–108.

Greenfield, N. J. (2007a). Using circular dichroism collected as a function of temperature to determine the thermodynamics of protein unfolding and binding interactions. *Nature Protocols, 1*, 2527–2535.

Greenfield, N. J. (2007b). Using circular dichroism spectra to estimate protein secondary structure. *Nature Protocols*, *1*, 2876–2890.

Haertlein, M., Moulin, M., Devos, J. M., Laux, V., Dunne, O., & Trevor Forsyth, V. (2016). Chapter five—Biomolecular deuteration for neutron structural biology and dynamics. In K. Zvi (Ed.), *Methods in enzymology: Vol. 566* (pp. 113–157). Cambridge, MA: Academic Press.

Haino, T., Fukuta, K., Iwamoto, H., & Iwata, S. (2009). Noncovalent isotope effect for guest encapsulation in self-assembled molecular capsules. *Chemistry*, *15*, 13286–13290.

Hammes-Schiffer, S. (2002). Impact of enzyme motion on activity. *Biochemistry*, *41*, 13335–13343.

Han, R., & Coleman, J. E. (1995). Dependence of the phosphorylation of alkaline phosphatase by phosphate monoesters on the pKa of the leaving group. *Biochemistry*, *34*, 4238–4245.

Hansen, P. E. (1988). Isotope effects in nuclear shielding. *Progress in Nuclear Magnetic Resonance Spectroscopy*, *20*, 207–255.

Hatcher, E., Soudackov, A. V., & Hammes-Schiffer, S. (2004). Proton-coupled electron transfer in soybean lipoxygenase. *Journal of the American Chemical Society*, *126*, 5763–5775.

Hattori, A., Crespi, H. L., & Katz, J. J. (1965). Effect of side-chain deuteration on protein stability. *Biochemistry*, *4*, 1213–1225.

Hay, S., Johannissen, L. O., Hothi, P., Sutcliffe, M. J., & Scrutton, N. S. (2012). Pressure effects on enzyme-catalyzed quantum tunneling events arise from protein-specific structural and dynamic changes. *Journal of the American Chemical Society*, *134*, 9749–9754.

Heinrich, F. (2016). Chapter seven—Deuteration in biological neutron reflectometry. In K. Zvi (Ed.), *Methods in enzymology: Vol. 566* (pp. 211–230): Academic Press.

Hoopes, J. T., Elberson, M. A., Preston, R. J., Reddy, P. T., & Kelman, Z. (2015). Chapter two—Protein labeling in Escherichia coli with ^{2}H, ^{13}C, and ^{15}N. In K. Zvi (Ed.), *Methods in enzymology: Vol. 565* (pp. 27–44). Cambridge, MA: Academic Press.

Hsieh, J., Walker, S. C., Fierke, C. A., & Engelke, D. R. (2009). Pre-tRNA turnover catalyzed by the yeast nuclear RNase P holoenzyme is limited by product release. *RNA*, *15*, 224–234.

Hull, W. E., Halford, S. E., Gutfreund, H., & Sykes, B. D. (1976). ^{31}P nuclear magnetic resonance study of alkaline phosphatase: The role of inorganic phosphate in limiting the enzyme turnover rate at alkaline pH. *Biochemistry*, *15*, 1547–1561.

Ikegami, A. (1986). The statistical nature of protein structural fluctuations. In G. R. Welch (Series Ed.) & I. Prigogine & G. Nicolis (Vol. Eds.), *The fluctuating enzyme: Vol. 5. Non-equilibrium problems in the physical sciences and biology* (pp. 1–191). Hoboken, NJ: John Wiley & Sons.

Johnson, K. A. (1992). 1 Transient-state kinetic analysis of enzyme reaction pathways. In S. S. David (Ed.), *The enzymes: Vol. 20* (pp. 1–61). Cambridge, MA: Academic Press.

Kipp, D. R., Silva, R. G., & Schramm, V. L. (2011). Mass-dependent bond vibrational dynamics influence catalysis by HIV-1 protease. *Journal of the American Chemical Society*, *133*, 19358–19361.

Klinman, J. P. (2015). Dynamically achieved active site precision in enzyme catalysis. *Accounts of Chemical Research*, *48*, 449–456.

Knapp, M. J., Rickert, K., & Klinman, J. P. (2002). Temperature-dependent isotope effects in soybean lipoxygenase-1: Correlating hydrogen tunneling with protein dynamics. *Journal of the American Chemical Society*, *124*, 3865–3874.

Kohen, A. (2015). Role of dynamics in enzyme catalysis: Substantial versus semantic controversies. *Accounts of Chemical Research*, *48*, 466–473.

Kohen, A., Cannio, R., Bartolucci, S., & Klinman, J. P. (1999). Enzyme dynamics and hydrogen tunnelling in a thermophilic alcohol dehydrogenase. *Nature*, *399*, 496–499.

Krzemińska, A., Moliner, V., & Świderek, K. (2016). Dynamic and electrostatic effects on the reaction catalyzed by HIV-1 protease. *Journal of the American Chemical Society*, *138*, 16283–16298.

Kuchitsu, K., & Bartell, L. S. (1962). Effect of anharmonic vibrations on the bond lengths of polyatomic molecules. II. Cubic constants and equilibrium bond lengths of methane. *The Journal of Chemical Physics*, *36*, 2470–2481.

Le, M. T., Brown, R. E., Simon, A. E., & Dayie, T. K. (2015). Chapter nineteen—*In vivo*, large-scale preparation of uniformly ^{15}N- and site-specifically ^{13}C-labeled homogeneous, recombinant RNA for NMR studies. In K. Zvi (Ed.), *Methods in enzymology: Vol. 565* (pp. 495–535). Cambridge, MA: Academic Press.

Lin, M. T., Fukazawa, R., Miyajima-Nakano, Y., Matsushita, S., Choi, S. K., Iwasaki, T., *et al.* (2015). Chapter three—Escherichia coli auxotroph host strains for amino acid-selective isotope labeling of recombinant proteins. In K. Zvi (Ed.), *Methods in enzymology: Vol. 565* (pp. 45–66). Cambridge, MA: Academic Press.

Liu, C. T., Francis, K., Layfield, J. P., Huang, X., Hammes-Schiffer, S., Kohen, A., et al. (2014a). Escherichia coli dihydrofolate reductase catalyzed proton and hydride transfers: Temporal order and the roles of Asp27 and Tyr100. *Proceedings of the National Academy of Sciences of the United States of America*, *111*, 18231–18236.

Liu, X., Hanson, B. L., Langan, P., & Viola, R. E. (2007). The effect of deuteration on protein structure: A high-resolution comparison of hydrogenous and perdeuterated haloalkane dehalogenase. *Acta Crystallographica Section D*, *63*, 1000–1008.

Liu, C. T., Layfield, J. P., Stewart, R. J., French, J. B., Hanoian, P., Asbury, J. B., et al. (2014b). Probing the electrostatics of active site microenvironments along the catalytic cycle for Escherichia coli dihydrofolate reductase. *Journal of the American Chemical Society*, *136*, 10349–10360.

Longbotham, J. E., Hardman, S. J., Gorlich, S., Scrutton, N. S., & Hay, S. (2016). Untangling heavy protein and cofactor isotope effects on enzyme-catalyzed hydride transfer. *Journal of the American Chemical Society*, *138*, 13693–13699.

Luk, L. Y., Ruiz-Pernia, J., Dawson, W. M., Roca, M., Loveridge, E. J., Glowacki, D. R., et al. (2013). Unraveling the role of protein dynamics in dihydrofolate reductase catalysis. *Proceedings of the National Academy of Sciences of the United States of America*, *110*, 16344–16349.

Luk, L. Y., Loveridge, E. J., & Allemann, R. K. (2014a). Different dynamical effects in mesophilic and hyperthermophilic dihydrofolate reductases. *Journal of the American Chemical Society*, *136*, 6862–6865.

Luk, L. Y., Ruiz-Pernia, J. J., Adesina, A. S., Loveridge, E. J., Tuñon, I., Moliner, V., et al. (2015). Chemical ligation and isotope labeling to locate dynamic effects during catalysis by dihydrofolate reductase. *Angewandte Chemie International Edition English*, *54*, 9016–9020.

Luk, L. Y., Ruiz-Pernia, J. J., Dawson, W. M., Loveridge, E. J., Tuñon, I., Moliner, V., et al. (2014b). Protein isotope effects in dihydrofolate reductase from Geobacillus stearothermophilus show entropic-enthalpic compensatory effects on the rate constant. *Journal of the American Chemical Society*, *136*, 17317–17323.

Lumry, R., & Gregory, R. B. (1986). Free-energy management in protein reactions: Concepts, complications, and compensation. In G. R. Welch (Series Ed.), I. Prigogine, & G. Nicolis (Vol. Eds.), *The fluctuating enzyme: Vol. 5. Nonequilibrium problems in the physical sciences and biology* (pp. 1–191). Hoboken, NJ: John Wiley & Sons.

Masterson, J. E., & Schwartz, S. D. (2015). Evolution alters the enzymatic reaction coordinate of dihydrofolate reductase. *The Journal of Physical Chemistry B*, *119*, 989–996.

Masterson, J. E., & Schwartz, S. D. (2013). Changes in protein architecture and subpicosecond protein dynamics impact the reaction catalyzed by lactate dehydrogenase. *The Journal of Physical Chemistry A*, *117*, 7107–7113.

Meilleur, F., Contzen, J., Myles, D. A. A., & Jung, C. (2004). Structural stability and dynamics of hydrogenated and perdeuterated cytochrome P450cam (CYP101). *Biochemistry*, *43*, 8744–8753.

Meilleur, F., Dauvergne, M.-T., Schlichting, I., & Myles, D. A. A. (2005). Production and X-ray crystallographic analysis of fully deuterated cytochrome P450cam. *Acta Crystallographica Section D, 61*, 539–544.

Michel, E., & Allain, F. H. T. (2015). Chapter sixteen—Selective amino acid segmental labeling of multi-domain proteins. In K. Zvi (Ed.), *Methods in enzymology: Vol. 565* (pp. 389–422). Cambridge, MA: Academic Press.

Mincer, J. S., & Schwartz, S. D. (2003). A computational method to identify residues important in creating a protein promoting vibration in enzymes. *The Journal of Physical Chemistry B, 107*, 366–371.

Northrop, D. B. (1981). The expression of isotope effects on enzyme-catalyzed reactions. *Annual Review of Biochemistry, 50*, 103–131.

Northrop, D. B. (1991). Intrinsic isotope effects in enzyme catalyzed reactions. In P. F. Cook (Ed.), *Enzyme mechanism from isotope effects* (pp. 181–202). Boca Raton: CRC Press.

Núñez, S., Wing, C., Antoniou, D., Schramm, V. L., & Schwartz, S. D. (2006). Insight into catalytically relevant correlated motions in human purine nucleoside phosphorylase. *The Journal of Physical Chemistry A, 110*, 463–472.

Perrin, C. L. (2010). Secondary equilibrium isotope effects on acidity. In P. R. John (Ed.), *Advances in physical organic chemistry: Vol. 44* (pp. 123–171). Cambridge, MA: Academic Press.

Perrin, C. L., & Dong, Y. (2007). Secondary deuterium isotope effects on the acidity of carboxylic acids and phenols. *Journal of the American Chemical Society, 129*, 4490–4497.

Peuker, S., Andersson, H., Gustavsson, E., Maiti, K. S., Kania, R., Karim, A., et al. (2016). Efficient isotope editing of proteins for site-directed vibrational spectroscopy. *Journal of the American Chemical Society, 138*, 2312–2318.

Pudney, C. R., Guerriero, A., Baxter, N. J., Johannissen, L. O., Waltho, J. P., Hay, S., et al. (2013). Fast protein motions are coupled to enzyme H-transfer reactions. *Journal of the American Chemical Society, 135*, 2512–2517.

Pudney, C. R., Johannissen, L. O., Sutcliffe, M. J., Hay, S., & Scrutton, N. S. (2010). Direct analysis of donor – acceptor distance and relationship to isotope effects and the force constant for barrier compression in enzymatic H-tunneling reactions. *Journal of the American Chemical Society, 132*, 11329–11335.

Quinn, D. M. (2005). Theory and practice of solvent isotope effects. In A. Kohen & H. H. Limbach (Eds.), *Isotope effects in chemistry and biology* (pp. 995–1018). Boca Raton: CRC Press.

Radkiewicz, J. L., & Brooks, C. L. (2000). Protein dynamics in enzymatic catalysis: Exploration of dihydrofolate reductase. *Journal of the American Chemical Society, 122*, 225–231.

Rechavi, D., Scarso, A., & Rebek, J., Jr. (2004). Isotopomer encapsulation in a cylindrical molecular capsule: A probe for understanding noncovalent isotope effects on a molecular level. *Journal of the American Chemical Society, 126*, 7738–7739.

Renger, T., & Marcus, R. (2002). On the relation of protein dynamics and exciton relaxation in pigment–protein complexes: An estimation of the spectral density and a theory for the calculation of optical spectra. *The Journal of Chemical Physics, 116*, 9997–10019.

Roca, M., Moliner, V., Tuñón, I., & Hynes, J. T. (2006). Coupling between protein and reaction dynamics in enzymatic processes: Application of Grote – Hynes theory to catechol *O*-methyltransferase. *Journal of the American Chemical Society, 128*, 6186–6193.

Rokop, S., Gajda, L., Parmerter, S., Crespi, H. L., & Katz, J. J. (1969). Purification and characterization of fully deuterated enzymes. *Biochimica et Biophysica Acta, 191*, 707–715.

Rose, I. A. (1980). The isotope trapping method: Desorption rates of productive E·S complexes. *Methods in Enzymology, 64*, 47–59.

Ruiz-Pernia, J. J., Luk, L. Y., Garcia-Meseguer, R., Marti, S., Loveridge, E. J., Tuñon, I., et al. (2013). Increased dynamic effects in a catalytically compromised variant of Escherichia coli dihydrofolate reductase. *Journal of the American Chemical Society, 135*, 18689–18696.

Saen-Oon, S., Quaytman-Machleder, S., Schramm, V. L., & Schwartz, S. D. (2008). Atomic detail of chemical transformation at the transition state of an enzymatic reaction. *Proceedings of the National Academy of Sciences of the United States of America, 105*, 16543–16548.

Schramm, V. L. (1998). Enzymatic transition states and transition state analog design. *Annual Review of Biochemistry, 67*, 693–720.

Schramm, V. L. (2015). Transition states and transition state analogue interactions with enzymes. *Accounts of Chemical Research, 48*, 1032–1039.

Schwartz, S. D., & Schramm, V. L. (2009). Enzymatic transition states and dynamic motion in barrier crossing. *Nature Chemical Biology, 5*, 551–558.

Silva, R. G., Murkin, A. S., & Schramm, V. L. (2011). Femtosecond dynamics coupled to chemical barrier crossing in a Born-Oppenheimer enzyme. *Proceedings of the National Academy of Sciences of the United States of America, 108*, 18661–18665.

Singh, P., Francis, K., & Kohen, A. (2015). Network of remote and local protein dynamics in dihydrofolate reductase catalysis. *ACS Catalysis, 5*, 3067–3073.

Singh, P., Islam, Z., & Kohen, A. (2016). Examinations of the chemical step in enzyme catalysis. *Methods in Enzymology, 577*, 287–318.

Somogyi, B., & Damjanovich, S. (1986). A microenvironmental approch to enzyme dynamics. In G. R. Welch (Series Ed.) & I. Prigogine & G. Nicolis (Vol. Eds.), *The fluctuating enzyme: Vol. 5. Nonequilibrium problems in the physical sciences and biology* (pp. 1–191). Hoboken, NJ: John Wiley & Sons.

Suarez, J., & Schramm, V. L. (2015). Isotope-specific and amino acid-specific heavy atom substitutions alter barrier crossing in human purine nucleoside phosphorylase. *Proceedings of the National Academy of Sciences of the United States of America, 112*, 11247–11251.

Świderek, K., Ruiz-Pernia, J. J., Moliner, V., & Tunon, I. (2014). Heavy enzymes—Experimental and computational insights in enzyme dynamics. *Current Opinion in Chemical Biology, 21*, 11–18.

Toney, M. D., Castro, J. N., & Addington, T. A. (2013). Heavy-enzyme kinetic isotope effects on proton transfer in alanine racemase. *Journal of the American Chemical Society, 135*, 2509–2511.

Truhlar, D. G., & Garrett, B. C. (1984). Variational transition state theory. *Annual Review of Physical Chemistry, 35*, 159–189.

Turowski, M., Yamakawa, N., Meller, J., Kimata, K., Ikegami, T., Hosoya, K., et al. (2003). Deuterium isotope effects on hydrophobic interactions: The importance of dispersion interactions in the hydrophobic phase. *Journal of the American Chemical Society, 125*, 13836–13849.

Wang, Z., Antoniou, D., Schwartz, S. D., & Schramm, V. L. (2016a). Hydride transfer in DHFR by transition path sampling, kinetic isotope effects, and heavy enzyme studies. *Biochemistry, 55*, 157–166.

Wang, Z., Chang, E. P., & Schramm, V. L. (2016b). Triple isotope effects support concerted hydride and proton transfer and promoting vibrations in human heart lactate dehydrogenase. *Journal of the American Chemical Society, 138*, 15004–15010.

Wang, Z., Singh, P., Czekster, C. M., Kohen, A., & Schramm, V. L. (2014). Protein mass-modulated effects in the catalytic mechanism of dihydrofolate reductase: Beyond promoting vibrations. *Journal of the American Chemical Society, 136*, 8333–8341.

Warshel, A. (1984). Dynamics of enzymatic reactions. *Proceedings of the National Academy of Sciences of the United States of America, 81*, 444–448.

Warshel, A., & Bora, R. P. (2016). Perspective: Defining and quantifying the role of dynamics in enzyme catalysis. *The Journal of Chemical Physics*, *144*, 180901.

Zhang, H., & van Ingen, H. (2016). Isotope-labeling strategies for solution NMR studies of macromolecular assemblies. *Current Opinion in Structural Biology*, *38*, 75–82.

Zhao, C., Parrish, R. M., Smith, M. D., Pellechia, P. J., Sherrill, C. D., & Shimizu, K. D. (2012). Do deuteriums form stronger CH – π interactions? *Journal of the American Chemical Society*, *134*, 14306–14309.

Zoi, I., Suarez, J., Antoniou, D., Cameron, S. A., Schramm, V. L., & Schwartz, S. D. (2016). Modulating enzyme catalysis through mutations designed to alter rapid protein dynamics. *Journal of the American Chemical Society*, *138*, 3403–3409.

CHAPTER FOUR

Extracting Kinetic Isotope Effects From a Global Analysis of Reaction Progress Curves

Sam Hay[1]

BBSRC/EPSRC Centre for Synthetic Biology of Fine and Speciality Chemicals (SYNBIOCHEM), Manchester Institute of Biotechnology and School of Chemistry, The University of Manchester, Manchester, United Kingdom

[1]Corresponding author: e-mail address: sam.hay@manchester.ac.uk

Contents

1. Introduction 86
2. (Global) Nonlinear Fitting 87
3. Global Fitting of a Time Course Described by Integrated Rate Equations 91
4. Modeling and Fitting Time Courses Using Numerical Integration of Rate Equations 94
5. Model Selection and Error Analysis 98
6. Specific Examples 102
   6.1 DNA Polymerase 102
   6.2 Ethanolamine Ammonia Lyase 104
   6.3 Model-Free Calculation of KIEs 107
7. Summary 109
Acknowledgments 109
References 109

Abstract

Enzyme reaction progress curves, or time course datasets, are often rich in information, yet their analysis typically reduces their information content to a single parameter, the initial velocity. An alternative approach is described here, where the time course is described by a model constructed from rate equations. In combination with global nonlinear regression, intrinsic rate and/or equilibrium constants can be directly obtained by fitting these data. This method can be greatly enhanced when combined with the measurement of (usually deuterium) isotope effects, which selectively perturb individual step(s) within the reaction, allowing better separation of fitted parameters and more robust model testing. This chapter focuses on practical considerations when using analytical and/or numerically integrated rate equations to model enzyme reactions. The emphasis is on the underlying methodology, which is demonstrated with specific examples alongside recommendations of suitable software.

Methods in Enzymology, Volume 596
ISSN 0076-6879
http://dx.doi.org/10.1016/bs.mie.2017.06.041

1. INTRODUCTION

At the heart of enzymology is the experimental measurement of enzyme kinetics, with the resulting kinetic parameters—rate and equilibrium constants—allowing *quantitative* comparison of different enzymes, substrates, and reaction conditions. The measurement of (kinetic) isotope effects extends this approach by allowing the selective perturbation of isotopically sensitive steps during the reaction. When analyzing multiple-turnover (steady-state) experiments, it is now common practice to only monitor the initial phase of the reaction time course and to extract initial velocities, v_0, by linear fitting of these data, the v_0 being the slope of product accumulation vs time. A number of experiments are performed often at different substrate concentrations, and the substrate concentration dependence of v_0 may then be analyzed using Michaelis–Menten theory to extract apparent k_{cat} and K_m values (Cornish-Bowden, 2015). Enzyme reaction progress curves need not be analyzed using the initial velocity approach, and both Henri (Henri, 1903) and Michaelis and Menten (Michaelis, Menten, Johnson, & Goody, 2011) analyzed their data using the more general integrated rate equation approach. However, it can be difficult to manually derive integrated rate equations to describe more complex enzyme mechanisms (multiple chemical steps, reversibility, inhibition, etc.) without making simplifying approximations; see, e.g., Boeker (1984, 1985), Jennings and Niemann (1955), Schwert (1969); Table 1; and Section 3. As advances were made in the analysis of initial velocity data (King & Altman, 1956), the analysis of enzyme kinetics became dominated by the initial velocity approach (Cornish-Bowden, 2015).

Table 1 Examples of Rate and Integrated Rate Laws

Reaction	Rate Law	Integrated Rate Law
$A \rightarrow B$	$(\partial/\partial t)[A] = -k_f[A]$	$[A] = [A]_0 \exp(-kt)$
$A + S \rightarrow B$	$(\partial/\partial t)[A] = -k_f[A][S]$	$[A] = [A]_0 \frac{[S]}{[S]_0} \exp\left(k_f\left([A]_0 - [S]_0\right)t\right)$ $\cong [A]_0 \exp\left(-k_f[S]_0 t\right)$ if $[S]_0 \gg [A]_0$
$A \rightleftharpoons B$	$(\partial/\partial t)[A] = k_r[B] - k_f[A]$	$[A] = \frac{1}{k_1 + k_2}\left\{\begin{array}{l}[A]_0(k_r + k_f \exp(-(k_f + k_r)t)) \\ +[B]_0(k_r - k_r \exp(-(k_f + k_r)t))\end{array}\right\}$

k_f and k_r are the forward and reverse rate constants and $[X]$ and $[X]_0$ denote the concentration of species X at times t and $t=0$, respectively.

With advances in scientific computing, by the 1970s it became practically feasible to use nonlinear regression (fitting) of reaction time courses to integrated forms of rate equations (Bates & Frieden, 1973a, 1973b; Duggleby & Morrison, 1977, 1978). With further advances in computation power and improvements in numerical algorithms (Press, Teukolsky, Vetterling, & Flannery, 2007), it is now possible to investigate an arbitrarily complex chemical reaction by modeling and fitting reaction time courses described directly in terms of their rate equations (cf. explicit integrated forms of these equations). These analyses offer a range of potential advantages over the initial velocity analysis, including the removal of any simplifying approximations regarding reaction conditions such as $[E]_0 \ll [S]_0$ (Duggleby, 2001; Johnson, 2009; Kuzmic, 2009).

This chapter will describe how we use rate equations, global nonlinear fitting, and KIE measurements to interrogate enzyme mechanisms. While the focus is on reaction time course data, a brief general introduction to global fitting is first given in Section 2, with other examples of its use in enzymology shown. While examples of suitable software will be mentioned at relevant points, the emphasis is on the underlying methodology and Mathematica (www.wolfram.com/mathematica) code snippets for key calculations are also included for demonstration purposes. Additionally, much of the initial discussion of time course modeling and fitting in Sections 3–5 makes use of simulated data so that the choice of fitting models (equations) can be considered in the context of the underlying data structure (information content). Section 6 considers three specific examples of global time course analysis from our laboratory. All examples make use of KIEs. In this chapter we will focus on H/D (deuterium) isotope effects due to the relatively large isotope effects often observed, and the practicality of uniform isotopic labeling with deuterium, which can be delivered via (per)deuterated substrates and/or exchange of ionizable protons (in the enzyme and/or substrate) by preequilibration in D_2O. However, as isotope effects are used in this work to selectively perturb chemical steps, this approach can be extended to the analysis of any isotope effect if the KIE is sufficiently large to observe above the experimental noise and fitting error.

2. (GLOBAL) NONLINEAR FITTING

Key to the analysis of much experimental data, including reaction time courses, is the use of (non)linear data fitting or regression. Regression typically relies on algorithms that attempt to minimize an objective function, *S*.

In many cases, S will be the sum of the squared residuals, essentially a measure of the total "distance" between the data and the fitted model. For a single dataset comprising n pairs of x,y values: $S = \sum_{i=1}^{n} r_i^2$, where x is the dependent variable (e.g., time, temperature, pH), r is the residuals: $r_i = y_i - f(x_i, \beta)$, and $f(x, \beta)$ is the fitting model with one or more adjustable (fitting) parameters, β (Bates & Watts, 2007). Of note is that the value of S is proportional to the number of data points, n, and the magnitude of the variance in the y data. Popular commercial fitting software includes Origin and OriginPro (www.originlab.com), SigmaPlot (www.sigmaplot.co.uk), and IgorPro (www.wavemetrics.com/products/igorpro/igorpro), which are all capable of employing user-defined fitting models and performing global fitting. It is also possible to analyze n-dimensional data with $n-1$ dependent variables, such as a series of spectra acquired over a continuous variable such as time, solution redox potential, or pH ($n=3$). In this case, the data are comprised of a matrix and may be first processed using singular value decomposition to reduce the dimensionality of the data (Press et al., 2007). This approach is popular in spectroscopy and will not be covered further here.

Global fitting uses a single fitting model, or a series of related fitting models, to simultaneously fit multiple datasets. The fitting model(s) may contain a mixture of local and global adjustable variables. When a parameter is global or "shared," a single value is calculated for all datasets over which it is shared (it may be a subset of the total datasets), whereas when a parameter is not shared, individual parameters are calculated for each dataset. In the simplest case, the datasets will share a common dependent variable (e.g., time), but this need not be the case, and as long as each dataset can be described by a model which shares at least one parameter with another model within the global optimization problem, then fitting is theoretically possible, although may not be supported by some software.

There are a number of classes of optimization algorithms including Bayesian, deterministic, stochastic (e.g., Monte Carlo sampling), and heuristic (e.g., Nelder–Mead (simplex), genetic and evolutionary), which can be used for global nonlinear least-squares optimization (Moles, Mendes, & Banga, 2003; Press et al., 2007). As the complexity of the fitting model increases, it can become increasingly difficult to identify whether the fitted set of parameters represents a global minimum. This has major implications for error estimation, which is discussed later in Section 5, and in general a good strategy is to use (or at least test) a combination of methods, e.g., a random search followed by steepest descent or Levenberg–Marquardt algorithm. Care must also be taken as the objective function for a global fitting problem is the sum of individual objective functions for each dataset: $S_{global} = S_1 + S_2 + \cdots + S_N$. As a result, if the magnitude of the variance in y data differs between datasets and/or if a particular dataset contains significantly different numbers of data points, then the objective function will be biased by those "bigger" datasets. If required (it may be that the bias is desirable), this bias can be avoided by scaling the data prior to fitting: variance in y data can be normalized and data can be resampled by interpolation so that all datasets contain the same number of data points.

Two examples of global fitting are given in Figs. 1 and 2, and the fitted parameters are compared with those obtained by local fitting in Table 2. Details are given in the relevant figure captions. The main advantage of global fitting is that it may allow better estimation of the uncertainty (error) in fitting parameters, as each dataset can act as constraint(s) on the fitting

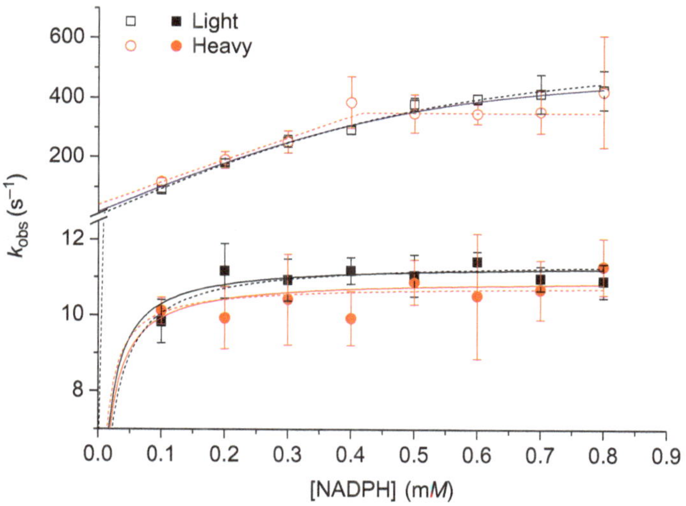

Fig. 1 An example of global fitting of rate constant data: the substrate (NADPH) concentration dependence of two rate constants (*open and filled symbols*) observed in stopped-flow experiments performed with wild-type (light) and isotopically labeled (heavy) flavoenzyme pentaerythritol tetranitrate reductase. The faster rate constant is fit to a quadratic binding model with all rate constants shared (*blue solid line*). The individual (nonglobal) fits are shown for comparison as *dashed lines*. The slower rate constant is expected to exhibit saturation behavior (Pudney, Hay, & Scrutton, 2009), but this was not observed due to experimental restraints (minimum usable NADPH concentration). Consequently, these rate constants were fit to a hyperbolic function both locally (*dashed lines*) and globally (shared K_S, local k_{red}; *solid lines*). Data are taken from Longbotham, Hardman, Gorlich, Scrutton, and Hay (2016) and fitted parameters are given in Table 2.

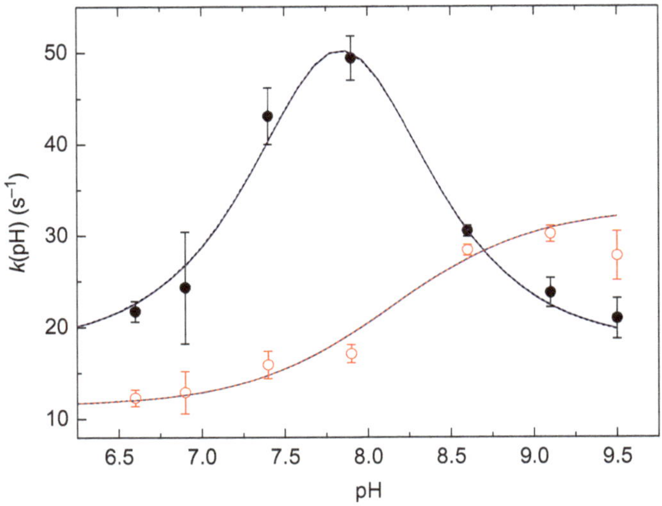

Fig. 2 An example of global fitting of rate constant data: the pH dependence of the apparent forward (*red open symbols*) and reverse (*black filled symbols*) rate constants for the formation of the quinonoid intermediate during the reaction of the H463F mutant of tryptophan indole lyase with L-Trp. The data are fit to a 2-pK_a expression with p$K_{a,1}$ values fit both locally (*dashed lines*) and shared (*solid blue line*). In this case, the fits are almost indistinguishable, but the fitting error associated with both pK_a values is significantly improved when the data are globally fit. Data are taken from Phillips, Kalu, and Hay (2012) and fitted parameters are given in Table 2.

Table 2 Comparison of Fitted Parameters Determined From Local and Global Fits of the Data in Figs. 1 and 2

Fig. 1

	Local Fitting		
Fitted Parameters	**Light Data**	**Heavy Data**	**Global Fitting**[a]
k_1, mM^{-1} s^{-1}	1127 ± 779	736 ± 173	975 ± 530
k_{-1}, s^{-1}	133 ± 338	43 ± 14	88 ± 172
k_2, s^{-1}	542 ± 384	0 ± 51	384 ± 460
k_{-2}, s^{-1}	30 ± 166	349 ± 37	124 ± 293
k_{red}, s^{-1}	11.46 ± 0.16	10.80 ± 0.27	11.36 ± 0.12 (light), 10.96 ± 0.23 (heavy)[a]
K_S, mM^{-1}	0.014 ± 0.05	0.007 ± 0.004	0.010 ± 0.003

Fig. 2

	Local Fitting		
Fitted Parameters	**k_f Data**	**k_r Data**	**Global Fitting**
pK_{a1}	8.11 ± 0.90	8.14 ± 0.16	8.14 ± 0.15
pK_{a2}	7.56 ± 0.76	ND[b]	7.54 ± 0.15

[a] k_{red} values, which were assumed to be the only isotopically sensitive step, were not shared during the global fitting and the given values are for the light and heavy enzymes, respectively.
[b] Not determined.

of all other datasets. This can make a significant difference when the information content (see Sections 3 and 4) of each dataset is lower than the number of adjustable parameters in the fitting model and/or when fitting parameters are partially correlated (correlated parameters have similar effects on the fit). Note that sharing parameters during global fitting does not always reduce their associated fitting errors (see Table 2), so the selection of which parameters to fit globally vs locally may need to be performed by trial and error. It is often good practice to present (for comparison) a range of fitting solutions where parameters of interest are fitted both locally and globally.

3. GLOBAL FITTING OF A TIME COURSE DESCRIBED BY INTEGRATED RATE EQUATIONS

As an example of the global fitting of time course data, consider the following mechanism for two sequential first-order reactions:

$$A \xrightarrow{k_1} B \xrightarrow{k_2} C \tag{1}$$

The time course is described by five parameters: two rate constants, k_1 and k_2, and the starting concentration of each species, $[A]_0$, $[B]_0$, and $[C]_0$. Fig. 3 shows two simulations of this reaction in which the change in concentration of the [A], [B], and [C] is plotted. The two simulations differ in the values of k_2 (1 and $0.2\,s^{-1}$) as would be seen if the second step has an intrinsic KIE = 5.0. In both cases $k_1 = 10\,s^{-1}$, so there is no intrinsic KIE on this step. Typically, the objective of analyzing such time course data is to extract the rate constant(s) that describe the reaction. This analysis can be difficult if the underlying mechanism is not (well) understood, but upon inspection of the transient accumulation of species B in Fig. 3, it is clear that at least two steps are involved in this reaction. However, if only species A and/or C are observed, which may be the case when the intermediate B is spectroscopically silent, is it still possible to determine k_1 and/or k_2 by fitting these data? By inspecting the integrated forms of the rate equations describing the time course in Fig. 3 we can, at least hypothetically, answer this question.

While it can be quite laborious to derive the integrated rate equations for multistep reactions, they are often easily determined using symbolic mathematical software. The rate equations are defined as a series of coupled ordinary differential equations (ODEs), and in the case of Eq. (1), these are simply:

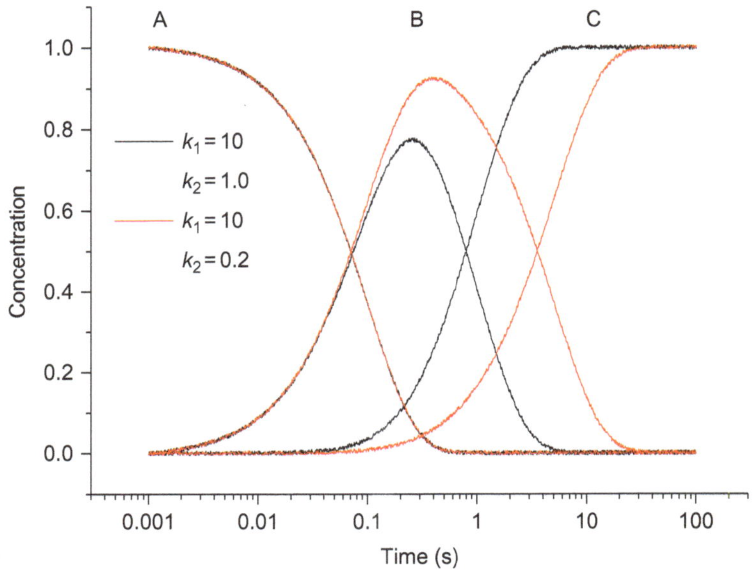

Fig. 3 A simulated time course for the reaction A → B → C as described in the main text. The isotope effect of 5.0 on k_2 is apparent in the perturbation of the time courses of species B and C.

$$\begin{aligned}(\partial/\partial t)[\mathrm{A}] &= -k_1[\mathrm{A}]\\(\partial/\partial t)[\mathrm{B}] &= k_1[\mathrm{A}] - k_2[\mathrm{B}]\\(\partial/\partial t)[\mathrm{C}] &= k_2[\mathrm{B}]\end{aligned} \tag{2}$$

Initial conditions are introduced such that the concentration of each species at some time, typically $t_i = 0$, is defined. The integrated forms of each rate equation can then be obtained by using the "DSolve" function in Mathematica:

```
DSolve[{
    A'[t] == -k1*A[t],          (* rate equations *)
    B'[t] == k1* A[t] - k2*B[t],
    C'[t] == k2*C[t],
    A[0] == A0,                  (* initial conditions *)
    B[0] == B0,
    C[0] == A0
  },
  {A, B, C},t]
```

The integrated rate equation can be written:

$$[\mathrm{A}]=[\mathrm{A}]_0\exp(-k_1 t)$$

$$[\mathrm{B}]=\frac{\exp(-(k_1+k_2)t)}{k_2-k_1}\left\{\begin{array}{l}[\mathrm{A}]_0 k_1(\exp(k_2 t)-\exp(k_1 t))\\ +[\mathrm{B}]_0(k_2-k_1)\exp(k_1 t)\end{array}\right\}$$

$$[\mathrm{C}]=\frac{\exp(-(k_1+k_2)t)}{k_2-k_1} \tag{3}$$

$$\left\{\begin{array}{l}[\mathrm{A}]_0\{k_1\exp(k_1 t)-k_2\exp(k_2 t)+(k_2-k_1)\exp((k_1+k_2)t)\}\\ +[\mathrm{B}]_0\{(k_1-k_2)\exp(k_1 t)+(k_2-k_1)\exp((k_1+k_2)t)\}\\ +[\mathrm{C}]_0(k_2-k_1)\exp((k_1+k_2)t)\end{array}\right\}$$

which clearly simplifies if $[\mathrm{B}]_0$ and/or $[\mathrm{C}]_0=0$. This allows the concentration of each species in the reaction to be defined at arbitrary time, t. Upon inspection of Eq. (3), it is clear that only k_1 can be obtained by fitting the concentration dependence of species A as this dataset does not have sufficient information content to determine any additional parameters. Conversely, both time courses for species B and C contain sufficient information, in principle, to determine both k_1 and k_2. However, the associated integrated rate equations are quite complex, which may be prone to getting "stuck" during regression as k_1 and k_2 have some correlation. Global fitting of two or three of the time course datasets (multiple species) with global/shared rate constants is likely to allow more robust fitting solutions. If species A data are included, k_1 should be well defined, which in turn will improve the fit of k_2. An alternative approach is to make use of the isotope effect on k_2 (or by extension to other rate constants). If only species B or C is observed, then global fitting of these time courses measured with both the light and heavy substrate isotopologues (i.e., S_H and S_D in Fig. 3) also allows some separation of variables, which may significantly improve the confidence in fitted k_1, k_2^H, and k_2^D values. In this case only k_1 would be a global/shared parameter and k_2^H and k_2^D are local parameters, fitted individually.

Note that the time courses are modeled in units of concentration vs t. For comparison to experiment, the concentrations can be transformed into absorbance units using the Beer–Lambert law with relevant extinction coefficients, or to fluorescence units via a scaling factor. If extinction coefficients or relevant scaling factors are not known, these can be added to the model as additional fitting parameters. Also, if the observed signal contains contributions from multiple species due to spectral overlap, the fitting function can be defined in terms of a linear combination of scaled species concentrations: $f = c_A[A] + c_B[B] + \cdots + c_N[N]$, where c are the scaling factors, which can be shared during global fitting. While this approach adds additional fitting parameters, it is generally preferable to normalizing the data (and setting $\Sigma c = 1$) as it allows more realistic error estimation.

4. MODELING AND FITTING TIME COURSES USING NUMERICAL INTEGRATION OF RATE EQUATIONS

It is often not possible or practical to derive analytical forms of integrated rate equations for a reaction of interest. In these cases, the time course can instead be obtained by directly solving the rate equation for each species in the reaction using numerical integration. Numerical integration algorithms are included in many popular mathematical software packages, including MATLAB (www.mathworks.com/products/matlab) and Mathematica, which are both commercial software requiring a license, as well as a range of open source software including GNU Octave (www.gnu.org/software/octave) and Sage (www.sagemath.org). A number of specialist software packages for modeling chemical kinetics also exist, and these can offer advantages in ease of use and/or performance. Examples include KinTek Explorer (kintekcorp.com (Johnson, 2009); commercial software requiring license), DynaFit (www.biokin.com/dynafit/ (Kuzmic, 2009); commercial software with free academic license), and systems biology pathway modelers such as COPASI (copasi.org (Hoops et al., 2006; Mendes, Messiha, Malys, & Hoops, 2009); open source with (free) artistic license 2.0). However, while analytical equations can be precompiled and/or only calculated once during a fitting calculation, in many cases numerical integration of the rate equations must be repeated every time a parameter is changed (e.g., during each optimization iteration). As a result, the use of numerically integrated rate equations may come with a significant computational overhead (Toney, 2013), and it is advisable to use integrated forms of the rate equations when possible.

To demonstrate how a reaction time course is modeled using numerical integration, the behavior of a "Michaelis–Menten" enzyme will be examined:

$$E + S \underset{k_r}{\overset{k_f}{\rightleftharpoons}} E \cdot S \overset{k_{cat}}{\rightarrow} E + P \tag{4}$$

The rate equation for each species in the reaction is again defined using the appropriate rate law, and in some software, rate equations may be generated automatically from a user-defined mechanism as "E+S=ES=E+P" in the notation used by KinTek explorer (Johnson, 2009). In the case of Eq. (4), the rate equations are:

$$\begin{aligned}(\partial/\partial t)[\mathrm{E}] &= (k_{\mathrm{r}} + k_{\mathrm{cat}})[\mathrm{E}\cdot\mathrm{S}] - k_{\mathrm{f}}[\mathrm{E}][\mathrm{S}]\\ (\partial/\partial t)[\mathrm{S}] &= k_{\mathrm{r}}[\mathrm{E}\cdot\mathrm{S}] - k_{\mathrm{f}}[\mathrm{E}][\mathrm{S}]\\ (\partial/\partial t)[\mathrm{E}\cdot\mathrm{S}] &= k_{\mathrm{f}}[\mathrm{E}][\mathrm{S}] - (k_{\mathrm{r}} + k_{\mathrm{cat}})[\mathrm{E}\cdot\mathrm{S}]\\ (\partial/\partial t)[\mathrm{P}] &= k_{\mathrm{cat}}[\mathrm{E}\cdot\mathrm{S}]\end{aligned} \tag{5}$$

Once the initial concentrations of each species (i.e., $[\mathrm{E}]_{t_i}$, $[\mathrm{S}]_{t_i}$, $[\mathrm{E{\cdot}S}]_{t_i}$, and $[\mathrm{P}]_{t_i}$) and the value for each rate constant (i.e., k_f, k_r, and k_{cat}) are defined, the coupled ODEs are solved (approximately) using numerical integration between t_i and some arbitrary final time point, t_f. As an example, the following Mathematica code will perform this calculation using the "NDSolve" function, which finds numerical solutions to ODEs:

```
ti = 0; tf = 1000;              (* integrate between ti and tf *)
species = {En, S, ES, P};       (* all species in the reaction *)
initialConc = {                 (* initial concentration of each species *)
   En[ti] == 10^-6,
   S[ti] == 10^-3,
   ES[ti] == 0,
   P [ti] == 0 };
kf = 10^4; kr = 1; kcat = 10;    (* rate constants *)
rateEq = {                        (* rate equations *)
   En'[t] == (kr + kcat)*ES[t]-kf*En[t]*S[t],
   S'[t] == kr*ES[t] - kf*En[t]*S[t],
   ES'[t] == kf*En[t]*S[t] - (kr + kcat)*ES[t],
   P'[t] == kcat*ES[t]
};
timecourse = NDSolve[{             (* the numerical integration step *)
     rateEq,
     initialConc
   },
   species,
   {t,ti,tf}]
```

Fig. 4 shows the time course of each species in the reaction shown in Eq. (4) using the following rate constants and initial conditions: $k_f = 10^4\,M^{-1}\,\mathrm{s}^{-1}$; $k_r = 1\,\mathrm{s}^{-1}$; $k_{cat} = 10\,\mathrm{s}^{-1}$; $[\mathrm{E}]_0 = 1\,\mu M$; $[\mathrm{S}]_0 = 1\,\mathrm{m}M$; $[\mathrm{E{\cdot}S}]_0 = 0$; $[\mathrm{P}]_0 = 0$. The Briggs–Haldane $K_M = (k_r + k_{cat})/k_f$ is 1.1 mM,

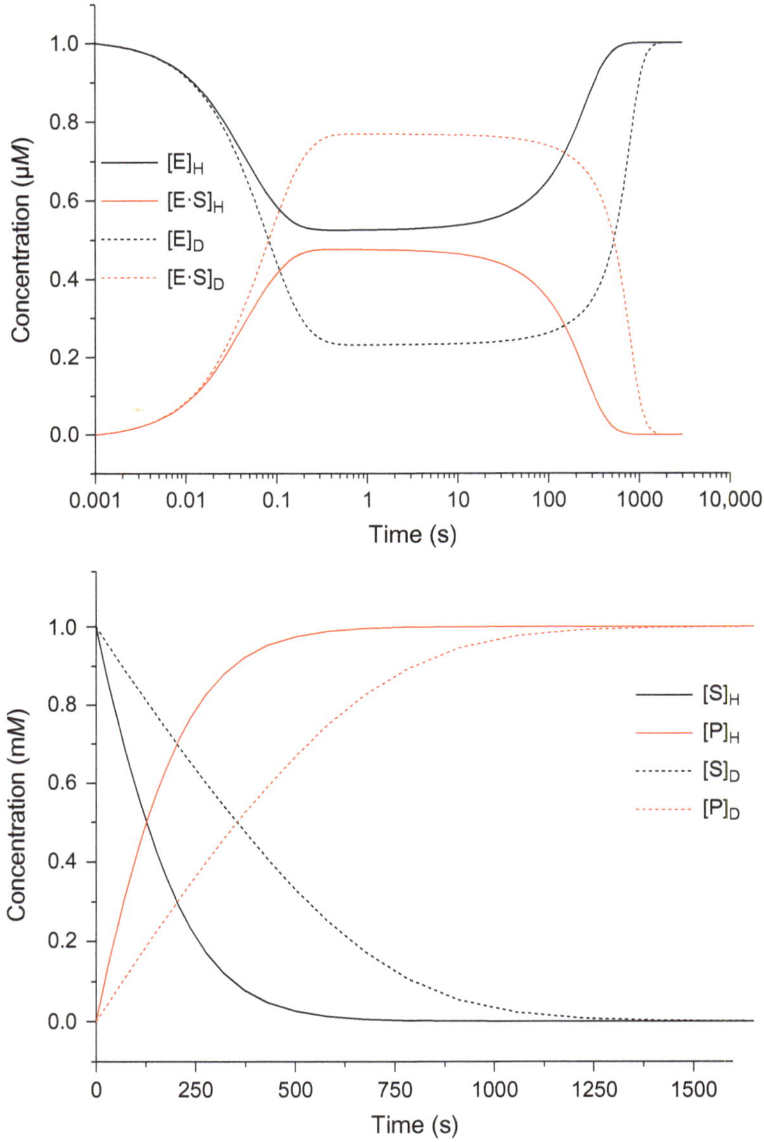

Fig. 4 Modeling the time course of each species involved in the reaction of a Michaelis–Menten enzyme, E, with substrates, S, and product, P, as described in Section 4. An intrinsic isotope effect of 5.0 on k_{cat} is demonstrated by comparing the reaction of isotopically light and heavy substrates S_H and S_D, respectively. Note the logarithmic timescale for the *top graph*.

which is similar to substrate concentration used in the simulation. Consequently, we see that [E] ~ [E·S] under conditions of steady state from ~0.3 to 30 s. The advantage of using isotope effects to probe mechanism again becomes apparent upon inspection of a similar reaction modeled with an isotopically labeled substrate, S_D. Here the specific perturbation of isotopically sensitive steps in a reaction is modeled by reducing k_{cat} to $2\,s^{-1}$ (KIE = 5.0), while leaving k_f and k_r unaffected. While binding isotope effects (see Chapter "Binding isotope effects for interrogating enzyme–substrate interactions" by Stratton et al. in this volume) will affect the values of k_f and k_r, the magnitude of these KIEs is often small enough to be lost within the noise of a typical kinetic experiment. Upon inspection of Fig. 4, the rate of substrate consumption and product formation is seen to be reduced by a factor of 5 for the reaction with S_D as evidenced by the greater accumulation of $[E-S]_D$ vs $[E-S]_H$ at steady state. Additionally, as a result of the reduction in k_{cat}, the period of steady state is increased fivefold to ~150 s and as the K_M is reduced to ~0.3 m*M*, [E] ~ ¼ × [E·S] under steady state. If one were able to directly monitor the free enzyme (e.g., oxidized flavin cofactor) and/or E·S complex (e.g., charge transfer complex), then fitting of these data over the full time course to a model that solves the rate equations in Eq. (5) would, in principle, allow good estimation of all three Michaelis–Menten parameters (k_f, k_r, and k_{cat}). Such an experiment would likely require a rapid mixing approach, such as is available with a stopped-flow spectrometer or rapid quench apparatus. Conversely, if the entire substrate and/or product time course is fitted to this model, it is possible to obtain good estimates of k_{cat} and K_M, but not individual k_f and k_r values as they are highly correlated. In either case, it is theoretically possible to extract (at least) k_{cat} and K_M values from a single reaction time course, thus highlighting one advantage of full time course fitting over initial velocity analysis of enzyme kinetics.

For those researchers who are accustomed to analyzing enzyme kinetics by means of the initial velocity approach, the claim that a K_M value can be obtained by fitting a single time course may require some further discussion. In an initial velocity experiment, v_0 is measured with a range of substrate concentrations, and subsequent plot of v_0 vs $[S]_0$ then gives the familiar Michaelis–Menten plot from which k_{cat} and K_M are obtained by fitting: $v_0 = (k_{cat}/[E]_0)[S]/(K_M + [S])$. Key to the good estimation of K_M is the measurement of v_0 at conditions where $[S]_0$ is of similar magnitude to the apparent K_M. In a full reaction time course analysis, free substrate [S] will vary from $[S] = [S]_0$ to $[S] = 0$ if the reaction is irreversible, not limited by the availability of additional substrate(s), etc. As a consequence, the instantaneous velocity, $v(t) = \partial[\mathrm{P}]/\partial t$, will vary from v_0 to 0 throughout the experiment. If the experiment is performed under conditions where $[\mathrm{S}]_0 > K_m$, the time course information content is at least equivalent to that found in a good Michaelis–Menten plot (Fig. 5).

To further illustrate this point, Fig. 5 shows additional simulations of the reaction shown in Fig. 4. Now, the reaction progress (related to amount of product formed) is determined over a range of substrate concentrations. As the enzyme concentration was not varied, the reaction takes longer to complete with more substrate (note the log timescale) and the general shape of the progress curve in the top panel is dependent on $[\mathrm{S}]_0$. The middle panel shows how the instantaneous velocity, $v(t)$, varies with time (note that $v(t) \rightarrow v_0$ as $t \rightarrow 0$). Of note is that the transition from maximal $v(t)$ to $v(t) = 0$ becomes progressively more abrupt as the initial substrate concentration, $[\mathrm{S}]_0$, is increased. If these time course data are fitted to a numerically integrated form of Eq. (5), reliable K_M values are only obtained under reaction conditions where $[\mathrm{S}]_0$ is below $\sim 10 \times K_M$. A Michaelis–Menten-type curve can be generated from these data by plotting $v(t)$ vs the instantaneous substrate concentration [S], and this is shown in the bottom panel of Fig. 5. This example highlights the importance of good experimental design: reaction conditions should be chosen to maximize the information content of the data collected. This almost always includes collecting data that are evenly sampled over *logarithmic* time (i.e., with a log time base) in order to prevent the relative oversampling of data associated with slower steps in the reaction. Further, time course analysis should be performed under a number of conditions in order to test the validity and accuracy of the fitting model and resulting fitted parameters and to ensure good data coverage; observe how those data collected with low $[\mathrm{S}]_0$ concentrations only sample the "bottom" of the Michaelis–Menten curve in Fig. 5. Approaches for model validation are discussed further in the next section.

5. MODEL SELECTION AND ERROR ANALYSIS

Perhaps the biggest caveat to using full time course analysis of enzyme kinetics is the absolute dependence of this method on the fitting model chosen. If the enzyme mechanism is not known, then the selection of an incorrect model (e.g., two-step vs three-step, ordered vs random binding) will lead to spurious fitting parameters. Fortunately, it is now relatively trivial to build and test a range of fitting models in silico, which collectively describe all reasonable reaction mechanisms. By fitting data to a range of mechanisms and comparing both the goodness of the fit (below) and the

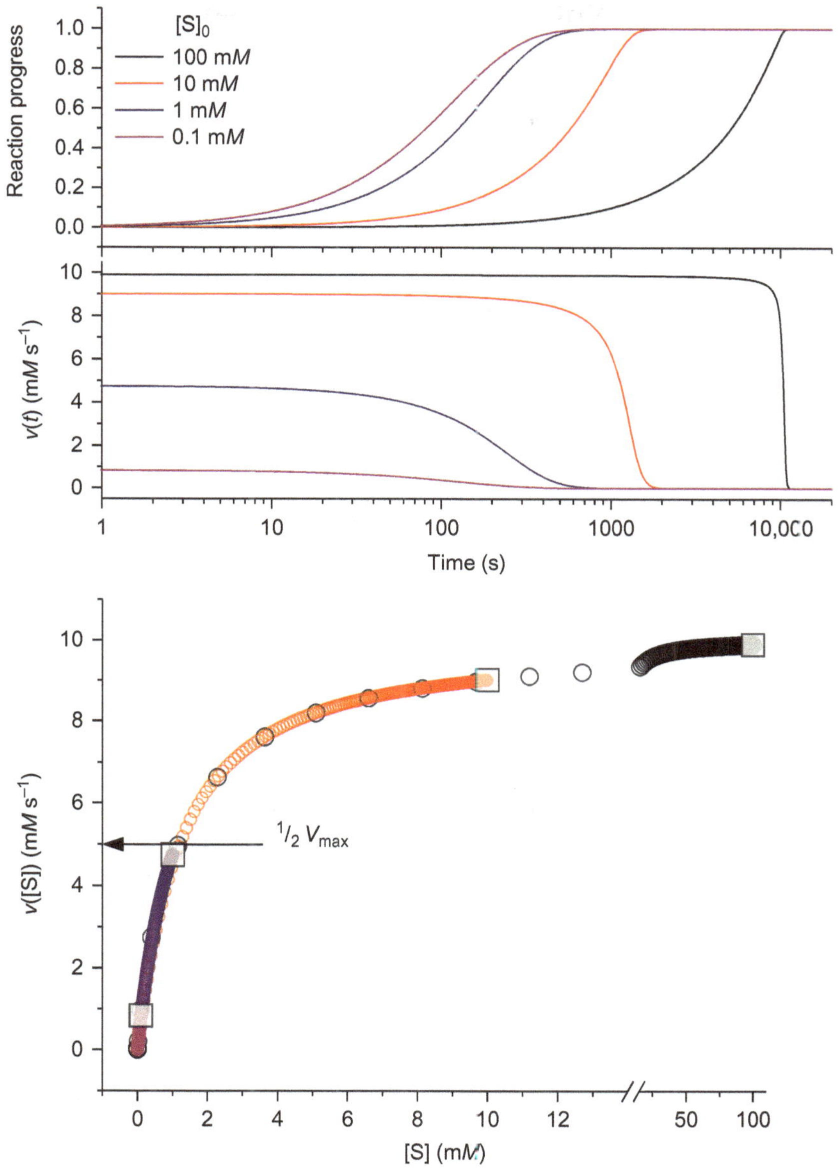

Fig. 5 Further modeling of the "Michaelis–Menten enzyme" in Fig. 4. Reaction progress is calculated by normalizing the product concentration, while *v*(*t*) is calculated by numerical differentiation using a three-point moving window approach. The Michaelis–Menten plot of these data shown in the *bottom panel* uses the same color scheme as the other panels. The initial velocities determined at each of the four $[S]_0$ conditions are shown as *black squares*.

fitted parameters, it may be possible to identify a single unique model (mechanism), which describes the reaction of interest. If not, Occam's razor dictates that one should chose the simplest model unless additional information is available. In either case, best practice is to describe all reasonable models when presenting the analysis.

When time course analysis is combined with global fitting, care must be taken in choosing whether parameters, particularly in the case of rate constants, are locally or globally fitted. In the specific case of the measurement of isotope effects, one may assume that isotopic substitution only measurably affects certain steps in the mechanism (e.g., the chemical step as is modeled in Fig. 4), and thus all other rate constants are identical and can be shared. This assumption should be tested by comparing fit results obtained when all rate constants are fitted locally vs where selected rate constants are fitted globally. Again, it is best practice to include fitting results in both cases when presenting these data. This is not limited to time course analysis and should be extended to all examples of global fitting.

When testing fitting models, one should aim to collect multiple time course datasets measured under different conditions as well as replicates measured under identical conditions. By varying substrate concentrations, it may be possible to isolate the second-order rate constant associated with the substrate binding step and/or test whether substrate and/or product inhibition is significant. Inhibition may be evident under conditions of higher $[S]_0$ if the reaction is run to completion, as higher $[S]_0$ will lead to the formation of higher product concentrations. Varying substrate concentration(s) can also have a knock-on effects as, for example, the second-order rate of substrate binding (sensitive to $[S]_0$) may be highly correlated with rate constants describing the reverse reaction (substrate dissociation) and/or neighboring steps. It has already been discussed how the use of isotope effects allows, in principle, the selective perturbation of certain steps within a reaction. To further increase the information available from such experiments, isotope effects can be combined; if solvent isotope effects are observed, these can be measured with both light and heavy substrate isotopologues. Such double-isotope fractionation experiments are also useful in testing whether a reaction is concerted or stepwise (Belasco, Albery, & Knowles, 1983; Hermes, Roeske, O'Leary, & Cleland, 1982). Additionally, proton inventory-style experiments can be performed with defined mixtures of protiated and deuterated substrate and/or solvent (Hay et al., 2008; Venkatasubban & Schowen, 1984). These data can in turn be analyzed globally, and the isotopically sensitive step(s) critically evaluated

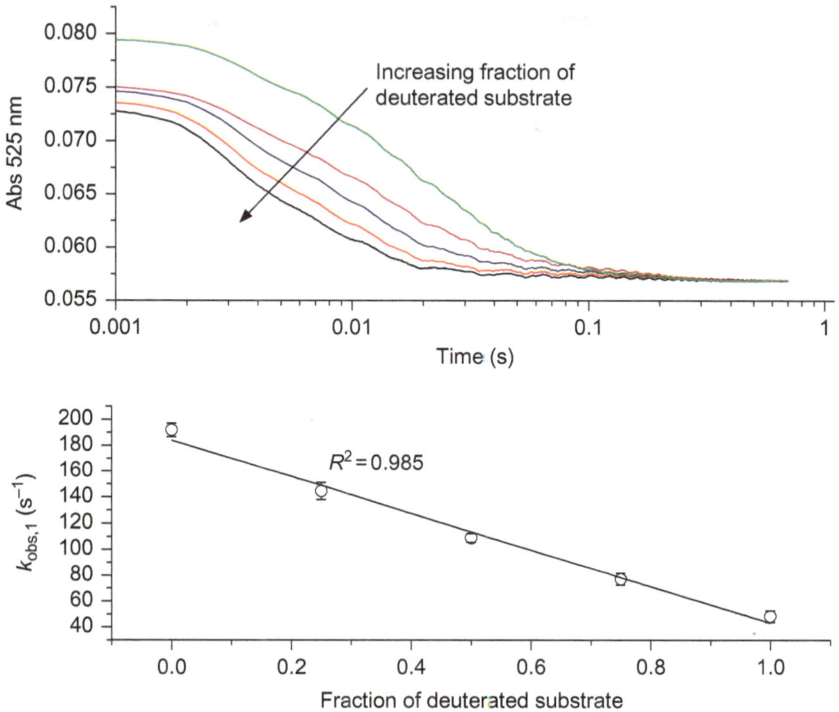

Fig. 6 Proton inventory-style stopped-flow experiments performed by mixing the enzyme ethanolamine ammonia lyase (EAL) with various mixtures of protiated and deuterated ethanolamine substrate. The *top panel* shows stopped-flow transients, while the *bottom panel* shows the apparently linear dependence of the faster rate constant determined from a double-exponential fit of the data. The reaction mechanism is described in Fig. 8 and discussed further in Section 6.2.

(Fig. 6 and Section 6.2). As no additional fitting parameters are added (the mole fraction of heavy isotope is a *dependent* variable), this type of experiment increases the information content of the experiment without increasing the complexity of the fitting model. Replicate experiments are important as these allow the fitting reproducibility (parameter estimation) to be independently tested; i.e., do you get the same fitting parameters, within fitting error, if you fit different time courses measured under identical conditions? Global fitting of these additional datasets will also potentially be useful if there is a high degree of experimental error, but care must be taken if these are combined with experiments measured under different conditions as biasing of the objective function will occur unless each experimental condition is represented by a similar number of data points (see Section 2).

As the complexity of the fitting model increases, the difficulty in realistically estimating the "goodness of fit" and the error on fitting parameters also increases. Nonlinear regression can grossly *underestimate* the error, so specialist software such as KinTek Explorer, DynaFit, and COPASI allow graphical analysis of fitting parameters via plots of confidence contours or intervals. This topic has been covered in depth elsewhere with respect to time course analysis of enzyme kinetic data (Johnson, 2009; Kuzmic, 2009), so will be kept brief here. Parameter error is often not normally distributed about some central best-fit value, so presenting error as ± fitting error can be misleading. Instead, it may be better practice to provide upper and lower confidence limits based on some cutoff criteria (see also Section 6.3). The success of the fitting can also be judged by some "goodness-of-fit" statistic such as R^2 for linear regression and χ^2 for nonlinear regression as these can be constructed from a sum of squared errors, which is directly related to the residuals. Graphical inspection of the fit (overlaid with the data) and the residuals often allows the experienced user to assess whether the fit is "good enough." In either case, this judgment should always be made in the context of the fitted parameters and their associated fitting errors; i.e., does the fit go through the data, are the residuals randomly distributed about 0, and do the fitted parameters seem reasonable?

6. SPECIFIC EXAMPLES

In this section, two recent examples will be examined, which have used global fitting of time course data to determine enzymatic H/D KIEs. (Jones, Rentergent, Scrutton, & Hay, 2015; Rentergent, Driscoll, & Hay, 2016). Finally, a method to calculate KIEs from nonnormally distributed rate constants will be described with reference to an enzymatic single-molecule KIE experiment (Pudney et al., 2013).

6.1 DNA Polymerase

Deriving useful kinetic parameters describing the processive DNA synthesis catalyzed by DNA polymerases has long proven to be a challenge. We recently set out to develop a strategy to assay and model the full reaction time course for polymerases with DNA templates of defined length (Driscoll, Rentergent, & Hay, 2014; Rentergent et al., 2016). We developed an assay, which uses the intercalating fluorophore PicoGreen to report on the formation of double-stranded DNA. Unfortunately, this must be performed as a discontinuous assay as we found PicoGreen (and other related dyes) to be potent inhibitors of DNA polymerization. Nevertheless, we were able to collect sufficient data to allow reproducible time course analysis. A two-step model was built where DNA binding to the enzyme is explicitly modeled, whereas the nucleotide binding and subsequent chemistry, which actually consists of five reversible steps, was condensed into a single reaction with Michaelis–Menten behavior (and associated k_{cat} and k_{cat}/K_M values; Fig. 7). This approximation is reasonable as only exo- mutants of DNA polymerase Klenow fragment were used, and in these mutants this condensed step is effectively irreversible. Crucially, the length, N, of the template DNA was explicitly accounted for by including $2N+2$ DNA species in the model: [DNA$_n$] and [E·DNA$_n$] for all $0 \leq n \leq N$. A range of DNA template lengths were used, up to a maximum length (excluding initial double-stranded primer region) of 100 nucleotides. At its largest, the model contained 205 species: 101 [DNA$_n$], 101 [E·DNA$_n$], free enzyme, free nucleotide, and pyrophosphate.

Model fitting was performed numerically using COPASI (Hoops et al., 2006). The data were fitted via a fluorescence scaling parameter, which operated on the pyrophosphate concentration (a convenient "product" species). PicoGreen was not included in the model as it was always added in saturating concentrations, and the inhibition was assumed to be rapid relative to other chemical steps. The model was calibrated by globally fitting three families of experiments where enzyme, DNA, and nucleotide concentrations were each varied (as dependent variables). Care was taken to scale the relative magnitude of each family of data to ensure approximately equal weighting.

The model in Fig. 7 allows DNA–enzyme dissociation and rebinding after each polymerization step to be described by k_{-1} and k_1, respectively. To reduce the number of fitting variables, the same values of k_1 and k_{-1} were used for every step except for the final full-length DNA product. Perhaps surprisingly, these k_1 and k_{-1} values were well defined by the model (standard error ~10% of value) and are in good agreement with literature, showing that some substrate binding kinetics can be obtained from multiple-turnover (steady-state-like) assays. The k_{cat} and K_M values, which were also shared across all polymerization steps, were also well defined, with similar standard errors to the k_1 and k_{-1} values. Solvent deuterium KIEs were measured on the apparent Michaelis–Menten parameters using two different lengths of DNA template. Similar values were obtained between experiments, and the KIEs on k_{cat}, K_M, and k_{cat}/K_M were found to be 3.0–3.2, 2.2–2.3, and 1.3–1.5, respectively. These suggest that either the

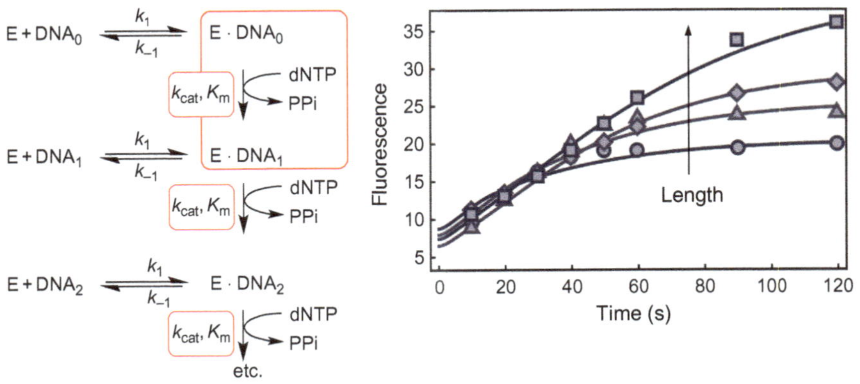

Fig. 7 *Left*, partial kinetic model of the processive synthesis of a DNA template with defined length. The model continues for N total vertical steps, producing DNA_N. In our study, DNA binding and dissociation (horizontal steps) was treated using rate equations with global k_1 and k_{-1} rate constants, while nucleotide incorporation (vertical steps; *red box*) was treated using a simple Michaelis–Menten model with global k_{cat} and K_m input parameters. *Right*, time course data measured with different lengths of DNA template. The data are globally fitted as described in the main text.

intrinsic KIE is inflated well above the semi-classical limit (k_{cat} and K_M are not intrinsic parameters), or the prevailing mechanism where the chemical step is rapid and flanked by slow conformational rearrangement steps is incorrect (Rentergent et al., 2016).

6.2 Ethanolamine Ammonia Lyase

The coenzyme B12-dependent enzyme ethanolamine ammonia lyase (EAL) catalyzes the conversion of ethanolamine to acetaldehyde and ammonia. A partial reaction scheme showing the steps immediately after substrate binding is shown in Fig. 8. The first chemical step (A to B) involves homolysis of the Co—C bond of the adenosyl cobalamin. The second step (B to C), which may be concerted with the previous step, involves hydrogen atom transfer from ethanolamine to the adenosyl radical. The following step involving substrate radical rearrangement is thought to be effectively irreversible. Rapid rotation of the adenosyl CH_3 group followed by transfer of the hydrogen back to the substrate radical (C to B′) will result in isotopic scrambling of the adenosyl methyl group, and under multiple-turnover conditions the 5′-deoxyadenosyl group becomes stably perdeuterated (Weisblat & Babior, 1971).

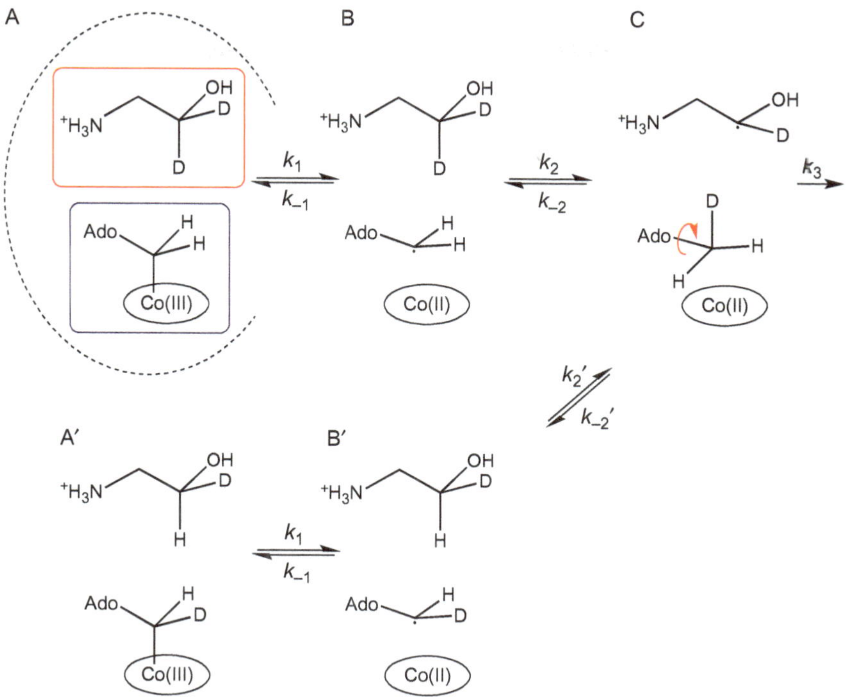

Fig. 8 A partial reaction scheme for the reaction of EAL (only the adenosyl cobalamin, shown in *blue square*) with deuterated ethanolamine substrate (*red square*). Rotation of the adenosyl CH_3 group is highlighted with a *red arrow*.

We investigated the initial hydrolysis steps in EAL by measuring KIEs with deuterated ethanolamine and/or adenosyl cobalamin (Jones et al., 2015). As these steps are rapid, experiments were performed with a stopped-flow spectrometer. Absorbance from the Co(III) oxidation state of the adenosyl cobalamin was monitored, which meant any steps following the irreversible substrate radical rearrangement are effectively silent, so only the first enzyme turnover was considered. In total, seven different experiments were performed, including proton inventory-style experiments with protiated and deuterated ethanolamine (shown in Fig. 6). Collectively, these data were globally fit to a model that contained 7 rate constants (shown in Fig. 8) and 16 chemical species (4 each of A, B, and C, and 2 each of A′ and B′). As an example, the experiment performed with protiated enzyme and deuterated ethanolamine (Fig. 8) is described by the following rate equations:

$$\begin{aligned}(\partial/\partial t)[\mathrm{A}] &= k_{-1}[\mathrm{B}](t) - k_1[\mathrm{A}](t)\\(\partial/\partial t)[\mathrm{A}'] &= k_{-1}[\mathrm{B}'](t) - k_1[\mathrm{A}'](t)\\(\partial/\partial t)[\mathrm{B}] &= k_1[\mathrm{A}](t) + (1/3)k_{-2}[\mathrm{C}](t) - (k_{-1}+k_2)[\mathrm{B}](t)\\(\partial/\partial t)[\mathrm{B}'] &= k_1[\mathrm{A}'](t) + (2/3)k'_{-2}[\mathrm{C}](t) - \left(k_{-1}+k'_2\right)[\mathrm{B}'](t)\\(\partial/\partial t)[\mathrm{C}] &= k_2[\mathrm{B}](t) + k'_2[\mathrm{B}'](t) - \left((1/3)k_{-2} + (2/3)k'_{-2}\right)[\mathrm{C}](t)\end{aligned} \tag{6}$$

which must be solved by numerical integration. Rate constants and a single extinction coefficient for the sum of A and A′ (Co(III)) species in each experiment were fitted globally. Additional individual fitting parameters to allow for error in enzyme concentration and the stopped-flow dead time were also included. To ensure even weighting of the objective function, replicate transients for each of the seven experiments were averaged and then resampled prior to fitting. Each averaged transient was linearly interpolated and 200 data points were taken at equal distances over a logarithmic time-scale between 1.6 ms (the instrument dead time) and 525 ms. Resampling was performed using Mathematica, while fitting was performed with COPASI (Hoops et al., 2006). Example fits are shown in Fig. 9.

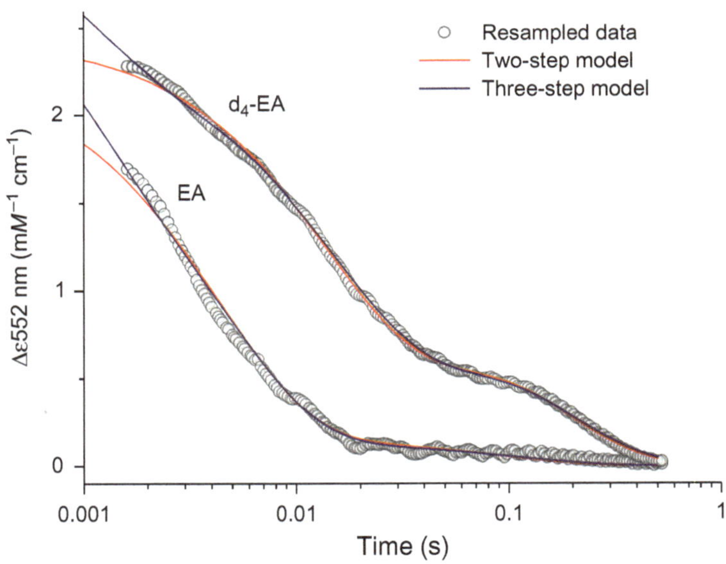

Fig. 9 Stopped-flow transients (*open circles*) measured with protiated EAL and both protiated and deuterated (d_4-)ethanolamine. Global fits to a two-step (A to B and B to C are concerted) and a three-step model are shown as *blue* and *red lines*, respectively.

During model testing we found that similar goodness-of-fit statistics were obtained with a simpler two-step mechanism where k_1 and k_2 steps are concerted, so we presented best-fit results for both models. In both models and for all rate constants and KIEs, except for the KIE on the k'_{-2} step, reasonable fitting values were obtained as judged by both their magnitude and standard error. The poor estimate of the KIE on the k'_{-2} step was rationalized as arising due to an oversimplification in the fitting model; this hydrogen "scrambling" step is likely to be preceded by a protein rearrangement step that was not included. Ultimately, this approach allowed the estimation of intrinsic rate constants and KIEs on the reversible H-transfer steps that precede substrate radical rearrangement in EAL. The KIEs were not found to be inflated above the semi-classical limit, and the "hump" observed in some stopped-flow transients was shown to arise from "scrambling" of the deuteration state of the 5′-deoxyadenosyl moiety (Jones et al., 2015).

6.3 Model-Free Calculation of KIEs

KIEs are usually calculated by $\text{KIE} = \overline{k}_{\text{l}}/\overline{k}_{\text{h}}$, where l and h denote the light and heavy isotope, respectively. However, as $\overline{k}_{\text{i}}$ is typically an average value and/or calculated with some associated fitting error, if its value is not normally distributed (as is the case with many fitted parameters) then this calculation will skew the KIE. One alternative is estimate the KIE from the distribution of every pairwise combination of individual k_{l} and k_{h} values. If there are N values of k_{l} and M values of k_{h}, then we construct an $N \times M$ matrix of all possible KIE values:

$$[\text{KIE}] = \begin{bmatrix} k_{\text{l},1}/k_{\text{h},1} & \cdots & k_{\text{l},N}/k_{\text{h},1} \\ \vdots & \ddots & \vdots \\ k_{\text{l},1}/k_{\text{h},M} & \cdots & k_{\text{l},N}/k_{\text{h},M} \end{bmatrix} \tag{7}$$

The representative KIE′ is then determined from the distribution of all values in the KIE matrix. The distribution can be examined by binning the [KIE] values and plotting these as a histogram (Fig. 10). If the k_{l} and/or k_{h} distributions are multimodal, this approach is still valid, but multiple KIE′ values will be observed as individual peaks in the [KIE] distribution, which may become quite complex.

As an example, upon inspection of a relatively large distribution of apparent rate constants ($1/\tau$) determined using single-molecule spectroscopy,

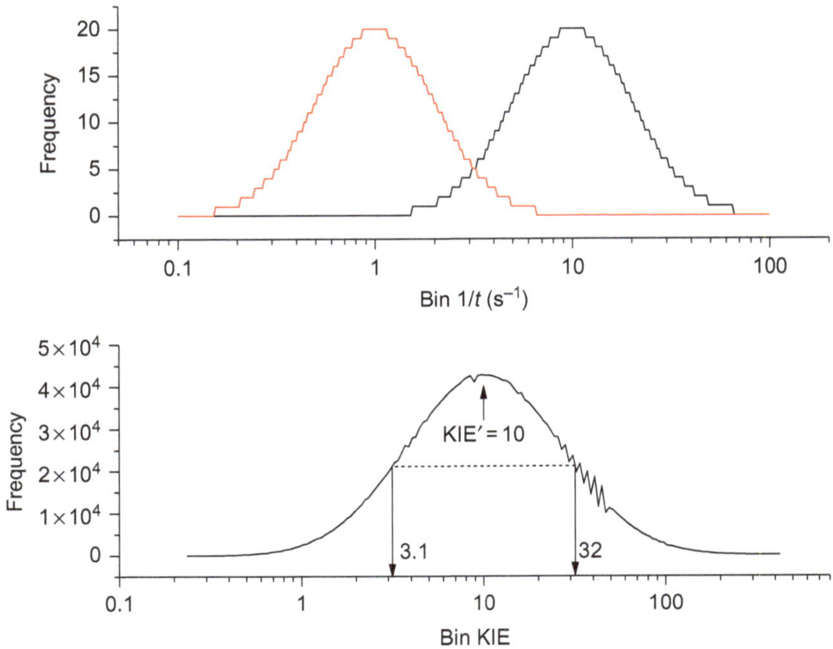

Fig. 10 Model-free calculation of KIEs using a pairwise method. The data are taken from a simulation in Pudney et al. (2013). The KIE distribution shows the representative KIE′ and the KIE values at the distribution's full width at half maximum values = 3.1 and 32, which may be used as pseudo-confidence limits for the representative KIE′.

we found these values to have an approximately log-normal distribution. Consequently, the distribution of pairwise log[KIE] values was also (approximately) log-normally distributed (Pudney et al., 2013). If the [KIE] values can be transformed such that their distribution is approximately normal (e.g., as is the case of the log[KIE] values in Fig. 10), the representative KIE′ can be determined from the arithmetic average of these values. The standard deviation and associated confidence limits of the distribution are also demonstrative of the uncertainty in KIE′. However, once these values are transformed into KIE values (e.g., by taking the exponential in this example), the error in the KIE must be presented as individual high and low confidence limits as they are no longer equally spaced either side of the KIE′ value (Fig. 10). If the KIE distribution is not normally distributed, a weighted average may allow the estimation of a representative KIE. Error estimation will be more complex, but again could be based on confidence limits after integrating the KIE distribution.

7. SUMMARY

The simulation of reaction time courses is useful both in the design of experiments and, when combined with nonlinear (global) fitting algorithms, to extract kinetic parameters of interest. With advances in both computing hardware and numerical algorithms, it is now possible to simultaneously fit large families of reaction time courses to models comprising many chemical species and rate constants using numerical integration. For example, consider the DNA polymerase model in Section 6.2, which contained as many as 205 chemical species. In some cases, specialist software is freely available (e.g., COPASI), and although calculations are quite computationally demanding, many fitting problems will converge within minutes when run on a contemporary desktop computer. Care must be taken when choosing model(s) and when choosing "shared" parameters during global nonlinear regression, but problems can usually be identified with adequate model testing and parameter analysis (confidence analysis). Isotope effect measurements can greatly enrich the analysis due to the selective perturbation of isotopically sensitive steps. Further, as *intrinsic* rate constants can be obtained from this analysis, intrinsic KIEs are also determined, further highlighting the advantages of this approach.

ACKNOWLEDGMENTS

This work was supported in part by the Biotechnology and Biological Sciences Research Council (BBSRC; BB/M007065/1 and BB/M017702/1).

REFERENCES

Bates, D. J., & Frieden, C. (1973a). Full time course studies on the oxidation of reduced coenzyme by bovine liver glutamate dehydrogenase. Use of computer simulation to obtain rate and dissociation constants. *The Journal of Biological Chemistry*, *248*, 7885–7890.

Bates, D. J., & Frieden, C. (1973b). Treatment of enzyme kinetic data. 3. The use of the full time course of a reaction, as examined by computer simulation, in defining enzyme mechanisms. *The Journal of Biological Chemistry*, *248*, 7878–7884.

Bates, D. M., & Watts, D. G. (2007). *Nonlinear regression analysis and its applications*. Chichester, NY: Wiley.

Belasco, J. G., Albery, W. J., & Knowles, J. R. (1983). Double isotope fractionation: Test for concertedness and for transition-state dominance. *Journal of the American Chemical Society*, *105*, 2475–2477.

Boeker, E. A. (1984). Integrated rate equations for enzyme-catalysed first-order and second-order reactions. *Biochemical Journal*, *223*, 15–22.

Boeker, E. A. (1985). Integrated rate equations for irreversible enzyme-catalysed first-order and second-order reactions. *Biochemical Journal*, *226*, 29–35.

Cornish-Bowden, A. (2015). One hundred years of Michaelis–Menten kinetics. *Perspectives in Science, 4*, 3–9.
Driscoll, M. D., Rentergent, J., & Hay, S. (2014). A quantitative fluorescence-based steady-state assay of DNA polymerase. *The FEBS Journal, 281*, 2042–2050.
Duggleby, R. G. (2001). Quantitative analysis of the time courses of enzyme-catalyzed reactions. *Methods (San Diego, Calif.), 24*, 168–174.
Duggleby, R. G., & Morrison, J. F. (1977). The analysis of progress curves for enzyme-catalysed reactions by non-linear regression. *Biochimica et Biophysica Acta, 481*, 297–312.
Duggleby, R. G., & Morrison, J. F. (1978). Progress curve analysis in enzyme kinetics: Model discrimination and parameter estimation. *Biochimica et Biophysica Acta, 526*, 398–409.
Hay, S., Pang, J., Monaghan, P. J., Wang, X., Evans, R. M., Sutcliffe, M. J., et al. (2008). Secondary kinetic isotope effects as probes of environmentally-coupled enzymatic hydrogen tunneling reactions. *Chemphyschem: A European Journal of Chemical Physics and Physical Chemistry, 9*, 1536–1539.
Henri, V. (1903). *General laws for the action of diastases (A. Cornish-Bowden, Trans.)*. Available at: http://bip.cnrs-mrs.fr/bip10/HenriEng.pdf. (accessed 26.07.2017).
Hermes, J. D., Roeske, C. A., O'Leary, M. H., & Cleland, W. W. (1982). Use of multiple isotope effects to determine enzyme mechanisms and intrinsic isotope effects. Malic enzyme and glucose-6-phosphate dehydrogenase. *Biochemistry, 21*, 5106–5114.
Hoops, S., Sahle, S., Gauges, R., Lee, C., Pahle, J., Simus, N., et al. (2006). COPASI—A complex pathway simulator. *Bioinformatics (Oxford, England), 22*, 3067–3074.
Jennings, R. R., & Niemann, C. (1955). The evaluation of the kinetic constants of enzyme-catalyzed reactions by procedures based upon integrated rate equations. *Journal of the American Chemical Society*, 77, 5432–5433.
Johnson, K. A. (2009). Fitting enzyme kinetic data with KinTek Global Kinetic Explorer. *Methods in Enzymology, 467*, 601–626.
Jones, A. R., Rentergent, J., Scrutton, N. S., & Hay, S. (2015). Probing reversible chemistry in coenzyme B12-dependent ethanolamine ammonia lyase with kinetic isotope effects. *Chemistry (Weinheim an der Bergstrasse, Germany), 21*, 8826–8831.
King, E. L., & Altman, C. (1956). A schematic method of deriving the rate laws for enzyme-catalyzed reactions. *The Journal of Physical Chemistry, 60*, 1375–1378.
Kuzmic, P. (2009). DynaFit—A software package for enzymology. *Methods in Enzymology, 467*, 247–280.
Longbotham, J. E., Hardman, S. J., Gorlich, S., Scrutton, N. S., & Hay, S. (2016). Untangling heavy protein and cofactor isotope effects on enzyme-catalyzed hydride transfer. *Journal of the American Chemical Society, 138*, 13693–13699.
Mendes, P., Messiha, H., Malys, N., & Hoops, S. (2009). Enzyme kinetics and computational modeling for systems biology. *Methods in Enzymology, 467*, 583–599.
Michaelis, L., Menten, M. L., Johnson, K. A., & Goody, R. S. (2011). The original Michaelis constant: Translation of the 1913 Michaelis-Menten paper. *Biochemistry, 50*, 8264–8269.
Moles, C. G., Mendes, P., & Banga, J. R. (2003). Parameter estimation in biochemical pathways: A comparison of global optimization methods. *Genome Research, 13*, 2467–2474.
Phillips, R. S., Kalu, U., & Hay, S. (2012). Evidence of preorganization in quinonoid intermediate formation from l-Trp in H463F mutant Escherichia coli tryptophan indole-lyase from effects of pressure and pH. *Biochemistry, 51*, 6527–6533.
Press, W. H., Teukolsky, S. A., Vetterling, W. T., & Flannery, B. P. (2007). *Numerical recipes: The art of scientific computing* (3rd ed.). Cambridge: Cambridge University Press.
Pudney, C. R., Hay, S., & Scrutton, N. S. (2009). Bipartite recognition and conformational sampling mechanisms for hydride transfer from nicotinamide coenzyme to FMN in pentaerythritol tetranitrate reductase. *The FEBS Journal, 276*, 4780–4789.

Pudney, C. R., Lane, R. S., Fielding, A. J., Magennis, S. W., Hay, S., & Scrutton, N. S. (2013). Enzymatic single-molecule kinetic isotope effects. *Journal of the American Chemical Society*, *135*, 3855–3864.

Rentergent, J., Driscoll, M. D., & Hay, S. (2016). Time course analysis of enzyme-catalyzed DNA polymerization. *Biochemistry*, *55*, 5622–5634.

Schwert, G. W. (1969). Use of integrated rate equations in estimating the kinetic constants of enzyme-catalyzed reactions. *The Journal of Biological Chemistry*, *244*, 1278–1284.

Toney, M. D. (2013). Common enzymological experiments allow free energy profile determination. *Biochemistry*, *52*, 5952–5965.

Venkatasubban, K. S., & Schowen, R. L. (1984). The proton inventory technique. *CRC Critical Reviews in Biochemistry*, *17*, 1–44.

Weisblat, D. A., & Babior, B. M. (1971). The mechanism of action of ethanolamine ammonia-lyase, a B 12-dependent enzyme. 8. Further studies with compounds labeled with isotopes of hydrogen: Identification and some properties of the rate-limiting step. *The Journal of Biological Chemistry*, *246*, 6064–6071.

CHAPTER FIVE

Natural Isotope Abundance in Metabolites: Techniques and Kinetic Isotope Effect Measurement in Plant, Animal, and Human Tissues

Illa Tea*,†,‡, Guillaume Tcherkez*,1

*Research School of Biology, Australian National University, Canberra, ACT, Australia
†Cancer Metabolism and Genetics Group, John Curtin School of Medical Research, Australian National University, Canberra, ACT, Australia
‡EBSI Team, CEISAM, University of Nantes-CNRS UMR 6230, Nantes, France
[1]Corresponding author: e-mail address: guillaume.tcherkez@anu.edu.au

Contents

1. Introduction 114
2. Compound-Specific ^{13}C and ^{15}N Analysis: History and Overview 120
   2.1 Brief History 120
   2.2 Overview of Sample Preparation and Analytical Strategies 121
3. Compound-Specific ^{15}N Analysis 124
   3.1 *N*-pivaloylation of the Amine Functional Group 126
   3.2 *N,O*-trifluoroacetylation (TFA and PFPA) 127
   3.3 *tert*-Butyltrimethylsilylation 128
4. Compound-Specific ^{13}C Analysis 129
   4.1 Principle and Precautions 129
   4.2 Silylation 130
   4.3 Acetylation/Methylboronation of Monosaccharides 130
   4.4 Methylation of Fatty Acids and Methods Not Requiring Derivatization 131
   4.5 LC-co-IRMS 131
5. KIE Measurement From LC-co-IRMS or GC-C-IRMS Data 133
   5.1 Principle 134
   5.2 Utilization of LC-co-IRMS and GC-C-IRMS Data 137
   5.3 Example 140
6. Conclusions and Perspectives 142
Acknowledgments 142
References 142

Methods in Enzymology, Volume 596
ISSN 0076-6879
http://dx.doi.org/10.1016/bs.mie.2017.07.020

Abstract

The natural isotope abundance in bulk organic matter or tissues is not a sufficient base to investigate physiological properties, biosynthetic mechanisms, and nutrition sources of biological systems. In fact, isotope effects in metabolism lead to a heterogeneous distribution of ^{2}H, ^{18}O, ^{13}C, and ^{15}N isotopes in metabolites. Therefore, compound-specific isotopic analysis (CSIA) is crucial to biological and medical applications of stable isotopes. Here, we review methods to implement CSIA for ^{15}N and ^{13}C from plant, animal, and human samples and discuss technical solutions that have been used for the conversion to CO_2 and N_2 for IRMS analysis, derivatization and isotope effect measurements. It appears that despite the flexibility of instruments used for CSIA, there is no universal method simply because the chemical nature of metabolites of interest varies considerably. Also, CSIA methods are often limited by isotope effects in sample preparation or the addition of atoms from the derivatizing reagents, and this implies that corrections must be made to calculate a proper δ-value. Therefore, CSIA has an enormous potential for biomedical applications, but its utilization requires precautions for its successful application.

1. INTRODUCTION

Compound-specific stable isotope analysis by ratio mass spectrometry (CSIA) enables precise measurement of naturally occurring stable isotopes (such as $^{13}C/^{12}C$, $^{15}N/^{14}N$, $^{2}H/^{1}H$, $^{18}O/^{17}O$) within metabolites from complex mixtures (e.g., biological extracts). Isotopes are quantified using the isotope composition or δ-value (usually expressed in per mil):

$$\delta = \frac{R}{R_{st}} - 1$$

where R is the isotope ratio heavy/light ($^{13}C/^{12}C$, $^{15}N/^{14}N$, etc.) of the sample (metabolite) of interest, and R_{st} that in the international reference (Pee Dee Belemnite for ^{13}C, atmospheric N_2 for ^{15}N, standard mean oceanic water for ^{2}H and ^{18}O). δ-values provide a convenient tool to trace the origin and the fate of organic matter in a wide range of application such as environmental monitoring (Elsner & Imfeld, 2016), archeological studies (Styring et al., 2015), plant sciences (Aranjuelo, Cabrera-Bosquet, Mottaleb, Araus, & Nogués, 2009; de Rijke, Truyols, & de Koster, 2016; Dumont et al., 2013; González-Pérez, Jiménez-Morillo, de la Rosa, Almendros, & González-Vila, 2016; Hansen, Fromberg, & Frandsen (2014); Hettmann, Gleixner, & Juchelka, 2008; Lehmann, Gamarra, Kahmen, Siegwolf & Saurer, 2017; Mauve et al., 2016; Molero, Aranjuelo, Teixidor, Araus, & Nogués, 2011; Paolini, Ziller, Laursen,

Husted, & Camin, 2015; Styring, Fraser, Bogaard, & Evershed, 2014; Zech et al., 2013; Zhou et al., 2016), human and animal nutrition (Cantalapiedra-Hijar et al., 2016; Egli et al., 2016; Ferchaud-Roucher, Albert, Champ, & Krempf, 2006; Godin, Richelle, Metairon, & Fay, 2004; Godin et al., 2013; Morrison, O'Hara, King, & Preston, 2011; Petzke, Boeing, & Metges, 2005; Recio, Martín, Raposo, & Raposo, 2013; Reichardt, Barclay, Weaver, & Morrison, 2011; Styring et al., 2015; Tea, Ferchaud-Roucher, Küster, Darmaun, & Robins, 2007; Tea et al., 2013, 2016; Thevis, Piper, Horning, Juchelka, & Schänzer, 2013; Verbeke et al., 2010; Wang et al., 2014; Webb et al., 2015). Furthermore, the analysis of natural isotope abundance in different metabolites from the same extract, especially $^{13}C/^{12}C$ and $^{15}N/^{14}N$, can be used to investigate biosynthetic mechanisms and metabolic pathways in plant, animal, and human tissues. CSIA can also be implemented to enzymatic reaction assays allowing determination of kinetic isotope effects (KIEs) or equilibrium isotope effects (EIEs). Conversely, EIE and KIE are important pieces of information because they allow one to interpret isotope abundances in metabolites by taking into account enzymatic effects in metabolism.

In this chapter, we will focus on techniques for CSIA and associated KIE/EIE measurements and briefly discuss examples of application in plant, animal, and biomedical science. Taken as a whole, there are two main techniques used for CSIA, based on gas chromatography (GC) or liquid chromatography (LC) coupled to isotope ratio mass spectrometry (IRMS). The fact that CSIA requires instrument coupling (chromatography and IRMS) is an important aspect that constraints the interface between GC/LC and IRMS (usually, combustion or pyrolysis furnace (denoted as "C") with the GC; chemical oxidizer (denoted as "co"); or high-temperature combustion (referred to as "HTC") with the LC) and the type of metabolite that can be analyzed (typically, volatile or not). There are other possible couplings, such as GC-ICP-MS (gas chromatography-inductively coupled plasma-mass spectrometry), that may allow the determination of isotope ratios with a multicollector MS, but this approach has not been implemented in biology and will not be discussed here.

There is now quite a few published works taking advantage of GC-C-IRMS or LC-co-IRMS in plant science (typical examples in Table 1) and in biomedicine (Table 2). Quite generally, it has been shown that it is possible to measure the compound-specific isotope composition of amino acids (and derivatives such as glutathione), sugars, organic acids, alkanes, sterols, and base (nucleic acid) derivatives. Most popular

Table 1 Summary of Compound-Specific Isotope Analyses Applied to Plant Tissues

Matrix	Metabolites	Detection Methods and Isotopes	Application	References
Arabidopsis	Glu, Gln	LC-co-IRMS ^{13}C, GC-C-IRMS ^{13}C and ^{15}N	KIE of glutamine synthesis using $^{14}N/^{15}N$ and solvent isotope effects	Mauve et al. (2016)
Leaves from french bean plants	Sugars: Suc, Fruc, β-Glc, o-Fru, α-Glc	GC-C-IRMS ^{13}C	Determination of kinetic $^{12}C/^{13}C$ isotope fractionation by invertase	Mauve et al. (2009)
Leaf water in C3 and C4 plants	Phytol and sterols	GC-Py-IRMS ^{2}H	Hydrogen isotopic differences between C3 and C4 land-plant lipids	Zhou et al. (2016)
Bell pepper	*n*-Alkanes	GC-Py-IRMS ^{2}H	Pepper classification according to geographical origin	de Rijke et al. (2016)
Leaf from *Pinus sylvestris*	Carbohydrates	GC-Py-IRMS ^{18}O	Development derivatization methods	Lehmann et al. (2017)
Wheat grain	Amino acids	GC-C-IRMS ^{13}C and ^{15}N	Verification of the fertilization history of wheat crops	Paolini et al. (2015)
C3 and C4 plant	Sucrose: furancarboxaldehyde (i.e., furfural); HMF	GC-C-IRMS ^{13}C	Isotopic signature of carbohydrate pyrolysis products from C3 and C4 plants: Py-CSIA of C3 and C4 plant sugars	González-Pérez et al. (2016)
Vanilla pods	Vanillin	GC-C-IRMS ^{13}C and ^{2}H	Authentication, traceability	Hansen et al. (2014)
Macroalga (Ulva sp.)	DNA, RNA	LC/IRMS ^{13}C	DNA and RNA biosynthesis	Moerdijk-Poortvliet, Stal, and Boschker (2014)

Needle extract	Soluble carbohydrate	GC-Py-IRMS $\delta(18)O$	$\delta(18)O$ sucrose of needles reflects the climatically controlled evaporative (18)O enrichment of leaf water	Zech et al. (2013)
Leguminous plants	aa	GC-C-IRMS ^{13}C	Biosynthetic pathways in plants	Styring et al. (2014)
Tobacco leaf	The wax and sterol fraction	2D (GC-C-IRMS)and high-temperature conversion (GC-HTC-IRMS)	Study of biosynthesis and delivery mechanisms of naturally occurring compounds in tobacco	Dumont et al. (2013)
Plant	aa	GC-C-IRMS ^{13}C	Study amino acid fluxes in a plant–microbe symbiotic association	Molero et al. (2011)
Plant	aa	GC-C-IRMS ^{13}C	Study carbon partitioning and dark respiration in cereals subjected to water stress	Aranjuelo et al. (2009)
Plant	Organics	LC/IRMS ^{13}C	Analysis of organic acids in plants	Hettmann et al. (2008)

Table 2 Summary of Compound-Specific Isotope Analyses Applied to Human and Animal Tissues

Matrix	Metabolites	Detection Methods and Isotopes	Application	References
Human plasma and urine	Short-chain fatty acids: acetate, propionate, butyrate	GC-C-IRMS ^{13}C	Influence of the type of indigestible carbohydrate on plasma and urine short-chain fatty acid	Verbeke et al. (2010) and Ferchaud-Roucher et al. (2006)
Human venous and arterial blood from the umbilical cord	Glutathione	GC-C-IRMS ^{13}C	In vitro measurement of fractional synthesis rate of glutathione	Tea et al. (2007)
Cancer culture cells and biopsie tissues	Amino acids, organic acids	GC-C-IRMS ^{13}C and ^{15}N LC-C-IRMS ^{13}C	Isotopic signature in cancer metabolism	Tea et al. (2016)
Milk and hair from infant	aa in proteins	GC-C-IRMS ^{15}N	Amino nitrogen transfer from human milk to infant hair protein	Tea et al. (2013) and Romek et al. (2013)
Lamb muscle	aa	GC-C-IRMS ^{15}N	Authentication and assessment of feed efficiency in ruminants	Cantalapiedra-Hijar et al. (2016)
Fat from pigs	FAME	GC-C-IRMS ^{13}C	Tool to ascertain the diet of Iberian pigs used	Recio et al. (2013)
Human and faunal from archeological samples	aa	LC-IRMS ^{13}C	Palaeodietary reconstructions	Webb et al. (2015)
Hair protein		GC-C-IRMS ^{15}N and ^{13}C	Vegetarians and omnivores are reflected in their hair protein ^{13}C and ^{15}N abundance	Petzke, Boeing, Klaus, and Metges (2005)

Human and animal tissues	aa	GC-C-IRMS ^{15}N	Refining human palaeodietary reconstruction	Styring et al. (2015)
Urine	Testosterone	GC-C-IRMS ^{13}C	IRMS detection of testosterone	Wang et al. (2014)
Urine		GC-Pyr-IRMS ^{2}H	Detecting target compounds for sports drug testing	Thevis et al. (2013)
Rat plasma and liver	Metabolites	GC-C-IRMS ^{13}C for lipids and LC-IRMS ^{13}C for glucose	Difference in ^{13}C of plasma metabolites and liver tissue between diabetic and nondiabetic Zucker diabetic fatty rats	Godin et al. (2013)
Human plasma	Cholesterol	GC-Pyr-IRMS ^{2}H	Determination of cholesterol absorption in normocholesterolemic volunteers	Godin et al. (2004)
	Glucose, fatty acids	GC-C-IRMS ^{13}C	Exercise performed immediately after fructose ingestion enhances fructose oxidation and suppresses fructose storage	Egli et al. (2016)
Plasma	Galactose, glucose	LC-IRMS ^{13}C	Quantitation of plasma ^{13}C-galactose and ^{13}C-glucose during exercise by liquid chromatography/isotope ratio mass spectrometry	Morrison et al. (2011)
Fecal sample	16s RNA	Anion-exchange LC-IRMS ^{13}C	Use of stable isotopes to measure the metabolic activity of the human intestinal microbiota	Reichardt et al. (2011)

applications are associated with sterols (doping control), amino acids (animal nutrition monitoring or plant ^{15}N labeling), or sugars (assessment of wine or honey adulteration). Up to now, ^{13}C and ^{15}N analyses in biology and biomedicine are more common than ^{18}O and ^{2}H (despite significant use for the analysis of volatile flavors, sometimes with coupling to headspace trapping, since the 2000s), simply because the latter cannot be performed by LC-co-IRMS and the use of the pyrolysis interface in GC-C-IRMS for oxygen and hydrogen isotopes can be difficult. For example, it is only recently that the $^{18}O/^{16}O$ ratio in plant sugars has been analyzed (Lehmann et al., 2017). Also, up to now, other isotopes such as sulfur ($^{34}S/^{32}S$ ratio) cannot be analyzed with the current techniques and require GC-ICP-MS. Thus, this chapter will focus on ^{13}C- and ^{15}N-CSIA.

2. COMPOUND-SPECIFIC ^{13}C AND ^{15}N ANALYSIS: HISTORY AND OVERVIEW

2.1 Brief History

Originally, CSIA was performed using off-line methods. They consisted of converting purified organic molecules to simple gases (CO_2 and N_2 for ^{13}C and ^{15}N analysis) in sealed quartz or metal tubes in the presence of oxidizing or reducing agents (de Groot, 2009). Gases were separated (isolated cryogenically) and then introduced to the IRMS via a dual-inlet system (Fig. 1A). This approach required tedious preparation samples and relatively large sample sizes. The concept of linking a GC with an IRMS was established in the mid- to late 1970s by Sano, Yotsui, Abe, and Sasaki (1976) and Matthews and Hayes (1978). In the early 1990s, the commercial production of the GC-C-IRMS provided a real breakthrough in CSIA. However, until 2000, the implementation of CSIA was very modest in biology as compared to geosciences. An alternative off-line technique consisted

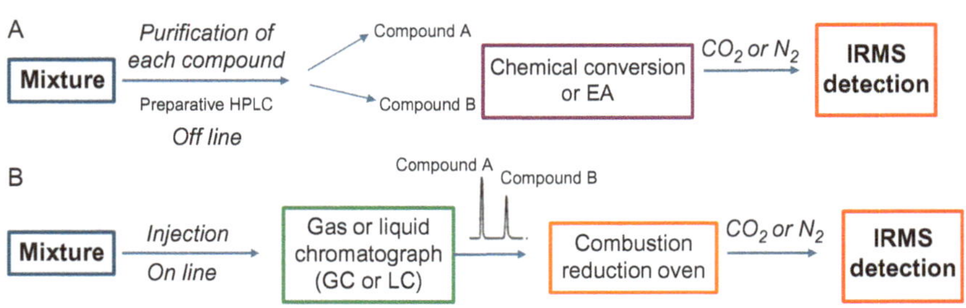

Fig. 1 Overview of instrumental coupling for CSIA, performed using off-line (A) and on-line methods (B).

of LC purification of compounds of interest using a preparative system (in which substantial amounts could be injected), and then isotope analysis by elemental analysis (EA) coupled to IRMS. This method has been used successfully for plant sugars and organic acids, for example (Duranceau, Ghashghaie, Badeck, Deleens, & Cornic, 1999). Then, LC-co-IRMS instruments were commercialized (in 2004) and enabled $^{13}C/^{12}C$ analysis. Recently, a system enabling both $^{13}C/^{12}C$ and $^{15}N/^{14}N$ analyses by LC coupled to IRMS via high-temperature combustion–desolvation (instead of chemical oxidation) has been proposed and is now available commercially (Federherr, Volders, Schmidt, Lange, & Sieper's, 2016). Both GC-C-IRMS and LC-co-IRMS on-line methods are relatively fast and allow the analysis of small quantities of sample (nmol of the compound of interest). Analytes are separated by GC or LC and converted quantitatively in gases in the interface, and finally transferred into the IRMS (principle summarized in Fig. 1B). These on-line techniques resulted in considerable increase in the number of studies using CSIA (Tables 1 and 2). However, it should be kept in mind that new developments remain relatively slow, because a critical step to carry out proper isotope analysis is sample preparation, from extraction to optimal derivatization and rigorous chromatographic conditions.

2.2 Overview of Sample Preparation and Analytical Strategies

In human and plant isotopic studies, advantage has been taken from other approaches such as metabolomics (including lipidomics) for standard sample extraction and preparation, and compound identification prior to isotope analysis. This is an important point: usual GC-C-IMRS or LC-co-IRMS instruments do not provide the identification of compounds separated by chromatography (IRMS only gives δ-values), and thus, biochemical identification should be performed beforehand, using GC-MS or LC-MS coupled to the same GC and LC conditions. Conversely, δ-values in individual metabolites provide information inaccessible with bulk stable isotope analyses (EA-IRMS) or mass spectrometers, and useful for metabolomics (e.g., to follow near natural abundance labeling). However, CSIA is more complex to undertake than bulk stable isotope analyses, simply because some isotope fractionation (see Box 1) may occur during the steps of sample preparation and/or chromatographic separation. Thus, to provide accurate δ-values in specific metabolites from biological tissues (i.e., leaves, roots, muscle) and fluids (plasma, urine, saliva), which both contain thousands of components (with highly variable chemical and physical properties),

BOX 1 Definitions

Analyte: compound analyzed by the instrument used that can be the metabolite itself (in LC systems) or a derivative of the metabolite of interest (in GC systems). A given metabolite can generate several analytes when the derivatization leads to several products. Conversely, a given analyte can originate from several metabolites, when the derivatization leads to identical products for distinct metabolites. For example, TMS-pyroglutamate is a shared analyte coming from both glutamine and glutamate upon silylation.

Derivatization: addition of chemical groups on metabolites of interest to improve their volatility (GC) or chromatographic separation (LC), or both.

δ-Value or isotope composition (usually expressed in ‰): abundance of the heavy isotope with respect to that of the light isotope, as compared to the international standard material. It is denoted as $\delta^{13}C$ for carbon and $\delta^{15}N$ for nitrogen.

Fractionation (Δ): isotopic difference (usually expressed in ‰) between the substrate and the product of a reaction. It is given by the deviation of the isotope effect (α) from 1, as $\Delta = \alpha - 1$. It can be shown that it can be calculated from δ-values in substrate and product as: $\Delta = (\delta_{substrate} - \delta_{product})/(\delta_{product} + 1)$.

Isotope effect (α): ratio of rate constants (k_{light}/k_{heavy}) or equilibrium constants (K_{light}/K_{heavy}) of the isotopologues of interest. For enzymatic reaction, it is the ratio of catalytic efficiency V/K: $\alpha = (V/K)_{light}/(V/K)_{heavy}$.

Isotopologue: isotopic analogue of a molecule, where one atom has been replaced by its isotope. For example, $^{13}C^{16}O_2$ is the ^{13}C-isotopologue of $^{12}C^{16}O_2$. Must not be confused with isotopomers, which refer to isotopic isomers (for example, $^{13}CH_3$–$^{12}CH_2OH$ and $^{12}CH_3$–$^{13}CH_2OH$ are two isotopomers of ethanol).

Isotope ratio mass spectrometer (IRMS): mass spectrometer based on a magnetic sector with (usually) fixed collectors (Faraday cups) adapted to quantify precisely the abundance of isotopic species of CO_2 (^{13}C analysis), N_2 (^{15}N analysis), CO (^{18}O analysis), H_2 (^{2}H analysis), or SO_2 (^{34}S analysis).

Quantitative reaction: chemical reaction that consumes all of the substrate molecules, thereby preventing any isotope fractionation. In nonquantitative reactions, substrate molecules left behind may have a δ-value different from the initial value because the reaction selects for an isotopic species (isotope effect).

it is crucial to pay attention to isotope behavior (i.e., possible isotope effects) at each step. These steps include: (i) extraction, (ii) derivatization (mostly, in GC-C-IRMS), and (iii) measurement itself. Fig. 2 shows an experimental workflow for the isolation of a great number of metabolites targeted for compound-specific stable isotope and structural analyses for plant, animal, and human applications. Before extraction, solid samples are preferably lyophilized and crushed (ground) to ensure homogeneity for bulk

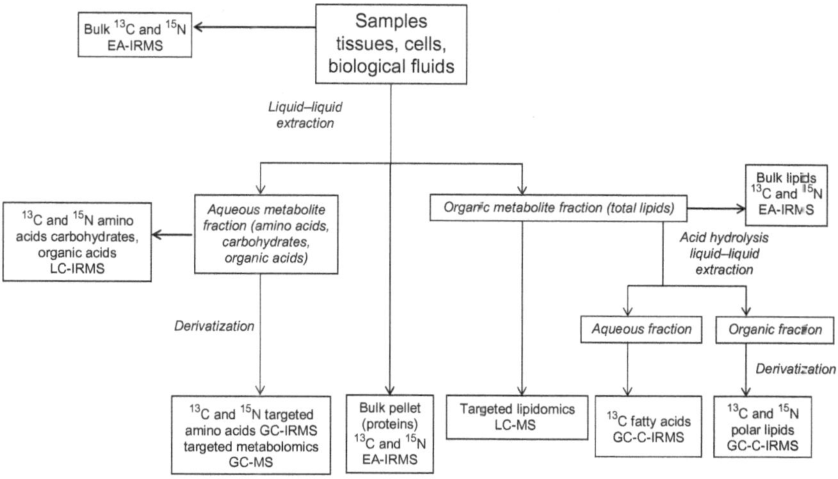

Fig. 2 Experimental workflow for the optimal isolation of a number of metabolites for CSIA for plant, animal, and human applications. This workflow integrates both bulk and compound-specific analyses as well as structural determination of metabolites so as to know the identity of compounds analyzed by GC-C-IRMS of LC-co-IRMS.

^{13}C and ^{15}N measurements and to increase the effectiveness of solvent penetration in the sample (extraction yield). The Bligh–Dyer method (Bligh & Dyer, 1959) is a well-known method originally used for lipids, and it can be used routinely for the liquid/liquid extraction of biological tissues. Two separated fractions are obtained: a total organic lipid fraction and an aqueous fraction (containing amino acids, carbohydrates, organic acids) to be analyzed by LC-co-IRMS and GC-MS, and then GC-C-IRMS after derivatization. The lipid fraction can be analyzed by LC-MS-targeted lipidomics, and bulk ^{13}C and ^{15}N abundance can be determined by EA-IRMS. The total lipid fraction includes polar lipids generally comprising fatty acids esterified to glycerol and polar headgroups (e.g., in phospholipids). Most commonly occurring polar lipids are phospholipids with choline or ethanolamine headgroups, i.e., phosphatidylcholine and phosphatidylethanolamine, while other polar lipids can include glycolipids (highly abundant in plant samples). Recently, CSIA of nitrogen-containing intact polar lipids by GC-C-IRMS has been published, for ^{15}N abundance in microbial phosphatidylethanolamine (Kayacelebi et al., 2015).

The step of functional group derivatization is inevitable in GC-C-IRMS just as in GC-MS. Indeed, metabolite classes found in biological samples are, for example, fatty acids, alcohol, monosaccharides, organic acids (carboxylic acids), and amino acids. Because of their functional groups (carboxylic, hydroxyl, and amino groups) which are polar and nonvolatile, these metabolites are potentially undetectable by GC. Derivatization reactions like esterification, acetylation, and silylation are thus employed to improve chromatographic detection and separation. Meier-Augenstein (1997, 1999) and Rieley et al. (1994) provide an extensive review of derivatizing reagents common in GC-C-IRMS. It is worth mentioning that derivatizing reagents must conform to several criteria in order to avoid isotope fractionation: (i) quantitative reaction with analytes, (ii) limited addition of the element (N or C) being analyzed (i.e., to avoid isotope dilution and important corrections, see Box 2), (iii) no chromatographic isotope effects, and (iv) no detrimental effects on the reaction interface (furnace or chemical oxidizer). For ^{15}N and ^{13}C analysis it is thus essential to verify whether the analytical protocol introduces an isotope effect in every instance. In this context, the use of two types of internal standards is recommended: first, a molecule of a similar chemical structure (but not requiring derivatization) and of known isotope composition should be added to the sample prior to sample preparation. This standard allows one to check the reliability of the instrument and to correct for possible drifts of chromatographic isotope effects in δ-values. Second, a standard of a similar chemical structure requiring derivatization and of known isotope composition should be added to the sample prior to sample preparation. This standard allows one to check the stability of the isotopic correction and/or potential undesirable isotope effects affecting the observed δ-values. Moreover, due to the addition of carbon atoms (exogenous carbons of the derivatizing reagent), it is recommended to measure the δ-value of the second standard before and after derivatization, so as to calculate the isotope composition of the added carbon group by mass balance (see also Section 5).

3. COMPOUND-SPECIFIC ^{15}N ANALYSIS

Nitrogen stable isotope analysis by GC-C-IRMS has been mostly used for amino acids and was first implemented by Merritt and Hayes (1994). They used *N*-acetyl-*n*-propyl ester derivatives for GC separation. Later, *N*-pivaloyl isopropyl esters became preferred over acetyl derivatization (Metges, Petzke, & Hennig, 1996), because of the better chromatographic resolution. *N*-trifluoroacetyl anhydride (TFA) or *N*-pentafluoropropyl anhydride (PFPA) have been employed as alternative reagents in several studies (Docherty, Jones, & Evershed, 2001; Hofmann, Gehre, & Jung, 2003). It is important to note that ^{15}N-CSIA requires excellent GC conditions

BOX 2 Isotopic corrections after derivatization

Principle: chemical derivatization used in GC-C-IRMS leads to the addition of carbon atoms, while the IRMS gives the δ-value of CO_2 generated in the interface from the whole derivative. Therefore, the contribution of the exogenous C-atoms of the derivatizing reagent must be subtracted to calculate the true value of the metabolite. There are two main methods to get the isotope signature of derivatizing chemical groups:

- *Direct measurement*: the derivatizing reagent is analyzed by EA-IRMS (as a liquid sample), and thus, its δ-value is measured directly. Then, it is assumed that this value is fully representative of the added C-atoms in metabolite derivatives. This assumption may be incorrect if (i) the reagent contains C-atoms that do not belong to derivatizing groups (such as a solvent) or (ii) there is some isotope effects during derivatization.
- *Indirect measurement*: pure standard metabolites of known $\delta^{13}C$ are derivatized, and the isotope composition of added carbon is calculated from the observed δ-value of the derivative. This technique has the advantage to integrate all of the effects during preparation and separation, in a compound-specific manner. A series of standards of different δ-values can also be used (see later).

Calculation: the calculation of the actual $\delta^{13}C$ of the metabolite of interest is carried out by mass balance as follows:

$$\delta_{obs} = p \cdot \delta_{dev} + (1 - p) \cdot \delta_{act} \quad \text{(B1)}$$

where δ_{obs}, δ_{dev}, and δ_{act} are the $\delta^{13}C$ of the derivative (observed), the derivatizing groups, and of the metabolite of interest, respectively. p is the proportion of exogenous (added) carbon in the derivative. p is a critical parameter to be determined. Classical chemical rules can be used to predict the number of C-atoms added, using the known number of functional group effectively derivatized. However, such a prediction is often invalid with silylation because some methyl groups can be lost (e.g., $-Si(CH_3)_2$ groups instead of $-Si(CH_3)_3$) as well as carbon atoms from the metabolite itself (typically, whole $-COOSi(CH_3)_3$ groups) in the furnace of the GC-C-IRMS. This process is not easily predictable and depends on the metabolite of interest. It may thus be necessary to determine p using ^{13}C-enriched and ^{13}C-depleted standards so as to draw a calibration curve and determine p and δ_{dev} by linear regression, applying Eq. (B1).

Precision of δ-values: a caveat of derivatization with addition of C-atoms is that the isotopic correction increases the standard deviation (σ) on actual $\delta^{13}C$ values. In fact, Eq. (B1) can be rewritten as: $\sigma(\delta_{act}) = (\sigma(\delta_{obs}) + p \cdot \sigma(\delta_{dev}))/(1 - p)$. This means that the imprecision on measured δ-values is additive. Also, when the proportion of added carbon (p) is large, the denominator is small, and thus, $\sigma(\delta_{act})$ is large. For example, for trimethylsilylation of an amino acid-like glutamate with three functionalized groups, C-atom addition represents 65% of total carbon, and a typical σ of 0.2‰ on observed δ-values translates into a 1‰ uncertainty in δ_{act}.

and thus judicious selection of the GC column used in order to obtain good peak resolution and avoid coelution of analytes. Moreover, ^{15}N-CSIA involves combustion followed by reduction of nitrous oxides (NO_x) to N_2. The oxidation–reduction conversion is usually done using a combustion reactor with metallic catalysts (CuO/Pt, 850°C or CuO/NiO/Pt, 940°C), sometimes followed by a second reduction reactor (elemental Cu, at 600°C), thereby reducing nitrous oxides to N_2. Another important point to take into account for N-isotopic analyses is that gases (CO_2, N_2) are not separated downstream of the combustion interface, and therefore, CO_2 may enter the IRMS and yield significant ion current at *m/z* 28 and 29 due to CO^+ fragments. Consequently, it is essential to remove CO_2 from the carrier gas stream, and this is generally done by cryogenic trapping (−100°C).

3.1 *N*-pivaloylation of the Amine Functional Group

Derivatizaton as *N*-pivaloyl-*O*-isopropyl esters appears to be the best-adapted method for compound-specific ^{15}N analysis techniques (for, e.g., amino acids), since it does not involve addition of nitrogen-containing groups and does not produce by-products of metabolites of interest. Moreover, this derivatization reaction does not introduce heteroatoms like fluorine, which could lead to a nonquantitative sample conversion in the combustion furnace of the GC-C-IRMS. In fact, the presence of fluorinated in derivatives allows the formation of extremely stable forms of fluorides with Cu and/or Ni in the combustion furnace (such as CuF_2 and NiF_2), reducing the amount of available CuO and NiO and thus the efficacy of combustion catalysis. Other advantages of this method include good chromatographic quality and stability of derivatives (derivatized metabolites remain stable for several weeks, in contrast with methods like silylation where the derivatives may be degraded after only 1–2 days). Also, the pivaloyl group makes the compound more volatile and nonpolar as compared to acetyl groups, thereby improving chromatographic separation in most cases.

Nevertheless, not all amino acids can be analyzed using this method, either because they cannot be derivatized satisfactorily, or because they are imperfectly resolved in the chromatogram. A few years ago, a good resolution in the GC chromatogram was obtained with a mixture of 15 amino acids using a polar column (60 m ZB WAX, 100% polyethylene glycol) (Romek et al., 2013; Tea et al., 2013; Fig. 3). However, the resolution

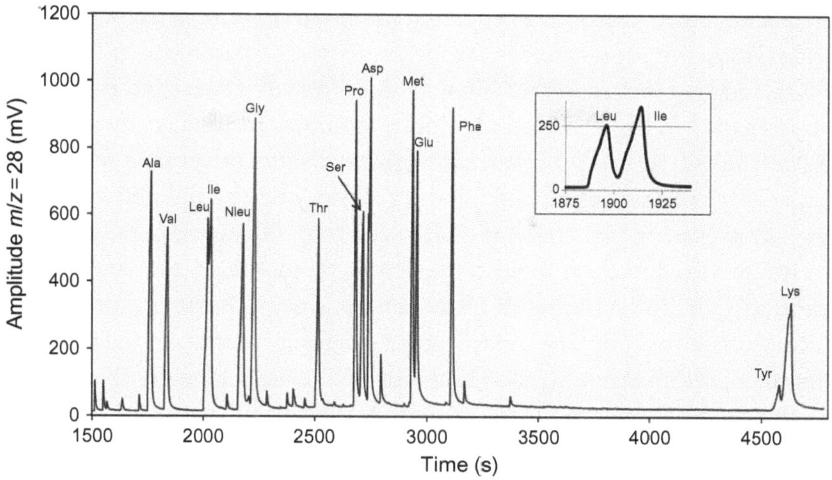

Fig. 3 Typical example of IRMS ^{15}N chromatogram (monitored by the signal from the collector of the major isotopic species at mass 28, $^{14}N_2$) showing the separation of 15 amino acids of a standard mixture in ethyl acetate (solvent), derivatized using the pivaloyl method. Each amino acid is present at 1.8 mg mL^{-1} (except for norleucine (internal standard) which is at 2 mg mL^{-1}). Inset shows the separation of Leu and Ile at 1 mg mL^{-1}.

of isomers of Leu and Ile and of the peaks of Tyr and Lys was not satisfactory. Amides like Asn or Gln, and Arg were not visible. Nevertheless, this derivatization had been effectively applied to the determination of $\delta^{15}N$ values in free amino acids (Petzke & Lemke, 2009), as well as amino acids from protein hydrolysates (Petzke & Lemke, 2009), hair samples (Petzke, Feist, Fleig, & Metges, 2006), and breast cancerous cells (Tea et al., 2016). Some amino acids such as arginine and glutamine, and polyamines like citrulline and ornithine are poorly detectable by this method, and thus, further work is needed to obtain their $\delta^{15}N$, possibly requiring an alternative derivatization approach.

3.2 *N,O*-trifluoroacetylation (TFA and PFPA)

Derivatizations to *N*-trifluoroacetyl isopropyl or *N*-pentafluoropropionyl esters are employed due to their excellent chromatographic properties compared with *N*-acetyl-isopropyl esters (NAIP). However, the compatibility with GC-C-IRMS is uncertain (i.e., δ-values are doubtful) due to the production of CuF_2 and NiF_2 in the combustion furnace of GC-C-IRMS (see earlier). In addition, it is necessary to reoxidize the combustion reactor every 30–40 runs to regenerate CuO, or even replace the furnace after 100–150 runs. The TFA method is suitable for most amino acids, but as for pivaloylation reaction, amides are poor reactants, leading to $\delta^{15}N$ values of Gln and Asn that represent mixed values from the acids and their amide counterparts. With classical methods, arginine is usually incompletely derivatized, since small amounts of ornithine and other by-products are formed. By contrast, the PFPA method allows one to fully derivatize arginine, as shown by classical GC-MS detection (Kayacelebi et al., 2015). Ornithine derivatization yields two products, in which the two primary amino groups are acylated differently, that is, one of the amino group gives a monosubstituted primary amide instead of a bisubstituted secondary amide. Our own preliminary results obtained by GC-C-IRMS show the presence of a unique peak of arginine derivative and two peaks for ornithine. However, $\delta^{15}N$ values including that of arginine were found to be strongly enriched (not shown here), simply reflecting the effect of fluorine, preventing a complete combustion reaction in the furnace and thus yielding CO that compromised $\delta^{15}N$ measurements. This analytical issue can be resolved by using a nonfluorinated analogue of PFPA derivatives, that is, NAIP.

Also, a polar chromatographic column can be used to improve resolution and avoid the problem of coelution observed with Lys, Cys, and Tyr, when using a non- or low-polar column. In fact, the choice of the GC column seems to be important to consider. It has been demonstrated that Chirasil-Val and DB-FFAP are unsuitable for such analyses, and that the HP-1MS column should be avoided to measure the $\delta^{15}N$ in amino acids (Chikaraishi, Takano, Ogawa, & Ohkouchi, 2010). This suggests that the stationary phase in these columns is too electrophilic, and therefore, the use of a DB-23, DB-35, HP-Chiral-20β, HPINNOWAX, and Ultra-2 columns should be preferred (Chikaraishi et al., 2010).

3.3 *tert*-Butyltrimethylsilylation

The *tert*-butyltrimethylsilylation (*t*-BDMS) reaction offers clear advantages in that the reaction is rapid and performed in one step only. The alkaline medium of the reaction assay also allows the simultaneous detection of glutamate and aspartate as well as their amide counterparts glutamine and asparagine. Beyond the ^{15}N analysis of amino acids, of particular interest in biomedical applications are other metabolites such as glutathione or taurine (a derivative of the sulfur-containing amino acid cysteine), which are involved in oxidative stress. The $\delta^{15}N$ in taurine has already been measured by GC-C-IRMS (Tea, Antheaume, Besnard, & Robins, 2010), whereby the amino ($-NH_2$) and sulfonic acid ($-SO_3H$) groups are derivatized using triethylorthoacetate, without any $^{14}N/^{15}N$ isotope effect. Triethylorthoacetylation thus appears as an interesting alternative to silylation. In fact, a $^{14}N/^{15}N$ isotope effect would have been observed with "classical" trimethylsilylation, because it would have given a mixture of trimethylsilylated and dimethylsilylated taurine and a number of other by-products (Meier-Augenstein, 1997). One obvious disadvantage of this derivatization is the instability of derivatized products. Furthermore, glutamate, glutamine, and arginine undergo an incomplete reaction and/or generate by-products. Note that this situation is even worse with trimethylsilylation where arginine is broken down in ornithine, citrulline, and other products, and glutamine is partially deamidated to glutamate and pyroglutamate.

4. COMPOUND-SPECIFIC ^{13}C ANALYSIS

4.1 Principle and Precautions

As for ^{15}N analysis, the conversion of metabolites (or their derivatives) to the simple gas analyzed by IRMS (here, CO_2) must be quantitative. Combustion also produces H_2O as a major by-product, as well as NO_x and SO_x, in the case of metabolite containing N and S. NO_x and SO_x are corrosive, and furthermore, NO_2 interferes with CO_2 analysis (the m/z of $^{14}N^{16}O_2$ is 46, identical to $^{12}C^{16}O^{18}O$). These species are easily eliminated by a reactor that includes a reduction section (Cu at 500–700°C). Water can also participate in proton-transfer reactions within the IRMS ion source, leading to the formation of $^{12}C^{16}O_2H^+$ which interferes with $^{13}C^{16}O_2$ at m/z 45 (Leckrone & Hayes, 1998). Water is removed via a counter-current drier gas passed onto a selectively permeable membrane of sulfonated fluoropolymer (Nafion®). A cryogenic trap at −70°C that is placed downstream of GC can also be used to eliminate water. In most cases where GC-C-IRMS is used for derivatized compounds (e.g., sugars, amino acids), derivatization methods introduce exogenous carbon atoms, which alter the $\delta^{13}C$ values (the number of added carbon atoms can be quite high, for example, in the case of silylation). That is, the δ-value obtained with the IRMS is that of the analyte (derivatized metabolite) and thus does include the contribution of exogenous carbons, and this contribution should be removed by mass-balance calculations (Box 2).

4.2 Silylation

Silylation methods are the most commonly employed for the derivatization of metabolites (sterols, alcohols, amino acids, monosaccharides). In the case of polyfunctionalized compounds such as amino acids and monosaccharides, the number of carbons added by the derivatization is proportionally larger than in, e.g., sterols, due to their molecular weight. In principle, the silylation reaction is rapid and quantitative, and this is not accompanied by KIEs (Rieley et al., 1994). Prevalent methods for derivatization of amino acids and monosaccharides use *t*-BDMS or trimethylsilyl (TMS) derivatives. Unfortunately, they can lead to the formation of several products from one compound (i.e., mono-, di-, and tri-TMS and isomeric derivatives), and some compounds are degraded during the reaction (Meier-Augenstein, 1997; and see Section 3.3). It has been also demonstrated that silicon in *t*-BDMS and TMS groups may form silicon carbide (SiC_x) in the oxidation reactor of GC-C-IRMS, leading to incomplete analyte conversion to CO_2, thus causing an isotope fractionation (Shinebarger, Haisch, & Matthews, 2002). In addition, C-atoms trapping in Si forms are not easily quantifiable, and thus, it is difficult to determine the final proportion in exogenous carbon atoms; therefore, the isotopic correction to apply is highly uncertain. Furthermore, silicon derivatives such as SiC can undergo sublimation at high temperature (in the furnace) and then may enter the mass spectrometer and dirty the ion source. For these reasons, silylation is not the preferred method for ^{13}C analyses.

4.3 Acetylation/Methylboronation of Monosaccharides

An alternative approach to silylation and trifluoroacetylation methods (mentioned earlier for ^{15}N) is the use of acetylation by alkylchloroformates, which has widely been employed for functionalized hydroxyl groups of monosaccharides and amino or thiol groups of amino acids. Despite the KIE associated with the reaction, the method has been shown to be reproducible (i.e., the KIE is rather constant) and a correction can be implemented to calculate proper δ^{13}C values. A clear advantage is that derivatives are very stable compared to silylation (*t*-BDMS). S-containing metabolites such as cysteine and glutathione can be derivatized, and both glutamine and glutamate derivatives can be detected (Montigon, Boza, & Fay, 2001).

The technique of acetylation of monosaccharides forming alditol acetates has been adapted for GC-C-IRMS analysis, but typical analytical issues encountered in that case are: (i) a KIE during the reaction and (ii) a high amount of added exogenous carbons in derivatives. To avoid the KIE associated with acetylation reactions, methylboronation (MBA) followed by trimethylsilylation can be employed. This approach leads to a lower carbon addition (+2 to 3 C-atoms per group) compared with alditol acetate derivatization (+10 to 12 C-atoms), but unfortunately, it has a low yield for several monosaccharides because it leads to an incomplete reaction with an associated KIE (van Dongen, Schouten, & Damsté, 2001).

4.4 Methylation of Fatty Acids and Methods Not Requiring Derivatization

Derivatization of fatty acids to fatty acid methyl esters (FAMEs) by the BF_3/methanol method is rapid and quantitative, thus avoiding isotopic fractionations. A correction is necessary to remove the contribution of the methyl C-atom introduced by the reagent. Instead, a simple procedure to avoid fatty acid methylation is the use of a DB-FFAP (nitroterephthalic acid-modified polyethyleneglycol) capillary GC column (typically 30 m, 0.32 mm i.d.). Fifteen fatty acids from C_{12} to C_{24} can be separated and analyzed by GC-C-IRMS with a total run time of 20 min. This method has already been validated in terms of specificity, linearity, precision, accuracy, sensitivity, and robustness, by using classical quantitative GC-FID (Zhang, Wang, & Liu, 2015), but has not been implemented for CSIA. However, this method is rapid and, importantly, does not add exogenous C-atoms.

4.5 LC-co-IRMS

The interface used to couple the LC system to the IRMS is now available commercially. It is based on the chemical wet oxidation, using sodium or ammonium peroxodisulfate (also simply referred to as "persulfate") and orthophosphoric acid as reactants to convert any metabolite containing carbon (organic molecule) to dissolved CO_2 within a reactor at 100°C. Then, the evolved CO_2 is extracted into a carrier gas by three gas-permeable membranes (one CO_2 separation unit and two water-extracting Nafion® membranes) and finally introduced into the IRMS (Fig. 4A). Recently, a new interface (Isochrom LC cube, manufactured by the company Elementar) has been commercialized and has the ability to perform both ^{13}C and ^{15}N analyses. In contrast to wet chemical oxidation used in previous systems, this new system uses a high-temperature combustion process with oxygen and platinum as a catalyst, oxidizing quantitatively metabolites into CO_2 and NO_x, subsequently reduced to N_2. It is thus referred to as

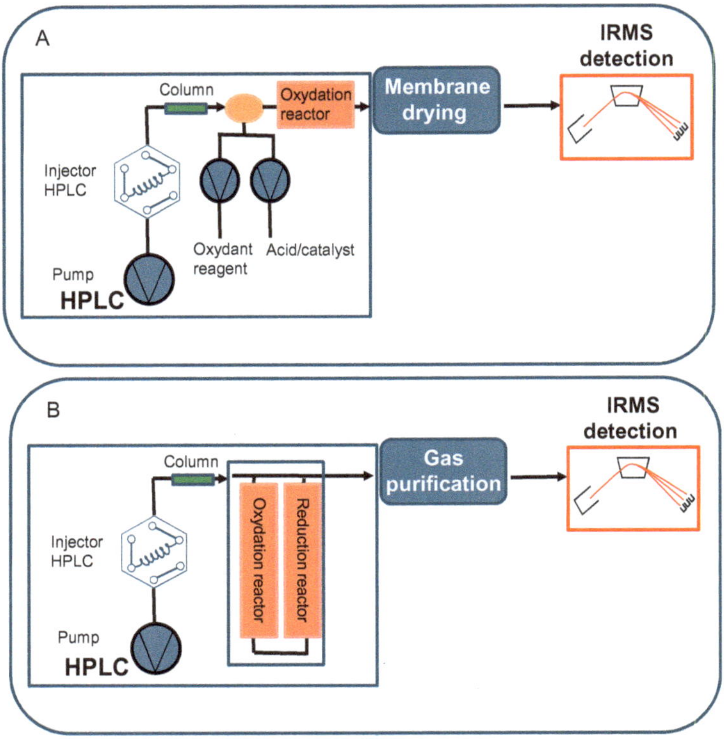

Fig. 4 Overview of the instrumental coupling in LC-C-IRMS systems, using (A) a flow reactor that mixes the aqueous sample with a strong oxidizing reagent and purges out carbon dioxide with a carrier gas or (B) the novel HTC interface technology based on catalyst-supported high-temperature combustion with subsequent gas purification/separation.

LC-HTC-IRMS. CO_2 and N_2 are purified by a selective trap or may be separated with a GC and then enter the IRMS (Fig. 4B). A proof of concept has been provided for the determination of both $\delta^{13}C$ and $\delta^{15}N$ in caffeine (Federherr et al., 2016).

LC-co- or LC-HTC-IRMS (collectively "LC-C-IRMS") is suitable for hydrosoluble molecules such as amino acids, nucleotides, sugars, peptides. The major advantage of LC-C-IRMS compared to GC-C-IRMS is that sample preparation does not require derivatization. Moreover, the isotopic precision of LC-C-IRMS is close to that of GC-C-IRMS. For example, the standard deviation of $\delta^{13}C$ values of underivatized amino acids in plasma was within 0.05‰–0.5‰, whereas the standard deviation of $\delta^{13}C$ values of the same amino acids derivatized and measured by GC-C-IRMS was within 0.2‰–0.6‰. However, there are critical analytical constraints associated with LC-co-IRMS. First, the mobile phase in the LC can only be made of water or a saline aqueous solution so as not to adulterate δ-values with exogenous nitrogen or carbon. Second, chemical oxidation of N-containing compounds like urea can generate nitric or nitrous acid that can disproportionate to NO_2 (monoisotopic m/z 46), which can compromise $\delta^{13}C$ determination. Third, chemical oxidation is sensitive to chemical conditions such as the presence of metallic ions or chlorate that can be found in environmental (soil and plant) samples. Therefore, LC-co-IRMS will probably be rapidly replaced by LC-HTC-IRMS in the near future. Still, the general use of LC-C-IRMS requires perfect LC conditions and thus an adapted LC column, as well as temperature and eluent conditions.

5. KIE MEASUREMENT FROM LC-CO-IRMS OR GC-C-IRMS DATA

A KIE is defined by the relative rates of isotopologues (v and v^* for light and heavy isotopologues, respectively) during a reaction $A \rightarrow B$ that converts a substrate A (the concentration of which is denoted as (italicized) A and A^* for the light and heavy isotopologue, respectively) into a product B as follows:

$$\alpha = \frac{A^*/A}{v^*/v} \quad (1)$$

Since the reaction rate v can be written (by definition) as $v = -dA/dt$, this gives the following differential expression, by substitution into Eq. (1):

$$\alpha = \frac{d\ln(A)}{d\ln(A^*)} \quad (2)$$

In practice, the KIE is never measured applying Eq. (2) with reaction rates obtained with the natural (light) and isotopic (heavy) forms separately, because they are far too close to be distinguished from experimental noise (with the notable exception of hydrogen and deuterium). By contrast, the method that uses natural abundance is perfectly adapted because it allows very sensitive and precise monitoring of the isotope ratio in substrates and products of the reaction. Nevertheless, the isotope competition method (upon isotopic labeling) can be used despite a lower precision that the use of natural isotope ratios. The isotope competition method has been used recently for the determination of $^{16}O/^{18}O$ KIE in nucleotide cleavage, using

^{13}C monitoring of labeled substrates by LC-co-IRMS (Du, Ferguson, Gregory, & Sprang, 2008).

Depending on the mathematical expression of the reaction rate v and the reaction considered, the significance of α may change: for a first-order inorganic reaction, $v = kA$ (where k is the rate constant) and thus $\alpha = k/k^*$; for an enzyme that follows Michaelis–Menten kinetics, $\alpha = (V/K)/(V^*/K^*)$. In what follows, we give essential information on KIEs and how to exploit GC- and LC-co-IRMS data to calculate them. Further details on experimental procedures, general calculations, and applications to dissect enzyme mechanisms are given in Cleland (1980, 1995), O'Leary (1980), and Northrop (1981). As will be apparent below, the advantage of the GC- and LC-co-IRMS systems is that they allow substrate and product to be analyzed within one instrument, possibly simultaneously. This is particularly useful to determine the atomic site (atom position) at which the KIE applies. For example, in the reaction catalyzed by the glutamine synthetase, which converts glutamate $+NH_3$ into glutamine, the $^{14}N/^{15}N$ isotope effects at the α-amino and δ-amino sites were differentiated by the simultaneous analysis of glutamate and glutamine by GC-C-IRMS (Mauve et al., 2016). Also, when several substrates (or products) can be consumed (or evolved, respectively) by a biochemical process, they can be monitored simultaneously by LC- or GC-C-IRMS and individual KIEs can be measured. For example, KIEs in biotransformation of short hydrocarbons in mixtures have been examined by GC-C-IRMS (Kinnaman, Valentine, & Tyler, 2007), and KIEs in the oxidation products of nitrobenzene (nitrobenzyl alcohol and catechol) by dioxygenase have been monitored by LC-co-IRMS (Pati, Kohler, Bolotin, Parales, & Hofstetter, 2014). It is worth noting that the principles given below also apply for EIEs (reversible reactions). In such a case, the isotope composition is substrate and products must be measured at equilibrium. For a recent example using GC-C-IRMS, see Skarpeli-Liati et al. (2011).

5.1 Principle

Isotopic data on substrates and products of a biochemical reaction obtained from LC- or GC-C-IRMS instruments can be used to calculate the KIE. The general procedure that applies is similar to that already described extensively including for carboxylation or decarboxylation reactions (where the $\delta^{13}C$ in evolved CO_2 is monitored) (O'Leary, 1980). Quite generally, it exploits the relationship describing the Rayleigh fractionation, in which the isotope composition in the substrate of the reaction of interest at time

t (R_A) is expressed as a function of the initial isotope composition of the substrate (R_0) at $t=0$ and the commitment of the reaction (percentage of substrate consumed or product formed):

$$\ln(R_A) = \ln(R_0) + \left(\frac{1}{\alpha} - 1\right) \ln(\xi) \tag{3}$$

where ξ is the ratio of the concentration of the substrate remaining at time t to that present initially ($=A/A_0$). Eq. (3) remains valid regardless of the expression of the reaction rate (i.e., whether it is a single-step reaction or an enzymatic reaction). Note that Eq. (3) reflects the fact that when a reaction fractionates against the heavy isotope ($\alpha > 1$), the substrate should progressively be enriched (increase in R_A), reflecting the production of depleted products by mass balance (Fig. 5A). Also note that the isotope effect α is a constant and thus does not change with time t or commitment ξ. What changes with time is the isotope composition of the substrate and thus the isotopic difference between total product and substrate pools. By contrast, the instantaneous isotope ratio of the product formed at time t (that is, in differential terms, between t and $t+dt$), R_{Bt}, relates to that of remaining substrate at time t as follows: $R_{Bt} = v/v^* = R_A/\alpha$. Note that Eq. (3) can be rearranged to:

$$\alpha = \frac{\ln(\xi)}{\ln(\xi) + \ln\left(\frac{R_A}{R_0}\right)} \tag{4}$$

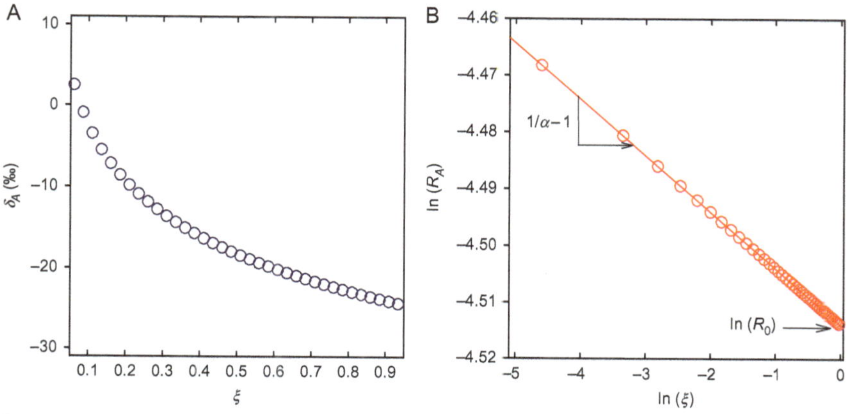

Fig. 5 Example of changes in δ-value of the substrate (δ_A) along a chemical reaction (A) and associated log–log plot (natural logarithm) to compute the KIE (B) using Eq. (3). The initial δ-value of the substrate (δ_0) was arbitrarily fixed at −25‰; note that the reaction proceeds from the *right* ($\xi=1$) to the *left* ($\xi=0$).

In practice, samples obtained from quenched reaction assays at different times (different values of ξ) are collected and the isotope composition in the substrate is measured. Then, the logarithm of the isotope ratio $\ln(R_A)$ is plotted against $\ln(\xi)$ and the slope gives $1/\alpha - 1$, using Eq. (3) (Fig. 5B). Alternatively, α can be calculated at each time point using Eq. (4).

At this stage, it is worth mentioning that the term "substrate" means the molecular species effectively consumed by the enzyme (or more generally, by the reaction). For example, the molecular species that binds to the enzyme can be involved in an equilibrium between different forms (e.g., between nonprotonated $-NH_2$ and protonated $-NH_3^+$), associated with an EIE (in this case, $^{15}\alpha_{eq}$ is about 1.015). This isotope effect has to be accounted for in calculations. An extensive description of equations associated with such calculations for ^{15}N has been provided in Mauve et al. (2016). Under certain circumstances, it is easier to carry out the isotopic analysis of the product (here denoted as B) rather than the substrate, because the substrate cannot be derivatized or chromatographically separated with available columns. This is typically the case for small compounds like NH_3 (NH_4^+) or HCO_3- that are not separable with most columns routinely used for organic metabolites. The isotope composition of the product at time t is:

$$\ln(R_B) = \ln(R_0) + \ln\left(\frac{1-\xi^{1/\alpha}}{1-\xi}\right) \tag{5}$$

Eq. (5) can be rearranged to yield a linear relationship with $\ln(\xi)$ as follows:

$$\ln(R_0 - R_B(1-\xi)) = \ln(R_0) + \frac{1}{\alpha}\ln(\xi) \tag{6}$$

which can be reformulated using the relative amount of formed product ($= B/A_0 = 1 - A/A_0 = 1 - \xi$), denoted as g:

$$\ln(R_0 - R_B g) = \ln(R_0) + \frac{1}{\alpha}\ln(1-g) \tag{6'}$$

The isotope effect is thus given by:

$$\alpha = \frac{\ln(1-g)}{\ln\left(1 - \frac{R_B}{R_0}g\right)} \tag{7}$$

As with the substrate, the isotope effect can be computed using the slope of the relationship between $\ln(R_0 - R_B(1-\xi))$ and $\ln(\xi)$ with Eq. (6) or a direct calculation with Eq. (7).

5.2 Utilization of LC-co-IRMS and GC-C-IRMS Data

LC-co-IRMS and GC-C-IRMS data provide the isotope composition in analytes of interest (substrate or product of a reaction or their derivatives obtained during sample preparation, see Section 2). Typically for $^{13}C/^{12}C$ isotopes, exogenous carbon atoms might have been added by chemical derivatization to allow analysis by GC-C-IRMS, and this has first to be corrected for before using δ-values to compute isotope effects (Box 2). Importantly, values obtained by LC-co-IRMS and GC-C-IRMS represent average isotope values over the whole molecule. In other words, these techniques do not, usually, provide positional isotope compositions (exceptions being differentiation using substrate and products as in the glutamine synthetase reaction, or the use of molecular fragmentation prior to analysis). By contrast, KIEs are position specific. Primary KIE is associated with atom positions directly involved in the reaction coordinate (i.e., in bonds being made or broken). Secondary KIEs arise due to changes in bonding at other atoms in the molecule. Thus, the isotope effect that is computed from LC-co-IRMS and GC-C-IRMS data is the average across positions. For example, if a molecule contains n carbon atoms and only one C-atom is involved in bonds being reshuffled during the reaction, the net observed $^{12}C/^{13}C$ isotope effect value will be:

$$\alpha = \frac{\alpha_{\text{int}} + (n-1)}{n} \tag{8}$$

where α_{int} is the intrinsic isotope effect at the C-atom position of interest. Of course, from a biochemical perspective, the relevant parameter is α_{int} because it gives information on the mechanism of catalysis. Eq. (8) assumes that there is no secondary isotope effect so that $(n-1)$ is not multiplied by anything in the numerator. This assumption is sometimes reasonable, because secondary isotope effects (i.e., on other, nondirectly involved atom positions) are generally small (i.e., close to 1.000). Eq. (8) also applies to other elements (N, H, O) when there are n atoms and a positional isotope effect α_{int}. It can even be the case for nitrogen, since some metabolites may have more than one N-atom (e.g., glutamine, arginine).

An important caveat of KIE measurements by these techniques is the precision that can be achieved. It is not unusual to observe standard deviations on standards or samples larger than 0.2‰, or even 0.5‰. By differentiating Eq. (4) at fixed ξ (i.e., at fixed time) and using the definition of the isotope composition ($\delta = R/R_{st} - 1$), we can easily obtain a relationship between the variation in the δ-value of the substrate, $\Delta\delta_A$, and the resulting change in the observed KIE, $\Delta\alpha$:

$$\Delta\alpha = -\frac{\Delta\delta_A}{\delta_A + 1} \cdot \frac{\ln(\xi)}{\left(\ln(\xi) + \ln\left(\frac{\delta_A + 1}{\delta_0 + 1}\right)\right)^2} \quad (9)$$

Eq. (9) shows that at the beginning of the reaction, ln(ξ) is small and the denominator of the right term does cause a large error in α. In other words, because we are at the beginning at the reaction, the actual change in δ_A as compared to δ_0 is rather small, and thus, a measurement error in δ_A, even modest, is proportionally very large compared to the difference $\delta_0 - \delta_A$. Conversely, near the end of the reaction, where ln(ξ) goes to very negative values, the right term tends to 1 and the error in α can be modest. This situation can nevertheless be problematic from an experimental point of view because under such circumstances, nearly all of the substrate is consumed, the concentration in A is small, and therefore measuring δ_A is potentially very difficult (meaning that $\Delta\delta_A$ can be large). This situation is reversed with the product (measurement of δ_B): at the beginning of the reaction, the product is present in small amounts and thus measuring its isotope composition can also be difficult. A numerical example of Eq. (9) is given in Fig. 6, where the actual isotope effect is fixed at $\alpha = 1.010$, the analytical error on δ_A is fixed at 0.2‰, and the initial substrate has an isotope composition (δ_0) of −25‰. At low reactional commitment ($\xi = 0.9$), the error made on observed α is about 2‰, while at high commitment ($\xi = 0.1$), the error is only 0.1‰. Again, in practice, $\Delta\delta_A$ is not constant and tends to increase at low concentration. It is therefore critical to ensure a good linearity of the IRMS and has isotopic standards (including internal standards) to have a reliable δ-value. It should be noted that a modest error in observed α of about 1‰ can be highly problematic because the resulting error in α_{int} is much larger (Eq. 8).

It is apparent in Eqs. (3)–(7) that the commitment of the reaction (measured as the fraction of remaining substrate (ξ) or the fraction of appeared

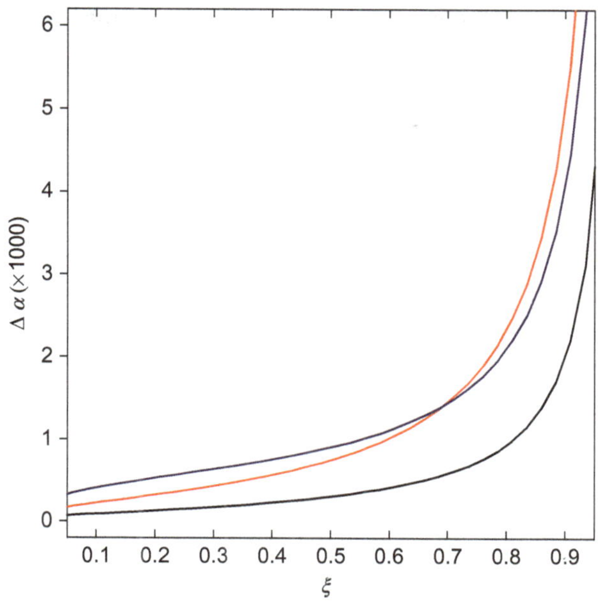

Fig. 6 Potential error in KIE determination ($\Delta\alpha$) as a function of the fraction of remaining substrate ξ when there is an uncertainty in measured δ-values of the substrate ($\Delta\delta$), as predicted by Eq. (9). *Black*, fixed $\Delta\delta$ of 0.2‰; *red*, fixed $\Delta\delta$ of 0.5‰; *blue*, varying $\Delta\delta$ (increasing from 0.35‰ to 1‰ as the reaction proceeds). In this example, the actual KIE of the reaction is 1.010; for example, at high ξ, $\Delta\alpha$ is 0.006 (6‰) or more, meaning that the observed value of the KIE ranges within 1.004 and 1.016.

product (*g*)) is an important parameter to be determined. GC-C-IRMS and LC-co-IRMS analyses may give the opportunity to measure precisely the amount of substrate and/or product when they are carried out in the split mode, i.e., when a fraction of the injected sample ends up to the LC or GC detector after the chromatography. Unfortunately, this method is compromised when the splitless mode is used to compensate for a low metabolite concentration and thus increase the signal in IRMS collectors. A compromise can be the use of a low split ratio (of, say, 5%) so that most of the injected sample ends up being analyzed by the IRMS. Alternatively, one may use the signal of the major mass (44 for carbon, 28 for nitrogen) to estimate the amount of material being analyzed (Tea et al., 2013). Of course, this requires a specific calibration curve to be done because the response of the collectors (in amperes or volts) is highly sensitive to analytical conditions such as the level of background signal.

5.3 Example

Here, we provide an example associated with the $^{12}C/^{13}C$ isotope effect in the catalysis of glycolate oxidase (GOX). This enzyme converts glycolate into glyoxylate, using O_2 as an oxidant and a flavine (FMN) as a cofactor. The KIE of the reaction has been determined recently by LC-co-IRMS, by following the $\delta^{13}C$ value of the substrate (glycolate) and the amount of remaining glycolate using refractometry (Dellero et al., 2015).

In that study, purified recombinant GOX was assayed in vitro and the reaction was quenched at different times using hydrochloric acid. Samples were lyophilized and resuspended in distilled water, and a volume of the resulting solution corresponding to 10–100 nmol of C in glycolate was injected into the LC-co-IRMS system. The LC allowed separation of the buffer (*tris*), glycolate, and glyoxylate by anion-exchange chromatography, with isocratic degassed water as the mobile phase. Detection was performed using a refractometer, and thus, glycolate amounts were determined with the LC signal (using a calibration curve). The split mode diverts one part of the sample to the LC detector, and the IRMS only sucks one fraction of the incoming sample (possibly via an interfacing needle valve), and therefore, the effective C amount entering the IRMS for isotopic analysis is much lower than 10 nmol (rather within the 100–1000 pmol range). The relationship obtained between reactional commitment and the logarithm of the isotope ratio is shown in Fig. 7, with the range of experimental values shown by gray shading. It is clear that at the beginning of the reaction, the potential error is relatively small, simply because the substrate is at high concentration (leading to a small analytical variability), and any variation of α (biological variability) does not translate into a big change in $\delta^{13}C$ of the substrate. However, as the reaction proceeds, both the lower concentration and the biological variability can lead to large uncertainty in the experimental points. This means that in the plot shown in Fig. 7B, the estimate of the intercept ($\ln(R_0)$) will be rather reliable (if not already known), while the slope (given by $1/\alpha - 1$, see Eq. 3) and thus the KIE may be associated with a large error. In that case, the average $^{12}C/^{13}C$ isotope effect is 1.019 and its standard deviation is 4‰, which is not negligible.

The $^{12}C/^{13}C$ isotope effect has also been determined using deuterated glycolate (Fig. 7, red curves). Under such circumstances, α-protonation of glycolate is slowed down and an increase in the $^{12}C/^{13}C$ isotope effect is expected. As a matter of fact, a value of 1.046 is found. When plotted together, it appears that the first time points (high $\ln(\xi)$ values) are close

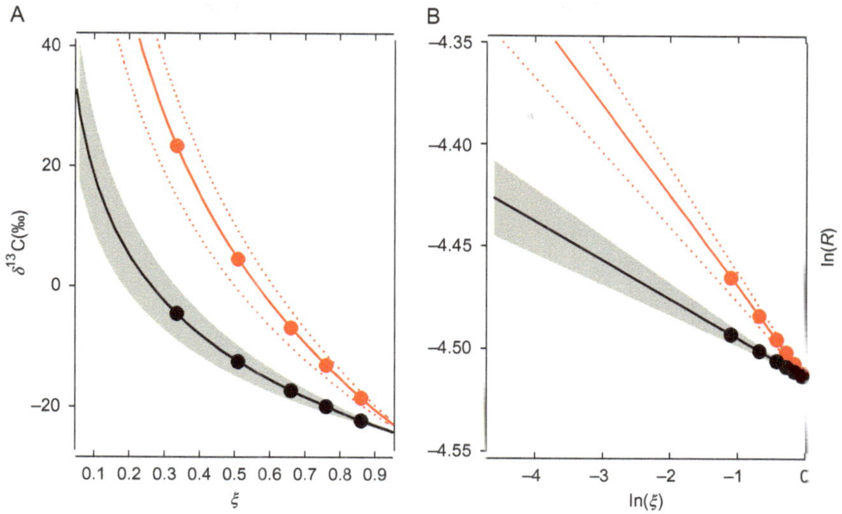

Fig. 7 Relationship between reaction commitment (fraction of remaining substrate, ξ) and the δ-value in substrate glycolate during the reaction catalyzed by *Arabidopsis* glycolate oxidase (A); and associated log–log relationship (natural logarithm) between ξ and the $^{13}C/^{12}C$ isotope ratio (B). *Black*, reaction with natural protonated glycolate; *red*, reaction with deuterated glycolate. *Solid lines, curves* driven by experimental points (means). *Dotted and shaded zones* represent regions of 80% confidence of measurement values (i.e., ± standard deviation). Note nearly indistinguishable data points on the *right hand side* of (B).

to those of the curve obtained with ordinary glycolate (Fig. 7B). This reflects the effect described by Eq. (9) whereby at low commitment, even very small variations in the δ-value of the substrate can cause large errors in observed α. That is, the relative change in δ_A (as compared to δ_0) is small and does not allow facile differentiation between distinct kinetic curves. In that respect, $\delta^{13}C$ values obtained at large commitment can be seen as essential, but unfortunately, they are associated with larger analytical errors.

Taken as a whole, accuracy and precision of GC-C-IRMS and LC-co-IRMS data are especially important to ensure good KIE determination. To do so, both replicating samples (assays) and running analyses within the same concentration range (e.g., using sample pooling for aliquots collected at low ξ) to avoid linearity bias (see Hartenbach et al., 2008 for an example of atrazine analysis by GC-C-IRMS) represent good practice to minimize potential errors in α.

6. CONCLUSIONS AND PERSPECTIVES

The potential of CSIA is enormous because of the variety of applications, from soil science to biomedicine. In fact, it is now well recognized that bulk isotope composition of tissues or organic matter is not a sufficient base to understand physiological mechanisms. For example, one of the most important applications of carbon isotopes worldwide is the use of $\delta^{13}C$ to select crop varieties with a better water use efficiency (Farquhar & Richards, 1984). Nonetheless, it is clear that this principle still needs refinements because of the multiplicity of isotope effects in metabolism and the nonhomogenous distribution of ^{13}C between plant metabolites (Tcherkez, Mahe, & Hodges, 2011). Presently, we believe that three lines of improvement warrant further work: First, CSIA extended to other elements such as oxygen and hydrogen (still in its infancy in biology and medicine) and sulfur (nearly nonexisting currently). Second, techniques other than GC-C-IRMS and LC-C-IRMS, such as high-resolution mass spectrometry (such as the Orbitrap® used in LC-MS that can potentially resolve ^{13}C and ^{15}N isotopologues, for example). Third, intramolecular isotope analysis is desirable, simply because the isotope composition within metabolites is heterogeneous due to position-specific isotope effects. In fact, considerable progress has been done on sugars using ^{13}C-NMR (for a review, see Gilbert, Silvestre, Robins, Remaud, & Tcherkez, 2012), but also GC-pyrolysis-IRMS in small organic acids (e.g., Dias, Freeman, & Franks, 2002).

ACKNOWLEDGMENTS

The authors thank the support of the initiative PLAISIR (Pays de la Loire Association for International Structure on "Isotopomic Research") funded by the French Regional International Strategy Grant Pays de la Loire, and the Australian Research Council through a Future Fellowship grant (under contract FT140100645).

REFERENCES

Aranjuelo, I., Cabrera-Bosquet, L., Mottaleb, S. A., Araus, J. L., & Nogués, S. (2009). $^{13}C/^{12}C$ isotope labeling to study carbon partitioning and dark respiration in cereals subjected to water stress. *Rapid Communications in Mass Spectrometry*, *23*, 2819–2828.

Bligh, E., & Dyer, J. (1959). A rapid method of total lipid extraction and purification. *Canadian Journal of Biochemistry and Physiology*, *37*, 911–917.

Cantalapiedra-Hijar, G., Ortigues-Marty, I., Schiphorst, A.-M., Robins, R. J., Tea, I., & Prache, S. (2016). Natural ^{15}N abundance in key amino acids from lamb muscle: Exploring a new horizon in diet authentication and assessment of feed efficiency in ruminants. *Journal of Agricultural and Food Chemistry*, *64*, 4058–4067. acs.jafc.6b00967.

Chikaraishi, Y., Takano, Y., Ogawa, N. O., & Ohkouchi, N. (2010). Instrumental optimization of compound-specific nitrogen isotope analysis of amino acids by gas chromatography/combustion/isotope ratio mass spectrometry. In N. Ohkouchi, I. Tayasu, & K. Koba (Eds.), *Earth, life, and isotopes*. Kyoto: Kyoto University Press, pp. 367–386.

Cleland, W. W. (1980). Measurement of isotope effects by the equilibrium perturbation technique. *Methods in Enzymology*, *64*, 104–125.

Cleland, W. W. (1995). Isotope effects: Determination of enzyme transition state structure. *Methods in Enzymology*, *249*, 341–373.

de Groot, P. A. (2009). *Handbook of stable isotope analytical techniques. Vol. 2*. Ham, Belgium: Elsevier.

Dellero, Y., Mauve, C., Boex-Fontvieille, E., Flesch, V., Jossier, M., Tcherkez, G., et al. (2015). Experimental evidence for a hydride transfer mechanism in plant glycolate oxidase catalysis. *The Journal of Biological Chemistry*, *290*, 1689–1698.

de Rijke, E., Truyols, V., & de Koster, C. G. (2016). UvA-DARE (digital academic repository) the use of ^{2}H and ^{18}O isotopic analyses combined with chemometrics as a traceability tool for the geographical origin of bell peppers. *Food Chemistry*, *204*, 122–128.

Dias, R. F., Freeman, K. H., & Franks, S. G. (2002). Gas chromatography–pyrolysis–isotope ratio mass spectrometry: A new method for investigating intramolecular isotopic variation in low molecular weight organic acids. *Organic Geochemistry*, *33*, 161–168.

Docherty, G., Jones, V., & Evershed, R. P. (2001). Practical and theoretical considerations in the gas chromatography/combustion/isotope ratio mass spectrometry $\delta^{13}C$ analysis of small polyfunctional compounds. *Rapid Communications in Mass Spectrometry*, *15*, 730–738.

Du, X., Ferguson, K., Gregory, R., & Sprang, S. R. (2008). A method to determine ^{18}O kinetic isotope effects in the hydrolysis of nucleotide triphosphates. *Analytical Biochemistry*, *372*, 213–221.

Dumont, E., Tienpont, B., Higashi, N., Mitsui, K., Ochiai, N., Kanda, H., et al. (2013). Heart-cutting two-dimensional gas chromatography in combination with isotope ratio mass spectrometry for the characterization of the wax fraction in plant material. *Journal of Chromatography A*, *1317*, 230–238.

Duranceau, M., Ghashghaie, J., Badeck, F., Deleens, E., & Cornic, G. (1999). $\delta^{13}C$ of CO_2 respired in the dark in relation to $\delta^{13}C$ of leaf carbohydrates in Phaseolus vulgaris L. under progressive drought. *Plant, Cell & Environment*, *22*, 515–523.

Egli, L., Lecoultre, V., Cros, J., Rosset, R., Marques, A.-S., Schneiter, P., et al. (2016). Exercise performed immediately after fructose ingestion enhances fructose oxidation and suppresses fructose storage. *American Journal of Clinical Nutrition*, *103*, 348–355.

Elsner, M., & Imfeld, G. (2016). Compound-specific isotope analysis (CSIA) of micropollutants in the environment—Current developments and future challenges. *Current Opinion in Biotechnology*, *41*, 60–72.

Farquhar, G., & Richards, R. (1984). Isotopic composition of plant carbon correlates with water-use efficiency of wheat genotypes. *Australian Journal of Plant Physiology*, *11*, 539.

Federherr, E., Volders, F., Schmidt, C., & Lange, L. (2016). A novel LC-IRMS interface for flexible compound-specific stable isotope analysis. *Rapid Communications in Mass Spectrometry*, *30*, 944–952.

Ferchaud-Roucher, V., Albert, C., Champ, M., & Krempf, M. (2006). Solid-phase microextraction method for carbon isotopic analysis of volatile carboxylic acids in human plasma by gas chromatography/combustion/isotope ratio mass spectrometry. *Rapid Communications in Mass Spectrometry*, *20*, 3573–3578.

Gilbert, A., Silvestre, V., Robins, R. J., Remaud, G. S., & Tcherkez, G. (2012). Biochemical and physiological determinants of intramolecular isotope patterns in sucrose from C_3, C_4 and CAM plants accessed by isotopic ^{13}C NMR spectrometry: A viewpoint. *Natural Product Reports*, *29*, 476.

Godin, J.-P., Richelle, M., Metairon, S., & Fay, L.-B. (2004). [2H/H] isotope ratio analyses of [2H5]cholesterol using high-temperature conversion elemental analyser isotope-ratio mass spectrometry: Determination of cholesterol absorption in normocholesterolemic volunteers. *Rapid Communications in Mass Spectrometry, 18*, 325–330.
Godin, J.-P., Ross, A. B., Cléroux, M., Pouteau, E., Montoliu, I., Moser, M., et al. (2013). Natural carbon isotope abundance of plasma metabolites and liver tissue differs between diabetic and non-diabetic Zucker diabetic fatty rats. C. Miele, Ed. *PLoS One, 8*. e74866.
González-Pérez, J. A., Jiménez-Morillo, N. T., de la Rosa, J. M., Almendros, G., & González-Vila, F. J. (2016). Compound-specific stable carbon isotopic signature of carbohydrate pyrolysis products from C3 and C4 plants. *Journal of the Science of Food and Agriculture, 96*, 948–953.
Hansen, A. M. S., Fromberg, A., & Frandsen, H. (2014). Authenticity and traceability of vanilla flavors by analysis of stable isotopes of carbon and hydrogen. *Journal of Agricultural and Food Chemistry, 62*(42), 10326–10331.
Hartenbach, A. E., Hofstetter, T. B., Tentscher, P. R., Canonica, S., Berg, M., & Schwarzenbach, R. P. (2008). Carbon, hydrogen, and nitrogen isotope fractionation during light-induced transformations of atrazine. *Environmental Science and Technology, 42*, 7751–7756.
Hettmann, E., Gleixner, G., & Juchelka, D. (2008). *irm-LC/MS: $\delta^{13}C$ analysis of organic acids in plants*. Thermo Scientific (Application note 30075).
Hofmann, D., Gehre, M., & Jung, K. (2003). Sample preparation techniques for the determination of natural $^{15}N/^{14}N$ variations in amino acids by gas chromatography-combustion-isotope ratio mass spectrometry (GC-C-IRMS). *Isotopes in Environmental and Health Studies, 39*, 233–244.
Kayacelebi, A. A., Knöfel, A.-K., Beckmann, B., Hanff, E., Warnecke, G., & Tsikas, D. (2015). Measurement of unlabeled and stable isotope-labeled homoarginine, arginine and their metabolites in biological samples by GC–MS and GC–MS/MS. *Amino Acids, 47*, 2023–2034.
Kinnaman, F. S., Valentine, D. L., & Tyler, S. C. (2007). Carbon and hydrogen isotope fractionation associated with the aerobic microbial oxidation of methane, ethane, propane and butane. *Geochimica et Cosmochimica Acta, 71*, 271–283.
Leckrone, K. J., & Hayes, J. M. (1998). Water-induced errors in continuous-flow carbon isotope ratio mass spectrometry. *Analytical Chemistry, 70*, 2737–2744.
Lehmann, M. M., Gamarra, B., Kahmen, A., Siegwolf, R. T. W., & Saurer, M. (2017). Oxygen isotope fractionations across individual leaf carbohydrates in grass and tree species. *Plant, Cell & Environment, 40*, 1658–1670.
Matthews, D. E., & Hayes, J. M. (1978). Isotope-ratio-monitoring gas chromatography-mass spectrometry. *Analytical Chemistry, 50*, 1465–1473.
Mauve, C., Bleton, J., Bathellier, C., Lelarge-Trouverie, C., Guerard, F., Ghashghaie, J., et al. (2009). Kinetic $^{12}C/^{13}C$ isotope fractionation by invertase: Evidence for a small *in vitro* isotope effect and comparison of two techniques for the isotopic analysis of carbohydrates. *Rapid Communications in Mass Spectrometry, 23*, 2499–2506.
Mauve, C., Giraud, N., Boex-Fontvieille, E. R. A., Antheaume, I., Tea, I., & Tcherkez, G. (2016). Kinetic commitment in the catalysis of glutamine synthesis by GS1 from Arabidopsis using $^{14}N/^{15}N$ and solvent isotope effects. *Plant Physiology and Biochemistry, 108*, 203–211.
Meier-Augenstein, W. (1997). The chromatographic side of isotope ratio mass spectrometry: Pitfalls and answers. *LCGC International, 15*, 244–253.
Meier-Augenstein, W. (1999). Applied gas chromatography coupled to isotope ratio mass spectrometry. *Journal of Chromatography A, 842*, 351–371.

Merritt, D. A., & Hayes, J. M. (1994). Nitrogen isotopic analyses by isotope-ratio-monitoring gas chromatography/mass spectrometry. *Journal of the American Society for Mass Spectrometry*, *5*, 387–397.

Metges, C. C., Petzke, K.-J., & Hennig, U. (1996). Gas chromatography/combustion/isotope ratio mass spectrometric comparison of N-acetyl- and N-pivaloyl amino acid esters to measure 15N isotopic abundances in physiological samples: A pilot study on amino acid synthesis in the upper gastro-intestinal tract of minipigs. *Journal of Mass Spectrometry*, *31*, 367–376.

Moerdijk-Poortvliet, T. C. W., Stal, L. J., & Boschker, H. T. S. (2014). LC/IRMS analysis: A powerful technique to trace carbon flow in microphytobenthic communities in intertidal sediments. *Journal of Sea Research*, *92*, 19–25.

Molero, G., Aranjuelo, I., Teixidor, P., Araus, J. L., & Nogués, S. (2011). Measurement of 13C and 15N isotope labeling by gas chromatography/combustion/isotope ratio mass spectrometry to study amino acid fluxes in a plant-microbe symbiotic association. *Rapid Communications in Mass Spectrometry*, *25*, 599–607.

Montigon, F., Boza, J. J., & Fay, L. B. (2001). Determination of13C- and15N-enrichment of glutamine by gas chromatography/mass spectrometry and gas chromatography/combustion/isotope ratio mass spectrometry after N(O,S)-ethoxycarbonyl ethyl ester derivatisation. *Rapid Communications in Mass Spectrometry*, *15*, 116–123.

Morrison, D. J., O'Hara, J. P., King, R. F. G. J., & Preston, T. (2011). Quantitation of plasma 13C-galactose and 13C-glucose during exercise by liquid chromatography/isotope ratio mass spectrometry. *Rapid Communications in Mass Spectrometry*, *25*, 2484–2488.

Northrop, D. B. (1981). The expression of isotope effects on enzyme-catalyzed reactions. *Annual Review of Biochemistry*, *50*, 103–131.

O'Leary, M. H. (1980). Determination of heavy-atom isotope effects on enzyme-catalyzed reactions. *Methods in Enzymology*, *64*, 83–104.

Paolini, M., Ziller, L., Laursen, K. H., Husted, S., & Camin, F. (2015). Compound-specific $\delta^{15}N$ $\delta^{13}C$ analyses of amino acids for potential discrimination between organically and conventionally grown wheat. *Journal of Agricultural and Food Chemistry*, *63*, 5841–5850.

Pati, S. G., Kohler, H.-P. E., Bolotin, J., Parales, R. E., & Hofstetter, T. B. (2014). Isotope effects of enzymatic dioxygenation of nitrobenzene and 2-nitrotoluene by nitrobenzene dioxygenase. *Environmental Science and Technology*, *48*, 10750–10759.

Petzke, K. J., Boeing, H., Klaus, S., & Metges, C. C. (2005). Carbon and nitrogen stable isotopic composition of hair protein and amino acids can be used as biomarkers for animal-derived dietary protein intake in humans. *Journal of Nutrition*, *135*(6), 1515–1520.

Petzke, K. J., Boeing, H., & Metges, C. C. (2005). Choice of dietary protein of vegetarians and omnivores is reflected in their hair protein13C and15N abundance. *Rapid Communications in Mass Spectrometry*, *19*, 1392–1400.

Petzke, K. J., Feist, T., Fleig, W. E., & Metges, C. C (2006). Nitrogen isotopic composition in hair protein is different in liver cirrhotic patients. *Rapid Communications in Mass Spectrometry*, *20*, 2973–2978.

Petzke, K. J., & Lemke, S. (2009). Hair protein and amino acid ^{13}C and ^{15}N abundances take more than 4 weeks to clearly prove influences of animal protein intake in young women with a habitual daily protein consumption of more than 1 g per kg body weight. *Rapid Communications in Mass Spectrometry*, *23*, 2411–2420.

Recio, C., Martín, Q., Raposo, C., & Raposo, C. (2013). GC-C-IRMS analysis of FAMEs as a tool to ascertain the diet of Iberian pigs used for the production of pork products with high added value. *Grasas y Aceites*, *64*, 181–190.

Reichardt, N., Barclay, A. R., Weaver, L. T., & Morrison, D. J. (2011). Use of stable isotopes to measure the metabolic activity of the human intestinal microbiota. *Applied and Environmental Microbiology*, *77*, 8009–8014.

Rieley, G., Brand, W., Mersch, F. J., Garritt, R., Fenwick, C. S., & Fallick, A. E. (1994). Derivatization of organic compounds prior to gas chromatographic–combustion–isotope ratio mass spectrometric analysis: Identification of isotope fractionation processes. *The Analyst*, *119*, 915–919.

Romek, K. M., Julien, M., Frasquet-Darrieux, M., Tea, I., Antheaume, I., Hankard, R., et al. (2013). Human baby hair amino acid natural abundance ^{15}N-isotope values are not related to the ^{15}N-isotope values of amino acids in mother's breast milk protein. *Amino Acids*, *45*, 1365–1372.

Sano, M., Yotsui, Y., Abe, H., & Sasaki, S. (1976). A new technique for the detection of metabolites labelled by the isotope ^{13}C using mass fragmentography. *Biomedical Mass Spectrometry*, *3*, 1–3.

Shinebarger, S. R., Haisch, M., & Matthews, D. E. (2002). Retention of carbon and alteration of expected ^{13}C-tracer enrichments by silylated derivatives using continuous-flow combustion-isotope ratio mass spectrometry. *Analytical Chemistry*, *74*, 6244–6251.

Skarpeli-Liati, M., Turgeon, A., Garr, A. N., Arnold, W. A., Cramer, C. J., & Hofstetter, T. B. (2011). pH-dependent equilibrium isotope fractionation associated with the compound specific nitrogen and carbon isotope analysis of substituted anilines by SPME-GC/IRMS. *Analytical Chemistry*, *83*, 1641–1648.

Styring, A. K., Fraser, R. A., Arbogast, R.-M., Halstead, P., Isaakidou, V., Pearson, J. A., et al. (2015). Refining human palaeodietary reconstruction using amino acid δ^{15}N values of plants, animals and humans. *Journal of Archaeological Science*, *53*, 504–515.

Styring, A. K., Fraser, R. A., Bogaard, A., & Evershed, R. P. (2014). Cereal grain, rachis and pulse seed amino acid δ^{15}N values as indicators of plant nitrogen metabolism. *Phytochemistry*, *97*, 20–29.

Tcherkez, G., Mahe, A., & Hodges, M. (2011). ^{12}C/^{13}C fractionations in plant primary metabolism. *Trends in Plant Science*, *16*, 499–506.

Tea, I., Antheaume, I., Besnard, J., & Robins, R. J. (2010). Determination of the natural abundance δ^{15}N of taurine by gas chromatography-isotope ratio measurement mass spectrometry. *Rapid Communications in Mass Spectrometry*, *24*, 3380–3386.

Tea, I., Ferchaud-Roucher, V., Küster, A., Darmaun, D., & Robins, R. J. (2007). Determination of ^{13}C isotopic enrichment of glutathione and glycine by gas chromatography/combustion/isotope ratio mass spectrometry after formation of the N- or N,S-ethoxycarbonyl methyl ester derivatives. *Rapid Communications in Mass Spectrometry*, *21*, 3245–3252.

Tea, I., Le Guennec, A., Frasquet-Darrieux, M., Julien, M., Romek, K., Antheaume, I., et al. (2013). Simultaneous determination of natural-abundance δ^{15}N values and quantities of individual amino acids in proteins from milk of lactating women and from infant hair using gas chromatography/isotope ratio mass spectrometry. *Rapid Communications in Mass Spectrometry*, *27*, 1345–1353.

Tea, I., Martineau, E., Antheaume, I., Lalande, J., Mauve, C., Gilard, F., et al. (2016). ^{13}C and ^{15}N natural isotope abundance reflects breast cancer cell metabolism. *Scientific Reports*, *6*, Article number 34251.

Thevis, M., Piper, T., Horning, S., Juchelka, D., & Schänzer, W. (2013). Hydrogen isotope ratio mass spectrometry and high-resolution/high-accuracy mass spectrometry in metabolite identification studies: Detecting target compounds for sports drug testing. *Rapid Communications in Mass Spectrometry*, *27*, 1904–1912.

van Dongen, B. E., Schouten, S., & Damsté, J. S. S. (2001). Gas chromatography/combustion/isotope-ratio-monitoring mass spectrometric analysis of methylboronic derivatives of monosaccharides: A new method for determining natural ^{13}C abundances of carbohydrates. *Rapid Communications in Mass Spectrometry*, *15*, 496–500.

Verbeke, K., Ferchaud-Roucher, V., Preston, T., Small, A. C., Henckaerts, L., Krempf, M., et al. (2010). Influence of the type of indigestible carbohydrate on plasma and urine short-chain fatty acid profiles in healthy human volunteers. *European Journal of Clinical Nutrition*, *64*, 678–684.

Wang, J., Yang, R., Yang, W., Liu, X., Xing, Y., & Xu, Y. (2014). IRMS detection of testosterone manipulated with ^{13}C labeled standards in human urine by removing the labeled ^{13}C. *Analytica Chimica Acta*, *852*, 146–152.

Webb, E. C., Honch, N. V., Dunn, P. J. H., Eriksson, G., Lidén, K., & Evershed, R. P. (2015). Compound-specific amino acid isotopic proxies for detecting freshwater resource consumption. *Journal of Archaeological Science*, *63*, 104–114.

Zech, M., Saurer, M., Tuthorn, M., Rinne, K., Werner, R. A., Siegwolf, R., et al. (2013). A novel methodological approach for $\delta^{18}O$ analysis of sugars using gas chromatography-pyrolysis-isotope ratio mass spectrometry. *Isotopes in Environmental and Health Studies*, *49*, 492–502.

Zhang, H., Wang, Z., & Liu, O. (2015). Development and validation of a GC–FID method for quantitative analysis of oleic acid and related fatty acids. *Journal of Pharmaceutical Analysis*, *5*, 223–230.

Zhou, Y., Grice, K., Stuart-Williams, H., Hocart, C. H., Gessler, A., & Farquhar, G. D. (2016). Hydrogen isotopic differences between C_3 and C_4 land plant lipids: Consequences of compartmentation in C_4 photosynthetic chemistry and C_3 photorespiration. *Plant, Cell & Environment*, *39*, 2676–2690.

CHAPTER SIX

Measurement of Kinetic Isotope Effects in an Enzyme-Catalyzed Reaction by Continuous-Flow Mass Spectrometry

Kenneth M. Roberts*, Paul F. Fitzpatrick†,1
*University of South Carolina Aiken, Aiken, SC, United States
†University of Texas Health Science Center, San Antonio, TX, United States
[1]Corresponding author: e-mail address: fitzpatrickp@uthscsa.edu

Contents

1. Introduction 149
2. Development of a Continuous-Flow Mass Spectrometry System 150
   2.1 Experimental Considerations 152
   2.2 Calibration of the System 153
3. Measurement of KIEs 158
   3.1 Equipment 159
   3.2 Buffers and Reagents 159
   3.3 Procedure 160
4. Conclusions 160
Acknowledgments 161
References 161

Abstract

Kinetic isotope effects (KIEs) provide powerful probes of the mechanisms of enzyme-catalyzed reactions. In this chapter, we describe the use of continuous-flow mass spectrometry to determine the deuterium KIE for the enzyme *N*-acetylpolyamine oxidase based on the ratio of labeled and unlabeled products in mass spectra of whole reaction mixtures.

1. INTRODUCTION

To fully determine the mechanism of an enzyme-catalyzed reaction, one must determine the structures of all intermediates and transition states. As discussed elsewhere in this volume, kinetic isotope effects (KIEs) provide arguably the most powerful method for analysis of mechanisms and

Methods in Enzymology, Volume 596
ISSN 0076-6879
http://dx.doi.org/10.1016/bs.mie.2017.07.001

determination of transition state structures. The methods for measuring KIEs are either competitive or noncompetitive, with all but deuterium isotopes generally being measured by competitive methods due to their higher precision. When the isotopes being used are not radioactive, the method of choice has been isotope ratio mass spectrometry (IRMS). While capable of impressive precision, IRMS has significant limitations. It is limited in the isotopes it can detect, with ^{15}N and ^{13}C being the most relevant for enzymological studies; this can be overcome by the use of a remote label, for example, by using a remote ^{15}N substituent to measure an ^{18}O effect (Cleland, 1990). IRMS can also require a significant amount of material, although improved instrumentation has decreased that problem. Finally, the method requires that the product or the remaining substrate be purified away from other compounds that contain the element of interest; this step can be the most demanding in that one must avoid any selective loss of one isotope vs another. For example, in isolating an amine product or a substrate for measurement of an ^{15}N effect, the use of an ion-exchange column can result in partial separation of ^{14}N- and ^{15}N-labeled material due to the isotope effect on the amine pK_a. Several of these problems are avoided if the reaction mixture is injected without any work-up directly into a mass spectrometer using a continuous-flow system. With this approach, the mass spectrometer effectively carries out the separation of products and substrates. Since there are no steps between the reaction and the analysis, adventitious fractionation is avoided. The sensitivity of modern electrospray mass spectrometers means that the requirements are small in terms of material. We describe here a method for the measurement of a deuterium isotope effect on an enzyme-catalyzed reaction. The method should also be applicable to measuring effects with heavier atoms.

2. DEVELOPMENT OF A CONTINUOUS-FLOW MASS SPECTROMETRY SYSTEM

The continuous-flow method, in which two reagents are mixed and then passed through a tube containing a detector, was the first method developed for observing chemical and enzyme-catalyzed reactions on a millisecond timescale (Gutfreund, 1999; Hartridge & Roughton, 1923). While this methodology has generally been replaced by stopped-flow methods, the continuous flow approach is much more versatile in the range of detection methods one can use. The basic components of a continuous-flow mass spectrometry (CFMS) system are pumps to drive solutions containing enzyme and substrate together, a mixer, and an electrospray mass spectrometer with a source connected to the mixer. The mass spectrometer thus serves as the detector, while the reaction time is varied by either changing the flow rate or changing the length of the tubing between the mixer and the source.

Our CFMS system is designed around a high-resolution mass spectrometer with a nanospray electrospray source incorporating a nanovolume chip mixer from Eksigent Technologies (Fitzpatrick, Chadegani, Zhang, Roberts, & Hinck, 2016; Gaweska, Roberts, & Fitzpatrick, 2012; Roberts, Tormos, & Fitzpatrick, 2014). Briefly, an HPLC pump pushes a sample containing enzyme and a sample containing substrate along separate lines to the chip mixer (Fig. 1). The chip mixer contains a Peltier temperature controller to permit regulation of the reaction temperature. The mixture passes out of the chip mixer to a nanospray emitter affixed directly to the exit port of the chip mixer. In reactions where the substrate is unaffected by the application of voltage (such as for our assays with *N*-acetylpolyamine oxidase, PAO) (Roberts et al., 2014), the voltage tee for the electrospray

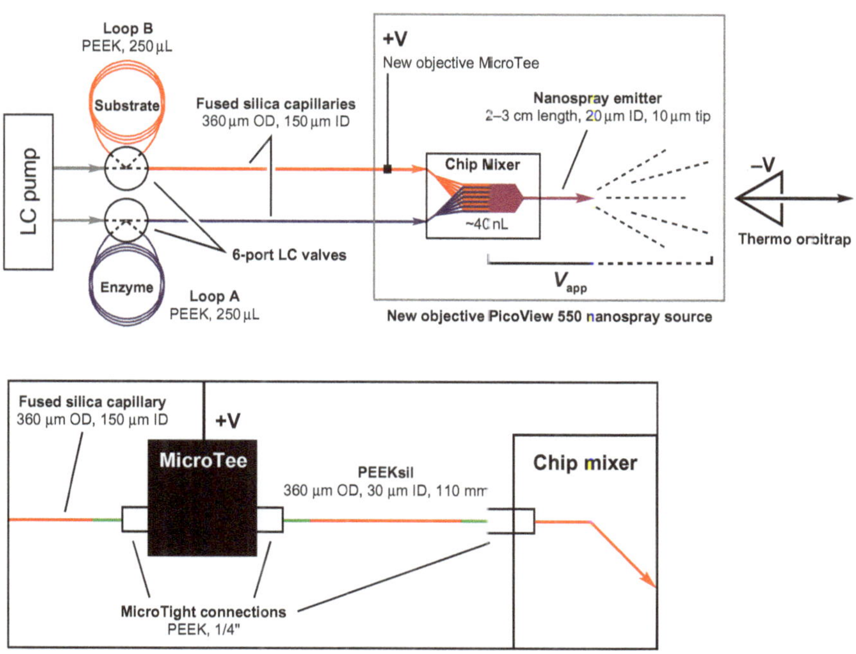

Fig. 1 Schematic of the continuous-flow mass spectrometry system used in our studies. The location of the voltage tee is indicated with +V. V_{app} represents the apparent delay volume.

is applied upstream of the chip mixer onto the line containing substrate. For applications in which the substrate is sensitive to the voltage field of the spray, the voltage tee can be applied between the chip mixer and the emitter. This does increase the delay volume significantly as the volume of the voltage tee is included in this arrangement.

Alternative systems have been described for CFMS of enzyme-catalyzed reactions. Anderson and coworkers (Paiva, Tilton, Crooks, Huang, & Anderson, 1997) utilized syringe pumps to drive reagents through a zero-volume mixer before injection into the electrospray mass spectrometer, although this system does not appear to have been used to measure isotope effects. Wilson and coworkers have described a continuous-flow mass spectrometer that uses a two-capillary system for mixing that varies the reaction time by altering the relative position of the capillaries and have recently applied it to measuring deuterium and ^{13}C isotope effects on enzyme-catalyzed reactions (Liuni, Olkhov-Mitsel, Orellana, & Wilson, 2013; Wilson & Konermann, 2003).

Our CFMS system is restricted to reaction times between 0.6 and 24 s. Different reaction times are achieved by varying the flow rate. A total flow rate of 0.25 μL/min yields a reaction time of 24 s; we have found that total flow rates less than 0.75 μL/min result in significant problems with electrospray quality when using emitters with 10 μm tips. This may be overcome using emitters with 5 μm tips, although the smaller tip diameter can result in greater tip heterogeneity and more potential for obstruction. An alternative for increased reaction times would be to insert fused-silica tubing between the mixer and the emitter. A total flow rate of 10 μL/min (5 μL/min for each sample line) is the maximum achievable with the NanoLC pump, giving a minimum reaction time of 0.6 s. While it would be possible to replace the pump with one capable of higher flow rates, we have found that total flow rates of 5 μL/min or greater result in increasingly heterogeneous electrosprays. Emitters with larger tips (20 or 50 μm) could be used instead at these higher flow rates; this will result in an increased delay volume.

2.1 Experimental Considerations

A significant limitation in analyzing enzymatic reactions by CFMS is that the buffers typically used for enzymes can be incompatible with electrospray mass spectrometry. Ammonium acetate has adequate buffer capacity at pH 8.5 and is a volatile salt. At pH 7 and 8, a buffer composed of

1,2-diammoniumethylene diacetate (EDDA) and ammonium chloride can be used (Fitzpatrick et al., 2016; Roberts et al., 2014). EDDA is a volatile salt with a pK_a of 7.6. The addition of 5% methanol to the reaction substantially increases the quality of the spray and can be required to obtain acceptable spectra; presumably other organic solvents miscible with water could be used, but methanol has the advantage of low cost. Typically, the enzyme is prepared in buffer without methanol, and the substrate is prepared in buffer plus 10% methanol. Mixing these solutions at equal volumes gives 5% methanol in the final reaction mixture. On the timescale of the reactions described here, the exposure of an enzyme to this concentration of methanol generally has little effect on structure or activity.

2.2 Calibration of the System

In continuous-flow assays, the reaction time is dependent on how long the reaction mixture takes to flow from the mixer to the detector (Gutfreund, 1999; Hartridge & Roughton, 1923). The reaction time (t), then, is dictated by the flow rate of the mixture (Q) and the volume of the delay line (V) (Eq. 1). The electrospray source reduces the droplet size of the solvent mixture until the individual ions are fully desolvated (Cole, 2000); the enzymatic reaction can take place within the droplets until the reaction components are separated immediately prior to detection. The "delay line" of the CFMS system, therefore, includes everything from the mixer (reported to us as 40 nL), on through the emitter (10 nL), and well into the electrospray, until all of the components are isolated and enter the mass spectrometer as individual ions. Consequently, it is necessary to determine an apparent delay volume (V_{app}) for a given mixer and source arrangement.

$$t = \frac{V}{Q} \quad (1)$$

The available reactions times were taken into consideration in selecting the enzyme-catalyzed reaction for initial characterization. Specifically, a reaction that would be 80%–90% complete in 10–20 s was desired. We had previously analyzed the reaction of PAO with several substrates (Henderson Pozzi, Gawandi, & Fitzpatrick, 2009). The reaction with one, N,N'-dibenzyl-1,4-diaminobutane (DBDB; Fig. 2), had appropriate kinetic parameters ($k_{cat} = 0.8\,s^{-1}$, $K_m = 15\,\mu M$) at pH 8.6 for the timescale we were interested in and consequently was selected.

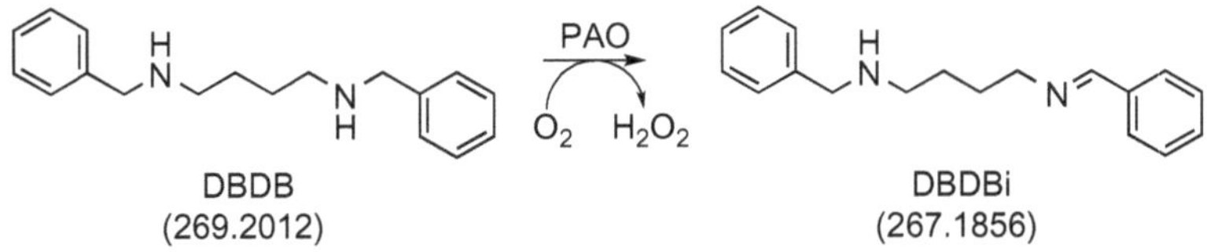

Fig. 2 Oxidation of DBDB by PAO.

Quantitative mass spectrometry requires the use of internal standards to account for the variation in the analyte signal with conditions. Critically, the ion intensity for a specific analyte varies with the total flow rate of the electrospray (Hopfgartner, Wachs, Bean, & Henion, 1993; Page, Kelly, Tang, & Smith, 2007); this flow-dependent variation is unique for each analyte. For the PAO reaction, it was thus necessary to find an analogue of DBDB that shows a similar flow-dependent signal profile. To do so, potential analogues and DBDB were mixed with buffer in the absence of enzyme and injected into the mass spectrometer. The relative intensities of the ions for DBDB and the analogue were then compared at multiple sample flow rates. We found that *N*-methylphenethylamine (MPEA) exhibited a change in intensity with flow a rate comparable to that observed for DBDB.

To determine the V_{app} of our CFMS system, the time course for the oxidation of DBDB by PAO was first determined directly under the buffer and temperature conditions to be used for CFMS. The concentrations of DBDB (40 μ*M*) and PAO (20 μ*M*) were chosen such that ~90% of the DBDB was consumed over the first 15 s of the reaction and the signal to noise for DBDB was acceptable. We used a Bio-Logic QFM-400 quenched-flow apparatus to mix 80 μ*M* DBDB plus 80 μ*M* MPEA, in 10% methanol, 50 m*M* ammonium acetate, pH 8.5, with 40 μ*M* PAO in 50 m*M* ammonium acetate, pH 8.5, at 30°C, quenching the reaction with 2 *M* HCl after 50 ms to 10 s. The amount of DBDB remaining at each time was then determined by HPLC. Since the initial concentration of DBDB was close to the K_m value, the change in the concentration with time was reasonably well described in Eq. (2), yielding the value for the first-order rate constant (k_{obs}) for the reaction under these conditions. The reaction was then repeated by the CFMS system under the same conditions, measuring the relative ion intensity of DBDB across a range of flow rates, using MPEA as an internal standard. The CFMS data were fit to Eq. (3) using the value of k_{obs} determined from the chemical quench analyses (Fig. 3) to obtain the V_{app} value. For two different chip mixers, we obtained values for V_{app} of 102 ± 7 and 210 ± 20 nL, respectively, when the CFMS system was set up as shown in Fig. 1.

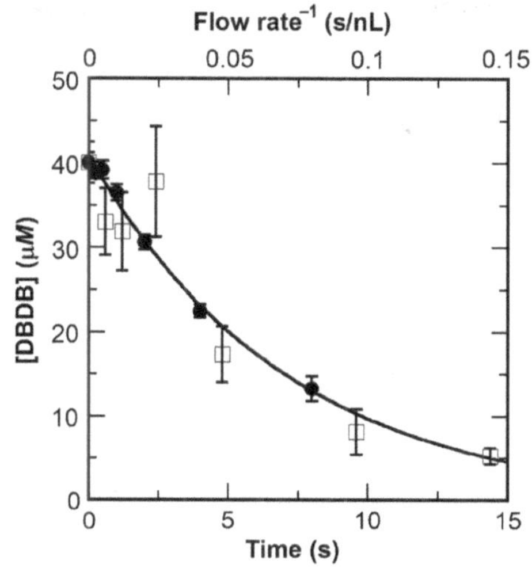

Fig. 3 Time dependence of the oxidation of DBDB by PAO by chemical quench (*closed circles*) and CFMS (*open squares*). Final reaction conditions were 40 μ*M* DBDB, 20 μ*M* PAO, 50 m*M* ammonium acetate, 5% methanol, pH 8.5, and 30°C. The *line* indicates the best-fit curve for the chemical quench data to Eq. (2) and the best-fit curve for the CFMS data to Eq. (3). *From Roberts, K. M., Tormos, J. R., & Fitzpatrick, P. F. (2014). Characterization of unstable products of flavin- and pterin-dependent enzymes by continuous-flow mass spectrometry.* Biochemistry, *53(16), 2672–2679. http://dx.doi.org/10.1021/bi500267c, http://pubs.acs.org/doi/full/10.1021/bi500267c.*

$$[\mathrm{DBDB}]_t = [\mathrm{DBDB}]_0 e^{-k_{\mathrm{obs}} t} \tag{2}$$

$$[\mathrm{DBDB}]_t = [\mathrm{DBDB}]_0 e^{-k_{\mathrm{obs}} \frac{V_{\mathrm{app}}}{Q}} \tag{3}$$

2.2.1 Equipment

- Thermo Fisher Scientific LTQ OrbiTrap Discovery
- New Objective PicoView nanospray source with Eksigent Technologies Nanovolume Chip Mixer
- Fused-silica tubing (360 μm OD, 150 μm ID)
- New Objective Picotip distal coated SilicaTip emitter (20 μm ID, 10 μm tip)
- Bio-Logic Science Instruments QFM-400 quenched-flow system
- Phenomenex Gemini-NX C18 HPLC column (3.0 μm, 1.0 × 150 mm) and HPLC with fluorescence detector
- Synergy Software KaleidaGraph graphing and data analysis software

2.2.2 Buffers and Reagents

- *N*,*N*′-Dibenzyl-1,4-diaminobutane
- *N*-Methylphenethylamine
- *N*-Acetylpolyamine oxidase
- Buffer A: 50 m*M* ammonium acetate, pH 8.5
- Buffer B: 10% methanol in 50 m*M* ammonium acetate, pH 8.5
- Sample Solvent: 0.1% trifluoroacetic acid in water
- Elution Solvent: 0.1% trifluoroacetic acid, 17.5% acetonitrile in water

2.2.3 Procedure—CFMS Assays

1. Set up the CFMS system as shown in Fig. 1:
   - **a.** Connect the LC pump separately to two six-port injection valves and the valves to the chip mixer using fused-silica tubing.
   - **b.** A few centimeters before the chip mixer, insert the nanospray voltage tee onto the line from Loop B.
   - **c.** Mount the sample loops onto the injection valve such that water from the pump may pass either directly through the valve to the chip mixer or through the sample loops to the chip mixer.
   - **d.** Cut nanospray emitter to appropriate length and connect directly to the chip mixer.
2. Start the LC pump at the chosen flow rate to establish steady, even flow throughout the valves, sample lines, and the chip mixer, generating a stable electrospray.
3. Load sample loops A and B with Buffer A (with or without enzyme) and 80 μ*M* DBDB, 80 μ*M* MPEA in Buffer B, respectively. Allow 2–5 min for the sample flows and pressure to reach an even, steady flow.
4. Once the sample flow has equilibrated, record 100 or more scans of the mass spectrum of the resulting reaction mixture in the range of m/z 130–300. Average the ratio of the ion intensities for DBDB ($[M+H]^+$ 269.2012) and MPEA ($[M+H]^+$ 136.1121) across all scans.

2.2.4 Notes

1. The voltage tee *should not* be placed in line with the protein sample upstream of the chip mixer. We found that this arrangement leads to immediate precipitation of the protein within the tee, resulting in complete blockage of the tee within minutes.
2. We use nanospray emitters with a distal conductive coating because the coating assisted with the formation of a successful electrospray at the low concentration of methanol used in these assays.

3. Successful electrospray conditions for reaction analyses included a spray voltage of 4–6 kV when the voltage tee was applied upstream of the chip mixer. Lower voltages resulted in no spray. Higher voltages were complicated by erratic spray and background electrochemistry involving the substrates.
4. When applied downstream of the chip mixer, spray voltages as low as 2 kV could be applied to the source for successful electrospray. However, this arrangement results in a significant increase in the delay volume of the system.
5. At higher flow rates (greater than 0.5 μL/min per line), the pressure in the two sample lines is high enough that varying from a 1:1 ratio of flow rates between the sample lines is possible. This can allow varying the concentrations of enzyme and substrate in the reaction.
6. At lower flow rates (less than 0.5 μL/min per line), the flow rates of the lines *must* be equal; otherwise the pressure of the slower line can become overpowered by the faster line, preventing adequate mixing within the mixer. This results in a lower concentration of the slower flow rate sample in the mixture than expected.
7. Allowing the system to reach a steady, even pressure/flow is imperative. It takes 2–5 min for the sample lines, mixer, and emitter to equilibrate such that the flow rate is equal throughout the system. Since the flow rate dictates the reaction time in this fixed-volume system, waiting until the flows stabilize ensures the mass spectra correlates to the correct reaction time.
8. To correct for the change in signal intensities at varying flow rates, samples include an internal standard (MPEA). The mass spectra of reaction samples are recorded for 100–200 scans and the ratio of the signal intensities of the analyte and the internal standard are averaged across all scans. Where possible, these ratios are compared to a standard curve.

2.2.5 Procedure—V_{app} Calibration

1. Using a quenched-flow apparatus, mix 50 μL 40 μ*M* PAO in Buffer A with 50 μL 80 μ*M* DBDB, 80 μ*M* MPEA in Buffer B at 30°C and quench with 50 μL 2 *M* HCl. Vary reaction times from 50 ms to 10 s.
2. Centrifuge the quenched samples at 12,000 × *g* for 60 s to remove the denatured protein. Add 100 μL of reaction supernatant to 900 μL Sample Solvent.
3. Quantify DBDB by injecting 20 μL of each sample onto a Phenomenex Gemini-NX C18 HPLC column with Elution Solvent as the mobile

phase at a flow rate of 0.25 mL/min. The fluorescence detector is set at 278 nm with an excitation wavelength of 255 nm. The concentration of DBDB is determined from the fluorescence intensity of DBDB relative to that of the MPEA internal standard.

4. Determine the observed rate constant, k_{obs}, from a fit of the concentration of DBDB vs time with Eq. (2).
5. Repeat the reaction on the CFMS system by mixing 40 μ*M* PAO in Buffer A with 80 μ*M* DBDB, 80 μ*M* MPEA in Buffer B at equal flow rates for a range of total flow rates (*Q*: 0.25–10 μL/min). For each flow rate (*Q*), calculate the concentration of DBDB using the MPEA internal standard across 100 scans and take the average of the values.
6. Determine the apparent delay volume (V_{app}) from a fit of the concentration of DBDB vs *Q* with Eq. (3).

2.2.6 Notes

1. An important aspect of this procedure is that the reaction is performed under the same conditions in both methods. This eliminates any condition-dependent variances and allows direct comparison of the time-dependent concentration of the analyte (DBDB) by both methods.
2. Only the quality of the fit is important for the success in determining V_{app}. The nature of that change is not important. Although a ratio of 2:1 for DBDB:PAO is not truly first order, the data were well fit with a first-order rate equation and thus the fit could be used as an appropriate representation of both reactions.

3. MEASUREMENT OF KIEs

The ability to follow the PAO reaction by CFMS provided an opportunity to determine isotope effects on the reaction by a competitive method without the need for synthesis of multiply labeled substrates or isolation of the product or remaining substrate at different reaction times. The product of the oxidation of DBDB by PAO is DBDB monoimine (DBDBi; Fig. 2). To determine the $D(k_{cat}/K_m)$ value, we used a mixture of DBDB and DBDB-d_{14} as substrate and followed the formation of the products from the nondeuterated ($m/z = 267.1856$) and deuterated ($m/z = 280.2672$) substrates simultaneously using CFMS (Fig. 4). The reactions contained DBDB-d_{14} and DBDB in a 5:1 ratio so that the intensities of the ions from the labeled and unlabeled substrates would be comparable. A separate

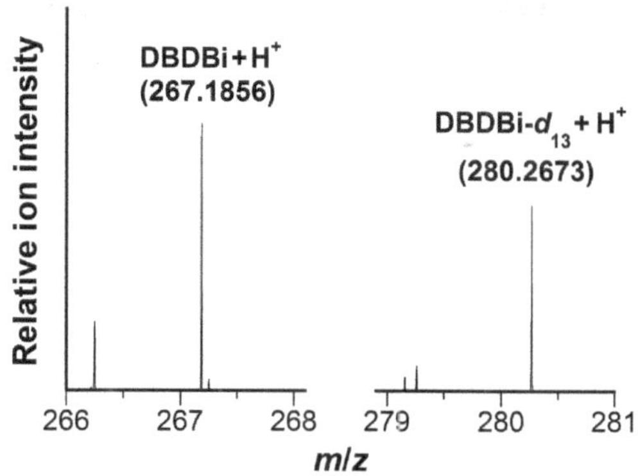

Fig. 4 Mass spectrum of the reaction of 40 μ*M* PAO with 6.7 μ*M* DBDB, 33 μ*M* DBD3-d_{14} at 30°C at a flow rate of 10 μL/min, for a reaction time of 1.3 s. The spectrum is an average of 100 scans.

analysis of the isotope effect using a noncompetitive approach yielded a value of 6.7 ± 0.6; in the absence of prior knowledge of the isotope effect, one could vary the ratio of the labeled and unlabeled substrate to optimize the result. No internal standard was required, since the isotope effect was determined from the ratio of the labeled and unlabeled products relative to the substrate ratio at the beginning of the reaction, and isotopic substitution is not expected to affect the ionization properties of the substrate and products. We tested this across a wide range of flow rates and found the relative ion intensities of the substrates to be independent of the flow rates. Fig. 5 shows the product isotope effect as a function of time. Extrapolation of the time-dependent isotope effect to t_0 using Eq. (4) yielded a value of 7.7 ± 0.3 for D(k_{cat}/K_m).

$$\frac{([\mathrm{DBDB}i]/[\mathrm{DBDB}i - d_{13}])_t}{([\mathrm{DBDB}]/[\mathrm{DBDB} - d_{14}])_0} = \mathrm{KIE}\, e^{-kt} \tag{4}$$

3.1 Equipment

- CFMS system

3.2 Buffers and Reagents

- *N,N'*-Di-(perdeuterobenzyl)-1,4-diaminobutane (DBDB-d_{14}) (Henderson Pozzi et al., 2009)

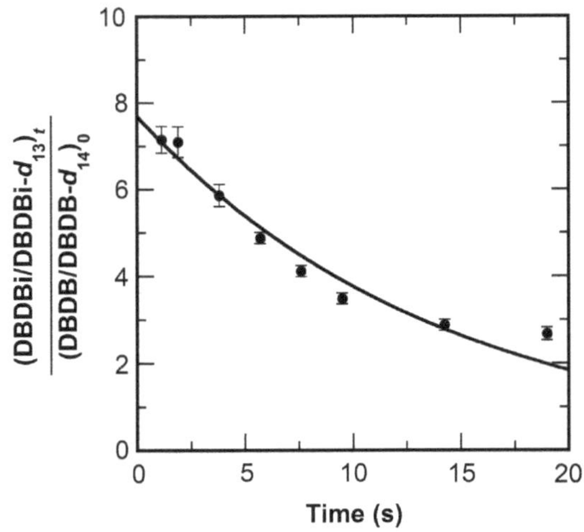

Fig. 5 Measurement of the deuterium isotope effect for oxidation of DBDB by PAO using continuous-flow mass spectrometry. The reaction contained 6.7 μ*M* DBDB, 33 μ*M* DBDB-d_{14}, and 40 μ*M* PAO in 5% methanol and 50 m*M* ammonium acetate (pH 8.5) at 30°C after mixing. The line is from a fit to Eq. (4). *From Roberts, K. M., Tormos, J. R., & Fitzpatrick, P. F. (2014). Characterization of unstable products of flavin- and pterin-dependent enzymes by continuous-flow mass spectrometry.* Biochemistry, *53(16), 2672–2679. http://dx.doi.org/10.1021/bi500267c, http://pubs.acs.org/doi/full/10.1021/bi500267c.*

3.3 Procedure

1. Perform the CFMS reaction by mixing 80 μ*M* PAO in Buffer A with 6.7 μ*M* DBDB, 33 μ*M* DBDB-d_{14} in Buffer B at 30°C at equal flow rates such that total flow rates yield reaction times from 0.6 to 24 s.
2. For each reaction time, record the mass spectrum (m/z 130–300) for 100 scans. Calculate the observed product isotope effect for each time point as the average of the ratio of the ion intensities for DBDBi ($[M+H]^+$ 267.1856) and DBDBi-d_{13} ($[M+H]^+$ 280.2672) across all 100 scans.
3. Fit the time dependence of the observed isotope effect to Eq. (4) to determine the isotope effect as the calculated value at time zero.

4. CONCLUSIONS

KIEs on an enzyme reaction can be measured using continuous-flow mass spectrometry. Important considerations for successful CFMS assays include a buffer system amenable to both an enzyme reaction and electrospray ionization mass spectrometry, quantification of analytes using an appropriate internal standard, accurate reaction times obtained from calibration of the flow rate and the apparent delay volume, and the starting conditions of the reaction.

ACKNOWLEDGMENTS

Development of the CFMS system was supported by grants from the National Institutes of Health (R01 GM058698) and The Welch Foundation (AQ1245).

REFERENCES

Cleland, W. W. (1990). Secondary ^{18}O isotope effects as a tool for studying reactions of phosphate mono-, di-, and triesters. *The FASEB Journal*, *4*(11), 2899–2905.

Cole, R. B. (2000). Some tenets pertaining to electrospray ionization mass spectrometry. *Journal of Mass Spectrometry*, *35*(7), 763–772. http://dx.doi.org/10.1002/1096-9888.

Fitzpatrick, P. F., Chadegani, F., Zhang, S., Roberts, K. M., & Hinck, C. S. (2016). Mechanism of the flavoprotein l-hydroxynicotine oxidase: Kinetic mechanism, substrate specificity, reaction product, and roles of active-site residues. *Biochemistry*, *55*(4), 697–703. http://dx.doi.org/10.1021/acs.biochem.5b01325.

Gaweska, H. M., Roberts, K. M., & Fitzpatrick, P. F. (2012). Isotope effects suggest a stepwise mechanism for berberine bridge enzyme. *Biochemistry*, *51*(37), 7342–7347. http://dx.doi.org/10.1021/bi300887m.

Gutfreund, H. (1999). Rapid-flow techniques and their contributions to enzymology. *Trends in Biochemical Sciences*, 24(11), 457–460. http://dx.doi.org/10.1016/S0968-0004(99)01468-1.

Hartridge, H., & Roughton, F. J. W. (1923). A method of measuring the velocity of very rapid chemical reactions. Proceedings of the Royal Society of London *Series A*, *104*(726), 376–394. http://dx.doi.org/10.1098/rspa.1923.0116.

Henderson Pozzi, M., Gawandi, V., & Fitzpatrick, P. F. (2009). Mechanistic studies of *para*-substituted N,N'-dibenzyl-1,4-diaminobutanes as substrates for a mammalian polyamine oxidase. *Biochemistry*, *48*(51), 12305–12313. http://dx.doi.org/10.1021/bi901694s.

Hopfgartner, G., Wachs, T., Bean, K., & Henion, J. (1993). High-flow ion spray liquid chromatography/mass spectrometry. *Analytical Chemistry*, *65*(4), 439–446. http://dx.doi.org/10.1021/ac00052a021.

Liuni, P., Olkhov-Mitsel, E., Orellana, A., & Wilson, D. J. (2013). Measuring kinetic isotope effects in enzyme reactions using time-resolved electrospray mass spectrometry. *Analytical Chemistry*, *85*(7), 3758–3764. http://dx.doi.org/10.1021/ac400191t.

Page, J. S., Kelly, R. T., Tang, K., & Smith, R. D. (2007). Ionization and transmission efficiency in an electrospray ionization-mass spectrometry interface. *Journal of the American Society for Mass Spectrometry*, 18(9), 1582–1590. http://dx.doi.org/10.1016/j.jasms.2007.05.018.

Paiva, A. A., Tilton, R. F., Crooks, G. P., Huang, L. Q., & Anderson, K. S. (1997). Detection and identification of transient enzyme intermediates using rapid mixing, pulsed-flow electrospray mass spectrometry. *Biochemistry*, *36*(49), 15472–15476. http://dx.doi.org/10.1021/bi971883i.

Roberts, K. M., Tormos, J. R., & Fitzpatrick, P. F. (2014). Characterization of unstable products of flavin- and pterin-dependent enzymes by continuous-flow mass spectrometry. *Biochemistry*, *53*(16), 2672–2679. http://dx.doi.org/10.1021/bi500267c.

Wilson, D. J., & Konermann, L. (2003). A capillary mixer with adjustable reaction chamber volume for millisecond time-resolved studies by electrospray mass spectrometry. *Analytical Chemistry*, *75*(23), 6408–6414. http://dx.doi.org/10.1021/ac0346757.

CHAPTER SEVEN

Primary Deuterium Kinetic Isotope Effects From Product Yields: Rationale, Implementation, and Interpretation

Tina L. Amyes, John P. Richard[1]
University at Buffalo, SUNY, Buffalo, NY, United States
[1]Corresponding author: e-mail address: jrichard@buffalo.edu

Contents

1. Introduction 164
2. Enzyme-Catalyzed Decarboxylation of OMP 166
3. Experimental Protocol 168
   3.1 A Comparison of Protocol for Determination of SDIE and PDIE 168
   3.2 Enzyme-Catalyzed Decarboxylation 169
4. Interpretation of Results 173
5. Other Enzymes 174
Acknowledgment 175
References 175
Further Reading 177

Abstract

A simple and convenient method is described to determine primary deuterium kinetic isotope effects (1°DKIEs) on reactions where the hydron incorporated into the reaction product is derived from solvent water. The 1°DKIE may be obtained by ^{1}H NMR analyses as the ratio of the yields of —H- and —D-labeled products from a reaction in 50:50 (v/v) HOH/DOD. The procedures for these ^{1}H NMR analyses are reviewed. This product deuterium isotope effect (PDIE) is defined as $1/\phi_{EL}$ for fractionation of hydrons between solvent and the transition state for the reaction examined. When the solvent is not the direct hydron donor, it is necessary to correct the PDIE for the fractionation factor ϕ_{EL} for partitioning of the hydron between the solvent and the direct donor EL. This method was used to determine the 1°DKIE on decarboxylation reactions catalyzed by wild-type orotidine 5′-monophosphate decarboxylase (OMPDC) and by mutants of OMPDC, and then in the determination of the 1°DKIE on the decarboxylation reaction catalyzed by 5-carboxyvanillate decarboxylase. The experimental procedures used in studies on OMPDC and the rationale for these procedures are described.

Methods in Enzymology, Volume 596
ISSN 0076-6879
http://dx.doi.org/10.1016/bs.mie.2017.06.043

1. INTRODUCTION

A simple and powerful experiment to characterize enzymatic reaction mechanisms is the determination of the effect of a change from HOH as a solvent to DOD on the reaction rate (Kresge, More O'Ferrall, & Powell, 1987; Schowen, 1972). The interpretation of solvent deuterium kinetic isotope effects (SDKIEs) requires a consideration of both the secondary solvent isotope effects on reactivity of the hydroxylic solvent water (Jencks, 1987) and the primary deuterium kinetic isotope effect (1°DKIE) on the rate of hydron transfer that arises from the difference in ground-state zero-point energy that is lost at the transition state for cleavage of X—H compared with X—D bonds (Westheimer, 1961).

These complications were examined and clarified in studies on the hydrolysis of vinyl ethers that proceed with rate-determining protonation of the substrate to form an oxocarbenium ion (Scheme 1) (Richard & Williams, 2007; Tsang & Richard, 2007, 2009). The ratio of second-order rate constants k_H/k_D determined for specific acid-catalyzed hydrolysis of vinyl ethers in HOH and DOD is smaller than the 1°DKIE on proton transfer (Kresge, Sagatys, & Chen, 1977). For example, the ratio $k_H/k_D = 4.0$ for specific acid-catalyzed hydrolysis of α-methoxystyrene in HOH and DOD is smaller than the 1°DKIE of 5.7 determined as described below, because of an inverse contribution from a secondary SDKIE of 0.70 (Tsang & Richard, 2009). This SDKIE arises because hydrogen bonds between solvent and —H at H_3O^+ are stronger than between solvent and —D at D_3O^+ (Jencks, 1987; Kresge et al., 1987; Schowen, 1972). The weakening of hydrogen bonds to lyonium ion at the transition state for specific acid-catalyzed reactions therefore results in a larger increase in the barrier for protonation by H_3O^+,

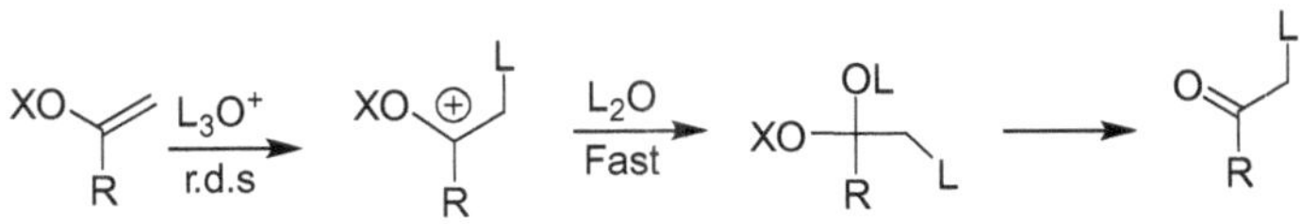

Scheme 1 The mechanism for hydrolysis of a simple vinyl ether to form the ketone. There is extensive evidence that protonation of the vinyl ether to form the oxocarbenium ion is effectively irreversible and the rate-determining step for the hydrolysis reaction, because the barrier to deprotonation of this intermediate to regenerate the vinyl ether is several kcal/mol higher than the barrier for addition of solvent to form the hemiacetal intermediate (Richard et al., 2000; Richard & Williams, 2007; Richard, Williams, & Amyes, 1999).

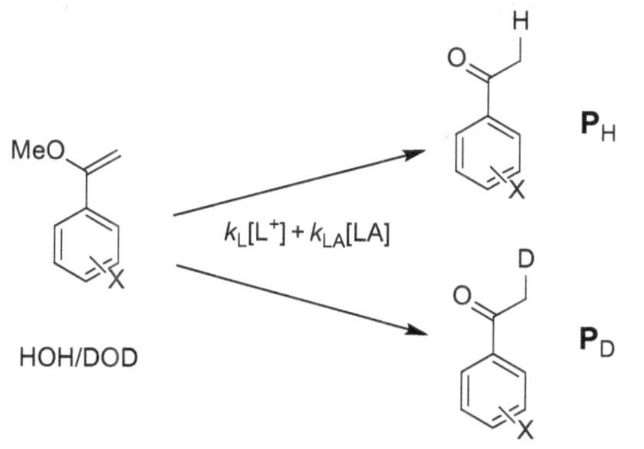

Scheme 2 The hydrolysis of a ring-substituted styrene in a mixed solvent to give a mixture of —H- and —D-labeled substituted acetophenones. The primary isotope effect on hydron transfer is calculated from the ratio of the yields of —H- and —D-labeled reaction products using Eq. (1).

where the hydrogen bonds are strongest, compared with D_3O^+: this is the underlying cause for the observed inverse SDKIE (Scheme 1).

$$1^{\circ}\text{DKIE} = \text{PDIE} = \frac{f_D[\mathbf{P}_H]}{f_H[\mathbf{P}_D]} = \left(\frac{f_D}{f_H}\right)\left(\frac{[A_{CH_3}]}{1.5[A_{CH_2D}]}\right) \tag{1}$$

The 1°DKIEs on protonation of ring-substituted α-methoxystyrenes may be obtained, free of any contribution from a SDKIE, by examining the reactions of —H and —D in a *common* mixed solvent. The 1°DKIE on the rate-determining proton transfer step is obtained from the product deuterium isotope effect (PDIE) determined using Eq. (1), where f_H and f_D are the mole fractions of H and D in the mixed solvent and $[\mathbf{P}_H]/[\mathbf{P}_D]$ is the ratio of the yield of the —H- and —D-labeled products (Scheme 2). This ratio is determined as the ratio of the area A_{CH3} of the peak for the singlet due to the $—CH_3$ group and $1.5A_{CH2D}$ for the upfield shifted triplet due to the $—CH_2D$ group.

This experimental protocol can be adapted to enzyme-catalyzed reactions of vinyl ethers and vinyl phosphates. In practice, experiments to determine the 1°DKIE for protonation of an enzyme-bound enol pyruvate intermediate by pyruvate kinase failed (Kuo, O'Connell, & Rose, 1979; Robinson & Rose, 1972), because proton transfer at the enzyme active site is reversible (Scheme 3) and results in formation of dideuterium-labeled pyruvate at early reaction times (W. Tsang & J. Richard, unpublished results).

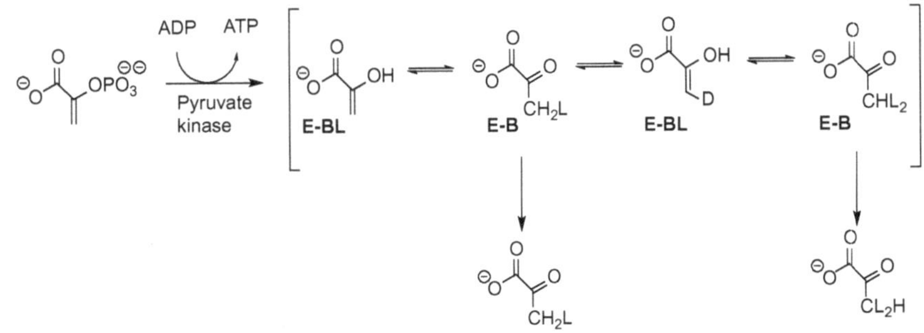

Scheme 3 Pyruvate kinase-catalyzed phosphorylation of ADP in 50/50 (v/v) HOH/DOD. This reaction gives a mixture of mono- and dideuterium-labeled pyruvate during the first enzyme turnover of substrate (W. Tsang & J. P. Richard, unpublished results).

2. ENZYME-CATALYZED DECARBOXYLATION OF OMP

A proton from the solvent is substituted for CO_2 in enzyme-catalyzed decarboxylation. Enzyme-catalyzed decarboxylation proceeds by a stepwise mechanism, through a carbanion reaction intermediate, when the negative charge at carbon is delocalized onto a second electron-deficient atom, such as an enolate oxygen (Richard, 1990). Orotidine 5′-monophosphate decarboxylase (OMPDC; Scheme 4) catalyzes the chemically difficult decarboxylation of orotidine 5′-monophosphate (OMP) (Miller & Wolfenden, 2002), where the negative charge at the vinyl carbanion reaction intermediate is largely localized at carbon-6 (Amyes et al., 2008; Tsang et al., 2012). There is considerable interest in understanding the mechanism by which OMPDC enables formation of this unstable vinyl carbanion reaction intermediate.

The SDKIE of 1.3 on k_{cat}/K_m for OMPDC-catalyzed decarboxylation in HOH and DOD shows that there are only small changes in bonding at solvent-derived hydrons on proceeding from free enzyme and substrate to the rate-determining transition state for OMPDC-catalyzed decarboxylation (Ehrlich, Hwang, Cook, & Blanchard, 1999). This SDKIE may be dominated by the effect of changes in the strength of hydrogen-bonding interactions to protein-bound hydrons (Chang et al., 1996; Kresge, 1973). The small observed SDKIE shows that there can be no more than a small 1°DKIE on the rate-determining step for enzyme-catalyzed decarboxylation, and is consistent with rate-determining loss of CO_2 to form a UMP carbanion reaction intermediate (Scheme 4). These results do not bear

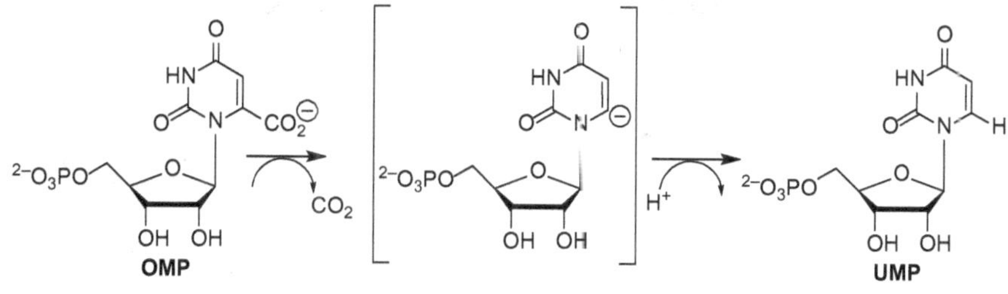

Scheme 4 OMPDC-catalyzed decarboxylation of OMP through a vinyl carbanion reaction intermediate (Amyes, Wood, Chan, Gerlt, & Richard, 2008; Tsang et al., 2012).

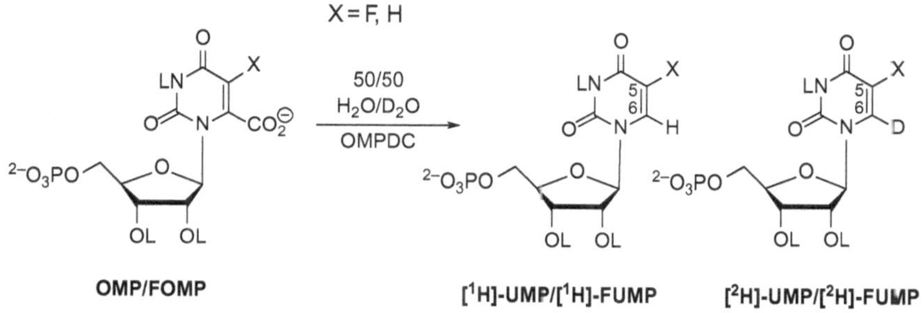

Scheme 5 The decarboxylation of OMP or FOMP in a mixed solvent catalyzed by OMPDC to give a mixture of —H- and —D-labeled reaction products. The primary isotope effect on hydron transfer is calculated from the ratio of the yields of —H ([$\mathbf{P}_H$])- and —D ([$\mathbf{P}_D$])-labeled products using Eq. (2).

directly on the magnitude of the 1°DKIE on the proton transfer reaction to C-6 of UMP for a concerted reaction mechanism (not shown) (Appleby, Kinsland, Begley, & Ealick, 2000; Begley & Ealick, 2004), or to C-6 of the UMP carbanion intermediate of a stepwise reaction (Scheme 4).

We were interested in characterizing the extent of bonding to the proton transferred from an acidic side chain of OMPDC to the C-6 carbon at the product-determining transition state for OMPDC-catalyzed decarboxylation. This is defined by the 1°DKIE on the hydron transfer reaction, which provides a measure for the fraction of the original zero-point energy of the original bond to hydrogen that is maintained at the transition state for the product-determining step (Westheimer, 1961). The 1°DKIE on OMPDC-catalyzed decarboxylation of OMP and 6-fluoroorotidine 5′-monophosphate (FOMP) was obtained, independent of contributions from solvent deuterium isotope effects, by determining the yields of —H- and —D-labeled products formed in a common solvent of 50/50 (v/v) HOH/DOD (Scheme 5) (Toth et al., 2007, 2010). We describe here the protocol for the determination of these product isotope effects.

3. EXPERIMENTAL PROTOCOL

3.1 A Comparison of Protocol for Determination of SDIE and PDIE

We first consider the strengths and limitations of the different protocol for determination of PDIEs for reactions in a mixed solvent of 50/50 (v/v) HOH/DOD and for determination of SDKIEs for separate reactions in HOH and DOD.

(1) The uncertainty in the value for the PDIE is small, because the transfer of —H and —D is examined under identical reaction conditions. The experimental error arises from the uncertainty in the yields of —H- and —D-labeled reaction products, calculated from the areas of the relevant NMR product peaks (Eq. 1). These areas can generally be determined to better than ±1% for each product, to give an uncertainty in the PDIE of better than ±2% (Tsang & Richard, 2009). A similar small uncertainty may possible for the SDIE, determined as the ratio of rate constants for separate experiments in HOH and DOD, provided the differences in the temperature of the runs are minimized by carrying the reactions out in parallel and with a single bath to control temperature.

(2) The determination of the SDIE on OMPDC-catalyzed decarboxylation requires an analytical method to continuously monitor conversion of substrate OMP to product UMP. It is not necessary to continuously monitor decarboxylation to obtain the PDIE, but an analytical method is needed to determine the yields of —H- and —D-labeled UMP products at the end of the reaction. The detection of reaction products by NMR analysis is generally less sensitive than detection by UV–Vis spectroscopy. This will complicate the determination of the PDIE, when the reactant is sparingly soluble in the aqueous solvent (Tsang & Richard, 2007, 2009).

$$\mathrm{PDIE} = \frac{[\mathbf{P}_\mathrm{H}]}{[\mathbf{P}_\mathrm{D}]} = \frac{1}{\phi_{\mathrm{TS}}} \tag{2}$$

$$\mathrm{KIE} = \left(\frac{k_{\mathrm{EH}}}{k_{\mathrm{ED}}}\right) = \left(\frac{K^{\ddagger}_{\mathrm{EH}}}{K^{\ddagger}_{\mathrm{ED}}}\right) = \frac{\phi_{\mathrm{EL}}}{\phi_{\mathrm{TS}}} = \phi_{\mathrm{EL}}\mathrm{PDIE} \tag{3}$$

(3) The value of the SDIE on OMPDC-catalyzed decarboxylation is a function of the complex secondary solvent deuterium isotope effects on enzyme reactivity (Kresge et al., 1987; Schowen, 1972, 2007), and of the intrinsic 1°DKIE on hydron transfer (see above). By comparison the PDIE depends only on the fractionation factor (ϕ_{TS}; Eq. 2) for partitioning of the transferred hydron between the mixed solvent and the transition state for enzyme-catalyzed decarboxylation. The hydron that is donated to form the product of OMPDC-catalyzed decarboxylation is derived from the cationic ammonium side chain of Lys93 (Miller, Hassell, Wolfenden, Milburn, & Short, 2000). The kinetic deuterium isotope effect on the reaction of the side chain labeled with —H (k_{EH}) and with —D (k_{ED}) is related to the product isotope effect by the term $\phi_{EL} \asymp 1.0$ (Eq. 3) for fractionation of the hydron between solvent and the ammonium side chain of Lys93 (Jarret & Saunders, 1986; Schowen & Schowen, 1982). The fractionation factor of $\phi_{EL} \asymp 1.0$ shows that the deuterium enrichment of the acidic side chain is essentially the same as water so that the product DIE is equal to the 1°DKIE on proton transfer from the —H- and —D-labeled side chains.

3.2 Enzyme-Catalyzed Decarboxylation

A single stock solution of 50/50 (v/v) HOH/DOD [1.00/1.00 molar ratio] is used for the preparation of all buffer solutions. Solutions of concentrated OMPDC were extensively dialyzed against the standard buffer of 50 m*M* MOPS buffer (50% free base) at pL 7.4 and an ionic strength of 0.1 (NaCl) in 50/50 (v/v) HOH/DOD. Standard assays described in earlier work were used to determine the activity of OMPDC for decarboxylation of OMP (Barnett, Amyes, Wood, Gerlt, & Richard, 2008). The enzyme concentrations used in these studies were sufficient to give decarboxylation of ca. 95% of the substrate in several hours.

Similar reaction conditions were used for OMPDC-catalyzed decarboxylation of OMP and FOMP. The decarboxylation of 2 m*M* substrate was carried out in 50/50 (v/v) HOH/DOD (2.0 mL) that contained 50 m*M* MOPS buffer (50% free base) at pL 7.4 and an ionic strength of 0.10 (NaCl). The reactions were initiated by addition of OMPDC and the reaction progress was determined periodically from the decrease in absorbance at 279 nm ($\Delta\varepsilon = 2400\,M\text{cm}^{-1}$) and 282 nm ($\Delta\varepsilon = 1380\,M\text{cm}^{-1}$) for enzyme-catalyzed decarboxylation of OMP and FOMP, respectively.

A wide range of enzyme concentrations (0.01–30 μ*M*) were used due to the large falloff in the activity for mutant forms of OMPDC examined (Barnett, Amyes, McKay Wood, Gerlt, & Richard, 2010; Barnett et al., 2008). Wildtype and mutant forms of OMPDC are stable for several days under these conditions.

When the reaction was judged to have reached 95% completion, generally after several hours (*x* min), an 800 μL aliquot was withdrawn and the enzyme was removed by ultrafiltration using Microcon® Centrifugal Filter Units (10 K molecular weight cutoff). After an additional reaction time of *x* min a second 800 μL aliquot was withdrawn and treated in the same manner. The filtered solutions were frozen and saved for NMR analysis. Prior to ^{1}H NMR analyses, the solutions of OMP were generally adjusted to 9.5–10 using NaOD, in order to increase the resolution of the signals for the 5-H and 1′-H of UMP. The solutions of FUMP were kept at the initial pH of 7.4 for ^{19}F NMR analysis.

NMR analyses of product yields. These experiments are designed to obtain a single experimental value, the deuterium enrichment of the product of the reaction in 50/50 HOH/DOD. This enrichment is obtained by integration of the appropriate NMR peaks for —H- and —D-labeled products, determined by averaging multiple Fourier-transform NMR spectra. Errors in peak areas may arise by failure of allow complete relaxation of nuclei between Fourier transform pulses. These errors were minimized by determining the relaxation times T_1 for all nuclei of interest, and then using a relaxation delay between pulses of $>7T_1$ for all nuclei.

The yields of [6-^{1}H]-UMP and [6-^{2}H]-UMP were determined by ^{1}H NMR analyses at 500 MHz. The spectra (ca. 30 transients) were recorded for a 6000 Hz spectral width, and with suppression of the water peak, using a 90 degree pulse angle, a 6 s acquisition time, and a 120 s relaxation delay between pulses ($>7T_1$). The yields of [6-^{1}H]-FUMP and [6-^{2}H]-FUMP were determined by ^{19}F NMR analyses at 470 MHz. The spectra (64 transients) were recorded for a 50,000 Hz spectral width using a 90 degree pulse angle, a 2.6 s acquisition time, and a 20 s relaxation delay between pulses ($>7T_1$).

The product spectra were analyzed in regions sensitive to the deuterium enrichment at carbon-6 of UMP or FUMP. Fig. 1 shows representative partial ^{1}H NMR spectra from decarboxylation of OMP catalyzed by OMPDC in 50/50 (v/v) HOH/DOD obtained at pH 7.4 (Fig. 1A) and 9.7 (Fig. 1B). In early studies the yield of UMP labeled with hydrogen at C-6 was obtained from the integrated area of the doublet due to the C-6 proton at 7.990 ppm (■, Fig. 1A) and the yield of [6-^{2}H]-UMP was obtained from the area of the

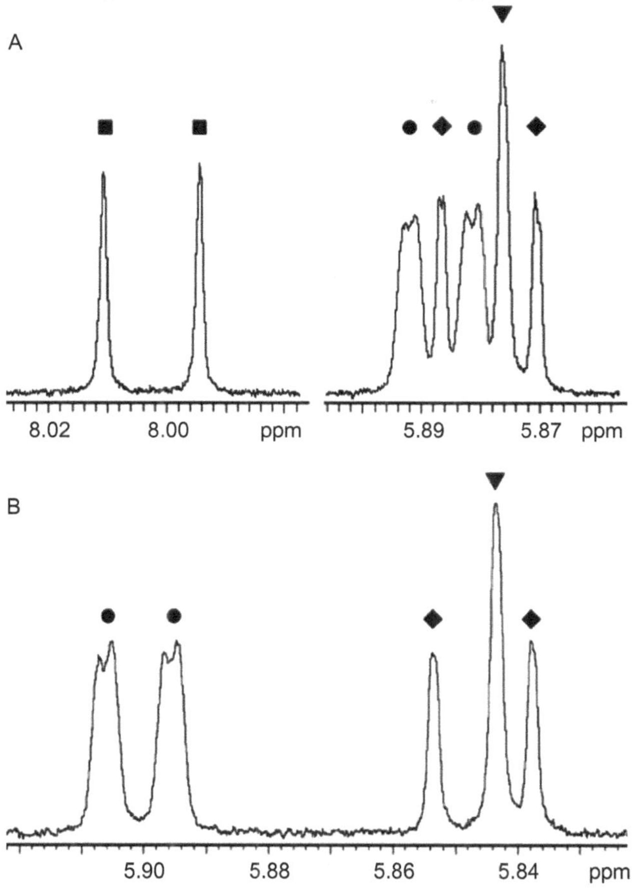

Fig. 1 (A) Partial ^{1}H NMR spectrum (500 MHz) at pL 7.4 of the UMP product of decarboxylation of OMP catalyzed by ScOMPDC (0.024 μ*M*) in 50/50 (v/v) HOH/DOD at 25°C. Key: (■), doublet due to the C-6 proton of [6-^{1}H]-UMP; (●), poorly resolved double doublets due to the anomeric protons of [6-^{1}H]-UMP and [6-^{2}H]-UMP; (◆), doublet due to the C-5 proton of [6-^{1}H]-UMP; (▼), collapsed singlet due to the C-5 proton of [6-^{2}H]-UMP. (B) Partial ^{1}H NMR spectrum determined at pL 9.7. Key: (●), double doublets due to the anomeric protons of [6-^{1}H]-UMP and [6-^{2}H]-UMP; (◆), doublet due to the C-5 proton of [6-^{1}H]-UMP; (▼), collapsed singlet due to the C-5 proton of [6-^{2}H]-UMP. *From Toth, K., Amyes, T. L., Wood, B. M., Chan, K., Gerlt, J. A., & Richard, J. P. (2010). Product deuterium isotope effects for orotidine 5′-monophosphate decarboxylase: Effect of changing substrate and enzyme structure on the partitioning of the vinyl carbanion reaction intermediate.* Journal of the American Chemical Society, 132*(20), 7018–7024 with permission. Copyright 2010 American Chemical Society.*

collapsed singlet due to the C-5 proton at 5.865 ppm (▼, Fig. 1A) (Toth et al., 2007). The spectrum at pL 7.4 shows overlap between the signals for the C-5 pyrimidine and the C-1′ anomeric ribosyl hydrogen. The spectral resolution at pL 9.7 is improved due to the effect on chemical shift of partial ionization of N-3 of UMP ($pK_a = 9.45$) (Massoud & Sigel, 1988) on chemical shift. In later studies the spectral analyses were carried out at pL 9.7 (Fig. 1B), and the PDIE calculated from the ratio of the integrated area of the doublet at 5.846 ppm for the C-5 hydrogen of [6-^{1}H]-UMP and of the collapsed singlet at 5.843 ppm for the C-5 hydrogen of [6-^{2}H]-UMP (Toth et al., 2010). The PDIE was determined as the ratio of these peak areas for UMP labeled with [^{1}H] ($\mathbf{A}_{\mathrm{H}}$) and [^{2}H] ($\mathbf{A}_{\mathrm{D}}$) at C-6 (Eq. 4).

$$\mathrm{PDIE} = \frac{[\mathbf{A}_{\mathrm{H}}]}{[\mathbf{A}_{\mathrm{D}}]} \tag{4}$$

Fig. 2 shows a representative ^{19}F NMR spectrum (470 MHz) at pL 7.4 of 5-FUMP obtained from the decarboxylation of 5-FOMP (2 m*M*) catalyzed

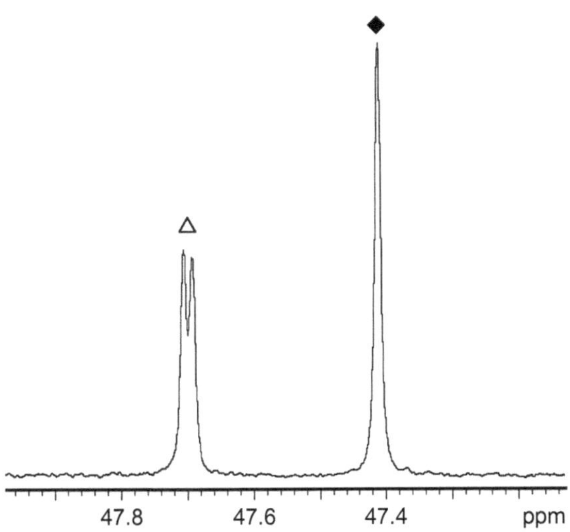

Fig. 2 The 470 MHz ^{19}F NMR spectrum of 5-FUMP, which is the product of decarboxylation of 5-FOMP catalyzed by R235A ScOMPDC (0.054 μ*M*) in 50/50 HOH/DOD (v/v) at pL 7.4. Key: (Δ), doublet at $\delta = 47.70$ ppm ($J_{\mathrm{HF}} = 6$ Hz) due to the fluorine at [6-^{1}H]-5-FUMP; (◆), downfield-shifted singlet at $\delta = 47.41$ ppm due to the fluorine at [6-^{2}H]-5-FUMP. *From Toth, K., Amyes, T. L., Wood, B. M., Chan, K., Gerlt, J. A., & Richard, J. P. (2010). Product deuterium isotope effects for orotidine 5′-monophosphate decarboxylase: Effect of changing substrate and enzyme structure on the partitioning of the vinyl carbanion reaction intermediate.* Journal of the American Chemical Society, 132*(20), 7018–7024 with permission. Copyright 2010 American Chemical Society.*

by R235A mutant ScOMPDC. Separate ^{19}F signals are observed for [6-^{1}H]-5-FUMP and [6-^{2}H]-5-FUMP because the C-6 deuterium perturbs the ^{19}F chemical shift of the C-5 fluorine by 0.29 ppm. The PDIE was determined as the ratio of the integrated area of the doublet at 47.70 ppm (J_{HF} = 6 Hz) for [6-^{1}H]-5-FUMP and of the downfield-shifted singlet at 47.41 for [6-^{2}H]-5-FUMP. In all cases the PDIEs (Eq. 4) were determined at two different reaction times, *x* min and 2*x* min.

4. INTERPRETATION OF RESULTS

The interpretation of the magnitude of the PDIE is straightforward. The large thermodynamic driving force for protonation of the vinyl carbanion reaction intermediate ensures that hydron transfer is irreversible (Tsang et al., 2012) so that the yields of —H- and —D-labeled products are determined entirely by the relative barriers for hydron transfer from OMPDC. We find that the decarboxylation of OMP in 50/50 (v/v) HOH/DOD catalyzed by OMPDC from *S. cerevisiae, M. thermautotrophicus*, and *E. coli* gives essentially equal yields of the corresponding —H- and —D-labeled products. This gives a value of 1.0 for the PDIE (Toth et al., 2007). The agreement from two different determinations at *x* and 2*x* min reaction times is generally ±2% or better.

The values of 1.0 and 0.9 for the PDIE for OMPDC-catalyzed decarboxylation of OMP and FOMP, respectively, do not change as k_{cat}/K_m is reduced 10^4-fold by site-directed mutations (Toth et al., 2010). This represents a rare example of a complete lack of selectivity for the transfer H and D from a hydroxylic solvent to the reaction product (Macciantelli, Seconi, & Eaborn, 1978), which requires that the zero-point energy associated with the bonds to the reacting hydrons not change as the hydron moves from the mixed HOH/DOD solvent to the product-determining transition state for OMPDC-catalyzed decarboxylation. The results show that proton transfer from OMPDC to C-6 of OMP does not provide electrophilic *push* to the loss of CO_2 in a concerted reaction (Begley, Appleby, & Ealick, 2000; Begley & Ealick, 2004), even when there is a very large barrier to reactions catalyzed by mutant OMPDC enzymes. The lack of selectivity between the reaction of —H and —D is consistent with restricted C—N bond rotation of the donor —CH_2—NL_3^+ group of the side chain of Lys-93. This ensures that protonation of the intermediate is faster than bond rotation that would allow the intermediate to select between reaction of —H and —D. We proposed that hydrogen bonds from the CH_2—NL_3^+ group of Lys-93 to the carboxylate groups of Asp-91 and Asp-96 restrict rotation about the carbon–nitrogen bond and directs the single unliganded hydron toward the C-6 UMP carbanion (Toth et al., 2007, 2010).

5. OTHER ENZYMES

The general protocol described here is suitable for the determination of the PDIE for any enzyme-catalyzed or nonenzymatic decarboxylation, and was recently applied in a study on the mechanism of the decarboxylation of 5-carboxyvanillate catalyzed by 5-carboxyvanillate decarboxylase (CVDB). This enzyme shows SDIEs on k_{cat} and k_{cat}/K_m of 2.9 and 2.0, respectively, and a PDIE of 4.6 that was determined as the ratio of —H- and —D-labeled products formed in a solvent of 50/50 (v/v) HOH/DOD (Scheme 6) (Vladimirova et al., 2016). The observation of a substantial kinetic SDIE for separate experiments in HOH and DOD, and of a PDIE for experiments in the mixed HOH/DOD solvents provides strong evidence that the rate and product-determining steps for CVDB-catalyzed decarboxylation are the same. The authors of this work provide strong support for the stepwise mechanism shown in Scheme 7, where effectively irreversible and rate-determining hydron transfer to form a cyclohexadiene reaction intermediate is followed by rapid loss of CO_2 (Vladimirova et al., 2016).

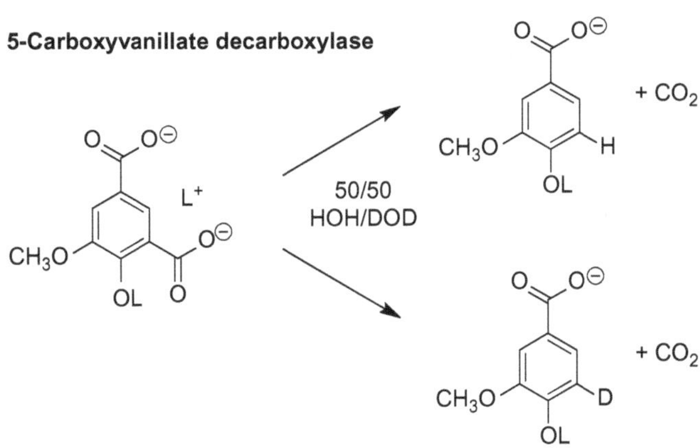

Scheme 6 The products of 5-carboxyvanillate decarboxylase-catalyzed reactions of 5-carboxyvanillate in 50/50 (v/v) HOH/DOD.

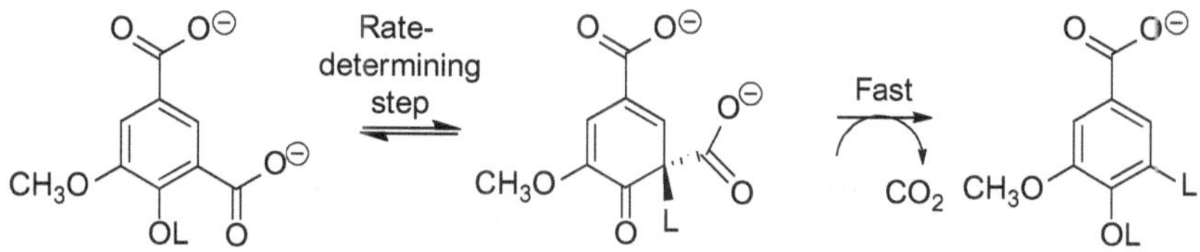

Scheme 7 The stepwise mechanism for the decarboxylation of 5-carboxyvanillate catalyzed by CVDB, where the rate-determining step is protonation of the substrate to form a cyclohexadiene intermediate.

ACKNOWLEDGMENT

We are grateful to the US National Institutes of Health Grants GM116921 and GM039754 for generous support of our work.

REFERENCES

Amyes, T. L., Wood, B. M., Chan, K., Gerlt, J. A., & Richard, J. P. (2008). Formation and stability of a vinyl carbanion at the active site of orotidine 5′-monophosphate decarboxylase: pK_a of the C-6 proton of enzyme-bound UMP. *Journal of the American Chemical Society, 130*(5), 1574–1575.

Appleby, T. C., Kinsland, C., Begley, T. P., & Ealick, S. E. (2000). The crystal structure and mechanism of orotidine 5′-monophosphate decarboxylase. *Proceedings of the National Academy of Sciences of the United States of America, 97*(5), 2005–2010.

Barnett, S. A., Amyes, T. L., McKay Wood, B., Gerlt, J. A., & Richard, J. P. (2010). Activation of R235A mutant orotidine 5′-monophosphate decarboxylase by the guanidinium cation: Effective molarity of the cationic side chain of Arg-235. *Biochemistry, 49*(5), 824–826.

Barnett, S. A., Amyes, T. L., Wood, B. M., Gerlt, J. A., & Richard, J. P. (2008). Dissecting the total transition state stabilization provided by amino acid side chains at orotidine 5′-monophosphate decarboxylase: A two-part substrate approach. *Biochemistry, 47*(30), 7785–7787.

Begley, T. P., Appleby, T. C., & Ealick, S. E. (2000). The structural basis for the remarkable catalytic proficiency of orotidine 5′-monophosphate decarboxylase. *Current Opinion in Structural Biology, 10*(6), 711–718.

Begley, T. P., & Ealick, S. E. (2004). Enzymatic reactions involving novel mechanisms of carbanion stabilization. *Current Opinion in Chemical Biology, 8*(5), 508–515.

Chang, T. K., Chiang, Y., Guo, H. X., Kresge, A. J., Mathew, L., Powell, M. F., et al. (1996). Solvent isotope effects in H2O-D2O mixtures (proton inventories) on serine-protease-catalyzed hydrolysis reactions. Influence of oxyanion hole interactions and medium effects. *Journal of the American Chemical Society, 118*(37), 8802–8807.

Ehrlich, J. I., Hwang, C.-C., Cook, P. F., & Blanchard, J. S. (1999). Evidence for a stepwise mechanism of OMP decarboxylase. *Journal of the American Chemical Society, 121*(29), 6966–6967.

Jarret, R. M., & Saunders, M. (1986). Use of various nuclei as probes in a new NMR method for obtaining proton/deuteron fractionation data. *Journal of the American Chemical Society, 108*(24), 7549–7553.

Jencks, W. P. (1987). *Catalysis in chemistry and enzymology*. New York: Dover Publications.

Kresge, A. J. (1973). Solvent isotope effects and the mechanism of chymotrypsin action. *Journal of the American Chemical Society*, *95*(9), 3065–3067.
Kresge, A. J., More O'Ferrall, R. A., & Powell, M. F. (1987). Solvent isotope effects, fractionation factors and mechanisms of proton transfer reactions. In E. Buncel (Ed.), *Isotopes in organic chemistry* (pp. 178–275). Amsterdam: Elsevier.
Kresge, A. J., Sagatys, D. S., & Chen, H. L. (1977). Vinyl ether hydrolysis. 9. Isotope effects on proton transfer from the hydronium ion. *Journal of the American Chemical Society*, *99*(22), 7228–7233.
Kuo, D. J., O'Connell, E. L., & Rose, I. A. (1979). Physical, chemical, and enzymological characterization of enolpyruvate. *Journal of the American Chemical Society*, *101*(17), 5025–5030.
Macciantelli, D., Seconi, G., & Eaborn, C. (1978). Further studies of solvent isotope effects in the cleavage of substituted benzyltrimethylsilanes by methanolic sodium methoxide. Intermediate kinetic isotope effects for reactions of carbanions with methanol. *Journal of the Chemical Society, Perkin Transactions*, *2*, 834–838.
Massoud, S. S., & Sigel, H. (1988). Metal ion coordinating properties of pyrimidine-nucleoside 5′-monophosphates (CMP, UMP, TMP) and of simple phosphate monoesters, including D-ribose 5′-monophosphate. Establishment of relations between complex stability and phosphate basicity. *Inorganic Chemistry*, *27*(8), 1447–1453.
Miller, B. G., Hassell, A. M., Wolfenden, R., Milburn, M. V., & Short, S. A. (2000). Anatomy of a proficient enzyme: The structure of orotidine 5'-monophosphate decarboxylase in the presence and absence of a potential transition state analog. *Proceedings of the National Academy of Sciences of the United States of America*, *97*(5), 2011–2016.
Miller, B. G., & Wolfenden, R. (2002). Catalytic proficiency: The unusual case of OMP decarboxylase. *Annual Review of Biochemistry*, *71*, 847–885.
Richard, J. P. (1990). *The chemistry of enols. The biochemistry of enols* (pp. 651–689). New York, NY: Wiley.
Richard, J. P., Amyes, T. L., Lin, S.-S., O'Donoghue, A. C. M., Toteva, M. M., Tsuji, Y., et al. (2000). How does organic structure determine organic reactivity? Partitioning of carbocations between addition of nucleophiles and deprotonation. *Advances in Physical Organic Chemistry*, *35*, 67–115.
Richard, J. P., & Williams, K. B. (2007). A Marcus treatment of rate constants for protonation of ring-substituted α-methoxystyrenes: Intrinsic reaction barriers and the shape of the reaction coordinate. *Journal of the American Chemical Society*, *129*(21), 6952–6961.
Richard, J. P., Williams, K. B., & Amyes, T. L. (1999). Intrinsic barriers for the reactions of an oxocarbenium ion in water. *Journal of the American Chemical Society*, *121*(36), 8403–8404.
Robinson, J. L., & Rose, I. A. (1972). Proton transfer reactions of muscle pyruvate kinase. *Journal of Biological Chemistry*, *247*(4), 1096–1105.
Schowen, R. L. (1972). Mechanistic deductions from solvent isotope effects. *Progress in Physical Organic Chemistry*, *9*, 275–332.
Schowen, R. L. (2007). The use of solvent isotope effects in the pursuit of enzyme mechanisms. *Journal of Labelled Compounds and Radiopharmaceuticals*, *50*(11 − 12), 1052–1062.
Schowen, K. B., & Schowen, R. L. (1982). Solvent isotope effects on enzyme systems. *Methods in Enzymology*, *87*(Enzyme Kinet. Mech., Pt. C), 551–606.
Toth, K., Amyes, T. L., Wood, B. M., Chan, K., Gerlt, J. A., & Richard, J. P. (2007). Product deuterium isotope effect for orotidine 5'-monophosphate decarboxylase: Evidence for the existence of a short-lived carbanion intermediate. *Journal of the American Chemical Society*, *129*(43), 12946–12947.
Toth, K., Amyes, T. L., Wood, B. M., Chan, K., Gerlt, J. A., & Richard, J. P. (2010). Product deuterium isotope effects for orotidine 5′-monophosphate decarboxylase: Effect of

changing substrate and enzyme structure on the partitioning of the vinyl carbanion reaction intermediate. *Journal of the American Chemical Society, 132*(20), 7018–7024.
Tsang, W.-Y., & Richard, J. P. (2007). A simple method to determine kinetic deuterium isotope effects provides evidence that proton transfer to carbon proceeds over and not through the reaction barrier. *Journal of the American Chemical Society, 129*(34), 10330–10331.
Tsang, W.-Y., & Richard, J. P. (2009). Structure-reactivity effects on primary deuterium isotope effects on protonation of ring-substituted α-methoxystyrenes. *Journal of the American Chemical Society, 131*(39), 13952–13962.
Tsang, W.-Y., Wood, B. M., Wong, F. M., Wu, W., Gerlt, J. A., Amyes, T. L., et al. (2012). Proton transfer from C-6 of uridine 5′-monophosphate catalyzed by orotidine 5′-monophosphate decarboxylase: Formation and stability of a vinyl carbanion intermediate and the effect of a 5-fluoro substituent. *Journal of the American Chemical Society, 134*(35), 14580–14594.
Vladimirova, A., Patskovsky, Y., Fedorov, A. A., Bonanno, J. B., Fedorov, E. V., Toro, R., et al. (2016). Substrate distortion and the catalytic reaction mechanism of 5-carboxyvanillate decarboxylase. *Journal of the American Chemical Society, 138*(3), 826–836.
Westheimer, F. H. (1961). The magnitude of the primary kinetic isotope effect for compounds of hydrogen and deuterium. *Chemical Reviews, 61*, 265–273.

FURTHER READING

Amyes, T. L., & Richard, J. P. (1992). Generation and stability of a simple thiol ester enolate in aqueous solution. *Journal of the American Chemical Society, 114*(26), 10297–10302.
Amyes, T. L., & Richard, J. P. (1996). Determination of the pK(a), of ethyl acetate: Bronsted correlation for deprotonation of a simple oxygen ester in aqueous solution. *Journal of the American Chemical Society, 118*(13), 3129–3141.

CHAPTER EIGHT

Measurement and Prediction of Chlorine Kinetic Isotope Effects in Enzymatic Systems

Lukasz Szatkowski[*,1], Rabindra Nath Manna[†,1], Anna Grzybkowska[‡,1], Rafał Kamiński[§], Agnieszka Dybala-Defratyka[§,2], Piotr Paneth[§,2]

[*]University of Arizona, Department of Chemistry and Biochemistry, Tucson, AZ, United States
[†]Indian Association for the Cultivation of Science, Physical Chemistry, Kolkata, West Bengal, India
[‡]Elucidation of Biosynthesis by Isotopic Spectrometry Group, CEISAM, University of Nantes—CNRS UMR6230, Nantes, France
[§]Lodz University of Technology, Faculty of Chemistry, Institute of Applied Radiation Chemistry, Lodz, Poland
[2]Corresponding authors: e-mail address: agnieszka.dybala-defratyka@p.lodz.pl; piotr.paneth@p.lodz.pl

Contents

1. Introduction 180
2. Experimental Determination of Cl-KIEs 181
   2.1 Chlorine Isotopic Composition 181
   2.2 Multiply Chlorinated Reactants 184
3. Theoretical Predictions of Cl-KIEs 186
   3.1 Cluster Model Approach 189
   3.2 Full Enzymatic Model Approaches 194
4. Summary 209
References 210

Abstract

Approaches to determine chlorine kinetic isotope effects (Cl-KIEs) on enzymatic dehalogenations are discussed and illustrated by representative examples. Three aspects are considered. First methodology for experimental measurement of Cl-KIEs, with stress being on FAB-IRMS technique developed in our laboratory, is described. Subsequently, we concentrate our discussion on the consequences of reaction complexity in the interpretation of experimental values, a problem especially important in cases of polychlorinated reactants. The most fruitful studies of enzymatic dehalogenations by Cl-KIEs require their theoretical evaluation, hence the computational focus of the second part of this chapter.

[1] These authors contributed equally to this work.

Methods in Enzymology, Volume 596
ISSN 0076-6879
http://dx.doi.org/10.1016/bs.mie.2017.07.021

1. INTRODUCTION

Understanding the mechanism of enzyme-catalyzed dehalogenations (Agarwal et al., 2017) is an important scientific issue since these reactions increasingly play a significant role in environmental remediation. Isotope effects provide the most sensitive kinetic tool for studying mechanisms of chemical and biological processes (Melander, 1960; Melander & Saunders, 1982; Wolfsberg, Van Hook, & Paneth, 2010) because isotopic substitution introduces the smallest perturbation to the studied system compared to other kinetic methods. Furthermore, kinetic isotope effects reflect directly on the properties of the transition states (TSs), thus allowing chemical insight not available from other methods. These merits of the studies of isotope effects are moderated by the fact that they are usually small and hard to determine experimentally. In the first approximation the magnitude of an isotope effect is proportional to the mass difference between isotopologues relative to atomic weight of the isotopic element. Thus, hydrogen isotope effects are by far the largest, while so-called heavy-atom isotope effects (carbon, nitrogen, oxygen, sulfur, chlorine, etc.) are one to two orders of magnitude smaller. Primary chlorine kinetic isotope effects (Cl-KIEs) are usually in the range of 0.5%–1.0% (percentage deviation from unity) although values as large as 2.5% have been predicted theoretically (Świderek & Paneth, 2012; Szatkowski & Dybala-Defratyka, 2013). In this chapter, we concentrate on Cl-KIEs on enzymatic dehalogenations. We analyze both experimental and theoretical approaches. Previous work has been summarized earlier (Paneth, 2003, 2006).

There are two other aspects of studies of chlorine isotope effects on enzyme-catalyzed reactions that require attention. First, usually modeling is performed for the dehalogenation step, while other steps might contribute to the overall reaction rate. Therefore, evaluation of reaction complexity needs to be taken into account (Paneth, 1985) and the relationship between the isotope effect on the chemical conversion within an enzyme active site (so-called intrinsic isotope effects) and the experimentally determined overall isotope effect (called apparent isotope effect) should be established. Second, many chlorinated compounds contain more than one chlorine atom per molecule. Since the heavier stable isotope of chlorine (^{37}Cl) is present in nature in large quantity, experimental analysis of polychlorinated compounds requires special attention. Herein, we present our approaches to these problems on the examples of our studies on chlorine isotope effects on enzymatic dehalogenations.

2. EXPERIMENTAL DETERMINATION OF Cl-KIEs

2.1 Chlorine Isotopic Composition

Cl-KIEs are small and therefore mostly primary kinetic isotope effects have been studied, while secondary effects are usually neglected (although it has been shown that even binding chlorine isotope effects can be sizable (Siwek et al., 2013). Due to this small size, experimental determination of chlorine isotope effects is by necessity limited to competitive methods where both isotopes are present simultaneously in the reaction mixture. In fact, since the natural isotopic abundance of ^{35}Cl and ^{37}Cl is close to 3:1, it would be tedious and costly to use a direct comparison of separate kinetic runs with each isotopologue anyway.

Precise measurements of chlorine isotopic composition are most frequently performed using isotope ratio mass spectrometry (IRMS) (Kaufmann, Long, Bentley, & Davis, 1984). Methodology varies for specific applications with major differences being the type of the ion source and inlet system. For example, for studies of chlorinated ethenes, persistent environmental pollutants, a method using continuous flow of the analyte (CF-IRMS) has been developed, in which collectors are set to detect isotopic molecular peaks of chlorinated ethanes (Shouakar-Stash, Drimmie, & Frape, 2005). For geological applications, methods using positive (Cs_2Cl^+) (Numata, Nakamura, & Gamo, 2001; Xiao & Zhang, 1992) or negative (Cl^-) (Fujitani, Yamashita, Numata, Kanazawa, & Nakamura, 2010; Vengosh, Chivas, & McCulloch, 1989) ions using thermal ionization (TIMS) were used. A promising methodology is provided by inductively coupled plasma (ICP-MS) mass spectrometers that are equipped with a multicollector for isotopic measurements. $^{37}Cl/^{35}Cl$ ratios have been determined using this technique (Van Acker, Shahar, Young, & Coleman, 2006).

In enzymatic studies, however, the "traditional" IRMS is routinely used. IRMS for chlorine isotopic composition uses a dual inlet system with reference gas, methyl chloride, of known isotopic composition, calibrated against the Standard Mean Ocean Chloride (SMOC) standard. Typically positive ions of a molecular cluster are analyzed although the most precise results (0.005‰ standard deviation) have been obtained in the negative ion mode analyzing the Cl^- peaks (Halas & Pelc, 2009). This technique is limited to volatile analytes and thus chlorinated material needs to be converted to gaseous form prior to the measurements. This can be achieved in several ways depending on the chemical nature of the studied compounds.

Since in a plethora of reactions chloride is an isolable product, a detailed protocol has been developed (Hill & Fry, 1962; Taylor & Grimsrud, 1969) in order to convert Cl^- to the volatile analyte, MeCl.

During the course of studying enzymatic dehalogenations, in order to avoid tedious conversion which is prone to isotopic fractionation without appropriate care, we have developed a new method of direct measurements. It involves isotopic analysis of Cl^- peaks using FAB-IRMS (fast-atom bombardment-isotope ratio mass spectrometer) in the negative ion mode (Westaway, Koerner, Fang, Rudziński, & Paneth, 1998).

2.1.1 Fast-Atom Bombardment-Isotope Ratio Mass Spectrometer

The ratio of chlorine isotopes was also measured (Westaway et al., 1998) using a modified, hybrid model FAB-IRMS (MI 1201E, PO Electron, Ukraine). It is equipped with two Faraday cups that can be focused, within a narrow range, on a particular beam, and one fixed electron multiplier with a pulse analyzer and a single-ion counter. While normally samples are introduced into the FAB source via a probe in suspension in a matrix, we have found that silver chloride can be deposited directly on the support—silver plate was used for this purpose and mounted on the tip of the probe after washing with nitric acid, water, and acetone. A few milligrams of silver chloride were deposited on the clean silver plate and the plate was heated gently (below 480°C) to melt the sample. Equally good results have been obtained with ironing the AgCl sample onto the silver support. Precautions were taken to keep the dimensions of the samples constant (about 0.2 mm thick and 4 mm in diameter). The silver plate with the adhered solid sample was mounted on the tip of the direct insertion probe. Xenon atoms (6 keV) hitting the surface of the probe at an incidence angle of 45 degree were used for ionization. In the negative ion mode, the ions were accelerated through a 5 kV potential and detected in a Faraday cup collector system. As can be seen in Fig. 1, spectra contained almost exclusively chloride peaks at m/z 35 and 37. After initial fluctuations, the ion current stayed constant for several hours (see Fig. 2) allowing for multiple determinations of isotopic ratios; typically the mean value of the isotopic ratio was obtained from between 30 and 50 separate determinations, each of which was an average of 10 individual measurements. It is worth noting that the very stable ion current allows one to measure the isotope ratio with the same precision with a single-ion counter alternating between m/z 35 and m/z 37.

The requirement for milligram quantities of AgCl is due to handling rather than the amount necessary for the measurement. In fact, after

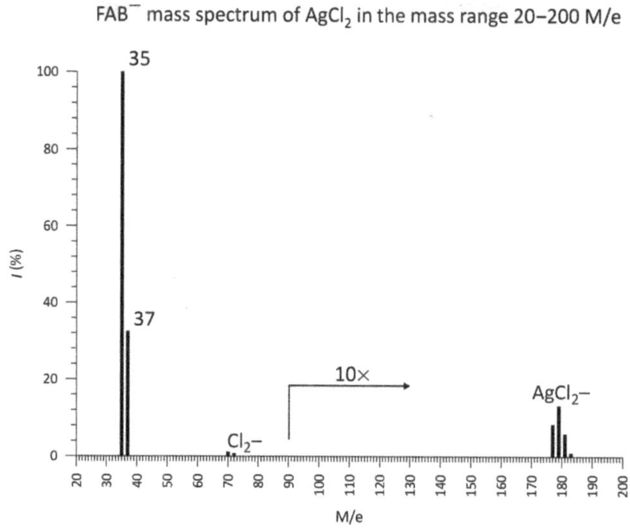

Fig. 1 FAB spectrum of AgCl in the negative ion mode.

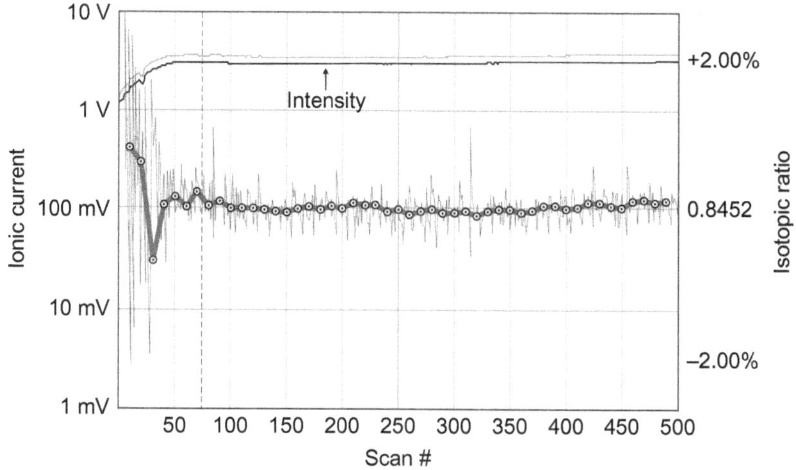

Fig. 2 Time dependence of the total ion current and the isotope ratio in FAB-IRMS of AgCl.

withdrawal from the source and removal of the silver from the surface the sample can be reused several times. We have managed to lower the required amount of silver chloride to about 0.1 mg, but further sample reduction was not possible. Thus in order to lower the required amount of sample, we have subsequently modified our methodology by switching to cesium chloride, which could be introduced from a solution. Initially, we anticipated using the positive ion mode with Cs_2Cl^+ ion detection, but its ion current was low, usually less than 10% of Cs^+. However, in the negative ion mode, Cl^- was again the dominant peak suitable for analysis (when thin, conducting layers were used—with thick samples the chloride ion current was significantly suppressed). This way much smaller amounts can be easily deposited on the silver plate. Furthermore, we have modified the FAB probe so that up to two samples could be simultaneously loaded into the FAB source, saving on the pumping time. This solution, however, did not give reproducible geometry of the probe and was consequently abandoned.

2.2 Multiply Chlorinated Reactants

Most reactions of chemical and biochemical interest involve chlorinated compounds only one chlorine atom in the molecule. Any experimental protocol in such situations is straightforward, and the only complication can arise from the reaction complexity (Paneth, 1985), where the isotope effect of the dehalogenation step may be masked by the reaction complexity or branching (Fisher, Palfey, Maniscalco, & Indyk, 2006; Gawlita-Kamzurska & Paneth, 1992; Mazurkiewicz-Łazowska & Paneth, 1991). Situation becomes more complicated when more than one chlorine atom is present in a studied molecule—a scenario quite frequent in environmental studies of chlorinated pollutants. In such cases, frequently compound-specific isotopic analysis (Bernstein et al., 2011) rather than isotope effects of each particular position is performed. Furthermore, complex kinetic trends have to be addressed mathematically using analytical or numerical methods (Hunkeler, Vanbreukelen, & Elsner, 2009) and presence of isotopologues carefully accounted for (Elsner & Hunkeler, 2008; Jin, Haderlein, & Rolle, 2013). We illustrate this problem on a relatively simple example of the reaction in which reactant contains two chlorine atoms.

2.2.1 Dechlorination of 1,2-Dichloroethane by the DhlA Dehalogenase

DhlA poses an interesting case of a reaction where the reactant has two chlorine atoms, one of which undergoes an S_N2 displacement in the rate-determining step. The other chlorine would contribute a secondary kinetic isotope effect, which can be neglected. However, due to large abundance of the heavy chlorine isotope, 1,2-dichloroethane (DCE) isotopologues $^{35}Cl–(CH_2)_2–^{35}Cl$, $^{35}Cl–(CH_2)_2–^{37}Cl$, and $^{37}Cl–(CH_2)_2–^{37}Cl$ are present in the ratio of about 9:6:1, respectively, and intramolecular competition in

$^{35}Cl\text{-}(CH_2)_2\text{-}^{37}Cl$ cannot be neglected when the product (chlorine anion) isotopic analysis is performed.

The relation between the original isotopic composition of DCE and the chloride product when only one chlorine atom is removed is thus complicated. The formal kinetics of this problem is the following:

$$
\begin{aligned}
&S_{LL} \xrightarrow{2k_L} P_L \\
&\quad\;\; \xrightarrow{k_L} P_L \\
&S_{LH} \\
&\quad\;\; \xrightarrow{k_H} P_H \\
&S_{HH} \xrightarrow{2k_H} P_H
\end{aligned}
\qquad (1)
$$

From the kinetic equation $\frac{-dS_{ij}}{dt} = (k_i + k_j) S_{ij}$, where i and j are either light (L) or heavy (H) isotopes and the subscript 0 indicates the initial concentration, the time dependence of each isotopologues reactant concentration can be written as

$$S_{ij} = S_{0ij} e^{-(k_i + k_j)t} \qquad (2)$$

when secondary isotope effects are neglected (which is an acceptable assumption in the case of Cl-KIEs). Substitution of function (2) into equations for the rates of product formation:

$$
\begin{aligned}
\frac{dP_L}{dt} &= 2k_L S_{LL} + k_L S_{LH} \\
\frac{dP_H}{dt} &= 2k_H S_{HH} + k_H S_{LH}
\end{aligned}
\qquad (3)
$$

and their integration leads to time dependence of concentrations of isotopologue products:

$$P_i = S_{0ii}\left(1 - e^{-2k_i t}\right) + \frac{k_i}{k_i + k_j} S_{0ij}\left[1 - e^{-(k_i + k_j)t}\right] \qquad (4)$$

P_H/P_L is the isotopic ratio (R) of the product (that contains only one isotopic chlorine atom). Simple rearrangements of the ratio of functions given by Eq. (4) for the isotopic product for the full conversion sample (time approaches infinity) yield mathematical expression:

$$R_{\infty} = \frac{\dfrac{S_{0HH}}{S_{0LH}} + \dfrac{1/\mathrm{KIE}}{1/\mathrm{KIE}+1}}{\dfrac{S_{0LL}}{S_{0LH}} + \dfrac{\mathrm{KIE}}{\mathrm{KIE}+1}} \quad (5)$$

where KIE (k_L/k_H) is the primary chlorine kinetic isotope effect. As can be seen, when S_{0HL} is approaching zero, R_{∞} approaches S_{0HH}/S_{0LL}, i.e., represents directly to isotopic composition of the reactant and corresponds to the intermolecular kinetic isotope effect. On the other extreme, when S_{0HL} is very large, R_{∞} approaches 1/KIE, and it reflects on the intrinsic isotope effect (these effects being equal under the assumption made that secondary isotope effects are negligible).

In the intermediate cases, when the abundance of molecules containing two different isotopes simultaneously (S_{HL} comparable to S_{LL} and S_{HH}) cannot be ignored, isotopic composition of the product bearing only one isotopic atom differs from that of the initial reactant, which is given by $S_0 = (S_{0HH}/S_{0LH} + 1/2)/(S_{0LL}/S_{0LH} + 1/2)$ and only if there is no isotope effect $S_0 = R_{\infty}$. Thus depending on the experimental protocol, results should be corrected accordingly.

We have used this approach in the studies on the DhlA-catalyzed dehalogenation of DCE. The chlorine isotope ratio was measured on AgCl precipitated after small and full conversion of the initial DCE by FAB-IRMS methodology described above and converted to the kinetic isotope effect by the following equation:

$$^{37}\mathrm{KIE_{obs}} = \mathrm{In}(1-f)/\mathrm{In}(1-fR_f/R_0) \quad (6)$$

where f, the fraction of reaction; and R, the isotopic ratio of the reactant (0) or product (f).

The experimentally obtained value of 1.0045 was initially interpreted as an indication of reaction complexity as the corresponding value obtained for the slow substrate, 1-chlorobutane, was 1.0066 (Lewandowicz, Rudziński, Tronstad, Widersten, Ryberg, Matsson et al., 2001). However, when the correction (5) has been applied, the Cl-KIE for DCE dechlorination by DhlA becomes 1.0068, in excellent agreement with the theoretical prediction of the intrinsic isotope effect, indicating that in both cases dehalogenation is a fully rate-determining step.

3. THEORETICAL PREDICTIONS OF Cl-KIEs

Experimental measurements do not always provide detailed information of a given chemical reaction; the reason may be technical or simply that, for instance, the reaction has several steps. In order to draw conclusions from such data, it is mandatory to understand the relationship between the isotope effect(s) on steps that occur during the conversion of the initial reactant to the final product and the overall isotope effect measured by the method. Isotope effects coming from theoretical calculations allow for mechanistic elucidation and give detailed information on TS structure. As they carry position-specific isotopic information, they also allow recognizing the contribution of each single atom of a molecule to the overall calculated effect and prevent erroneous interpretations. In addition, if reactions are composed of several steps, computations make it possible to calculate isotope effects for each of them so that the observable "net" isotope fractionation can be interpreted as a weighted sum of the separate steps. Computational chemistry tools can therefore fill the gaps needed for such interpretation since they provide information on molecular positions and reaction steps which is not experimentally accessible. In order to accurately predict the values of KIEs, reliable geometries of the involved species (reactants and products for equilibrium IEs and reactants and transition state for KIEs) and a reliable approximation for the second derivatives of the potential energy with respect to the nuclear coordinates of the system (Hessian) calculations are needed. Since a KIE reflects differences in bonding between the reactants and transition states, it is necessary to use quantum mechanical (QM) methods to describe them. For small systems (up to 100 atoms), for which the chemical change can be successfully mimicked using gas-phase approximation, modern computational chemists are equipped with a whole palette of electronic structure methodologies to optimize the coordinates of a system and calculate its energy. For systems of this size, it is also feasible to calculate the Hessian at the same selected quantum level of theory. If reactions in the condensed phase are of interest, like the ones taking place in enzymatic active sites, then one needs to keep in mind that most likely a different treatment will be required as the systems involving well-defined environment are simply large. This has natural consequences including both the necessity of lowering the level of theory and finding the best approximation to include the environmental effects. Although the incorporation of the environment into the overall picture is essential, in general, the most traditional method is to exclude any averaging from fluctuating environment or effects arising from anharmonicities of low vibrational modes due to the ease of usage and reasonable computational cost. Despite the obvious limitations, such approaches proved to help the interpretation of experimental KIEs measured for many chemical systems. There are, however, numerous examples of systems (even model ones) for which different combinations of QM method and basis set have not reproduced all experimentally observed KIEs for a particular system (Adamczyk, Dybala-Defratyka, & Paneth, 2011; Fang et al., 2003; Glad & Jensen, 1996). One possible reason for the apparent failure of theory was relying on KIE predictions based only on individual TSs instead of taking into account multiple configurations that are visited by enzymes during the reaction they catalyze and thus proper averaging over them (Ruiz-Pernía & Williams, 2012). For example, QM/MM molecular dynamics simulations can in principle provide an ensemble-averaged approach for computing KIEs, apart from proposing an appropriate scheme for the systems where specific reaction center–environment interactions are important. Nevertheless, an ensemble approach like this involves significant computational effort and in many cases own "group/researcher"-developed/modified software; therefore currently this approach is not accessible for a wider group of chemists. An ensemble of structures resulting from such simulations has to be subsequently optimized to respective transition and corresponding reactant states. Those optimized geometries have to be of very high quality in order to compute accurate force constants based on them. Furthermore, it is either not always absolutely necessary or feasible to use full explicit solvation models whenever environmental effects are of interest (Grzybkowska, Kaminski, & Dybala-Defratyka, 2014). Once energy optimized, determining the force constant matrix (Hessian) comprises another challenge. While using continuum solvation model does not increase the size of a system, any approach involving explicit treatment of surrounding media surely does, and if a process of interest requires including many solvent/enzyme molecules, then a choice of method to perform normal mode analysis to obtain Hessian and partition functions, and thence isotope effects, is limited to protocols allowing computations only for a subset of atoms (isotopically sensitive, so to speak) (Ghysels et al., 2010; Williams, 2012). Quality of calculated vibrational frequencies and subsequently isotope effects will, thus, depend largely on a selected approximation. In larger systems, when only a subset of atoms can be used for the Hessian calculation, the way of treating low-frequency modes can be also crucial. Recently, Williams has highlighted the importance of including six external degrees of freedom in computation of a Hessian as they may constitute a response of a system to its environment (when represented explicitly), and in larger systems these eigenvalues are usually nonzero (Williams, 2012). The choice of this subset should not be arbitrary, as pointed out by the same researcher, as KIEs are not necessarily local properties of the system.

Cl-KIEs as well as computational chemistry tools have been shown before to provide important information on the mechanism of reaction catalyzed by dehalogenating enzymes and hence could have been successfully applied to a number of biochemical transformations (Lewandowicz, Rudziński, Luo, Dunaway-Mariano, & Paneth, 2002; Lewandowicz et al., 2001; Manna & Dybala-Defratyka, 2013, 2014; Manna, Zinovjev, Tuñón, & Dybala-Defratyka, 2015; Papajak et al., 2006; Schürner et al., 2015; Siwek et al., 2013; Szatkowski & Dybala-Defratyka, 2013; Szatkowski, Thompson, Kaminski, Franzen, & Dybala-Defratyka, 2011).

A more detailed description of these transformations at the molecular level may provide meaningful insight into the existing knowledge on strategies employed by environmental systems and leading to the elimination of substituents responsible for toxicity of organic compounds. Furthermore, such analysis may constitute a basis for an interpretation tool of trends observed during isotopic analysis of environmental samples, in particular, when the results coming from isotopic ratio measurements are supported by modeling these effects using most advanced methods of computational chemistry.

3.1 Cluster Model Approach

Theoretical modeling of enzymatic reactions is very challenging and troublesome. However, normally the essential reaction takes places in a well-defined area of the enzyme (i.e., active site) and the reaction usually involves a few functional groups only. Therefore, in the cluster model or the big-QM approach, the most important parts of the enzyme and substrates are included to capture the main features of reaction mechanisms. It is important to keep the size of the model relatively small to reduce the computational cost. Himo, Siegbahn, and their coworkers have studied a large number of enzymatic reactions belonging to different enzyme families using the QM cluster modeling approach to date (Blomberg, Borowski, Himo, Liao, & Siegbahn, 2014; Himo & Siegbahn, 2003; Noodleman, Lovell, Han, Li, & Himo, 2004). In principle, cluster models are prepared from the enzyme–substrate complexes resulted from available X-ray structures of enzymes with ligands bound or from docking studies, and subsequent model construction including initial equilibration of the system using various simulation techniques in order to have a system similar to the one present in solution instead of crystal. During the QM cluster model calculations, usually tight convergence and optimization criteria are applied. All atoms are either allowed to move freely during the optimization or some atoms belonging to outside sphere of the model are kept fixed to their crystal structure positions. Molecular geometries of all species are fully optimized, and vibrational analysis is carried out to confirm the nature of the stationary points ($3n-6$ real vibrations in the case of reactants and one imaginary frequency corresponding to the desired reaction coordinate in the case of TSs). Additionally, the intrinsic reaction coordinate (IRC) (Hratchian & Schlegel, 2005) calculations are performed to ascertain the true nature the TSs. The geometries obtained as the end-points on either side of the IRC trajectories are subjected to further optimization to obtain the reactant and product states, such that their connection to the TS is established. Cluster model calculations are either performed in the gas phase or some simplified forms of including environment using background charges originating from the larger model of the system and applied force field or continuum solvation models employing a very low dielectric constant to mimic the protein environment surrounding the active site. In the latter case, care has to be taken since the polarity of the environment can strongly affect the location of the TS (Sicińska, Truhlar, & Paneth, 2001). Having stationary states located, in particular reactants and transition states, and relevant sets of force constants for each of them, the conventional TS theory within the Born–Oppenheimer and the harmonic oscillator approximations is usually used to predict KIEs. For his purpose either expression of KIE in terms of partition function (7) or vibrations (8) is used:

$$\mathrm{KIE}=\frac{k_\mathrm{L}}{k_\mathrm{H}}=\frac{\left(Q^{\mathrm{TS}}/Q^{\mathrm{RS}}\right)_\mathrm{L}}{\left(Q^{\mathrm{TS}}/Q^{\mathrm{RS}}\right)_\mathrm{H}}=\frac{\left(Q_\mathrm{H}/Q_\mathrm{L}\right)^{\mathrm{RS}}}{\left(Q_\mathrm{H}/Q_\mathrm{L}\right)^{\mathrm{TS}}}=\frac{f_\mathrm{RS}}{f_\mathrm{TS}} \tag{7}$$

where k_L and k_H are the rate constants for the reactions of light (L) and heavy (H) isotopically substituted reactants, respectively. Q^{RS} and Q^{TS} are molecular partition functions for the reactant state (RS) and transition state (TS) structures, respectively. The isotopic partition function ratio (IPFR) is expressed as a ratio of heavy isotope molecular partition over light isotope partition function (Ruiz-Pernía & Williams, 2012; Williams, 2012). Furthermore, f_RS and f_TS represent IPFRs of RS and TS structures, respectively:

$$\mathrm{KIE}=\left[\frac{\vartheta_\mathrm{L}^{\neq}}{\vartheta_\mathrm{H}^{\neq}}\right]\prod_i^{3N_s-6}\left[\frac{\vartheta_{i\mathrm{H}}}{\vartheta_{i\mathrm{L}}}\right]_\mathrm{RS}\left[\frac{\sinh(u_{i\mathrm{L}}/2)}{\sinh(u_{i\mathrm{H}}/2)}\right]_\mathrm{RS}\prod_i^{3N_s-7}\left[\frac{\vartheta_{i\mathrm{L}}}{\vartheta_{i\mathrm{H}}}\right]_\mathrm{TS}\left[\frac{\sinh(u_{i\mathrm{H}}/2)}{\sinh(u_{i\mathrm{L}}/2)}\right]_\mathrm{TS} \tag{8}$$

where $u=hc\vartheta/k_\mathrm{B}T$, with the frequency ($\mathrm{s}^{-1}$) replaced by wavenumber (ϑ, cm^{-1}), and h, k_B, c, and T are the Planck constant, Boltzmann constant, velocity of light, and absolute temperature, respectively. $3N_s-6$ and $3N_s-7$ are pure vibrational eigenvalues for RS and TS, respectively. Eq. (8) was proposed by Bigeleisen-Mayer and implemented in the ISOEFF program (Anisimov & Paneth, 1999). Sometimes keeping a number of atoms fixed during the geometry optimization results in a certain number of small imaginary frequencies which are usually removed prior to isotope effect calculations as they do not contribute significantly to the zero-point energy (Manta, Raushel, & Himo, 2014). However, before neglecting them, one always needs to analyze their contribution first as they may be a nonnegligible source of the environment influence (Williams, 2012). Two illustrative examples are presented later.

3.1.1 Dechlorination of 1,2-Dichloroethane and n-Butyl Chloride by the DhLA Dehalogenase

In early calculations of chlorine KIEs, only limited computational resources were available and therefore mostly cluster models were used, sometimes with inclusion of a continuum solvent model with a reduced dielectric constant to take into account the low polarity of the active site. Studies of the dehalogenation step in the DhlA-catalyzed dechlorination of 1,2-dichloroethane and *n*-butyl chloride provide comparison between small cluster model and hybrid/multilevel calculations (Lewandowicz et al., 2001).

DhlA catalyzes dechlorination in two consecutive S_N2 steps. In the first and rate-limiting one, a carboxylate group of the enzyme acts as a nucleophile and attacks a chlorine-bearing carbon atom, yielding chloride and a reactant-enzyme intermediate. This ester intermediate is subsequently hydrolyzed by a neighboring water molecule, restoring the enzyme and yielding the product alcohol.

The active site of the enzyme was constructed from the Protein Data Bank 2DHC (Verschueren, Seljee, Rozeboom, Kalk, & Dijkstra, 1993) structure and truncated to include Asp124, Trp125, Trp175 residues, one water molecule, and the reactant. N-terminals were capped with hydrogen atoms and C-terminals were capped with the amide groups yielding 79-atom and 85-atom models for the DhlA active site with $(ClCH_2)_2$ or *n*BuCl, respectively. Fig. 3 shows major components of the models.

The first step of the reaction was modeled. In hybrid calculations the model was divided into two quantum parts, one of which was treated at the DFT level and the more remote atoms from the incipient O—Cl bond semiempirically. The B1LYP functional was used to describe atoms in the upper QM part. Three sizes of the DFT level were considered. The main results, collected in Table 1, illustrate that in this case the size of the QM part, basis set, and frequency scaling have only a marginal influence on the final calculations of chlorine KIEs.

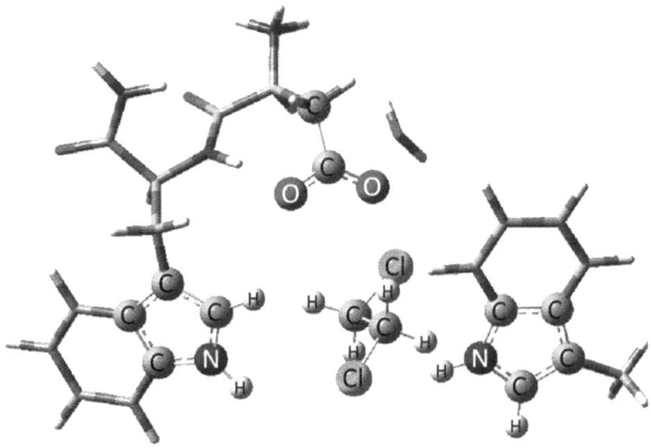

Fig. 3 Active site model used in hybrid/cluster calculations of Cl-KIE on DhlA-catalyzed reaction.

Table 1 Theoretical Values of Cl-KIEs on DhlA-Catalyzed Reaction

Theory Level	Trp Model	*n*BuCl	$(CH_2Cl)_2$
B1LYP/6-31+G(d):PM3	NH_3		1.0068
B1LYP/6-31G(d):PM3	NH_3	1.0063	1.0065 (1.0063 freq. scaled)
	Pyrrole rings	1.0061	1.0063
	Indole rings	1.0065	

3.1.2 The Aerobic Dehydrochlorination of HCH Isomers by the LinA Enzyme

Hexachlorocyclohexane (HCH) isomers are the most prevalent halogenated organic insecticides that have been mainly used for crop protection (Bala et al., 2012; Lal et al., 2010). They pose a health hazard due to their persistent toxicity effects on humans, wildlife, plants, soil, water, and the atmosphere (Willett, Ulrich, & Hites, 1998). There have been a growing number of international projects with the objective to study the biodegradation processes of HCH isomers under various environmental conditions (Nijenhuis & Richnow, 2016; Zhang et al., 2014). Dehalogenases from the Lin family of enzymes have been shown to metabolize HCH isomers to less toxic compounds (Oakley et al., 2004; Okai et al., 2010). Catalyzed by LinA, one H/Cl axial pair elimination from α-, γ-, δ-, and β-HCH leads to a respective pentachlorocyclohexene formation (Scheme 1; Brittain et al., 2011).

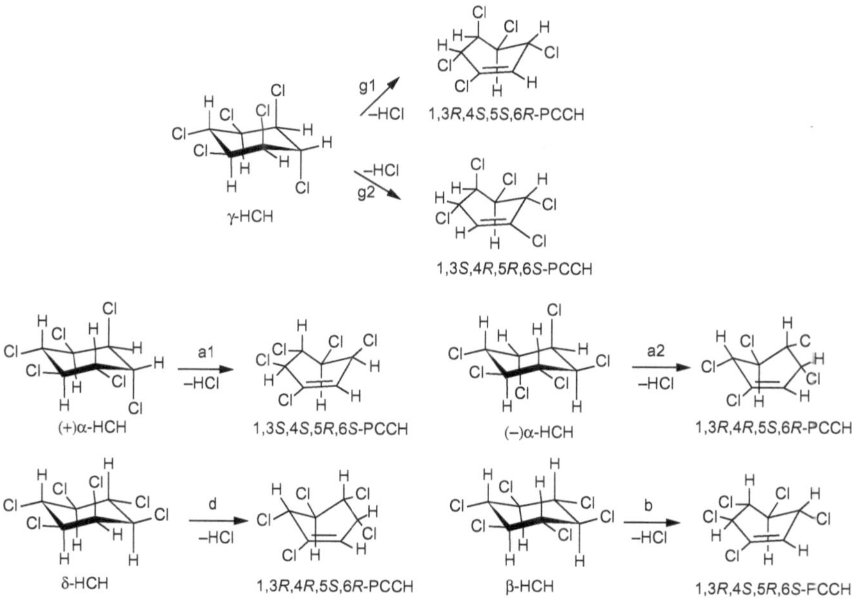

Scheme 1 Dehydrochlorination reaction of γ-, α-, δ-, and β-HCH isomers catalyzed by LinA.

A small model of the HCH dehydrochlorinase LinA active site comprising its catalytic dyad His73-Asp25 was constructed based on the receptor–ligand complexes suggested by docking studies (Manna & Dybala-Defratyka, 2013). This enzymatic cluster model is comprised of the respective HCH molecule, the imidazole ring of His73, and the carboxylate group of Asp25.

The DFT studies on enzymatic cluster models indicated that dehydrochlorination of all the HCH isomers occurs by an E2 mechanism (Manna & Dybala-Defratyka, 2013). β-HCH exhibits the highest free energy of activation among all four HCH isomers. The values obtained for α-, γ-, and β-HCH elimination reflected the trend observed experimentally ($\Delta G^{\neq}$ of 14.9, 16.0, and 21.5 kcal mol^{-1}, respectively) (Brittain et al., 2011). The QM study allowed the elucidation of distinctive features of β-HCH vs the other three isomers (Manna & Dybala-Defratyka, 2013). The TS structure for the β-HCH was found to be different from the TSs of the other three isomers. Furthermore, β-HCH was shown to react by a *syn-* rather than an *anti* E2 mechanism. The primary ^{37}Cl-KIEs of α-, δ-, and γ-HCH were 1.0030, 1.0034, and 1.0034, respectively, at 300 K, which were lower than the ^{37}Cl-KIEs obtained for β-HCH (1.0107). Although all isomers were predicted to undergo a concerted E2 reaction, the contribution of proton ($H_{\delta 1}$) transfer and carbon–chlorine-leaving group bond stretch in the TSs differed for the different isomers giving rise to different values of predicted Cl-KIEs. However, Cl-KIE patterns calculated in that study indicated that they could be used as a mechanistic indicator for these systems (β-HCH vs the rest of isomers).

3.2 Full Enzymatic Model Approaches

3.2.1 One TS Structure Protocol

For the study of the reaction mechanism of small systems (up to 200–300 atoms) a full QM model can be used with an IEF-PCM (Miertus, Scrocco, & Tomasi, 1981) solvent representation as implemented, for example, in the Gaussian09 software (Frisch et al., 2009). Unfortunately, QM models have not always been found appropriate for studying reaction mechanisms of enzymatic systems or systems with explicit solvent as they provide a very limited picture of the environment influence and for some systems that turned out not to be sufficient (Lonsdale, Harvey, & Mulholland, 2012; Warshel, 2003). Molecular mechanics (MM) approaches could be a good alternative to handle systems with a large number of atoms; however, they do not allow for a description of the process of forming and breaking of chemical bonds. Therefore, a combination of both methods QM and MM is the most common approach used in the studies of enzymatic reactions (Lonsdale et al., 2012; Warshel, 2003). Atoms that are directly involved in the change of electronic structure of the system and their nearest surrounding atoms are treated quantum mechanically, while the remaining enzyme atoms and/or solvent are represented by an appropriate force field (MM). The correct description of the energy of such systems, especially the interaction energies between QM and MM subsystems, is the key issue in the QM/MM methods. It is especially important if both subsystems are directly connected to each other. However in the simplest case, when the MM subsystem does not have direct connections with the QM subsystem, all electrostatic interactions between the subsystems can be limited to Coulombic interactions between MM and QM subsystems, where nonbonded interactions between them can be limited to van der Waals interactions (Gao & Xia, 1992). The problem is much more complicated when the QM and MM subsystems have direct bonded interactions to each other. In this case the simplest method to address this problem was developed by Morokuma and coworkers and was implemented in the ONIOM model (Dappricha, Komáromia, Byuna, Morokuma, & Frisch, 1999). The total energy of the whole system is described as the difference between the sum of the energy of the QM subsystem at the QM level and the MM subsystem at the MM level minus the MM energy of the QM subsystem. The second type of method uses a link atom common to both subsystems to correctly describe the valence terms and electrostatics on the boundary between the QM and MM subsystems (Eurenius, Chatfield, Brooks, & Hodoscek, 1996). Finally, the third approach is to use localized frozen QM orbitals to create an interface between the QM and MM subsystems (Philipp & Friesner, 1999), which, for example, has been implemented in the Schrödinger QSite software (QSite, version 5.0, Schrödinger, 2007). The total energy ($E^{QM/MM}$) in QSite is described according to Eq. (9):

$$E^{QM/MM} = E^{QM} + E^{MM} + E^{int} \tag{9}$$

where E^{QM} is the QM subsystem energy, E^{MM} is the energy of the MM subsystem calculated with the usage of a selected force field for biological system and solvent molecules, and E^{int} is an interaction energy between the QM and MM subsystems.

In general the first step in predicting KIEs for enzymatic reactions is construction of its energetic profile. Taking into account energetic barriers of each step of the reaction KIEs can be predicted and analyzed. One of the possible approaches to construct the energy profile is conducting classical molecular dynamics simulation and then exploring the potential energy surface along reaction coordinates using the QM/MM scheme. The structure of the equilibrated system is used as the initial configuration for the PES map construction. Initial TS structures are located on obtained PES maps and subsequently optimized using an optimization algorithm, as for instance micro–macro iteration method (Martí, Moliner, & Tuñón, 2005; Turner, Moliner, & Williams, 1999) implemented in f-Dynamo library (Field, Albe, Bret, Proust-De Martin, & Thomas, 2000). The Hessian matrix is then calculated only for the QM atoms in the presence of background charges representing the environment. For each optimized TS the IRC calculations are performed to trace down the system toward the reactant and product valleys in mass-weighted Cartesian coordinates. Based on the optimized structures and calculated frequencies, position-specific Cl-KIEs are calculated using normal modes generated for light and heavy isotopologues of the Cl atom of interest using the Bigeleisen equation. Proposed potential energy profiles allow pinpointing the rate-determining steps and thus propose the steps that will govern the magnitude of the observed (measured) isotope effects.

In the following part, we show two illustrative examples of the potential of various QM/MM approaches in the study of dehalogenation reactions, predictions, and validations of Cl-KIEs. These include a chlorophenol dehalogenation reactions catalyzed by two peroxidases (horseradish peroxidase, HRP, and dehaloperoxidase, DHP), and hydrolysis of atrazine catalyzed by wild type and mutant (Glu241Gln) TrzN chlorohydrolase.

Plant *peroxidases*, like HRP (Nicell, 2001) or DHP from the marine bristle worms *Amphitrite ornata*, which has a peroxidase activity (Chen, Woodin, Lincoln, & Lovell, 1996; Klibanov, Alberti, Morris, & Felshin, 1980), can remove halogens from phenolic rings through oxidative dehalogenation. The oxidative pathways toward degradation of chlorophenols can lead to various products, from quinones to polymerized species (Dec & Bollag, 1995) via two one-electron transfers (Ferrari, Laurenti, & Trotta, 1999). Both enzymes need the presence of H_2O_2 as a cosubstrate in both these steps. It ensures that during enzymatic transformation, chlorophenols could be converted to the phenoxy radicals as products of the first one-electron transfer, while a product of the second one-electron transfer could vary from phenoxy radicals (if after the first one-electron transfer, the phenoxy radical had been released from enzyme active site and replaced by a second chlorophenol molecule) to cationic form of chlorophenol ring (if the phenoxy radical remained inside the enzyme active center). Therefore, two different pathways could be drawn: direct dehalogenation—via direct water attack on the aromatic carbon of the substrate whether in its anionic, radical–anionic, or cationic form (Wiese, Chang, Lloyd, Freeman, & Samokyszyn, 1997); or indirect dehalogenation—by the coupling of two chlorophenol rings (radical anion form of chlorophenol with another chlorophenol whether in its anionic, radical anionic, or cationic form) with simultaneous expulsion of the halogen substituent from the attacked ring (Dec & Bollag, 1994).

To study this problem, we compared a reference reaction using 2,4,6-trichlorophenol (TCP) dehalogenation in an aqueous environment (explicit model, QM/MM study) with a dehalogenation of 4-chlorophenol (4-CP) modeled using the enzymatic environment (QM/MM study) (Szatkowski et al., 2011), and a QM model of chlorophenol coupling (Szatkowski & Dybala-Defratyka, 2013). For the description of the reference reaction and enzymatic dehalogenation a QM/MM model implemented in the QSite program (QSite, version 5.0, Schrödinger, 2007) was used. The QM part included the substrate molecule 4-CP (for HRP system) or TCP (for a reference reaction) and selected water molecules and was treated with the B3LYP (Becke, 1993; Lee, Yang, & Parr, 1988) functional (for HRP system) with lacvp* basis set (Hay & Wadt, 1985), while the MM part with TIP3P model water molecules and the enzyme was described by the OPLS-AA force field (Jorgensen, Maxwell, Tirado-Rives, & Turner, 1996; Kaminski, Friesner, Tirado-Rives, & Jorgensen, 2001). Meanwhile, a QM approach implemented in the Gaussian09 software (Frisch et al., 2009) together with the IEF-PCM solvation model (Miertus et al., 1981) was used for the description of the dimerization process.

A theoretical reference model was built with the substrate molecule (TCP) and 201 surrounding water molecules. The TCP molecule with the four closest water molecules was defined as the QM region, while the rest of the system was treated as an MM region (positions of all water molecules beyond 8 Å away from the solute were fixed). In the case of HRP-catalyzed model, the 4-CP molecule along with six water molecules was defined as the QM part (30 atoms), while the rest of the system was assigned to the MM part. Positions of the atoms beyond 15 Å from the substrate molecule were kept fixed. In both cases the reaction coordinate was defined as the decreasing distance between the attacked C atom in the phenol ring and the attacking O atom in the water molecule.

For the reference reaction, three different dehalogenation pathways comprising three distinct forms of substrate undergoing the reaction were studied for the chlorine substituent at the *para* position: phenolate (anion), phenoxy radical, and cation. For anionic and phenoxy radical forms, large Cl-KIE values were predicted (Table 2). However, the energetic data allowed the anionic pathway to be disregarded as unrealistic. Moreover, in the case of the cationic pathway no appropriate TS was found, as there was no barrier for the water attack on aromatic carbon; therefore the Cl-KIE for this pathway was unity. Instead, a formation of the low energy Meisenheimer complex (MC) ($-48.3\,\text{kcal mol}^{-1}$, respectively, *para* position

Table 2 Theoretical Values of Cl-KIE and Saddle Point QM/MM Barriers for Dehalogenation of the TCP in Anionic and Phenoxy Radical Form in Water at 300 K and for HRP-Catalyzed Phenoxy Radical Model of 4-CP Dehalogenation

	Water Environment		HRP-Catalyzed Reaction
	Phenolate	**Phenoxy Radical**	**Phenoxy Radical**
Cl-KIE	1.0037	1.0085	1.0074
QM/MM barrier (kcal mol^{-1})	57.3	31.3	47.0

model) was observed. A similar MC complex was also found for a direct water attack on TCP chlorine substituent at *ortho* positions. Further analysis of the expulsion of the chlorine ion from MC showed the existence of two saddle points. However, neither from these saddle points could be reoptimized to the corresponding TS.

The energy profiles for the dehalogenation of the cationic phenoxy radical form of 4-CP catalyzed by HRP were modeled. It turned out that the phenoxy radical pathway catalyzed by HRP proceeds in a similar manner to the uncatalyzed dehalogenation of the TCP phenoxy radical by direct water attack. Cl-KIE values as well as computed QM/MM energy barriers were similarly large for both cases (see Table 2). Furthermore, the formation of an enzyme-Meisenheimer complex ($-52\,\text{kcal}\,\text{mol}^{-1}$ stabilization energy, EMc) was also observed in the case of the cationic pathway catalyzed by HRP, with a similar nature as the reference reaction. Once again the studies of the subsequent chloride ion departure process from the EMc complex (similar to those for the uncatalyzed reaction) revealed a lack of any TS. A very similar mechanism, through an MC (with Cl-KIE equal to unity), was also suggested for dehalogenation of chlorophenols catalyzed by DHP (Szatkowski et al., 2011).

Three different mechanistic scenarios for a coupling of two identical rings of 2-CP, 4-CP, TCP, and DMCP were investigated: radical–anion (R–A), radical–radical (R–R) both in triplet and singlet states, and radical–cation (R–C) (Szatkowski & Dybala-Defratyka, 2013). The reaction coordinate was defined as an attack of the oxygen radical (from one chlorophenol molecule) onto respective carbon atom with a chlorine substituent (from the second chlorophenol molecule) either at *ortho* or at *para* position. Therefore, with respect to the coupling position in the attacked ring, five different systems were studied: 2-CP (*ortho*), 4-CP (*para*), TCP-*o* (*ortho*), TCP-*p* (*para*), and DMCP (*para*). Only in the case of R–A and R–R triplet models was a carbon–chlorine bond cleavage observed. However, the R–R triplet pathway was found to be energetically unfavorable (energy barrier higher by more than $25\,\text{kcal}\,\text{mol}^{-1}$ compared to other pathways). Predicted Cl-KIE values for all chlorophenols for R–A pathway were close to unity: 2-CP 0.9999, 4-CP 1.0002; TCP-*o* 0.9997, TCP-*p* 1.0001; and DMCP 1.0002. Other pathways were identified as two-step reactions: dimer formation and intramolecular chlorine atom transfer (between two carbon atoms in the attacked ring) and, because of lack of chlorine expulsion, the Cl-KIEs on the coupling step for those pathways could not be directly compared with the experimental data.

Atrazine, the most popular representative of *s*-triazine compounds (herbicides used in agricultural production), is a serious source of ground- and surface-water contamination. Its widespread presence and environmental accumulation lead to conditions toxic to other species present in the shared contaminated area. Removal of the halide substituent as the first step in the metabolism of chlorinated *s*-triazine herbicides, leading to nontoxic hydroxylated analogs, is catalyzed by atrazine *chlorohydrolases TrzN*. Hydrolysis of atrazine catalyzed by wild type of TrzN is a complex reaction consisting of multiple proton transfer leading to *s*-triazine ring protonation, nucleophilic attack, a Meisenheimer-like complex formation, and leaving group departure with its concurrent protonation (Scheme 2). In the mutant enzyme (Glu241Gln) glutamic acid is replaced by glutamine at position 241. In that case the triazine ring protonation occurs via direct proton transfer to nitrogen in the vicinity of Glu68 from the nucleophilic water.

Hydrolysis of atrazine catalyzed by wild type and mutant (Glu241Gln) form of TrzN enzyme was investigated utilizing the construction of numerous PESs. The equilibrated systems were divided into the QM subsystem consisting of the substrate—atrazine, the zinc ion along with its coordination sphere (His63, His65, and His238, and nucleophile water molecule), the catalytically important residues Glu241, His274, Glu68, Asp300, and another crystal water, while the MM subsystem comprised of the rest of the protein, water molecules, and sodium ions. Due to the presence of the Zn^{2+} ion the QM part was described by the PM3ZnB Hamiltonian (Brothers, Suarez, Deerfield, & Merz, 2004), while the MM subsystem was treated using the OPLS-AA force field (Jorgensen et al., 1996; Kaminski et al., 2001) and TIP3P3 for protein and sodium ions, and water, respectively. To construct PES maps key distances were monitored. To investigate aromatic ring protonation in the wild-type form, $H1_{Nu}$–O_{Nu} (Nu–nucleophile) and $N_{H274}\cdots H1_{Nu}$, N_{H274}–$H1_{Nu}$ and $O_{E241}\cdots H1_{Nu}$, and O_{E241}–$H1_{Nu}$ and $N_{ATZ}\cdots H1_{Nu}$ interatomic distances were used as reaction coordinates subsequently for proton transfer to histidine 274 from nucleophilic water, its shuttle to glutamic acid and its transfer to aromatic ring nitrogen. Then, the PES for the nucleophile attack, in which the $C6_{ATZ}\cdots O_{109}$ and $C6_{ATZ}$–Cl_{ATZ} distances were monitored, was explored. Next, a combination of bond breaking ($H2_{Nu}$–O_{Nu}) and bond forming ($H2_{Nu}$–N_{H274}) and a combination of bond breaking (C_{ATZ}–Cl_{ATZ}) and bond forming (C_{ATZ}–O_{Nu}) were used to describe the proton transfer to His274 from the attached OH group. Subsequently, a combination of bond breaking ($H2_{Nu}$–N_{H274}) and bond forming ($H2_{Nu}$–Cl_{ATZ}) and

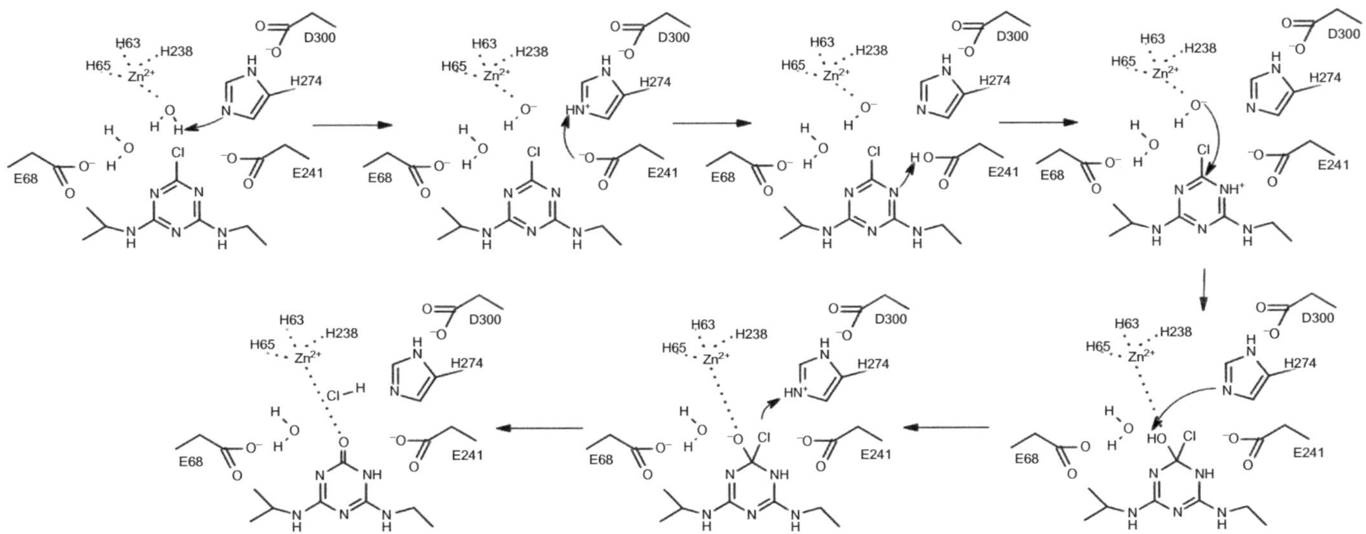

Scheme 2 Proposed mechanism for the atrazine hydrolysis catalyzed by the wild-type TrzN.

C_{ATZ}–Cl_{ATZ} bond was used as reaction coordinates to monitor proton shuttle to chlorine from His274. In the case of the reaction catalyzed by the Glu241Gln form, $H1_{Nu}$–O_{Nu} and $N_{ATZ}\cdots H1_{Nu}$ distances were used to construct the PES map of direct proton transfer to *s*-triazine ring from the nucleophile. Nucleophilic attack was investigated in the same manner as in the wild-type enzyme. Next, N_{H274}–$H2_{Nu}$ and $H2_{Nu}\cdots N_{H274}$, and N_{H274}–$H2_{Nu}$ and $H2_{Nu}\cdots Cl_{ATZ}$ distances were used to define reaction coordinates for proton transfer to His274 and its shuttle to chlorine, respectively. Obtained potential energy profiles for both reactions are presented in Fig. 4. Hydrolysis of atrazine seemed to be a three-step reaction in which ring protonation, nucleophilic attack, and chloride protonation can be distinguished.

Based on the proposed potential energy profiles the rate-determining step of the reactions could be pinpointed. In the case of wild-type TrzN–Atz the nucleophilic attack leading to the formation of Meisenheimer-like complex was a rate-determining step with the energetic barrier of 25.0 kcal mol^{-1} (black plot, Fig. 4). However, ring protonation was a rate-determining step (20.5 kcal mol^{-1}) for the Glu241Gln–Atz reaction (gray plot, Fig. 4), and therefore a Cl-KIE near unity was expected. The Cl-KIEs were predicted for the nucleophilic attack (18.6 kcal mol^{-1}) as its energetic barrier is three times higher than that for the chloride protonation (6.3 kcal mol^{-1}). The very small inverse Cl-KIE (0.9996) and small normal Cl-KIE (1.0003) predicted for the wild-type and Glu241Gln TrzN–atrazine complexes, respectively, indicate that chloride departure is not a rate-determining step, which agrees with the findings from the energetic profile for the atrazine hydrolysis. Although Cl-KIEs on hydrolysis of atrazine by TrzN were not

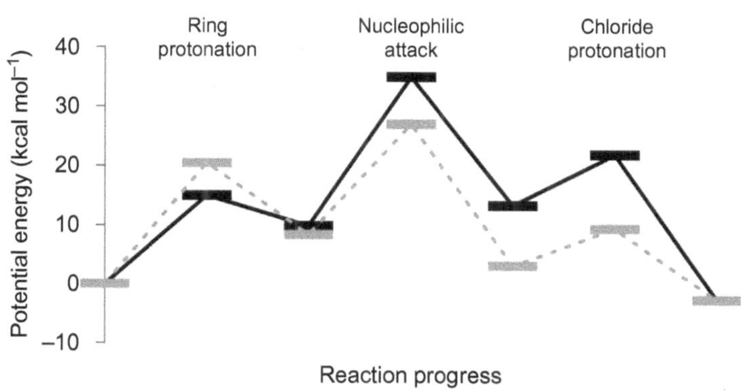

Fig. 4 Potential energy profiles for wild-type TrzN–Atz (*black line*) and mutant TrzN–Atz (*gray line*) obtained at MPW1K/6-31+G(d,p)/MM//PM3ZnB/MM.

measured, we assessed our model of the reaction by comparing carbon and nitrogen KIE determined experimentally and theoretically for the same reaction. The predicted C-KIE equal to 1.0434 for hydrolysis occurring in the wild-type enzyme is in good agreement with the experimental value of1.041 (Meyer, Penning, & Elsner, 2009; Schürner et al., 2015). The magnitude of 1.0450 obtained for the reaction catalyzed by the mutant is a bit larger than the measured value (1.035), but it represents the same trend (normal isotope effect). In contrast, inverse nitrogen isotope effects, which generally occur if the bonding environment in the TS is tighter/more restricted than in the reactant, were measured. N-KIEs values equal to 0.988 and 0.989 (Meyer et al., 2009; Schürner et al., 2015) for hydrolysis occurring in the wild type and Glu241Gln, respectively, indicated that aromatic ring has to be protonated prior to nucleophilic attack. Theoretically predicted N-KIEs equal to 0.9983 and 0.9990 for reaction catalyzed by wild type and mutant TrzN, respectively, are overestimated with respect to the experimental values; however, the same trend (inverse isotope effects) was reproduced which along with the data for the carbon fractionation confirms a good quality of the model and applied computational approach.

3.2.2 Ensemble of TS Structures Protocol

In order to provide more detail for the mechanistic scenarios as well as to confirm findings resulting from cluster models and include the fact that enzymes are quite flexible molecules by nature so instead of describing a single chemical reaction they catalyze by a single TS structure, one, if possible, should rather consider an approach that allows providing an ensemble of structures. For these purposes, free energy profiles of studied reactions should be constructed using various available computational techniques, for instance, umbrella sampling technique (Kumar, Bouzida, Swendsen, Kollman, & Rosenberg, 1992; Kumar, Rosenberg, Bouzida, Swendsen, & Kollman, 1995; Sugita, Kitao, & Okamoto, 2000), within which a biasing potential is applied in terms of a harmonic potential that enhances the sampling in the neighborhood of chosen values of a reaction coordinate. From the probability distribution of the system along the reaction coordinate, the change in free energy in each window is calculated.

The windows are then combined by the weighted histogram analysis method (WHAM) (Grossfield, 2011) and free energy profiles are constructed. The so-called potential of mean force (PMF) is the free energy change along the reaction coordinate (or path) in the configuration space. The PMF ($W(\xi)$) related to the normalized probability of finding ($\rho(r)$)

the system at particular value of chosen one-dimensional reaction coordinate (ξ) at temperature T can be written as

$$W(\xi) = C - k_b T \ln \int \rho(r^N)\delta(\xi(r^N) - \xi)\,dr^{N-1} \tag{10}$$

where k_b is the Boltzmann constant and constant (C) term is for the biased ensemble and thus varies between differently biased simulations. The activation free energy can be described as

$$\Delta G^{\neq}(\xi) = W(\xi^{\neq}) - \left[W(\xi^{R}) + G_{\xi}(\xi^{R})\right] \tag{11}$$

where the superscripts indicate the value of the reaction coordinate at the reactants (R) and transition state ($\neq$). Furthermore, $G_{\xi}(\xi^{R})$ is the free energy associated with setting the reaction coordinate to a specific value at the reactant state. This last term makes a small contribution and the activation free energy is directly estimated from the PMF change between the maximum of the profile ($W(\xi^{\neq})$) and the reactant's minimum ($W(\xi^{R})$) energy using the following equation:

$$\Delta G^{\neq}(\xi) \asymp W(\xi^{\neq}) - W(\xi^{R}) = \Delta W^{\neq}(\xi) \tag{12}$$

The selection of the reaction coordinate is trivial when the reaction mechanism can be driven by a single-internal coordinate. However, this is not the case when the reaction involves several internal coordinates. Then, a two-dimensional PMF is computed using two coordinates: ξ_1 and ξ_2. Furthermore, each reaction coordinate can be defined as a simple combination or an antisymmetric combination of interatomic distances. 2D-PMF related to the probability of finding the system at particular values of these two coordinates can be expressed as

$$W(\xi) = C' - k_b T \ln \int \rho(r^N)\delta(\xi_1(r^N) - \xi_1)\delta(\xi_2(r^N) - \xi_2)\,dr^{N-2} \tag{13}$$

The conventional free energy calculation techniques, such as umbrella sampling technique and metadynamics (Barducci, Bonomi, & Parrinello, 2011), are limited by the so-called curse of dimensionality, as they become prohibitively expensive when the dimensionality of the necessary free energy surface becomes larger than two. Other techniques, such as string method and path collective variables (CVs) (Branduardi, Gervasio, & Parrinello, 2007; Zinovjev, Martí, & Tuñón, 2012; Zinovjev, Ruiz-Pernía, & Tuñón, 2013; Zinovjev & Tuñón, 2014), were shown to be a successful treatment for complex reactions catalyzed by various enzymes (Aranda, Zinovjev, Roca, & Tuñón, 2014; Manna et al., 2015; Sanchez-Martinez, Field, & Crehuet, 2015; Zhang et al., 2015). Briefly, the main idea lies in focusing on the minimum free energy path (MFEP) on the free energy surface instead of calculating the entire surface and a one-dimensional reaction coordinate is defined along this path. This framework reduces the multidimensional process to a single dimension, and hence this process becomes tractable by any conventional free energy sampling technique.

The on-the-fly string method (Maragliano, Fischer, Vanden-Eijnden, & Ciccotti, 2006; Maragliano & Vanden-Eijnden, 2007) allows MFEPs between metastable states to be found. A number of CVs are defined to locate these MFEPs. Such a multidimensional space is usually sufficient to cover all different mechanistic scenarios for a system of interest. Once MFEPs are obtained, a metric-corrected path CV (Zinovjev et al., 2012, 2013; Zinovjev & Tuñón, 2014) is defined to obtain a PMF using the umbrella sampling technique.

PMFs obtained at lower levels of theory with respect to the QM region of the entire system can be subsequently corrected using some interpolation techniques (Manna & Dybala-Defratyka, 2014; Manna et al., 2015; Ruiz-Pernía, Silla, Tuñón, & Martí, 2006; Ruiz-Pernía, Silla, Tuñón, Martí, & Moliner, 2004). All aforementioned calculations aimed at obtaining PMFs for studied enzymatic reactions were carried out with the AMBER simulation package, f-Dynamo libraries, and its interface with Gaussian 09.

QM/MM methods often have been used over the last decades for KIE calculations for the condensed-phase reactions (Barnes & Williams, 1996; Kanaan et al., 2011; López-Canut, Ruiz-Pernía, Tuñón, Ferrer, & Moliner, 2009). But computation of a full ($3N \times 3N$) Hessian matrix for a large number of atoms in an enzymatic system was not found practical. Therefore, subset ($3N_s \times 3N_s$) Hessians of only the QM region atoms are determined by means of QM/MM calculations in which the QM (N_s) atoms experience the influence of the surrounding atoms of the environment. Williams has (Ruiz-Pernía & Williams, 2012; Williams, 2012) suggested that it is better to include all vibrational modes ($3N_s$) of subset QM (N_s) atoms with respect to its environment in the QM/MM schemes. Therefore, the IPFR of the RS structure over $3N_s$ vibrational modes is given by

$$f_{\mathrm{RS}} = \left[\prod_{j}^{N_s} \frac{m_{j\mathrm{L}}}{m_{j\mathrm{H}}}\right]^{3/2} \prod_{i}^{3N_s} \left[\frac{\sinh(u_{i\mathrm{L}}/2)}{\sinh(u_{i\mathrm{H}}/2)}\right]_{\mathrm{RS}} \tag{14}$$

where m_j is the atomic mass of jth atom. The products of all $3N_s$ of isotopic mass ratios for the isotopically substituted atoms are identical for both RS and TS, and thence, these factors cancel from Eq. (7).

IPFR of TS structure over $3N_s - 1$ vibrational modes is defined as

$$f_{TS} = \left[\prod_{j}^{N_s} \frac{m_{jL}}{m_{jH}}\right]^{3/2} \prod_{i}^{3N_s-1} \left[\frac{\sinh(u_{iL}/2)}{\sinh(u_{iH}/2)}\right]_{TS} \tag{15}$$

Ensemble averaging is essential for KIE calculations for condensed-phase reactions in order to incorporate the influence of the environment fluctuations of the protein and solvent, respectively.

Thus, the average IPFRs of RS and TS configurations are described as

$$\langle f_{RS}\rangle = \frac{1}{M_{RS}} \sum_{i}^{M_{RS}} (f_{RS})_i \tag{16}$$

$$\langle f_{TS}\rangle = \frac{1}{M_{TS}} \sum_{i}^{M_{TS}} (f_{TS})_i \tag{17}$$

Hence, the ensemble average KIE is defined using the following equation:

$$\langle \text{KIE}\rangle = \frac{\langle f_{RS}\rangle}{\langle f_{TS}\rangle} \tag{18}$$

Therefore, the steps for the ensemble-averaged KIE calculations can be as follows: QM/MM free simulations are performed to construct the PMF profile, then QM/MM optimization calculations are carried out to optimize RS and TS structures using the micro–macro iteration scheme along all the possible pathways from the PMFs, the Hessians matrices are computed for a subset of atoms (N_s) from the QM region within the full enzymatic system, and finally Cl-KIEs were calculated using the UJISO program (Ruiz-Pernía & Williams, 2012; Williams, 2012).

Below, we illustrate the use of selected ensemble of TS structures protocols for two dehalogenases from the Lin family of enzymes: LinB and LinA.

3.2.2.1 Dechlorination of β-HCH Catalyzed by LinB

LinB, a haloalkane dehalogenase from *Sphingomonas paucimobilis* UT26, is known to metabolize halohydrocarbons to halide ions and the respective alcohols (Heeb, Zindel, Geueke, Kohler, & Lienemann, 2012;

Koudelakova et al., 2011; Stucki & Thuer, 1995; Swanson, 1999). During the dechlorination reaction of β-HCH, Asp108 of LinB enzyme nucleophilically attacks the C_1 carbon atom of β-HCH, which results in the C_1–Cl_1 bond cleavage and yields an acyl-enzyme ester complex (Fig. 5).

Dechlorination of β-HCH catalyzed by LinB was investigated using hybrid QM/MM methodologies in order to incorporate the effects of the protein environment and its flexibility. Free energy surfaces were calculated using 1D- and 2D-PMF profiles obtained at the PM3/MM level of theory (Manna & Dybala-Defratyka, 2014). The overestimated energetic barriers by the PM3 Hamiltonian were corrected at the M06-2X/6-31+G(d,p) level of theory. The resulting activation energies (16 and 19 kcal mol^{-1} from 1D- and 2D-PMF profiles, respectively) for the dechlorination reaction of β-HCH in the active site of LinB were in very good agreement with the experimentally determined value of 17 kcal mol^{-1} (Brittain et al., 2011). Furthermore, both 1D- and 2D-PMF indicated that the formation of the $O_{\delta1}$–C_1 bond and the cleavage of the C_1–Cl_1 bond occur via a concerted mechanism. In the TS the key interatomic distances (such as $O_{\delta1}$–C_1 and C_1–Cl_1) are quite loose (above 2.2 Å) and a carbocation develops on the central carbon atom. The departing chloride ion is stabilized via hydrogen-bonding interactions with halide-stabilizing residues (mainly Asn38 and Trp109). The predicted ^{37}Cl-KIEs for the dechlorination reaction of β-HCH were calculated over 10 optimized configurations of the RS and the TS structures at the M06-2X/6-31+G(d,p)/OPLS-AA level of theory. The computed ensemble average ^{37}Cl-KIEs are 1.0081 ± 0.0005 and 1.0079 ± 0.0004 from 1D- and 2D-PMF, respectively, at 300 K. The predicted Cl-KIEs for the dechlorination reaction of β-HCH are comparable to the data reported for other dehalogenases catalyzing S_N2 dechlorination reactions (Devi-Kesavan & Gao, 2003; Lewandowicz et al., 2001; Martí et al., 2005).

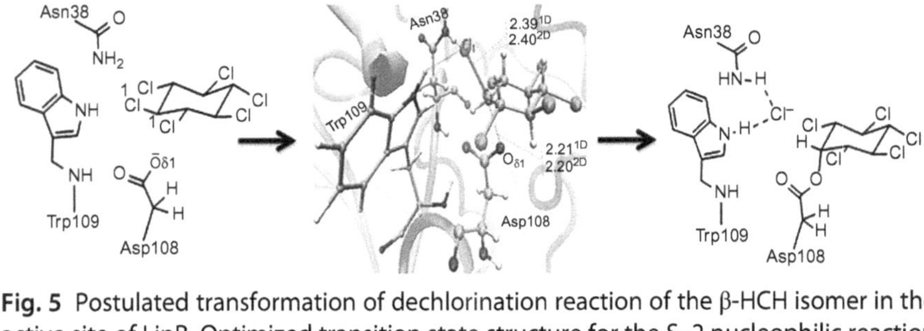

Fig. 5 Postulated transformation of dechlorination reaction of the β-HCH isomer in the active site of LinB. Optimized transition state structure for the S_N2 nucleophilic reaction between β-HCH and Asp108 in the active site of LinB obtained at the M06-2X/6-31+G(d,p)/OPLS-AA level of theory. 1D and 2D represent the data obtained from 1D- and 2D-PMF calculations, respectively.

3.2.2.2 Dehydrochlorination Reaction of γ-HCH by LinA

The key intermediates of the dehydrochlorination reactions of HCH isomers have to be properly understood in order to evaluate the fate of HCH isomers in the environment. Reduced models of the LinA showed that the dehydrochlorination reaction of HCH followed a concerted mechanism in which these two proton ($H_{\delta 1}$, H_1) transfers, the C_2–Cl_2 bond cleavage, and the C_1–C_2 double bond formation occur within a single reaction step (Brittain et al., 2011; Manna & Dybala-Defratyka, 2013). Moreover, the dehydrochlorination reactions of HCHs may involve several chemical steps. However, the sequence in which the transfers of these protons ($H_{\delta 1}$, H_1) occur remained unresolved (Scheme 3). Additionally different findings called for inclusion of the full enzymatic environment in the

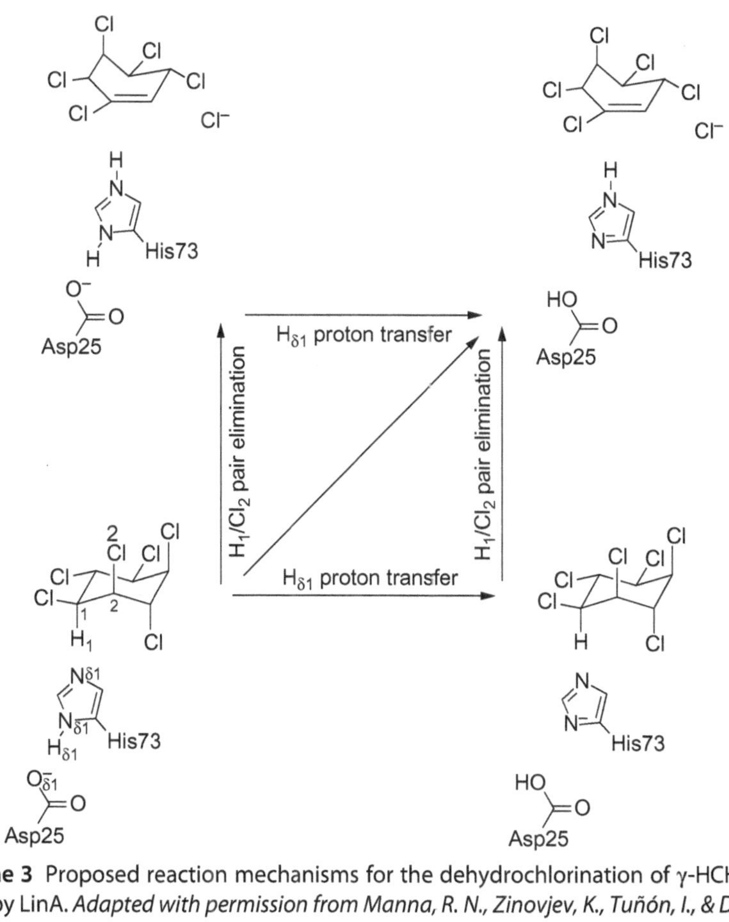

Scheme 3 Proposed reaction mechanisms for the dehydrochlorination of γ-HCH catalyzed by LinA. *Adapted with permission from Manna, R. N., Zinovjev, K., Tuñón, I., & Dybala-Defratyka, A. (2015). Dehydrochlorination of hexachlorocyclohexanes catalyzed by the LinA dehydrohalogenase. A QM/MM study,* Journal of Physical Chemistry B, *119, 15100–15109. Copyright 2015 American Chemical Society.*

constructed models as it was not present in those two previous studies. Taking into account the multidimensionality of the dehydrochlorination reaction of γ-HCH the path CV PMF calculations were performed to calculate a free energy profile along the mean free energy profiles.

The QM/MM free simulation studies provided two distinct minimum free energy pathways (concerted and stepwise) for γ-HCH (Fig. 6; Manna et al., 2015). It turned out that the reaction pathways for the dehydrochlorination reactions of γ-HCH depend on the initial mutual orientations of the HCH substrates in the active site of LinA and the protonation state of His73 residue. The sequential pathway comprised the transfer of the proton between His73 and Asp25, and subsequently the H_1/Cl_2 pair elimination from the HCH molecule. The concerted mechanism of the dehydrochlorination reaction of γ-HCH was initiated with neutral His73. The free energy of activation along path A and path B for the dehydrochlorination reaction of γ-HCH from M05-2X/6-31 + G(d,p) corrected path coordinate PM3/MM PMF were 13.2 and 17.2 kcal mol^{-1}, respectively (Fig. 6), which are in good agreement with the experimental values of 16 kcal mol^{-1} (Brittain et al., 2011). Based on these findings both pathways (concerted and stepwise) seemed to be feasible for the dehydrochlorination reaction of γ-HCH.

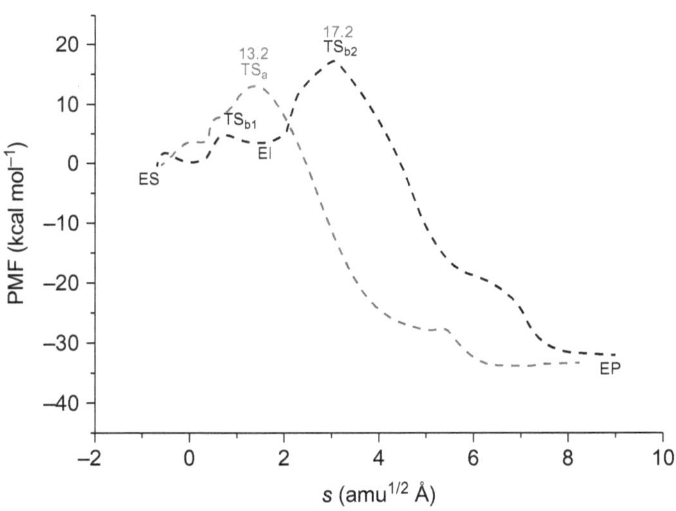

Fig. 6 M05-2X/6-31 + G(d,p) corrected PMFs obtained as a function of path collective variable (*s*) for the dehydrochlorination reactions of γ-HCH isomers. The path collective variable (*s*) measures the advance of the system along the minimum free energy path (MFEP).

Table 3 ^{37}Cl-KIEs Predicted for Dehydrochlorination of γ-HCH by LinA Using QM/MM Ensemble and QM Models of the Reaction

QM/MM		
Path A	**Path B**	**QM Cluster Model**
1.004 ±0.000	1.003 ±0.001	1.003

QM/MM optimization calculations were subsequently carried out to generate Hessian matrices for the QM subset of atoms in each model. Then, ^{37}Cl-KIEs (i.e., at the Cl_2 position) were calculated over 10 optimized structures for each considered pathway for the dehydrochlorination of γ-HCH at 300 K using the UJISO program. The path coordinate PMF revealed that path B follows a stepwise mechanism via two TS structures TS_{b1} and TS_{b2} (Manna et al., 2015; Fig. 6). The H_1/Cl_2 pair elimination step was found to determine the overall rate of the reaction. The TS_{b1} and enzyme intermediate along path B were not located at the M05-2X/6-31+G(d,p)/OPLS-AA level of theory either for γ-HCH. Therefore, the KIEs for path B could be only calculated using the optimized structures of ES and TS_{b2}. The predicted ensemble average ^{37}Cl-KIEs for path A (1.0038 ± 0.0004) for γ-HCH are slightly larger than for path B (1.0029 ± 0.0007). These differences were related to different C_2–Cl_2 (1.96 and 1.92 Å from path A and path B) and C_1–H_1 (1.57 and 1.46 Å from path A and path B) bond stretches in TS_a and TS_{b2}. The concerted pathway described in this section is similar to our findings (Manna & Dybala-Defratyka, 2013) using the cluster model of LinA active site (Section 3.1). The possibility of two reaction mechanisms for the γ-HCH isomer could only have been demonstrated in the presence of the full enzymatic environment in the model. However, they did not result in different isotopic fractionation patterns (Table 3).

4. SUMMARY

Establishing molecular mechanisms of enzyme-catalyzed dehalogenations is of great importance currently, and Cl-KIEs play an important role in their evaluation. However, neither their experimental determination nor theoretical prediction is an easy task. Furthermore, due to their small values and complexity of enzymatic processes, their interpretation is not straightforward. We believe that the best approach is an interplay between theory and experiment, which usually allows valid mechanistic conclusions to be reached. Selecting from a plethora of experimental techniques is frequently a result of available instrumentation. In theoretical studies the shift toward QM/MM methodology from cluster models is observed in recent years and a majority of Cl-KIEs have been evaluated theoretically using this approach. Although the small value of these isotope effects is a constraint, the local character of isotopic perturbation allows nowadays for their reliable predictions. Finally, the mechanistic complexity of a reaction must be carefully considered, especially when interpretation of results obtained with polychlorinated compounds are being considered.

REFERENCES

Adamczyk, P., Dybala-Defratyka, A., & Paneth, P. (2011). DFT study of trichloroethene reaction with permanganate in aqueous solution. *Environmental Science & Technology, 45*, 3006–3011.

Agarwal, V., Miles, Z. D., Winter, J. M., Eustaquio, A. S., El Gamal, A. A., & Moore, B. S. (2017). Enzymatic halogenation and dehalogenation reactions: Pervasive and mechanistically diverse. *Chemical Reviews, 117*, 5619–5674.

Anisimov, V., & Paneth, P. (1999). ISOEFF98. A program for studies of isotope effects using hessian modifications. *Journal of Mathematical Chemistry, 26*, 75–86.

Aranda, J., Zinovjev, K., Roca, M., & Tuñón, I. (2014). Dynamics and reactivity in Thermus aquaticus N6-adenine methyltransferase. *Journal of the American Chemical Society, 136*, 16227–16239.

Bala, K., Geueke, B., Miska, M. E., Rentsch, D., Poiger, T., Dadhwal, M., et al. (2012). Enzymatic conversion of ε-hexachlorocyclohexane and a heptachlorocyclohexane isomer, two neglected components of technical hexachlorocyclohexane. *Environmental Science & Technology, 46*, 4051–4058.

Barducci, A., Bonomi, M., & Parrinello, M. (2011). Metadynamics. *WIREs Computational Molecular Science, 1*, 826–843.

Barnes, J. A., & Williams, I. H. (1996). Theoretical modelling of kinetic isotope effects for glycoside hydrolysis in aqueous solution by a hybrid quantum-mechanical/molecular-mechanical method. *Chemical Communications*, 193–194.

Becke, A. D. (1993). Density-functional thermochemistry. III. The role of exact exchange. *The Journal of Chemical Physics, 98*, 5648–5652.

Bernstein, A., Shouakar-Stash, O., Ebert, K., Laskov, C., Hunkeler, D., Jeannottat, S., et al. (2011). Compound-specific chlorine isotope analysis: A comparison of gas chromatography/isotope ratio mass spectrometry and gas chromatography/quadrupole mass spectrometry methods in an interlaboratory study. *Analytical Chemistry, 83*, 7624–7634.

Blomberg, M. R. A., Borowski, T., Himo, F., Liao, R.-Z., & Siegbahn, P. E. M. (2014). Quantum chemical studies of mechanisms for metalloenzymes. *Chemical Reviews, 114*, 3601–3658.

Branduardi, D., Gervasio, F. L., & Parrinello, M. (2007). From A to B in free energy space. *The Journal of Chemical Physics, 126*, 054103.

Brittain, D. R. B., Pandey, R., Kumari, K., Sharma, P., Pandey, G., Lal, R., et al. (2011). Competing SN2 and E2 reaction pathways for hexachlorocyclohexane degradation in the gas phase, solution and enzymes. *Chemical Communications, 47*, 976–978.

Brothers, E. N., Suarez, D., Deerfield, D. W., & Merz, K. M. J. (2004). PM3-compatible zinc parameters optimized for metalloenzyme active sites. *Computers & Chemistry, 25*, 1677–1692.

Chen, Y. P., Woodin, S. A., Lincoln, D. E., & Lovell, C. R. (1996). An unusual dehalogenating peroxidase from the marine terebellid polychaete *Amphitrite ornata*. *The Journal of Biological Chemistry, 271*, 4609–4612.

Dappricha, S., Komáromia, I., Byuna, K. S., Morokuma, K., & Frisch, M. J. (1999). A new ONIOM implementation in Gaussian98. Part I. The calculation of energies, gradients, vibrational frequencies and electric field derivatives. *Journal of Molecular Structure: THEOCHEM, 461–462*, 1–21.

Dec, J., & Bollag, J.-M. (1994). Dehalogenation of chlorinated phenols during oxidative coupling. *Environmental Science & Technology, 28*, 484–490.

Dec, J., & Bollag, J.-M. (1995). Effect of various factors on dehalogenation of chlorinated phenols and anilines during oxidative coupling. *Environmental Science & Technology, 29*, 657–663.

Devi-Kesavan, L. S., & Gao, J. (2003). Combined QM/MM study of the mechanism and kinetic isotope effect of the nucleophilic substitution reaction in haloalkane dehalogenase. *Journal of the American Chemical Society, 125*, 1532–1540.

Elsner, M., & Hunkeler, D. (2008). Evaluating chlorine isotope effects from isotope ratios and mass spectra of polychlorinated molecules. *Analytical Chemistry, 80*, 4731–4740.

Eurenius, K. P., Chatfield, D. C., Brooks, B. R., & Hodoscek, M. (1996). Enzyme mechanisms with hybrid quantum and molecular mechanical potentials. I. Theoretical considerations. *International Journal of Quantum Chemistry, 60*, 1189–1200.

Fang, Y., Gao, Y., Ryberg, P., Eriksson, J., Kołodziejska-Huben, M., Dybała-Defratyka, A., et al. (2003). Experimental and theoretical multiple kinetic isotope effects for an SN2 reaction. An attempt to determine transition-state structure and the ability of theoretical methods to predict experimental kinetic isotope effects. *Chemistry—A European Journal, 9*, 2696–2709.

Ferrari, R. P., Laurenti, E., & Trotta, F. (1999). Oxidative 4-dechlorination of 2,4,6 trichlorophenol catalyzed by horseradish peroxidase. *Journal of Biological Inorganic Chemistry, 4*, 232–237.

Field, M. J., Albe, M., Bret, C., Proust-De Martin, F., & Thomas, A. (2000). The dynamo library for molecular simulations using hybrid quantum mechanical and molecular mechanical potentials. *Journal of Computational Chemistry, 21*, 1088–1100.

Fisher, H. F., Palfey, B. A., Maniscalco, S. J., & Indyk, L. (2006). Relationship between the time-dependence of a transient-state kinetic isotope effect and the location of complexes in a reaction sequence. *The Journal of Physical Chemistry. A, 110*, 4465–4472.

Frisch, M. J., Trucks, G. W., Schlegel, H. B., Scuseria, G. E., Robb, M. A., Cheeseman, J. R., et al. (2009). *Gaussian 09, revision E.01.* Wallingford, CT: Gaussian, Inc.

Fujitani, T., Yamashita, K., Numata, M., Kanazawa, N., & Nakamura, N. (2010). Measurements of chlorine stable isotopic composition by negative thermal ionization mass spectrometry using total evaporation technique. *Geochemical Journal, 44*, 241–246.

Gao, J., & Xia, X. (1992). A priori evaluation of aqueous polarization effects through Monte Carlo QM-MM simulations. *Science, 258*, 631–635.

Gawlita-Kamzurska, E., & Paneth, P. (1992). Time-dependence of kinetic isotope effects of parallel reactions. *Polish Journal of Chemistry, 66*, 209–213.

Ghysels, A., Van Speybroeck, V., Pauwels, E., Catak, S., Brooks, B. R., Van Neck, D., et al. (2010). Comparative study of various normal mode analysis techniques based on partial Hessians. *Journal of Computational Chemistry, 31*, 994–1007 [and references cited therein].

Glad, S. S., & Jensen, F. (1996). Basis set and correlation effects on transition state geometries and kinetic isotope effects. *The Journal of Physical Chemistry, 100*, 16892–16898.

Grossfield, A. (2011). *WHAM: The weighted histogram analysis method, version 2.0.5.*

Grzybkowska, A., Kaminski, R., & Dybala-Defratyka, A. (2014). Theoretical predictions of isotope effects versus their experimental values on an example of uncatalyzed hydrolysis of atrazine. *Physical Chemistry Chemical Physics, 16*, 15164–15172.

Halas, S., & Pelc, A. (2009). New isotope ratio mass spectrometric method of precise δ^{37}Cl determinations. *Rapid Communications in Mass Spectrometry, 23*, 1061–1064.

Hay, P. J., & Wadt, W. R. (1985). Ab initio effective core potentials for molecular calculations. Potentials for K to Au including the outermost core orbitals. *The Journal of Chemical Physics, 82*, 299–310.

Heeb, N. V., Zindel, D., Geueke, B., Kohler, H.-P. E., & Lienemann, P. (2012). Biotransformation of hexabromocyclododecanes (HBCDs) with LinB-An HCH-converting bacterial enzyme. *Environmental Science & Technology, 46*, 6566–6574.

Hill, J. W., & Fry, A. (1962). Chlorine isotope effects in the reactions of benzyl and substituted benzyl chlorides with various nucleophiles. *Journal of the American Chemical Society, 84*, 2763–2769.

Himo, F., & Siegbahn, P. E. M. (2003). Quantum chemical studies of radical-containing enzymes. *Chemical Reviews, 103*, 2421–2456.

Hratchian, H. P., & Schlegel, H. B. (2005). Using hessian updating to increase the efficiency of a hessian based predictor-corrector reaction path following method. *Journal of Chemical Theory and Computation, 1*, 61–69.

Hunkeler, D., Vanbreukelen, B., & Elsner, M. (2009). Modeling chlorine isotope trends during sequential transformation of chlorinated ethenes. *Environmental Science & Technology, 43*, 6750–6756.

Jin, B., Haderlein, S. B., & Rolle, M. (2013). Integrated carbon and chlorine isotope modeling: Applications to chlorinated aliphatic hydrocarbons dechlorination. *Environmental Science & Technology, 47*, 1443–1451.

Jorgensen, W. L., Maxwell, D. S., Tirado-Rives, J., & Turner, A. (1996). Development and testing of the OPLS all-atom force field on conformational energetics and properties of organic liquids. *Journal of the American Chemical Society, 118*, 11225–11236.

Kaminski, G. A., Friesner, R. A., Tirado-Rives, J., & Jorgensen, W. L. (2001). Evaluation and reparametrization of the OPLS-AA force field for proteins via comparison with accurate quantum chemical calculations on peptides. *The Journal of Physical Chemistry. B, 105*, 6474–6487.

Kanaan, N., Ferrer, S., Martí, S., Garcia-Viloca, M., Kohen, A., & Moliner, V. (2011). Temperature dependence of the kinetic isotope effects in thymidylate synthase. A theoretical study. *Journal of the American Chemical Society, 133*, 6692–6702.

Kaufmann, R. S., Long, A., Bentley, H., & Davis, S. (1984). Natural chlorine isotope variations. *Nature, 309*, 338–340.

Klibanov, A. M., Alberti, B. N., Morris, E. D., & Felshin, L. M. (1980). Enzymatic removal of toxic phenols and anilines from waste waters. *Journal of Applied Biochemistry, 2*, 414–421.

Koudelakova, T., Chovancova, E., Brezovsky, J., Monincova, M., Fortova, A., Jarkovsky, J., et al. (2011). Substrate specificity of haloalkane dehalogenases. *The Biochemical Journal, 435*, 345–354.

Kumar, S., Bouzida, D., Swendsen, R. H., Kollman, P. A., & Rosenberg, J. M. (1992). The weighted histogram analysis method for free-energy calculations on biomolecules. I. The method. *Journal of Computational Chemistry, 13*, 1011–1021.

Kumar, S., Rosenberg, J. M., Bouzida, D., Swendsen, R. H., & Kollman, P. A. (1995). Multidimensional free-energy calculations using the weighted histogram analysis method. *Journal of Computational Chemistry, 16*, 1339–1350.

Lal, R., Pandey, G., Sharma, P., Kumari, K., Malhotra, S., Pandey, R., et al. (2010). Biochemistry of microbial degradation of hexachlorocyclohexane and prospects for bioremediation. *Microbiology and Molecular Biology Reviews, 74*, 58–80.

Lee, C., Yang, W., & Parr, R. G. (1988). Development of the Colle-Salvetti correlation-energy formula into a functional of the electron density. *Physical Review B, 37*, 785–789.

Lewandowicz, A., Rudziński, J., Luo, L., Dunaway-Mariano, D., & Paneth, P. (2002). Determination of the chlorine kinetic isotope effect on the 4-chlorobenoyl-CoA dehalogenase-catalyzed nucleophilic aromatic substitution. *Archives of Biochemistry and Biophysics, 398*, 249–252.

Lewandowicz, A., Rudziński, J., Tronstad, L., Widersten, M., Ryberg, P., Matsson, O., et al. (2001). Chlorine kinetic isotope effects on the haloalkane dehalogenase reaction. *Journal of the American Chemical Society, 123*, 4550–4555.

Lonsdale, R., Harvey, J. N., & Mulholland, A. J. (2012). A practical guide to modelling enzyme-catalysed reactions. *Chemical Society Reviews*, *41*, 3025–3038.

López-Canut, V., Ruiz-Pernía, J. J., Tuñón, I., Ferrer, S., & Moliner, V. (2009). Theoretical modeling on the reaction mechanism of *p*-nitrophenylmethylphosphate alkaline hydrolysis and its kinetic isotope effects. *Journal of Chemical Theory and Computation*, *5*, 439–442.

Manna, R. N., & Dybala-Defratyka, A. (2013). Insights into the elimination mechanisms employed for the degradation of different hexachlorocyclohexane isomers using kinetic isotope effects and docking studies. *Journal of Physical Organic Chemistry*, *26*, 797–804.

Manna, R. N., & Dybala-Defratyka, A. (2014). A computational study of the dechlorination of β-hexachlorocyclohexane (β-HCH) catalyzed by the haloalkane dehalogenase LinB. *Archives of Biochemistry and Biophysics*, *562*, 43–50.

Manna, R. N., Zinovjev, K., Tuñón, I., & Dybala-Defratyka, A. (2015). Dehydrochlorination of hexachlorocyclohexanes catalyzed by the LinA dehydrohalogenase. A QM/MM study. *The Journal of Physical Chemistry. B*, *119*, 15100–15109.

Manta, B., Raushel, F. M., & Himo, F. (2014). Reaction mechanism of zinc-dependent cytosine deaminase from Escherichia coli: A quantum-chemical study. *The Journal of Physical Chemistry. B*, *118*, 5644–5652.

Maragliano, L., Fischer, A., Vanden-Eijnden, E., & Ciccotti, G. (2006). String method in collective variables: Minimum free energy paths and isocommittor surfaces. *The Journal of Chemical Physics*, *125*, 024106.

Maragliano, L., & Vanden-Eijnden, E. (2007). On-the-fly string method for minimum free energy paths calculation. *Chemical Physics Letters*, *446*, 182–190.

Martí, S., Moliner, V., & Tuñón, I. (2005). Improving the QM/MM description of chemical processes: A dual level strategy to explore the potential energy surface in very large systems. *Journal of Chemical Theory and Computation*, *1*, 1008–1016.

Mazurkiewicz-Łazowska, H., & Paneth, P. (1991). Numerical evaluation of time-dependence of concentrations, rates and kinetic isotope effects. *Computers & Chemistry*, *15*, 347–349.

Melander, L. (1960). *Isotope effect on reaction rates*. New York: The Ronald Press.

Melander, L., & Saunders, W. H. (1982). *Reaction rates of isotopic molecules*. New York: John Wiley & Sons.

Meyer, A. H., Penning, H., & Elsner, M. (2009). C and N isotope fractionation suggests similar mechanisms of microbial atrazine transformation despite involvement of different enzymes (AtzA and TrzN). *Environmental Science & Technology*, *43*, 8079–8085.

Miertus, S., Scrocco, E., & Tomasi, J. (1981). Electrostatic interaction of a solute with a continuum. A direct utilization of ab initio molecular potentials for the prevision of solvent effects. *Chemical Physics*, *55*, 117–129.

Nicell, J. A. (2001). Environmental applications of enzymes. *Interdisciplinary Environmental Review*, *3*, 14–41.

Nijenhuis, I., & Richnow, H. H. (2016). Stable isotope fractionation concepts for characterizing biotransformation of organohalides. *Current Opinion in Biotechnology*, *41*, 108–113.

Noodleman, L., Lovell, T., Han, W.-G., Li, J., & Himo, F. (2004). Quantum chemical studies of intermediates and reaction pathways in selected enzymes and catalytic synthetic systems. *Chemical Reviews*, *104*, 459–508.

Numata, M., Nakamura, N., & Gamo, T. (2001). Precise measurement of chlorine stable isotopic ratios by thermal ionization mass spectrometry. *Geochemical Journal*, *35*, 89–100.

Oakley, A. J., Klvana, M., Otyepka, M., Nagata, Y., Wilce, M. C. J., & Damborsky, J. (2004). Crystal structure of haloalkane dehalogenase LinB from Sphingomonas paucimobilis UT26 at 0.95 Å resolution: Dynamics of catalytic residues. *Biochemistry*, *43*, 870–878.

Okai, M., Kubota, K., Fukuda, M., Nagata, Y., Nagata, K., & Tanokura, M. (2010). Crystal structure of γ-hexachlorocyclohexane dehydrochlorinase LinA from Sphingobium japonicum UT26. *Journal of Molecular Biology*, *403*, 260–269.

Paneth, P. (1985). On the application of the steady state to kinetic isotope effects. *Journal of the American Chemical Society*, *107*, 7070–7071.

Paneth, P. (2003). Chlorine kinetic isotope effects on enzymatic dehalogenations. *Accounts of Chemical Research*, *36*, 120–126.

Paneth, P. (2006). Chlorine isotope effects in biological systems. In A. Kohen & H. H. Limbach (Eds.), *Isotope effects in chemistry and biology* (pp. 875–891). Baton Rouge: CRC Press. Chapter 35.

Papajak, E., Kwiecień, R. A., Kamiński, R., Sicińska, D., Rudziński, J., Szatkowski, Ł., et al. (2006). Mechanism of DL-2-haloacid dehalogenase reactions from kinetic isotope effects studies. *Biochemistry*, *45*, 6012–6017.

Philipp, D. M., & Friesner, R. A. (1999). Mixed ab initio QM/MM modeling using frozen orbitals and tests with alanine dipeptide and tetrapeptide. *Journal of Computational Chemistry*, *20*, 1468–1494.

QSite, version 5.0, Schrödinger, LLC, New York, 2007.

Ruiz-Pernía, J. J., Silla, E., Tuñón, I., & Martí, S. (2006). Hybrid quantum mechanics/molecular mechanics simulations with two-dimensional interpolated corrections: Application to enzymatic processes. *The Journal of Physical Chemistry. B*, *110*, 17663–17670.

Ruiz-Pernía, J. J., Silla, E., Tuñón, I., Martí, S., & Moliner, V. (2004). Hybrid QM/MM potentials of mean force with interpolated corrections. *The Journal of Physical Chemistry. B*, *108*, 8427–8433.

Ruiz-Pernía, J. J., & Williams, I. H. (2012). Ensemble-averaged QM/MM kinetic isotope effects for the S_N2 reaction of cyanide anions with chloroethane in DMSO solution. *Chemistry—A European Journal*, *18*, 9405–9414 [and references cited therein].

Sanchez-Martinez, M., Field, M., & Crehuet, R. (2015). Characterization of a dynamic string method for the construction of transition pathways in molecular reactions. *The Journal of Physical Chemistry. B*, *119*, 1103–1113.

Schürner, H. K. V., Seffernick, J. L., Grzybkowska, A., Dybala-Defratyka, A., Wackett, L. P., & Elsner, M. (2015). Characteristic isotope fractionation patterns in s-triazine degradation have their origin in multiple protonation options in the s-triazine hydrolase TrzN. *Environmental Science & Technology*, *49*, 3490–3498.

Shouakar-Stash, O., Drimmie, R. J., & Frape, S. K. (2005). Determination of inorganic chlorine stable isotopes by continuous flow isotope ratio mass spectrometry. *Rapid Communications in Mass Spectrometry*, *19*, 121–127.

Sicińska, D., Truhlar, D. G., & Paneth, P. (2001). Solvent-dependent transition states for decarboxylations. *Journal of the American Chemical Society*, *123*, 7683–7686.

Siwek, A., Omi, R., Hirotsu, K., Jitsumori, K., Esaki, N., Kurihara, T., et al. (2013). Modeling of binding modes and chlorine isotope effects on DL-2-haloacid dehalogenase-catalyzed reactions. *Archives of Biochemistry and Biophysics*, *540*, 26–32.

Stucki, G., & Thuer, M. (1995). Experiences of a large-scale application of 1,2-dichloroethane degrading microorganisms for groundwater treatment. *Environmental Science & Technology*, *29*, 2339–2345.

Sugita, Y., Kitao, A., & Okamoto, Y. (2000). Multidimensional replica-exchange method for free-energy calculations. *Journal of Computational Chemistry*, *113*, 6042–6051.

Swanson, P. (1999). Dehalogenases applied to industrial-scale biocatalysis. *Current Opinion in Biotechnology*, *10*, 365–369.

Świderek, K., & Paneth, P. (2012). Extending limits of chlorine kinetic isotope effects. *The Journal of Organic Chemistry*, 77, 5120–5124.

Szatkowski, L., & Dybala-Defratyka, A. (2013). A computational study on enzymatically driven oxidative coupling of chlorophenols: An indirect dehalogenation reaction. *Chemosphere*, *91*, 258–264.

Szatkowski, L., Thompson, M. K., Kaminski, R., Franzen, S., & Dybala-Defratyka, A. (2011). Oxidative dechlorination of halogenated phenols catalyzed by two distinct enzymes: Horseradish peroxidase and dehaloperoxidase. *Archives of Biochemistry and Biophysics, 505*, 22–32.

Taylor, J. W., & Grimsrud, E. P. (1969). Chlorine isotopic ratios by negative ion mass spectrometry. *Analytical Chemistry, 41*, 805–810.

Turner, A., Moliner, V., & Williams, I. H. (1999). Transition-state structural refinement with GRACE and CHARMM: Flexible QM/MM modelling for lactate dehydrogenase. *Physical Chemistry Chemical Physics, 1*, 1323–1331.

Van Acker, M. R. M. D., Shahar, A., Young, E. D., & Coleman, M. L. (2006). GC/multiple collector ICPMS method for chlorine stable isotope analysis of chlorinated aliphatic hydrocarbons. *Analytical Chemistry, 78*, 4663–4667.

Vengosh, A., Chivas, A. R., & McCulloch, M. T. (1989). Direct determination of boron and chlorine isotopic compositions in geological materials by negative thermal-ionization mass spectrometry. *Chemical Geology: Isotope Geoscience, 79*, 333–343.

Verschueren, K. H. G., Seljee, F., Rozeboom, H. J., Kalk, K. H., & Dijkstra, B. W. (1993). Crystallographic analysis of the catalytic mechanism of haloalkane dehalogenase. *Nature, 363*, 693–698.

Warshel, A. (2003). Computer simulations of enzyme catalysis: Methods, progress, and insights. *Annual Review of Biophysics and Biomolecular Structure, 32*, 425–443.

Westaway, K. C., Koerner, T., Fang, Y.-R., Rudziński, J., & Paneth, P. (1998). A new method of determining chlorine kinetic isotope effects. *Analytical Chemistry, 70*, 3548–3552.

Wiese, F. W., Chang, H. C., Lloyd, R. V., Freeman, J. P., & Samokyszyn, V. M. (1997). Peroxidase-catalyzed oxidation of 2,4,6-trichlorophenol. *Archives of Environmental Contamination and Toxicology, 34*, 217–222.

Willett, K. L., Ulrich, E. N., & Hites, R. A. (1998). Differential toxicity and environmental fates of hexachlorocyclohexane isomers. *Environmental Science & Technology, 32*, 2197–2207.

Williams, I. H. (2012). Kinetic isotope effects from QM/MM subset hessians: "Cut-off" analysis for SN2 methyl transfer in solution. *Journal of Chemical Theory and Computation, 8*, 542–553.

Wolfsberg, M., Van Hook, W. A., & Paneth, P. (2010). *Isotope effects in the chemical, geological and bio sciences*. London: Springer.

Xiao, Y. K., & Zhang, C. G. (1992). High precision isotopic measurement of chlorine by thermal ionization mass spectrometry of Cs_2Cl^+ ion. *International Journal of Mass Spectrometry and Ion Processes, 116*, 183–192.

Zhang, N., Bashir, S., Qin, J., Schindelka, J., Fischer, A., Nijenhuis, I., et al. (2014). Compound specific stable isotope analysis (CSIA) to characterize transformation mechanisms of α-hexachlorocyclohexane. *Journal of Hazardous Materials, 280*, 750–757.

Zhang, S., Ganguly, A., Goyal, P., Bingaman, J. L., Bevilacqua, P. C., & Hammes-Schiffer, S. (2015). Role of the active site guanine in the *glmS* ribozyme self-cleavage mechanism: Quantum mechanical/molecular mechanical free energy simulations. *Journal of the American Chemical Society, 137*, 784–798.

Zinovjev, K., Martí, S., & Tuñón, I. (2012). A collective coordinate to obtain free energy profiles for complex reactions in condensed phases. *Journal of Chemical Theory and Computation, 8*, 1795–1801.

Zinovjev, K., Ruiz-Pernía, J. J., & Tuñón, I. (2013). Toward an automatic determination of enzymatic reaction mechanisms and their activation free energies. *Journal of Chemical Theory and Computation, 9*, 3740–3749.

Zinovjev, K., & Tuñón, I. (2014). Exploring chemical reactivity of complex systems with path-based coordinates: Role of the distance metric. *Journal of Computational Chemistry, 35*, 1672–1681.

CHAPTER NINE

Kinetic Deuterium Isotope Effects in Cytochrome P450 Reactions

Frederick P. Guengerich[1]
Vanderbilt University School of Medicine, Nashville, TN, United States
[1]Corresponding author: e-mail address: f.guengerich@vanderbilt.edu

Contents

1. Introduction and History of Cytochrome P450 Kinetic Isotope Effects 218
2. Types of Experiments 219
   2.1 Noncompetitive Intramolecular Experiment 221
   2.2 Competitive Intermolecular Experiment 222
   2.3 Competitive Intramolecular Experiment 222
   2.4 Noncompetitive Intermolecular Experiment 223
   2.5 Estimation of the Intrinsic KIE 223
3. Choice of Enzymatic Parameters 225
4. KIEs for Various P450 Reactions 226
5. Examples of P450 KIE Studies 228
6. In Vivo KIEs With P450s 231
7. Consideration of P450 KIEs in Drug Development 232
8. Summary and Conclusions 233
Acknowledgment 234
References 234

Abstract

Cytochrome P450 (P450, CYP) research provides many opportunities for the application of kinetic isotope effect (KIE) strategies. P450s collectively catalyze oxidations of more substrates than any other group of enzymes, and C—H bond cleavage is a major feature in a large fraction of these reactions. The presence of a significant primary deuterium KIE is evidence that hydrogen abstraction is at least partially rate-limiting in the reactions, and this appears to be the case in many P450 reactions. The first report of a KIE in (P450-linked) drug metabolism appeared in 1961 (for morphine *N*-demethylation), and in a number of cases, it has been possible to modulate the in vivo metabolism or toxicity of chemicals by deuterium substitution. A number of efforts are in progress to utilize deuterium substitution to alter the metabolism of drugs in an advantageous manner.

Methods in Enzymology, Volume 596
ISSN 0076-6879
http://dx.doi.org/10.1016/bs.mie.2017.06.036

1. INTRODUCTION AND HISTORY OF CYTOCHROME P450 KINETIC ISOTOPE EFFECTS

The concept and theory of the kinetic isotope effect (KIE) are long-established and treated elsewhere in this monograph. This chapter will be restricted to deuterium KIEs on catalysis by cytochrome P450 (P450, CYP) enzymes. In the general mechanism of P450 catalysis (Fig. 1), there is a single C—H bond-breaking step that could be affected by isotopic substitution, and all KIEs must be interpreted in this context. (D_2O KIEs have been reported for P450 reactions, but the interpretation of these effects is complicated and their meaning is less well understood (Fersht, 1999), and they will not be considered because of space limitations.)

P450 reactions are well suited to the application of KIE approaches, in that a large fraction of the reactions catalyzed by these enzymes involve the breaking of C—H bonds (Guengerich, 2001; Guengerich &

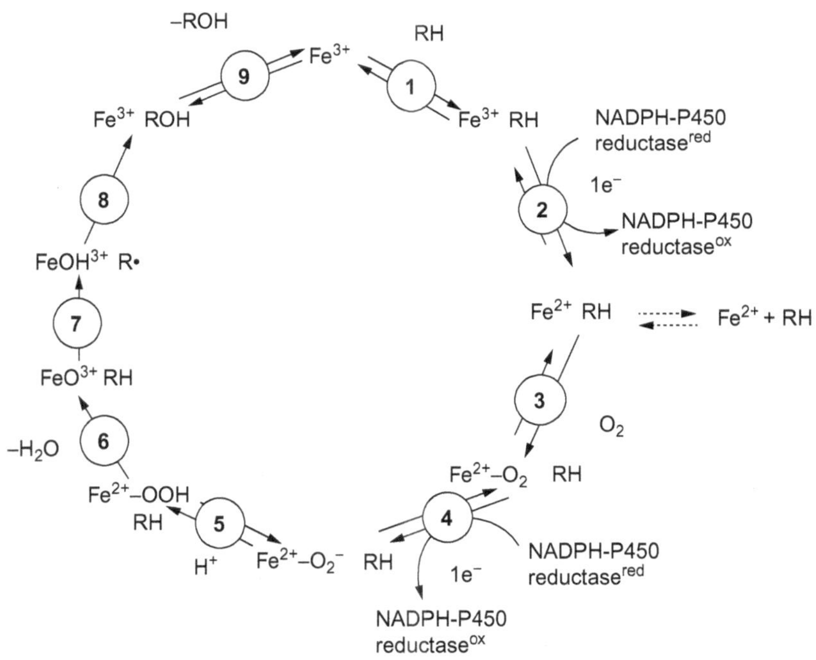

Fig. 1 Catalytic cycle of microsomal P450s. RH is the substrate, with a C—H bond to be broken. With the seven mitochondrial P450s and many bacterial P450s, the electrons are delivered from a ferredoxin (including adrenodoxin) instead of the diflavin NADPH-P450 reductase. In some cases, cytochrome b_5 (b_5) delivers an electron in step 4, but not always. Note that the only step in which C—H bond breaking occurs in step 7.

Macdonald, 1984; Ortiz de Montellano, 2015). Thus, the ability to determine the extent to which this step is rate-limiting in a complex cycle is very useful, and therefore many KIE studies have been done. The first report on a P450 KIE appears to be that of Elison, Rapoport, Laursen, and Elliott (1961), who showed that substitution of deuterium in an *N*-methyl group of morphine slowed its metabolism and increased its potency in rats.

The interpretation of P450 KIE studies is, in one sense, simple. If a deuterium KIE is observed, the rate of C—H bond breaking has affected the overall rate of catalysis. However, enzyme mechanisms are complex and exactly how such a change is manifested is a function of the experimental design.

2. TYPES OF EXPERIMENTS

One of the most common P450 reactions is the conversion of a methyl, methylene, or methine group to an alcohol. Other common C—H bond-breaking reactions are heteroatom cleavages (e.g., N, O, halogen), where a formal hydroxylation is followed by a breakdown of a carbinolamine, hemiacetal, halohydrins, etc.

There are myriad P450 reactions (Rendic & Guengerich, 2012, 2015). Most of them can be readily analyzed by modern chromatography methods, i.e., HPLC (or UPLC) and GC. With both methods, mass spectrometry (electrospray for charged molecules, atmospheric pressure chemical ionization for neutral ones) generally provides the best sensitivity. With HPLC/UPLC, UV–visible and fluorescence methods also provide excellent sensitivity in many cases. Mass spectrometry, done in a full-scan mode, has the advantages of (i) verifying the identity of products and (ii) screening out overlap between molecules that are not well resolved by chromatography. Internal standards are useful for quantitation and, in the case of GC, critical with split injection methods.

An extensive discussion of synthetic methods for generating substrate isotopologues is beyond the scope of this chapter. When *N*- and *O*-alkyl derivatives (which are often P450 substrates) are under consideration, the synthesis is generally straightforward with Gabriel methods (i.e., the reaction of amines or alcohols with alkyl halides) or (for *N*-alkylamines) hydride reduction of amides prepared with coupling an amine with an acyl halide or alkylchlorformate (e.g., see Miwa, Walsh, Kedderis, & Hollenberg, 1983; Okazaki & Guengerich, 1993). The synthesis of deuterated (or tritiated) derivatives of some hydrocarbon chains can be accomplished by treatment of halides with $NaBH_4$ (Bell, Vanderslice, & Spehar, 1969; Kim, Cha,

Nagy, Yun, & Guengerich, 2014). Other molecules can be much more problematic, depending on the precursor molecules available (Gelb, Heimbrook, Mälkönen, & Sligar, 1982; Krauser & Guengerich, 2005; Yoshimoto & Guengerich, 2014).

The use of multiply deuterated substrates is sometimes involved. For instance, the oxidation of —CD_3 in a molecule is common for P450 substrates. In some cases, a perdeuterated substrate may be commercially available, e.g., medium- and long-chain fatty acids. These compounds can be problematic: (i) there is potential for *vicinal* secondary KIEs and (ii) interpretations of any metabolic switching may be more complex. However, synthesis of deuterated materials is not trivial, and the placement of deuterium may often be a matter of practicality.

An issue to consider in any experimental design (and synthesis) is prochirality. This is often an issue in labeling methylene groups, in that only one of the two hydrogens may be abstracted by a P450 (Krauser & Guengerich, 2005; White, Miller, Favreau, & Bhattacharyya, 1986).

Finally, atomic purity is a serious issue. The experiments discussed under Sections 2.1–2.3 are not very sensitive to impurities because both the "protium" and "deuterium" reactions occur with the same enzyme. The noncompetitive intermolecular experiment (Section 2.4) is, however, very sensitive to impurities. If the deuterated substrate contains an enzyme inhibitor, the apparent $^{D}(V/K)$ value will be too high. (*Note*: the conventions of Northrop, 1982 regarding ^{D}V, $^{D}(V/K)$, ^{D}k, etc., are used in this chapter.) If the deuterated substrate has low atomic enrichment, then (depending on some aspects of the reaction) the reaction may preferentially occur with the protiated material and yield an apparent $^{D}(V/K)$ that is too low (an experiment intended to be of the 2.4 type (Fig. 2D) may partially become an experiment of type 2.2 (Fig. 1B)). For these reasons, deuterated substrates should have as high an atomic excess as possible (preferably >98%), as analyzed by ^{1}H-NMR (in our experience this is a more accurate method than mass spectrometry).

In that several of the experiments involve only analysis by mass spectrometry (Sections 2.1–2.3; Fig. 2A–C), some caution is in order. A critical problem is the ^{13}C spillover (1.1% × number of carbons), which can be quite problematic. In principle these can be corrected using formulae (Biemann, 1962), although a more accurate approach is to utilize high-resolution mass spectrometry. At a resolution of 60,000–100,000, the difference between M + 1 ions due to ^{13}C and ^{2}H is readily distinguished (Yoshimoto & Guengerich, 2014).

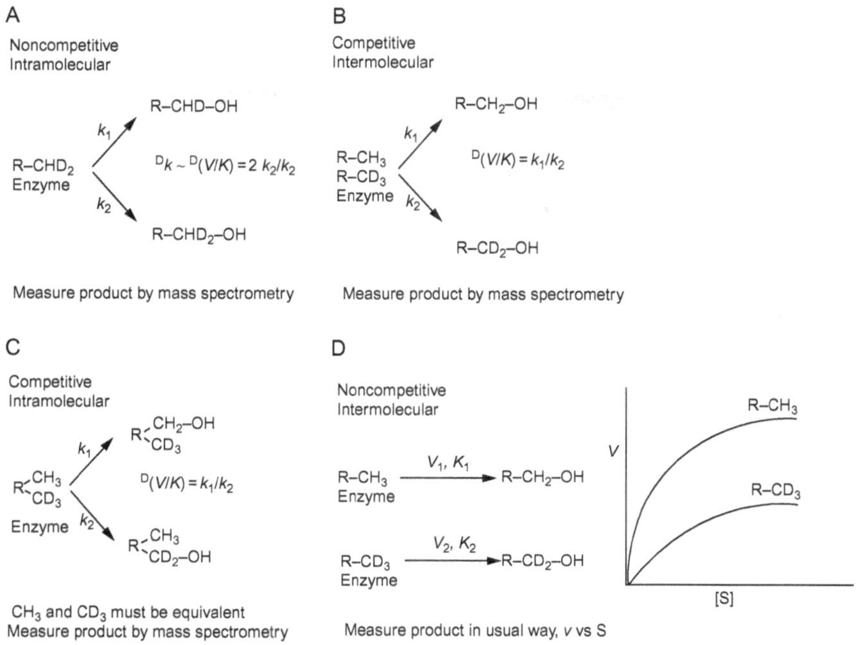

Fig. 2 Types of deuterium KIE studies used in P450 research. (A) Noncompetitive intramolecular. The factor of 2 is a statistical correction, in that there are two deuteriums and one protium available here. The factor would be ½ if the substrate were —CH_2D. (B) Competitive intermolecular. (C) Competitive intramolecular. (D) Noncompet tive intermolecular. In the "k_1" pathways, a C—H bond is broken; in "k_2" pathways a C—D bond is broken.

What a KIE means will depend on the type of experiment used to generate the data. Following are five approaches used with P450 experiments (Fig. 2).

2.1 Noncompetitive Intramolecular Experiment

The noncompetitive intramolecular experiment (Fig. 2A) works well with hydroxylations of methyl groups by P450 enzymes, which is a fairly ccmmon reaction that generates a primary alcohol. The reaction is noncompetitive because a single isotopologue is being analyzed, and the deuterium atoms to be removed are equivalent with each other. The ccmparison of C—H bonds being broken is intramolecular. The KIE should be independent of issues of substrate concentration and only a single (mass spectrometry) measurement is required (preferably replicates).

In principle, the KIE in this approach is an estimate of the intrinsic KIE (designated ^{D}k), i.e., the KIE for the actual C—H bond-breaking step, as in a purely chemical reaction. There are three caveats about a KIE measured using this approach:

(a) The method works well with methyl groups but probably will not with a *pro*-chiral methylene, in that only one of the two hydrogen atoms may be abstracted (Krauser & Guengerich, 2005).

(b) The KIE may not be the ^{D}k, due to possible contribution of a secondary KIE, coming from either the *geminal* or *vicinal* deuterium substitution (which can be up to 1.4 (Fersht, 1999)), and these are multiplicative (Matsson & Westaway, 1998).

(c) A KIE measured by this approach provides no information about how rate-limiting the C—H bond-breaking step is. The competition is between a protium and a deuterium (or tritium) atom at the same carbon, but no insight into which step in Fig. 1 limits the overall rate is provided.

2.2 Competitive Intermolecular Experiment

In a competitive intermolecular experiment, there is competition between two kinds of molecules, the protiated- and the deuterium-labeled (Fig. 2B). In this reaction two different substrate isotopologues are used that compete for enzyme binding and the reaction is considered intermolecular relative to Section 2.1. The experiment shown in Fig. 2B has two methyl hydroxylations, but in principle the comparison can be done at any position on the substrate molecule. The results, as in Section 2.1, can easily be determined by mass spectrometry with assays done at a single-substrate concentration to provide an estimate of $^{D}(V/K)$.

As in Section 2.1, the resulting KIE is not a direct reflection of the rate-limiting nature of the C—H bond-breaking step. A high value, based on this argument, would indicate that the chemical step is at least partially rate-limiting and also that the enzyme has fast exchange of substrates. If this value is low compared to ^{D}k, the intrinsic KIE, this is evidence that the substrate is "locked in" the active site (low k_{off} rate for substrate) and there is "high commitment to catalysis" (Northrop, 1982).

2.3 Competitive Intramolecular Experiment

This is a rather specialized experiment and can only be done with substrates that have equivalent moieties within them, e.g., *N*,*N*-dimethyl groups or methyl groups on alkanes (*n*-octane). The experiments cannot be done if *pro*-chirality is an issue. As in Sections 2.1–2.3, the analysis is done using mass spectrometry and yields an estimate of $^{D}(V/K)$.

The results are a measure of the ability of the substrate to rearrange (tumble) in the active site. A value much lower than the ^{D}k for the intermolecular reaction can be interpreted in this context. However, the results can also be a reflection of the substrate k_{off} (release and rebinding), so this behavior may not be discernable in the absence of other information such as an isotope dilution experiment using a product or nonreactive substrate trap (Gonzalez & Guengerich, 2017).

2.4 Noncompetitive Intermolecular Experiment

This experiment requires more effort than any of the three already described (Sections 2.1–2.3), in that assays must be done at multiple substrate concentrations with both protiated and deuterated substrates. The experiments yield both ^{D}V and $^{D}(V/K)$, and noncompetitive experiments are the *only* method to determine ^{D}V. This type of experiment is also subject to more error because of the multiplicity of measurements and the propagation of error in quotients because they are ratios, which multiply error in the numerator and divisor. Finally, those experiments are subject to two problems with impurities:

(a) If the deuterated substrate has any inhibitory contaminant, the enzyme activity will be reduced and the observed KIE will be artificially high.

(b) If the deuterated substrate is contaminated with protiated substrate, then (if substrate exchange is rapid) a high KIE may be attenuated because the enzyme can "pick" the protiated substrate out of the deuterated pool and use it.

Despite these inherent problems, the approach is very powerful and provides information to address the question of how rate-limiting the C—H bond-breaking step is. If the KIE (^{D}V or $^{D}(V/K)$) is >1, then the C—H bond-breaking step is at least partially rate-limiting. If ^{D}V or $^{D}(V/K)$ approaches the ^{D}k, then the C—H bond-breaking step is very rate-limiting.

2.5 Estimation of the Intrinsic KIE

As already mentioned, ^{D}k is the estimate of the KIE for the C—H bond-breaking step, as in a strictly chemical reaction. There are at least three approaches that have been used, with two used in P450 systems, to determine the extent to which the observed KIE reflects the intrinsic KIE on the chemical step.

(a) Northrop's Method. This approach, described in more detail elsewhere (Northrop, 1975, 1977, 1982), involves comparison of the deuterium and tritium *V*/*K* results.

$$\frac{\dfrac{{}^{\mathrm{D}}(V/K)-1}{{}^{\mathrm{T}}(V/K)-1}={}^{\mathrm{D}}k-1}{{}^{\mathrm{T}}k-1}$$

$$y=\frac{(k_{\mathrm{H}}/k_{\mathrm{D}})-1}{(k_{\mathrm{H}}/k_{\mathrm{T}})-1}$$

The procedure used in Section 2.4 will not generate ${}^{\mathrm{T}}(V/K)$ because complete tritium substitution is impractical and one of the other experiments is necessary. The value y can be estimated from tables (Northrop, 1977). This approach has been used for some P450 reactions (Krauser & Guengerich, 2005; Miwa, Walsh, & Lu, 1984).

(b) The "multiple isotope effect" method utilizes a combination of ^{2}H and ^{13}C isotope effects (Cook & Cleland, 2007). Only a few ^{13}C KIEs have been reported with P450 reactions, and apparently this approach to determining ${}^{\mathrm{D}}k$ has not been applied.

(c) The "isotopically sensitive branching" method. This was developed by Jones, Korzekwa, Rettie, and Trager (1986) and Jones, Rettie, and Trager (1990) and can be applied when a reaction is irreversible, and a minor product is also formed. In the work of Jones et al. (1986, 1990) the method is based on ${}^{\mathrm{D}}V$ results (actually only single-substrate concentration results were used in the original papers). Thus, using the Northrop nomenclature, where the major product (P) comes from deuterated substrate but the minor product does not,

$${}^{D}V\text{observed}=\frac{\left[{}^{D}k+\left(\dfrac{[\mathrm{P}_{\mathrm{major}}]}{[\mathrm{P}_{\mathrm{minor}}]}\right)\right]}{1+\left(\dfrac{[\mathrm{P}_{\mathrm{major}}]}{[\mathrm{P}_{\mathrm{minor}}]}\right)}$$

The approach can be used, although it becomes more problematic in cases in which there are either (a) multiple alternate products or (b) no alternate products. It is still possible for a P450 reaction to show a KIE, in that the activated complex has an alternative to productive oxygenation—the formation of reduced oxygen products, e.g., water (Gorsky, Koop, & Coon, 1984).

3. CHOICE OF ENZYMATIC PARAMETERS

In considering what KIE results to measure, it should be noted that several approaches measure only $^{D}(V/K)$ values (Fig. 2A–C). Many of the P450 KIE results in the literature report only KIE determinations done at a single-substrate concentration, often a high one that approximates k_{cat} conditions. However, considerably more information can be obtained about what is rate-limiting by measuring k_{cat} and K_m and thus ^{D}V and $^{D}(V/K)$. Exactly which steps of a model mechanism are reflected by ^{D}V, $^{D}(V/K)$, ^{D}k, etc., are a function of the microscopic rate constants for the multiple turnover reaction and are discussed elsewhere (Northrop, 1982; Walsh, 1979).

In general, most P450 papers, when they have considered obtaining results varying substrate concentrations, have relied on ^{D}V. However, consider the case of oxidation of ethanol to acetaldehyde by human P450 2E1 (Bell & Guengerich, 1997; Bell-Parikh & Guengerich, 1999).

$$CH_3CH_2OH \rightarrow CH_3CH(OH)_2 \leftrightarrows CH_3CHO + H_2O$$
$$CH_3CD_2OH \rightarrow CH_3CD(OH)_2 \leftrightarrows CH_3CDO + H_2O$$

The rate-limiting step occurs after product formation.

Simplified, the reaction is depicted as (note that we are changing the convention of step numbers now):

$$E \underset{k_{-1}}{\overset{k_1 S}{\rightleftarrows}} ES \xrightarrow{k2} EP \xrightarrow{k3} E + P$$

where S is the substrate and P the product. Then

$$k_{cat} = \frac{k_2 k_3}{k_2 + k_3}[E]_T,$$

and K_m has units of molarity. Therefore,

$$\frac{^{D}V = {}^{D}k2 + k_2/k_3}{1 + k_2/k_3}$$

and

$$ {}^{D}(V/K) = \frac{{}^{D}k2 + k_2/k_{-1}}{1 + k_2/k_{-1}} $$

where ${}^{D}V$ is the KIE on k_{cat} and ${}^{D}(V/K)$ is the KIE on the ratio of k_{cat}/K_m, according to the convention of Northrop (1982).

The value of ${}^{D}(V/K)$ is ~3–5 (Bell & Guengerich, 1997). A plot of product vs time indicates "burst kinetics"—rate-limiting step is after product formation. The ${}^{D}k$ for C—H bond breaking was found to be 3.8, and the isotopic effect on product formation measured using pre-steady-state kinetic studies was estimated to be 3.2. Since ${}^{D}V \approx 1$ due to the fact that a slow step after catalysis is limiting for V, then ${}^{D}k_2 + k_2/k_3 \approx 1 + k_2/k_3$. This is valid only if $k_2 \gg k_3$. From the results presented here, k_2 (the rate of product formation) is much faster than k_3 (the rate of product release).

The above expressions for k_{cat} and K_m are reduced to

$$ K_m \cong \frac{k_3 k_{-1}}{k_1 k_2} $$

and

$$ k_{cat} = \frac{k_2 k_3}{k_2 + k_3} \quad [E]_T $$

Thus, this paradigm includes rate-limiting product release (or something else following product formation) in the overall reaction. K_m increases as k_2 decreases, consistent with the observed effect of isotopic substitution and is a general case (Walsh, 1979). Further, K_m is a function of k_{cat} (and definitely not equivalent to K_d).

4. KIEs FOR VARIOUS P450 REACTIONS

In our own experience, almost every deuterated analog of a compound hydroxylated by P450s (at a particular site that has then been deuterated) shows at least some KIE. As discussed elsewhere (Guengerich, 2013), there are some general predictions that can be made about the tendency for P450 reactions to show kinetic deuterium isotope effects. For instance, amine *N*-dealkylation and aromatic hydroxylation reactions have low KIEs for mechanistic reasons and are generally not good candidates for resolving mechanistic detail (Figs. 3 and 4). However, dealkylations of ethers and

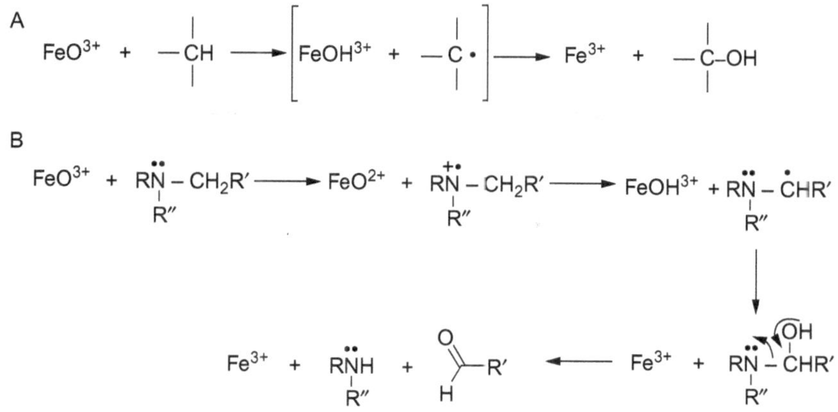

Fig. 3 Comparisons of mechanisms of carbon hydroxylation (A) and *N*-dealkylation (B) (Guengerich & Macdonald, 1984).

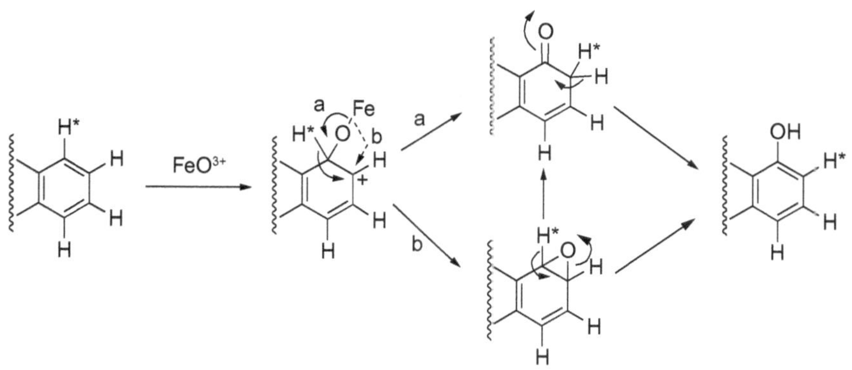

Fig. 4 General mechanisms of aryl hydroxylation (Guengerich, 2001; Ortiz de Montellano, 2015). *H indicates a labeled hydrogen atom, i.e., deuterium or tritium. Alternate pathways (a, *solid line*; b, *broken line*) for the collapse of the tetrahedral iron–oxygen intermediate lead to the carbonyl and to the epoxide, respectively, and both can form the phenol.

amides (Hall & Hanzlik, 1990) are different, due to the much higher oxidation potentials, and high KIEs are often observed. Alkyl groups, i.e., methylenes, methines, and methyls, are possibilities, depending on what is rate-limiting in the particular P450 reaction. However, care is necessary about the issue of *pro*-chirality of methylenes, in that the abstraction of hydrogen and oxygen rebound may, or may not be stereoselective. For example, P450 3A4 abstracts only the β-hydrogen at the C-6 carbon of testosterone (Krauser & Guengerich, 2005), so that this information must be considered in evaluating any results.

Miwa et al. (1983) reported high KIEs (8.6–10.1) for amine *N*-dealkylation reactions catalyzed by peroxidases. However, the KIEs for such reactions are low for P450s, usually ≤2 (Guengerich, 1990). An explanation was provided in that amine dealkylations by both P450s and peroxidases proceed via initial 1-electron abstraction (Fig. 3), but P450s use base catalysis with FeO^{2+} for α-proton transfer while peroxidases do not (instead utilizing heme edge electron transfer), leaving α C—H bond breaking uncatalyzed and allowing the aminium radicals to accumulate (Guengerich, Yun, & Macdonald, 1996; Okazaki & Guengerich, 1993).

For an aryl C—H bond, the enzyme does an addition but does not break the C—H/C—D bond (Fig. 4). Therefore, one expects to see KIEs due to deuterium substitution in alkyl groups but not aryl groups. In this regard, Tomaszewski, Jerina, and Daly (1975) concluded that isomerization to phenols proceeds without detectable isotope effects (Kasperek, Bruice, Yagi, & Jerina, 1972). Arene oxide formation from deuterated aromatic substrates also proceeds without a KIE because C—H bonds are not broken in the oxidation. The general lack of KIEs in formation of phenols in aromatic substrates (Daly & Jerina, 1969; Guroff et al., 1967; Perel, Dayton, Tauriello, Brand, & Mark, 1967; Tanabe, Yasuda, Tagg, & Mitoma, 1967) is consistent with (but does not necessarily prove) the formation of arene oxides as intermediates.

Tanabe, Tagg, Yasuda, LeValley, and Mitoma (1970) reported an in vitro KIE of 1.47 for the aryl oxidation of the drug zoxazolamine. However, this result could not be repeated under the same conditions by Tomaszewski et al. (1975), i.e., the KIE was not greater than unity. Other studies have consistently shown low KIE values (little effect of deuterium substitution) for aryl oxidations, interpreted as evidence for the proposed mechanism shown in Fig. 4 (Bush & Trager, 1982; Hanzlik, Hogberg, & Judson, 1984; Korzekwa, Swinney, & Trager, 1989; Ortiz de Montellano, 1995; Ortiz de Montellano & DeVoss, 2005; Tomaszewski et al., 1975). Typical KIE values for some substituted aryl molecules have been reported in the range of 0.95–1.27 (Korzekwa et al., 1989; Ortiz de Montellano, 1995).

5. EXAMPLES OF P450 KIE STUDIES

The concept of inhibiting rates of P450 oxidations by C—D substitution has been largely the same since the field of P450 KIE research began (Elison et al., 1961) (which is actually 1 year before P450s were formally discovered (Omura & Sato, 1962)). KIE measurements have been made in many systems, but what new information has resulted? Many of the earliest studies were of the intermolecular competition type, and several investigators concluded that KIEs were absent or modest (Björkhem, 1972; Tanabe et al., 1967; Ullrich, 1969).

However, some early results with intramolecular reactions showed very high observed KIEs (Foster, Jarman, Stevens, Thomas, & Westwood, 1974; Hjelmeland, Aronow, & Trudell, 1977) and possibly another study as well (McMahon, Sullivan, Craig, & Pereira, 1969), although the results were later reinterpreted (Hjelmeland et al., 1977). Groves, McClusky, White, and Coon (1978) considered the high intramolecular competitive KIE observed using rabbit P450 2B4 (11.5), plus earlier work, as the first definitive evidence for the radicaloid, Compound I, oxygen-rebound mechanism that is generally accepted today for hydroxylation (Fig. 3A).

Another concept in the field of P450 KIEs is that of "metabolic switching," which Baillie (1981) first attributed to Mitoma, Dehn, and Tanabe (1971), followed by Horning et al. (1975) and others. The concept has been reviewed (Miwa & Lu, 1987). Basically, a KIE on a reaction at one site of a substrate results in *enhanced* activity at another site of the substrate, i.e., yielding an alternate product. For instance, deuteration at the 6β-position of testosterone attenuated the formation of 6β-hydroxytestosterone but enhanced the production of 2β-hydroxytestosterone (Krauser & Guengerich, 2005). The concept is that the enzyme is capable of moving to a reactive intermediate (i.e., Compound I) with the substrate but then "turns" the substrate for catalysis. The amount of total product formed from the deuterated substrate may be as great as with the protiated analog, or it may be less.

Metabolic switching is of interest in both basic and applied contexts. The existence of switching indicated that formation of activated iron–oxygen complex (i.e., FeO^{3+}) is not reversible and, in a sense, contributes to a strong commitment to catalysis (Harada, Miwa, Walsh, & Lu, 1984). However, it should be pointed out that although the Compound I form of P450 (formally FeO^{3+}) is formed irreversibly, it will break down to H_2O if not used to catalyze substrate oxidation (Gorsky et al., 1984). In addition, it is possible to exploit metabolic switching in estimating the intrinsic KIE, ${}^{D}k$ (Section 2.5).

There are also practical aspects of metabolic switching, which are relevant to the issue of utilizing deuterium substitution in drug development (Section 7, vide infra). If the issue is attenuating rates of metabolism to improve drug half-life or lower drug clearance, then the existence of metabolic switching would abrogate a positive effect of deuteration (but only if the parent drug has pharmacological activity). Another issue is that deuteration could produce a metabolic switch to a toxic product. An example is the analgesic phenacetin (in hamsters), where deuterium substitution in the ethoxymethylene decreased liver necrosis but raised blood methemoglobin levels by 50% (Nelson et al., 1978). The authors rationalized the results as meaning that *O*-deethylation was blocked, so that deacetylation by another enzyme, an amidase, was favored (Nelson & Trager, 2003) and gave rise to the more toxic molecule *p*-phenetidine (4-ethoxyaniline) (Ross et al., 1985).

Another aspect is consideration of when KIEs are observed in P450 reactions. As already pointed out, deuterium KIEs should not be seen for P450 reactions in which no C—H bonds are broken during enzyme catalysis, e.-g., epoxidation, heteroatom oxygenation, aryl hydroxylation (Fig. 4). As indicated earlier, in our experience we have found generally low KIEs for amine *N*-dealkylations, which we have rationalized in terms of initial one-electron oxidation followed by proton transfer-linked rearrangement and oxygen rebound (Fig. 3B), at least in some cases (Guengerich, 1990; Guengerich et al., 1996; Okazaki & Guengerich, 1993).

The P450 reactions for which we have measured KIEs all involve mammalian enzymes and are summarized in Fig. 5. In a previous analysis of this type, there was an inverse correlation between the reaction rate and the

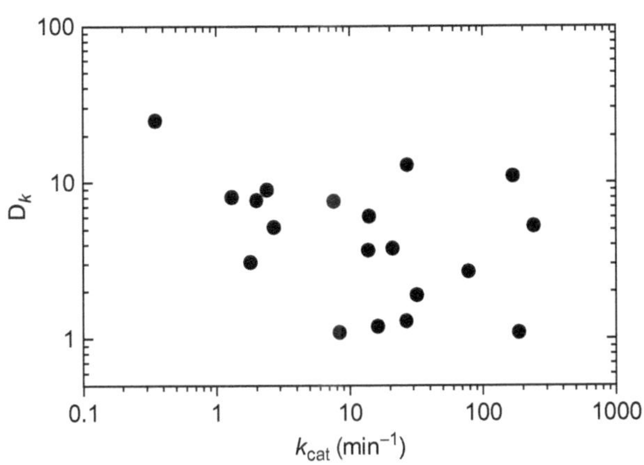

Fig. 5 Relationship between KIE (intermolecular, noncompetitive) and k_{cat} for a number of P450 reactions, mainly with human P450s. Data points are from the indicated references (Bell & Guengerich, 1997; Bell-Parikh & Guengerich, 1999; Chowdhury, Calcutt, & Guengerich, 2010; Chowdhury, Calcutt, Nagy, & Guengerich, 2012; Guengerich, Krauser, & Johnson, 2004; Guengerich, Miller, Hanna, Sato, & Martin, 2002; Krauser & Guengerich, 2005; Pallan et al., 2015; Shinkyo & Guengerich, 2011; Yun, Miller, & Guengerich, 2000).

intermolecular noncompetitive KIE, with an r^2 value of 0.62 (Guengerich, 2013). Even though the correlation coefficient suggested that only ~60% of the variability could be attributed to such a relationship, the idea that ease of C—H bond-breaking limits overall reaction rates among a plethora of P450s has some appeal. Since the earlier publication appeared (4 years ago), we have added our work with P450 21A2, which has a relatively fast substrate turnover rate and also high KIEs (for both substrates, progesterone and 17α-hydroxyprogesterone) (Pallan et al., 2015; Wang et al., 2017). Now any relationship is completely missing ($r^2 = 0.015$). Thus, the absolute rate of a P450 reaction does not appear to provide any insight into how rate-limiting the C—H bond-breaking step is. In more recent studies we have used site-directed mutagenesis to make a series of naturally occurring P450 21A2 variants identified in the clinic (Wang et al., 2017). Some of these have only $1/10^6$ the catalytic activities of the wild-type enzyme but the high KIEs were seen with all. The mutations are unlikely to increase substrate-binding commitments, which we found only limited evidence for, and such a result provides additional evidence that the chemical step is rate-limiting. Therefore, even with the wild-type enzyme, with a specificity constant (k_{cat}/K_m) of $10^7 M^{-1} s^{-1}$, C—H bond-breaking is at least partially rate-limiting. This could be viewed as a less-than-perfect fit of the FeO^{3+} entity and the C-21 hydroxylation site or simply a reflection of the symmetry of the transition state (Melander & Saunders, 1980; Westheimer, 1961).

6. IN VIVO KIEs WITH P450s

Although the first application of KIEs to P450 drug metabolism in 1961 involved in vivo work (Elison et al., 1961), far fewer studies have been done in vivo than in vitro. Another example involves (methylene) dideuteration of the barbiturate butethal, which doubled the sleeping time when the compound was administered to mice (Tanabe, Yasuda, LeValley, & Mitoma, 1969). Early work in the in vivo area was reviewed by Baillie (1981).

One example of a large KIE was seen in our own work on the further metabolism of a product of the calcium channel blocker nifedipine (Funaki, Soons, Guengerich, & Breimer, 1989). P450s catalyze the oxidative demethylation of methyl esters, with high KIE values (Guengerich, 1987; Guengerich, Peterson, & Bocker, 1988). The strong effect of deuteration on pharmacokinetic parameters provides evidence that the in vitro ester cleavage is due to oxidation and not hydrolysis.

In several cases large KIEs had been observed in cell culture or in vivo (in animals), including procarbazine, tamoxifen, *N*-nitrosodimethylamine, pulegone, methylene chloride, chloroform, bromoform, ethylene dibromide, furosemide, *N*-methylformamide, 3,3′-iminodiproprionitrile, *N*-(3,5-didhlorophenyl)succinimide, 3-methylcholanthrene, butylated hydroxytoluene, phenacetin, and 3-methylindole, and these are all related to altered biological effects such as toxicity, as reviewed by Baillie (1981) and Nelson and Trager (2003). As an example, deuterated *N*,*N*-dimethylnitrosamine is threefold less carcinogenic (than the protiated version) in rats (Keefer, Lijinsky, & Garcia, 1973). The role of P450 oxidation in the phenomenon was established by in vitro (microsomal) *N*-demethylation kinetics (Wade et al., 1987) and in vivo measurement of methylated DNA adducts (Swann, Mace, Angeles, & Keefer, 1983). Another cancer-related example is ethylene dibromide (1,2-dibromethane), which is oxidized by P450s. However, its genotoxicity is related to another pathway, glutathione conjugation (Guengerich, 2005). Deuteration of ethylene dibromide blocks oxidation and increases the genotoxicity in rats (White, Gandolfi, Bowden, & Sipes, 1983), as does administration of the P450 2E1/2A6 inhibitor disulfiram (Wong, Winston, Hong, & Plotnick, 1982).

7. CONSIDERATION OF P450 KIEs IN DRUG DEVELOPMENT

The first application of KIEs to P450 drug metabolism was that of Elison et al. (1961) on the *N*-demethylation of morphine, which is recognized today as a cytochrome P450 reaction. Subsequent work on the P450-catalyzed *O*-demethylation of anisoles was published by Mitoma, Yasuda, Tagg, and Tanabe (1967) and Foster et al. (1974). In a review article Foster (1984) concluded that "Two promising areas involve the attenuation of the metabolism pathways which generate the active metabolites for prodrugs and, where overall metabolism cannot be retarded, the deflection of metabolism away from pathways leading to metabolites with toxic properties or other undesirable biological activity toward innocuous pathways." As pointed out in 2009 (Yarnell, 2009), "The strategy is far from new." That 2009 article also cites a 2006 paper by Sepracor on tramadol (Shao, Abolin, Hewitt, Koch, & Varney, 2006). In the 56 years since the reports of Elison et al. (1961), there have been numerous theoretical and fundamental studies of KIEs as well as efforts to modify in vivo pharmacokinetics.

In 2009–10 there was a renewed interest in deuterated analogs of known pharmacologically active compounds. Deuteration strategy was highlighted in 2009 as the "latest retro rage in the pharmaceutical world" (Yarnell, 2009) and several companies were established (e.g., Concert, Dueteria, Auspex (now part of Teva), Protia). Concert advertised that this approach allows it to "rapidly create novel, differentiated compounds with substantially reduced R&D risk, time and expense." The concept is that of routine optimization by deuterating at known metabolic hotspots. For a 2016 review of progress in the commercialization, see Halford (2016). In April 2017 the US Food and Drug Administration (FDA) approved deutetrabenazine (Austedo, from Auspex-Teva) for the treatment of chorea associated with Huntington's disease.

8. SUMMARY AND CONCLUSIONS

P450s are very appropriate for KIE studies because of their proclivity for C—H bond-breaking reactions in the oxidations that they catalyze. KIE studies have provided important insights into how P450s catalyze reactions, and today there is considerable interest in utilizing deuterium substitution (and KIEs) in drug development. What does the future hold?

Many of the reported KIEs for P450 reactions are larger than the classical limit of ~7 (Fig. 5) but to the author's knowledge no reports of serious efforts at studying tunneling have appeared. Some of the approaches to studying tunneling are technically difficult, but there are probably some good candidates. Archebacterial P450s would seem to be good candidates because of their broader temperature range for comparing activation energies for protiated and deuterated substrates (Klinman, 2013), although only limited information is available regarding their substrates and redox partners.

In principle, one could examine thousands of more P450 reactions with KIE studies to determine how rate-limiting the C—H bond-breaking step is. This would not be very practical nor would the results be that enlightening. Nevertheless, many individual reactions catalyzed by P450s are of inherent interest, and KIE analysis is a relatively rapid method (assuming ease of deuterated synthesis) of establishing if C—H bond breaking is a rate-limiting step. In some cases, this information is important in the development of P450 systems for industrial catalysis. The potential for development of drugs with altered metabolism properties has already been discussed.

ACKNOWLEDGMENT

Research in this area in the author's laboratory is supported by National Institutes of Health Grants R01 GM118122 and R01 GM103937. Thanks are extended to M.J. Reddish and to M.E. Harris for comments on the draft and initial submission.

REFERENCES

Baillie, T. A. (1981). The use of stable isotopes in pharmacological research. *Pharmacological Reviews*, *33*(2), 81–132.

Bell, L. C., & Guengerich, F. P. (1997). Oxidation kinetics of ethanol by human cytochrome P450 2E1. Rate-limiting product release accounts for effects of isotopic hydrogen substitution and cytochrome b_5 on steady-state kinetics. *The Journal of Biological Chemistry*, *272*(47), 29643–29651.

Bell, H. M., Vanderslice, C. W., & Spehar, A. (1969). The reduction of organic halogen compounds by sodium borohydride. *Journal of Organic Chemistry*, *34*, 3923–3926.

Bell-Parikh, L. C., & Guengerich, F. P. (1999). Kinetics of cytochrome P450 2E1-catalyzed oxidation of ethanol to acetic acid via acetaldehyde. *The Journal of Biological Chemistry*, *274*(34), 23833–23840.

Biemann, K. (1962). *Mass spectrometry, organic chemical applications* (pp. 223–225). New York: McGraw-Hill.

Björkhem, I. (1972). On the rate-limiting step in microsomal hydroxylation of steroids. *European Journal of Biochemistry*, *27*(2), 354–363.

Bush, E. D., & Trager, W. F. (1982). Evidence against an abstraction or direct insertion mechanism for cytochrome P-450 catalysed *meta* hydroxylations. *Biochemical and Biophysical Research Communications*, *104*, 626–632.

Chowdhury, G., Calcutt, M. W., & Guengerich, F. P. (2010). Oxidation of *N*-nitrosoalkylamines by human cytochrome P450 2A6: Sequential oxidation to aldehydes and carboxylic acids and analysis of reaction steps. *The Journal of Biological Chemistry*, *285*(11), 8031–8044.

Chowdhury, G., Calcutt, M. W., Nagy, L. D., & Guengerich, F. P. (2012). Oxidation of methyl and ethyl nitrosamines by cytochrome P450 2E1 and 2B1. *Biochemistry*, *51*(50), 9995–10007.

Cook, P. F., & Cleland, W. W. (2007). *Enzyme kinetics and mechanism* (pp. 277–280). New York: Garland Science Publishing.

Daly, J., & Jerina, D. (1969). Migration of deuterium during aryl hydroxylation. 3. Effect of *ortho*- and *meta*-substituents. *Archives of Biochemistry and Biophysics*, *134*(1), 266–268.

Elison, C., Rapoport, H., Laursen, R., & Elliott, H. W. (1961). Effect of deuteration of N-CH_3 group on potency and enzymatic N-demethylation of morphine. *Science*, *134*(3485), 1078–1079.

Fersht, A. (1999). *Structure and mechanism in protein science* (pp. 98–99). New York: Freeman.

Foster, A. B. (1984). Deuterium isotope effects in studies of drug metabolism. *Trends in Pharmacological Sciences*, *5*, 524–527.

Foster, A. B., Jarman, M., Stevens, J. D., Thomas, P., & Westwood, J. H. (1974). Isotope effects in *O*- and *N*-demethylations mediated by rat liver microsomes: An application of direct insertion electron impact mass spectrometry. *Chemico-Biological Interactions*, *9*(5), 327–340.

Funaki, T., Soons, P. A., Guengerich, F. P., & Breimer, D. D. (1989). In vivo oxidative cleavage of a pyridine-carboxylic acid ester metabolite of nifedipine. *Biochemical Pharmacology*, *38*(23), 4213–4216.

Gelb, M. H., Heimbrook, D. C., Mälkönen, P., & Sligar, S. G. (1982). Stereochemistry and deuterium isotope effects in camphor hydroxylation by the cytochrome $P450_{cam}$ monooxygenase system. *Biochemistry*, *21*, 370–377.

Gonzalez, E., & Guengerich, F. P. (2017). Kinetic processivity of the two-step oxidations of pregnenolone and progesterone to androgens by human cytochrome P450 17A1. *The Journal of Biological Chemistry*, *292* [in press].

Gorsky, L. D., Koop, D. R., & Coon, M. J. (1984). On the stoichiometry of the oxidase and monooxygenase reactions catalyzed by liver microsomal cytochrome P-450: Products of oxygen reduction. *The Journal of Biological Chemistry*, *259*, 6812–6817.

Groves, J. T., McClusky, G. A., White, R. E., & Coon, M. J. (1978). Aliphatic hydroxylation by highly purified liver microsomal cytochrome P-450: Evidence for a carbon radical intermediate. *Biochemical and Biophysical Research Communications*, *81*, 154–160.

Guengerich, F. P. (1987). Oxidative cleavage of carboxylic esters by cytochrome P-450. *The Journal of Biological Chemistry*, *262*(18), 8459–8462.

Guengerich, F. P. (1990). Low kinetic hydrogen isotope effects in the dehydrogenation of 1,4-dihydro-2,6-dimethyl-4-(2-nitrophenyl)-3,5-pyridinedicarboxylic acid dimethyl ester (nifedipine) by cytochrome P-450 enzymes are consistent with an electron/proton/electron transfer mechanism. *Chemical Research in Toxicology*, *3*(1), 21–26.

Guengerich, F. P. (2001). Common and uncommon cytochrome P450 reactions related to metabolism and chemical toxicity. *Chemical Research in Toxicology*, *14*(6), 611–650.

Guengerich, F. P. (2005). Activation of alkyl halides by glutathione transferases. *Methods in Enzymology*, *401*, 342–353.

Guengerich, F. P. (2013). Kinetic deuterium isotope effects in cytochrome P450 oxidation reactions. *Journal of Labelled Compounds and Radiopharmaceuticals*, *56*(9–10), 428–431.

Guengerich, F. P., Krauser, J. A., & Johnson, W. W. (2004). Rate-limiting steps in oxidations catalyzed by rabbit cytochrome P450 1A2. *Biochemistry*, *43*(33), 10775–10788.

Guengerich, F. P., & Macdonald, T. L. (1984). Chemical mechanisms of catalysis by cytochromes P-450: A unified view. *Accounts of Chemical Research*, *17*, 9–16.

Guengerich, F. P., Miller, G. P., Hanna, I. H., Sato, H., & Martin, M. V. (2002). Oxidation of methoxyphenethylamines by cytochrome P450 2D6. Analysis of rate-limiting steps. *The Journal of Biological Chemistry*, *277*(37), 33711–33719.

Guengerich, F. P., Peterson, L. A., & Bocker, R. H. (1988). Cytochrome P-450-catalyzed hydroxylation and carboxylic acid ester cleavage of Hantzsch pyridine esters. *The Journal of Biological Chemistry*, *263*(17), 8176–8183.

Guengerich, F. P., Yun, C. H., & Macdonald, T. L. (1996). Evidence for a 1-electron oxidation mechanism in *N*-dealkylation of *N,N*-dialkylanilines by cytochrome P450 2B1. Kinetic hydrogen isotope effects, linear free energy relationships, comparisons with horseradish peroxidase, and studies with oxygen surrogates. *The Journal of Biological Chemistry*, *271*(44), 27321–27329.

Guroff, G., Daly, J. W., Jerina, D. M., Renson, J., Witkop, B., & Udenfriend, S. (1967). Hydroxylation-induced migration: The NIH shift. *Science*, *157*, 1524–1530.

Halford, B. (2016). Deuterium switcheroo breathes life into old drugs. *Chemical and Engineering News*, *94*(27), 32–36.

Hall, L. R., & Hanzlik, R. P. (1990). Kinetic deuterium isotope effects on the *N*-demethylation of tertiary amides by cytochrome P-450. *The Journal of Biological Chemistry*, *265*, 12349–12355.

Hanzlik, R. P., Hogberg, K., & Judson, C. M. (1984). Microsomal hydroxylation of specifically deuterated monosubstituted benzenes. Evidence for direct aromatic hydroxylation. *Biochemistry*, *23*, 3048–3055.

Harada, N., Miwa, G. T., Walsh, J. S., & Lu, A. Y. H. (1984). Kinetic isotope effects on cytochrome P-450-catalyzed oxidation reactions: Evidence for the irreversible formation of an activated oxygen intermediate of cytochrome P-448. *The Journal of Biological Chemistry, 259*, 3005–3010.

Hjelmeland, L. M., Aronow, L., & Trudell, J. R. (1977). Intramolecular determination of primary kinetic isotope effects in hydroxylations catalyzed by cytochrome P-450. *Biochemical and Biophysical Research Communications, 76*, 541–549.

Horning, M. G., Haegele, K. D., Sommer, K. R., Nowlin, J., Stafford, M., & Thenot, J.-P. (1975). In E. R. Klein & P. D. Klein (Eds.), *Metabolic switching of drug pathways as a consequence of deuterium substitution. Proceedings of the 2nd international conference on stable isotopes. October 20–23, 1975, Oak Brook, IL,* Washington/Springfield, VA: Energy Research and Development Administration/National Technical Information Service.

Jones, J. P., Korzekwa, K. R., Rettie, A. E., & Trager, W. F. (1986). Isotopically sensitive branching and its effect on the observed intramolecular isotope effects in cytochrome P-450 catalyzed reactions: A new method for the estimation of intrinsic isotope effects. *Journal of the American Chemical Society, 108*, 7074–7078.

Jones, J. P., Rettie, A. E., & Trager, W. F. (1990). Intrinsic isotope effects suggest that the reaction coordinate symmetry for the cytochrome P-450 catalyzed hydroxylation of octane is isozyme independent. *Journal of Medicinal Chemistry, 33*(4), 1242–1246.

Kasperek, G. J., Bruice, T. C., Yagi, H., & Jerina, D. M. (1972). Differentiation between the concerted and stepwise mechanisms for aromatization (NIH-shift) of arene epoxides. *Journal of the Chemical Society, Chemical Communications*, (13), 784–785.

Keefer, L. K., Lijinsky, W., & Garcia, H. (1973). Deuterium isotope effect on the carcinogenicity of dimethylnitrosamine in rat liver. *Journal of the National Cancer Institute, 51*, 299–302.

Kim, D., Cha, G. S., Nagy, L. D., Yun, C. H., & Guengerich, F. P. (2014). Kinetic analysis of lauric acid hydroxylation by human cytochrome P450 4A11. *Biochemistry, 53*(39), 6161–6172.

Klinman, J. P. (2013). Importance of protein dynamics during enzymatic C-H bond cleavage catalysis. *Biochemistry, 52*(12), 2068–2077.

Korzekwa, K. R., Swinney, D. C., & Trager, W. F. (1989). Isotopically labeled chlorobenzenes as probes for the mechanism of cytochrome P-450 catalyzed aromatic hydroxylation. *Biochemistry, 28*, 9019–9027.

Krauser, J. A., & Guengerich, F. P. (2005). Cytochrome P450 3A4-catalyzed testosterone 6β-hydroxylation stereochemistry, kinetic deuterium isotope effects, and rate-limiting steps. *The Journal of Biological Chemistry, 280*(20), 19496–19506.

Matsson, O., & Westaway, K. C. (1998). Secondary deuterium kinetic isotope effects and transition state structure. *Advances in Physical Organic Chemistry, 31*, 143–248.

McMahon, R. E., Sullivan, H. R., Craig, J. C., & Pereira, W. E., Jr. (1969). The microsomal oxygenation of ethyl benzene: Isotopic, stereochemical, and induction studies. *Archives of Biochemistry and Biophysics, 132*(2), 575–577.

Melander, L., & Saunders, W. H. (1980). *Reaction rates of isotopic molecules*. New York: John Wiley & Sons.

Mitoma, C., Dehn, R. L., & Tanabe, M. (1971). In vitro metabolic studies on propyl *p*-nitrophenyl ether. *Biochimica et Biophysica Acta, 237*(1), 21–27.

Mitoma, C., Yasuda, D. M., Tagg, J., & Tanabe, M. (1967). Effect of deuteration of the O-CH_3 group on the enzymic demethylation of *o*-nitroanisole. *Biochimica et Biophysica Acta, 136*(3), 566–567.

Miwa, G. T., & Lu, A. Y. H. (1987). Kinetic isotope effects and 'metabolic switching' in cytochrome P450-catalyzed reactions. *BioEssays*, 7, 215–219.

Miwa, G. T., Walsh, J. S., Kedderis, G. L., & Hollenberg, P. F. (1983). The use of intramolecular isotope effects to distinguish between deprotonation and hydrogen atom abstraction mechanisms in cytochrome P-450- and peroxidase-catalyzed *N*-demethylation reactions. *The Journal of Biological Chemistry, 258*, 14445–14449.

Miwa, G. T., Walsh, J. S., & Lu, A. Y. H. (1984). Kinetic isotope effects on cytochrome P-450-catalyzed oxidation reactions: The oxidative *O*-dealkylation of 7-ethoxycoumarin. *The Journal of Biological Chemistry, 259*, 3000–3004.

Nelson, S. D., Garland, W. A., Mitchell, J. R., Vaishnav, Y., Statham, C. N., & Buckpitt, A. R. (1978). Deuterium isotope effects on the metabolism and toxicity of phenacetin in hamsters. *Drug Metabolism and Disposition, 6*, 363–367.

Nelson, S. D., & Trager, W. F. (2003). The use of deuterium isotope effects to probe the active site properties, mechanism of cytochrome P450-catalyzed reactions, and mechanisms of metabolically dependent toxicity. *Drug Metabolism and Disposition, 31*(12), 1481–1498.

Northrop, D. B. (1975). Steady-state analysis of kinetic isotope effects in enzymic reactions. *Biochemistry, 14*, 2644–2651.

Northrop, D. B. (1977). Determining the absolute magnitude of hydrogen isotope effects. In W. W. Cleland, M. H. O'Leary, & D. B. Northrop (Eds.), *Isotope effects on enzyme-catalyzed reactions. Proceedings of the sixth annual Harry Steenbock symposium* (pp. 122–152). Baltimore, London, and Tokyo: University Park Press.

Northrop, D. B. (1982). Deuterium and tritium kinetic isotope effects on initial rates. *Methods in Enzymology, 87*, 607–625.

Okazaki, O., & Guengerich, F. P. (1993). Evidence for specific base catalysis in *N*-dealkylation reactions catalyzed by cytochrome P450 and chloroperoxidase. Differences in rates of deprotonation of aminium radicals as an explanation for high kinetic hydrogen isotope effects observed with peroxidases. *The Journal of Biological Chemistry, 268*(3), 1546–1552.

Omura, T., & Sato, R. (1962). A new cytochrome in liver microsomes. *The Journal of Biological Chemistry, 237*, 1375–1376.

Ortiz de Montellano, P. R. (1995). Oxygen activation and reactivity. In P. R. Ortiz de Montellano (Ed.), *Cytochrome P450: Structure, mechanism, and biochemistry* (2nd ed., pp. 245–303). New York: Plenum Press.

Ortiz de Montellano, P. R. (2015). Substrate oxidation. In P. R. Ortiz de Montellano (Ed.), *Cytochrome P450: Structure, mechanism, and biochemistry* (4th ed., pp. 111–176). New York: Springer.

Ortiz de Montellano, P. R., & DeVoss, J. J. (2005). Substrate oxidation by cytochrome P450 enzymes. In P. R. Ortiz de Montellano (Ed.), *Cytochrome P450: Structure, mechanism, and biochemistry* (3rd ed., pp. 183–245). New York: Plenum Publishers.

Pallan, P. S., Wang, C., Lei, L., Yoshimoto, F. K., Auchus, R. J., Waterman, M. R., et al. (2015). Human cytochrome P450 21A2, the major steroid 21-hydroxylase: Structure of the enzyme-progesterone substrate complex and rate-limiting C-H bond cleavage. *The Journal of Biological Chemistry, 290*(21), 13128–13143.

Perel, J. M., Dayton, P. G., Tauriello, C. L., Brand, L., & Mark, L. C. (1967). Metabolic studies with deuterated phenobarbital. *Journal of Medicinal Chemistry, 10*(3), 371–374.

Rendic, S., & Guengerich, F. P. (2012). Contributions of human enzymes in carcinogen metabolism. *Chemical Research in Toxicology, 25*(7), 1316–1383.

Rendic, S., & Guengerich, F. P. (2015). Survey of human oxidoreductases and cytochrome P450 enzymes involved in the metabolism of xenobiotic and natural chemicals. *Chemical Research in Toxicology, 28*(1), 38–42. http://dx.doi.org/10.1021/tx500444e.

Ross, D., Larsson, R., Andersson, B., Nilsson, U., Lindquist, T., Lindeke, B., et al. (1985). The oxidation of *p*-phenetidine by horseradish peroxidase and prostaglandin synthase and the fate of glutathione during such oxidations. *Biochemical Pharmacology, 34*, 343–351.

Shao, L., Abolin, C., Hewitt, M. C., Koch, P., & Varney, M. (2006). Derivatives of tramadol for increased duration of effect. *Bioorganic & Medicinal Chemistry Letters, 16*(3), 691–694.

Shinkyo, R., & Guengerich, F. P. (2011). Cytochrome P450 7A1 cholesterol 7α-hydroxylation: Individual reaction steps in the catalytic cycle and rate-limiting ferric iron reduction. *The Journal of Biological Chemistry, 286*(6), 4632–4643.

Swann, P. F., Mace, R., Angeles, R. M., & Keefer, L. K. (1983). Deuterium isotope effect on metabolism of *N*-nitrosodimethylamine in vivo in rat. *Carcinogenesis*, *4*, 821–825.
Tanabe, M., Tagg, J., Yasuda, D., LeValley, S. E., & Mitoma, C. (1970). Pharmacologic and metabolic studies with deuterated zoxazolamine. *Journal of Medicinal Chemistry*, *13*(1), 30–32.
Tanabe, M., Yasuda, D., LeValley, S., & Mitoma, C. (1969). The pharmacologic effect of deuterium substitution on 5-*n*-butyl-5-ethylbarbituric acid. *Life Sciences*, *8*(19), 1123–1128.
Tanabe, M., Yasuda, D., Tagg, J., & Mitoma, C. (1967). Absence of isotope effects in the microsomal hydroxylation of acetanilide. *Biochemical Pharmacology*, *16*(11), 2230–2233.
Tomaszewski, J. E., Jerina, D. M., & Daly, J. W. (1975). Deuterium isotope effects during formation of phenols by hepatic monoxygenases. Evidence for an alternative to arene oxide pathway. *Biochemistry*, *14*(9), 2024–2031.
Ullrich, V. (1969). On the hydroxylation of cyclohexane in rat liver microsomes. *Hoppe-Seyler's Zeitschrift fur Physiologische Chemie*, *350*, 357–365.
Wade, D., Yang, C. S., Metral, C. J., Roman, J. M., Hrabie, J. A., Riggs, C. W., et al. (1987). Deuterium isotope effect on denitrosation and demethylation of N-nitrosodimethylamine by rat liver microsomes. *Cancer Research*, *47*, 3373–3377.
Walsh, C. (1979). *Enzymatic reaction mechanisms* (pp. 118–121). San Francisco: W. H. Freeman Co.
Wang, C., Pallan, P. S., Zhang, W., Lei, L., Yoshimoto, F. K., Waterman, M. R., et al. (2017). Functional analysis of human cytochrome P450 variants involved in congenital adrenal hyperplasia. *The Journal of Biological Chemistry*, *292*, 10767–10778.
Westheimer, F. H. (1961). The magnitude of the primary kinetic isotope effect for compounds of hydrogen and deuterium. *Chemical Reviews*, *61*(3), 265–273.
White, R. D., Gandolfi, A. J., Bowden, G. T., & Sipes, I. G. (1983). Deuterium isotope effect on the metabolism and toxicity of 1,2-dibromoethane. *Toxicology and Applied Pharmacology*, *69*, 170–178.
White, R. E., Miller, J. P., Favreau, L. V., & Bhattacharyya, A. (1986). Stereochemical dynamics of aliphatic hydroxylation by cytochrome P-450. *Journal of the American Chemical Society*, *108*, 6024–6031.
Wong, L. C. K., Winston, J. M., Hong, C. B., & Plotnick, H. (1982). Carcinogenicity and toxicity of 1,2-dibromoethane in the rat. *Toxicology and Applied Pharmacology*, *63*, 155–165.
Yarnell, A. T. (2009). Heavy-hydrogen drugs turn heads, again. *Chemical & Engineering News*, *87*(25), 36–39.
Yoshimoto, F. K., & Guengerich, F. P. (2014). Mechanism of the third oxidative step in the conversion of androgens to estrogens by cytochrome P450 19A1 steroid aromatase. *Journal of the American Chemical Society*, *136*(42), 15016–15025.
Yun, C.-H., Miller, G. P., & Guengerich, F. P. (2000). Rate-determining steps in phenacetin oxidations by human cytochrome P450 1A2 and selected mutants. *Biochemistry*, *39*(37), 11319–11329.

CHAPTER TEN

Use of Isotopes and Isotope Effects for Investigations of Diiron Oxygenase Mechanisms

Rahul Banerjee, Anna J. Komor, John D. Lipscomb[1]
University of Minnesota, Minneapolis, MN, United States
[1]Corresponding author: e-mail address: lipsc001@umn.edu

Contents

1. Introduction 240
2. General Methods 240
   2.1 Preparation of sMMO Components 240
   2.2 Preparation of CmlI 246
   2.3 Preparation of Reduced Enzymes 247
3. Methane Monooxygenase 249
   3.1 Mass Properties That Allow Substrate-to-Product Conversions to Be Tracked 249
   3.2 Atomic Properties That Allow Specialized Spectroscopies 249
   3.3 Mass Properties That Impact Primarily Vibrational Spectroscopies 256
   3.4 BDE Shifts That Permit Dynamic Isotope Effect Studies 263
4. CmlI Oxygenase 273
   4.1 Mass Properties That Allow Substrate-to-Product Conversions to Be Tracked 273
   4.2 Atomic Properties That Allow Specialized Spectroscopies 273
   4.3 Mass Properties That Impact Primarily Vibrational Spectroscopies 276
   4.4 Mass Properties That Allow the Mechanism of the Oxidative Pathway to Be Probed 279
5. Conclusions 286
References 286

Abstract

Isotope effects of four broad and overlapping categories have been applied to the study of the mechanisms of chemical reaction and regulation of nonheme diiron cluster-containing oxygenases. The categories are: (a) mass properties that allow substrate-to-product conversions to be tracked, (b) atomic properties that allow specialized spectroscopies, (c) mass properties that impact primarily vibrational spectroscopies, and (d) bond dissociation energy shifts that permit dynamic isotope effect studies of many types. The application of these categories of isotope effects is illustrated using the soluble methane monooxygenase system and CmlI, which catalyzes the multistep arylamine to arylnitro conversion in the biosynthetic pathway for chloramphenicol.

Methods in Enzymology, Volume 596
ISSN 0076-6879
http://dx.doi.org/10.1016/bs.mie.2017.07.016

1. INTRODUCTION

Enzymes that activate molecular oxygen for incorporation of one or both oxygen atoms into a substrate (oxygenases) are common in nature and utilize a variety of cofactors (Denisov, Makris, Sligar, & Schlichting, 2005; Kovaleva & Lipscomb, 2008; Solomon et al., 2014). Oxygenases that employ a nonheme dinuclear iron cluster in the active site generate powerful oxygenating intermediates and exhibit a remarkable versatility in the types of reactions that they catalyze (Fox, Shanklin, Somerville, & Münck, 1993; Pikus et al., 1996; Sazinsky, Bard, Di Donato, & Lippard, 2004; Tinberg & Lippard, 2011; Wallar & Lipscomb, 1996; Xing et al., 2006). Isotopes and isotope effects have been broadly employed to investigate the mechanisms of diiron oxygenases, and these studies have resulted in some of the most detailed descriptions of reaction cycle intermediates among metalloenzymes (Doan, Shanmugam, Stubbe, & Hoffman, 2015; Krebs et al., 2013; Ling, Sahlin, Sjöberg, Loehr, & Sanders-Loehr, 1994; Lipscomb, 2006; Moenne-Loccoz, Baldwin, Ley, Loehr, & Bollinger, 1998; Shanklin, Achim, Schmidt, Fox, & Münck, 1997; Wallar & Lipscomb, 1996). Generally, four properties of isotopes have been exploited with considerable overlap between classes: (1) mass properties that allow substrate-to-product conversions to be tracked, (2) atomic properties that allow specialized spectroscopies, (3) mass properties that impact primarily vibrational spectroscopies, and (4) bond dissociation energy (BDE) shifts that permit dynamic isotope effect studies of many types. In this chapter, the methods used to apply isotopes to probe the reaction mechanism of the diiron cluster-containing soluble form of methane monooxygenase (sMMO) and the *N*-oxygenase CmlI are presented (Makris et al., 2015; Tinberg & Lippard, 2011; Wallar & Lipscomb, 1996). These enzymes catalyze distinct classes of oxygenase reactions, and they serve to illustrate the power of isotope-based methods in mechanistic investigations of reaction cycles (sMMO) and reaction pathways (CmlI). Each of the four classes of isotope applications is encompassed in one or both of these studies.

2. GENERAL METHODS

2.1 Preparation of sMMO Components

2.1.1 Purification of the sMMO Hydroxylase Component (sMMOH)

Equipment

1. Anaerobic glove box (Coy)
2. Stirring cold plate (Stir-Kool SK-12, Thermoelectrics Unlimited, Inc.)

3. Sonifier (Branson S-450A)
4. Floor centrifuge (Beckman), centrifuge tubes
5. DEAE-Sepharose CL6B and Q-Sepharose HR (Sigma Q1754) anion-exchange resins
6. Chromatographic columns (Kimble Kontes Chromaflex)
7. Peristaltic pump and gradient mixer (Teledyne Isco, Biorad Econo pump)
8. Stirred cell concentrator, 200–500 mL
9. Amicon Ultra-15 centrifugal filter device MWCO: 30 kDa
10. UV–vis spectrophotometer

Buffers and Reagents

1. Buffer A: 25 m*M* MOPS, pH 7.0. 200 μ*M* $Fe(NH_4)_2SO_4{\cdot}6H_2O$ and 2 m*M* L-cysteine are added to the buffer after it has been made anaerobic and immediately before use.
2. Buffer B: 25 m*M* MOPS, pH 7.0, saturated with methane at 4°C.
3. Buffer C: 25 m*M* MOPS, pH 7.0, saturated with oxygen at 4°C.
4. Buffer D: 50 m*M* MOPS, pH 7.0.
5. Buffer E: 25 m*M* MOPS, pH 6.8.
6. Buffer F: 50 m*M* MOPS, pH 7.6.
7. Nitrobenzene stock solution—1.6 μL of nitrobenzene (99%, Sigma) is added to 1 mL of buffer F in a 5-mL serum vial. The vial is crimp cap sealed and vortexed vigorously to make a fine suspension of the organic nitrobenzene liquid in buffer. The concentration of this saturated solution of nitrobenzene is 16 m*M* (Merck Index).
8. Ethanol-free NADH.

Assay Methods

The activity of sMMOH in the presence of the regulatory protein MMOB and the reductase MMOR can be assayed by polarography, gas chromatography, or colorimetric NADH consumption assays that were detailed in a previous *Methods in Enzymology* volume (Fox, Froland, Jollie, & Lipscomb, 1990). A convenient colorimetric assay is based on the conversion of nitrobenzene to nitrophenol and is detailed here (Brazeau & Lipscomb, 2003).

Nitrobenzene Assay Procedure

1. The nitrobenzene oxidation activity assay of the sMMO enzyme relies on the measurement of the rate of production of the chromophoric *p*-nitrophenol product at 404 nm ($\varepsilon_{404} = 15\ \mathrm{m}M^{-1}/\mathrm{cm}$ at pH 7.6).
2. The assay can be used to either characterize purified sMMOH or locate and approximate the relative activity of sMMOH containing fractions

during enzyme purification. For the former application, the sMMO protein complex is reconstituted in the following ratio: sMMOH:MMOB: MMOR = 1:2:4 in buffer F. In the latter application, a solution of MMOB (1.0 μ*M*) and MMOR (0.4 μ*M*) in buffer F is mixed with 10 μL of the chromatography-derived fractions to generate the assay solutions.

3. A calculated volume of the nitrobenzene stock solution is added to provide a working concentration of 1 m*M*.
4. The addition of 0.5 m*M* NADH initiates the reaction. The specific activity of the sMMO enzyme is calculated from an initial rate measurement of the absorbance increase at 404 nm.
5. In the protein characterization mode, the specific activity of sMMO for nitrobenzene oxidation is described in terms of nanomoles of *p*-nitrophenol product formed per minute per mg of the sMMOH.

Notes

The regiospecificity of nitrobenzene oxidation is an important factor in this activity assay. As the extinction coefficients of *p*-nitrophenol ($\varepsilon_{404} = 15\,\text{m}M^{-1}/\text{cm}$ at pH 7.6) and *m*-nitrophenol ($\varepsilon_{404} = 0.4\,\text{m}M^{-1}/\text{cm}$ at pH 7.6) differ significantly at 404 nm, it is vital to measure the isomeric distribution of the nitrophenol products by HPLC when assaying an sMMO protein variant. For example, the regiospecificity of nitrobenzene oxidation for the wild-type sMMOH:MMOB:MMOR reaction is 89% *p*-nitrophenol and 11% *m*-nitrophenol, but when using the N107A and S110A MMOB variant, the ratio changes to 69% *p*-nitrophenol and 31% *m*-nitrophenol.

sMMOH Purification Procedure

1. The hydroxylase protein component of sMMO (sMMOH) is purified from the obligate aerobic and methane-oxidizing bacterium *Methylosinus trichosporium* OB3b, the growth procedure for which has been described previously (Fox et al., 1990). A frozen cell pellet (200 g) is thawed in a warm water bath (50°C) and transferred into an anaerobic glove bag. The thawed cells are resuspended in 400 mL of cold (4°C), anaerobic buffer A.
2. The cell slurry is poured into a stainless steel beaker wrapped with frozen dry ice packs. A stirring cold plate provides further cooling. The cells are ruptured by sonication for 10 min (3/4 in. disrupter horn, 80% power output, 80% duty cycle). The temperature of the cell suspension varies between 10°C and 15°C.

3. The sonicated cell suspension is aliquoted into sealable centrifuge tubes inside the glove box, and the cell debris spun down by centrifugation at 45,000 × *g* for 60 min.
4. The sealed centrifuge tubes are brought back into the glove bag, and the cell-free extract is carefully decanted. It is diluted with 200 mL of cold (4°C), anaerobic buffer A.
5. The cell-free extract is batch bound with 250 mL of DEAE-Sepharose resin that has been equilibrated with anaerobic buffer A inside the glove box. The batch binding process is allowed to proceed for 30 min at 25°C.
6. The cell-free extract and DEAE resin slurry are poured into a sealable chromatography column (4.5 × 13 cm) and packed with the aid of a peristaltic pump.
7. The sealed chromatography column is taken out of the glove box and transferred to a cold room. The resin is washed with 50 mL of anaerobic buffer A (not containing the Fe(II) and L-cysteine additive) followed by 600 mL of a 1:1 mixture of cold (4°C) buffer B and C at a linear flow rate of 38 cm/h (ml/h/cross-sectional area of the column).
8. The bound proteins are eluted with a 2.0-L salt gradient from 0 to 0.4 *M* NaCl in buffer D at a linear flow rate of 19 cm/h. The sMMOH protein elutes at a salt concentration of 0.08 *M* NaCl. Fractions containing sMMOH are identified with the colorimetric nitrobenzene to nitrophenol oxidation assay.
9. Fractions possessing activity that are within 20% of the activity of the peak fraction are pooled and concentrated by ultrafiltration with a YM-30 membrane to a final volume of 35 mL.
10. The concentrated protein pool is applied to the top of a Sephadex G-25 column (3 × 22 cm) under a buffer head. This desalting chromatographic column is equilibrated and developed with buffer E. The protein pool is eluted from the column at a linear flow rate of 70 cm/h. The protein eluent is easily identified by a coeluting colored heme protein contaminant.
11. The desalted protein pool is subsequently loaded on a high-resolution Q-Sepharose column (3 × 12 cm) that has been equilibrated with buffer E at a linear flow rate of 26 cm/h. The column is washed with 50 mL of buffer E at the same flow rate.
12. The bound proteins are eluted from the resin with a 900-mL gradient from 0 to 0.08 *M* NaCl in buffer E at a linear flow rate of 33 cm/h.

sMMOH elutes at a salt concentration of 0.035 *M* NaCl in buffer E. The fractions containing a nitrobenzene oxidation activity greater than 20% of the maximum are pooled and concentrated by ultrafiltration.

13. Glycerol is added to a final concentration of 5% (v/v) to the sMMOH protein pool as a cryoprotectant. The concentrated protein is frozen in liquid N_2 and stored at −80°C.

Notes

1. The $Fe(NH_4)_2SO_4 \cdot 6H_2O$ and L-cysteine solution acts as a stabilizer for preserving sMMOH activity by preventing the loss of active site iron (Fox, Froland, Dege, & Lipscomb, 1989). However, the Fe(II) additive leads to Fenton chemistry in the presence of oxygen in the purification buffers, which adversely affects sMMOH activity. The first few steps of the purification are conducted in an anaerobic manner to prevent the accumulation of hydroxyl radicals.
2. The sMMOH protein is present in the reduced diferrous state in the cell-free extract. After the sMMOH protein has been bound to the first DEAE column, the wash with methane and oxygen-saturated buffer converts sMMOH to the oxidized diferric state. The sMMOH is separated from the reductase MMOR and NADH present in cell-free extract; consequently, it is no longer converted back to the diferrous state. The iron atoms of the diiron cluster are more tightly bound to the ligands in the Fe(III) state. Subsequent to this diiron cluster oxidation, the $Fe(NH_4)_2SO_4 \cdot 6H_2O$ and L-cysteine stabilizer is not required, thus obviating the need for anaerobic conditions.

2.1.2 Purification of the sMMO Regulatory Component (MMOB)

Equipment

1. Sonifier (Branson S-450A)
2. Floor centrifuge (Beckman), centrifuge tubes
3. Q-Sepharose HR (Sigma Q1754) anion-exchange resins
4. Chromatographic columns (Kimble Kontes Chromaflex)
5. Peristaltic pump and gradient mixer (Teledyne Isco, Biorad Econo pump)
6. Stirred cell concentrator, 200–500 mL
7. Amicon Ultra-15 centrifugal filter device MWCO: 10 kDa

Buffers and Reagents

1. Buffer G: 25 m*M* MOPS, pH 7.0.
2. Buffer H: 25 m*M* MOPS, pH 7.0 containing 8 *M* urea.

Procedure

1. The MMOB protein is obtained from a heterologous overexpression system in *Escherichia coli* BL21(DE3) cells containing the pBWJ400 plasmid. The MMOB protein is found in the inclusion body fraction.
2. 20–30 g of cell paste are resuspended in buffer G (5 mL/g of cell paste) followed by sonication to rupture the cells (1/2 in. disrupter horn, 80% power output, 80% duty cycle).
3. DNase I (5 mg), Triton X-100 (1%, v/v), and NaCl (0.5 *M*) are added to the ruptured cell suspension, which is then stirred at 4°C for 15 min.
4. This mixture is centrifuged at 43,000 × *g* for 30 min to obtain the inclusion bodies containing MMOB.
5. The pellet is resuspended in buffer G in order to remove traces of the Triton X-100 detergent. It is spun down again by centrifugation at 43,000 × *g* for 15 min.
6. The pellet is dissolved in 25 mL of buffer H (with constant stirring at room temperature for 30 min), following which the insoluble fraction is removed by centrifugation at 43,000 × *g* for 20 min.
7. The urea denatured MMOB protein is refolded by dropwise addition of the urea solution into a stirred solution (600 mL) of buffer G at 4°C.
8. The refolded protein solution is centrifuged at 43,000 × *g* for 15 min in order to remove insoluble material.
9. The remaining soluble protein solution is loaded onto a high-resolution Q-Sepharose column (3 × 12 cm) that has been equilibrated with buffer G at a linear flow rate of 26 cm/h. The column is washed with 50 mL of buffer G containing 0.08 *M* NaCl at the same flow rate.
10. The bound proteins are eluted off the column with a 900-mL gradient from 0.08 to 0.3 *M* NaCl in buffer G at a linear flow rate of 33 cm/h. The MMOB protein elutes at a salt concentration of 0.18 *M* NaCl at the pH of this buffer. Fractions containing MMOB are identified on the basis of its electronic absorption spectrum (see note) and SDS-PAGE gels.
11. The pooled fractions are concentrated with a 10-kDa cutoff ultrafiltration membrane in Amicon stirred cells and centrifugal filter units. The concentrated protein is diluted with buffer G to remove NaCl. The concentrated protein is frozen in liquid N_2 and stored at −80°C.

Notes

The eluent fractions from ion-exchange chromatography column are carefully monitored by electronic absorption spectroscopy at 280 and 290 nm. Only fractions containing protein with an A_{280}/A_{290} ratio of less than 1.45 are pooled. This threshold ensures that an eluent protein peak containing truncated MMOB protein missing its N- and C-terminal ends (ESI-MS molecular weight = 10,123 Da) is avoided.

2.1.3 Purification of the sMMO Reductase Component (MMOR)

1. The reductase protein MMOR is purified from the native organism, and its purification protocol has been described previously (Fox et al., 1990). The following note describes a modification of the original protocol that reduces the number of chromatographic separation steps.
2. The eluent pool containing MMOR from the first DEAE-Sepharose column is loaded onto a high-resolution Q-Sepharose column (3 × 12 cm) that has been equilibrated with 25 m*M* MOPS, pH 6.5, with 5 m*M* sodium mercaptoacetate added immediately before use (buffer I) at a linear flow rate of 26 cm/h. The column is washed with 100 mL of buffer I containing 0.15 *M* NaCl. The bound protein is eluted with a 500-mL salt gradient from 0.15 to 0.35 *M* NaCl in buffer I at a linear flow rate of 26 cm/h. This chromatography step provides sufficient resolution to obtain pure MMOR and removes the need for additional chromatographic separation steps.

2.2 Preparation of CmlI

Equipment

1. Sonifier (Branson S-450A)
2. Thermometer, preferably electric with a metal probe
3. Floor centrifuge (Beckman), centrifuge tubes
4. 25 mL Ni-NTA agarose (Qiagen) or another affinity chromatography matrix appropriate for binding His-tagged proteins and column such as Econo-Column (Bio-Rad)
5. Biorad Biologic LP chromatography system with fraction collector
6. Stirred cell concentrator, 200–500 mL (Amicon)
7. Dialysis tubing (Spectra/Por molecular porous membrane tubing, MWCO: 6–8 kDa, Spectrum Labs) and dialysis tubing clips
8. Amicon Ultra-15 centrifugal filter device MWCO: 10 kDa

Buffers and Reagents

1. Buffer A (Lysis buffer): 50 m*M* NaH_2PO_4, 300 m*M* NaCl, 10 m*M* imidazole, pH 8
2. Buffer B (Elution buffer): 50 m*M* NaH_2PO_4, 300 m*M* NaCl, 300 m*M* imidazole, pH 8
3. Buffer C (Reaction buffer): 50 m*M* Bicine, pH 9

Procedure

1. The gene for CmlI is cloned from genomic DNA isolated from *Strepto-myces venezuelae* mycelia and inserted into plasmid pVP91A (DNASU plasmid repository) to allow expression of a His_8-tagged enzyme in *E. coli* BL21(DE3) as previously described (Makris et al., 2015).
2. Cells are recovered by centrifugation and resuspended in ~3 mL of buffer A per gram of cell paste by stirring at room temperature.
3. Cells are lysed via sonication at 4°C while stirring, with frequent checks to ensure that the solution temperature does not exceed 10°C (1/2 in. disruptor horn, maximum power, 50% duty cycle, 20 min).
4. The lysed cell suspension is loaded into centrifuge tubes and spun at 17,000 × *g* for 45–60 min.
5. The resulting supernatant is gravity loaded onto the prepacked Ni-NTA column. The pellet from the centrifugation can be discarded.
6. The Biorad Biologic is set up with buffer A and buffer B, and a 80-min program in which the first 10 min are 100% buffer A and the next 70 min are a gradient from 0% to 100% buffer B. The fraction collector is set up to collect 1 min (~2.5 mL) fractions. Fractions are assayed by SDS-PAGE gel for the presence of CmlI (monomer ~34 kDa) and then pooled and concentrated to ~50 mL in a stirred cell concentrator under N_2 gas.
7. The protein is then loaded into dialysis tubing and dialyzed overnight into buffer C. The following day, protein is concentrated to 1–2 m*M* using centrifugal spin filters, flash frozen in liquid N_2, and stored at −80°C. Protein concentration is determined using the monomer extinction coefficient, $\varepsilon_{280} = 50\,\text{m}M^{-1}/\text{cm}$.

2.3 Preparation of Reduced Enzymes

In their natural reaction cycles, sMMOH and CmlI are reactive with O_2 in their two-electron reduced states (Lee, Nesheim, & Lipscomb, 1993; Makris et al., 2015). Thus, all experiments designed to probe a single turnover reaction of the enzyme require the preparation of the diferrous state of the diiron cluster found in these enzymes.

Equipment

1. Anaerobic glove box (Coy)
2. Sephadex G-25 PD-10 column, 8.3 mL (GE Healthcare)

Chemicals and Reagents

1. Sodium hydrosulfite
2. Methyl viologen

Procedure

1. The sMMOH or CmlI protein is made anaerobic by incubation at 4°C for 30–60 min under an argon headspace while gently stirring the protein.
2. The anaerobic protein is transferred into an anaerobic glove box and reduced with a onefold excess of sodium hydrosulfite and a catalytic amount of methyl viologen (10% of enzyme active site concentration) at room temperature for a minimum of 30 min. The sodium hydrosulfite solution must be made in the anaerobic chamber as it is highly reactive with oxygen.
3. The chemical reductants are separated from the sMMOH or CmlI protein in the anaerobic glove box by desalting through a PD-10 desalting column, which has been equilibrated with the anaerobic buffer specific to the enzyme and the experiment in which it is to be used.
4. A small aliquot of the reduced enzyme is taken out of the glove box to measure the protein concentration based upon its electronic absorbance at 280 nm (sMMOH $\varepsilon_{280} = 539\,\mathrm{m}M^{-1}/\mathrm{cm}$; CmlI $\varepsilon_{280} = 100\,\mathrm{m}M^{-1}/\mathrm{cm}$; in both cases, these values refer to the dimeric protein with two active sites).

Notes

1. In experiments where the working concentration of the protein is greater than 60 mg/mL, the high viscosity of the protein solution disrupts the flow rate and resolution of the desalting column. For these applications, the protein is reduced with a stoichiometric equivalent amount of sodium hydrosulfite. This ensures that there is no chemical reductant left in the reduced state at the end of the process. As the concentration of the chemical reductant is reduced, more time is provided for the reaction to reach completion (1 h at room temperature). Alternatively, the protein can be chemically reduced at a lower working concentration than desired, desalted on the PD-10 column, and then reconcentrated with spin-concentrator devices in a centrifuge housed in the anaerobic glove box.
2. As MMOB decreases the midpoint reduction potential of the diiron cluster of sMMOH and impedes the reduction process, MMOB is added to sMMOH only after the reduction is complete.
3. Diferrous sMMOH is unstable in comparison to diferric sMMOH and tends to precipitate at high protein concentrations (>0.5 mM) at room temperature.

3. METHANE MONOOXYGENASE

3.1 Mass Properties That Allow Substrate-to-Product Conversions to Be Tracked

3.1.1 Source of Oxygen in Product Methanol

sMMO is isolated from methanotrophic bacteria and has three protein components (Wallar & Lipscomb, 1996). The growth of methanotrophs, the preparation of the protein components, and the methods to assay activity have been described in previous *Methods in Enzymology* chapters (Fox et al., 1990; Pilkington & Dalton, 1990). Some modifications in these procedures have been made in recent years (Banerjee, Meier, Münck, & Lipscomb, 2013) as described in Section 2.1. The active site diiron cluster is found in the alpha subunit of sMMOH, which has an overall $(\alpha\beta\gamma)_2$ quaternary structure. Even though the reaction catalyzed by sMMO is the oxidation of methane, O_2 activation requires the input of two electrons. These are derived from NADH and delivered by a reductase protein (MMOR). The overall reaction of sMMO is regulated by the third component termed MMOB, which forms a complex with sMMOH and controls the binding of both O_2 and methane (Liu, Nesheim, Lee, & Lipscomb, 1995; Wallar & Lipscomb, 2001). This regulatory activity allows sMMO to be specific for methane oxidation despite the fact that it is the most difficult to oxidize abundant alkane. Methanotrophs can only use methane as an efficient growth substrate, and so the ability to selectively oxidize methane is tantamount to survival. The reaction carried out by sMMO was characterized by the use of the ^{18}O isotopes of O_2 and H_2O. The methods for the determination of the source and stoichiometry of oxygen in the products are described in Section 3.4. The characterization of the sMMO reaction using isotopically enriched O_2 and substrates showed that overall reaction is:

$$CH_4 + O_2 + NADH + H^+ \rightarrow CH_3OH + H_2O + NAD^+$$

3.2 Atomic Properties That Allow Specialized Spectroscopies

3.2.1 Mössbauer Spectroscopy

Mössbauer spectroscopy has been one of the most useful spectroscopies for iron-containing enzymes (Münck, 2000). In its normal mode of application, a ^{57}Co source decays by electron capture to ^{57}Fe in a nuclear excited state. Subsequent return to the ground state occurs by the recoilless (and therefore nearly monochromatic) emission of a 14.4-keV gamma ray. The energy of the gamma ray is modulated by oscillation of the ^{57}Co source at 5–10 Hz and then passed through a sample containing ^{57}Fe. When the Doppler-shifted frequency of the gamma ray matches the energy splitting of the ^{57}Fe atomic nucleus in the sample, absorption occurs and is detected by a counter mounted behind the sample. The energy splitting of the ^{57}Fe nucleus in the sample is modulated primarily by the oxidation state of the iron, the symmetry of the electronic environment, and coupling phenomena involving other metals or ligands. Sophisticated techniques have been developed to fit the resulting spectra over a range of applied magnetic fields to characterize the electronic structure of the iron atom in exquisite detail as described in previous *Methods in Enzymology* chapters (Münck, 1978; Münck, Surerus, & Hendrich, 1993).

Incorporation of ^{57}Fe into sMMOH

1. The ^{57}Fe is obtained commercially either as a fine metal powder or as a metal strip.
2. A measured amount of the metal is dissolved in a small amount (~1 mL for 100 mg of ^{57}Fe metal) of aqua regia (3 parts HCl to 1 part HNO_3) overnight at room temperature. The amount of acid required to solubilize the iron is minimized in order to prevent the pH of the growth media from being altered after iron supplementation.
3. The acid solution is diluted with distilled water to prepare a 40-m*M* stock solution of ^{57}Fe. This solution is filter sterilized (0.2 μm, Gelman), and a calculated amount is added directly to the growth media to achieve a working concentration of 80 μ*M*. While this working concentration is significantly larger than the average iron concentration of 10 μ*M* in Luria Bertani media, the methanotroph culture is pushed to very high optical densities ($A_{540}=15$–25) in the fermentor where iron tends to limit the methanotroph growth.

The Mössbauer spectra of resting and reduced sMMOH were obtained by flash freezing the respective samples in Mössbauer cups. The Mössbauer spectrum of resting sMMOH appeared as the superposition of two quadrupole doublets, one from each iron of the diiron cluster (Fig. 1; Fox, Hendrich, et al., 1993). The parameters of the doublets showed that each iron is in the high-spin ($S=5/2$) Fe^{3+} oxidation state, but the appearance as doublets rather than the characteristic six-line "magnetic" spectrum of high-spin Fe^{3+} showed that the irons are antiferromagnetically exchange coupled through bridging ligands. The X-ray crystal structure of diferric sMMOH later revealed these ligands to be a bridging 1,3-carboxylate from Glu144 and two bridging hydroxides or waters (Elango et al., 1997;

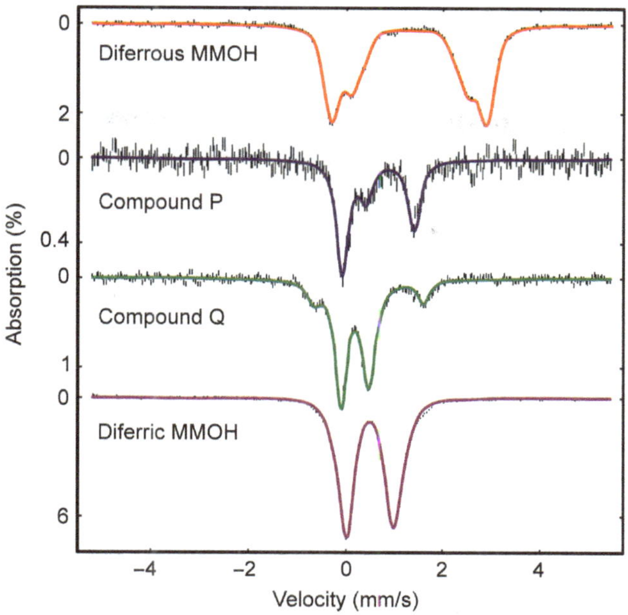

Fig. 1 The Mössbauer spectra for the states of sMMOH indicated on the figure. Data are shown as *hash marks* and fit as overlaid *solid lines*. *Adapted from Shu, L., Nesheim, J. C., Kauffmann, K., Münck, E., Lipscomb, J. D., & Que, L., Jr. (1997). An $Fe^{IV}_2O_2$ diamond core structure for the key intermediate Q of methane monooxygenase.* Science, 275, *515–518.*

Rosenzweig, Frederick, Lippard, & Nordlund, 1993). Reduction of the diiron cluster required two electrons, and so each iron becomes Fe^{2+}. The crystal structure of the reduced protein showed that one or both bridging solvents are lost, and one of the two carboxylate oxygens from Glu243 becomes bridging (Rosenzweig, Nordlund, Takahara, Frederick, & Lippard, 1995). This change in bridging structure generated adjacent, easily occupied ligand sites on the irons to allow O_2 binding. Mössbauer spectroscopy showed that the irons become ferromagnetically coupled with an inverted zero-field splitting during reduction (Fig. 1; Fox, Hendrich, et al., 1993). This configuration accounted for the appearance of an EPR resonance near $g = 16$ (Fox et al., 1989; Hendrich, Münck, Fox, & Lipscomb, 1990) that allowed the reaction with O_2 and subsequent formation of early reaction cycle intermediates to be followed in a single turnover cycle. Stopped-flow and rapid freeze-quench (RFQ) methods were used to detect, kinetically characterize, and trap the intermediates for spectroscopic studies.

3.2.2 Method for Single Turnover Reactions of sMMOH (or CmlI) in the Stopped-Flow and RFQ Instruments for Studies of Isotopically Enriched Intermediates

A stopped-flow analysis of optical changes during a single turnover of the enzyme is used to identify intermediates of interest and the time regime in which they appear. This is a prerequisite to using RFQ methods to generate cryogenically trapped intermediates for spectroscopic analysis. Stopped flow is also used directly to analyze differences in reaction rate constants arising from isotopic incorporation into substrates and solvent.

Equipment

1. Stopped-flow instrument (Applied Photophysics)
2. RFQ instrument (Update Instruments System1000 or BioLogic QFM-4000) and freezing device (custom design)
3. Gastight syringes (Hamilton, SGE)

Procedure

1. For stopped-flow experiments, the diferrous oxygenase (sMMOH or CmlI) sample is loaded into a gastight syringe inside the anaerobic glove box. This syringe is then used to load the reduced protein in the drive syringe of the stopped-flow instrument, which had been scrubbed of O_2 with consecutive 10 mL washes of 20–50 mM sodium hydrosulfite solution and anaerobic buffer. Typically, a 1.5–2-mL solution of sMMOH or CmlI is prepared at a concentration of 60 μM (120 μM active sites).
2. The other drive syringe is loaded with O_2-containing buffer. The O_2 concentration may be saturating for experiments designed to detect and trap intermediates, or subsaturating for experiments designed to evaluate O_2-concentration dependence. For sMMOH reactions, the second syringe also contains a stoichiometric equivalent amount of MMOB relative to the sMMOH active sites. For some sMMOH experiments, a substrate is added in this solution as well. For experiments with CmlI, the substrate may be added with the O_2-saturated buffer or anaerobically in the glove box to the diferrous protein to generate an ES complex. Experiments with sMMOH indicate that it does not form a kinetically relevant ES complex prior to formation of the reactive intermediate compound Q.
3. Intermediates with spectral features in the visible are detected in the stopped-flow instrument with single wavelength, diode array, or spectra-kinetic multiple single wavelength detection methods. Substrate concentration, pH, and temperature dependence of the rate constants of

individual kinetic steps in the single turnover reaction cycle are also measured.

4. Kinetic time courses for intermediate formation and decay are analyzed with single wavelength nonlinear fitting and/or multiple-wavelength global fitting methods that are included in the software package provided with the instrument or through third-party suppliers such as KinTek Inc. This enables an informed decision of reaction time points where best to trap intermediates using the RFQ instrument for characterization by Mössbauer, EPR, or a variety of other spectroscopic techniques.
5. The RFQ instrument must feature drive syringes that can be dismounted and loaded with the reduced protein sample inside the anaerobic glove box. For the high sMMOH concentration RFQ experiments, the MMOB protein (anaerobic) is added to the reduced sMMOH protein inside the glove box and the protein complex is then loaded into the RFQ syringe. This procedure significantly decreases the viscosity of the protein solution due to sMMOH–MMOB interaction, but it also greatly increases the rate of reaction with any O_2 inadvertently introduced during sample handling. Typically, a 2.5-mL solution of sMMOH is prepared at a concentration of 500 μ*M* (1 m*M* active sites) for RFQ experiments.
6. Traditionally, RFQ samples are frozen by dispensing samples into an isopentane bath at −150°C. An improved and far less hazardous method for freezing the samples has been developed by freezing the dispensed solution on counter-rotating aluminum wheels cooled to liquid N_2 temperature (Fig. 2A and B; Lin, Gerfen, Rousseau, & Yeh, 2003; Mbughuni et al., 2010). The large heat capacity of the wheels causes very rapid freezing, and the counter rotation of the wheels allows the frozen solution to be ground to a coarse powder to facilitate packing into a sample container.
7. The sample reaction time in a RFQ experiment is controlled by two factors: (1) solution velocity through the mixer and (2) the length of tubing between the mixer and a nozzle that dispenses the reaction solution onto the rotating wheels. A calibrated set of tubing is provided with the instrument.
8. The freeze-quench powder samples are packed into EPR tubes and Mössbauer cups with custom-designed adapters (Fig. 2C and D).

The intermediates that have been discovered in the sMMOH reaction cycle are shown in Fig. 3. Reaction of diferrous sMMOH with O_2 occurred

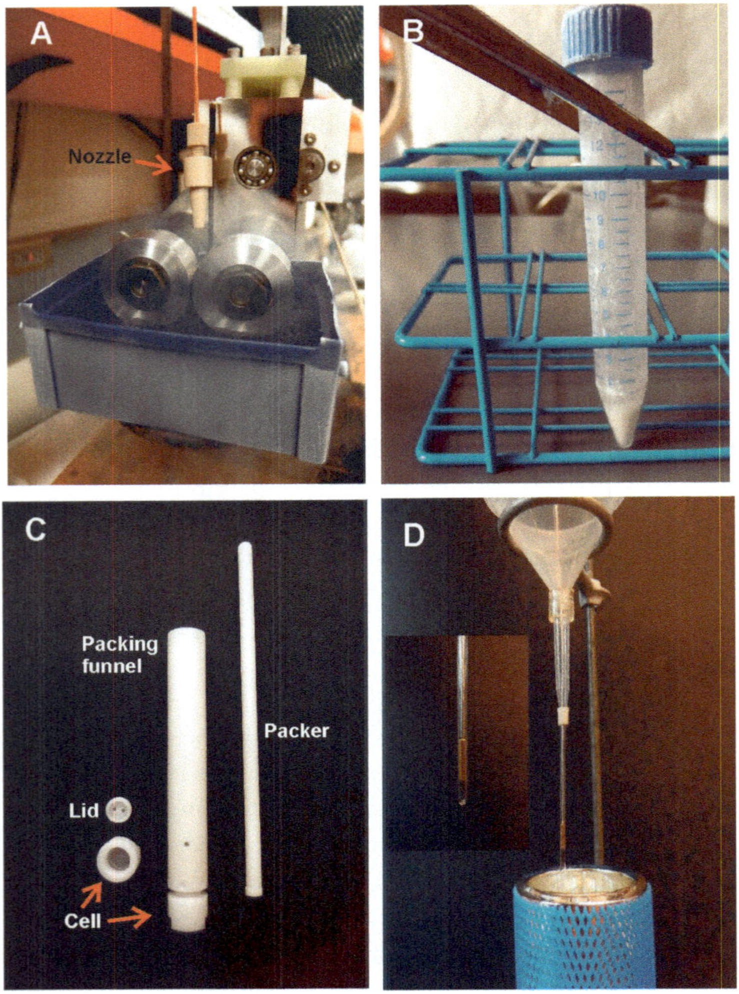

Fig. 2 (A) RFQ freezing device. The outlet nozzle from the RFQ instrument sprays the sample on the surface of the counter-rotating aluminum wheels, which are partially immersed in liquid nitrogen. (B) Freeze-quenched powder recovered from the freezing bath is stored under liquid nitrogen in 15 mL conical tubes. (C) Apparatus for packing the RFQ powder into Mössbauer cells, which are shown both attached and detached from the packing funnel. (D) Apparatus for packing the RFQ powder into EPR tubes under liquid nitrogen conditions. *Inset*—a packing rod machined to the inside diameter of the EPR tube is used to pack the frozen powder.

rapidly only in the presence of MMOB. The $g=16$ EPR spectrum disappeared and the optical spectrum of an intermediate termed P appeared between 600 and 700 nm (Lee, Nesheim, et al., 1993). The Mössbauer spectrum of P showed that the irons are exchange-coupled high-spin Fe^{3+}, but

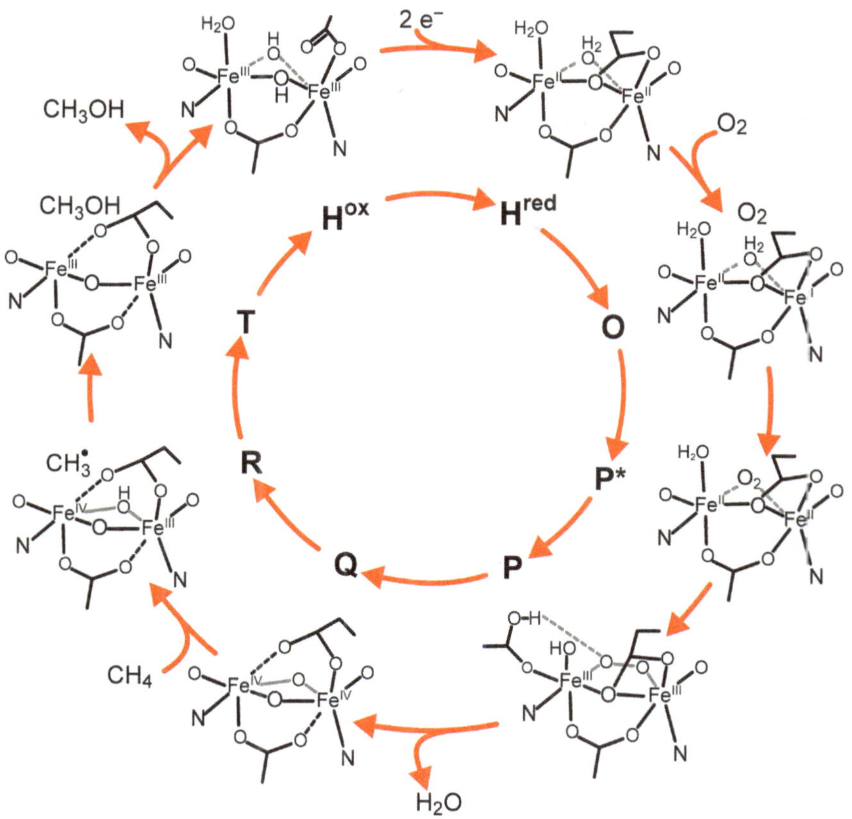

Fig. 3 Proposed catalytic cycle of sMMO. The reaction starts from the resting diferric form of sMMOH (H^{ox}) by reduction with two electrons from MMOR to generate the diferrous state (H^{red}). The sMMOH:MMOB complex in this state is primed to react with O_2 and generate intermediates O, P*, P, Q, R, and T in the catalytic cycle.

with parameters similar to those of model complexes in which a μ-1,2-peroxo complex had formed (Fig. 1; Banerjee et al., 2013). In essence, O_2 had occupied a bridging position between the irons and the electrons had moved almost completely from the metals to the oxygens. The rate constant for the disappearance of the $g=16$ signal was more than double the rate constant of appearance of intermediate P based upon its absorption at 700 nm, showing that another intermediate, termed P*, occurs in between (Banerjee et al., 2013; Lee & Lipscomb, 1999). Moreover, the rate constant for $g=16$ disappearance was found to be independent of O_2 concentration, meaning that an intermediate, termed O, must be irreversibly formed in between diferrous sMMOH and P* (Liu et al., 1995). The presence of the $g=16$ signal and the ^{57}Fe Mössbauer spectra of RFQ samples from the earliest time points after O_2 addition showed that intermediate O has a diferrous cluster indistinguishable from that of diferrous sMMOH. This finding suggested that O_2 binds in the active site but not to the diiron cluster. Due to the loss of the $g=16$ signal during the formation of P*, it was assumed that this intermediate had a diferric cluster as found in P. However, carefully coordinated optical stopped-flow and RFQ ^{57}Fe Mössbauer experiments showed that P* still has a diferrous cluster, but a change in the nature of the exchange coupling eliminated the $g=16$ signal (Banerjee et al., 2013). Thus, O_2 is likely to bind weakly to the cluster in way that does not allow transfer of the electrons from the metals to stabilize the complex. Subsequent proton transfer in the active site detected by pH-dependent kinetics resulted in the formation of the diferric peroxo intermediate P (Lee & Lipscomb, 1999). The weak optical features and lack of EPR spectra of P* and P meant that isotopic substitution of ^{57}Fe to allow Mössbauer spectroscopy was critical for the structural characterization for these intermediates.

Intermediate P converts spontaneously in another pH-dependent reaction to an intensely yellow intermediate termed Q (Lee, Nesheim, et al., 1993). In contrast to O, P*, and P, intermediate Q was shown in stopped-flow experiments to react rapidly with methane and adventitious substrates to yield a product complex termed T. Thus, Q is the reactive intermediate capable of breaking the 105-kcal per mole bond of methane and inserting an atom of oxygen. Mössbauer spectra of an sample obtained by RFQ of ^{57}Fe-enriched Q revealed a novel antiferromagnetically coupled, high-spin dinuclear Fe^{4+} cluster (Fig. 1; Lee, Fox, Froland, Lipscomb, & Münck, 1993). No similar cluster had been observed in a biological system, and no synthetic, biomimetic complex had been synthesized at the time of this finding. The structure of Q was initially proposed based on RFQ extended X-ray absorption fine structure (EXAFS) measurements, which showed an unusually short Fe–Fe distance of 2.46 Å, a distance that could only be achieved by breaking the O–O bond and creating a *bis*-μ-oxo "diamond core" structure (Shu et al., 1997).

3.3 Mass Properties That Impact Primarily Vibrational Spectroscopies

3.3.1 Time-Resolved Resonance Raman Spectroscopy

The intense yellow color of Q suggested that it would be a good candidate for characterization by vibrational spectroscopies, in particular resonance Raman spectroscopy (rRaman). This technique greatly enhances the intensity of Stokes-shifted, scattered radiation when the sample is illuminated at a wavelength near that of an absorbance maximum of the chromophoric species (i.e., an electronic transition). The shift reveals the characteristic vibrational modes of bonds within the species giving rise to the chromophore, in this case, the diiron cluster complex. Unfortunately, rRaman of frozen biological samples poses experimental problems and transient intermediates such as Q do not generally have a sufficient lifetime in solution to obtain an interpretable spectrum. They may also be photolabile in the intense laser light used in the rRaman experiment. One method to address these problems is continuous flow time-resolved resonance Raman spectroscopy (TR^3) perfected in the laboratory of Denis Proshlyakov at Michigan State University (Banerjee, Proshlyakov, Lipscomb, & Proshlyakov, 2015; Proshlyakov, Henshaw, Monterosso, Ryle, & Hausinger, 2004). In brief, the reduced enzyme and O_2-saturated buffer are actively mixed in a 1:1 ratio, followed by flow through an observation cell (Fig. 4). The aging time of the sample prior to data collection is controlled by the flow rate. In this way, probing the enzyme reaction at a certain distance from the mixing point enables the resonance Raman spectrum for a specific intermediate to be determined. A continuous flow of the reactants ensures that fresh reaction sample is constantly being probed by the laser beam and avoids data collection from the decay products of photolabile enzyme intermediates.

Equipment

1. High-pressure syringe infusion pumps (Harvard Apparatus)
2. Gastight syringes (Hamilton, SGE)
3. Computer-controlled actuated HPLC valves (Upchurch Scientific, Rheodyne, IDEX Health & Science)
4. Temperature controller (Model 34, Cryocon)
5. Quartz flow cuvette—25 mm long, internal cross section 0.15×1.5 mm (Starna)
6. Translation stage, actuator motor (model ZST25, Thorlabs Inc.), motion control board (KFLOP, DynoMotion)
7. Near-UV laser (351 nm Argon laser, Coherent model Innova 300)

Reagents and Buffers

1. Buffer A: 75 m*M* MOPS, pH 7.0 containing 5% glycerol, 0.1% Triton X-100, 200 μ*M* $Fe(NH_4)_2SO_4 \cdot 6H_2O$, and 2 m*M* L-cysteine. For experiments with D_2O, the buffer is prepared with D_2O and the pH meter reading adjusted according to the equation pD = pH + 0.4.

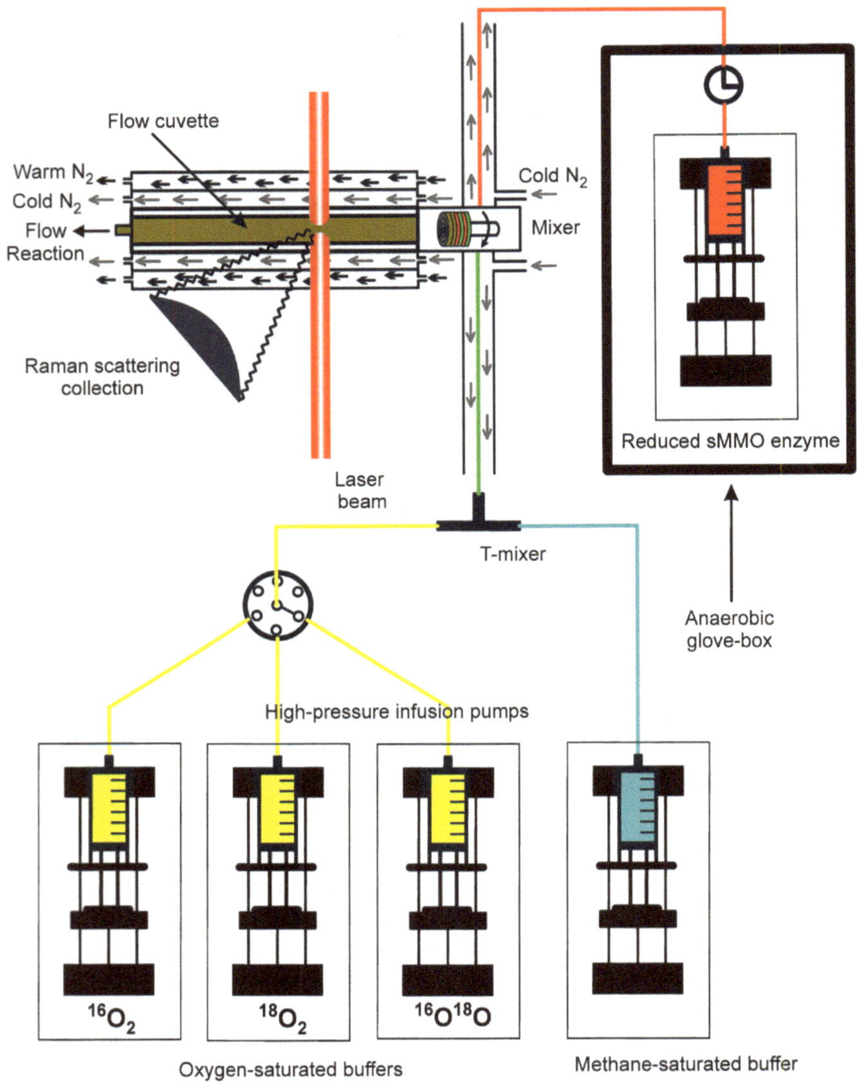

Fig. 4 Schematic diagram of the continuous flow time-resolved rRaman experimental setup. The reduced sMMOH:MMOB complex is combined sequentially with isotopic oxygen-saturated buffers using an active mixer to initiate the reaction. In some experiments, methane-saturated buffer solution is also mixed. The mixed enzyme solution flows through a cuvette where it is probed by the laser beam, which leads to Raman scattering.

2. Buffer B: 75 m*M* MOPS pH 7.0 containing 5% glycerol.
3. Oxygen ($^{16}O_2$, $^{18}O_2$, and $^{16}O^{18}O > 99\%$ purity; Cambridge Isotope Labs or ICON Isotopes: $^{16}O^{18}O$ was synthesized by E. H. Appelman using techniques detailed in Grzyska, Appelman, Hausinger, & Proshlyakov, 2010).
4. Methane (CH_4, C^2H_4, (CD_4) > 98% purity; Cambridge Isotope Labs or ICON Isotopes).

Procedure

Protein, isotopic oxygen-saturated, and methane-saturated buffer preparation

1. The sMMOH protein is chemically reduced as described in Section 2.3. The protein solution is buffer exchanged into buffer A during the PD-10 desalting step.
2. A stoichiometric amount of anaerobic MMOB relative to sMMOH active sites is added to the reduced sMMOH. The working concentration of sMMOH is 240 μ*M* (480 μ*M* active sites). The protein concentration is selected based upon the extinction coefficient of the intermediate at the probed wavelength and the desired absorbance in the flow cuvette.
3. The sMMOH:MMOB protein complex is loaded into a gastight syringe which is installed on a computer-controlled high-pressure syringe infusion pump inside an anaerobic glove box.
4. Isotopic O_2-saturated buffers are prepared by adding 1 mL of O_2 gas ($^{16}O_2$, $^{18}O_2$, $^{16}O^{18}O$) to 6 mL of anaerobic buffer B in separate 10 mL gastight syringes followed by vigorous agitation with a stir bar on a stir plate for 30 min. 5 mL of the isotopic O_2-saturated buffer is withdrawn into individual 5 mL gastight syringes while taking care to avoid air bubbles. These three syringes ($^{16}O_2$, $^{18}O_2$, $^{16}O^{18}O$) are installed on computer-controlled high-pressure syringe infusion pumps.
5. Methane-saturated buffers in two isotopic forms (CH_4, CD_4) are prepared and used in an identical manner as the oxygen-saturated buffers.

TR^3 instrument setup

1. The core of the TR^3 instrument (Fig. 4) is largely the same as that designed for cryogenic studies (Grzyska et al., 2010).
2. The reactant syringes are connected to the TR^3 system through HPLC tubing and computer-controlled actuator valves. This enables an automated and consistent switching of the isotopic O_2-saturated buffer from one to the other without stopping the flow of the reduced enzyme and/or methane-saturated buffer. This minimizes the nonisotopic variation in the Raman scattering data.

3. The temperature of the reactant solutions, the mixer, and the quartz flow cuvette is maintained at 4°C with cold nitrogen gas flowing within insulation jackets enclosing the reactant solution lines. The nitrogen gas is chilled to 4°C by adjusting the temperature of the nitrogen stream coming from a dry ice/ethanol bath with a 25-W temperature controller. The temperature of the mixer is additionally controlled with miniature Peltier cells.
4. The enzymatic reaction is probed with the laser beam at a distance of 5 mm from the mixer outlet in the flow cuvette. The dead volume is ~3 μL. Accordingly, varying the reactant flow rate between 100 and 2.5 μL/min per reactant line leads to reaction aging times of 1–36 s at the probe point.
5. Owing to laminar flow, the reaction solution close to the walls of the cuvette flows slower than the intended flow rate and represents a longer aged sample. To avoid Raman scattering from this sample, the laser beam is focused so that a small subset of the reaction solution close to the center of the flow cuvette is sampled.
6. The slow-moving protein solution also precipitates on the walls of the cuvette at the probe point of the laser. To avoid this, the cuvette is constantly laterally translated in a reciprocal manner along the direction of sample flow with the help of a translation stage driven by an actuator motor controlled by a motion control board (distance = 1 mm; speed 0.02 mm/s).

Resonance Raman data collection and analysis

1. A laser line at 351 nm is utilized in order to maximize the resonance enhancement from the electronic absorption spectrum of Q ($\varepsilon_{350} = 20{,}000\,M^{-1}$/cm). The laser power is maintained at 60 mW at the sample probe point.
2. The Raman shifts are calibrated with acetone in the flow cell before data collection. An extensive purge with anaerobic buffer removes the acetone from the flow setup to prevent reaction with Q.
3. The Raman scattering data were collected by alternating between the O_2 isotopes 4–12 times during the course of a single experiment. A short pulse at a higher flow rate (200 μL/min) of the O_2-saturated buffers is delivered following every switch to purge the previous oxygen isotope from the lines.
4. When testing for the effect of substrate, the methane-saturated buffer is mixed in a 2:1 ratio with the oxygen-saturated buffer through a T-junction mixer before the combined solution is mixed with the reduced enzyme in the TR^3 flow cuvette.

5. A quantitative analysis of the Raman spectra collected over multiple runs with different batches of protein was enabled through normalization to the intensity of a protein vibrational band at 756 cm^{-1}.
6. The absolute rRaman spectrum of the sMMO reaction is overwhelmed by nonresonant vibrations of the buffer solution. However, the extreme care taken in the instrumental design to minimize nonisotopic variations allows subtraction of the absolute spectra to create difference spectra that highlight the Raman mode involving motion of the labeled oxygen atom(s) associated with the chromophore of Q.

The vibrational mode of Q was identified in the $^{16}O_2$–$^{18}O_2$ difference spectrum as a feature that rapidly disappeared when CH_4 was added to the mixed solutions, but largely retained upon addition of CD_4 (see discussion of the kinetic isotope effect (KIE) below) (Fig. 5). The vibrational mode was found at 690 cm^{-1} with an ^{18}O downshift of 36 cm^{-1} (Banerjee et al., 2015). The energy of the vibrational mode and the magnitude of the downshift are characteristic of a tetraatomic vibration as observed for *bis*-μ-oxo "diamond core" model complexes (Wilkinson et al., 1998). This finding definitively

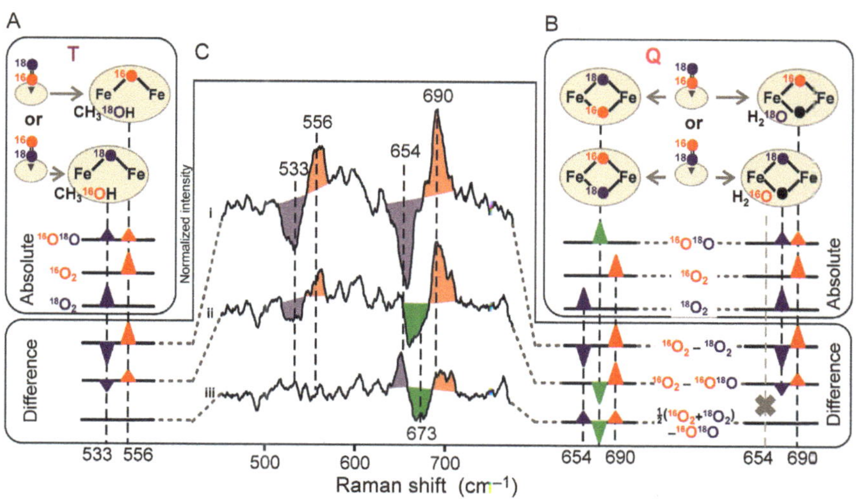

Fig. 5 The rRaman spectra expected from the different isotopic oxygen incorporation patterns for species T (A) and Q (B) are shown. (C) The experimentally observed rRaman difference spectra are compared with the theoretical difference spectra derived from the spectra in (A) and (B). The comparisons indicate that both atoms from O_2 are retained in a diamond core structure in Q. One of these is transferred to product as T is formed. *Adapted from Banerjee, R., Proshlyakov, Y., Lipscomb, J. D., & Proshlyakov, D. A. (2015). Structure of the key species in the enzymatic oxidation of methane to methanol.* Nature, 518, *431–434.*

identified the structure of the diiron cluster of Q. Importantly, a single vibrational mode with exactly half the downshift was observed when $^{16}O^{18}O$ was used, showing that both oxygens from O_2 are retained in the diamond core. If this were not true, one of the oxygen atoms would originate from solvent ($H_2{}^{16}O$) and the other randomly from $^{16}O^{18}O$ so that two vibrational bands from $Fe_2{}^{16}O_2$ and $Fe_2{}^{18}O^{16}O$ diamond cores would be observed. The finding that both O_2 oxygens are retained in the diamond core is strongly suggestive of a homolytic O–O bond cleavage mechanism, in sharp contrast to the heterolytic cleavage typically proposed for the formation of the electronically equivalent compound I of heme enzymes (Ogliaro, de Visser, & Shaik, 2002; Poulos, 2014).

A second vibrational mode was observed in the TR^3 spectra at $556\,cm^{-1}$ with an ^{18}O downshift of $23\,cm^{-1}$. In contrast to the Q vibrational mode, the $556\,cm^{-1}$ vibrational band increased in amplitude when substrates were added to the reaction mixture, suggesting that it arises from the product complex T. The frequency and oxygen isotopic downshift combined with a lack of isotopic sensitivity to solvent deuteration are consistent with a μ-oxo-bridged diiron cluster, suggesting that one oxygen atom had been transferred to the substrate to form the product. The same two vibrational frequencies at half the intensity were observed when the mixed isotope $^{16}O^{18}O$ was employed as expected if only one oxygen atom from Q is retained in the diiron cluster structure. This finding also showed that the product is not incorporated into the cluster as a terminal or bridging ligand as it is formed.

3.3.2 Other Spectroscopies

Another method that has great potential application in the study of enzyme intermediates such as Q combines the Mössbauer effect with vibrational spectroscopy. Nuclear resonance vibrational spectroscopy (NRVS) utilizes synchrotron radiation near the frequency of the monochromatic ^{57}Fe nuclear transition of the Mössbauer effect (Scheidt, Durbin, & Sage, 2005). All vibrational modes associated with the iron ligation are detected, thereby yielding more information than rRaman spectroscopy, which in principle only detects modes arising from symmetric vibrational modes (although the low symmetry of enzyme active sites often allows asymmetric modes to be observed). Also, the technique can be used with frozen RFQ samples of intermediates such as Q. To date, no results of NRVS studies of Q formed with ^{57}Fe-enriched sMMOH and various O_2 isotopes have been reported.

3.4 BDE Shifts That Permit Dynamic Isotope Effect Studies

3.4.1 Tracking the Origin of the Oxygen Atom in Methanol

The ^{18}O isotope of oxygen found in $^{18}O_2$ and $H_2{}^{18}O$ has traditionally been used to identify the source of the oxygen atom in the oxidized product (Hayaishi, Katagiri, & Rothberg, 1955; Mason, Fowlks, & Peterson, 1955). Oxygenase enzymes can also be unambiguously classified as either monooxygenases or dioxygenases by tracking the fate of the isotopically labeled oxygen atoms. In sMMO, a single turnover reaction with $^{18}O_2$ in the presence of $H_2{}^{16}O$ buffer demonstrated that 100% of the methanol product was isotopically labeled with ^{18}O (Nesheim & Lipscomb, 1996).

Equipment

1. Reaction vial (Reacti-Vial, Thermo Scientific)
2. Gastight syringes (Hamilton, SGE)

Reagents

1. $^{18}O_2$ (Cambridge Isotope Labs or ICON Isotopes, >99% purity)
2. Methane (CH_4, CD_4; Cambridge Isotope Labs or ICON Isotopes, >98% purity)

Procedure

Single turnover reaction

1. The sMMOH protein is chemically reduced as described in Section 2.3. A stoichiometric equivalent amount of anaerobic MMOB as sMMOH active sites (120 μ*M* active sites) is added to the reduced sMMOH protein in a 3-mL septum-sealed reaction vial inside the anaerobic glove box.
2. The $^{18}O_2$-saturated buffer (50 m*M* MOPS, pH 7.0) is prepared by adding 1 mL of $^{18}O_2$ gas to 6 mL of anaerobic buffer contained in a 6-mL crimp cap sealed vial. 1 mL of buffer is removed from the vial to maintain the sealed vial at atmospheric pressure. The solution is vigorously agitated for 30 min with a stir bar and stir plate at 4°C. The solubility of oxygen in water at 4°C is 1.8 m*M*.
3. The methane-saturated buffer is prepared in an identical manner as the oxygen-saturated buffer. The solubility of methane in water at 4°C is 2.0 m*M*.
4. The methane and the oxygen-saturated buffers are withdrawn into a gastight syringe in volumetric amounts to enable a working concentration of 1.4 m*M* methane and 0.4 m*M* oxygen. This solution is mixed in a 1:1 ratio with the reduced sMMOH:MMOB solution to initiate the reaction.

Methanol derivatization

1. As methanol is miscible with water, it is hard to detect with great sensitivity on a gas chromatograph instrument. Accordingly, it is esterified with the Schotten–Baumann procedure. The sMMO single turnover reaction solution (2 mL) is added to 50 mL of dichloromethane and esterified using 3,5-dinitrobenzoyl chloride (2 mmol) and NaOH (2 mmol). The reaction is allowed to proceed for 17 h at room temperature with constant stirring.
2. Multiple washes with 1.0 *M* HCl (3 × 100 mL) and a 5% $NaHCO_3$, 2.0 *M* NaCl solution (3 × 100 mL) is applied to the organic layer. Anhydrous sodium sulfate is then used to dry the organic layer. The contents of the organic layer are concentrated by evaporation until dry, followed by dissolution in 2 mL dichloromethane.
3. The methyl ester product is isolated (R_f = 0.5–0.7) by thin-layer chromatography on silica gel plates with chloroform as the mobile phase.
4. The region containing the product is scrapped off the plate and its contents dissolved in dichloromethane. The solution is concentrated by evaporation.
5. The methyl ester product is quantified on a capillary DB-5 column-equipped gas chromatograph instrument with direct injection into a Kratos MS-25 double-focusing mass spectrometer set at an ionization potential of 70 eV. The retention times and concentration are estimated with the help of authentic standards.

Notes

The use of $^{18}O_2$ and/or deuterated or ^{13}C-labeled methane shifts the mass of the enzyme reaction-derived product from the trace levels of contaminating methanol in the organic solvents used for the derivatization reaction.

3.4.2 Chiral Ethane

The mechanism of O–O and C–H bond cleavage and subsequent oxygen insertion into methane by sMMO has benefited from isotope effect studies of several types. Important early insight into the mechanism of the sMMO reaction derived from the synthesis of the alternative substrate ethane with a chiral methyl functional group (Priestley et al., 1992). Through the use of chiral borohydride reagents, acetaldehyde was reduced to yield *R*- or *S*-[1-^{2}H,1-^{3}H]-ethane. Subsequent reaction in a multiple turnover sMMO system resulted in the formation of ethanol in which the large tritium kinetic isotope effect allowed oxygen to be preferentially inserted in only the C–^{1}H or C–^{2}H bonds. The product ethanol enantiomers were derivatized to form the (2*R*)-2-acetoxy-2-phenylethanoates and then analyzed by tritium NMR to determine the degree to which chirality had been retained. It was observed that both *R*- and *S*-chiral ethane became about 35% racemized in the product, showing that an intermediate is formed that can rotate around the C–C bond before the final product is made (Fig. 6). Two conclusions could be drawn: (a) the intermediate is likely to be a radical formed by hydrogen atom abstraction in a reaction reminiscent of that proposed for P450 compound I reaction with substrates and (b) the intermediate has a lifetime on the order of a bond rotation, $\sim 10^{-12}$ s since incomplete racemization was observed. This lifetime proved to be very difficult to evaluate by nonisotope-based techniques such as radical clock chemistry (Jin & Lipscomb, 1999; Liu, Johnson, Newcomb, & Lippard, 1993; Valentine, LeTadic-Biadatti, Toy, Newcomb, & Lippard, 1999). The conclusions from the chiral ethane experiment also launched numerous computational studies that evoked many new models for the reaction mechanism (Basch, Mogi, Musaev, & Morokuma, 1999; Rinaldo, Philipp, Lippard, & Friesner, 2007; Siegbahn,

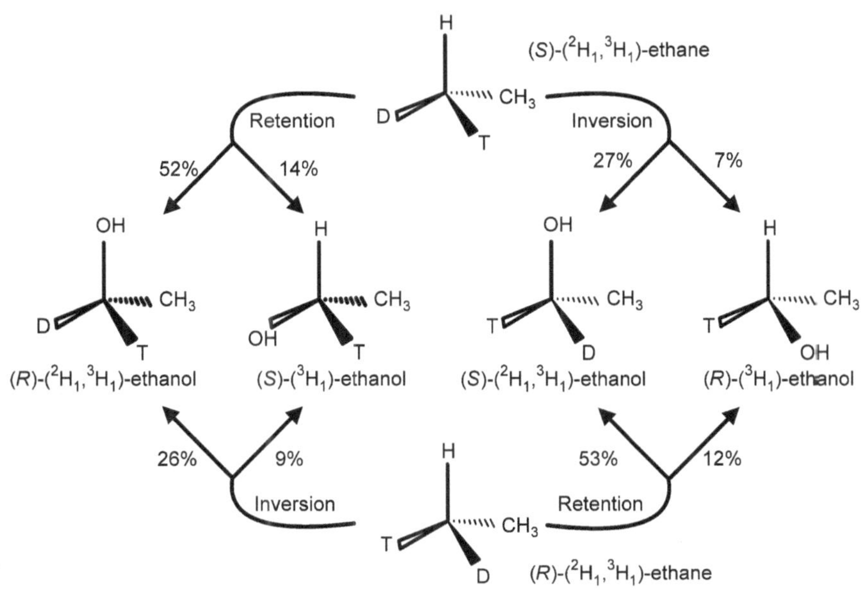

Fig. 6 Stereochemical distribution of ethanol products obtained through the sMMO oxidation of chiral *R*- or *S*-[1-^{2}H,1-^{3}H]-ethane. *Adapted from Priestley, N. D., Floss, H. G., Froland, W. A., Lipscomb, J. D., Williams, P. G., & Morimoto, H. (1992). Cryptic stereospecificity of methane monooxygenase.* Journal of the American Chemical Society, 114, *7561–7562.*

2001; Yoshizawa, 2000). Currently, a mechanism is favored in which a P450-like hydrogen atom abstraction and rebound step is envisioned with the caveat that the radical intermediate must be "caged" so that no diffusion is required during the rebound step (Gherman, Dunietz, Whittington, Lippard, & Friesner, 2001). This mechanism would account for only partial racemization of chiral ethane despite that high rate constant of C–C bond rotation.

Equipment

1. Specialized equipment for manipulating radioactive gas, radiolabeled reagents, substrates, and products.
2. 320 MHz NMR instrument with a $^{3}H/^{1}H$ 5 mm dual probe.

Reagents

1. Tritium gas (97.9% $^{3}H_2$—Oak Ridge National Laboratories)
2. (*R*)- or (*S*)-alpine borane (Sigma-Aldrich)

Procedure

Synthesis of (*R*)- and (*S*)-[1-$^{2}H_1$,$^{3}H_1$]ethane

1. The synthesis is initiated by the reduction of [1-^{2}H]acetaldehyde (99% ^{2}H) with either (*R*)- or (*S*)-alpine borane to give (*R*)- or (*S*)-[1-^{2}H]ethanol, respectively, with approximately 40%–50% yield. Following distillation, the enantiomeric purities of the ethanol products are determined by ^{2}H NMR analysis of their (2*R*)-2-acetoxy-2-phenylethanoate derivatives (88% and 96% ee for (*R*)- and (*S*)-[1-$^{2}H_1$]ethanol, respectively).
2. The ethanol enantiomers are converted to ethyl tosylates using 4-methyl benzenesulfonyl (tosyl) chloride without loss of enantiomeric enrichment.
3. The next step involves the preparation of carrier-free $LiEt_3B^{3}H$ by a reaction of *n*-butyllithium (*n*-BuLi) and $^{3}H_2$ on a 0.2-mmol scale (5.7 Ci per synthesis). A solution of *n*-BuLi in hexane (0.2 mmol) is rapidly stirred in the presence of an atmosphere of carrier-free tritium gas. Tetramethylethylenediamine (TMEDA; 0.250 mmol) is added, leading to the precipitation of LiT. After 1 h, a stoichiometric amount of triethylborane (0.2 mmol) in tetrahydrofuran or diglyme is added to the reaction vial to form the supertritide, $LiEt_3B^{3}H$.
4. The (*R*)- and (*S*)-[1-$^{2}H_1$,$^{3}H_1$]ethanes are finally generated from the (*R*)- and (*S*)-[1-$^{2}H_1$]ethyl tosylates, in ca. 45% yield, by the S_N2 displacement of the tosylate group with carrier-free $LiEt_3B^{3}H$. The yield is calculated from the increase in pressure in the reaction vial, typically increasing from 0 to approximately 210–320 mm Hg at 25°C in the 10-mL round bottom reaction vial.

Enzyme reaction

1. The chiral ethane is prepared in a vial attached to a Schlenk line. The ethane gas is then transferred to another vial on the Schlenk line containing activated charcoal at low temperature. This vial is connected to another vial containing the sMMO enzyme components, and the ethane gas is transferred to the headspace of the latter vial by warming the activated charcoal.
2. The sMMO enzyme (160 μ*M* sMMOH, 320 μ*M* MMOB, 30 μ*M* MMOR) reaction with the (*R*)- or (*S*)-chiral ethane substrates (90 μmol) is initiated by the addition of >90 μmol O_2 and 22 μmol NADH in 50 m*M* MOPS buffer at pH 7.5.
3. After completion of the reaction (30 min), the volatile components of the reaction solution, i.e., water and ethanol are recovered by lyophilization (yield ~120 mCi ^3H).

Product derivatization and analysis

1. The ethanol products are derivatized with (2*R*)-2-acetoxy-2-phenylacaetic acid (acetylmandelic acid). Acetylmandelic acid (230 mg) and *N*,*N*-dimethyl-4-aminopyridine (DMAP) (5 mg) are dissolved in 8.0 mL of dichloromethane and cooled to −20°C. Dicyclohexylcarbodiimide (265 mg dissolved in 1.0 mL dichloromethane) is added dropwise to the stirred mandelate/DMAP solution, thus forming a white precipitate.
2. The ethanol/water lyophilisate from the sMMO reaction is added dropwise to the activated acetylmandelate solution at −20°C, and the reaction is subsequently warmed to room temperature and filtered.
3. The solution phase is then frozen and the dichloromethane evaporated under vacuum. The derivatized (2*R*)-2-acetoxy-2-phenylethanoate products are then transferred to 5 mm glass NMR tubes with screw caps.
4. ^3H NMR spectra are recorded using a $^3H/^1H$ 5 mm dual probe, over approximately 12 ppm using a 320-MHz NMR instrument.

Notes

The procedures described here invoke the use of carrier-free tritium gas, which is radioactive, and require specialized equipment and exceptional precautions. However, the general synthetic approach can be used to synthesize chiral ethane with any desired level of tritium enrichment.

3.4.3 Rate Constant for Q Reaction With Deuterated Methane Isotopomers

Another approach to the characterization of the mechanism of sMMO was to examine the effects of isotopic substitution on the rate constant for reaction of intermediate Q with substrates (Nesheim & Lipscomb, 1996). sMMO differs from most oxygenases in that the rate constants for the interconversion of reaction cycle intermediates prior to the reactive species (Q in this case) are independent of substrate concentration (Lee, Nesheim, et al., 1993). Thus, the reaction can be initiated by rapidly mixing diferrous sMMOH plus stoichiometric MMOB with a solution containing excess O_2 and a variable concentration of a substrate. Q forms with the same rate constant in all cases, but the decay of Q depends on substrate type and concentration (Lee, Nesheim, et al., 1993). As a result, the specific reaction cycle step in which oxygen is transferred into the unactivated C–H bond of methane or adventitious substrates can be directly observed by following the decay of the yellow chromophore of Q. A linear substrate concentration dependence was observed for the rate constant of Q decay for all alkanes (<C8) and most other substrates.

Interestingly, a KIE of approximately 50 was observed when CD_4 was used in place of CH_4 (Fig. 7A and B). The rate constants for the reactions with the other methane isotopomers decreased linearly with the number of

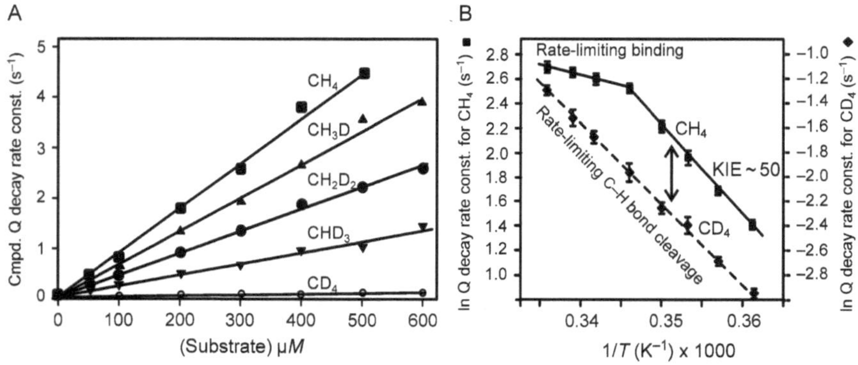

Fig. 7 (A) The rate constant of Q decay displays a large substrate KIE and decreases as the fraction of deuterium increases in the methane substrate. (B) The Arrhenius plots for Q decay during reaction with CH_4 and CD_4 show a distinct behaviors. The breakpoint in the plot with CH_4 arises from a change in the rate-limiting step from substrate binding in the active site to C–H bond breaking. In contrast, the C–D bond-breaking step is always rate limiting in the tested temperature range.

deuterium atoms in the molecule. The magnitude of the isotope effect is most consistent with a tunneling mechanism for the C–H bond-breaking reaction (Zheng & Lipscomb, 2006). Moreover, similarly large KIE values have since been associated experimentally and computationally with hydrogen atom abstraction mechanisms in model systems and other oxygenases (Carr & Klinman, 2014; Kaizer et al., 2004; Meyer, Tomchick, & Klinman, 2008; Mukherjee, Angeles-Boza, Huff, & Roth, 2011; Rittle & Green, 2010).

3.4.4 Detection of Size Selectivity and Quantum Tunneling in Q Reactions

The reactions of methane and ethane with Q occur with similar apparent second-order rate constants despite the 5-kcal per mole difference in BDE. If this difference were the only contributor to the activation energies of these reactions, a ~4000-fold difference in rate would be expected for Q decay with methane and ethane. Interestingly, there is no detectable deuterium KIE for the reaction of Q with ethane (Brazeau & Lipscomb, 2000). It was proposed that this startling difference in KIE for similar molecules arose from different types of rate-limiting steps. Structural studies of the sMMOH–MMOB interface together with kinetic studies using site-directed MMOB variants, suggested that the entry route for substrates into the active site is carefully regulated (Chang, Wallar, Lipscomb, & Mayo, 1999; Wallar & Lipscomb, 2001). It appears to be size selective for methane. Consequently, the reaction with Q is rate limited by C–H bond cleavage for methane, but the rate-limiting step becomes substrate entry into the active site for ethane. This effect is also seen in a comparison of the temperature dependence of the rate constants for CH_4 and CD_4. The former exhibits a break in the Arrhenius plot, whereas no break is seen for the latter. This suggests that the stronger C–D bond makes bond-breaking rate limiting throughout the accessible temperature range, whereas binding becomes rate limiting for CH_4 at sufficiently low temperature (Fig. 7B). By increasing the size of the entry channel using MMOB variants, a KIE of 2 for ethane was induced, presumably by accelerating the entry rate so that C–H bond cleavage becomes rate limiting (Brazeau, Wallar, & Lipscomb, 2001). However, the KIE for ethane remains in the classical range (<7), so quantum tunneling appears to be unique to the methane C–H bond cleavage reaction. Another MMOB variant removed the barriers to substrate entry, and its reactions exhibited KIE values in the classical range for both methane and ethane

(Brazeau & Lipscomb, 2003). The rate constants for the reactions using this variant gave the expected 4000-fold difference predicted by the difference in BDEs. Thus, examination of isotope effects revealed the mechanism of C–H bond cleavage used by sMMO as well as the remarkable means by which the enzyme uses size exclusion and quantum tunneling to enhance specificity for methane as a substrate (Zheng & Lipscomb, 2006). The latter characteristic is necessary because all other potential substrates are toxic for methanotrophs.

Equipment

1. Stopped-flow UV/vis spectrometer (Applied Photophysics) or similar
2. Gastight syringes (Hamilton, SGE)

Reagents

1. Methane (CH_4, CH_3D, CH_2D_2, CHD_3, CD_4; Cambridge Isotope Labs or ICON Isotopes, >98% purity)
2. Ethane (C_2H_6, C_2D_6; Cambridge Isotope Labs or ICON Isotopes, >98% purity)

Procedure

1. The sMMOH protein (120 μ*M* active sites) is chemically reduced and loaded into the drive syringe of the stopped-flow instrument as described in Section 2.3.
2. The other drive syringe of the stopped-flow instrument is loaded with a solution containing MMOB (120 μ*M*) in O_2-saturated buffer (50 m*M* MOPS, pH 7.0). If substrate is to be added to the reaction, it is mixed with this solution. This step is accomplished by addition of a calculated volume of methane- or ethane-saturated buffer (chosen to maintain pseudo-first order conditions for substrate concentration) into the same gastight syringe as the MMOB and oxygenated buffer solution. The solubility of methane and ethane in water at 25°C is 1.5 and 1.9 m*M*, respectively.
3. The single turnover sMMO reaction is followed at 430 nm, which is the λ_{max} for Q. The kinetic time course is analyzed by nonlinear regression and fits to a sum of 2-exponentials. One of the reciprocal relaxation times from this analysis is independent of the type and concentration of substrate, while the second reciprocal relaxation time depends on both. The concentration-dependent plot is linear to the limit of solubility of methane or ethane and represents the reaction of Q with the substrate. The slope of this plot yields the apparent second-order rate constant for the reaction of Q with each isotopolog of methane or ethane. The concentration-independent reciprocal relaxation time gives the rate constant for Q formation from intermediate P.

Notes

1. The substrate H/D KIE for the overall sMMO reaction can also be estimated by measuring the product ratios with methane and its deuterated isotopologs. The methanol product is quantitated as described in Section 3.4.1 and the isotopic content determined from the mass spectrum.
2. The substrate concentration-dependent reciprocal relaxation time is smaller than the substrate-independent reciprocal relaxation time at low substrate concentration, but becomes much larger at high substrate concentrations. As a result, the rate constant by which Q *appears* to form in the time course at high substrate concentration is actually reflective of the rate constant for Q decay and vice versa. Thus, care must be taken in correctly plotting the reciprocal relaxation times as described earlier.
3. Commercially supplied methane gas varies in the types and amounts of contaminants present. Many of these are either substrates (ethane, ethylene, propane, propylene, etc.) or mechanism-based irreversible inhibitors (acetylene). The true H/D KIE for methane is so large that these alternative reactions can introduce significant error in the observed KIE.

3.4.5 Measuring the Number of Protons Required During P and Q Formation

In order to decipher the role of protons in the sMMOH reaction, a pH profile was prepared for each kinetic step in the sMMO catalytic cycle (Lee & Lipscomb, 1999). Only the chemical steps of P and Q formation were found to be pH sensitive as the rate constants of both kinetic steps increased with a drop in pH from 9.0 to 6.5 ($pK_a = 7.6$ for both steps) (Fig. 8A). These proton-dependent steps were further probed by generating proton inventory plots for each (Fig. 8B). These plots depict the variation of the kinetic rate constant as a function of the deuterium mole fraction in the aqueous solvent (Schowen & Schowen, 1982). These plots can provide an estimation of the number of proton transfer events involved in the transition state of a reaction. For the P and Q formation steps, a solvent KIE value $k_H/k_D = 1.3 \pm 0.1$ and 1.4 ± 0.1 were obtained, respectively. The proton inventory plots were linear for both kinetic steps, indicating the presence of a single proton in flight during the transition stage of each conversion. Combined with the pK_a of 7.6 for each protonation event, this information indicated that neither conserved active site residues Asp242, Thr213, and Cys151 nor a localized water molecule that is part of a hydrogen bond

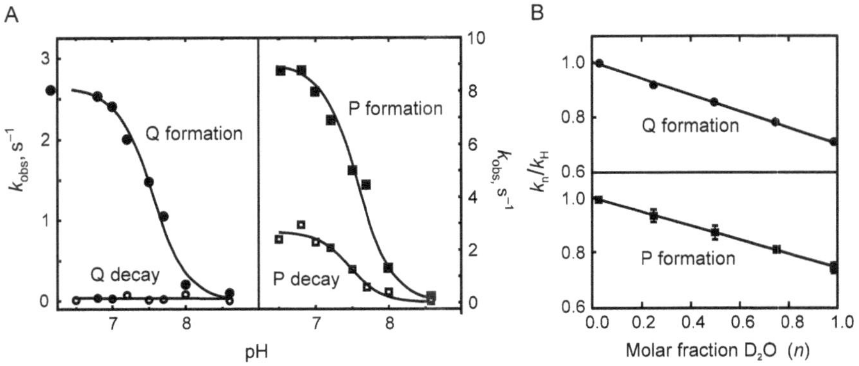

Fig. 8 (A) pH dependence of the rate constants of the formation of P, its decay to generate Q, and Q decay. (B) Proton inventory plots for the pH-dependent steps of P and Q formation. *Adapted from Lee, S. K., & Lipscomb, J. D. (1999). Oxygen activation catalyzed by methane monooxygenase hydroxylase component: Proton delivery during the O-O bond cleavage steps.* Biochemistry, 38, *4423–4432.*

network are the likely proton donors. Instead, an iron atom-bound water molecule that has been observed in X-ray crystal structures was suggested to be the crucial donor molecule during the steps of P and Q formation.

Equipment

1. Stopped-flow instrument (Applied Photophysics) or similar
2. Gastight syringes (Hamilton, SGE)

Reagents

1. 100 m*M* MOPS buffer pH = 7.0. The buffer is prepared with H_2O and D_2O. NaOD is used for pD adjustment of the D_2O buffer and the pH meter reading corrected according to the equation pD = pH + 0.4.

Procedure

1. The sMMOH protein is chemically reduced as described in Section 2.3. The sMMOH protein is buffer exchanged into the pure H_2O, D_2O, and mixtures of the H_2O/D_2O buffers on the PD-10 desalting column, where the column is equilibrated with the chosen buffer. Alternatively, a concentrated aliquot of sMMOH protein can also be diluted into the respective buffers and the protein chemically reduced with a stoichiometric equivalent of sodium hydrosulfite (see Section 2.3).
2. The reduced sMMOH, MMOB protein, and O_2-saturated buffers are prepared and loaded onto the stopped-flow instrument as described in Section 3.2.2.
3. The steps of formation and decay of intermediate P and Q are followed at 700 nm (λ_{max} for P) and 430 nm, respectively.
4. Plotting the rate constants for the formation and decay of P and Q as a function of the mole fraction of deuterium in the buffer enables a measurement of the proton inventory of the individual chemical steps.

Note

The sMMOH protein exchanges slowly in D_2O so that an incubation time of 8 h or more is required for complete exchange. The kinetics of intermediate formation are significantly altered if the exchange is not complete.

4. CmlI OXYGENASE

4.1 Mass Properties That Allow Substrate-to-Product Conversions to Be Tracked

CmlI is a diiron cluster-containing enzyme that catalyzes the final step in the biosynthesis of chloramphenicol (CAM) in *S. venezuelae* (Fig. 9; Lu, Chanco, & Zhao, 2012; Makris et al., 2015). An arylamine group of the precursor molecule is converted to the arylnitro moiety of the active antibiotic in a reaction that requires O_2. Interestingly, sMMO cannot carry out this reaction and CmlI cannot catalyze hydrocarbon oxidation, so the comparison of these enzymes provides insight into the mechanistic versatility of the diiron cluster. CmlI was definitively shown to be an oxygenase based on ^{18}O incorporation from $^{18}O_2$ into the product during a single turnover reaction (Komor et al., 2016). The procedure for monitoring isotopic incorporation into the reaction pathway intermediates and the final product CAM is described in Section 4.4.

4.2 Atomic Properties That Allow Specialized Spectroscopies

4.2.1 Mössbauer Spectroscopy

X-ray crystal structures show that the diiron clusters of sMMOH and CmlI are similar in that the supporting ligand structures consist of only Glu and His ligands, although CmlI employs one more His ligand in the site that is occupied by a water molecule in sMMOH (Knoot, Kovaleva, & Lipscomb, 2016; Rosenzweig et al., 1993). In solution at pH 9, the irons in the diferric CmlI cluster are bridged by a μ-1,3 Glu carboxylate and an oxo moiety from solvent (Makris et al., 2015). Interestingly, the X-ray crystal structure at pH 6.8 shows that the oxo bridge is replaced by one of the Glu ligands shifting to an

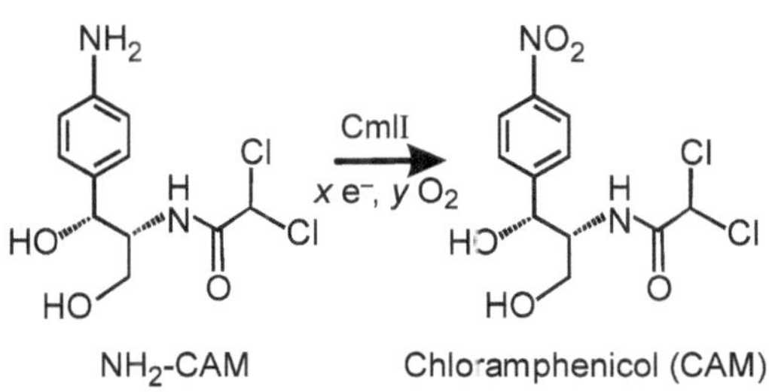

Fig. 9 Reaction catalyzed by CmlI to form chloramphenicol.

unusual μ-1,3 (η^1:η^2) bridging mode, suggesting that there are at least two conformations of the cluster ligand structure. As in the case with sMMOH, isotopic replacement of the active site iron with ^{57}Fe allowed the use of Mössbauer spectroscopy to detect the presence of the cluster and then characterize its electronic properties (Makris et al., 2015). ^{57}Fe CmlI was made by addition of 3 mg/L (53 μ*M*) ^{57}Fe in aqua regia to the growth media immediately prior to IPTG induction. The Mössbauer spectra also showed that there are at least two conformations of the cluster that change relative population with pH. The ^{57}Fe Mössbauer spectra of these two conformations become essentially indistinguishable when the diiron cluster is reduced to the diferrous state in preparation for reaction with O_2.

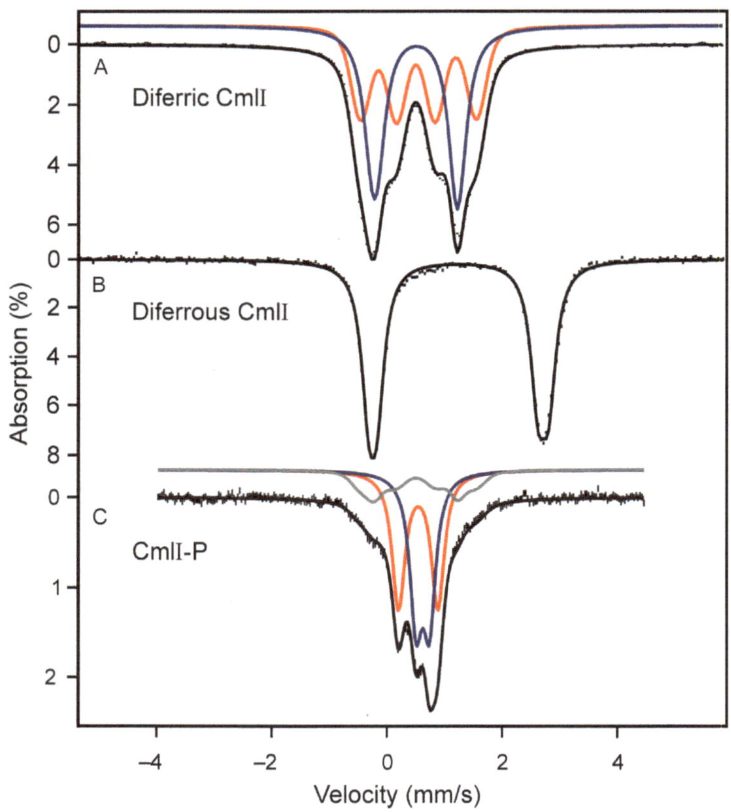

Fig. 10 Mössbauer spectra of (A) diferric, (B) diferrous, and (C) diferric peroxo CmlI. (A) The *black hash marks* show the raw data. The *blue* and *red solid lines* model two structures of the diferric cluster, the sum of which (*black solid line*) fit the experimental data. (B) The raw data are fit well by the spectrum of a single homogeneous diferrous cluster (*black solid line*). (C) The *black hash marks* show the raw data, which can be fit by the *(Continued)*

4.2.2 Preparation of the Reactive Peroxo Intermediate

Exposure of diferrous CmlI to O_2 results in the rapid formation of an intermediate with a new, broad optical chromophore that can be monitored without interference at 500 nm (Makris et al., 2015). This new species is remarkably stable in the absence of an arylamine substrate, decaying with a half-life of 3 h at 4°C, pH 9. When exposed to substrate, the species decays 100,000 times faster and produces the arylnitro product. The ^{57}Fe Mössbauer spectrum of this intermediate shows that the irons remain exchange coupled, but are each oxidized to the high-spin Fe(III) state (Fig. 10). Together, these findings suggest that a peroxo intermediate (CmlI-P) is formed and that this is the substrate reactive species rather than a dinuclear Fe(IV) species analogous to Q in sMMO. The Mössbauer spectra also show that the iron atoms of the cluster are in markedly different ligand environments, which has a bearing on the assignment of the peroxo structure as described later.

Equipment

1. 5 mL Reacti-Vials (Thermo Fisher) with lid, septum, and triangle stir bar
2. UV–vis spectrophotometer

Reagents

1. O_2 isotopes ($^{16}O_2$, $^{18}O_2$, $^{16}O^{18}O$)
2. 50 m*M* Bicine buffer, pH 9 or 50 m*M* Bicine buffer in D_2O, pD 8.6

Procedure

1. Working in an anaerobic glove box, the desired amount of diferrous CmlI (prepared as described in Section 2.3) is dispensed into a Reacti-Vial containing a stir bar. The sealed vial is removed from the anaerobic glove box and cooled on ice for 5 min. While still on ice, the vial is placed on a stir plate and O_2 gas is introduced into the headspace for 2–3 min. The vial may be manually tipped several times to ensure complete equilibration (do not invert).
2. CmlI-P concentration is determined by absorbance at 500 nm ($\varepsilon_{500} = 500\,M^{-1}$/cm) under the assumption of nearly stoichiometric formation of the intermediate.

Fig. 10—Cont'd spectrum of a single diferric peroxo species with two electronically different iron centers (*red* and *blue lines*) and a 20% diferric contaminant (*gray line*). The sum of the spectra of these species is shown as the *black line* over the data. *Adapted from Makris, T. M., Vu, V. V., Meier, K. K., Komor, A. J., Rivard, B. S., Münck, E., et al. (2015). An unusual peroxo intermediate of the arylamine oxygenase of the chloramphenicol biosynthetic pathway.* Journal of the American Chemical Society, 137, *1608–1617.*

Notes

1. CmlI-P can also be prepared in D_2O buffer by exchanging the enzyme into D_2O buffer in the diferrous state while still in the anaerobic chamber. Be aware that the solution may freeze while on ice at low enzyme concentrations. It is best to keep the enzyme as cool as possible without freezing. However, no deleterious effects from freezing and subsequent thawing have been noted.
2. The only buffer in which CmlI-P is known to be highly stable is 50 m*M* Bicine, pH 9.

4.3 Mass Properties That Impact Primarily Vibrational Spectroscopies

4.3.1 Resonance Raman Spectroscopy

The stability of CmlI-P allowed rRaman spectra to be obtained from enzyme in solution for samples prepared using $^{16}O_2$, $^{18}O_2$, and mixed-label O_2 (Fig. 11; Makris et al., 2015). A vibrational mode from an O–O stretch was observed at 791 cm^{-1} that downshifted 43 cm^{-1} when the sample was prepared with $^{18}O_2$. This value is close to the expected downshift value of 45 cm^{-1} computed using Hooke's Law for a diatomic O–O oscillator. An intermediate downshift of 23 cm^{-1} was observed when CmlI-P was prepared with the mixed-label O_2, showing that both oxygens are present in CmlI-P. Together, these observations are consistent with the assignment of CmlI-P as a diiron(III)-peroxo species. However, the observed vibrational mode is at least 60 cm^{-1} lower in energy than that of any other characterized diiron peroxo complex reported for either biological or chemical systems. Thus, the CmlI-P peroxo ligand is not bound in the usual μ-1,2-bridging mode. A tentative assignment of the structure of the novel CmlI-P intermediate was made based on rRaman spectra of diatomic mixed-metal peroxo complexes that exhibit O–O vibrational modes with similar energies. These peroxo complexes containing heme and copper, or chelated iron and yttrium or scandium have been shown to bind the oxygen as a bridging ligand side-on to one or both metals (Halime et al., 2010; Li, Van, Meier, Münck, & Que, 2013). The μ-η^1:η^2 binding mode with both oxygen atoms coordinated to one of the iron atoms and only one to the second metal exhibited the ν(O–O) values the most similar to those observed in CmlI-P and was the basis for a preliminary assignment (Makris et al., 2015). Subsequent rRaman measurements of CmlI-P formed in $^{16}OH_2$- or

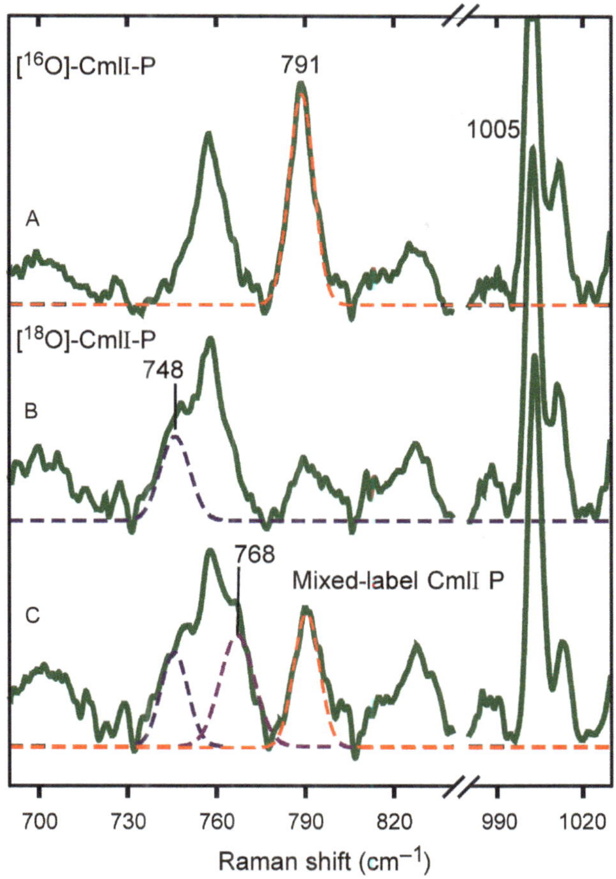

Fig. 11 Resonance Raman spectra of oxygenated CmlI, pH 9.0, prepared with (A) $^{16}O_2$, (B) $^{18}O_2$, and (C) mixed-labeled O_2. *Dashed lines* represent Gaussian fits to the peaks of interest. A peak corresponding to the vibration of an O_2-bonded species is identified due to its downshift using progressively heavier O_2 isotopes. *Adapted from Makris, T. M., Vu, V. V., Meier, K. K., Komor, A. J., Rivard, B. S., Münck, E., et al. (2015). An unusual peroxo intermediate of the arylamine oxygenase of the chloramphenicol biosynthetic pathway.* Journal of the American Chemical Society, 137, *1608–1617.*

$^{18}OH_2$-enriched buffers revealed the characteristic symmetric and antisymmetric stretching modes of an Fe–O–Fe μ-oxo-bridging ligand (Jasniewski, Komor, Lipscomb, & Que, 2017). Moreover, coordinated EXAFS and rRaman studies revealed an unprotonated μ-1,1-peroxo ligand instead of the μ-η^1:η^2 peroxo originally proposed. The resulting (μ-oxo)(μ-1,

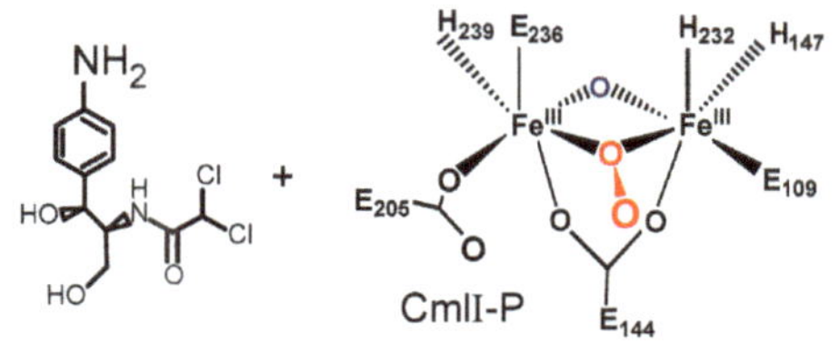

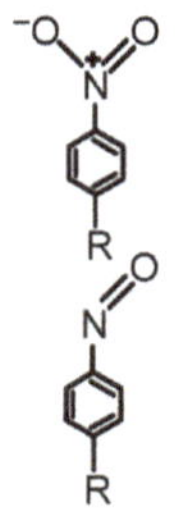

Product	Label	^{16}O-peroxo $^{16}O_2$-headspace ^{18}O-solvent	^{18}O-peroxo $^{18}O_2$-headspace ^{16}O-solvent	^{18}O-peroxo $^{16}O_2$-headspace ^{16}O-solvent
		%	%	%
NO_2-CAM	^{16}O,^{16}O	100	0.4	9
	^{16}O,^{18}O	0	7	44
	^{18}O,^{18}O	0	93	47
NO-CAM	^{16}O	98	4	8
	^{18}O	2	96	92

Fig. 12 The reaction between CmlI-P and NH_2-CAM carried out under three different isotope combinations (*top row* of table) reveals the source of oxygen incorporated into the NO-CAM and NO_2-CAM products. *Adapted from Komor, A. J., Rivard, B. S., Fan, R., Guo, Y., Que, L., Jr., & Lipscomb, J. D. (2016). Mechanism for six-electron aryl-N-oxygenation by the non-heme diiron enzyme CmlI.* Journal of the American Chemical Society, 138, *7411–7421.*

1-peroxo)diferric core, pictured in Fig. 12 below, is unprecedented and is apparently formed to increase the reactivity of the normally rather inert μ-1,2-peroxo bridge.

Equipment

1. Flat bottom NMR quartz tubes
2. Syringe with needle long enough to reach the bottom of the NMR tube
3. Acton AM-506 spectrophotometer (1200-groove grating)
4. Princeton Instruments LN_CCD-1100-PB_UVAR detector cooled to −120°C with liquid nitrogen
5. Spectra-Physics BeamLok 2060-KR-RS Argon ion laser
6. Kaiser Optical holographic super notch filter for appropriate wavelength
7. Data processing software such as GRAMS/AI (Thermo Galactic, Salem, NH), Origin, Spectrum, or Spekwin32

Procedure

1. This experiment requires CmlI-P made with $^{16}O_2$, $^{18}O_2$, and $^{16}O^{18}O$, which may be prepared immediately before the experiment or thawed from frozen CmlI-P preparations.

2. CmlI-P samples of 70–100 μL are loaded into a flat bottom NMR tube and then placed into the sample holder. Temperature is maintained at ~4°C by a steady stream of liquid N_2-cooled N_2 gas.
3. rRaman spectra are collected using the 488 nm laser line at 100 mW power using a 15 s exposure and a total collection time of 10–20 min per sample. Spectra are collected in 90 degree scattering geometry at resolution of 4 cm^{-1} and referenced to indene.
4. O_2-sensitive peaks are identified by determining which peaks shift in approximate accordance with Hooke's Law (see note) when the different O_2 isotopes are used. The absolute peak position and shifts are compared to analogous values of known complexes to generate structural hypotheses.

Notes

Hooke's Law prediction for diatomic O–O stretch: the expected resonance Raman peak shift due to isotopic substitution can be calculated using the following equation, where ν_i is the vibrational frequency of the initial peak, ν_f is that of the new peak after isotopic substitution, and μ is the reduced mass of the diatomic oscillator:

$$\nu_f = \nu_i \sqrt{\frac{\mu_f}{\mu_i}}, \ \mu = \frac{m_1 m_2}{m_1 + m_2}$$

4.4 Mass Properties That Allow the Mechanism of the Oxidative Pathway to Be Probed

4.4.1 Source of Oxygen in the Product CAM

A fundamental question in mechanistic studies of CmlI is the source of the oxygen atoms introduced to form the nitro group of CAM. The conversion of an arylamine to arylnitro is a six-electron oxidation reaction. As shown for sMMO, diiron clusters typically form one- or two-electron oxidants, so the arylamine to arylnitro conversion must occur in several steps. These steps might all involve O_2 activation and insertion, or some might proceed by a different route that results in oxygen incorporation from water. This question was addressed directly by forming CmlI-P using either $^{16}O_2$ or $^{18}O_2$ in ^{18}O-enriched or unenriched buffer, and then reacting it with substrate in a single turnover reaction (Fig. 12; Komor et al., 2016). The stability of CmlI-P allows it to be formed with $^{16}O_2$ or $^{18}O_2$, degassed, and transferred into environments with a different O_2 isotope in the headspace. Both

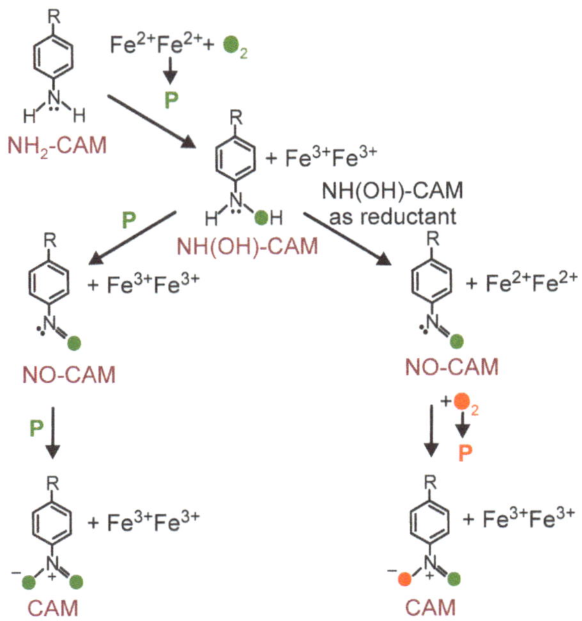

Fig. 13 Comparison of two mechanisms for the six electron oxidation of NH_2-Ar to NO_2-Ar by a diiron cluster. (*Left*) transformation is carried out by three sequential oxidation reactions performed by CmlI-P (P). (*Right*) the three steps of the mechanism are oxygenation of NH_2-CAM by P, oxidation of the resulting NH(OH)-CAM with concomitant reformation of diferrous cluster, and oxygenation of NO-CAM by P formed from the diferrous cluster by reaction with headspace O_2. *Adapted from Komor, A. J., Rivard, B. S., Fan, R., Guo, Y., Que, L., Jr., & Lipscomb, J. D. (2016). Mechanism for six-electron aryl-N-oxygenation by the non-heme diiron enzyme CmlI.* Journal of the American Chemical Society, 138, *7411–7421.*

oxygens in the CAM nitro group were labeled with oxygen derived from O_2, showing that at least two of the oxidation steps proceed by an oxygenase mechanism of some type. As illustrated in Fig. 13 (left scheme), the simplest model for the overall reaction in the native host would involve three steps in which an activated form of CmlI carries out a similar reaction at each step. Such a process would require six electrons from an external reductant to activate CmlI. Detailed isotope incorporation studies showed that this is not the case as described later.

Equipment

1. Lyophilizer (Labconco)
2. 5 mL Reacti-Vials (Thermo Fisher) with lid, septum, and triangle stir bar
3. Benchtop centrifuge that accommodates 2 mL epitubes

4. UPLC/MS^E system (Waters Acquity UPLC coupled to a Waters Synapt G2 HDMS quadrupole orthogonal acceleration time of flight mass spectrometer Waters Corp., Milford, MA, USA)
5. Waters HSS T3 C_{18} 2.1 mm × 100 mm column (1.7 μm diameter particles)

Reagents

1. $H_2^{18}O$
2. 10% Trifluoroacetic acid in water
3. UPLC buffers: 0.1% formic acid in water and 0.1% formic acid in acetonitrile

Procedure

1. To test for incorporation from O_2 derived from P, CmlI-P-$^{18}O_2$ is reacted with one equivalent of NH_2-CAM substrate while either stirring on ice on the benchtop (i.e., exposed to an $^{16}O_2$ environment) or under a positive pressure of $^{18}O_2$ gas in the headspace of a sealed vial.
2. To test for the incorporation of oxygen from water, the reactions in step 1 are repeated using CmlI-P-$^{16}O_2$ in buffer prepared with $H_2{}^{18}O$ (see next step).
3. Preparation of buffer in $H_2{}^{18}O$: 2 mL of 50 m*M* Bicine, pH 9 prepared in $H_2{}^{16}O$ is lyophilized to powder overnight and brought into the anaerobic glove box. 2 mL of $H_2{}^{18}O$ is added to the powder, and diferrous CmlI is exchanged into the $H_2{}^{18}O$ buffer by repeated concentration and dilution using a microcentrifuge and centrifugal filters located inside the anaerobic glove box.
4. The reactions between various CmlI-P preparations and NH_2-CAM are run for 10 min, at which point they are quenched with an equal volume of 10% TFA, centrifuged to remove the precipitated protein, and transferred to LCMS vials and stored at −20°C.
5. UPLC/MS is used to determine the percentage of ^{18}O incorporation under each reaction condition.

Notes

Analysis of the MS data is complicated by the fact that this experiment seeks to monitor a 2 Da mass difference (16 vs 18), and the CAM product also contains two Cl atoms. Cl has two isotopes in natural abundance that are also shifted by 2 Da (35 and 37). Thus, this experiment was first performed with the ^{16}O isotope (CmlI-P-$^{16}O_2$, $^{16}O_2$ atmosphere, and $H_2{}^{16}O$ buffer) to set a standard for the expected isotope distribution due to the Cl atoms. From this data, an Excel spreadsheet was generated that automatically determined which higher-molecular-weight products were due to ^{18}O incorporation, and which were due to ^{37}Cl incorporation. An example spreadsheet is available in the Supporting Information on http://dx.doi.org/10.1016/bs.mie.2017.07.016.

4.4.2 Testing for Ligand Exchange in the Active Oxidant

Mixed-labeled CAM product ($N^{16}O^{18}O$-CAM) was observed when one isotope of O_2 was used to prepare CmlI-P and a second was present in the headspace. One possible explanation for the mixed-labeled product was that CmlI-P could exchange its bound O_2 ligand with O_2 in the headspace. This was tested as shown in Fig. 14.

Equipment

1. 5 mL Reacti-Vials (Thermo Fisher) with lid, septum, and triangle stir bar
2. Benchtop centrifuge that fits 2-mL epitubes
3. UPLC/MSE system (Waters Acquity UPLC coupled to a Waters Synapt G2 HDMS quadrupole orthogonal acceleration time of flight mass spectrometer Waters Corp., Milford, MA, USA)
4. Waters HSS T3 C_{18} 2.1 mm × 100 mm column (1.7 μm diameter particles)

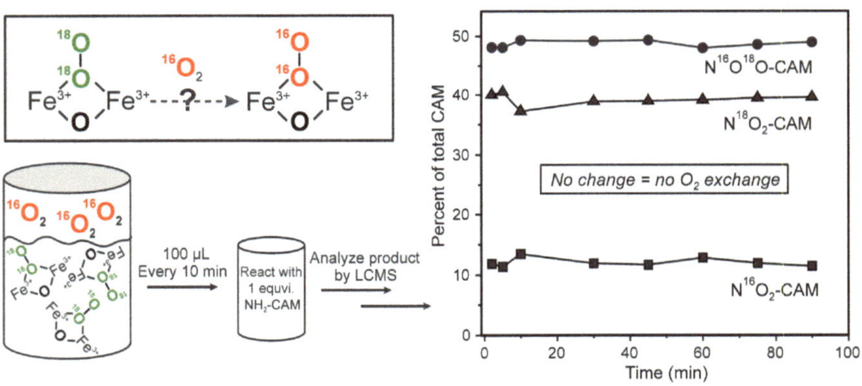

Fig. 14 Demonstration that CmlI-P does not exchange with O_2. Tracking the fate of oxygen from O_2 through the steps in the pathway requires that exchange of CmlI-P with O_2 in solution be evaluated. This is accomplished by forming CmlI-P with $^{18}O_2$ and then changing the headspace oxygen to $^{16}O_2$. Samples withdrawn and reacted with NH_2-CAM over the course of 90 min show no change in the isotope distribution in the product, indicating no O_2 exchange into CmlI-P (*right frame*). *Adapted from Komor, A. J., Rivard, B. S., Fan, R., Guo, Y., Que, L., Jr., & Lipscomb, J. D. (2016). Mechanism for six-electron aryl-N-oxygenation by the non-heme diiron enzyme CmlI.* Journal of the American Chemical Society, 138, *7411–7421.*

Reagents

1. $^{18}O_2$, $^{16}O_2$
2. 10% Trifluoroacetic acid in water
3. UPLC buffers: 0.1% formic acid in water and 0.1% formic acid in acetonitrile

Procedure

1. CmlI-P-$^{18}O_2$ (~1.5 mL) is prepared as described earlier in a 5-mL Reacti-Vial and then brought into the anaerobic chamber and allowed to stir for ~10 min until the solution is degassed. The solution is kept cold by placing the vial in a precooled metal block that can accommodate 5 mL Reacti-Vials.
2. The vial is removed from the anaerobic chamber and placed on a stir plate under a steady stream of $^{16}O_2$.
3. Periodically, 100 μL of the solution is withdrawn and reacted with one equivalent of NH_2-CAM substrate. The reaction with substrate is allowed to proceed for 10 min and then quenched. A total of 12 aliquots are taken and reacted over the course of 90 min.
4. Samples are analyzed for isotope incorporation by UPLC–MS as described earlier.

CmlI-P made with $^{18}O_2$ did not exchange with $^{16}O_2$ in the headspace or with ^{16}O from water over the period of several hours as shown by a constant level of ^{18}O incorporation into NO_2-CAM for samples reacted over this period of time. However, it was noted that the number of labeled oxygens in the nitro group of CAM did depend on the isotopic composition of the headspace O_2 during the single turnover reaction of CmlI-P with NH_2-CAM (Fig. 12). When CmlI-P made with $^{18}O_2$ was reacted with NH_2-CAM in an $^{18}O_2$ atmosphere, both nitro oxygens of CAM were labeled, but in an $^{16}O_2$ atmosphere, a substantial fraction of the CAM contained only one ^{18}O.

The possible series of oxidized intermediates shown in Fig. 13 includes hydroxylamino-CAM (NH(OH)-CAM) and nitroso-CAM (NO-CAM). These two potential intermediates were synthesized by established methods (Eyer & Schneller, 1983; Komor et al., 2016). Both intermediates proved to be substrates for CmlI-P, each producing CAM. Only NO-CAM could be detected as an intermediate during the single turnover reaction using NH_2-CAM as the substrate because NH(OH)-CAM is converted readily to NO-CAM during product extraction and workup. The oxygen in NO-CAM was shown to be derived from CmlI-P independent of the oxygen isotope in the solvent water or headspace O_2 (Fig. 12).

4.4.3 Method for Determining Oxidation State of Diiron Cluster Intermediates

The study exploring isotopic oxygen incorporation into substrates of CmlI suggested that the diiron cluster was converted to a diferrous state by a substrate-mediated reduction step (Komor et al., 2016). To determine the oxidation state of the diiron cluster, samples were taken at different points in the reaction and analyzed by Mössbauer spectroscopy. Reduction of CmlI by the pathway intermediate NH(OH)-CAM was demonstrated.

Equipment

1. 10 mL Reacti-Vial (Thermo Fisher) with lid and septum
2. Tongs
3. Face shield
4. Liquid nitrogen Dewar flask ~5 in. deep
5. Cold-protectant gloves (commercially available or layer nitrile gloves over anticut Kevlar laboratory gloves)

Procedure

1. To test the reaction between diferric resting state CmlI and NH(OH)-CAM, diferric resting state CmlI (560 μ*M*) is degassed first by purging with argon gas for 20 min, and then by transferring to the anaerobic glove bag and stirring for 10 min in the anaerobic atmosphere.
2. One equivalent of NH(OH)-CAM (freshly prepared in the anaerobic glove bag, 10% MeOH, 90% water) is added and allowed to stir for 5 min. A 100-μL aliquot of this sample is removed and quenched in the anaerobic glove bag with 100 μL of 5% TFA for HPLC analysis. The remaining enzyme substrate mixture is transferred to a Mössbauer cup that sits securely in a 10-mL Reacti-Vial and then sealed.
3. The Reacti-Vial containing the Mössbauer cup is brought out of the anaerobic chamber and carefully lowered into a shallow liquid nitrogen Dewar with just enough liquid nitrogen to cover the Reacti-Vial. *A face shield, cold-protectant gloves, and full PPE should be worn for this step and the next due to the risk of explosion.*
4. After 10 min of cooling, the top of the vial is removed, and Mössbauer cup is removed for analysis.
5. To test the reaction between CmlI-P and NH_2-CAM, CmlI-P (300 μ*M*, prepared as described earlier with $^{16}O_2$) is degassed in the anaerobic chamber, and then one equivalent NH_2-CAM (freshly prepared in the anaerobic chamber, 10% MeOH, 90% water) is added and allowed

to react, stirring, for 5 min. Mössbauer samples are prepared as described for the previous examples.

4.4.4 Mechanism

The mechanism advanced to account for these results (Fig. 15) was based on reformation of CmlI-P during the single turnover reaction, a process which would require reduction of diferric CmlI formed as a result of CmlI-P reaction (Komor et al., 2016). The substrate-based reducing agent was identified by demonstrating that NH(OH)-CAM reacts with diferric CmlI in the presence of O_2 to form CAM (Fig. 13, right pathway). The anaerobic combination of NH(OH)-CAM with diferric CmlI was definitively shown to form diferrous CmlI using Mössbauer spectroscopy with ^{57}Fe-enriched enzyme. Thus, the oxidation pathway begins by reaction of CmlI-P with NH_2-CAM to form NH(OH)-CAM and diferric CmlI. The NH(OH)-CAM resident in the same active site as the CmlI diferric cluster allows facile reduction of the cluster and oxidation of NH(OH)-CAM to NO-CAM, the next intermediate in the pathway. Finally, CmlI-P is reformed using the O_2 from the headspace and reacts with the

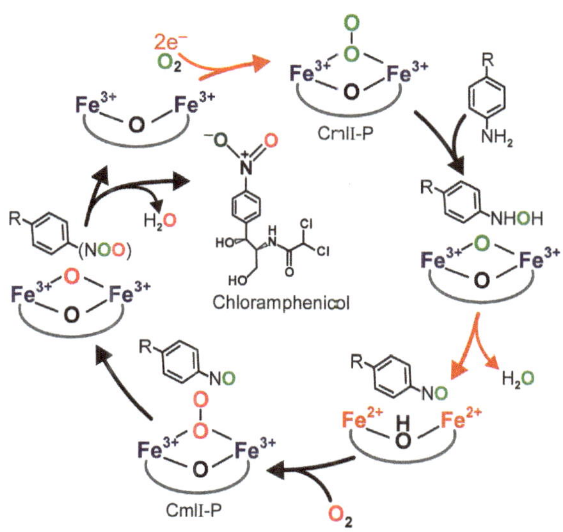

Fig. 15 Proposed CmlI reaction cycle. The *red cycle arrows* represent reduction steps. *Adapted from Komor, A. J., Rivard, B. S., Fan, R., Guo, Y., Que, L., Jr., & Lipscomb, J. D. (2016). Mechanism for six-electron aryl-N-oxygenation by the non-heme diiron enzyme CmlI.* Journal of the American Chemical Society, 138, *7411–7421.*

NO-CAM to form CAM. This process allows CAM to be formed from NH_2-CAM using only two external electrons and without the need to dissociate the progressively oxidized intermediates from the active site. It also accounts for the mixed isotope found in the nitro group of CAM when the oxygen isotope in CmlI-P and the headspace oxygen differ. The discovery of this complex mechanism was only possible through the use of isotopic-labeling techniques.

5. CONCLUSIONS

The studies described here demonstrate the power of the application of isotopes and isotope effects in the study of complex enzyme mechanisms. Indeed, the level of understanding that we now have regarding the chemical steps of the mechanisms of sMMO and CmlI, as well as the remarkable strategies these enzymes employ for regulation of activated oxygen species, is due largely to the application of isotopically enriched materials. It should be emphasized that in the broader scope of the study of enzyme mechanisms, there are many more subcategories to the four broad categories we have defined. In particular, the application of primary and secondary isotope effects in steady-state kinetic and isotope exchange studies has made seminal contributions to our understanding of enzyme catalysis. There is a continuing need for specialized facilities to enhance the use of isotopes in mechanistic studies. While techniques such as mass spectroscopy and NRVS have made stunning advances in recent years, techniques developed in the past such as isotope ratio mass spectrometry, steady-state oxygen isotope exchange, and carrier-free tritium labeling (to name a few) should be sustained for future applications.

REFERENCES

Banerjee, R., Meier, K. K., Münck, E., & Lipscomb, J. D. (2013). Intermediate P* from soluble methane monooxygenase contains a diferrous cluster. *Biochemistry*, *52*, 4331–4342.

Banerjee, R., Proshlyakov, Y., Lipscomb, J. D., & Proshlyakov, D. A. (2015). Structure of the key species in the enzymatic oxidation of methane to methanol. *Nature*, *518*, 431–434.

Basch, H., Mogi, K., Musaev, D. G., & Morokuma, K. (1999). Mechanism of the methane → methanol conversion reaction catalyzed by methane monooxygenase: A density functional study. *Journal of the American Chemical Society*, *121*, 7249–7256.

Brazeau, B. J., & Lipscomb, J. D. (2000). Kinetics and activation thermodynamics of methane monooxygenase compound Q formation and reaction with substrates. *Biochemistry*, *39*, 13503–13515.

Brazeau, B. J., & Lipscomb, J. D. (2003). Key amino acid residues in the regulation of soluble methane monooxygenase catalysis by component B. *Biochemistry, 42*, 5618–5631.

Brazeau, B. J., Wallar, B. J., & Lipscomb, J. D. (2001). Unmasking of deuterium kinetic isotope effects on the methane monooxygenase compound Q reaction by site-directed mutagenesis of component B. *Journal of the American Chemical Society, 123*, 10421–10422.

Carr, C. A. M., & Klinman, J. P. (2014). Hydrogen tunneling in a prokaryotic lipoxygenase. *Biochemistry, 53*, 2212–2214.

Chang, S. L., Wallar, B. J., Lipscomb, J. D., & Mayo, K. H. (1999). Solution structure of component B from methane monooxygenase derived through heteronuclear NMR and molecular modeling. *Biochemistry, 38*, 5799–5812.

Denisov, I. G., Makris, T. M., Sligar, S. G., & Schlichting, I. (2005). Structure and chemistry of cytochrome P450. *Chemical Reviews, 105*, 2253–2277.

Doan, P. E., Shanmugam, M., Stubbe, J., & Hoffman, B. M. (2015). Composition and structure of the inorganic core of relaxed intermediate X(Y122F) of *Escherichia coli* ribonucleotide reductase. *Journal of the American Chemical Society, 137*, 15558–15566.

Elango, N., Radhakrishnan, R., Froland, W. A., Wallar, B. J., Earhart, C. A., Lipscomb, J. D., et al. (1997). Crystal structure of the hydroxylase component of methane monooxygenase from *Methylosinus trichosporium* OB3b. *Protein Science, 6*, 556–568.

Eyer, P., & Schneller, M. (1983). Reactions of the nitroso analogue of chloramphenicol with reduced glutathione. *Biochemical Pharmacology, 32*, 1029–1036.

Fox, B. G., Froland, W. A., Dege, J. E., & Lipscomb, J. D. (1989). Methane monooxygenase from *Methylosinus trichosporium* OB3b. Purification and properties of a three-component system with high specific activity from a type II methanotroph. *The Journal of Biological Chemistry, 264*, 10023–10033.

Fox, B. G., Froland, W. A., Jollie, D. R., & Lipscomb, J. D. (1990). Methane monooxygenase from *Methylosinus trichosporium* OB3b. *Methods in Enzymology, 188*, 191–202.

Fox, B. G., Hendrich, M. P., Surerus, K. K., Andersson, K. K., Froland, W. A., Lipscomb, J. D., et al. (1993). Mössbauer, EPR, and ENDOR studies of the hydroxylase and reductase components of methane monooxygenase from *Methylosinus trichosporium* OB3b. *Journal of the American Chemical Society, 115*, 3688–3701.

Fox, B. G., Shanklin, J., Somerville, C., & Münck, E. (1993). Stearoyl-acyl carrier protein Δ^9 desaturase from *Ricinus communis* is a diiron-oxo protein. *Proceedings of the National Academy of Sciences of the United States of America, 90*, 2486–2490.

Gherman, B. F., Dunietz, B. D., Whittington, D. A., Lippard, S. J., & Friesner, R. A. (2001). Activation of the C-H bond of methane by intermediate Q of methane monooxygenase: A theoretical study. *Journal of the American Chemical Society, 123*, 3836–3837.

Grzyska, P. K., Appelman, E. H., Hausinger, R. P., & Proshlyakov, D. A. (2010). Insight into the mechanism of an iron dioxygenase by resolution of steps following the $Fe^{IV}=O$ species. *Proceedings of the National Academy of Sciences of the United States of America, 107*, 3982–3987.

Halime, Z., Kieber-Emmons, M. T., Qayyum, M. F., Mondal, B., Gandhi, T., Puiu, S. C., et al. (2010). Heme-copper-dioxygen complexes: Toward understanding ligand-environmental effects on the coordination geometry, electronic structure, and reactivity. *Inorganic Chemistry, 49*, 3629–3645.

Hayaishi, O., Katagiri, H., & Rothberg, S. (1955). Mechanism of the pyrocatechase reaction. *Journal of the American Chemical Society, 77*, 5450–5451.

Hendrich, M. P., Münck, E., Fox, B. G., & Lipscomb, J. D. (1990). Integer-spin EPR studies of the fully reduced methane monooxygenase hydroxylase component. *Journal of the American Chemical Society, 112*, 5861–5865.

Jasniewski, A. J., Komor, A. J., Lipscomb, J. D., & Que, L. (2017). An unprecedented (μ-1,1-peroxo)diferric structure for the ambiphilic orange peroxo intermediate of the nonheme N-oxygenase CmlI. *Journal of the American Chemical Society, 139*, 10472–10485.

Jin, Y., & Lipscomb, J. D. (1999). Probing the mechanism of C-H activation: Oxidation of methylcubane by soluble methane monooxygenase from *Methylosinus trichosporium* OB3b. *Biochemistry*, *38*, 6178–6186.

Kaizer, J., Klinker, E. J., Oh, N. Y., Rohde, J.-U., Song, W. J., Stubna, A., et al. (2004). Nonheme $Fe^{IV}O$ complexes that can oxidize the C-H bonds of cyclohexane at room temperature. *Journal of the American Chemical Society*, *126*, 472–473.

Knoot, C. J., Kovaleva, E. G., & Lipscomb, J. D. (2016). Crystal structure of CmlI, the arylamine oxygenase from the chloramphenicol biosynthetic pathway. *Journal of Biological Inorganic Chemistry*, *21*, 589–603.

Komor, A. J., Rivard, B. S., Fan, R., Guo, Y., Que, L., Jr., & Lipscomb, J. D. (2016). Mechanism for six-electron aryl-N-oxygenation by the non-heme diiron enzyme CmlI. *Journal of the American Chemical Society*, *138*, 7411–7421.

Kovaleva, E. G., & Lipscomb, J. D. (2008). Versatility of biological non-heme Fe(II) centers in oxygen activation reactions. *Nature Chemical Biology*, *4*, 186–193.

Krebs, C., Dassama, L. M. K., Matthews, M. L., Jiang, W., Price, J. C., Korboukh, V., et al. (2013). Novel approaches for the accumulation of oxygenated intermediates to multi-millimolar concentrations. *Coordination Chemistry Reviews*, *257*, 234–243.

Lee, S.-K., Fox, B. G., Froland, W. A., Lipscomb, J. D., & Münck, E. (1993). A transient intermediate of the methane monooxygenase catalytic cycle containing a $Fe^{IV}Fe^{IV}$ cluster. *Journal of the American Chemical Society*, *115*, 6450–6451.

Lee, S. K., & Lipscomb, J. D. (1999). Oxygen activation catalyzed by methane monooxygenase hydroxylase component: Proton delivery during the O-O bond cleavage steps. *Biochemistry*, *38*, 4423–4432.

Lee, S. K., Nesheim, J. C., & Lipscomb, J. D. (1993). Transient intermediates of the methane monooxygenase catalytic cycle. *The Journal of Biological Chemistry*, *268*, 21569–21577.

Li, F., Van, H. K. M., Meier, K. K., Münck, E., & Que, L., Jr. (2013). Sc^{3+}-triggered oxoiron (IV) formation from O_2 and its non-heme iron(II) precursor via a Sc^{3+}-peroxo-Fe^{3+} intermediate. *Journal of the American Chemical Society*, *135*, 10198–101201.

Lin, Y., Gerfen, G. J., Rousseau, D. L., & Yeh, S.-R. (2003). Ultrafast microfluidic mixer and freeze-quenching device. *Analytical Chemistry*, *75*, 5381–5386.

Ling, J., Sahlin, M., Sjöberg, B. M., Loehr, T. M., & Sanders-Loehr, J. (1994). Dioxygen is the source of the μ-oxo bridge in iron ribonucleotide reductase. *The Journal of Biological Chemistry*, *269*, 5595–5601.

Lipscomb, J. D. (2006). Catalysis and regulation in the soluble methane monooxygenase system: Applications of isotopes and isotope effects. In A. Kohen & H. H. Limbach (Eds.), *Isotope effects in chemistry and biology, Ch. 38* (pp. 931–954). Boca Raton: CRC Press.

Liu, K. E., Johnson, C. C., Newcomb, M., & Lippard, S. J. (1993). Radical clock substrate probes and kinetic isotope effect studies of the hydroxylation of hydrocarbons by methane monooxygenase. *Journal of the American Chemical Society*, *115*, 939–947.

Liu, Y., Nesheim, J. C., Lee, S.-K., & Lipscomb, J. D. (1995). Gating effects of component B on oxygen activation by the methane monooxygenase hydroxylase component. *The Journal of Biological Chemistry*, *270*, 24662–246625.

Lu, H. G., Chanco, E., & Zhao, H. M. (2012). CmlI is an N-oxygenase in the biosynthesis of chloramphenicol. *Tetrahedron*, *68*, 7651–7654.

Makris, T. M., Vu, V. V., Meier, K. K., Komor, A. J., Rivard, B. S., Münck, E., et al. (2015). An unusual peroxo intermediate of the arylamine oxygenase of the chloramphenicol biosynthetic pathway. *Journal of the American Chemical Society*, *137*, 1608–1617.

Mason, H. S., Fowlks, W. L., & Peterson, E. (1955). Oxygen transfer and electron transport by the phenolase complex. *Journal of the American Chemical Society*, *77*, 2914–2915.

Mbughuni, M. M., Chakrabarti, M., Hayden, J. A., Bominaar, E. L., Hendrich, M. P., Münck, E., et al. (2010). Trapping and spectroscopic characterization of an Fe^{III}-superoxo intermediate from a nonheme mononuclear iron-containing enzyme. *Proceedings of the National Academy of Sciences of the United States of America*, *107*, 16788–16793.

Meyer, M. P., Tomchick, D. R., & Klinman, J. P. (2008). Enzyme structure and dynamics affect hydrogen tunneling: The impact of a remote side chain (I553) in soybean lipoxygenase-1. *Proceedings of the National Academy of Sciences of the United States of America*, *105*, 1146–1151.

Moenne-Loccoz, P., Baldwin, J., Ley, B. A., Loehr, T. M., & Bollinger, J. M., Jr. (1998). O_2 activation by non-heme diiron proteins: Identification of a symmetric μ-1,2-peroxide in a mutant of ribonucleotide reductase. *Biochemistry*, *37*, 14659–14663.

Mukherjee, A., Angeles-Boza, A. M., Huff, G. S., & Roth, J. P. (2011). Catalytic mechanism of a heme and tyrosyl radical-containing fatty acid α-(di)oxygenase. *Journal of the American Chemical Society*, *133*, 227–238.

Münck, E. (1978). Mössbauer spectroscopy of proteins: Electron carriers. *Methods in Enzymology*, *54*, 346–379.

Münck, E. (2000). Aspects of ^{57}Fe Mössbauer spectroscopy. In L. Que, Jr. (Ed.), *Physical methods in bioinorganic chemistry* (pp. 287–319). Sausalito, CA: University Science Books.

Münck, E., Surerus, K. K., & Hendrich, M. P. (1993). Combining Mössbauer spectroscopy with integer spin electron paramagnetic resonance. *Methods in Enzymology*, *227*, 463–479.

Nesheim, J. C., & Lipscomb, J. D. (1996). Large isotope effects in methane oxidation catalyzed by methane monooxygenase: Evidence for C−H bond cleavage in a reaction cycle intermediate. *Biochemistry*, *35*, 10240–10247.

Ogliaro, F., de Visser, S. P., & Shaik, S. (2002). The 'push' effect of the thiolate ligand in cytochrome P450: A theoretical gauging. *Journal of Inorganic Biochemistry*, *91*, 554–567.

Pikus, J. D., Studts, J. M., Achim, C., Kauffmann, K. E., Münck, E., Steffan, R. J., et al. (1996). Recombinant toluene-4-monooxygenase: Catalytic and Mössbauer studies of the purified diiron and Rieske components of a four-protein complex. *Biochemistry*, *35*, 9106–9119.

Pilkington, S. J., & Dalton, H. (1990). Soluble methane monooxygenase from *Methylococcus capsulatus* (Bath). *Methods in Enzymology*, *188*, 181–190.

Poulos, T. L. (2014). Heme enzyme structure and function. *Chemical Reviews*, *114*, 3919–3962.

Priestley, N. D., Floss, H. G., Froland, W. A., Lipscomb, J. D., Williams, P. G., & Morimoto, H. (1992). Cryptic stereospecificity of methane monooxygenase. *Journal of the American Chemical Society*, *114*, 7561–7562.

Proshlyakov, D. A., Henshaw, T. F., Monterosso, G. R., Ryle, M. J., & Hausinger, R. P. (2004). Direct detection of oxygen intermediates in the non-heme Fe enzyme taurine/alpha-ketoglutarate dioxygenase. *Journal of the American Chemical Society*, *126*, 1022–1023.

Rinaldo, D., Philipp, D. M., Lippard, S. J., & Friesner, R. A. (2007). Intermediates in dioxygen activation by methane monooxygenase: A QM/MM study. *Journal of the American Chemical Society*, *129*, 3135–3147.

Rittle, J., & Green, M. T. (2010). Cytochrome P450 compound I: Capture, characterization, and C-H bond activation kinetics. *Science*, *330*, 933–937.

Rosenzweig, A. C., Frederick, C. A., Lippard, S. J., & Nordlund, P. (1993). Crystal structure of a bacterial non-haem iron hydroxylase that catalyses the biological oxidation of methane. *Nature*, *366*, 537–543.

Rosenzweig, A. C., Nordlund, P., Takahara, P. M., Frederick, C. A., & Lippard, S. J. (1995). Geometry of the soluble methane monooxygenase catalytic diiron center in two oxidation states. *Chemistry & Biology*, *2*, 409–418.

Sazinsky, M. H., Bard, J., Di Donato, A., & Lippard, S. J. (2004). Crystal structure of the toluene/o-xylene monooxygenase hydroxylase from *Pseudomonas stutzeri* OX1. Insight into the substrate specificity, substrate channeling, and active site tuning of multicomponent monooxygenases. *The Journal of Biological Chemistry*, *279*, 30600–30610.

Scheidt, W. R., Durbin, S. M., & Sage, J. T. (2005). Nuclear resonance vibrational spectroscopy—NRVS. *Journal of Inorganic Biochemistry*, *99*, 60–71.
Schowen, K. B., & Schowen, R. L. (1982). Solvent isotope effects on enzyme systems. *Methods in Enzymology*, *87*, 551–606.
Shanklin, J., Achim, C., Schmidt, H., Fox, B. G., & Münck, E. (1997). Mössbauer studies of alkane omega-hydroxylase—Evidence for a diiron cluster in an integral-membrane enzyme. *Proceedings of the National Academy of Sciences of the United States of America*, *94*, 2981–2986.
Shu, L., Nesheim, J. C., Kauffmann, K., Münck, E., Lipscomb, J. D., & Que, L., Jr. (1997). An $Fe_2^{IV}O_2$ diamond core structure for the key intermediate Q of methane monooxygenase. *Science*, *275*, 515–518.
Siegbahn, P. E. M. (2001). O − O bond cleavage and alkane hydroxylation in methane monooxygenase. *Journal of Biological Inorganic Chemistry*, *6*, 27–45.
Solomon, E. I., Heppner, D. E., Johnston, E. M., Ginsbach, J. W., Cirera, J., Qayyum, M., et al. (2014). Copper active sites in biology. *Chemical Reviews*, *114*, 3659–3853.
Tinberg, C. E., & Lippard, S. J. (2011). Dioxygen activation in soluble methane monooxygenase. *Accounts of Chemical Research*, *44*, 280–288.
Valentine, A. M., LeTadic-Biadatti, M.-H., Toy, P. H., Newcomb, M., & Lippard, S. J. (1999). Oxidation of ultrafast radical clock substrate probes by the soluble methane monooxygenase from *Methylococcus capsulatus* (Bath). *The Journal of Biological Chemistry*, *274*, 10771–10776.
Wallar, B. J., & Lipscomb, J. D. (1996). Dioxygen activation by enzymes containing binuclear non-heme iron clusters. *Chemical Reviews*, *96*, 2625–2657.
Wallar, B. J., & Lipscomb, J. D. (2001). Methane monooxygenase component B mutants alter the kinetics of steps throughout the catalytic cycle. *Biochemistry*, *40*, 2220–2233.
Wilkinson, E. C., Dong, Y. H., Zang, Y., Fujii, H., Fraczkiewicz, R., Fraczkiewicz, G., et al. (1998). Raman signature of the Fe_2O_2 "diamond" core. *Journal of the American Chemical Society*, *120*, 955–962.
Xing, G., Diao, Y., Hoffart, L. M., Barr, E. W., Prabhu, K. S., Arner, R. J., et al. (2006). Evidence for C-H cleavage by an iron-superoxide complex in the glycol cleavage reaction catalyzed by myo-inositol oxygenase. *Proceedings of the National Academy of Sciences of the United States of America*, *103*, 6130–6135.
Yoshizawa, K. (2000). Two-step concerted mechanism for methane hydroxylation on the diiron active site of soluble methane monooxygenase. *Journal of Inorganic Biochemistry*, *78*, 23–34.
Zheng, H., & Lipscomb, J. D. (2006). Regulation of methane monooxygenase catalysis based on size exclusion and quantum tunneling. *Biochemistry*, *45*, 1685–1692.

CHAPTER ELEVEN

Characterization of Substrate, Cosubstrate, and Product Isotope Effects Associated With Enzymatic Oxygenations of Organic Compounds Based on Compound-Specific Isotope Analysis

Sarah G. Pati*[,†], Hans-Peter E. Kohler*, Thomas B. Hofstetter*[,†,1]
*Eawag, Swiss Federal Institute of Aquatic Science and Technology, Dübendorf, Switzerland
†Institute of Biogeochemistry and Pollutant Dynamics (IBP), ETH Zürich, Zürich, Switzerland
[1]Corresponding author: e-mail address: thomas.hofstetter@eawag.ch

Contents

1. Introduction 292
2. Deriving Apparent Kinetic Isotope Effects From Measurements of Stable Isotope Ratios 294
   2.1 Instrumental Approaches and Observable Quantitites 294
   2.2 Substrate Isotope Fractionation 295
   2.3 Product Isotope Fractionation 300
   2.4 Method Comparison for Deriving Isotope Effects 303
   2.5 Multidimensional Isotope Fractionation Analysis 305
3. Experimental Approaches for Determining Isotope Fractionation During Oxygenation Reactions 307
   3.1 Experiment Design and Sampling Strategies 307
   3.2 Enzyme Assays for Isotope Analysis of Substrates 308
   3.3 Whole Cell Assays for C Isotope Analysis of Organic Reaction Products 311
   3.4 Enzyme Assays for O Isotope Analysis of Aqueous O_2 312
4. Instrumentation for Stable Isotope Analysis by Isotope Ratio Mass Spectrometry 313
   4.1 Instrumental Strategies 313
   4.2 Substrate Isotope Analysis by GC/IRMS 315
   4.3 Oxygen Isotope Analysis of Aqueous O_2 by GC/IRMS 318
   4.4 Product Isotope Analysis by LC/IRMS 321
5. Summary and Conclusion 323
Acknowledgments 323
References 323

Methods in Enzymology, Volume 596
ISSN 0076-6879
http://dx.doi.org/10.1016/bs.mie.2017.06.044

Abstract

Enzymatic oxygenations are among the most important biodegradation and detoxification reactions of organic pollutants. In the environment, however, such natural attenuation processes are extremely difficult to monitor. Changes of stable isotope ratios of aromatic pollutants at natural isotopic abundances serve as proxies for isotope effects associated with oxygenation reactions. Such isotope fractionations offer new avenues for revealing the pathway and extent of pollutant transformation and provide new insights into the mechanisms of catalysis by Rieske non-heme ferrous iron oxygenases. Based on compound-specific C, H, N, and O isotope analysis, we present a comprehensive methodology with which isotope effects can be derived from the isotope fractionation measured in substrates, the cosubstrate O_2, and organic oxygenation products. We use dioxygenation of nitrobenzene and 2-nitrotoluene by nitrobenzene dioxygenase as illustrative examples to introduce different mathematical procedures for deriving apparent substrate and product isotope effects. We present two experimental approaches to control reactant and product turnover for isotope fractionation analysis in experimental systems containing purified enzymes, *E. coli* clones, and pure strains of environmental microorganisms. Finally, we present instrumental procedures and sample treatment instructions for analysis of C, H, and N isotope analysis in organic compounds and O isotope analysis in aqueous O_2 by gas and liquid chromatography coupled to isotope ratio mass spectrometry.

1. INTRODUCTION

Enzymatic oxygenations are among the most important biodegradation and detoxification reactions of organic pollutants (Gibson & Parales, 2000; Kohler, 2011; Neilson & Allard, 2008). The oxidation of organic pollutants through the insertion of one or two oxygen atoms from molecular O_2 makes these compounds more polar and more bioavailable. The oxygenated organic products may be channeled into common metabolic pathways, which ultimately enable the mineralization of the pollutants (Boyd & Bugg, 2006; Bugg & Winfield, 1998; Schwarzenbach, Gschwend, & Imboden, 2016). Such processes enable natural attenuation of pollutants but are inherently difficult to monitor in contaminated soil and water. In the environment, oxidative biodegradation typically occurs through multiple different reaction pathways and takes place over timescales of years to decades. Moreover, the concentration dynamics of organic pollutants are often determined by nonreactive processes such as dilution and phase transfers, making it almost impossible to assess biodegradation quantitatively (Schwarzenbach et al., 2006).

Compound-specific isotope analysis (CSIA) of organic pollutants has become an alternative avenue to reveal the type of degradation reaction and to quantify the extent of transformation by measuring changes of pollutant isotope ratios at *natural isotopic abundances* (Aelion, Höhener, Hunkeler, & Aravena, 2010; Elsner, 2010; Elsner et al., 2012; Hofstetter & Berg, 2011; Hofstetter, Schwarzenbach, & Bernasconi, 2008). CSIA relies on the phenomenon of kinetic isotope effects, KIE (Eq. 1), that arises from small differences in activation free energies of reactant molecules with different isotopic substitution (Kohen & Limbach, 2006; Wolfsberg, Hook, Paneth, & Rebelo, 2010)

$$\mathrm{KIE} = \frac{{}^{l}k}{{}^{h}k} \tag{1}$$

where ${}^{l}k$ and ${}^{h}k$ denote reaction rate constants of light and heavy isotopologues, respectively. As a consequence, bonding differences between the ground state and transition state, especially at sites undergoing bond cleavage reactions, give rise to changes of stable isotope ratios, the so-called isotope fractionation, in the reacting pollutant. Isotope fractionation associated with pollutant biodegradation serves as a proxy for isotope effects of reactive processes, from which both the pathway and extent of pollutant degradation can be derived. As will be shown in Section 2, the link between pollutant isotope fractionation and the mechanisms of biodegradation through oxygenation reactions can be made in a number of different ways. The link can be established most rigorously in laboratory experiments when isotope fractionation of several elements in the pollutant, such as C, H, and N, are considered simultaneously together with the O isotope fractionation of aqueous O_2.

However, a widespread application of CSIA to assess enzymatic oxygenation reactions of organic pollutants is currently compromised. At present, we lack comprehensive knowledge of the mechanisms of mono- and dioxygenations of organic pollutants and the magnitude of isotope effects that are associated with the initial steps of the reactions. Moreover, the kinetic complexity of enzymatic O_2 activation and its consequences for the expression of isotope fractionation in the organic substrate is poorly understood (Wijker, Pati, Zeyer, & Hofstetter, 2015). The isotope fractionation observed during enzyme-catalyzed oxygenations of organic pollutants can vary substantially even for identical reactions because of multiple rate-limiting steps. We recently observed highly variable kinetic isotope effects of dioxygenations of different nitroarenes by nitrobenzene and nitrotoluene dioxygenases (Pati, Kohler, Bolotin, Parales, & Hofstetter, 2014; Pati,

Kohler, et al., 2016). In these works, we hypothesized that different steps of the pathway for activation of molecular O_2 contribute in a substrate-specific manner to the catalytic cycle of Rieske non-heme ferrous iron oxygenases. Based on these observations, we developed methodologies for assessing isotope effects of oxygenation reactions (1) in aromatic pollutants, (2) their oxygenated organic products, as well as (3) in aqueous O_2 based on measuring isotope fractionation.

The approach presented here allows a comprehensive characterization of isotope effects in substrates, cosubstrates, and products. Our examples focus on oxygenations, but similar procedures may be used to study isotope effects on other enzymatic processes based on isotope fractionation observed at natural isotopic abundances and measured by gas and liquid chromatography in combination with isotope ratio mass spectrometry. First, we summarize the different mathematical procedures with which isotope effects can be derived from data on the fractionation of stable isotopes of organic substrates and their reaction products. Thereafter, we describe experiments in laboratory model systems at different levels of biological complexity covering pure cultures of pollutant-degrading microorganisms, *E. coli* clones expressing enzyme system of interest, crude cell extracts thereof, as well as enzyme assays of purified nitroarene dioxygenases. In Section 4, we illustrate customized analytical methods for quantification of C, H, N, and O isotope ratios in different analytes before concluding with a summary.

2. DERIVING APPARENT KINETIC ISOTOPE EFFECTS FROM MEASUREMENTS OF STABLE ISOTOPE RATIOS

2.1 Instrumental Approaches and Observable Quantitites

Isotope ratios of individual compounds present in a solution or gas mixture are typically measured with continuous flow isotope ratio mass spectrometry by coupling gas or liquid chromatography with an isotope ratio mass spectrometer (GC/IRMS and LC/IRMS, Elsner et al., 2012; Jochmann & Schmidt, 2012; Sessions, 2006; Schmidt & Jochmann, 2012). Such measurements at natural isotopic abundances involve, in most cases, the conversion of analytes to small molecules such as CO_2, N_2, and H_2 with a limited number of stable isotopologues. Measured isotope ratios therefore reflect the averages of all atoms of the studied element in a molecule. Recent instrumental developments extend the scope to molecules with multiple heavy isotopic substitutions (clumped isotopes) and different isotopomers, but those techniques are not discussed here (Bernstein et al., 2011; Eiler, 2013; Magyar, Orphan, & Eiler, 2016; Piasecki et al., 2016).

Due to these instrumental boundary conditions, one observes changes of isotope ratios as isotope fractionation of all atoms of one element in a molecule. Isotope effects can only be inferred indirectly by making explicit assumptions with regard to the mechanisms of reaction and the number and position of reactive atoms in a molecule. The mathematical procedures for obtaining such "observed" or "apparent" isotope effects have been established for a while (Melander & Saunders, 1980; Singleton & Thomas, 1995) and are now applied extensively for CSIA based on a formalism introduced by Elsner, Zwank, Hunkeler, and Schwarzenbach (2005) and Elsner (2010). The evaluation of isotope fractionation with CSIA is now used for two purposes. (i) In laboratory model systems, quantification of isotope fractionation of (model) pollutants enables one to assess the magnitude and variability of apparent kinetic isotope effects of (bio)degradation pathways. In combination with theoretical analyses, for example, through computational investigation of reaction sequences, such data are key to assess the extent, to which intrinsic isotope effects of bond cleavage reactions are observable as isotope fractionation because isotope effects may be masked by the kinetic complexity in the chemical or enzymatic reaction (i.e., the isotopically sensitive step is not rate limiting, Cook & Cleland, 2007; Northrop, 1981) or through phase transfer processes (Aeppli et al., 2009; Eckert, Qiu, Elsner, & Cirpka, 2013). (ii) Derivation of apparent kinetic isotope effects from isotope fractionation measured in (contaminated) environments and its comparison with data from laboratory investigations is one of the most promising means to identify if, how, and to what extent pollutants are transformed in complex systems.

In the following sections, we illustrate the mathematical procedures for evaluating isotope fractionation data and deriving apparent kinetic isotope effects for enzymatic oxygenations based on two examples, in which the pollutants nitrobenzene and 2-nitrotoluene function as substrates for nitrobenzene dioxygenase. The two data sets from Pati et al. (2014) show how different evaluation procedures for substrate and product isotope fractionation data can be applied to obtain insights into the mechanisms of pollutant oxygenation and catalytic cycle of a Rieske non-heme ferrous iron oxygenase.

2.2 Substrate Isotope Fractionation

Measurements of substrate isotope fractionation are the most widespread application of CSIA. Here, we consider the enzymatic dioxygenation of nitrobenzene, compound **1** in Fig. 1, by nitrobenzene dioxygenase

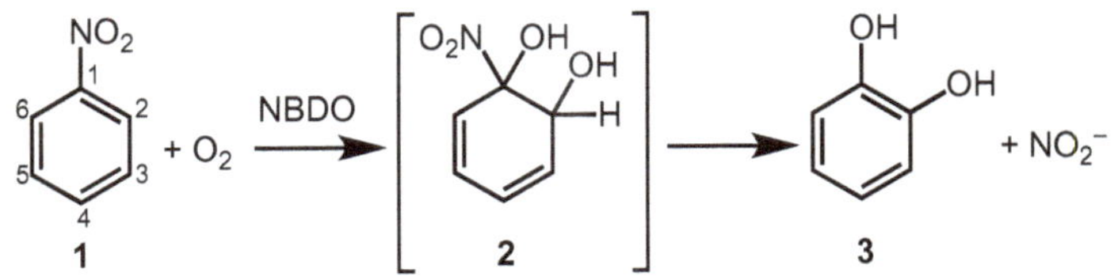

Fig. 1 Oxygenation of nitrobenzene (**1**) by nitrobenzene dioxygenase (NBDO) leading to a *cis*-dihydrodiol intermediate (**2**) and two products, that is, catechol (**3**) and nitrite (NO_2^-).

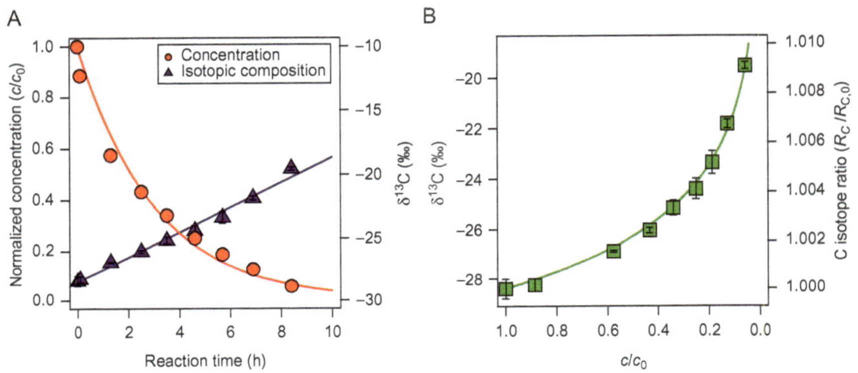

Fig. 2 (A) Time-course of nitrobenzene concentration and change in C isotope signature during the dioxygenation by NBDO. (B) Nonlinear correlation between change in C isotope signature and carbon isotope ratios and fraction of remaining nitrobenzene concentration (c/c_0). Note that the two *y*-axes, $\delta^{13}C$, and R/R_0, in panel (B) are equivalent.

(NBDO). This reaction leads to a *cis*-dihydrodiol intermediate (**2**), which undergoes spontaneous elimination of nitrite (NO_2^-) to catechol (**3**). In laboratory experiments, we determined the dynamics of substrate and product concentrations by conventional methods, that is, gas chromatography/mass spectrometry (GC/MS) for **1**, high-performance liquid chromatography (HPLC) for **3**, and a colorimetric assay for nitrite (Pati et al., 2014). Isotope ratios of **1** and **3** are measured by GC/IRMS and LC/IRMS, respectively. A typical time-course of concentrations and $^{13}C/^{12}C$ ratios of nitrobenzene during its dioxygenation by NBDO is illustrated in Fig. 2A.

2.2.1 Isotope Ratios and Isotope Signatures

We express the isotope ratios of an element E as the ratio of heavy (^{h}E) and light (^{l}E) element concentrations, R_E, and as isotope signature, $\delta^{h}E$. Eqs. (2) and (3) show the relationship between R_E and $\delta^{h}E$. $\delta^{h}E$ is derived from the ratio of ${}^{h}\mathrm{E}/{}^{l}\mathrm{E}$ for the analyte and internationally accepted reference material (Brand, Coplen, Vogl, Rosner, & Prohaska, 2014; Coplen, 2011). Eq. (4) illustrates this relationship for C isotope signatures, $\delta^{13}\mathrm{C}$, of an analyte and the international standard (Vienna Pee Dee Belemnite, VPDB).

$$R_E = \frac{{}^{h}\mathrm{E}}{{}^{l}\mathrm{E}} \tag{2}$$

$$\delta^{h}\mathrm{E} = \frac{({}^{h}\mathrm{E}/{}^{l}\mathrm{E})_{\mathrm{analyte}}}{({}^{h}\mathrm{E}/{}^{l}\mathrm{E})_{\mathrm{reference}}} - 1 \tag{3}$$

$$\delta^{13}\mathrm{C} = \frac{({}^{13}\mathrm{C}/{}^{12}\mathrm{C})_{\mathrm{analyte}}}{({}^{13}\mathrm{C}/{}^{12}\mathrm{C})_{\mathrm{VPDB}}} - 1 \tag{4}$$

Referencing of isotope ratios to standard materials not only matters for the comparison of measurements across studies and quality assurance of stable isotope laboratories but also for the quantification of isotope fractionation (Coleman & Meier-Augenstein, 2014). Different initiatives make organic reference materials available for CSIA (Schimmelmann et al., 2009, 2016) and simple spreadsheet templates facilitate correct referencing (Dunn, Hai, Malinovsky, & Goenaga Infante, 2015). The use of isotope signatures simplifies handling of data for isotope fractionation at natural isotopic abundances. For example, nitrobenzene used in this study exhibits an initial $\delta^{13}\mathrm{C}$-value of -0.0284 or -28.4‰. However, nitrobenzene has a $\delta^{13}\mathrm{C}$-value of -0.0196 or -19.6‰ after 94% conversion by NBDO. This difference of ${}^{13}\mathrm{C}/{}^{12}\mathrm{C}$ ratio is small and more difficult to keep track with the R_C notation ($R_{\mathrm{C},0} = 0.011\mathbf{047}$ vs $R_{\mathrm{C},94\%} = 0.011\mathbf{147}$). Note, however, that isotope signatures and isotope ratios are interconvertible and despite general recommendations by Coplen (2011), terminologies may vary slightly among different scientific disciplines.

2.2.2 Quantifying Isotope Fractionation

We derive kinetic isotope effects from the fractionation of stable isotopes in the substrate and the reaction progress, which is quantified as fraction of remaining substrate, c/c_0. The calculations are shown here for C isotope fractionation of nitrobenzene during its transformation to catechol but procedures apply equally to H, N, and O isotope fractionation data. Changes of $\delta^{13}\mathrm{C}$-values are related to c/c_0 of nitrobenzene through the C isotope enrichment factor, ϵ_C, as shown in Eq. (5) for the experimental data shown in Fig. 2B. We calculate the ϵ_C-value associated with the dioxygenation of nitrobenzene by NBDO by two procedures; (a) the observational data ($\delta^{13}\mathrm{C}$

vs c/c_0) are modeled directly by the function given in Eq. (5) and best fit parameters are obtained through nonlinear regression analysis. (b) The same data ($\delta^{13}C$ vs c/c_0) are transformed according to Eq. (6) and parameters are derived by linear regression analysis. Note that normalization of isotope fractionation as $R_C/R_{C,0}$ is used extensively in CSIA even though more accurate estimates of parameter uncertainties are obtained through linear regression of the nonnormalized form (Eq. (7), Scott, Lu, Cavanaugh, and Liu (2004), application examples in Tobler, Hofstetter, and Schwarzenbach (2007) and Hofstetter, Neumann, et al. (2008)). Applying linear and nonlinear regressions using data shown in Fig. 2 results in identical ϵ_C-values within uncertainty, that is, -3.28 ± 0.15‰ and -3.25 ± 0.13‰ using Eqs. (5) and (6), respectively. Note that all uncertainties shown in this chapter represent 95% confidence intervals.

$$\frac{R_C}{R_{C,0}} = \frac{\delta^{13}C + 1}{\delta^{13}C_0 + 1} = \left(\frac{c}{c_0}\right)^{\epsilon_C} \tag{5}$$

$$\ln\left(\frac{R_C}{R_{C,0}}\right) = \ln\left(\frac{\delta^{13}C + 1}{\delta^{13}C_0 + 1}\right) = \epsilon_C \cdot \ln\left(\frac{c}{c_0}\right) \tag{6}$$

$$\ln\left(\delta^{13}C + 1\right) = \epsilon_C \cdot \ln(c) + \ln\frac{\delta^{13}C_0 + 1}{c_0^{\epsilon_C}} \tag{7}$$

2.2.3 Apparent Kinetic Isotope Effects

Apparent kinetic isotope effects, AKIEs, are derived from isotope enrichment factors with an explicit assumption made regarding the reaction mechanism. In most cases, those assumptions are made to account for the presence of atoms that do not participate in the reaction (isotopic dilution). If a substrate exhibits multiple reactive positions, heavy isotopologues will react with both rates for light and heavy isotopes (Eq. 1) because the heavy isotope is located only in one of the several equivalent reactive sites (intramolecular isotopic competition). The resulting isotope fractionation is then smaller than predicted from the intrinsic KIE necessitating correction for the intramolecular isotopic competition effect (Elsner, 2010). The procedure leading to AKIEs shown in Eq. (8) apportions the observed isotope fractionation to reactions, in which bonds are broken and formed and results in estimates for primary KIEs. Because secondary isotope effects due to isotopic substitution at atoms that are not at the reactive site of a molecule are neglected, AKIE-values may slightly exceed those determined by alternative means (e.g., through experiments with site-specifically labeled compounds or computed KIEs).

For the dioxygenation of nitrobenzene leading to the *cis*-dihydrodiol (**2** in Fig. 1), we assume that changes of C hybridization and formation of C–O bonds reflect the rate-limiting reaction step. The ^{13}C-AKIE is calculated based on the ϵ_C-value according to Eq. (8).

$$^{13}\text{C-AKIE} \approx \frac{1}{1 + n/x \cdot z \cdot \epsilon_C} \tag{8}$$

where ϵ_C is the isotope enrichment factor derived with Eqs. (5)–(7), n is the total number of C atoms of nitrobenzene, x is the number of reactive sites, and z denotes the number of reactive sites that are subject to intramolecular isotopic competition. For dioxygenation of nitrobenzene n is 6 while x and z depend on whether one or both C–O bonds are considered to form in the rate-limiting reaction step. If we assume that the two C–O bonds are formed simultaneously in a synchronous dioxygenation (DelMonte et al., 1997; Houk & Strassner, 1999), the number of reactive sites, x, is 2 (C-1 and C-2 or C-1 and C-6 in Fig. 1), z is 1, and the ^{13}C-AKIE equals 1.011 ± 0.001. If, however, the two C–O bonds form in subsequent step as asynchronous dioxygenation, of which only one is rate-limiting, then $x = z = 2$ and the ^{13}C-AKIE amounts to 1.023 ± 0.002 (Pati et al., 2014). The larger ^{13}C-AKIE has indeed been confirmed by hybrid quantum mechanical/molecular mechanical (QM/MM) calculations (Pati, Kohler, et al., 2016). ^{15}N-AKIE-values calculated according to Eq. (8) with $n = x = z = 1$ reflects a secondary isotope effect regardless of the mechanistic assumption because the N atom is not involved directly in the reaction. The ^{15}N-AKIE-value (1.001 ± 0.001) is not significantly different from unity. A primary ^{2}H KIE arises if one assumes an asynchronous dioxygenation initiated at C-2. In this case, the primary ^{2}H-AKIEs of 1.029 ± 0.007 was obtained using the parameters $n = 5$, $x = 2$, $z = 2$.

Note that Eq. (8) is a good approximation of the relationship between AKIE and ϵ, which does not hold for large H isotope fractionation due to large ^{2}H-KIEs (Dorer, Höhener, Hedwig, Richnow, & Vogt, 2014; Elsner, 2010). Wijker, Adamczyk, Bolotin, Paneth, and Hofstetter (2013) proposed a kinetic model, in which a set of ordinary differential equation account explicitly for all isotopomers of singly substituted heavy isotopologues of the substrate and thereby accounts for the phenomena of isotopic dilution and intramolecular isotopic competition.

2.3 Product Isotope Fractionation

Apparent kinetic isotope effects of enzymatic oxygenations may not only be derived from analysis of the *substrate* but are also reflected in the isotope fractionation of the oxygenated organic reaction *product(s)*. This analysis by CSIA is often based on the implicit assumptions that the isotope fractionation of the catalytic step exhibiting bonding changes is not obscured by the following steps of the catalytic cycle (e.g., product release from the enzyme) before isotopic analysis of the dissolved reaction *product* can be carried out. Here, we illustrate the information that can be obtained from product isotope fractionation for two simple cases, in which oxygenation of a nitroaromatic substrate leads to one or two oxygenated organic products. These two cases are encountered, for example, when NBDO transforms nitrobenzene and 2-nitrotoluene to catechol and a mixtures of 3-methylcatechol and 2-nitrobenzyl alcohol, respectively (Figs. 1 and 3). We focus our discussions on the derivation of C isotope enrichment factors, ϵ_C, from measured isotope signatures, $\delta^{13}C$. All ϵ_C-values can be converted to AKIEs according to Eq. (8) bearing in mind the same assumptions made for the analysis of substrate isotope fractionation.

2.3.1 Simple Substrate to Product Relationships

The C isotope fractionation of catechol formed through dioxygenation of nitrobenzene is shown in Fig. 4A. All C atoms of nitrobenzene end up in catechol and the total number of heavy and light C atoms is conserved during any state of the reaction. Therefore, the initial C isotope ratio of the substrate should match the one measured in the product after complete substrate conversion. This requirement is almost met for the dioxygenation of nitrobenzene to catechol with an initial $\delta^{13}C$-value of nitrobenzene of $-28.4 \pm 0.4‰$ as compared to a final $\delta^{13}C$-value of $-29.6 \pm 0.2‰$ measured for catechol. The difference, however, is small and in the range of the total instrumental uncertainty of C isotope ratio measurements (Sherwood Lollar, Hirschorn, Chartrand, & Lacrampe-Couloume, 2007).

The ϵ_C-value for the transformation of nitrobenzene to catechol is determined by Eq. (9). ϵ_C can then be derived by nonlinear regression analysis. In Eq. (9), $\delta^{13}C_P$ is the product isotope signature and $\delta^{13}C_{S,0}$ is the initial isotope signature of the substrate. The ϵ_C determined in this way for the data set shown in Fig. 4B was $-4.5 \pm 0.1‰$ (Pati et al., 2014), which is slightly

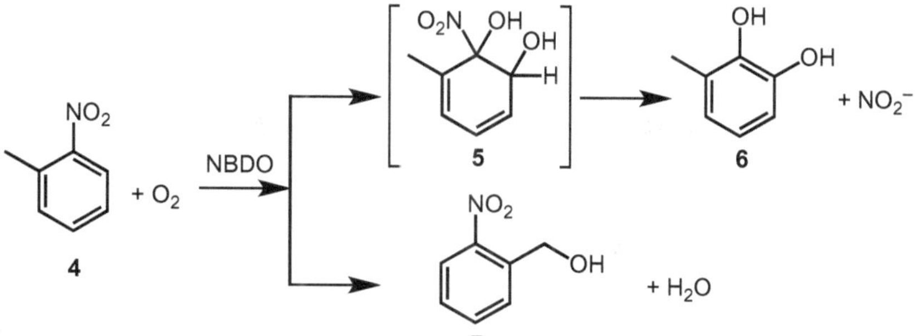

4 5 6 7

Fig. 3 Dioxygenation (*top*) and CH_3-group oxidation (*bottom*) of 2-nitrotoluene (**4**) catalyzed simultaneously by NBDO with the products 3-methylcatechol (**6**) and NO_2^-, which form spontaneously from a *cis*-dihydrodiol intermediate (**5**), and 2-nitrobenzyl alcohol (**7**), respectively.

larger (i.e., more negative) than the value determined by substrate isotope fractionation. A comparison of approaches follows in Section 2.4.

$$\frac{\delta^{13}C_P + 1}{\delta^{13}C_{S,0} + 1} = \frac{1 - (c/c_0)^{(\epsilon_C + 1)}}{1 - (c/c_0)} \tag{9}$$

$$\lim_{c/c_0 \to 1} \epsilon_C \approx \delta^{13}C_{P,0} - \delta^{13}C_{S,0} \tag{10}$$

Eq. (9) can be simplified to Eq. (10) for low substrate conversion when c/c_0 is close to unity. In this case, ϵ_C equals the difference between the isotope signature of product at early stages of the reaction, $\delta^{13}C_{P,0}$, and the initial substrate isotope signature, $\delta^{13}C_{S,0}$. In practice, this approach is applied to $\delta^{13}C_P$ measured for substrate turnovers below 10% when $\delta^{13}C_P$ is still close to $\delta^{13}C_{P,0}$ (Melander & Saunders, 1980). In the example of catechol formation from nitrobenzene, we determined a ϵ_C-value of $-$ 3.76 ± 0.24‰.

2.3.2 Multiple Reaction Products

Here, we describe a procedure based on a modified form of Eq. (9) that can be used for the analysis of isotope fractionation of multiple oxygenated organic products (Eq. 11). Note that this procedure does not enable one to distinguish reaction mechanisms in which two (or more products) form directly from the substrate molecule from cases, in which the reaction products result from a common intermediate. An in-depth discussion of these cases can be found in Pati, Kohler, et al. (2016) and Elsner, Chartrand, VanStone, Lacrampe Couloume, and Sherwood Lollar (2008).

Fig. 3 shows the NBDO-catalyzed oxygenation of 2-nitrotoluene (**4**) to 3-methylcatechol (**6**) and 2-nitrobenzyl alcohol (**7**). The two products are formed in almost equal amounts (see data in Pati et al., 2014). The ϵ_C-value determined for substrate isotope fractionation of 2-nitrotoluene according to Eq. (5) was $-1.3 \pm 0.1‰$ (Pati et al., 2014), which is much smaller than the one for nitrobenzene. The difference is caused by two different KIEs that are responsible for the substrate isotope fractionation of 2-nitrotoluene; one KIE for the dioxygenation of aromatic C atoms and one for the oxidation of the CH_3-group. The ^{13}C-AKIE for each of the two independent reactions cannot be derived from analysis of the ϵ_C-value of the substrate. Instead, isotope enrichment factors for dioxygenation and CH_3-group oxidation must necessarily be obtained from the isotope fractionation measured in the two products (Fig. 4B).

We calculate a *concentration-weighted average* of the isotope signatures of 3-methylcatechol and 2-nitrobenzyl alcohol according to Eq. (11), which is a modified form of Eq. (9) (Elsner et al., 2008). Fig. 4B shows that the "average" product isotope signature calculated with Eq. (11) follows the same trend as the isotope signatures of the two products, but with a constant offset. In Eq. (11), ϵ_C denotes the isotope enrichment factor of the "average"

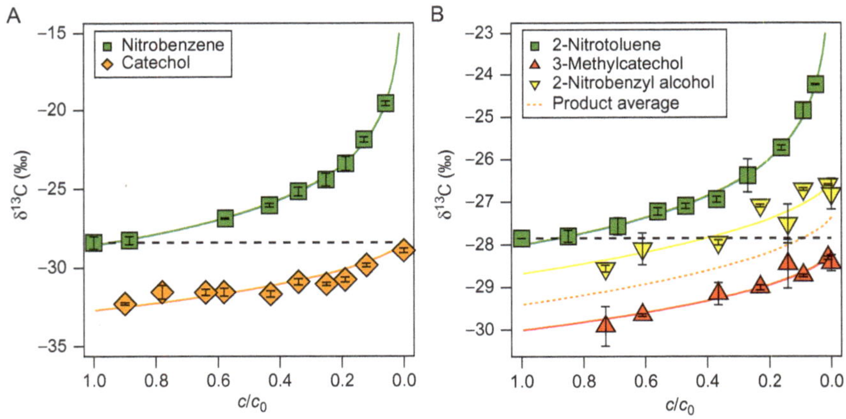

Fig. 4 (A) $\delta^{13}C$ of nitrobenzene (substrate) and catechol (final product) vs fraction of remaining substrate, c/c_0, during dioxygenation by NBDO. (B) $\delta^{13}C$ of 2-nitrotoluene, 3-methylcatechol, and 2-nitrobenzyl alcohol vs fraction of remaining substrate, c/c_0, during simultaneous dioxygenation and CH_3-group oxidation by NBDO. *Reprinted with permission from Pati, S. G., Kohler, H.-P. E., Bolotin, J., Parales, R. E., & Hofstetter, T. B. (2014). Isotope effects of enzymatic dioxygenation of nitrobenzene and 2-nitrotoluene by nitrobenzene dioxygenase.* Environmental Science & Technology, 48*(18), 10750–10759. http://dx.doi.org/10.1021/es5028844. Copyright 2014, American Chemical Society.*

product isotope signature and $D_{3\text{-MC}}$ is the offset between the δ^{13}C-value of 3-methylcatechol and the "average" product isotope signature

$$\frac{\delta^{13}C_{3\text{-MC}}+1}{\delta^{13}C_{2\text{-NT},0}+1}=(1+D_{3\text{-MC}})\frac{1-(c/c_0)^{(\epsilon_C+1)}}{1-(c/c_0)} \tag{11}$$

where $\delta^{13}C_{3\text{-MC}}$ is the C isotope signature of 3-methylcatechol, $\delta^{13}C_{2\text{-NT},0}$ is the initial isotope signature of 2-nitrotoluene, c/c_0 is the fraction of remaining substrate.

Assuming that the dioxygenation and CH_3-group oxidation of 2-nitrotoluene are two independent reactions that do not share a common intermediate, the reaction-specific isotope enrichment factor for the dioxygenation reaction of 2-nitrotoluene to 3-methylcatechol ($\epsilon_C^{2NT\to3-MC}$) is derived from $D_{3\text{-MC}}$ according to Eq. (12).

$$\epsilon_C^{2\text{-NT}\to3\text{-MC}}=D_{3\text{-MC}}+\epsilon_C \tag{12}$$

The identical procedure can be applied to the data for C isotope signatures of 2-benzyl alcohol instead of 3-methylcatechol. The offset of fitted δ^{13}C-values relative to the average product signatures would then correspond to $D_{2\text{-NBA}}$ and enable one to calculate $\epsilon_C^{2\text{-NT}\to2\text{-NBA}}$ with δ^{13}C-values of 2-nitrobenzyl alcohol with the same outcome. Using the data shown in Fig. 4B, one obtains $\epsilon_C^{2\text{-NT}\to3\text{-MC}}$ and $\epsilon_C^{2\text{-NT}\to2\text{-NBA}}$-values of -2.0 ± 0.4‰ and -0.2 ± 0.4‰, respectively (Pati et al., 2014).

It follows from the reasoning in Section 2.3.1 that the ϵ_C-values for each reaction pathway can also be obtained using the approximation at low substrate turnover (Eq. 10) introduced above. This procedure leads to $\epsilon_C^{2\text{-NT}\to3\text{-MC}}$ and a $\epsilon_C^{2\text{-NT}\to2\text{-NBA}}$ values of -3.2 ± 1.0‰ and -0.2 ± 0.9‰, respectively (Pati et al., 2014). Eqs. (10) and (11) lead to identical numbers but uncertainties are larger when the low substrate turnover approximation of Eq. (10) is used.

2.4 Method Comparison for Deriving Isotope Effects

The different approaches for calculation of ϵ and AKIE-values are best benchmarked against a set of ordinary differential equations that describe the system of interest (13). In its most simple form, two differential equations for species with light and heavy isotopic substitution, respectively, are used for deriving Eq. (5) and calculating ϵ and AKIE-values as shown above (Hunkeler & Elsner, 2010; Melander & Saunders, 1980). More comprehensive numerical models include several chemical species and isotopic elements as well as multiple isotopologues and isotopomers thereof (e.g., Maggi & Riley, 2010; Wijker, Adamczyk, et al., 2013; Jin, Haderlein, & Rolle, 2013; Höhener & Atteia, 2014). Those models do not require corrections for isotopic dilution and intramolecular isotopic competition, but applying Eqs. (5) and (8) for interpretation of isotope fractionation seems more popular in large parts of the stable isotope community.

A combined evaluation of substrate and product isotope fractionation for the dioxygenation and CH_3-group oxidation of different substrates by NBDO was carried out with Eqs. (13) and (14).

$$\frac{dc_\kappa}{dt} = \sum_j \nu \cdot k_j \cdot c_\kappa \quad (13)$$

$$\epsilon_j = \frac{{}^h k_j}{{}^l k_j} - 1 \quad (14)$$

where c_κ is the concentration of species κ. Here, we considered one heavy and one light isotopologue for each substrate and product. k_j is the rate constant for reaction j, and ν is the stoichiometric coefficient of reaction j. The ratio of reaction rate constants for heavy and light species, which relates to the enrichment factor of reaction j (Eq. 14), was estimated by fitting the model to measured concentrations and isotope signatures of all available species. We used Aquasim (Reichert, 1994) to implement the model and estimate enrichment factors but many other software packages (e.g., R (R Core Team, 2014), Matlab) are equally suited.

Table 1 shows a compilation of all ϵ_C-values derived for nitrobenzene dioxygenation as well as 2-nitrotoluene dioxygenation and CH_3-group oxidation by NBDO (Figs. 1–4; Pati et al., 2014). In the case of nitrobenzene dioxygenation to catechol, four differential equations according to Eq. (13) were considered for light and heavy C isotopologues of nitrobenzene as well as light and heavy C isotopologues of catechol. The resulting ϵ_C-value derived according to Eq. (14) was $-4.1 \pm 0.2‰$ (Pati et al., 2014), which is in between the ϵ_C-values derived with Eqs. (6) and (9), respectively (see Table 1). For the simultaneous dioxygenation and CH_3-group oxidation of 2-nitrotoluene to 3-methylcatechol and 2-nitrobenzyl alcohol, the number of differential equations according to Eq. (13) was 6 and the model resulted in reaction-specific C isotope enrichment factors of $\epsilon_C^{2\text{-NT}\rightarrow 3\text{-MC}} = -2.5 \pm 0.2‰$ and $\epsilon_C^{2\text{-NT}\rightarrow 2\text{-NBA}} = -0.4 \pm 0.2‰$ (see Table 1, Pati et al., 2014).

Table 1 Compilation of C Isotope Enrichment Factors Derived With Different Evaluation Methods and Isotope Fractionation Data for NBDO-Catalyzed Reactions of Nitrobenzene (NB) and 2-Nitrotoluene (2-NT) to Catechol (CAT), 3-Methylcatechol (3-MC), and 2-Nitrobenzyl Alcohol (2-NBA), Respectively[a]

Isotope Fractionation of	ϵ_C^{NB}	$\epsilon_C^{NB\rightarrow CAT}$	$\epsilon_C^{2\text{-}NT}$	$\epsilon_C^{2\text{-}NT\rightarrow 3\text{-}MC}$	$\epsilon_C^{2\text{-}NT\rightarrow 2\text{-}NBA}$
Substrate (Eq. 6)	−3.7 ± 0.2		−1.3 ± 0.1		
Product (Eq. 9)		−4.5 ± 0.1			
Product (Eq. 10)		−3.6 ± 0.6		−3.2 ± 1.0	−0.2 ± 0.9
Product (Eqs. 11 and 12)				−2.0 ± 0.4	−0.2 ± 0.4
Substrate and product (Eqs. 13 and 14)		−4.1 ± 0.2		−2.5 ± 0.2	−0.4 ± 0.2

[a]All values given in ‰ with uncertainties as 95% confidence intervals.
Original data can be found in Pati et al. (2014).

All approaches for deriving enrichment factors, and thus AKIE-values, gave consistent results for both nitrobenzene and 2-nitrotoluene dioxygenation by NBDO and the estimated parameters match within their 95% confidence intervals (see Table 1). Besides providing a general and powerful internal check on AKIE accuracy, there are, however, two advantages for combining substrate and product isotope fractionation for parameter estimation. First, uncertainties associated with reaction-specific enrichment factors were smaller. Second, in the case of 2-nitrotoluene, the concentration-weighted average of the two reaction-specific enrichment factors (i.e., of $\epsilon_C^{2\text{-NT}\rightarrow 3\text{-MC}}$ and $\epsilon_C^{2\text{-NT}\rightarrow 2\text{-NBA}}$) was in best agreement with the substrate isotope enrichment factor ($\epsilon_C^{2\text{-NT}}$).

2.5 Multidimensional Isotope Fractionation Analysis

If intrinsic KIEs of enzymatic bond cleavage reactions are masked by non-isotopic steps of catalytic cycles and independent information, such as from computational theory, is unavailable, ambiguities of AKIE interpretations can be circumvented with CSIA of two (or more) isotopic elements. This so-called multielement isotope fractionation analysis is based on the linear correlation of the observed isotope fractionation of different elements of a compound. Slopes of such correlations will be independent of masking because all elements in the breaking bond will be affected by masking in the same way. The formalism is shown for C and H isotope fractionation in Eq. (15) and illustrates how the correlation slope, $\Lambda_{H/C}$, is related to the intrinsic KIEs (Elsner, 2010).

$$\Lambda_{H/C} = \frac{\Delta\delta^2 H}{\Delta\delta^{13}C} \approx \frac{\epsilon_H}{\epsilon_C} \approx \frac{(n/x)_H}{(n/x)_C} \cdot \frac{^2H-KIE-1}{^{13}C-KIE-1} \cdot \frac{1+{}^{13}C-KIE\cdot(z_C-1)}{1+{}^2H-KIE\cdot(z_H-1)} \tag{15}$$

An example for multielement isotope fractionation analysis is shown in Fig. 5 for the δ^2H vs $\delta^{13}C$-values of 2-, 3-, and 4-nitrotoluene during their transformation by NBDO (Pati, Kohler, et al., 2016). Both ϵ-values and AKIEs differ significantly for these three substrates despite a common reaction mechanism. For example, ϵ_H-values for 3- and 4-nitrotoluene dioxygenation by NBDO were -2.6 ± 1.6‰ and -5.5 ± 2.3‰, respectively. Nevertheless, slopes of the linear regressions of δ^2H vs $\delta^{13}C$, $\Lambda_{H/C}$, are identical within uncertainties (1.5 ± 2.2‰ for 3-nitrotoluene and 2.1 ± 0.6‰ for 4-nitrotoluene). Identical $\Lambda_{H/C}$-values confirm that the observable C and H isotope fractionation originate, for both substrates, from the

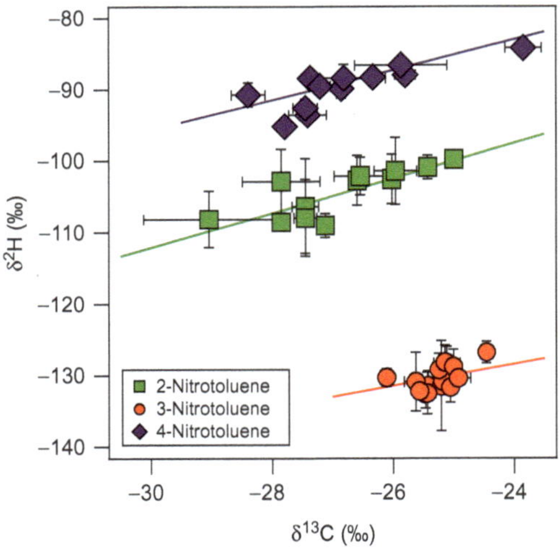

Fig. 5 δ^2H vs $\delta^{13}C$ values of 2-, 3-, and 4-nitrotoluene during dioxygenation by NBDO. Similar correlation slopes of C and H isotope fractionation (Eq. 15) are independent evidence for a common reaction mechanism. *Reprinted with permission from Pati, S. G., Kohler, H.-P. E., Pabis, A., Paneth, P., Parales, R. E., & Hofstetter, T. B. (2016). Substrate and enzyme specificity of the kinetic isotope effects associated with the dioxygenation of nitroaromatic contaminants.* Environmental Science & Technology, 50*(13) 6708–6716. http://dx.doi.org/10.1021/acs.est .5b05084, Copyright 2016, American Chemical Society.*

dioxygenation of aromatic carbon atoms (Fig. 1). In fact, our most recent work suggests that masking occurs through rate-limiting O_2 activation and that contributions of masking are substrate dependent (Pati, Kohler, et al., 2016). Note that Eq. (15) is a good approximation but not valid for reactions associated with large primary ^{2}H-KIE (Dorer et al., 2014; Wijker, Adamczyk, et al., 2013).

3. EXPERIMENTAL APPROACHES FOR DETERMINING ISOTOPE FRACTIONATION DURING OXYGENATION REACTIONS

3.1 Experiment Design and Sampling Strategies

The purpose of the experiments described below is to obtain samples for concentration and isotope ratio measurements of substrate, products, and cosubstrate (O_2). In contrast to experiments for assessment of concentration dynamics, such as those used for determining kinetic parameters of enzymatic reactions, samples for stable isotope analysis at natural abundance requires (i) large amounts of analytes, and (ii) an adequate extent of substrate conversion and product formation concomitant with strong isotopic enrichment (or depletion) to determine isotope fractionation parameters reliably. For substrate isotope fractionation, large extents of conversion of >90% are preferred whereas substrate conversion <10% is suited for quantifying product isotope fractionation.

To obtain sufficient amounts of analytes in a sample, one can either use high initial substrate concentrations provided that no substrate inhibition occurs or large sample volumes for subsequent automated analyte enrichment by solid-phase microextraction (SPME, see Section 4). We identified two separate approaches, with which samples with >90% substrate conversion can be generated. (a) In "continuous transformation assays," the reaction progress is controlled by stopping a reaction, for example, through acidification or analyte extraction. Those experiments can be conducted in one large volume reactor (>100 mL), from which adequate sample volumes are withdrawn at different time points. Alternatively, experiments are run in multiple reactors of smaller volumes (<30 mL), which are sacrificed at specific time points. (b) In the "limited turnover assays," the extent of transformation is controlled by limiting the amount of available cofactors (e.g., NADH). Both approaches were used to study the oxygenation of nitroaromatic compounds and three example procedures are illustrated below for isotopic analyses of substrates, products, and cosubstrate (O_2).

3.2 Enzyme Assays for Isotope Analysis of Substrates

Isotope effects on the oxygenation of (nitro)aromatic substrates, such as nitrobenzene, 2,6-dinitrotoluene, and naphthalene, by nitrobenzene dioxygenase (NBDO) can be obtained in assays with the purified, three-component enzyme system, as illustrated here. Those experiments follow the "limited turnover" approach illustrated in Fig. 6 (left-hand side, "1. Substrate isotope analysis") but alternative approaches with "continuous transformation assays" (Fig. 7) are also possible when working with whole cell cultures and cell extracts (Pati et al., 2014; Pati, Kohler, et al., 2016). Using purified enzymes instead of whole cells of *E. coli* clones can be required to avoid interferences, for example, from substrate loss through sorption to cell material when working at high substrate concentrations (up to 1 m*M*) and high cell densities. The activity of NBDO depends on the type of buffer used, concentration of dissolved Fe^{2+}, and initial substrate concentration. Consequently, the most practical approach to ensure adequate substrate conversion are experiments at high enzyme concentrations (0.3 μ*M* oxygenase) and addition of dissolved Fe^{2+}. Because substrate dioxygenation is too fast to be quenched at selected time points under these conditions, substrate turnover was limited by adding different substoichiometric amounts of NADH to different reactors. Purification procedures of the three enzyme components of NBDO (reductase, ferredoxin, and oxygenase) are reported elsewhere (Parales et al., 2005).

3.2.1 Procedure

1. Mix the three purified enzyme components of NBDO in 50 m*M* MES buffer at pH 6.8 containing 100 μ*M* $(NH_4)_2Fe(SO_4)_2$ to final concentrations of 0.3 μ*M* reductase, 3.6 μ*M* ferredoxin, and 0.3 μ*M* oxygenase.
2. Dissolve the nitroaromatic substrate (e.g., nitrobenzene) in MES buffer and add an appropriate volume to the enzyme mixture to give an initial substrate concentration of 200 μ*M*.
3. Fill nine 10-mL serum vials with 5 mL buffer solution containing enzymes and substrate and cap them with Viton rubber stoppers. The stoppers limit the loss of volatile substrate, however, the rubber material should be tested to avoid sorptive losses of the substrate into the stoppers.
4. Amend all but one vial, which is used as a control as is, with different amounts of the cofactor NADH ranging from 100 to 980 μ*M*.

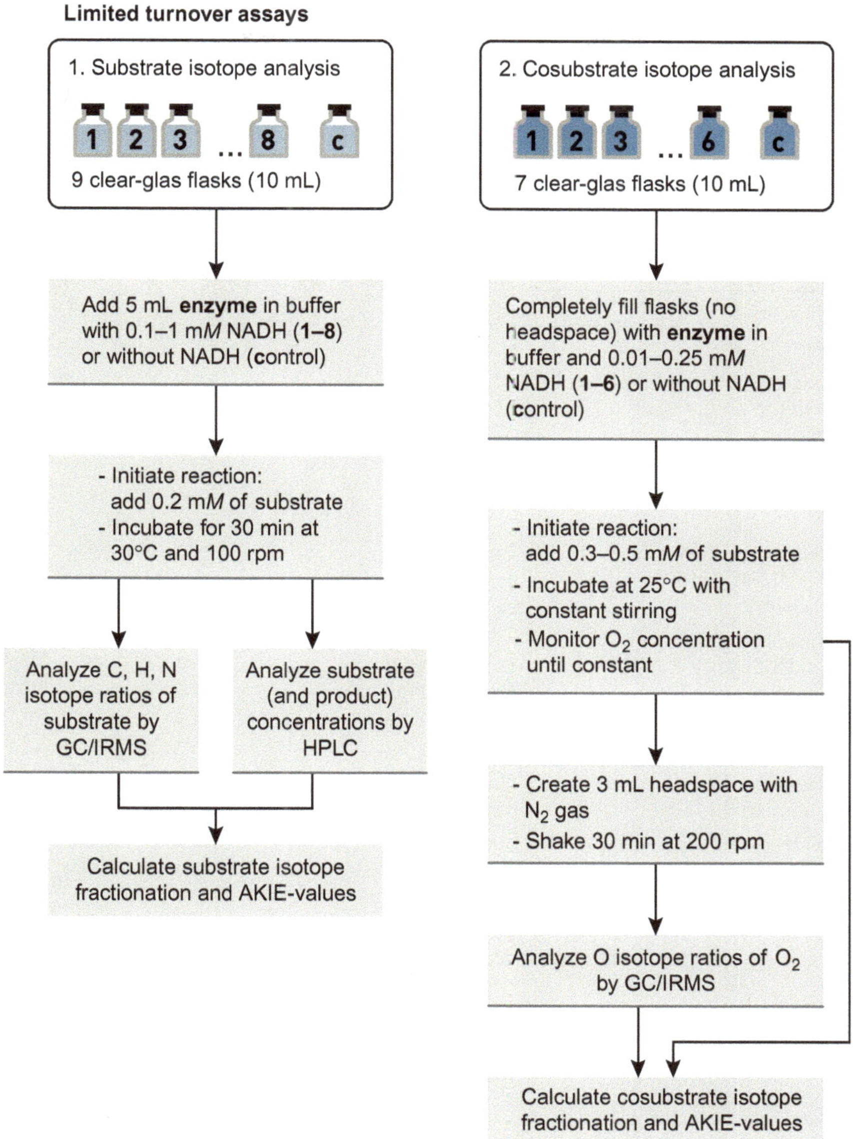

Fig. 6 Schematic view of experimental procedures for "limited turnover assays" where the extent of substrate turnover and product formation is controlled through the substoichiometric addition of NADH. Procedures for *substrate* isotope analysis, for example, of nitrobenzene, and for the *cosubstrate*, that is, aqueous O_2, are shown on the left- and right-hand side, respectively.

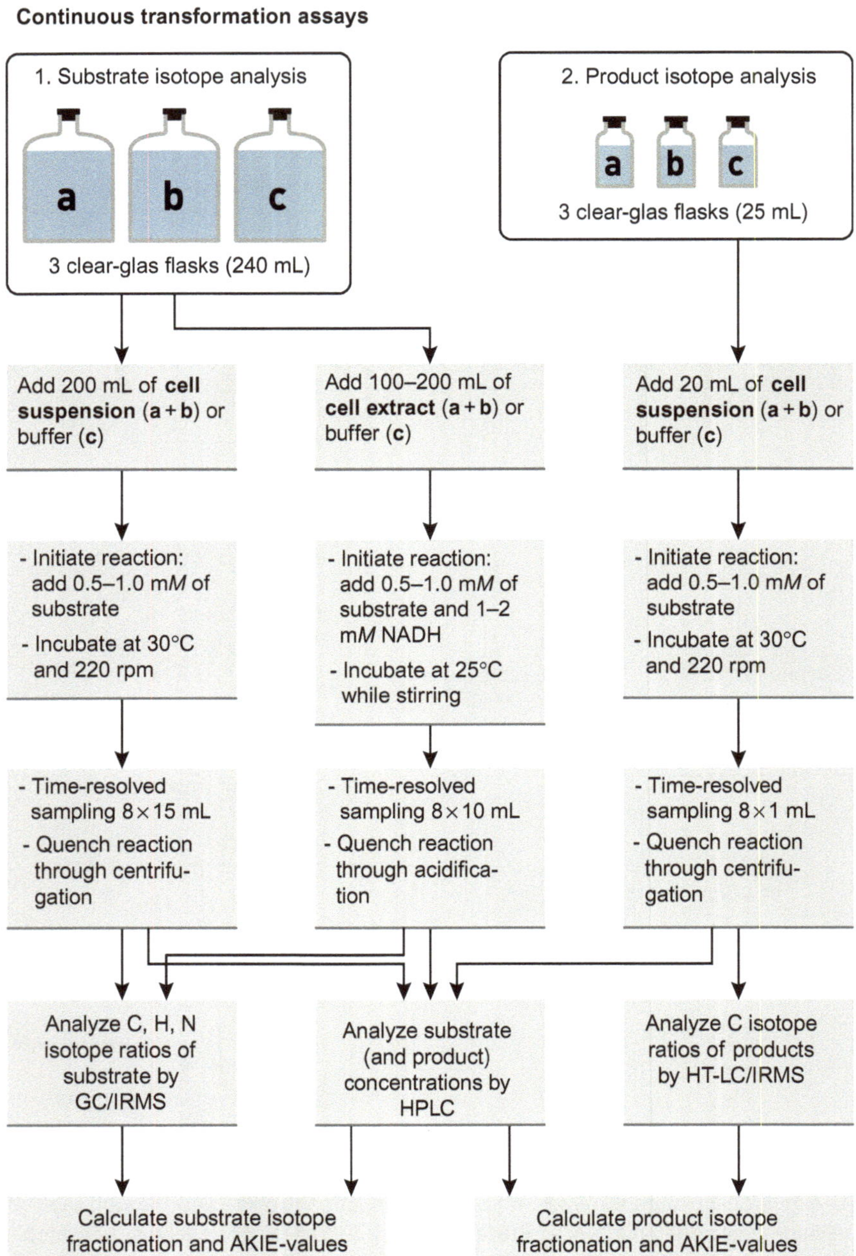

Fig. 7 Schematic view of experimental procedures for "continuous transformation assays." The extent of substrate turnover and product formation is controlled by withdrawing samples from large reactors at predefined time points. Procedures for *substrate* and *product* isotope analysis are shown on the left- and right-hand side, respectively.

5. Incubate all reactions at 30°C while shaking at 100 rpm for 30 min. Thereafter, all NADH should be oxidized and the transformation of the substrate terminated.
6. Store vials at 4°C until chemical and isotopic analyses. The concentrations of substrate and reaction products should be determined immediately by high-performance liquid chromatography (HPLC). Analysis of C, H, and N isotopic ratios of the substrate by GC/IRMS is described in Section 4.

3.3 Whole Cell Assays for C Isotope Analysis of Organic Reaction Products

Procedures for the quantification of isotope fractionation in the organic products of enzymatic oxygenation reactions take into account that these polar compounds are often only amenable to isotopic analysis by LC/IRMS. Isotopic analysis by LC/IRMS was restricted to C isotope analysis until recently (Godin & McCullagh, 2011; Zhang, Kujawinski, Jochmann, & Schmidt, 2011) but may be extended to N in the near future (Federherr et al., 2016). For these purposes, suitable experimental systems include enzyme assays and whole cells assays of *E. coli* clones because in both systems the products are not transformed further. Products of substituted nitroaromatic compound oxygenation, that is, substituted catechols, phenols, and benzyl alcohols, however, are not necessarily stable in the different experimental systems. In particular catechols, are easily oxidized abiotically. We found that oxygenation products were stable in whole cell assays with *E. coli* clones expressing NBDO (Pati et al., 2014). Note that addition of acid or base to stop reactions always resulted in artifacts and product concentration were substoichiometric. The procedure described below follows the "continuous transformation assay" with few reactors (Fig. 7) and can also be applied to determine the substrate isotope fractionation in assays with whole cells of *E. coli* clones as well as with pure strains of environmental microorganisms and their crude cell extracts.

There are two major differences between assays for product and substrate analysis shown in Fig. 7. First, in assays for nitroaromatic compound analysis, reactor and sample volumes are usually 4–10 times larger than for analysis of the hydroxylated products. The difference is due to the requirements for instrumental analysis (see GC/IRMS vs LC/IRMS in Section 4). Second, samples for analysis of nitroaromatic compounds can be acidified and thus stored at 4°C for days to a few weeks, whereas samples for analysis of the hydroxylated products cannot be preserved and need to be analyzed immediately.

3.3.1 Procedure

1. Grow and induce cultures of *E. coli* clones according to standard methods (overnight growth in nutrient rich medium) and dilute cells to appropriate densities (3–9 g/L) with phosphate buffer (40 m*M*, pH 7.0).
2. Fill two 25-mL serum flasks with 20 mL cell suspension and cap them with Viton rubber stoppers. An additional serum flask, which only contains phosphate buffer, should be run simultaneously as a control.
3. Initiate reactions by adding small volumes of a nitroaromatic substrate in a methanolic solution resulting in initial substrate concentrations of 0.5–1.0 m*M*.
4. After vigorous shaking, withdraw the first sample of 1.5 mL immediately with a glass syringe. Subsequently, incubate the reactors at 30°C while shaking at 200 rpm.
5. Take additional 6–8 samples after time intervals that enable sufficient isotope fractionation of the target analyte. Withdraw samples with gas tight glass syringes through the stopper after injecting an equivalent volume of air into the reactors. In most experiments, 90% substrate conversion was achieved after 8–10 h. However, in reactors with larger volumes reactions were often slower due to O_2 limitation.
6. Transfer samples into 1.5-mL plastic tubes and centrifuge them at 13,000 rpm for 5 min. The supernatant should be transferred into HPLC vials and stored at 4°C until concentration (by HPLC) and isotope analysis (see LC/IRMS in Section 4).

3.4 Enzyme Assays for O Isotope Analysis of Aqueous O_2

Oxygen isotope effects for the activation of O_2 by NBDO can be obtained from the analysis of O isotope fractionation in aqueous O_2. Pati, Bolotin, et al. (2016) developed a procedure for "limited turnover assays" with purified enzyme component similar to the one described in the first example for substrate isotope fractionation (Fig. 6, right-hand side). The major differences between enzyme assay for cosubstrate vs substrate isotope analysis are that (i) vials need to be filled with solution without headspace, (ii) one 10-mL reactor is sacrificed for every measurement of O isotope composition (no replicate measurements possible), and (iii) O_2 concentrations are monitored continuously during the reaction with an optical micro-sensor. Details on the sampling procedure and reaction progress monitoring will be shown in Section 4.3.

3.4.1 Procedure

1. Completely fill 5–10 10-mL glass vials with MES buffer (50 m*M*, pH 6.8) containing 0.15 μ*M* reductase, 1.8 μ*M* ferredoxin, 0.15 μ*M* oxygenase, 100 μ*M* $(NH_4)_2Fe(SO_4)_2$, and 300–500 μ*M* of a nitroaromatic substrate. Enzymes were purified according to Pati et al. (2014).
2. Close the vials with butyl rubber stoppers and aluminum crimp caps and ensure that no air bubbles are entrapped within the reactors.
3. Insert a fiber-optic oxygen microsensors (PreSens Precision Sensing GmbH, Germany; Pati, Bolotin, et al., 2016) with a stainless needle through the stopper for monitoring O_2 concentrations.
4. Initiate the reaction by adding different amounts of NADH (50–250 μ*M*) with a syringe through the stopper to all but one reactor (control). The reaction solution should be stirred slightly during the reaction with a magnetic stir bar inside the vial.
5. Monitor O_2 concentrations until a stable value is reached. The decrease in O_2 concentration was usually proportional to the amount of NADH added.
6. Remove the microsensor from the vessel and prepare the reactor for analysis. Due to potential leaking of ambient O_2 into the vials, measurements of O isotope composition by GC/IRMS should be performed on the same day the experiment is conducted (see Section 4.2 for details).

4. INSTRUMENTATION FOR STABLE ISOTOPE ANALYSIS BY ISOTOPE RATIO MASS SPECTROMETRY

4.1 Instrumental Strategies

CSIA relies on specialized isotope ratio mass spectrometers with sector field mass analyzers for the measurement of stable isotope ratios at natural isotopic abundance. Details on instrumentation and functioning of isotope ratio mass spectrometers can be found in many compilations such as Eiler (2013), Elsner et al. (2012), Sessions (2006), Jochmann and Schmidt (2012), Amrani, Sessions, and Adkins (2010), Said Ahmad et al. (2017), Bernstein et al. (2011), Zakon, Halicz, Lev, and Gelman (2016), Zakon, Halicz, and Gelman (2014), and Gelman and Halicz (2010). Here we highlight only the most important features of particular relevance for studying isotope effects of enzyme-catalyzed reactions.

Due to the instrumental requirement to focus isotopic ratio measurements on a few small molecules and the fact that analytes are present as mixture of compounds, analytes need to be converted, in most cases, into sample gases through the use of chemical reaction interfaces under continuous flow conditions. GC/IRMS devices are most versatile and enable one to measure C, H, N, and O isotope ratios of (semi-)volatile organic compound because analytes can be converted to CO_2, H_2, N_2, and CO through oxidation, reduction, and pyrolysis. A simplified scheme for this instrumental setup is shown in Fig. 8A. Most applications are based on commercially available standard instrumentations, but customized reactors may be required for the chemical conversion of highly oxidized compounds (Gehre et al., 2015; Nijenhuis, Renpenning, Kümmel, Richnow, & Gehre,

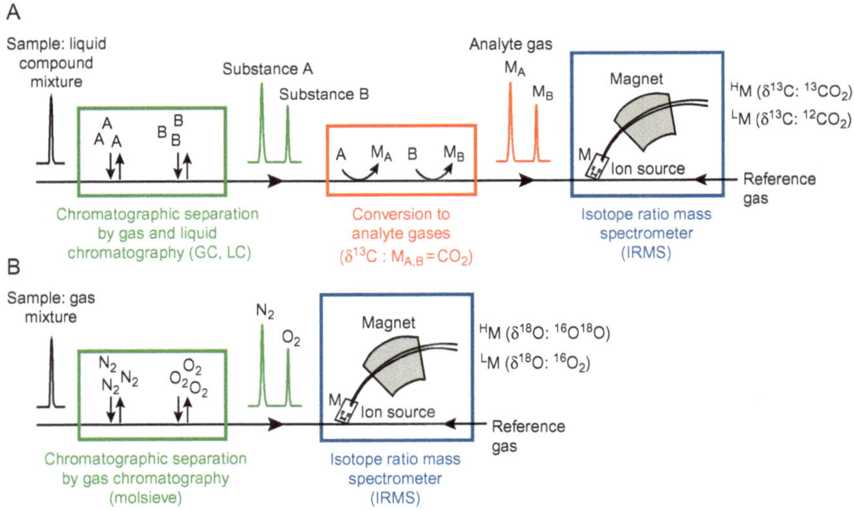

Fig. 8 Simplified schematic view of instrumentation for compound-specific isotope analysis by gas and liquid chromatography/isotope ratio mass spectrometry. (A) Instrumental setup for measurement of C, H, N, O isotope ratios in organic compounds after chemical conversion to small molecule gases by oxidation, reduction, and pyrolysis. Examples are shown here for determination of $\delta^{13}C$-values. "M" would correspond to H_2, N_2, and CO for measurements of δ^2H, $\delta^{15}N$, $\delta^{18}O$, respectively. (B) Instrumental setup for isotope ratio measurements without chemical conversion interface as exemplified for $\delta^{18}O$ of O_2. Note that gas purification devices used to remove, for example, H_2O and S-containing gases, are not shown here for simplicity. *Adapted from Elsner, M., Jochmann,M., Hofstetter, T., Hunkeler, D., Bernstein, A., Schmidt, T., & Schimmelmann, A. (2012). Current challenges in compound-specific stable isotope analysis of environmental organic contaminants.* Analytical and Bioanalytical Chemistry, 403, *2471–2491. http://dx.doi.org/10.1007/s00216-011 -5683-y.*

2016; Renpenning, Hitzfeld, et al., 2015; Renpenning, Kümmel, Hitzfeld, Schimmelmann, & Gehre, 2015) or complexing agents (Spahr et al., 2013). On the other hand, LC/IRMS devices apply wet chemical oxidation interfaces of analytes in aqueous solutions (Federherr et al., 2016; Godin & McCullagh, 2011; Krummen et al., 2004; Zhang et al., 2011). This approach is largely restricted to measuring C, and very recently, N isotope ratios.

The instrumental setup is selected primarily based on the physical–chemical properties of the analyte molecules. GC/IRMS is suitable for (semi-)volatile organic compounds extracted from water or injected in organic solvents as well as for gases. Conversely, LC/IRMS enables one to analyze polar and ionic organic compounds exclusively from aqueous solutions. Because enzyme assays are generally conducted in aqueous solution, direct injection of aqueous sample to an LC/IRMS system is most straightforward. For GC/IRMS measurements, analytes need to be extracted into organic solvents, or adsorbed to solid phases, for example, by solid phase (micro)extraction, and by purge and trap concentrators (Zwank, Berg, Schmidt, & Haderlein, 2003). Many of these steps can be automated. Details for sample preparation, instrumental settings, and data evaluation are illustrated below in three practical examples related to the three assays described in Section 3.

Finally, the isotope ratios of a series of gases and organic vapors are accessible by continuous-flow isotope ratio mass spectrometry without the use of chemical conversion interfaces. This approach is chosen here for measurement of $^{18}O/^{16}O$ ratios of O_2 as shown in Fig. 8B because O_2 isotopologues can be ionized and measured directly in an isotope ratio mass spectrometer (Pati, Bolotin, et al., 2016). The same strategy is pursued, for example, for analysis of halogen isotopes in organic compounds where isotopologue ratios are determined in fragments of the target analyte (Bernstein et al., 2011; Elsner & Hunkeler, 2008; Zakon et al., 2016).

4.2 Substrate Isotope Analysis by GC/IRMS

C, H, and N isotope ratios of nitroaromatic compounds can be measured by GC/IRMS after solid phase microextraction (SPME) from different matrices and similar analytical procedures apply for aminoaromatic compounds and alkylated and halogenated benzenes and phenols (Berg, Bolotin, & Hofstetter, 2007; Ratti, Canonica, McNeill, Bolotin, & Hofstetter, 2015; Ratti, Canonica, McNeill, Erickson, et al., 2015; Skarpeli-Liati et al., 2011; Wijker, Bolotin, Nishino, Spain, & Hofstetter, 2013; Wijker, Kurt, et al., 2013; Wijker, Zeyer, & Hofstetter, 2017). The range of analyte concentrations (3–200 μ*M*) in samples from enzyme assays makes SPME the ideal method for introducing analytes into the gas chromatograph. SPME fibers with different sorbate materials are commercially available for a range of organic compounds. Analytes are desorbed from the solid phase during a bake-out in the heated injector of the GC resulting in the transfer of analytes onto an analytical column. A GC compatible autosampler device can perform the whole extraction and desorption procedure automatically. The analyte(s) of interest are separated on an appropriate GC column from any other compounds that were extracted onto the SPME fiber so that for each compound a separate CO_2, N_2, or H_2 peak can be measured. The isotopic composition of the nitroaromatic substrates is derived from peak area ratios of heavy and light isotopologues of the corresponding analyte gas peaks. It is best practice to dilute samples for GC/IRMS analysis to a range of concentrations in which isotope ratios can be measured without mass bias from nonlinear detector responses (Jochmann, Blessing, Haderlein, & Schmidt, 2006; Sherwood Lollar et al., 2007).

4.2.1 Sample Preparation Procedure

1. Dilute samples from enzyme assays with different degree of substrate conversion with buffer to a specific substrate concentration that is ideal for GC/IRMS analysis. Such ranges depend on the element studied, e.g., 2 μ*M* for C and 10 μ*M* for H isotope analysis of nitrobenzene. Note that different instrumental runs are required for the various elements.
2. Adjust sample pH to 7.0 with NaOH or HCl to prolong the lifetime of the SPME fibers and to obtain analytes as neutral species.
3. Filter samples with precipitates or particulate matter to prevent clogging and distortion of the SPME fibers. Additional test should be performed to ensure that filtration does not change the isotopic composition of the analyte.
4. For each sample taken from an assay, prepare three identical analysis samples by mixing 1.3 mL diluted sample solution with 0.303 mg NaCl (4.0 *M* final concentration) to increase adsorption efficiency of nitroaromatic compounds onto the SPME fiber.
5. In addition to assay samples, prepare control samples containing the analyte(s) of interest in buffered solutions or water at the same analyte and NaCl concentrations. Analyzing control samples can reveal whether isotope fractionation occurred during sample preparation.
6. Additional samples containing in-house isotope standards, compounds with known isotopic composition that have been calibrated against

certified standard material, need to be measured to ensure accuracy (Werner & Brand, 2001).

4.2.2 Instrumental Parameters

The automated SPME extraction method is performed by a GC autosampler. Analysis vials are 1.5-mL glass vials with magnetic screw or crimp caps. The first sample is transported from the sample rack into an agitator that heats and agitates the vials during analyte extraction. For SPME of nitroaromatic compounds the fiber is immersed into the sample solution for 45 min at an agitator temperature of 40°C (Berg et al., 2007; Wijker, Bolotin, et al., 2013). Thereafter, nitroaromatic compounds are desorbed from the SPME fiber in the injector of the gas chromatograph for 5 min at 270°C. Nitroaromatic compounds are separated by means of gas chromatography using standard columns and temperature programs. Complete base-line separation of the analyte(s) is a prerequisite for GC/IRMS analysis. For C isotope analysis, the interface between GC and IRMS consists of a combustion oven operated at 1000°C. The setup for N isotope analysis requires an additional reduction step for conversion of NO_x species to N_2 at 650°C (Berg et al., 2007; Hartenbach, Hofstetter, Berg, Bolotin, & Schwarzenbach, 2006; Hartenbach et al., 2008). H isotope analysis is carried out with a pyrolysis reactor at 1200°C (Wijker, Adamczyk, et al., 2013). The ion masses that are recorded by the Faraday cups in the mass analyzer are m/z 2 and 3 for H isotope analysis, m/z 28, 29, and 30 for N isotope analysis, as well as m/z 44, 45, and 46 for C isotope analysis.

4.2.3 Sample Sequence and Data Evaluation

To compensate for signal drifts and offsets, control samples containing unreacted analyte should be dispersed throughout the sequence of measurements. Ideally, the analyte of interest is available as a calibrated in-house standard. If the isotopic composition of the analyte has not been calibrated externally, control samples should be amended with an additional compound of known isotopic composition. A typical sequence includes three replicate vials with control samples followed by three replicate vials of 2–3 assay samples (total of 6–9 injections). Thereafter, another three replicate vials with control samples are measured followed by three replicate vials of 2–3 assay samples, and, finally, another three replicate vials with control samples. With a run time of approximately 50 min per sample, this sequence takes 18–23 h. An experiment with nine assay samples requires six full days of GC/IRMS instrument time to determine C, H, and N isotope fractionation. As a quality control within and between measurement sequences, the control samples allow to detect signal drift over time as well as offsets, e.g., due to the SPME extraction procedure. Data can be reconciled using spreadsheet templates such as described by Dunn et al. (2015).

4.3 Oxygen Isotope Analysis of Aqueous O_2 by GC/IRMS

For determining O isotope ratios in aqueous O_2 a similar GC/IRMS setup as for substrate isotope analysis is used. The major differences are (i) mode of injection, (ii) operation of the conversion interface, and (iii) data evaluation strategy. O_2-containing gases can be analyzed with the isotope ratio mass spectrometer without chemical conversion. Major challenge for using a GC/IRMS are potential contamination of samples before and during analysis with atmospheric O_2. Because atmospheric O_2 has a constant isotopic composition (Barkan & Luz, 2005), small amounts of contamination from ambient O_2 can be corrected for with a blank sample subtraction. With this approach, accurate $\delta^{18}O$-values can be determined for aqueous O_2 concentrations ranging from 20 to 250 μM (0.6–8 mg/L) (Pati, Bolotin, et al., 2016).

4.3.1 Sample Preparation Procedure

1. Create a headspace in every 10-mL assay vial by manually replacing 3 mL reaction solution with N_2 gas while holding the vials upside down (see Fig. 9). The 3 mL excess solution can be used to quantify substrate consumption and product formation to establish reaction stoichiometries.
2. Once all vials contain a headspace, place them upside down on an orbital shaker for 30 min at 200 rpm to facilitate the transfer of O_2 into the gas phase.
3. Afterward, place samples onto the autosampler and automatically inject aliquots of the headspace from the same vials, in which the enzyme assays were run. Because of a danger of small contaminations of the vials with ambient air, only one injection should be made per vial.
4. Prepare control samples containing air-equilibrated water and blank samples containing O_2-free water in the same way as assay samples (make headspace manually and shake for phase transfer, see Fig. 9). O_2-free water can be produced by purging water with N_2 gas for 3 h and subsequently filling 10-mL vials headspace free in an anaerobic glove box under a N_2 atmosphere ($O_2 < 0.1$ ppm). Blanks should undergo the same sample preparation scheme as assay and control samples (including headspace creation).

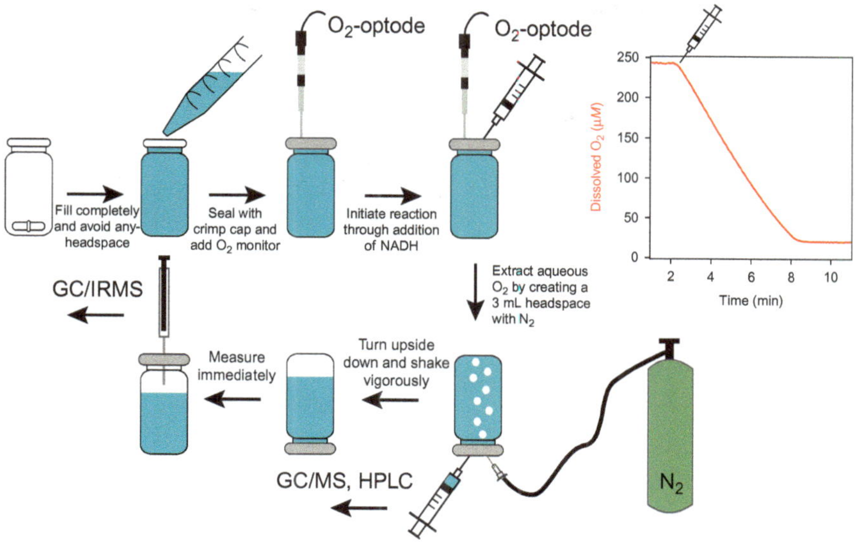

Fig. 9 Procedure for extraction of aqueous O_2 from "limited turnover assays" containing NBDO in aqueous solution (see Sections 3.4 and 4.3.1 for details). An O_2 microsensor (optode) is used for continuous monitoring of O_2 consumption during the experiment. $\delta^{18}O$ is measured in O_2 of the N_2 headspace.

4.3.2 Instrumental Parameters

After equilibration of gas and water phases in the vial, the O_2 concentration in the headspace should range between 50 and 600 μ*M* (2–20 mg/L) or 0.12–1.5 vol%. The autosampler injects 250 μL of the headspace with a large-volume headspace syringe into the split/splitless injector of the gas chromatograph. Before piercing the septum, the syringe should be flushed for 1 min with N_2 gas and automated procedures can be programmed with an autosampler. Setting the injector temperature up to 200°C will ensure a better reproducibility and precision of measured isotope ratios compared to having the injector operated at ambient temperature (Pati, Bolotin, et al., 2016). Separation of O_2 from other gases, in particular N_2, is achieved by means of gas chromatography with a molecular sieve PLOT column (5 Å, 30 m × 0.32 μm I.D.) at a constant oven temperature of 30°C. The combustion reactors in the interface between gas chromatograph and mass spectrometers are turned off (ambient temperature) and O_2 isotopologues can be analyzed directly with recording ion masses *m*/*z* of 32, 33, and 34. $^{18}O/^{16}O$ isotope ratios of O_2 are derived from peak area ratios at *m*/*z* 32 and 34. $^{17}O/^{16}O$ ratios can, in principle, be derived from peak are ratios at m/z 32 and 33, however, special multipliers for the middle Faraday cup are required to do so accurately.

4.3.3 Sample Sequence and Data Evaluation

The measurement sequence for O isotope ratio analysis is similar to the one described for substrate isotope analysis above with the exception that only one measurement per assay sample is performed. As an example, the fractionation of O isotopes during dioxygenation of nitrobenzene by NBDO is shown in Fig. 10. Measurements lasted 6 min per sample. Triplicates of control samples containing air-equilibrated water should be injected at the beginning and the end of the sequence and, depending on the sequence length, also in regular intervals within the sequence. Three blank samples (O_2-free water) should be injected at the end of the sequence to ensures that the maximum amount of potential contamination with ambient air is captured for the blank correction (Eq. 16). Control samples allow for corrections of signal drifts within a sequence as well as between sequences. One O_2 reference gas was used for calibrating isotope signatures of O_2 and the use of a second isotope standard with a different O isotope signature could improve accuracy of measurements. The procedure for correcting for

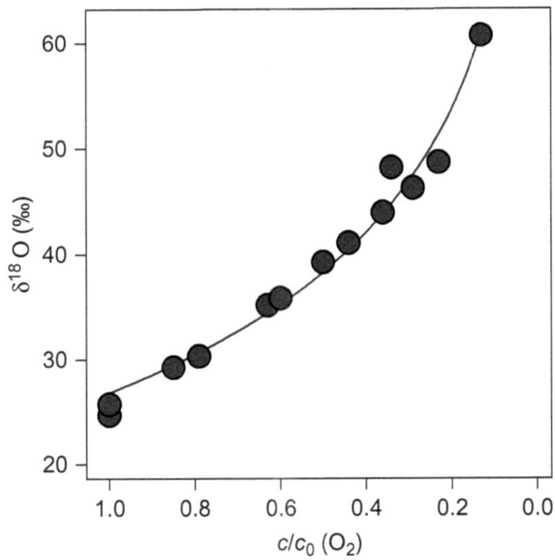

Fig. 10 δ^{18}O of aqueous O_2 after partial O_2 conversion during the dioxygenation of nitrobenzene by NBDO using "limited turnover assays" using the procedure shown in Fig. 9.

ambient O_2 contamination is described in detail in Pati, Bolotin, et al. (2016) and is performed according to Eq. (16).

$$\delta^{18}O_{corr} = \frac{\delta^{18}O_{meas} \cdot A_{meas} - \delta^{18}O_{blank} \cdot A_{blank}}{A_{meas} - A_{blank}} \quad (16)$$

where $\delta^{18}O_{meas}$ and A_{meas} are the measured isotope signature of O_2 and m/z 32 peak area in an assay sample, while $\delta^{18}O_{blank}$ and A_{blank} are the averages of the same parameter in the three blank samples. $\delta^{18}O_{corr}$ is the blank-corrected isotope signature for the given assay sample, which is used for further data evaluation.

4.4 Product Isotope Analysis by LC/IRMS

Measuring isotope ratios of oxygenation products by LC/IRMS circumvents the need for derivatization. Injection volumes are larger than with a GC/IRMS. With a 100 μL syringe, the lowest concentration for reliable isotope analysis for catechols and nitrobenzyl alcohols was approx. 30 μ*M* and thus 10 times higher than for C isotope analysis of nitroaromatic compounds by GC/IRMS.

4.4.1 Sample Preparation Procedure

1. Filter assay samples containing particles or precipitates prior to analysis by LC/IRMS to protect the column.
2. Adjust the pH of assay samples to $\leq$7.0 to ensure that the hydroxylated analytes are protonated (Pati et al., 2014).
3. Fill 500 μL of each assay sample into analysis vials so that multiple injections can be made from the same vial.
4. Prepare control samples by dissolving analytes in an appropriate buffer solution.

As with GC/IRMS, measurements should be performed with constant amounts of analytes in all samples (including controls). With LC/IRMS, however, constant amounts of injected analytes can be achieved by adjusting injection volumes instead of diluting samples for analysis by GC/IRMS. If multiple analytes are present in the same sample, the compounds need to be present at comparable concentrations or repeated measurements with different injection volumes should be performed. Measurement sequences and strategies for isotopic calibration and referencing are analogous to those for substrate isotope analysis by GC/IRMS.

4.4.2 Instrumental Parameters

Organic analytes are separated on an appropriate column by LC and converted to compound-specific CO_2-peaks in the wet oxidation interface before entering the isotope ratio mass spectrometer. The principle of the chemical conversion interface requires that organic solvent cannot be used as eluents for chromatographic separation and eluents consist of aqueous solution or inorganic buffers. Ion exchange chromatography and high-temperature reversed-phase chromatography are valuable alternatives (Godin & McCullagh, 2011; Zhang et al., 2011). In fact, high-temperature reversed-phase chromatography led to equal or better separation of substituted catechols and nitrobenzyl alcohols from each other and the corresponding nitroaromatic substrates than conventional HPLC with organic solvents.

Temperature gradients from 30°C to 160°C can be applied to change the retention of analytes on the analytical column (Zhang et al., 2011). Here, we used a 10 m*M* phosphate buffer solution at pH 2.5 to minimize the amount of CO_2 dissolved in our solutions. The effluent of the LC column (0.5 mL/min) is channeled into a wet chemical oxidation interface. The reactor in the oxidation interface is heated to 100°C and supplied continuously with phosphoric acid and $Na_2S_2O_8$. Analytes need to be base-line separated from all interfering compounds, so that compound-specific CO_2 peaks are formed in the interface. The CO_2 is subsequently stripped from the eluent when flowing through gas-permeable membranes and introduced with a He stream into the isotope ratio mass spectrometer. The IRMS settings are identical to those of C isotope analysis by GC/IRMS, CO_2 background levels, however, are considerably higher with LC/IRMS. It is therefore very important that CO_2 background levels do not change during elution of the target analyte peak. This issues can be circumvented by evaluating different temperature programs accordingly and testing reference compounds with known isotopic composition as mixtures in the LC/IRMS and through direct injection of the pure compounds into the oxidation interface. Note that the organic matter contents of the sample, for example, enzymes, cell components, can compromise the analysis if that material was converted into CO_2. It may be advantageous to use columns that retain the analytes sufficiently long so that for the first 1–2 min after injection the eluent with high organic backgrounds can bypass the wet oxidation interface.

5. SUMMARY AND CONCLUSION

Enzymatic oxygenations are among the most important biodegradation reactions in the environment, and they also contribute to initial transformation of numerous organic pollutants through cometabolic oxygenations. Despite its relevance, knowledge of enzymatic mechanisms of O_2 activation and oxygenation of organic pollutants are scarce (e.g., Wijker et al., 2015). It is therefore very challenging to assess whether such processes will happen unless the stable isotope fractionation of soil and water pollutants is understood in greater detail. The theoretical, experimental, and instrumental procedures illustrated here enable a researcher to investigate isotope enrichment factors and kinetic isotope effects that come along with biodegradation of many persistent organic pollutants. Because oxygenases are widely involved in oxidative biodegradation and biotransformation and exhibit a wide spectrum of possible substrates, the presented methods can be applied beyond the compounds and enzymes discussed here.

ACKNOWLEDGMENTS

This work was supported by the Swiss National Science Foundation (grants no. 206021-139'111/1) and the Swiss-Polish Research Collaboration (PSRP-025/200).

REFERENCES

Aelion, C. M., Höhener, P., Hunkeler, D., & Aravena, R. (2010). *Environmental isotopes in biodegradation and bioremediation*. Boca Raton, FL: Taylor & Francis Group.

Aeppli, C., Berg, M., Cirpka, O. A., Holliger, C., Schwarzenbach, R. P., & Hofstetter, T. B. (2009). Influence of mass-transfer limitations on carbon isotope fractionation during microbial dechlorination of trichloroethene. *Environmental Science & Technology*, *43*(23), 8813–8820. http://dx.doi.org/10.1021/es901481.

Amrani, A., Sessions, A. L., & Adkins, J. F. (2010). Compound-specific δ^{34}S analysis of volatile organics by coupled GC/multicollector-ICPMS. *Analytical Chemistry*, *81*(21), 9027–9034. http://dx.doi.org/10.1021/ac9016538.

Barkan, E., & Luz, B. (2005). High precision measurements of $^{17}O/^{16}O$ and $^{18}O/^{16}O$ ratios in H_2O. *Rapid Communications in Mass Spectrometry*, *19*(24), 3737–3742. http://dx.doi.org/10.1002/rcm.2250.

Berg, M., Bolotin, J., & Hofstetter, T. B. (2007). Compound-specific nitrogen and carbon isotope analysis of nitroaromatic compounds in aqueous samples using solid-phase microextraction coupled to GC/IRMS. *Analytical Chemistry*, *79*(6), 2386–2393. http://dx.doi.org/10.1021/ac0622577.

Bernstein, A., Shouakar-Stash, O., Ebert, K. A., Laskov, C., Hunkeler, D., Jeannottat, S., … Elsner, M. (2011). Compound-specific chlorine isotope analysis: A comparison of gas chromatography/isotope ratio mass spectrometry and gas chromatography/

quadrupole mass spectrometry methods in an interlaboratory study. *Analytical Chemistry*, *83*(20), 7624–7634. http://dx.doi.org/10.1021/ac200516c.

Boyd, D. R., & Bugg, T. D. H. (2006). Arene *cis*-dihydrodiol formation: From biology to application. *Organic & Biomolecular Chemistry*, *4*(2), 181–192. http://dx.doi.org/10.1002/chin.200618269.

Brand, W. A., Coplen, T. B., Vogl, J., Rosner, M., & Prohaska, T. (2014). Assessment of international reference materials for isotope-ratio analysis (IUPAC Technical Report). *Pure and Applied Chemistry*, *86*(3), 425–467. http://dx.doi.org/10.1515/pac-2013-1023.

Bugg, T. D. H., & Winfield, C. J. (1998). Enzymatic cleavage of aromatic rings: Mechanistic aspects of the catechol dioxygenases and later enzymes of bacterial oxidative cleavage pathways. *Natural Product Reports*, *15*(5), 513–530. http://dx.doi.org/10.1039/a815513y.

Coleman, M., & Meier-Augenstein, W. (2014). Ignoring IUPAC guidelines for measurement and reporting of stable isotope abundance values affects us all. *Rapid Communications in Mass Spectrometry*, *28*(17), 1953–1955. http://dx.doi.org/10.1002/rcm.6971.

Cook, P. F., & Cleland, W. W. (2007). *Enzyme kinetics and mechanism*. New York: Garland Science.

Coplen, T. B. (2011). Guidelines and recommended terms for expression of stable-isotope-ratio and gas-ratio measurement results. *Rapid Communications in Mass Spectrometry*, *25*(17), 2538–2560. http://dx.doi.org/10.1002/rcm.5129.

DelMonte, A. J., Haller, J., Houk, K. N., Sharpless, K. B., Singleton, D. A., Strassner, T., & Thomas, A. A. (1997). Experimental and theoretical kinetic isotope effects for asymmetric dihydroxylation. Evidence supporting a rate-limiting "(3 + 2)" cycloaddition. *Journal of the American Chemical Society*, *119*(41), 9907–9908.

Dorer, C., Höhener, P., Hedwig, N., Richnow, H.-H., & Vogt, C. (2014). Rayleigh-based concept to tackle strong hydrogen fractionation in dual isotope analysis—the example of ethylbenzene degradation by aromatoleum aromaticum. *Environmental Science & Technology*, *48*(10), 5788–5797. http://dx.doi.org/10.1021/es404837g.

Dunn, P. J. H., Hai, L., Malinovsky, D., & Goenaga Infante, H. (2015). Simple spreadsheet templates for the determination of the measurement uncertainty of stable isotope ratio delta values. *Rapid Communications in Mass Spectrometry*, *29*(22), 2184–2186. http://dx.doi.org/10.1002/rcm.7376.

Eckert, D., Qiu, S., Elsner, M., & Cirpka, O. A. (2013). Model complexity needed for quantitative analysis of high resolution isotope and concentration data from a toluene-pulse experiment. *Environmental Science & Technology*, *47*(13), 6900–6907.

Eiler, J. M. (2013). The isotopic anatomies of molecules and minerals. *Annual Review of Earth and Planetary Sciences*, *41*(1), 411–441. http://dx.doi.org/10.1146/annurev-earth-042711-105348.

Elsner, M. (2010). Stable isotope fractionation to investigate natural transformation mechanisms of organic contaminants: Principles, prospects and limitations. *Journal of Environmental Monitoring*, *12*(11), 2005–2031. http://dx.doi.org/10.1039/c0em00277a.

Elsner, M., Chartrand, M., VanStone, N., Lacrampe Couloume, G., & Sherwood Lollar, B. (2008). Identifying abiotic chlorinated ethene degradation: Characteristic isotope patterns in reaction products with nanoscale zero-valent iron. *Environmental Science & Technology*, *42*(16), 5963–5970. http://dx.doi.org/10.1021/es8001986.

Elsner, M., & Hunkeler, D. (2008). Evaluating chlorine isotope effects from isotope ratios and mass spectra of polychlorinated molecules. *Analytical Chemistry*, *80*(12), 4731–4740. http://dx.doi.org/10.1021/ac702543y.

Elsner, M., Jochmann, M., Hofstetter, T., Hunkeler, D., Bernstein, A., Schmidt, T., & Schimmelmann, A. (2012). Current challenges in compound-specific stable isotope analysis of environmental organic contaminants. *Analytical and Bioanalytical Chemistry*, *403*, 2471–2491. http://dx.doi.org/10.1007/s00216-011-5683-y.

Elsner, M., Zwank, L., Hunkeler, D., & Schwarzenbach, R. P. (2005). A new concept linking observable stable isotope fractionation to transformation pathways of organic pollutants. *Environmental Science & Technology*, *39*(18), 6896–6916. http://dx.doi.org/10.1021/es0504587.

Federherr, E., Willach, S., Roos, N., Lange, L., Molt, K., & Schmidt, T. (2016). A novel high-temperature combustion interface for compound-specific stable isotope analysis of carbon and nitrogen via high-performance liquid chromatography/isotope ratio mass spectrometry. *Rapid Communications in Mass Spectrometry*, *30*(7), 2537–2544. http://dx.doi.org/10.1002/rcm.7524.

Gehre, M., Renpenning, J., Gilevska, T., Qi, H., Coplen, T. B., Meijer, H. A. J., … Schimmelmann, A. (2015). On-line hydrogen-isotope measurements of organic samples using elemental chromium: An extension for high temperature elemental-analyzer techniques. *Analytical Chemistry*, *87*(10), 5198–5205. http://dx.doi.org/10.1021/acs.analchem.5b00085.

Gelman, F., & Halicz, L. (2010). High precision determination of bromine isotope ratio by GC-MC-ICPMS. *International Journal of Mass Spectrometry*, *289*(2–3), 167–169. http://dx.doi.org/10.1016/j.ijms.2009.10.004.

Gibson, D. T., & Parales, R. E. (2000). Aromatic hydrocarbon dioxygenases in environmental biotechnology. *Current Opinion in Biotechnology*, *11*(3), 236–243. http://dx.doi.org/10.1016/s0958-1669(00)00090-2.

Godin, J. P., & McCullagh, J. S. O. (2011). Review: Current applications and challenges for liquid chromatography coupled to isotope ratio mass spectrometry (LC/IRMS). *Rapid Communications in Mass Spectrometry*, *25*(20), 3019–3028. http://dx.doi.org/10.1002/rcm.5167.

Hartenbach, A., Hofstetter, T. B., Berg, M., Bolotin, J., & Schwarzenbach, R. P. (2006). Using nitrogen isotope fractionation to assess abiotic reduction of nitroaromatic compounds. *Environmental Science & Technology*, *40*(24), 7710–7716. http://dx.doi.org/10.1021/es061074z.

Hartenbach, A. E., Hofstetter, T. B., Aeschbacher, M., Sander, M., Kim, D., Strathmann, T. J., … Schwarzenbach, R. P. (2008). Variability of nitrogen isotope fractionation during the reduction of nitroaromatic compounds with dissolved reductants. *Environmental Science & Technology*, *42*(22), 8352–8359. http://dx.doi.org/10.1021/es801063u.

Hofstetter, T. B., & Berg, M. (2011). Assessing transformation processes of organic contaminants by compound-specific stable isotope analysis. *TrAC-Trends Analytical Chemistry*, *30*(4), 618–627. http://dx.doi.org/10.1016/j.trac.2010.10.012.

Hofstetter, T. B., Neumann, A., Arnold, W. A., Hartenbach, A. E., Bolotin, J., Cramer, C. J., & Schwarzenbach, R. P. (2008). Substituent effects on nitrogen isotope fractionation during abiotic reduction of nitroaromatic compounds. *Environmental Science & Technology*, *42*(6), 1997–2003. http://dx.doi.org/10.1021/es702471k.

Hofstetter, T. B., Schwarzenbach, R. P., & Bernasconi, S. M. (2008). Assessing transformation processes of organic compounds using stable isotope fractionation. *Environmental Science & Technology*, *42*(21), 7737–7743. http://dx.doi.org/10.1021/es801384j.

Höhener, P., & Atteia, O. (2014). Rayleigh equation for evolution of stable isotope ratios in contaminant decay chains. *Geochimica et Cosmochimica Acta*, *126*, 70–77. http://dx.doi.org/10.1016/j.gca.2013.10.036.

Houk, K. N., & Strassner, T. (1999). Establishing the (3 + 2) mechanism for the permanganate oxidation of alkenes by theory and kinetic isotope effects. *The Journal of Organic Chemistry*, *64*(3), 800–802.

Hunkeler, D., & Elsner, M. (2010). Principles and mechanisms of isotope fractionation. In C. M. Aelion, P. Höhener, D. Hunkeler, & R. Aravena (Eds.), *Environmental isotopes in biodegradation and bioremediation* (pp. 43–77). Boca Raton, FL: Taylor & Francis Group.

Jin, B., Haderlein, S. B., & Rolle, M. (2013). Integrated carbon and chlorine isotope modeling: Applications to chlorinated aliphatic hydrocarbons dechlorination. *Environmental Science & Technology*, *47*(3), 1443–1451. http://dx.doi.org/10.1021/es304053h.

Jochmann, M. A., Blessing, M., Haderlein, S. B., & Schmidt, T. C. (2006). A new approach to determine method detection limits for compound-specific isotope analysis of volatile organic compounds. *Rapid Communications in Mass Spectrometry*, *20*(24), 3639–3648. http://dx.doi.org/10.1002/rcm.2784.

Jochmann, M. A., & Schmidt, T. C. (2012). *Compound-specific stable isotope analysis*. RSC Publishing.

Kohen, A., & Limbach, H.-H. (2006). *Isotope effects in chemistry and biology*. New York: CRC Press/Taylor & Francis.

Kohler, H.-P. E. (2011). Biochemistry of environmental contaminant transformation: Nonylphenolic compounds and hexachlorocyclohanes - Two case studies. In B. Xing, N. Senesi, & P. M. Huang (Eds.), *Biophysico-chemical processes of anthropogenic organic compounds in environmental systems* (pp. 465–481). Hoboken, NJ: John Wiley & Sons, Inc.

Krummen, M., Hilkert, A. W., Juchelka, D., Duhr, A., Schlüter, H.-J., & Pesch, R. (2004). A new concept for isotope ratio monitoring liquid chromatography/mass spectrometry. *Rapid Communications in Mass Spectrometry*, *18*(19), 2260–2266. http://dx.doi.org/10.1002/rcm.1620.

Maggi, F., & Riley, W. J. (2010). Mathematical treatment of isotopologue and isotopomer speciation and fractionation in biochemical kinetics. *Geochimica et Cosmochimica Acta*, *74*(6), 1823–1835. http://dx.doi.org/10.1016/j.gca.2009.12.021.

Magyar, P. M., Orphan, V. J., & Eiler, J. M. (2016). Measurement of rare isotopologues of nitrous oxide by high-resolution multi-collector mass spectrometry. *Rapid Communications in Mass Spectrometry*, *30*(17), 1923–1940. http://dx.doi.org/10.1002/rcm.7671.

Melander, L. C. S., & Saunders, W. H. (1980). *Reaction rates of isotopic molecules*. New York: John Wiley & Sons.

Neilson, A. H., & Allard, A.-S. (2008). *Environmental degradation and transformation of organic chemicals*. Boca Raton, FL: CRC Press.

Nijenhuis, I., Renpenning, J., Kümmel, S., Richnow, H.-H., & Gehre, M. (2016). Recent advances in multi-element compound-specific stable isotope analysis of organohalides: Achievements, challenges and prospects for assessing environmental sources and transformation. *Trends in Environmental Analytical Chemistry*, *11*, 1–8. http://dx.doi.org/10.1016/j.teac.2016.04.001.

Northrop, D. B. (1981). The expression of isotope effects on enzyme-catalyzed reactions. *Annual Review of Biochemistry*, *50*, 103–131. http://dx.doi.org/10.1146/annurev.bi.50.070181.000535.

Parales, R. E., Huang, R., Yu, C.-L., Parales, J. V., Lee, F. K. N., Lessner, D. J., … Gibson, D. T. (2005). Purification, characterization, and crystallization of the components of the nitrobenzene and 2-nitrotoluene dioxygenase enzyme systems. *Applied and Environmental Microbiology*, *71*(7), 3806–3814. http://dx.doi.org/10.1128/aem.71.7.3806-3814.2005.

Pati, S. G., Bolotin, J., Brennwald, M. S., Kohler, H.-P. E., Werner, R. A., & Hofstetter, T. B. (2016). Measurement of oxygen isotope ratios ($^{18}O/^{16}O$) of aqueous O_2 in small samples by gas chromatography/isotope ratio mass spectrometry. *Rapid Communications in Mass Spectrometry*, *30*(6), 684–690. http://dx.doi.org/10.1002/rcm.7481.

Pati, S. G., Kohler, H.-P. E., Bolotin, J., Parales, R. E., & Hofstetter, T. B. (2014). Isotope effects of enzymatic dioxygenation of nitrobenzene and 2-nitrotoluene by nitrobenzene dioxygenase. *Environmental Science & Technology*, *48*(18), 10750–10759. http://dx.doi.org/10.1021/es5028844.

Pati, S. G., Kohler, H.-P. E., Pabis, A., Paneth, P., Parales, R. E., & Hofstetter, T. B. (2016). Substrate and enzyme specificity of the kinetic isotope effects associated with the dioxygenation of nitroaromatic contaminants. *Environmental Science & Technology*, *50*(13), 6708–6716. http://dx.doi.org/10.1021/acs.est.5b05084.

Piasecki, A., Sessions, A. L., Lawson, M., Ferreira, A. A., Neto, E. V. S., & Eiler, J. M. (2016). Analysis of the site-specific carbon isotope composition of propane by gas source isotope ratio mass spectrometer. *Geochimica et Cosmochimica Acta*, *188*, 58–72. http://dx.doi.org/10.1016/j.gca.2016.04.048.

R Core Team. (2014). *R:A language and environment for statistical computing*. Retrieved from http://www.R-project.org/. (R Foundation for Statistical Computing).

Ratti, M., Canonica, S., McNeill, K., Bolotin, J., & Hofstetter, T. B. (2015). Isotope fractionation associated with the indirect photolysis of substituted anilines in aqueous solution. *Environmental Science & Technology*, *49*(21), 12766–12773. http://dx.doi.org/10.1021/acs.est.5b03119.

Ratti, M., Canonica, S., McNeill, K., Erickson, P. R., Bolotin, J., & Hofstetter, T. B. (2015). Isotope fractionation associated with the direct photolysis of 4-chloroaniline. *Environmental Science & Technology*, *49*(7), 4263–4273. http://dx.doi.org/10.1021/es505784a.

Reichert, P. (1994). Aquasim—A tool for simulation and data-analysis of aquatic systems. *Water Science and Technology*, *30*, 21–30.

Renpenning, J., Hitzfeld, K. L., Gilevska, T., Nijenhuis, I., Gehre, M., & Richnow, H.-H. (2015). Development and validation of an universal interface for compound-specific stable isotope analysis of chlorine ($^{37}Cl/^{35}Cl$) by GC-high-temperature conversion (HTC)-MS/IRMS. *Analytical Chemistry*, *87*(5), 2832–2839. http://dx.doi.org/10.1021/ac504232u.

Renpenning, J., Kümmel, S., Hitzfeld, K. L., Schimmelmann, A., & Gehre, M. (2015). Compound-specific hydrogen isotope analysis of heteroatom-bearing compounds via gas chromatography-chromium-based high-temperature conversion (Cr/HTC)-isotope ratio mass spectrometry. *Analytical Chemistry*, *87*(18), 9443–9450. http://dx.doi.org/10.1021/acs.analchem.5b02475.

Said Ahmad, W., Wong, K., Mcnall, M., Shawar, L., Jacksier, T., Turich, C., … Amrani, A. (2017). Compound-specific sulfur isotope analysis of petroleum gases. *Analytical Chemistry*, *89*(5), 3199–3207. http://dx.doi.org/10.1021/acs.analchem.6b05131.

Schimmelmann, A., Albertino, A., Sauer, P. E., Qi, H., Molinie, R., & Mesnard, F. (2009). Nicotine, acetanilide and urea multi-level ^{2}H-, ^{13}C-and ^{15}N-abundance reference materials for continuous-flow isotope ratio mass spectrometry. *Rapid Communications in Mass Spectrometry*, *23*(22), 3513–3521. http://dx.doi.org/10.1002/rcm.4277.

Schimmelmann, A., Qi, H., Coplen, T. B., Brand, W. A., Fong, J., Meier-Augenstein, W., … Werner, R. A. (2016). Organic reference materials for hydrogen, carbon, and nitrogen stable isotope-ratio measurements: Caffeines, n-alkanes, fatty acid methyl esters, glycines, L-valines, polyethylenes, and oils. *Analytical Chemistry*, *88*(8), 4294–4302. http://dx.doi.org/10.1021/acs.analchem.5b04392.

Schmidt, T. C., & Jochmann, M. A. (2012). Origin and fate of organic compounds in water: Characterization by compound-specific stable isotope analysis. *Annual Review of Analytical Chemistry*, *5*(1), 133–155. http://dx.doi.org/10.1146/annurev-anchem-062011-143143.

Schwarzenbach, R. P., Escher, B. I., Fenner, K., Hofstetter, T. B., Johnson, C. A., von Gunten, U., & Wehrli, B. (2006). The challenge of micropollutants in aquatic systems. *Science*, *313*(5790), 1072–1077. http://dx.doi.org/10.1126/science.1127291.

Schwarzenbach, R. P., Gschwend, P. M., & Imboden, D. M. (2016). *Environmental organic chemistry* (2nd ed.). Hoboken, NJ: John Wiley & Sons Inc.

Scott, K. M., Lu, X., Cavanaugh, C. M., & Liu, J. S. (2004). Optimal methods for estimating kinetic isotope effects from different forms of the Rayleigh distillation equation. *Geochimica et Cosmochimica Acta*, *68*(3), 433–442. http://dx.doi.org/10.1016/s0016-7037(03)00459-9.

Sessions, A. L. (2006). Isotope-ratio detection for gas chromatography. *Journal of Separation Science, 29*(12), 1946–1961. http://dx.doi.org/10.1002/jssc.200600002.

Sherwood Lollar, B., Hirschorn, S. K., Chartrand, M. M. G., & Lacrampe-Couloume, G. (2007). An approach for assessing total instrumental uncertainty in compound-specific carbon isotope analysis: Implications for environmental remediation studies. *Analytical Chemistry, 79*(9), 3469–3475. http://dx.doi.org/10.1021/ac062299v.

Singleton, D. A., & Thomas, A. A. (1995). High-precision simultaneous determination of multiple small kinetic isotope effects at natural-abundance. *Journal of the American Chemical Society, 117*(36), 9357–9358.

Skarpeli-Liati, M., Turgeon, A., Garr, A. N., Arnold, W. A., Cramer, C. J., & Hofstetter, T. B. (2011). pH-Dependent equilibrium isotope fractionation associated with compound-specific nitrogen and carbon isotope analysis by SPME-GC/IRMS. *Analytical Chemistry, 83*(5), 1641–1648. http://dx.doi.org/10.1021/ac102667y.

Spahr, S., Huntscha, S., Bolotin, J., Maier, M. P., Elsner, M., Hollender, J., & Hofstetter, T. B. (2013). Compound-specific isotope analysis of benzotriazole and its derivatives. *Analytical and Bioanalytical Chemistry, 405*(9), 2843–2856. http://dx.doi.org/10.1007/s00216-012-6526-1.

Tobler, N. B., Hofstetter, T. B., & Schwarzenbach, R. P. (2007). Assessing iron-mediated oxidation of toluene and reduction of nitroaromatic contaminants in anoxic environments using compound-specific isotope analysis. *Environmental Science & Technology, 41*(22), 7773–7780. http://dx.doi.org/10.1021/es071129c.

Werner, R. A., & Brand, W. A. (2001). Referencing strategies and techniques in stable isotope ratio analysis. *Rapid Communications in Mass Spectrometry, 15*(7), 501–519. http://dx.doi.org/10.1002/rcm.258.

Wijker, R. S., Adamczyk, P., Bolotin, J., Paneth, P., & Hofstetter, T. B. (2013). Isotopic analysis of oxidative pollutant degradation pathways exhibiting large H isotope fractionation. *Environmental Science & Technology, 47*(23), 13459–13468. http://dx.doi.org/10.1021/es403597v.

Wijker, R. S., Bolotin, J., Nishino, S. F., Spain, J. C., & Hofstetter, T. B. (2013). Using compound-specific isotope analysis to assess biodegradation of nitroaromatic explosives in the subsurface. *Environmental Science & Technology, 47*(13), 6872–6883. http://dx.doi.org/10.1021/es3051845.

Wijker, R. S., Kurt, Z., Spain, J. C., Bolotin, J., Zeyer, J., & Hofstetter, T. B. (2013). Isotope fractionation associated with the biodegradation of 2-and 4-nitrophenols via monooxygenation pathways. *Environmental Science & Technology, 47*(24), 14185–14193. http://dx.doi.org/10.1021/es403876u.

Wijker, R. S., Pati, S. G., Zeyer, J., & Hofstetter, T. B. (2015). Enzyme kinetics of different types of flavin-dependent monooxygenases determine the observable contaminant stable isotope fractionation. *Environmental Science & Technology Lett, 2*(11), 329–334. http://dx.doi.org/10.1021/acs.estlett.5b00254.

Wijker, R. S., Zeyer, J., & Hofstetter, T. B. (2017). Isotope fractionation associated with the simultaneous biodegradation of multiple nitrophenol isomers by *Pseudomonas putida* B2. *Environmental Science: Processes & Impacts, 74*(19), 775–784. http://dx.doi.org/10.1039/C6EM00668J.

Wolfsberg, M., Hook, W. A. V., Paneth, P., & Rebelo, L. P. N. (2010). *Isotope effects in chemical, geological, and bio sciences*. Heidelberg: Springer.

Zakon, Y., Halicz, L., & Gelman, F. (2014). Isotope analysis of sulfur, bromine, and chlorine in individual anionic species by ion chromatography/multicollector-ICPMS. *Analytical Chemistry, 86*(13), 6495–6500. http://dx.doi.org/10.1021/ac5010025.

Zakon, Y., Halicz, L., Lev, O., & Gelman, F. (2016). Compound-specific bromine isotope ratio analysis using gas chromatography/quadrupole mass spectrometry. *Rapid Communications in Mass Spectrometry, 30*(17), 1951–1956. http://dx.doi.org/10.1002/rcm.7672.

Zhang, L., Kujawinski, D. M., Jochmann, M. A., & Schmidt, T. C. (2011). High-temperature reversed-phase liquid chromatography coupled to isotope ratio mass spectrometry. *Rapid Communications in Mass Spectrometry*, *25*(20), 2971–2980. http://dx.doi.org/10.1002/rcm.5069.

Zwank, L., Berg, M., Schmidt, T. C., & Haderlein, S. B. (2003). Compound-specific carbon isotope analysis of volatile organic compounds in the low-microgram per liter range. *Analytical Chemistry*, *75*(20), 5575–5583. http://dx.doi.org/10.1021/ac034230i.

CHAPTER TWELVE

pH-Free Measurement of Relative Acidities, Including Isotope Effects

Charles L. Perrin[1]
University of California, San Diego, La Jolla, CA, United States
[1]Corresponding author: e-mail address: cperrin@ucsd.edu

Contents

1. Introduction to NMR Titration 332
2. Multicomponent NMR Titration 333
   2.1 Data Analysis 333
   2.2 A Typical Experimental Procedure 335
   2.3 Electrodeless pH Titrations 335
3. Motivation: Applications to the Reverse Anomeric Effect 337
4. Applications of Other NMR Titrations to Relative Acidities 341
5. Measurement of Binding Constants 344
   5.1 Single-Point Comparisons 346
   5.2 Intramolecular Comparisons 347
   5.3 Ratio K_1/K_2 349
6. Heavy-Atom Isotope Effects on Acidity 350
7. Secondary Deuterium Isotope Effects on Acidity 351
8. Secondary Deuterium Isotope Effects on Binding 355
9. Further Studies of Secondary Deuterium Isotope Effects, Toward Elucidating Their Origin 356
10. Advantages of NMR Titration 359
11. Some Practical Considerations 360
    11.1 Pitfalls 363
12. Summary 364
Acknowledgment 364
References 364

Abstract

A powerful pH-free multicomponent NMR titration method can measure relative acidities, even of closely related compounds, with excellent accuracy. The history of the method is presented, along with details of its implementation and a comparison with earlier NMR titrations using a pH electrode. Many of its areas of applicability are described, especially equilibrium isotope effects. The advantages of the method, some practical considerations, and potential pitfalls are considered.

Methods in Enzymology, Volume 596
ISSN 0076-6879
http://dx.doi.org/10.1016/bs.mie.2017.07.012

1. INTRODUCTION TO NMR TITRATION

NMR pH titration was an early NMR method for measurement of acidity constants. It depends on the fact that proton transfer among acid and base forms is very fast. Consequently the observed chemical shift δ of an acid is the average of the chemical shifts δ_A and δ_B of its acid and base forms, but weighted by their respective concentrations, as in Eq. (1). Solving this equation for [B]/[A] and substituting into the Henderson–Hasselbalch equation leads to Eq. (2). Solving Eq. (2) for δ then leads to Eq. (3), expressing the dependence of the chemical shift on pH. This relationship has the appearance of a familiar titration curve, as sketched in Fig. 1. If the limiting chemical shifts are extrapolated to δ_{max} at low pH and δ_{min} at high pH, a linear least-squares fit of the left-hand side of Eq. (4) vs pH will give a slope near 1 and an intercept equal to $-pK_a$.

$$\delta = \frac{\delta_B[B] + \delta_A[A]}{[B] + [A]} \tag{1}$$

$$pH = pK_a + \log\frac{\delta_A - \delta}{\delta - \delta_B} \tag{2}$$

$$\delta = \frac{\delta_A 10^{-pH} + \delta_B 10^{-pK_a}}{10^{-pH} + 10^{-pK_a}} \tag{3}$$

$$\log[(\delta_{max} - \delta)/(\delta - \delta_{min})] = pH - pK_a \tag{4}$$

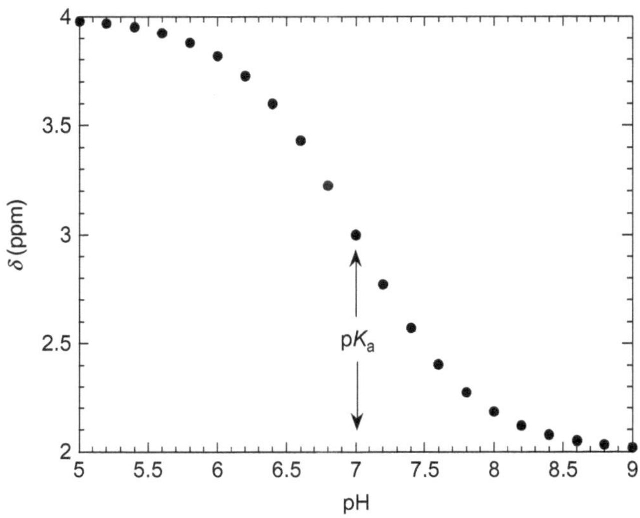

Fig. 1 Simulation of Eq. (3).

2. MULTICOMPONENT NMR TITRATION

The above NMR titration can be extended to more than one substance, because the sensitivity of NMR chemical shifts to environment allows the separate measurement of each substance's chemical shift. Consequently an NMR titration can be adapted to simultaneously measure the difference in acid dissociation constants (ΔpK_a) of two or more structurally similar compounds (Fabian, Perrin, & Sinnott, 1994; Perrin, Fabian, & Armstrong, 1994). The NMR titration is capable of high precision and accuracy, and it can be carried out without the necessity of measuring pH. From the variations of the NMR chemical shifts during the titration of a single solution, the ratio of acidity constants of two or more compounds can be determined. During a titration, the spectral characteristics of the most basic or acidic compound will change at lower proton concentration than those of the other compounds. A linear plot created from the chemical shifts of the neutral and ionic forms of each compound vs the observed chemical shifts after each addition of titrant gives the desired ratio and thus ΔpK_a, the difference between the pK_as of the compounds. This method differs from the closest previous method (Rabenstein & Mariappan, 1993), in that it relies explicitly only on chemical shifts and on a linear plot.

One key feature that makes the NMR titration feasible is that proton exchange is rapid on the NMR timescale. Second-order rate constants for thermoneutral exchange are at least $3\times10^8\,M^{-1}\,s^{-1}$ in water and are reduced only to $\sim 5\times10^6\,M^{-1}\,s^{-1}$ for some hindered amines in $CDCl_3$ (Grunwald & Cocivera, 1965; Perrin & Wang, 1982). These values mean that peak broadening from insufficiently rapid exchange is negligible even at millimolar concentrations, although this may not be the case for some proteins in solutions lacking sufficient buffer to mediate proton transfer (Bachovchin & Roberts, 1978; Sudmeier, Evelhoch, & Jonsson, 1980).

2.1 Data Analysis

Certain NMR chemical shifts of an acidic or basic compound undergo a change upon protonation or deprotonation sufficient to permit the NMR determination of the extent of ionization. If a mixture of two compounds, A and B, is titrated with acid, the chemical shifts of the more basic compound will change at lower proton concentration during the titration than those of the less basic one. Likewise, the chemical shifts of the more acidic compound will change at lower concentrations of base. The ratio of acidity constants of the different compounds, called K in Eq. (5), can readily be measured from the variations in chemical shifts, as derived below.

$$K = \frac{K_a^{AH^+}}{K_a^{BH^+}} = \frac{[A][BH^+]}{[AH^+][B]} \quad (5)$$

The observed chemical shifts are weighted averages of the chemical shifts for the protonated and deprotonated species. The observed chemical shift of base A is given by Eq. (6), where δ_{A^o} and δ_{AH^+} are the limiting chemical shifts of the neutral and protonated forms, respectively. A similar equation holds for the competing base B.

$$\delta_a = \frac{\delta_{A^o}[A] + \delta_{AH^+}[AH^+]}{[A] + [AH^+]} \quad (6)$$

Algebraic manipulation of these equations leads to Eq. (7), which is nonlinear whenever $K \neq 1$. The desired ratio, K, can be obtained from a nonlinear least-squares fit to Eq. (7). However, further manipulations lead to the linearized form in Eq. (8), relating observed chemical shifts to the desired K. This trivial linearization has been dignified as the Perrin–Fabian method. Then from the observed chemical shifts δ_a and δ_b during the titration of a mixture of compounds, the ratio of acidity constants can be calculated from a plot of $(\delta_b - \delta_{B^o})(\delta_{AH^+} - \delta_a)$ vs $(\delta_a - \delta_{A^o})(\delta_{BH^+} - \delta_b)$, which ought to be a straight line, with slope K and zero intercept (Perrin & Fabian, 1996). Therefore, the beginning chemical shifts of the neutral compounds (δ_{B^o} and δ_{A^o}) and the ending chemical shifts of the ions (δ_{BH^+} and δ_{AH^+}) must be known to accurately determine the ratio of acidity constants using this methodology.

$$\delta_a = \delta_{A^o} + \frac{(\delta_{AH^+} - \delta_{A^o})(\delta_b - \delta_{B^o})}{(1-K)(\delta_b - \delta_{B^o}) + K(\delta_{BH^+} - \delta_{B^o})} \quad (7)$$

$$(\delta_b - \delta_{B^o})(\delta_{AH^+} - \delta_a) = K(\delta_a - \delta_{A^o})(\delta_{BH^+} - \delta_b) \quad (8)$$

It must be acknowledged that many laboratories have undertaken to use NMR to measure acidity constants. Consequently there have been various approaches, all rather similar, but each with its own set of parameters and equations. Rather than incorporate one consistent set of parameters and symbols, this review preserves the original presentations.

2.2 A Typical Experimental Procedure (Reprinted With Permission From Perrin & Fabian, 1996, Copyright (1996) American Chemical Society)

A stock solution of DCl was prepared from 0.66 g CD_3OD, 0.08 g D_2O, and 0.27 g of 35% DCl/D_2O solution. A similar stock solution of NaOD was prepared from 0.66 g CD_3OD, 0.09 g D_2O, and 0.25 g of 40% NaOD/D_2O. For titrations in DMSO-d_6 a stock solution of trifluoroacetic acid (TFA) was prepared from 1.00 g DMSO-d_6 and 0.20 g TFA. A similar stock solution of *t*-BuOK was prepared from 1.00 g DMSO-d_6 and 0.19 g *t*-BuOK.

The NMR samples prepared for each titration contained 1.00 mL solvent, 5 μL tetramethylsilane (TMS) or 1,4-dioxane as internal standard, and 0.05, 0.10, or 0.20 mmol mixture of substances to be compared. An initial ^{1}H or ^{13}C NMR spectrum was taken of the sample, which is listed as the "0" titration point. Aliquots of 5 μL of the appropriate stock acid or base solution were then continually added until the chemical shifts no longer changed. The ^{1}H or ^{13}C NMR chemical shifts were recorded after each addition of the stock solution.

Amines titrated in CD_3OD/D_2O required two separate titrations, because a small percent of the amine becomes protonated in the solution. Therefore, two samples were prepared by splitting a solution containing 1.33 g CD_3OD, 0.50 g D_2O, 10 μL TMS, and 0.20 mmol mixture of amines. A 5-μL aliquot of the NaOD stock solution was added to the first sample. A ^{1}H NMR spectrum was taken of the sample, which is recorded as the "−1" titration point. Further addition of base gave no further change in the chemical shifts. A ^{1}H NMR spectrum was taken of the second sample, which is recorded as the "0" titration point. Aliquots of 5 μL of the DCl stock solution were then continually added until the chemical shifts no longer changed, indicating that the endpoint of the titration had been passed. The ^{1}H NMR chemical shifts were recorded after each addition of the stock solution.

NMR spectra were recorded on a Varian Unity-500 spectrometer (499.8 MHz ^{1}H, 125.7 MHz ^{13}C) using an indirect probe. Chemical shifts are referenced to TMS (δ 0.00) for ^{1}H spectra and 1,4-dioxane (δ 66.5) for ^{13}C spectra.

2.3 Electrodeless pH Titrations

An earlier variant on the pH-free NMR titration was a pH titration, measuring how NMR chemical shifts depend on pH. This has been especially common in biochemical systems, where the first applications were simply to measure pH from the observed chemical shift of various indicators

(Baryshnikova, Williams, & Sykes, 2008; Szakács, Kraszni, & Noszál, 2004; Tynkkynen, Tiainen, Soininen, & Laatikainen, 2009; Valcour & Woodworth, 1986). An example of such a series, which spans much of the acidic pH range, is dichloroacetic acid, chloroacetic acid, acetic acid, sodium dihydrogen phosphate, and tris(hydroxymethyl)aminomethane (TRIS) (Szakács, Hägele, & Tyka, 2004), but it was noted that extension of the scale to the stronger acids histidine (pK_a 1.83) and dichloroacetic acid (pK_a 0.93) required careful control of ionic strength and extrapolation to the chemical shift of unionized dichloroacetic acid (Szakács & Hägele, 2004).

In an early study the acidities of linear and branched carboxylic acids could be evaluated, relative to acetic acid, from the variation with the extent of protonation of the measured ^{13}C NMR shifts of either the carboxyl or α-CH_2 carbons (Pehk, Kiirend, Lippmaa, & Ragnarsson, 1996).

Another challenge is that measurement of pH in strongly basic solutions is also subject to error. Therefore, a series of weakly acidic substances was designed: trimethylammonium ion, *tert*-butylammonium ion, 4-hydroxypyridinium ion, cytosine, acetone oxime, acetamidinium ion, and methylguanidinium ion. These seven were anchored by sarcosine, whose pK_a was evaluated as 10.15 from a plot of chemical shift vs measured pH, and then pairwise to less acidic values by NMR titration using a nonlinear fit of the chemical shift of the decreasingly acidic substances vs the chemical shift of a substance whose pK_a had been evaluated (Orgován & Noszál, 2011).

Once such a series of pH indicators became available, with a known dependence of chemical shift on pH, it became possible to carry out an "electrodeless NMR titration" (Szakács, Hägele, et al., 2004). Thus the acidity constants of six fluoroquinolone antibacterials could be determined from ^{1}H NMR chemical shift vs pH, but the pK_a for the most acidic proton is near 0, so this was evaluated by NMR titration in a mixture with dichloroacetic acid, of known pK_a = 1.14 (Rusu et al., 2012).

As another example, the pH dependences of ^{13}C chemical shifts for a series of carbamoylcholine homologues with various amino groups were measured (Jensen et al., 2004). The data were fit to Eq. (9), where δ_i is the observed chemical shift of the *i*th carbon at a given pH, and δ_i^0 and δ_i^+ are the chemical shifts of the *i*th carbon in nonprotonated and protonated amine, respectively. Nonlinear regression analysis with simultaneous iteration on all carbons of the molecule (except for the *N*-methyls of the amine) then yielded pK_a.

$$\delta_i = \frac{\delta_i^0 \cdot 10^{-pK_a} + \delta_i^+ \cdot 10^{-pH}}{10^{-pK_a} + 10^{-pH}} \tag{9}$$

Among other examples of the use of electrodeless NMR titration are the comparisons of the pK_as of *cis* and *trans* rotamers of *N*-acetyl amino acids (Kubyshkin, Durkin, & Budisa, 2016), and of acetylsarcosine, glycylsarcosine, glycylglycylsarcosine, glycylproline, glycylhydroxyproline, alanylproline, sarcosine, proline, and hydroxyproline (Evans & Rabenstein, 1974).

An automated pH–NMR titration setup, consisting of a Bruker 250-MHz NMR instrument and a potentiometric titration unit, allows pH titrations with simultaneous recording of ^{1}H and ^{31}P NMR spectra at each titration point (Peters, Siegfried, & Kaden, 1999). This setup was applied to protonation of three diazacrown ethers carrying dangling phosphonate groups. From the fitting of the pH dependences of the chemical shifts of the ^{31}P and ^{1}H signals the protonation constants as well as the chemical shifts of the individual protonated species were obtained. The main advantages of this setup are that relatively small amounts of substance (0.05 mmol) are needed for a single titration and that the system needs no operator intervention during the entire experiment (about 20 h).

3. MOTIVATION: APPLICATIONS TO THE REVERSE ANOMERIC EFFECT

Our entry into this topic arose through a consideration of the reverse anomeric effect. The (normal) anomeric effect is the preference of an electronegative substituent for the axial position on a tetrahydrofuran (Juaristi & Cuevas, 1992; Kirby, 1983; Thatcher, 1993; Thibaudeau & Chattopadhyaya, 1999). The reverse anomeric effect is the preference of a cationic substituent for the equatorial position on a tetrahydropyran. In connection with the claim of a reverse anomeric effect in *N*-xylosylimidazolium ions (Paulsen, Gyorgydeak, & Friedmann, 1974), we wanted to compare the steric bulk of a cationic substituent with that of the corresponding neutral one, specifically a comparison of imidazolium with imidazolyl. A measure of the steric bulk of a substituent X is its *A* value, the extra energy required to place that substituent axial, as expressed in Eq. (10). In an attempt to compare the steric bulk of cationic with neutral, the conformational equilibria of *cis*-*N*-(4-methylcyclohexyl)imidazole (**1cis**, R = CH_3), *cis*-*N*-(4-phenylcyclohexyl) imidazole (**2cis**, R = Ph), and their conjugate acids were measured by low-temperature NMR (Perrin et al., 1994). The *A* value of an imidazolyl group was thereby found to be 2.2 ± 0.1 kcal/mol. The *A* value of an *N*-protonated imidazolyl group was also found to be 2.2 ± 0.1 kcal/mol, which is the same within experimental error.

$$A_X = G^\circ(\text{axial X}) - G^\circ(\text{equatorial X}) \tag{10}$$

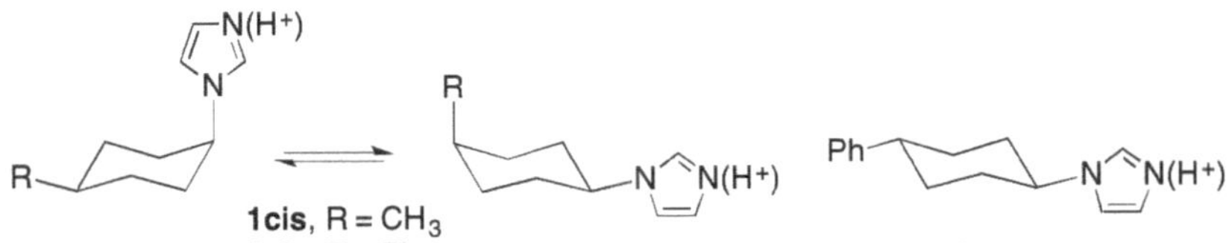

1cis, R = CH_3
2cis, R = Ph

2trans

To compare the two sizes more precisely, an NMR titration method, applicable to a mixture of isomers, was developed. The ratio of acidity constants of *cis*- and *trans*-*N*-(4-phenylcyclohexyl)imidazoles, as expressed in Eq. (11), could be determined from the variation of H1 chemical shifts as a 1:1 mixture was titrated with DCl. The observed chemical shift, δ_c, of the *cis* isomer **2cis** is given by Eq. (12), where δ_{CH^+} and δ_C are limiting chemical shifts of protonated and unprotonated forms, respectively. A similar equation relates δ_t, δ_{TH^+}, and δ_T, the chemical shift of the *trans* isomer, **2trans**, and its limiting chemical shifts. Algebraic manipulation of Eq. (11), Eq. (12), and the equivalent equation for δ_t leads to Eq. (13), which is nonlinear whenever $K \neq 1$. Indeed, a plot of δ_c against δ_t, as shown in Fig. 2, shows slight but systematic upward curvature, indicating that the *trans* isomer is more readily protonated. (To clarify the behavior, the diagonal dotted line corresponds to the hypothetical situation of identical basicity for *cis* and *trans*, and the horizontal and vertical dotted lines correspond to the hypothetical extreme where the *trans* is infinitely more basic than the *cis* and is protonated first during the titration.)

$$K = \frac{K_a^{CH^+}}{K_a^{TH^+}} = \frac{[C][TH^+]}{[CH^+][T]} \tag{11}$$

$$\delta_c = \frac{\delta_C[C] + \delta_{CH^+}[CH^+]}{[C] + [CH^+]} \tag{12}$$

$$\delta_c = \delta_C + \frac{(\delta_{CH^+} - \delta_C)(\delta_t - \delta_T)}{(1-K)(\delta_t - \delta_T) + K(\delta_{TH^+} - \delta_T)} \tag{13}$$

Further manipulations lead to the linearized form (Eq. 14), relating observed chemical shifts to the desired K. In particular, a plot of $(\delta_t - \delta_T)(\delta_{CH^+} - \delta_c)$ vs $(\delta_c - \delta_C)(\delta_{TH^+} - \delta_t)$ ought to be a straight line, with slope K and zero intercept. Fig. 3 shows such a plot. In aqueous acetone the

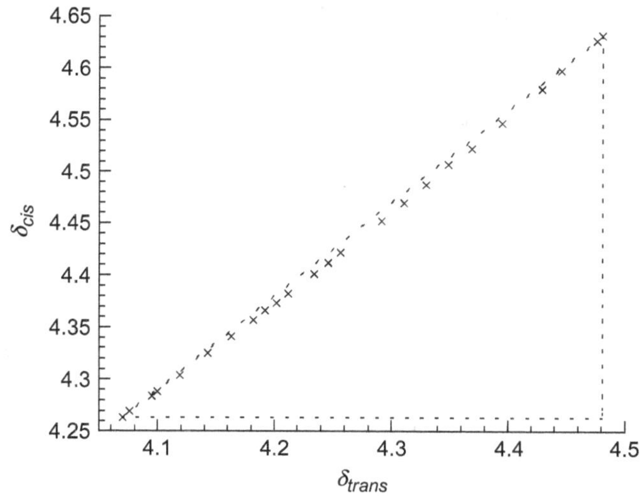

Fig. 2 Chemical shifts δ_{cis} and δ_{trans} during titration of a 1:1 mixture of *cis*- and *trans-N*-(4-phenylcyclohexyl)imidazoles (**2**) with DCl in 1:l acetone-d_6/D_2O. *Adapted with permission from data in table 3 of Perrin, C. L., Fabian, M. A., & Armstrong, K. B. Solvation effect on steric bulk of ionic substituents: Imidazolium vs imidazole.* Journal of Organic Chemistry, *59, 5246–5253. Copyright (1994) American Chemical Society.*

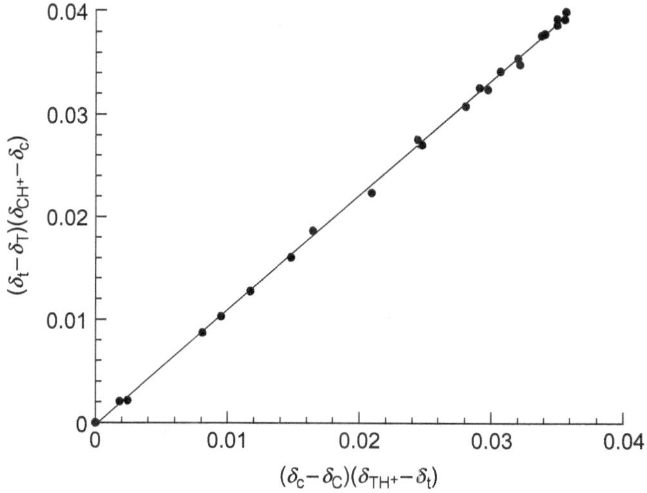

Fig. 3 Linearized plot of chemical shifts (Eq. 14) during titration of a 1:1 mixture of *cis*- and *trans-N*-(4-phenylcyclohexyl)imidazoles (**2**) with DCl in 1:l acetone-d_6/D_2O. *Reprinted with permission from Perrin, C. L., Fabian, M. A., & Armstrong, K. B. Solvation effect on steric bulk of ionic substituents: Imidazolium vs imidazole.* Journal of Organic Chemistry, *59, 5246–5253. Copyright (1994) American Chemical Society.*

cis isomer was found to be $0.048 \pm 0.002\,\mathrm{p}K$ units less basic. Through the application of a thermodynamic cycle, this corresponds to $\Delta A = A_{N\text{-imidazolylH}^+} - A_{N\text{-imidazolyl}} = 0.089 \pm 0.004$ kcal/mol, with the protonated imidazolyl slightly but detectably larger.

$$(\delta_t - \delta_T)(\delta_{CH^+} - \delta_c) = K(\delta_c - \delta_C)(\delta_{TH^+} - \delta_t) \quad (14)$$

This NMR titration method was used similarly to measure the effects of *N*-protonation on the anomeric equilibria in *N*-(D-glucopyranosy1)imidazole (**3**) and its tetra-*O*-acetyl derivative **4** (Fabian et al., 1994). In principle, the magnitude of any reverse anomeric effect could be measured as the increase in the proportion of the β anomer on protonation of an equilibrating mixture of anomers. However, glycosylimidazoles are configurationally stable, so that the two anomers do not equilibrate. Nevertheless, the reverse anomeric effect could be measured instead from the difference in $\mathrm{p}K_a$ between α and β anomers, as evaluated from the slope K_a^α/K_a^β in Eq. (15), analogous to Eq. (14), but where δ_α and δ_β are observed chemical shifts and δ_α^+, δ_α°, δ_β^+, and δ_β° are limiting chemical shifts of protonated and unprotonated α and β. It was found that *N*-protonation does not shift the equilibrium toward an equatorial imidazolyl, but rather shifts the equilibrium toward axial, which is opposite to what was claimed for the reverse anomeric effect.

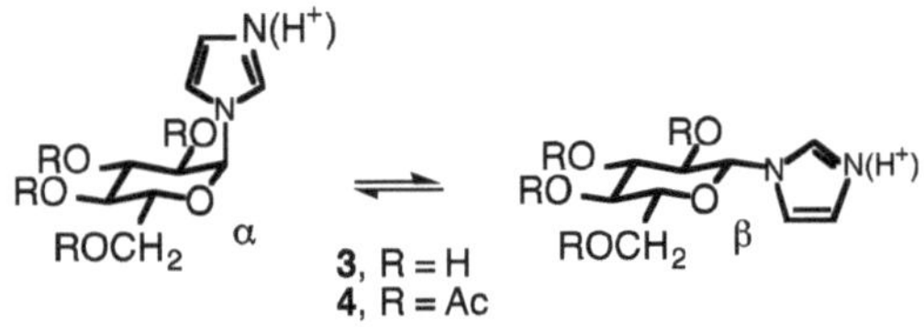

$$(\delta_\beta - \delta_{\beta^\circ})(\delta_\alpha^+ - \delta_\alpha) = (K_a^\alpha/K_a^\beta)(\delta_\alpha - \delta_{\alpha^\circ})\left(\delta_\beta^+ - \delta_\beta\right) \quad (15)$$

This study of the anomeric equilibrium was extended to a wider range of *N*-(glycopyranosyl)imidazoles (Perrin, Fabian, Brunckova, & Ohta, 1999). The NMR titration data indicate a $\Delta\Delta G^\circ_{\beta\to\alpha}$ that is almost always negative, corresponding to a greater preference for the axial position of a protonated imidazolyl group than of an unprotonated. This preference counters the small steric effect, arising from hindrance to ionic solvation. These results are exactly opposite to what is expected from the reverse anomeric effect and led to the conclusion that there is no firm evidence for this effect.

The ability to measure small changes in acidity next permitted the comparison of *N*-protonation-induced shifts of the anomeric equilibrium in a series of *N*-(tetra-*O*-methylglucopyranosyl)anilines (**5**, R = CH_3) and of the *cis/trans* equilibrium in *N*-(4-*tert*-butylcyclohexyl)anilines (**6**) (Perrin & Kuperman, 2003). In both series the equilibrium shifts toward equatorial upon *N*-protonation, consistent with steric hindrance to ionic solvation. The shift is smaller for the glucosylanilines than for the cyclohexylanilines, consistent with an enhancement of the normal anomeric effect that counters the steric hindrance and reduces the shift toward the equatorial β anomer. Moreover, the shift of the equilibrium toward equatorial increases slightly but detectably with electron-withdrawing substituents on the aromatic ring, which fine-tune the steric hindrance to ionic solvation. In contrast, the shift of the equilibrium toward equatorial decreases for the glucosylanilines. This result is consistent with an enhancement of the normal anomeric effect due to a more localized positive charge, rather than with a reverse anomeric effect.

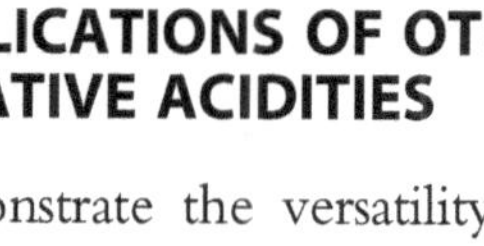

5

6

4. APPLICATIONS OF OTHER NMR TITRATIONS TO RELATIVE ACIDITIES

To demonstrate the versatility of the NMR titration method, the ΔpK_a between the conjugate acids of the two stereoisomers of 4-*t*-butylcyclohexylamine (**7**) was determined in a single ^1H NMR titration experiment in CD_3OD/D_2O and also in DMSO-d_6 (Perrin & Fabian, 1996). The ΔpK_a between the two stereoisomers of 4-*t*-butylcyclohexanecarboxylic acid (**8**) was determined in both of these solvents using both ^1H and ^{13}C NMR. To further demonstrate the scope of this method, the ΔpK_as of the conjugate acids of all four (racemic) stereoisomers of 2-decalylamine (**9**) were determined in a single ^1H NMR titration experiment.

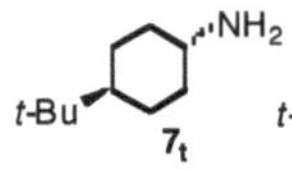

7_t

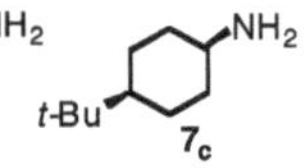

7_c

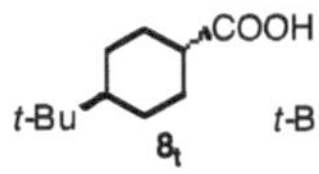

8_t

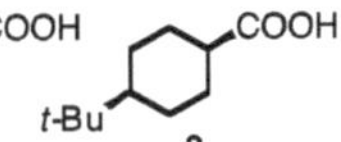

8_c

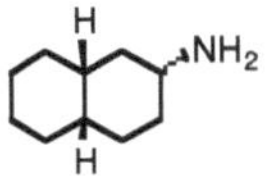

9_{cc}

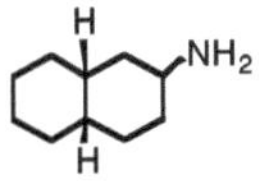

9_{ct}

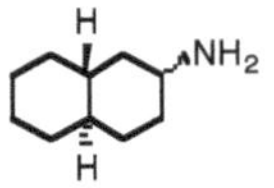

9_{tc}

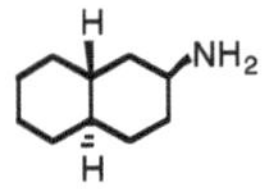

9_{tt}

The ^{1}H NMR titration method was applied to measure the basicities of cycloalkylamines (**10**), all relative to isopropylamine (Perrin, Fabian, & Rivero, 1999). Cyclopropylamine ($n=3$) and cyclobutylamine ($n=4$) are less basic than cyclopentylamine ($n=5$) and cyclohexylamine ($n=6$), owing to the hybridization of the carbon atoms in the small rings. Medium-ring amines ($n=7$, 8, 9) are even more basic, with cyclooctylamine ($n=8$) the most basic, again owing to hybridization. Surprisingly, large-ring amines ($n=12$, 15, 16, 21) are slightly less basic than cyclohexylamine, even though they all have the same hybridization. This difference may be due to conformational restrictions in the large-ring ammonium ions. This subtle effect required NMR titration to detect reliably.

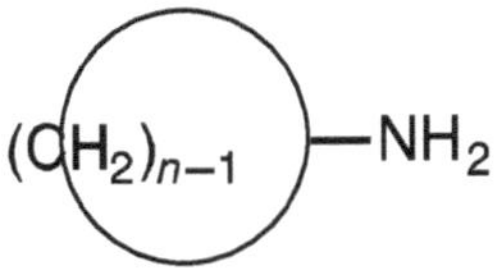

10, n = 3–12,15–16, 21

To investigate the effect of molecular curvature on basicity, the basicity of a pyrrolidine fused to a coronene (**11**) was compared to that of a pyrrolidine fused to a fullerene (**12**) (Steffens, Baldridge, & Siegel, 2000).

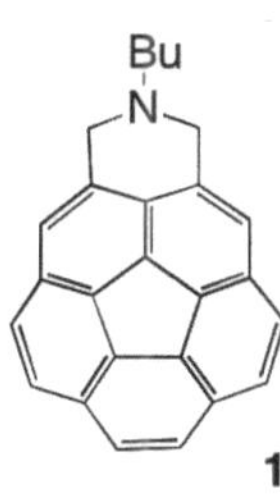

11

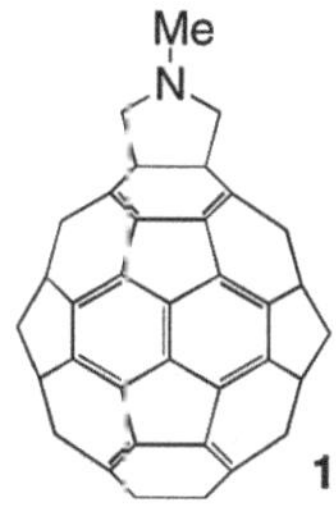

12

^{1}H NMR titration in DMSO-d_6 was used to probe polar–π interactions in two pyridinocyclophanes (**13**, X = H, F)). It was found that the pyridine is more basic when it faces a benzene ring, compared to when it faces a tetrafluorinated benzene ring, owing to the difference in electron density, which repels the N lone pair. Surprisingly, 2,6-lutidine, without the cyclophane, is still more basic, even though it lacks the electron density of the benzene ring (Baldridge, Cozzi, & Siegel, 2012).

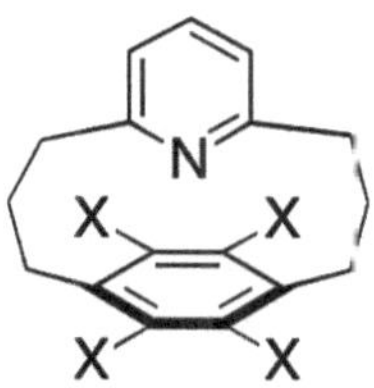

13

NMR titration was used to measure ΔpK for a series of amines in DMSO, including 1,8-bis(dimethylamino)naphthalene (Proton sponge, **14**) and the chiral *N*,*N*′-bis(2-quinolinyl)-*trans*-1,2-cyclohexanediamine (**15**), which is effective as a chiral Brønsted acid to catalyze an enantioselective reaction (Hess, Yoder, & Johnston, 2006).

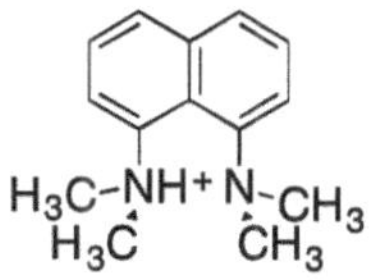

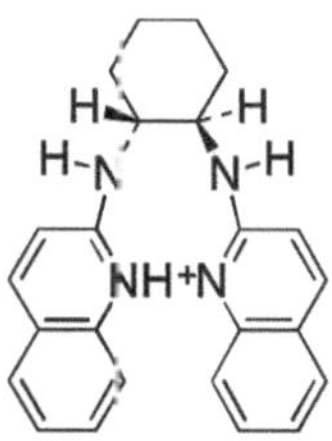

14 **15**

^{1}H NMR titrations were used to measure the relative basicities of *cis* and *trans* 1,5-oxazaspiro[5,5]undecanes (**16**, R = methyl or *t*-butyl) (Guerrero-Alvarez, Mas-Ku, Garcías-Morales, & Ariza-Castolo, 2010).

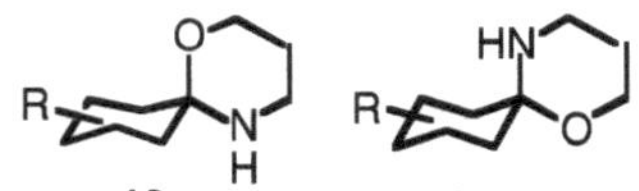
16_{cis} 16_{trans}

The pK_a of *N*-Cbz-2,5-diazabicyclo[4.1.0]heptane (**17a**) was determined to be 6.74±0.05 and that of *N*-Cbz piperazine·HCl (**17b**) to be 8.09±0.02 (Taylor et al., 2010). The latter pK_a is in accord with the reported pK_a of *N*-carbethoxypiperazine and that of *N*-benzoylpiperazine.

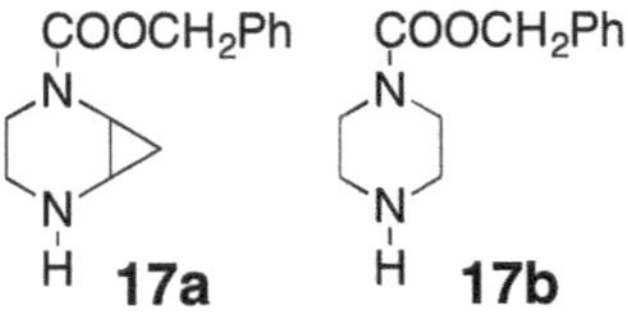
17a **17b**

The ΔpK_{a1} values for the first ionization of arylphosphonic acids, ArP(=O)(OH)$_2$, relative to ionization of dichloroacetic acid as pH indicator, were determined by electrodeless NMR titration (Keglevich et al., 2012). The ^{1}H chemical shifts of the aromatic protons of the compounds were fit to Eq. (16), where δL_{obs} and δI_{obs} are the observed chemical shifts of the phosphonic acid and of dichloroacetic acid, δHL and δL are the limiting chemical shifts of the acidic and basic forms of the phosphonic acid, and δHI and δI are the limiting chemical shifts of the acidic and basic forms of the indicator.

$$\delta L_{obs} = \delta L + \frac{(\delta HL - \delta L)(\delta I_{obs} - \delta I)}{(1 - 10^{\Delta pK})(\delta I_{obs} - \delta I) + 10^{\Delta pK}(\delta HI - \delta I)} \tag{16}$$

5. MEASUREMENT OF BINDING CONSTANTS

Two detailed guides to measurement of binding constants are available, applicable to both NMR titration and UV–vis (Hirose, 2001; Thordarson, 2011). Both include much practical advice, including precautions, choice of concentrations, and data analysis.

NMR titration can be used to measure binding constants for formation of binary complexes in multicomponent mixtures. A challenging test is the analysis of the binding of 9-ethyladenine (**18**) to host dicarboxylic acid (**19**) in the presence of 2-aminopyrimidine (**20a**) and imidazolidone (**20b**) (Wilcox, Adrian, Webb, & Zawacki, 1992). Each species is in equilibrium with its dimer and with all binary complexes with the other components. Dimerization constants were evaluated by fitting the NH chemical shifts for each single component to its concentration dependence. Then binding constants for all 1:1 complexes were evaluated by fitting the NH chemical shifts for each binary mixture. Finally the binding constant for **18** to host **19** was evaluated as $(5.05 \pm 1.0) \times 10^5$ from analysis of chemical shifts in three-component mixtures, where the other binding constants do affect the observed chemical shifts but represent only small corrections.

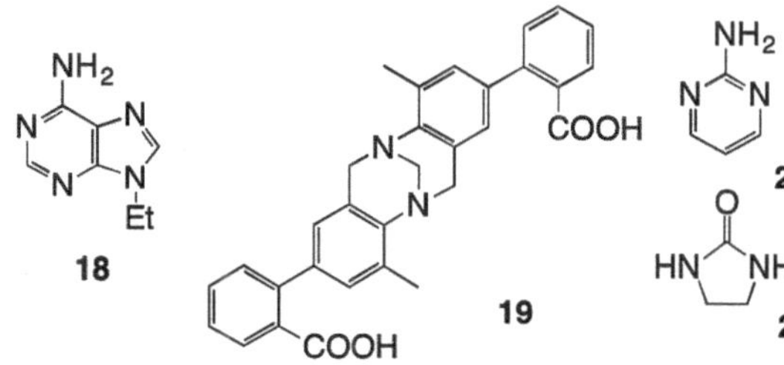

Binding constants could be measured in toluene-d_8 for the complexes between $(CpSiMe_3)_3M$ and Cp*Ga ($Cp = C_5H_5$, $Cp^* = C_5(CH_3)_5$, M=U or Nd) (Minasian et al., 2009). To the solution of $(CpSiMe_3)_3M$ varying amounts of Cp*Ga were added, and X_{cplx}, the mole fraction of the complex, was evaluated as $(\delta_{obs} - \delta_M)/(\delta_{MGa} - \delta_M)$, where δ_{obs}, δ_M, and δ_{MGa} are observed chemical shift and chemical shifts of free and bound $(CpSiMe_3)_3M$, not only at methyl but also at the two different CH of the Cp ring. Next, the average of these three values was converted to the binding constant, $K = [(CpSiMe_3)_3M\text{-}GaCp^*]/[(CpSiMe_3)_3M]([Cp^*Ga]_{total} - [(CpSiMe_3)_3M\text{-}GaCp^*])$. It may be noted that the signals are broadened by the paramagnetic metals, but paramagnetic shifts can be large, and the change of chemical shift on complexation is >10 ppm at two of the sites, providing high accuracy. Enthalpy and entropy of binding could then be evaluated from the temperature dependence of *K*.

NMR titration is also applicable to relative acidities of Lewis acids. Thus, reaction of azines L (L=pyridine, 3-picoline, 3,5-lutidine, 3-chloropyridine, pyridazine, pyrimidine, pyrazine, 3,5-dimethylpyrazine, and *s*-triazine) with metallocenes $[M(C_5H_4R)_3]$ (M=Ce, U; R=tBu, $SiMe_3$) in toluene gave the Lewis base adducts $[M(C_5H_4R)_3(L)]$ (Mehdoui, Berthet, Thuéry, & Ephritikhine, 2004). The equilibria between $[M(C_5H_4R)_3]$, L, and $[M(C_5H_4R)_3L)]$ were studied by ^1H NMR spectroscopy, even though the metal complexes are paramagnetic.

5.1 Single-Point Comparisons

Instead of a complete NMR titration and fitting of all the chemical shifts to Eq. (14), it is possible to abbreviate the experiment to a single point near the middle of the titration, along with the endpoints. The slope in Eq. (14) is most sensitive to that point, where the equilibrium is most closely balanced, so that the other points only reinforce the slope. Little accuracy is sacrificed by this simplification, because the slope of a line is determined by its extremities.

A single-point NMR titration was used to compare the acidities of *o*- and *p*-aroylaminobenzoic acids ($ArCONHC_6H_4COOH$), in order to evaluate the effect of the intramolecular hydrogen bond between amide NH and carboxyl, which is present only in the ortho isomer (Emenike, Liu, Naveo, & Roberts, 2013). Titrations were carried out in DMSO by ^1H or ^{13}C NMR of an equimolar mixture of a test acid AH and the tetrabutylammonium salt of B^-, the conjugate base of a standard acid BH (*p*-bromobenzoic, *p*-nitrobenzoic, or salicylic) of known acidity constant K_a^{BH}, or else an equimolar mixture of a standard acid BH and the tetrabutylammonium salt of A^-, the conjugate base of the test acid. It should be noted that it is not necessary to mix the two components in exactly equimolar proportions. The resulting chemical shifts of a reporter nucleus (^1H or ^{13}C, not specified further) in each species, along with those chemical shifts in acids AH and BH alone and in conjugate bases A^- and B^- alone, were used to evaluate the position of equilibrium between the two acids and relate K_a^{AH} to the known K_a^{BH}. Eq. (17) describes the relationship among those chemical shifts and the ionization constants of the two acids. It was further found that substituent effects on the acidities could be described well by a Hammett plot. It was also claimed that the best results are obtained when AH and BH are separated by $\sim 2\,pK_a$ units, but this is misleading and is discussed later in Section 11.

$$K_a^{AH}(\delta_a - \delta_A)(\delta_{BH} - \delta_b) = K_a^{BH}(\delta_b - \delta_B)(\delta_{AH} - \delta_a) \tag{17}$$

Single-point NMR titrations are particularly convenient for investigating the temperature dependence of a ΔpK_a. Titrations performed at multiple temperatures could be carried out using a single sample (Perrin & Fabian, 1996). During each point along the titration the sample was heated to each of five temperatures and allowed to equilibrate within the probe for 15 min at each temperature. Temperatures were calibrated using a separate CD_3OD/D_2O sample with an internal ethylene glycol reference.

It may be noted that the binding constants for the complexes between $(CpSiMe_3)_3M$ and Cp*Ga (Cp = C_5H_5, Cp* = $C_5(CH_3)_5$, M = U or Nd) were measured by single-point NMR titrations at 10°C intervals between 21.5 and 72.0°C or between −33.2 and 19.2°C, and then converted to enthalpy and entropy of binding (Minasian et al., 2009).

5.2 Intramolecular Comparisons

There are two different limiting cases of intramolecular comparisons, one where there are two or more functional groups in the molecule with similar acidities, and the other where two or more functional groups differ substantially in acidity. This latter case often applies to the determination of K_1/K_2, the ratio of first and second dissociation constants of a diprotic acid. The equations below are for the case of two functional groups, but the extension to more than two is straightforward mathematically.

Although this NMR titration method was developed initially for intermolecular comparisons, it is often desirable to measure relative pK_as of similar functional groups in the same molecule. For example, one often wants to assess the role of the local environment on acidity constants of residues in biological macromolecules. A pH titration provides only the macroscopic (overall) pK_as. The ability of NMR to distinguish different nuclei makes it possible to measure microscopic pK_as associated with individual functional groups.

The difference between intermolecular and intramolecular comparisons is that in the former the chemical shift of one functional group is independent of the state of protonation of the other one. To apply the NMR titration to an intramolecular case, it is necessary that this independence also hold, as would be true if the groups are noninteracting. One recommendation is that the differences between the pK_as must be >2, or else the different sites must not influence one another (Shivapurkar & Jeannerat, 2011). Alternatively, the dependence of chemical shift on state of protonation may be known, or may be deduced from fitting the data. When these conditions are not fulfilled, the chemical shift is not simply the average of only two forms, and one has to use more sophisticated models and specialized software to determine the pK_a by fitting the experimental data.

Two classic cases studied by measuring chemical shifts vs pH, *N,N*-dimethylpropanediamine ($H_2NCH_2CH_2CH_2N(CH_3)_2$) and histidylhistidine, illustrate the difficulties. In the first case it was assumed that chemical shifts depend only on the state of protonation of the attached nitrogen, and are independent of the state of protonation of the other nitrogen (Onasch, Schwartz, Aikens, & Bunce, 1991). This cannot be true for nitrogens that are four bonds away from each other. In the second example, the data were fit so as to estimate how chemical shifts of one imidazole depend on the state of protonation of the other one, but the results were quite inaccurate (Bradbury & Sheraga, 1966; Shrager, Cohen, Heller, Sachs, & Schechter, 1972).

A successful ^{1}H NMR method was developed for determining residue-specific acid-dissociation constants for peptides that contain more than one residue of the same acidic or basic amino acid (Rabenstein, Hari, & Kaerner, 1997). In particular, this method was applied to the side-chain $-NH_3^+$ groups of the four lysine residues in the pentadecapeptide Lys-Asn-Asn-Gln-Lys-Ser-Glu-Pro-Leu-Ile-Gly-Arg-Lys-Lys-Thr-NH_2. The resonances for the $C_\varepsilon H_2$ protons are sensitive to the state of protonation of the adjacent $-NH_3^+$, but these show an extensive overlap of signals because their environments are so similar. In contrast, the backbone $C_\alpha H$ chemical shifts vary from residue to residue, even for lysines, owing to nearest-neighbor residues and possibly to secondary structure. Therefore, the four lysine signals could be resolved using two-dimensional total correlation spectroscopy (TOCSY) spectra, which show resolved $C_\alpha H—C_\varepsilon H_2$ cross peaks. Thus, chemical shift-pH titration data were obtained for each lysine side-chain $-NH_3^+$ group, measured as a function of pH. Each pK_a could then be evaluated by fitting the observed chemical shifts for each acidic group to Eq. (18), where δ_{HA} and δ_A are fitting parameters describing the chemical shifts of acidic and basic forms, respectively. The pK_as of the Lys1, Lys5, Lys13, and Lys14 side-chain ammonium groups were thus found to be 11.14 ± 0.01, 10.95 ± 0.01, 10.96 ± 0.02, and 11.09 ± 0.02, respectively, showing how it is possible to distinguish even such similar acids. Likewise, the pK_as of the Cys3 and Cys13 thiol groups were determined to be 9.21 ± 0.07 and 8.60 ± 0.06, respectively. Alternatively, the relative acidities of the four lysines could be evaluated with greater accuracy, and without the necessity of measuring pH, directly from the chemical-shift data, which were subjected to a nonlinear least-squares fit to Eq. (19) of Δ, the difference in chemical shifts between the CH signals of two imidazoles. In Eq. (19) $\Delta_A = \delta_a^1 - \delta_A^x$, $\Delta^1 = \delta_A^1 - \delta_{HA}^1$, $\Delta^x = \delta_A^x - \delta_{HA}^x$, $R_{1x} = K_a^1/K_a^x$, and $n = (\delta_{obs} - \delta_A)(\delta_{HA} - \delta_A)$, the fraction of protonation of Lys_1. It may be noted that one of the acid groups is chosen as a reference, designated as "1", and the acidities for other acid groups, designated as "x", are evaluated relative to "1", whereas in Eq. (14)

or (17) all acids are treated equivalently. This is a necessary arbitrariness when there are more than two acid groups to be compared, as was also necessary with the four decalylamine stereoisomers (**9**).

$$\delta_{obs} = \frac{[H^+]\delta_{HA} + K_a\delta_A}{[H^+] + K_a} \tag{18}$$

$$\Delta = \Delta_A + \frac{R_{1x} n \Delta^x}{R_{1x} n - n + 1} - n\Delta^1 \tag{19}$$

In the study of the above pentadecapeptide there is an implicit assumption that protonation of one Lys does not affect the chemical shift of another Lys. Because the $-NH_3^+$ groups are well separated in space, this assumption is valid to a high degree of approximation. Likewise, this method was applied to peptides containing >1 histidyl residue, and ΔpK_a could be evaluated by a nonlinear fit of chemical shifts for the separated histidine signals (Kállay et al., 2007).

An instructive example of both intermolecular and intramolecular comparisons was achieved by ^{13}C NMR titration of a mixture of 2-methyl- and 3-methyl-adipic acids (**21**, **22**) and analyzed by Eq. (19) (Pehk et al., 1996). There is no problem in relating the acidity of each of the carboxyl groups of **22** to the acidity of the C6 carboxyl of **21**. It is a greater challenge to compare the acidity of the C1 carboxyl of **21** to that of its C6 carboxyl. Although the chemical shift of one carboxyl is not independent of the state of protonation of the other that is separated by five C—C bonds, this long-range variation was found to be more than an order of magnitude smaller than the variation of chemical shift of a carboxyl group with the state of its own protonation. Therefore this intramolecular comparison is valid.

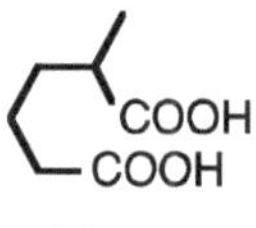

21

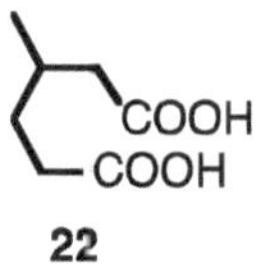

22

5.3 Ratio K_1/K_2

This method can be extended to diacids H_2A in order to evaluate K_1/K_2, the ratio of first and second acidity constants (Choi, Petterson, & Roberts, 2002). If $K_1 \gg K_2$ this simplifies to the separate evaluations of two acidity constants, by using Eq. (17), whereby K_1 or K_2 is compared to K_a^{BH}, the acidity constant of a reference acid. In this context, Eq. (17) becomes Eq. (20) or (21), where the chemical shifts refer to H_2A, HA^-, and A^-, and a single-point measurement suffices. If K_1 is not much greater than K_2, then the full equation, (Eq. 22), must be used, and two measurements are required, to provide two equations in the two unknowns. The method was successfully applied to fumaric, maleic, malonic, diethylmalonic, phthalic, furan-2,3-dicarboxylic, and bicyclo[2.2.1]-2,5-heptadiene-2,3-dicarboxylic acids.

$$\delta_a - \delta_{HA-} = \left(K_a^{BH}/K_1\right)(\delta_b - \delta_B)(\delta_{H_2A} - \delta_a)/(\delta_{BH} - \delta_b) \qquad (20)$$

$$\delta_a - \delta_{A=} = \left(K_a^{BH}/K_2\right)(\delta_b - \delta_B)(\delta_{HA-} - \delta_a)/(\delta_{BH} - \delta_b) \qquad (21)$$

$$(\delta_a - \delta_{A=})\frac{\delta_{BH} - \delta_b}{\delta_b - \delta_B}\left(K_2/K_a^{BH}\right) + (\delta_a - \delta_{HA-}) + (\delta_a - \delta_{H_2A})\frac{\delta_B - \delta_b}{\delta_b - \delta_{BH}}\left(K_a^{BH}/K_1\right) = 0 \qquad (22)$$

6. HEAVY-ATOM ISOTOPE EFFECTS ON ACIDITY

A demanding test of the NMR titration method is its applicability to secondary isotope effects, which is the focus of this volume of *Methods in Enzymology*. Primary isotope effects are those where the bond to the isotope is broken, whereas secondary isotope effects are those where the bond to the isotope remains intact. Therefore secondary isotope effects are quite small, even for deuterium relative to protium, and much smaller for other isotopic substitutions. To measure them, a method of high precision and accuracy is needed.

Imagine an NMR titration applied to a mixture of two acids, HA and HA′, each of whose chemical shifts, δ and δ', follows Eq. (1). If $\delta - \delta'$, the difference between those two chemical shifts, is designated as Δ, it is then possible to eliminate pH from the equation and express it in terms of n, the number of equivalents of titrant added, leading to Eq. (23), where $\Delta^- = \delta^- - \delta'^-$, $\Delta^\circ = \delta^\circ - \delta'^\circ$, and $R = K_a/K_a'$ (Ellison & Robinson, 1963). Moreover, that n can be evaluated as $(\delta - \delta^-)/(\delta^\circ - \delta^-)$, from the variation of the chemical shift of HA during the titration. Then, by using the three variables Δ^-, Δ°, and R as parameters in a nonlinear fitting routine, it is possible to evaluate the desired R, the ratio of acidity constants.

$$\Delta = \Delta^- + \frac{Rn(\delta^- - \delta^\circ + \Delta^\circ - \Delta^-)}{(Rn - n + 1)} - n(\delta^- - \delta^\circ) \qquad (23)$$

Eq. (23) is general, so that it applies to measurement of isotope effects on acidity. The titration of a mixture of formic acid and $^{18}O_2$-formic acid thus permitted the evaluation of the ^{18}O isotope effect on acidity from the ^{13}C NMR chemical shifts δ and δ' of the carboxyl carbons. For one ^{18}O ΔpK_a was found to be 0.00623 ± 0.00005. This same equation was used with ^{31}P chemical shifts to evaluate the ^{18}O isotope effects on the acidities of phosphoric acid and alkyl phosphates (Knight, Weiss, & Cleland, 1986). For four ^{18}O $K_a^{16}/K_a^{18} = 1.019 \pm 0.001$, and proportionally lower for fewer ^{18}O. The ability to accurately measure such small isotope effects is a tribute to the method.

The ^{15}N isotope effect on the first pK_a of glycine was measured from the NMR separations in the carboxyl ^{13}C NMR spectrum (Rabenstein & Mariappan, 1993). Although this carbon is more distant from the reactive $-NH_3^+$ than is the CH_2, the CH_2 signal is complicated by spin–spin coupling to the ^{15}N. The equilibrium isotope effect is related to the chemical shifts by Eq. (24), where $\Delta_d = \delta_d^{14} - \delta_p^{15}$, $\Delta^{14} = \delta_d^{14} - \delta_p^{14}$, $\Delta^{15} = \delta_d^{15} - \delta_p^{15}$, $R = K_a^{14}/K_a^{15}$, and $n = (\delta_{obs} - \delta_d)(\delta_p - \delta_d)$. This is analogous to Eq. (19), except that here the term Δ_A is ignored because there is no resolvable isotope effect on the chemical shifts of the acidic and basic forms. It may again be noted that the ^{14}N acid was chosen as the reference, rather than treating both acids equivalently.

$$\Delta = \frac{Rn\Delta^{15}}{Rn - n + 1} - n\Delta^{14} \tag{24}$$

Isotope effects were determined by ^{13}C NMR titration for a wide variety of carboxylic acids, including the ^{18}O effect on formic acid, acetic acid, and glycine acidities, the ^{15}N isotope effect on glycine acidity, and the ^{13}C isotope effect on acetic acid and glycine acidities (Pehk, Kiirend, Lippmaa, Ragnarsson, & Grehn, 1997).

According to DFT calculations, the secondary ^{18}O isotope effect on the acidity of a carboxylic acid is greater than the primary (Perrin & Burke, 2015). This is counterintuitive, because the H atom that is lost is closer to the ^{18}O atom that is responsible for the primary isotope effect. This is a result that could not be verified experimentally, because the proton exchanges too rapidly between the two oxygens.

7. SECONDARY DEUTERIUM ISOTOPE EFFECTS ON ACIDITY

It had been found that deuteration increases the basicity of amines, but the effects are small. For benzylamine-α-d_2 the isotope effect, expressed as ΔpK_a $(= -\log_{10}K_a^D + \log_{10}K_a^H$, where K_a is the acidity constant of the conjugate acid), was found by direct titration of the two amines to be 0.054 ± 0.001, which was subsequently revised to 0.032 ± 0.001 (Bary, Gilboa, & Halevi, 1979; Halevi, Nussim, & Ron, 1963). These error estimates are measures only of precision, not necessarily of accuracy, and may be overoptimistic if systematic error arises from an impurity in one of the samples. Similarly, ΔpK_a is 0.056 for methylamine-d_3 and 0.12 for dimethylamine-d_6, but surprisingly these were found to be temperature independent (Van der Linde & Robertson, 1964), as would be consistent with impurities. For trimethylamine-d_9 ΔpK_a is 0.185 (Northcott & Robertson, 1969), well beyond experimental error, but this could be due to steric repulsions, which flatten trimethylamine-h_9.

These cases were puzzling. They differ from the more familiar solvolyses in that there is no rehybridization, neither of the carbon bearing the isotope nor of the nitrogen, which remains nominally sp^3 on deprotonation. The isotope effect was therefore attributed to an inductive effect arising from an electrostatic interaction between the positive charge on the NH^+ and the dipole moment of the CH or CD bond (Halevi, 1963). Since dipole moment is charge times distance, and since the average CH bond is longer than CD, owing to anharmonicity, it is not impossible that deuterium could be effectively electron donating. This sort of inductive effect is consistent with the Born–Oppenheimer approximation, and it was generally accepted as the source of these isotope effects. Yet both the charge separation in a CH bond, which is quite nonpolar, and the bond-length difference due to anharmonicity, are small. Thus the isotope effect ought to be vanishingly small. Yet they were detected. Since such isotope effects were crucial to a proposed study to assess the symmetry of NHN hydrogen bonds (Perrin & Ohta, 2001, 2003), we sought to confirm them.

The NMR titration method permits precise measurement of relative basicities. When successive aliquots of acid are added to a mixture of bases, the acid will preferentially protonate the more basic species, whose NMR chemical shift will move ahead of that of the less basic one. The acidity constants K_a and chemical shifts δ can be related through Eq. (25), where δ^+ or δ^0 is for the protonated or deprotonated form, measured at the beginning or end of the titration. A plot of the quantity on the left vs $(\delta_1 - \delta_1^0)(\delta_2^+ - \delta_2)$ should be linear, with zero intercept, and with a slope equal to the ratio of acidity constants.

$$\left(\delta_1^+ - \delta_1\right)\left(\delta_2 - \delta_2^0\right) = \left(K_a^1/K_a^2\right)\left(\delta_1 - \delta_1^0\right)\left(\delta_2^+ - \delta_2\right) \tag{25}$$

Isotope effects were thereby measured for dimethylamine-d_3, benzylamine-d, and N,N-dimethylaniline-d_3 (Perrin, Ohta, & Kuperman, 2003). Mixing each of these with unlabeled material, or taking a methylamine-$d_{0,1,2,3}$ mixture, produces a ^{1}H NMR spectrum with resolvable signals due to different isotopologues (isomers that differ in the number of isotopic substitutions). Table 1 presents the results, which verify that there really is a secondary deuterium isotope effect on amine basicities, even though there is no rehybridization.

Isotope effects, K_a^H/K_a^D, on the acidities of N-methyl-d_3 (**23-*d*$_3$**) and N,N-dimethyl-d_6–4-fluoroanilinium (**24-*d*$_6$**) ions in aqueous dioxane were measured by ^{1}H-decoupled ^{19}F NMR titration, with a resolution-enhancing weighting function (Forsyth & Yang, 1986). At a pH far from the pK_a only a small intrinsic shift is observed. Near the pK_a there is a larger isotope shift, reaching a maximum of ~1 ppm, as shown for **24** in Fig. 4. The downfield shift (deshielding) of the deuterated isotopologue indicates that the labeled aniline is more basic and hence more fully protonated. It may be noted that the intrinsic isotope shift of deuteration is upfield (shielded, negative), so that the signals cross at intermediate pH values. From an analysis of those isotope shifts, K_a^H/K_a^D was found to be 1.117 ± 0.011 for **23** and 1.252 ± 0.011 for **24**.

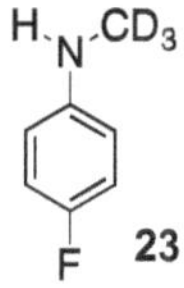
23

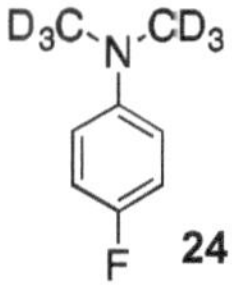
24

The method was also applied to secondary deuterium isotope effects on the acidities of a wide variety of carboxylic acids, including formic, acetic,

Table 1 Secondary Deuterium Isotope Effects on Amine Basicities

Amine	K_H/K_D	ΔpK_a	$\Delta\Delta G^{0a}$
Methylamine-d	1.040 ± 0.006	0.017 ± 0.003	23.2 ± 3.4
Methylamine-d_2	1.081 ± 0.004	0.034 ± 0.002	23.1 ± 1.1
Dimethylamine-d_3	1.144 ± 0.005	0.058 ± 0.002	26.6 ± 0.9
Benzylamine-d	1.0419 ± 0.0009	0.0178 ± 0.0004	24.4 ± 0.5
N,N-dimethylaniline-d_3	1.1051 ± 0.0018	0.0434 ± 0.0007	19.7 ± 0.3

[a]cal/mol per D.

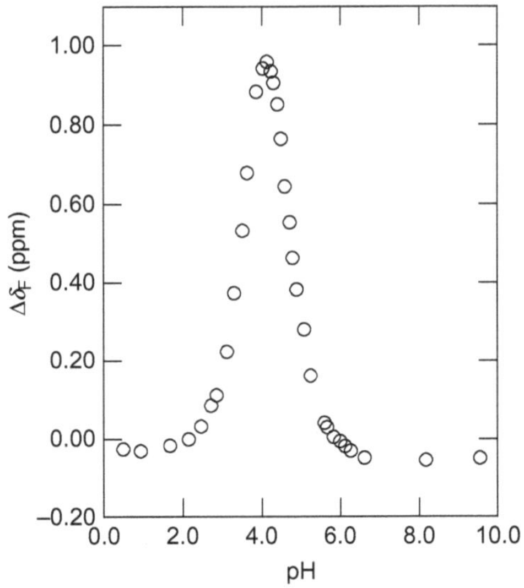

Fig. 4 pH dependence of the ^{19}F isotope shift between **24-*d*$_6$** and **24** in 50% aqueous dioxane. *Reprinted with permission from Forsyth, D. A., & Yang, J.-R. Intrinsic and equilibrium NMR isotope shift evidence for negative hyperconjugation.* Journal of the American Chemical Society, 108, *2157–2161. Copyright (1986) American Chemical Society.*

propionic, butyric, caproic, and also glycine and alanine (Pehk et al., 1997). The pK_a ratios could be evaluated from the variation with the degree of protonation of the measured ^{13}C NMR chemical-shift differences at atoms close to the site of isotopic substitution in a mixture of isotopologues. The data were fit by nonlinear least squares to Eq. (26), where δ_H and δ_D are the observed chemical shifts of undeuterated and deuterated isotopologues, δ_H° and δ_D° are those chemical shifts in the deprotonated form, δ_H^+ and δ_D^+ are those chemical shifts in the protonated form, $R = K_a^H/K_a^D$, and n is the degree of protonation of the undeuterated material. This is the same equation as Eq. (14), but adapted to deuteration, and again n is evaluated as $(\delta_H^\circ - \delta_H)/(\delta_H^\circ - \delta_H^+)$.

$$\delta_H - \delta_D = \delta_H^\circ - \delta_D^\circ - n\left(\delta_H^\circ - \delta_H^+\right) + \frac{Rn}{1+(R-1)n}\left(\delta_D^\circ - \delta_D^+\right) \qquad (26)$$

Trifluoroethanol-d_2 is less acidic than trifluoroethanol, according to ^{13}C NMR pH titration, with a ΔpK_a of 0.056 (Gawlita et al., 2000).

Additional examples of secondary isotope effects on acidity, some measured by NMR titration, are included in a review (Perrin, 2010).

8. SECONDARY DEUTERIUM ISOTOPE EFFECTS ON BINDING

^{31}P NMR titration was used to measure the H/D isotope effects on binding of various deuterium-containing $PhCH_2P(CH_3)_3^+$ isotopologues to the exterior of the supramolecular host tetrahedral $[Ga_4L_6]^{12-}$ (L = 1,5-bis(2,3-dioxybenzamido)naphthalene^{-4}, **25**) containing a strongly bound Et_4N^+ guest (Mugridge, Bergman, & Raymond, 2010). For the linearized plot the correlation coefficient was an impressive 0.99999. It was found that the binding-constant ratios are $K_{d0}/K_{d7} = 1.0302 \pm 0.0004$ and $K_{d0}/K_{d9} = 1.047 \pm 0.002$, and these two values could be corroborated with $K_{d7}/K_{d9} = 1.017 \pm 0.003$. Similar results, but of greater magnitude, were obtained for the H/D isotope effects on binding of those deuterium-containing $PhCH_2P(CH_3)_3^+$ isotopologues to the interior of the same $[Ga_4L_6]^{12-}$ (Mugridge, Bergman, & Raymond, 2012).

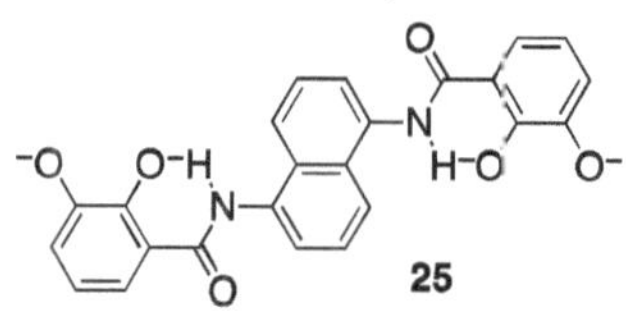

25

The equilibrium deuterium isotope effect for Cl^- binding to anion receptor (**26**) in DMSO-d_6 was measured by ^{13}C NMR titration (Tresca, Brueckner, Haley, Cheong, & Johnson, 2017). From the signals of the indicated alkyne C in the two isotopologues the linearized equation gave $K^H/K^D = 1.019 \pm 0.010$, where the value >1 means that binding is favored with ^1H. This value agrees with a calculated 1.020, whereas the calculated IEs for fragments is inverse (e.g., 0.989 for 2,6-ethynyl-4-nitro), which implies that the contributions from the separate fragments are nonadditive. The precision possible with this competitive titration may be compared with the much lower precision from the direct evaluation of the binding constant in $CHCl_3$, for which log $K = 4.39 \pm 0.62$.

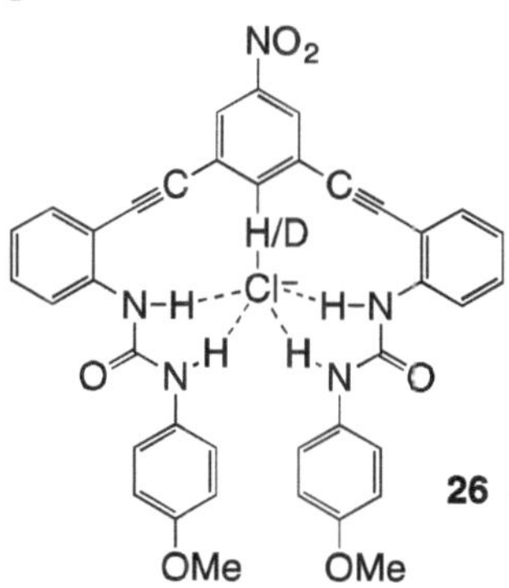

26

9. FURTHER STUDIES OF SECONDARY DEUTERIUM ISOTOPE EFFECTS, TOWARD ELUCIDATING THEIR ORIGIN

The data in Table 1 show that there is indeed a secondary deuterium isotope effect on the basicity of amines. An inductive or electrostatic origin for this isotope effect still seemed unlikely. Therefore we sought to investigate them further to learn their origin.

A key result was the ratio of acidity constants for the isotopomers of 1-benzyl-4-methylpiperidine-d_3 (**27**) (Perrin et al., 2003). This was an especially demanding measurement, because isotopomers are isomers (here stereoisomers) that differ only in the position of an isotope, as distinguished from isotopologues, which differ in the number of isotopic substitutions. It was found that the isotopomer with deuterium axial and protium equatorial is more basic, with $K_a^{eq}/K_a^{ax} = 1.060 \pm 0.006$. Therefore the isotope effect can be attributed to delocalization of the nitrogen lone pair into the antiperiplanar C—H bond, which decreases its zero-point energy and causes that isotopomer to resist protonation more than an antiperiplanar C—D does. Moreover, for 2-methyl-2-azanorbornane-3,3-d_2 (**28**, R = CH_3) and for pyrrolizidine-d (**29**) deuteration can exert an isotope effect when synperiplanar to a lone pair, but smaller in magnitude (Perrin, Ohta, Kuperman, Liberman, & Erdélyi, 2005). These conclusions could be confirmed by computations of the dependence of C—D stretching frequency in DCH_2NH_2 on dihedral angle. In summary, these isotope effects are attributed to a lower zero-point energy of a C—H bond adjacent to an amine nitrogen, arising from delocalization of either a syn or an anti lone pair, and with no detectable angle-independent inductive effect.

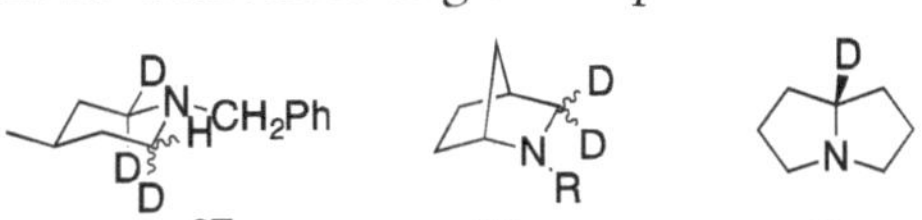

27 **28** **29**

Secondary deuterium isotope effects on acidities of carboxylic acids (**30–33**, **37–38**) and phenols (**34–36**) were also measured by ^1H, ^{13}C, and ^{19}F NMR titrations (Perrin & Dong, 2007). Deuteration definitely decreases the acidity of carboxylic acids and phenols, by up to 0.031 in the ΔpK per D. For aliphatic acids the isotope effects decrease as the site of deuteration becomes more distant from the OH, as expected, but a surprising result is that isotope effects in both phenol and benzoic acid do not decrease as the site of deuteration moves from ortho to meta to para. The experimental data are supported by ab initio computations which, however, substantially overestimate the isotope effects, perhaps because of inaccuracies in contributions due to solvation. The isotope effects originate in isotope-sensitive vibrations whose frequencies and zero-point energies are lowered on deprotonation. In the simplest case, formate, the key vibration can be recognized as the C—H stretch, which is weakened by delocalization of the antiperiplanar lone pairs on the oxygens. For the aromatic acids delocalization cannot account for the near constancy of isotope effects from ortho, meta, and para deuteriums, but the observed isotope effects are consistent with calculated vibrational frequencies and electron densities. Moreover, the ability of the frequency analysis to account for the isotope effects is evidence against an inductive origin.

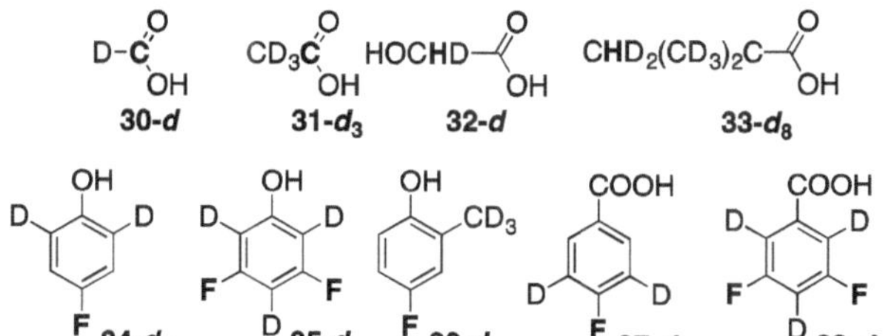

30-*d* **31-*d*$_3$** **32-*d*** **33-*d*$_8$**

34-*d*$_2$ **35-*d*$_3$** **36-*d*$_3$** **37-*d*$_2$** **38-*d*$_3$**

Secondary deuterium isotope effects on basicities of various deuterated pyridine isotopologues and of 2,6-lutidine-2,6-$(CD_3)_2$ were accurately measured in aqueous solution (Perrin & Karri, 2010). Deuteration at any position of pyridine increases the basicity, but the isotope effect per deuterium is largest for substitution at the 3-position, and smallest for the 2-position that is closest to the site of *N*-protonation, smaller even than for 2-CD_3 substitution. DFT computations at the B3LYP/cc-pVTZ level overestimate the magnitude of the measured isotope effects, but largely reproduce the variability with isotopic position. Because the calculated isotope effects are based on changes in vibrational frequencies on *N*-protonation, the correspondence between calculated and experimental isotope effects implies that they arise from zero-point energies, rather than from inductive effects.

In a remarkable demonstration of the power of the NMR titration method, secondary deuterium isotope effects on the basicities of isotopologues of trimethylamine were measured (Perrin & Dong, 2008). Deuteration definitely increases the basicity, by ~0.021 per D. Fig. 5 shows a linear plot from the NMR titration of a mixture of tri(methyl-*d*)ammonium and tri(methyl-d_2)ammonium hydrochlorides with NaOD in

D_2O. An impressive result is that the correlation coefficient is 0.999999. The isotope effect is attributed to the lowering of the CH stretching frequency and zero-point energy by delocalization of the nitrogen lone pair into the C—H antibonding orbital. Because this depends on the dihedral angle between the lone pair and the C—H, a further consequence is a preference for conformations with H antiperiplanar to the lone pair, and D gauche. This leads to a predicted nonadditivity of isotope effects, which could be confirmed experimentally. Specifically, it was found that the decrease in basicity, per deuterium, increases with the number of deuteriums.

Further insight into the origin of these isotope effects comes from their temperature dependence (Perrin & Flach, 2011). The secondary deuterium isotope effects on the acidities of formic-*d* acid, acetic-*d* and acetic-d_2 acids, and 3,5-difluorophenol-2,4,6-d_3 were measured by the NMR titration method, using ^{13}C NMR, ^{1}H NMR, and ^{19}F NMR, respectively. Instead of a complete titration for each sample, a single-point titration, as described above, used only three titration points, obtained from three samples prepared at room temperature. Two of the points correspond to the fully protonated and fully deprotonated samples, and the third point corresponds to a 50% neutralized sample.

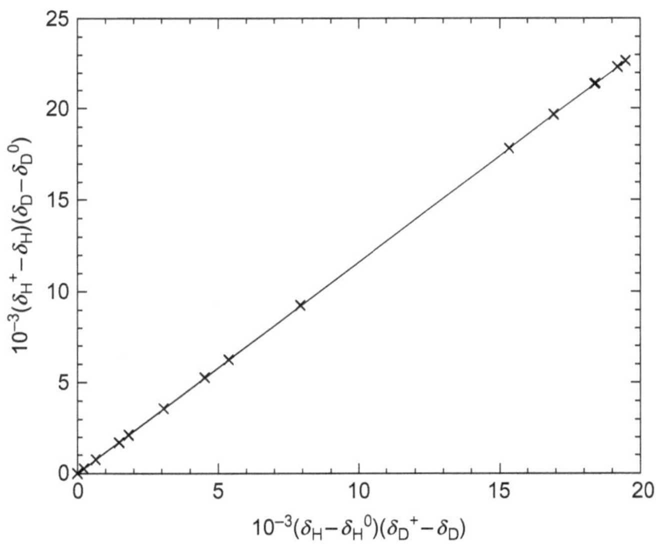

Fig. 5 NMR Titration of a mixture of $(CH_2D)_3NH^+$ and $(CH_2D)_3NH^+$. *Reprinted with permission from Perrin, C. L., & Dong, Y. (2008). Nonadditivity of secondary deuterium isotope effects on basicity of trimethylamine.* Journal of the American Chemical Society, 130, *11,143–11,148. Copyright (2008) American Chemical Society.*

Table 2 Enthalpy and Entropy Contributions to Secondary Deuterium Isotope Effects on Acidities

Acid	$\Delta\Delta H^{\circ}$ (cal/mol)	$\Delta\Delta S^{\circ}$ (cal/mol-K)	$T\Delta\Delta S^{\circ}$, 20°C
Formic-d	-37.4 ± 1.2	0.009 ± 0.004	2.6 ± 1.2
Acetic-d	-7.9 ± 0.4	-0.006 ± 0.001	-1.65 ± 0.4
Acetic-d_2	-15.2 ± 0.6	-0.010 ± 0.002	-2.9 ± 0.6
3,5-F_2Phenol-d_3	-27.9 ± 1.2	-0.0085 ± 0.004	-2.5 ± 1.2

The temperature dependence of the isotope effects could then be analyzed by fitting the data to Eq. (27), where $\Delta\Delta S^{\circ}$, obtained from the intercept vs $1/T$, is the entropic contribution to the isotope effect, $(S^{\circ}_{H^-} - S^{\circ}_{H^0}) - (S^{\circ}_{D^-} - S^{\circ}_{D^0})$, and similarly for $\Delta\Delta H^{\circ}$. The data are presented in Table 2. The data show that these isotope effects lie entirely in the enthalpy, as expected if their origin lies in vibrational frequencies and zero-point energies, whereas inductive effects generally manifest themselves in entropy.

$$\ln\left(\frac{K_a{}^{H}}{K_a{}^{D}}\right) = -\frac{\Delta\Delta H^{\circ}}{R}\frac{1}{T} + \frac{\Delta\Delta S^{\circ}}{R} \tag{27}$$

10. ADVANTAGES OF NMR TITRATION

This method is capable of exquisite precision, since it is based only on chemical-shift measurements. With modern NMR spectrometers, chemical shifts can be measured with remarkable accuracy and precision. In contrast, the classical pH titration requires accurate measurements of both volume and molarity. The data in Table 1 show that ΔpK_a can be measured with an error of as little as ± 0.0004, which is nearly two orders of magnitude better than can commonly be achieved with a pH electrode.

Since the titration is performed on a mixture of the two bases, under conditions guaranteed identical for both, it avoids systematic error due to possible impurities.

An NMR titration can be performed in any solvent, including ones like dichloromethane or chloroform, where a pH electrode would be useless. The only requirement is that the acidic and basic forms must remain soluble, even if they are ionic.

An important advantage of this approach is that it relies solely on chemical shifts. The concentrations of the components of the mixture are not involved, meaning that sample preparation requires no special care and does not need to produce stoichiometric or precisely known amounts. The same is true for the titrant, which can be added to the sample with a low-precision or uncalibrated addition device.

Although the initial development of NMR titration used ^{1}H NMR, other nuclei are applicable: Certainly ^{13}C, ^{15}N, ^{19}F, and ^{31}P NMR titrations benefit from the broader range of chemical shifts, so that signal overlap is rare and protonation shifts are larger than with ^{1}H NMR. Those chemical shifts are likely to be more sensitive to long-range effects such as protonation on distant groups, but that may be an advantage. It is often feasible to use ^{13}C in natural abundance, but enrichment is probably necessary for ^{15}N. The abundant ^{14}N nucleus is also sensitive to chemical environment, but its nuclear spin I of 1 often leads to signal broadening by nuclear quadrupole effects, which may compromise NMR titration.

NMR titration provides only relative acidities, although these are often more relevant than absolute acidities. Of course, relative acidities can be converted to absolute acidities by anchoring them to an acid of known pK_a, but then the absolute acidities are only as accurate as that known pK_a.

Isotope effects on acidity can be used to guide chromatographic separations on an acid–base column to separate isotopologues (Tanaka et al., 1989).

11. SOME PRACTICAL CONSIDERATIONS

Because NMR chemical shifts of the substrates vary with the solution environment, it is advisable to choose an internal standard for NMR that is as similar to the substrate as possible. Thus an alkane- or arene-sulfonate may be more suitable than tetramethylsilane for titration of a carboxylate, and a quaternary ammonium ion may be more suitable for titration of an amine.

In principle any nucleus that reports on the state of protonation of the molecule is useful. However, the greatest precision is obtained from monitoring those NMR chemical shifts that undergo the largest change. Therefore, those nuclei closest to the site of protonation/deprotonation are best suited for this method. These are also especially favorable, since they are often downfield of (more deshielded than) other signals and are readily discerned in the NMR spectrum.

Because the procedure involves the repeated addition to the sample of small aliquots of acid or base, an NMR titration can be quite time-consuming, especially if it is necessary to remove the sample from the NMR probe, add the aliquot, and mix the solution thoroughly. This is even more time-consuming if it is also necessary to allow the sample to equilibrate to the probe temperature after each addition. To streamline the procedure, it is convenient to add the aliquot directly into the sample in the probe. A device was designed to facilitate this experiment (Perrin & Rivero, 1999), and manufacturers offer various systems for flow NMR (Hentschel, Holtin, Steinhauser, & Albert, 2012), and for monitoring reactions (Gomez & de la Hoz, 2017).

The researcher should decide whether to carry out an NMR titration from low pH to high or from high to low. Because chemical shifts are often sensitive to the medium, and specifically to its ionic strength, it may be advisable to maintain constant ionic strength during the NMR titration. Instead of adding anionic base (such as NaOH) to neutral acid (such as RCOOH, Eq. 28) or cationic acid (such as HCl, Eq. 29), to neutral base (such as an amine), it is possible to first add a slight excess of anionic base or cationic acid to the substrate and then back-titrate with cationic acid or anionic base, respectively. In the first case the anionic conjugate base of the neutral substrate is gradually replaced during the titration by the anionic conjugate base of the cationic acid (Eq. 30). In the second case the cationic conjugate acid of the neutral substrate is gradually replaced during the titration by the cationic component of the anionic base (Eq. 31). Alternatively, the titration might start with the carboxylate salt or with the amine hydrochloride, if available. It can be seen that this technique maintains constant ionic strength during the titration.

$$OH^- + RCOOH \rightarrow RCO_2^- + H_2O \tag{28}$$

$$H^+ + R_3N \rightarrow R_3NH^+ \tag{29}$$

$$HCl + RCO_2^- \rightarrow RCOOH + Cl^- \tag{30}$$

$$R_3NH^+ + NaOH \rightarrow R_3N + Na^+ + H_2O \tag{31}$$

To take maximum advantage of accurate chemical-shift measurements, it may be necessary to output them in Hz, rather than ppm, because some instruments provide an extra significant figure for the frequencies.

To apply Eq. (8) or the other, equivalent version, it is crucial to have accurate values for the limiting chemical shifts, δ_{A°, δ_{AH^+}, δ_{B°, and δ_{BH^+} of the fully protonated and deprotonated species. Errors resulting from uncertainties in those limits have been presented graphically (Shivapurkar & Jeannerat, 2011). It is not advisable to simply take the chemical shifts at the beginning and end of the titration. Instead the pH of the starting solution should be adjusted by adding a small amount of acid or base so that the protonation or deprotonation of the species of interest starts only after two or three additions of the titrant. Similarly the titration should continue for a few points after the species of interest have been completely protonated or deprotonated. These additional points can then provide an estimate of how chemical shifts vary with medium, independently of the state of protonation. They may permit an extrapolation to the correct endpoint chemical shifts. Still another possibility is to use the endpoints as additional parameters in the nonlinear fit to Eq. (7).

Although H_2O is the standard solvent for pH measurements, ^1H NMR titrations in H_2O are plagued by the intense water signal, which limits the dynamic range of the spectrum. Therefore such titrations are best carried out in D_2O. This condition has two significant consequences: (1) Signals of the NH groups of amides are often missing because the NH protons exchange into the solvent. (2) A correction of 0.40 must be added to the pH-meter readings in D_2O to get pD values (Glasoe & Long, 1960).

A concentrated solution of titrant should be used, to minimize dilution, which reduces sensitivity and may change the solvent conditions.

Because this NMR titration method depends on a comparison between two acids, one of which serves as a reference, it is advisable to choose the reference wisely and not randomly. Of course, if the comparison is of two isomers (or isotopologues or isotopomers), there is no choice. However, if the reference is of known acidity, it is advisable to choose an acid of similar acidity. This is apparent from consideration of Eq. (23). If the acidities differ widely, one acid is almost completely protonated while the other is almost completely deprotonated, so that one factor on each side is always the difference between two large numbers. In contrast, if the acidities are similar, all factors become of similar magnitude at the midpoint of the titration, which contributes most strongly to the slope. Thus it was misleading advice to claim that the best results are obtained when AH and BH are separated by $\sim 2\,pK_a$ units (Emenike et al., 2013).

If signals overlap, the NMR titration method can be applied with two-dimensional NMR. For example, the intramolecular comparison of the relative acidities of the side-chain $-NH_3^+$ groups of the four lysine residues in a pentadecapeptide could be resolved with TOCSY (Rabenstein et al., 1997). Similarly, 2D ^1H–^{13}C HSQC (Homonuclear Single-Quantum

Correlation Spectroscopy) NMR was used for the titration with DCl of a six-component mixture of carboxylates, with acetic acid as a reference (Shivapurkar & Jeannerat, 2011).

11.1 Pitfalls

Because of the need for accurate chemical shifts, in practice it is necessary that the signal of the relevant reporter nucleus show baseline separation from other signals. Overlapping signals are a warning that data will be badly behaved.

The NMR titration method was initially developed for monoprotic acids or bases, and difficulties may arise in connection with intramolecular comparisons, where there are two or more sites of protonation. An example of such a difficulty can be seen in the ^{1}H NMR titration of two 8,8′-biquinolyls, where a linearized plot (Eq. 14) shows a correlation coefficient of only 0.9715, and the points can be seen to follow an ellipse. This was attributed to the complication that one of those substrates can be doubly protonated, but in general the ellipse is a signal that an NMR titration has a problem (Blakemore, Milicevic, Perera, Shvarev, & Zakharov, 2008). Thus special care was taken in the ^{13}C NMR titration of a mixture of glycine and glycine-d_2 (Romero, Reyes, & Wist, 2010). In order to determine the deuterium isotope effect on the first deprotonation of the NH_3^+, to produce a zwitterion, the titration was carried out only until the isoelectric point was reached (although it is unclear how that was ascertained). Anyhow, it was found that including an extra point or discarding one from the analysis modifies the result only within the experimental error.

The pK_as of substituted salicyl *N*-benzylaldimines, **39H**, were determined from pH titration and fitting to Eq. (14) (Ortegón-Reyna et al., 2014). It was further claimed that the tautomeric equilibrium constant K_{taut} could be obtained from the slope of a plot of δ_{CH_2} vs δ_{ortho} according to an adaptation of Eq. (14). Unfortunately, this analysis is faulty. While K_a for **39** is $[H^+][\mathbf{39}^-]/([\mathbf{39H}_{imino}] + [\mathbf{39H}_{keto}])$, where $\mathbf{39}^-$ is the common anion, so that K_{taut} is the ratio of the separate (microscopic) acidity constants $\mathbf{39H}_{imino}$ and $\mathbf{39H}_{keto}$, those separate acidity constants cannot be measured. This differs from the intramolecular comparisons above, where the chemical shift of one reporter nucleus must be independent of the state of protonation of the other. Here both δ_{CH_2} and δ_{ortho} follow Eq. (14), governed by the same pK_a, while the position of the tautomeric equilibrium is independent of pH. The information provided by a plot of δ_{CH_2} and δ_{ortho} according to Eq. (14) is only of the relative sensitivity of those two chemical shifts to protonation. This same fallacy applies to the attempt to determine by ^{1}H NMR titration the relative basicities of the two nitrogens in 6-substituted 2,3-dimethylquinoxalines (**40**) (Peralta-Cruz, Díaz-Fernández, Avila-Castro, Ortegón-Reyna, & Ariza-Castolo, 2016).

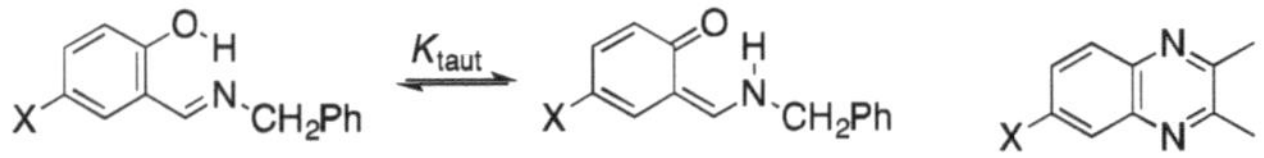

39H$_{imino}$ **39H**$_{keto}$ **40**

12. SUMMARY

NMR titration is a valuable technique for measuring relative acidities and basicities with exceptional accuracy. It can be applied to similar substances, including isotopically substituted ones, which thereby provides equilibrium isotope effects. With care it can be applied to diprotic acids and to intramolecular comparisons. It has the advantage that it is not necessary to measure pH, sample volumes, or molarities. It has wide applicability.

ACKNOWLEDGMENT

Preparation of this chapter was supported by NSF Grant CHE11-48992.

REFERENCES

Bachovchin, W. W., & Roberts, J. D. (1978). Nitrogen-15 nuclear magnetic resonance spectroscopy. The state of histidine in the catalytic triad of alpha-lytic protease. Implications for the charge-relay mechanism of peptide-bond cleavage by serine proteases. *Journal of the American Chemical Society*, *100*, 8041–8047.

Baldridge, K. K., Cozzi, F., & Siegel, J. S. (2012). Basicity of (2,6-Pyridino)paracyclophanes: Lone pair–π, cation–π, and solvation effects. *Angewandte Chemie International Edition*, *2012*(51), 2903–2906.

Bary, Y., Gilboa, H., & Halevi, E. A. (1979). Secondary hydrogen isotope effects. Corrigendum and addendum. *Journal of the Chemical Society, Perkin Transactions*, *2*, 938.

Baryshnikova, O. K., Williams, T. C., & Sykes, B. D. (2008). Internal pH indicators for biomolecular NMR. *Journal of Biomolecular NMR*, *41*, 5–7.

Blakemore, P., Milicevic, S. D., Perera, H., Shvarev, A., & Zakharov, L. N. (2008). Determination of pK_a values for diether derivatives of 7,7'-dihydroxy-8,8'-biquinolyl: Dependence of basicity on interannular dihedral angle. *Synthesis*, 2271–2277.

Bradbury, J. H., & Sheraga, H. A. (1966). Structural studies of ribonuclease. XXIV. The application of nuclear magnetic resonance spectroscopy to distinguish between the histidine residues of ribonuclease1. *Journal of the American Chemical Society*, *88*, 4240.

Choi, P. J., Petterson, K. A., & Roberts, J. D. (2002). Ionization equilibria of dicarboxylic acids in dimethylsulfoxide as studied by NMR. *Journal of Physical Organic Chemistry, 15*, 278.

Ellison, S. L. R., & Robinson, M. J. T. (1963). A precise N.M.R. method for measuring isotope effects on acid-base equilibria: Effects of solvent, solvent isotopic composition, and ionic strength on hydrogen and oxygen isotope effects in formic acid. *Journal of the Chemical Society, Chemical Communications*, 745–746.

Emenike, B. U., Liu, A. T., Naveo, E. P., & Roberts, J. D. (2013). Substituent effects on energetics of peptide-carboxylate hydrogen bonds as studied by ^{1}H NMR spectroscopy: Implications for enzyme catalysis. *Journal of Organic Chemistry, 78*, 11765–11771.

Evans, C. A., & Rabenstein, D. L. (1974). Nuclear magnetic resonance studies of the acid-base chemistry of amino acids and peptides. 11. Dependence of the acidity of the C-Terminal carboxyl group on the conformation of the C-terminal peptide bond. *Journal of the American Chemical Society, 96*, 312–7317.

Fabian, M. A., Perrin, C. L., & Sinnott, M. L. (1994). Absence of reverse anomeric effect: Conformational analysis of glucosylimidazolium and glucosylimidazole. *Journal of the American Chemical Society, 116*, 8398–8399.

Forsyth, D. A., & Yang, J.-R. (1986). Intrinsic and equilibrium NMR isotope shift evidence for negative hyperconjugation. *Journal of the American Chemical Society, 108*, 2157–2161.

Gawlita, E., Lantz, M., Paneth, P., Bell, A. F., Tonge, P. J., & Anderson, V. E. (2000). H-bonding in alcohols is reflected in the C_αH bond strength: Variation of C–D vibrational frequency and fractionation factor. *Journal of the American Chemical Society, 122*, 11660–11669.

Glasoe, P. K., & Long, F. A. (1960). Use of glass electrodes to measure acidities in deuterium oxide. *Journal of Physical Chemistry, 64*, 188–190.

Gomez, M. V., & de la Hoz, A. (2017). NMR reaction monitoring in flow synthesis. *Beilstein Journal of Organic Chemistry, 13*, 285–300.

Grunwald, E., & Cocivera, M. (1965). Solvent participation in proton transfer reactions of amines and their conjugate acids. *Discussions of the Faraday Society, 39*, 105–111.

Guerrero-Alvarez, J. A., Mas-Ku, W. P., Garcías-Morales, C., & Ariza-Castolo, A. (2010). Relative acidities and structure analysis of *cis* and *trans* isomers of 1,5-oxazaspiro[5.5] undecane derivatives by multinuclear magnetic resonance. *Magnetic Resonance in Chemistry, 48*, 356–361.

Halevi, E. A. (1963). Secondary isotope effects. *Progress in Physical Organic Chemistry, 1*, 109.

Halevi, E. A., Nussim, M., & Ron, A. (1963). Secondary hydrogen isotope effects. Part I. Strengths of α-deuterated carboxylic acids and amines. *Journal of the Chemical Society*, 866.

Hentschel, P., Holtin, K., Steinhauser, L., & Albert, K. (2012). Monitoring the on-line titration of enantiomeric omeprazole employing continuous-flow capillary microcoil ^{1}H NMR spectroscopy. *Chirality, 24*, 1074–1076.

Hess, A. S., Yoder, R. A., & Johnston, J. N. (2006). Chiral proton catalysis: pK_a determination for a BAM-HX Brønsted acid. *Synlett*, 147–149.

Hirose, K. (2001). A practical guide for the determination of binding constants. *Journal of Inclusion Phenomena and Macrocyclic Chemistry, 39*, 193–209.

Jensen, A. A., Mikkelsen, I., Frølund, B., Frydenvang, K., Brehm, L., Jaroszewski, J. W., et al. (2004). Carbamoylcholine homologs: Synthesis and pharmacology at nicotinic acetylcholine receptors. *European Journal of Pharmacology, 497*, 125–137.

Juaristi, E., & Cuevas, G. (1992). Recent studies of the anomeric effect. *Tetrahedron, 48*, 5019–5087.

Kállay, C., Ösz, K., Dávid, A., Valastyán, Z., Malandrinos, G., Hadjiliadis, N., et al. (2007). Zinc(II) binding ability of tri-, tetra- and penta-peptides containing two or three histidyl residues. *Dalton Transactions*, 4040–4047.

Keglevich, C., Grün, A., Bölcskei, A., Drahos, L., Kraszni, M., & Balogh, G. T. (2012). Synthesis and proton dissociation properties of arylphosphonates: A microwave-assisted catalytic Arbuzov reaction with aryl bromides. *Heteroatom Chemistry*, *23*, 574–582.

Kirby, A. J. (1983). *The anomeric and related stereoelectronic effects at oxygen*. Springer: Berlin.

Knight, W. B., Weiss, P. M., & Cleland, W. W. (1986). Determination of equilibrium ^{18}O isotope effects on the deprotonation of phosphate and phosphate esters and the anomeric effect on deprotonation of glucose-6-phosphate. *Journal of the American Chemical Society*, *108*, 2759–2761.

Kubyshkin, V., Durkin, P., & Budisa, N. (2016). Energetic contribution to both acidity and conformational stability in peptide models. *New Journal of Chemistry*, *40*, 5209.

Mehdoui, T., Berthet, J.-C., Thuéry, P., & Ephritikhine, M. (2004). Lanthanide(III)/actinide (III) differentiation in coordination of azine molecules to tris(cyclopentadienyl) complexes of cerium and uranium. *Dalton Transactions*, 579–590.

Minasian, S. G., Krinsky, J. L., Rinehart, J. D., Copping, R., Tyliszczak, T., Janousch, M., et al. (2009). A comparison of 4*f vs* 5*f* metal-metal bonds in $(CpSiMe_3)_3M$-ECp* (M = Nd, U, & E = Al, Ga, & Cp* = C_5Me_5): Synthesis, thermodynamics, magnetism, and electronic structure. *Journal of the American Chemical Society*, *131*, 13767–13783.

Mugridge, J. S., Bergman, R. G., & Raymond, K. N. (2010). High-precision measurement of isotope effects on noncovalent host-guest interactions. *Journal of the American Chemical Society*, *132*, 1182–1183.

Mugridge, J. S., Bergman, R. G., & Raymond, K. N. (2012). Equilibrium isotope effects on noncovalent interactions in a supramolecular host-guest system. *Journal of the American Chemical Society*, *134*, 2057–2066.

Northcott, D., & Robertson, R. E. (1969). Secondary deuterium isotope effect in the ionization of trimethylamine-d_9 in water. *Journal of Physical Chemistry*, *73*, 1559.

Onasch, O. F., Schwartz, H. M., Aikens, D. A., & Bunce, S. C. (1991). The microscopic protonation constants of *N*,*N*-dimethyl-1,3-propanediamine: An NMR titration experiment. *Journal of Chemical Education*, *68*, 791.

Orgován, G., & Noszál, B. (2011). Electrodeless, accurate pH determination in highly basic media using a new set of ^{1}H NMR pH indicators. *Journal of Pharmaceutical and Biomedical Analysis*, *54*, 958–964.

Ortegón-Reyna, D., Garcías-Morales, C., Padilla-Martínez, I., García-Báez, E., Aríza-Castolo, A., Peraza-Campos, A., et al. (2014). NMR structural study of the prototropic equilibrium in solution of Schiff bases as model compounds. *Molecules*, *19*, 459–481.

Paulsen, H., Gyorgydeak, Z., & Friedmann, M. (1974). Einfluß des anomeren und inversen anomeren effektes auf konformationsgleichgewichte von *N*-substituierten *N*-pentopyranosiden. *Chemisches Berichte*, *107*, 1590–1613.

Pehk, T., Kiirend, E., Lippmaa, E., & Ragnarsson, U. (1996). General approach to measurements of pKa differences by C NMR spectroscopy. *Journal of the Chemical Society, Perkin Transactions*, *2*, 2351–2357.

Pehk, T., Kiirend, E., Lippmaa, E., Ragnarsson, U., & Grehn, L. (1997). Determination of isotope effects on acid–base equilibria by ^{13}C NMR spectroscopy. *Journal of the Chemical Society, Perkin Transactions*, *2*, 445–450.

Peralta-Cruz, J., Díaz-Fernández, M., Avila-Castro, A., Ortegón-Reyna, D., & Ariza-Castolo, A. (2016). An experimental and theoretical study of intramolecular regioselective oxidations of 6-substituted 2,3-dimethylquinoxaline derivatives. *New Journal of Chemistry*, *40*, 5501–5515.

Perrin, C. L. (2010). Secondary equilibrium isotope effects on acidity. In J. P. Richard (Ed.), *Advances in Physical Organic Chemistry, Chapter 3: Vol. 44* (pp. 123–171).

Perrin, C. L., & Burke, K. D. (2015). Can a secondary isotope effect be larger than a primary? *Journal of Physical Chemistry A*, *119*, 5009–5018.

Perrin, C. L., & Dong, Y. (2007). Secondary deuterium isotope effects on the acidity of carboxylic acids and phenols. *Journal of the American Chemical Society, 129*, 4490–4497.
Perrin, C. L., & Dong, Y. (2008). Nonadditivity of secondary deuterium isotope effects on basicity of trimethylamine. *Journal of the American Chemical Society, 130*, 11143–11148.
Perrin, C. L., & Fabian, M. A. (1996). Multicomponent NMR titration for simultaneous measurement of relative pK_as. *Analytical Chemistry, 68*, 2127.
Perrin, C. L., Fabian, M. A., & Armstrong, K. B. (1994). Solvation effect on steric bulk of ionic substituents: Imidazolium vs imidazole. *Journal of Organic Chemistry, 59*, 5246–5253.
Perrin, C. L., Fabian, M. A., Brunckova, J., & Ohta, B. K. (1999). Absence of reverse anomeric effect in glycosylimidazoles. *Journal of the American Chemical Society, 121*, 6911–6918.
Perrin, C. L., Fabian, M. A., & Rivero, I. A. (1999). Basicities of cycloalkylamines: Baeyer strain theory revisited. *Tetrahedron, 55*, 5773–5780.
Perrin, C. L., & Flach, A. (2011). No contribution of an inductive effect to secondary deuterium isotope effects on acidity. *Angewandte Chemie International Edition, 50*, 7674–7676.
Perrin, C. L., & Karri, P. (2010). Position-specific secondary deuterium isotope effects on basicity of pyridine. *Journal of the American Chemical Society, 132*, 12145–12149.
Perrin, C. L., & Kuperman, J. (2003). Anomeric effects versus steric hindrance to ionic solvation in protonated glucosylanilines and cyclohexylanilines. *Journal of the American Chemical Society, 125*, 8846–8851.
Perrin, C. L., & Ohta, B. K. (2001). Symmetry of N-H-N hydrogen bonds in 1,8-bis(dimethylamino)naphthalene·H^+ and 2,7-dimethoxy-1,8-bis(dimethylamino)naphthalene·H^+. *Journal of the American Chemical Society, 123*, 6520.
Perrin, C. L., & Ohta, B. K. (2003). Symmetry of NHN hydrogen bonds in solution. *Journal of Molecular Structure, 644*, 1.
Perrin, C. L., Ohta, B. K., & Kuperman, J. (2003). β-Deuterium isotope effects on amine basicity, "inductive" and stereochemical. *Journal of the American Chemical Society, 125*, 15008–15009.
Perrin, C. L., Ohta, B. K., Kuperman, J., Liberman, J., & Erdélyi, M. (2005). Stereochemistry of β-deuterium isotope effects on amine basicity. *Journal of the American Chemical Society, 127*, 9641–9647.
Perrin, C. L., & Rivero, I. A. (1999). Apparatus for direct addition of reagents into a nuclear magnetic resonance (NMR) sample in the NMR probe. *Review of Scientific Instruments, 70*, 2173–2174.
Perrin, C. L., & Wang, W.-H. (1982). Nitrogen to nitrogen transfer. Significance of large negative entropies of activation. *Journal of the American Chemical Society, 109*, 2325–2326.
Peters, M., Siegfried, L., & Kaden, T. A. (1999). A fully automated pH–NMR titration set-up for protonation studies. *Journal of the Chemical Society, Dalton Transactions*, 1603–1607.
Rabenstein, D. L., Hari, S. P., & Kaerner, A. (1997). Determination of acid dissociation constants of peptide side-chain functional groups by two-dimensional NMR. *Analytical Chemistry, 69*, 4310–4316.
Rabenstein, D. L., & Mariappan, S. V. S. (1993). Determination of ^{15}N isotope effects on the acid-base equilibria of amino groups in amino acids by ^{13}C NMR. *Journal of Organic Chemistry, 58*, 4487–4489.
Romero, J., Reyes, A., & Wist, J. (2010). Secondary deuterium isotope effects on the acidity of glycine. *Spectrochimica Acta A, 77*, 845–848.
Rusu, A., Tóth, G., Szöcs, L., Kökösi, J., Kraszni, M., Gyéresi, A., et al. (2012). Triprotic site-specific acid–base equilibria and related properties of fluoroquinolone antibacterials. *Journal of Pharmaceutical & Biomedical Analysis, 66*, 50–57.

Shivapurkar, R., & Jeannerat, D. (2011). Determination of the relative pKa's of mixtures of organic acids using NMR titration experiments based on aliased ^{1}H–^{13}C HSQC spectra. *Analytical Methods*, *3*, 1316.
Shrager, R. I., Cohen, J. S., Heller, S. R., Sachs, D. H., & Schechter, A. N. (1972). Nuclear magnetic resonance titration curves of histidine ring protons. *Biochemistry*, *11*, 541.
Steffens, R. J., Baldridge, K. K., & Siegel, J. S. (2000). Synthesis and properties of 2,3-dihydro-1 h-corannuleno[2,3-*cd*]pyridine (2,3-dihydro-1 *h*-dibenzo[1,10:6,7] fluorantheno[3,4-*cd*]pyridine). derivatives: Heterocyclic *peri*-annellated corannulenes. *Helvetica Chimica Acta*, *83*, 2644–2654.
Sudmeier, J. L., Evelhoch, J. L., & Jonsson, N. B.-H. (1980). Dependence of NMR lineshape analysis upon chemical rates and mechanisms: Implications for enzyme histidine titrations. *Journal of Magnetic Resonance*, *40*, 377–390.
Szakács, Z., & Hägele, G. (2004). Accurate determination of low p*K* values by ^{1}H NMR titration. *Talanta*, *62*, 819–825.
Szakács, Z., Hägele, G., & Tyka, R. (2004). $^{1}H/^{31}P$ NMR pH indicator series to eliminate the glass electrode in NMR spectroscopic pK_a determinations. *Analytica Chimica Acta*, *522*, 247–258.
Szakács, Z., Kraszni, M., & Noszál, B. (2004). Determination of microscopic acid–base parameters from NMR–pH titrations. *Analytical and Bioanalytical Chemistry*, *378*, 1428–1448.
Tanaka, N., et al. (1989). Separation of nitrogen and oxygen isotopes by liquid chromatography. *Nature*, *341*, 727–728.
Taylor, R. R. R., Twin, H. C., Wen, W. W., Mallot, R. J., Lough, A. J., Gray-Owen, S. D., et al. (2010). Substituted 2,5-diazabicyclo[4.1.0]heptanes and their application as general piperazine surrogates: Synthesis and biological activity of a Ciprofloxacin analogue. *Tetrahedron*, *66*, 3370–3377.
Thatcher, G. R. J. (Ed.), (1993). The Anomeric Effect and Associated Stereoelectronic Effects. In *ACS symposium series*. Washington, DC: American Chemical Society.
Thibaudeau, C., & Chattopadhyaya, J. (1999). *Stereoelectronic effects in nucleosides and nucleotides and their structural implications*. Uppsala University Press.
Thordarson, P. (2011). Determining association constants from titration experiments in supramolecular chemistry. *Chemical Society Reviews*, *40*, 1305–1323.
Tresca, B. W., Brueckner, A. C., Haley, M. M., Cheong, P. H.-Y., & Johnson, D. W. (2017). Computational and experimental evidence of emergent equilibrium isotope effects in anion receptor complexes. *Journal of the American Chemical Society*, *139*, 3962–3965.
Tynkkynen, T., Tiainen, M., Soininen, P., & Laatikainen, R. (2009). From proton nuclear magnetic resonance spectra to pH. Assessment of ^{1}H NMR pH indicator compound set for deuterium oxide solutions. *Analytica Chimica Acta*, *648*, 105–112.
Valcour, A., & Woodworth, R. C. (1986). Internal proton magnetic resonance probes for pH titration of proteins. *Journal of Magnetic Resonance*, *66*, 536–541.
Van der Linde, W., & Robertson, R. E. (1964). K_D/K_H for base ionization of mono- and dimethylamine. *Journal of the American Chemical Society*, *86*, 4505.
Wilcox, C. S., Adrian, J. C., Jr., Webb, T. H., & Zawacki, F. J. (1992). Approaches to quantitative supramolecular chemistry. Hydrogen-bond-based molecular recognition phenomena and sigmoidal behavior in multicomponent mixtures. *Journal of the American Chemical Society*, *114*, 10189–10197.

CHAPTER THIRTEEN

Isotope Ratio Monitoring ^{13}C Nuclear Magnetic Resonance Spectrometry for the Analysis of Position-Specific Isotope Ratios

Richard J. Robins[1], Gérald S. Remaud, Serge Akoka
University of Nantes—CNRS UMR6230, Nantes, France
[1]Corresponding author: e-mail address: richard.robins@univ-nantes.fr

Contents

1. Introduction 370
2. Generic Aspects of the Methodology 371
   2.1 Comments About the Quality and Suitability of the Sample 371
   2.2 Comments About the NMR Spectrometer and "Absolute" vs "Relative" Values 373
   2.3 Comments About Sample Preparation for Optimal Spectral Acquisition 375
   2.4 Comments About Internal vs External Standards 376
   2.5 Comments About Line-Fitting and *f*/*F* to Delta 378
3. Examples of Methods 381
   3.1 Application of irm-^{2}H NMR to the Study of Enzymes Involved in the Biosynthesis of Long-Chain Fatty Acids 381
   3.2 Interconversion of Glucose and Fructose: The Exploitation of irm-^{13}C NMR to Explain Differences in the ^{13}C Positional Distribution in These Two Sugars 387
   3.3 Differentiating Between Different Pathways of Glucose Fermentation by Isotopic Fractionation Measured by irm-^{13}C NMR 390
   3.4 irm-^{13}C NMR to Probe the Mechanism of the Biotransformation of Ferulic Acid to Vanillin 393
4. Summary and Conclusions 397
References 398

Abstract

Studies of changes in isotope ratios in a given molecule caused by an enzyme reaction can give considerable insight into mechanism. Similarly, cascades of enzymes—pathways—can also introduce fractionation that can be used to characterize the type of metabolism being exploited. In both cases, classical studies have used enrichment techniques that have allowed the determination of isotope fractionation. More recent approaches, however, allow such studies to be carried out on substrates in which the

Methods in Enzymology, Volume 596
ISSN 0076-6879
http://dx.doi.org/10.1016/bs.mie.2017.07.023

isotope ratios are in the natural abundance range. This has the advantage of avoiding potentially demanding synthesis of specifically labeled compounds as well as allowing multipositional isotope ratio determinations. The downside can be that considerable quantities of analyte are required. In this chapter, we present the use of isotope ratio monitoring by NMR spectrometry as a means to access positional isotope ratios, potentially for all positions in the target molecule(s). It should be noted that, while the approach is generic, in that the general conditions for obtaining the required data follow the same protocol, no single protocol exists for all applications. The chapter is therefore split into two parts: general comments pertinent to the approach followed by a number of examples illustrating how different questions can be approached.

1. INTRODUCTION

The analysis of enzyme mechanism by the exploitation of isotope effects has been recognized as an efficient and informative approach for many years (Cook, 1991). The majority of studies have involved specifically labeled substrates in which heavy isotope (stable or radioactive) is introduced into one or more specific positions in the molecule and the change in the heavy/light ratio is monitored by a suitable analytical technique. This is, largely, mass spectrometry. However, it was always accepted that high enrichment in a heavy isotope could cause a modified response of the enzyme to the presence/absence of label, leading to aberrant values for the change in isotope ratio relative to those occurring at natural abundance. In the 1980s and 1990s, the approach of using heavy isotopes at natural abundance—mostly of ^{2}H or ^{13}C—was introduced, at least partly as a way of avoiding both this possibility and the need to synthesize specifically enriched substrates. The downside of this approach was the need to use a methodology that was sufficiently accurate to determine the changes in the isotope ratio at the permil level (‰) for ^{13}C. While isotope ratio monitoring by mass spectrometry (irm-MS) was exploitable, the disadvantage of this technique is that the effect on the isotope ratio at a given position is diluted by the isotope ratios at all the other positions in the compound. For compounds of more than 8–10 carbons, the effect is likely to be below the level of detection (Bouchard, Hunkeler, & Höhener, 2008). It also cannot indicate fractionation simultaneously at two or more positions, as can occur (see Section 3.4 for an example).

An alternative way of measuring the ^{2}H/^{1}H or ^{13}C/^{12}C ratios—that of isotope ratio monitoring by nuclear magnetic resonance spectrometry (irm-NMR)—was put forward in the early 1980s (Martin, Martin, Mabon, & Bricout, 1982; Martin, Sun, Guillou, & Martin, 1985) and applications to reaction mechanism (Singleton & Thomas, 1995) or metabolic causes of fractionation were subsequently developed (Martin, 1995; Martin, Zhang, Naulet, & Martin, 1986; Zhang, Yunianta, & Martin, 1995). This has the great advantage that, potentially, the isotope ratios at each resolved atom can be measured and, under the right conditions, the measurement can achieve a precision in the order of 1‰, required for ^{13}C. The disadvantage of irm-NMR is the relative insensitivity of the technique, leading to a need for a large amount of transformation to occur: in the order of hundreds of mg. Nevertheless, the application of this method has led to the measurement of isotope ratio modifications in relation to biotransformations in a number of cases.

In this chapter, we present examples of this approach, which illustrate strategies to be adopted to monitor isotope ratios related to enzymatic modifications. It must be borne in mind that the methodology needs to be adapted to each individual case study. The approach of irm-NMR is generic, but the details need adapting to the extant circumstances.

2. GENERIC ASPECTS OF THE METHODOLOGY

The application of irm-NMR has a number of features in common, which are to some extent independent of the nucleus being studied, and which will be described before embarking on a series of applications.

2.1 Comments About the Quality and Suitability of the Sample

2.1.1 Purity

Paramount to this technique is to know the sample. Pure compound is desirable, preferably at a purity >98%–99%. In many cases this can be achieved by the use of commercial analyte. In others, considerable effort may be required to attain the necessary purity.

In some circumstances, complete purity is not required: this depends on the question being asked. For example, in an early application we wanted to examine by irm-^{2}H NMR the molecule capsaicin, the hot pungent principle of *Capsicum frutescens*. Capsaicin is an amide between a unit of vanillylamine and a C10 branched-chain carboxylic acid and occurs naturally with a considerable proportion of 6,7-dihydrocapsaicin. In the commercial sample studied, these compounds were in an approximate 70:30 ratio. When this mixture was studied by irm-^{2}H NMR, a degree of overlap in the signals was found, as anticipated (Markai et al., 2002). The distribution of the $(^{2}H/^{1}H)_i$ values obtained allowed certain observations to be made on their biosynthesis, including the biosynthetic relationship between 6,7-dihydrocapsaicin and capsaicin (Fig. 1). The $(^2H/^1H)_7$ values for the two molecules show closely similar values, whereas the $(^2H/^1H)_6$ values differ by 50 ppm. This can be interpreted as an indication of an isotope effect in ^2H comparable to that observed for the fatty acid desaturases. Hence, indications as to the potential mechanism of this as-yet-undescribed enzyme can be indirectly obtained by irm-^2H NMR analysis even though the sample is mixed and there is some line overlap (Markai et al., 2002).

2.1.2 Line Overlap

In other circumstances, line overlap can be highly detrimental, leading to inaccurate measurement of the surfaces under the peak. This is especially

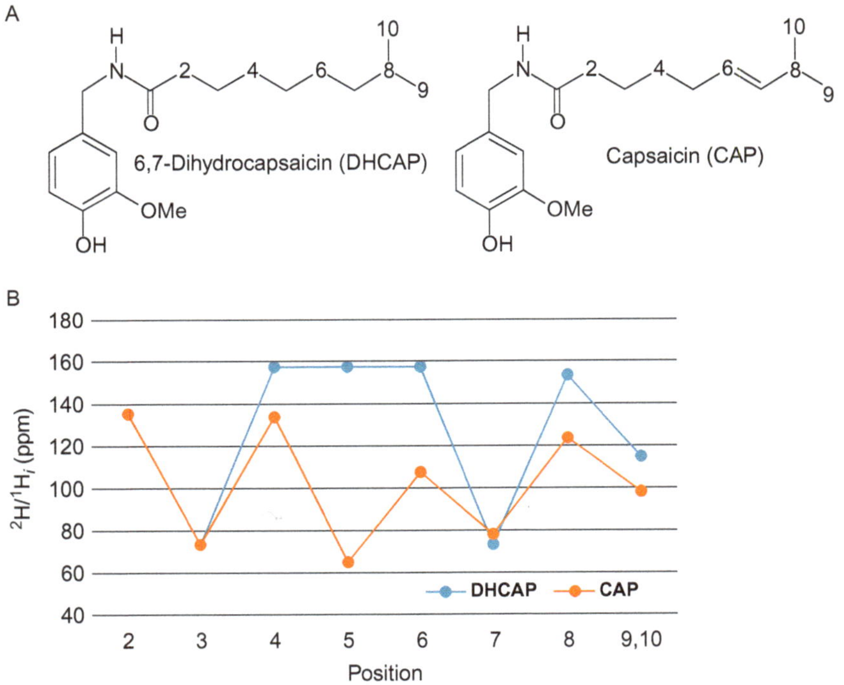

Fig. 1 Analysis of the distribution of ^2H in natural capsaicin and 6,7-dihydrocapsaicin from *Capsicum frutescens* extract. (A) Structures of the compounds. (B) The ^2H_i values for the side-chain moieties of 6,7-dihydrocapsaicin (DHCAP) and capsaicin (CAP). Note that the values shown for DHCAP for C4, C5, C6 are the mean for these positions which resonate at the same frequency in the ^2H NMR spectrum. *Figure adapted from Markai, S., Marchand, P. A., Mabon, F., Baguet, E., Billault, I., & Robins, R. J. (2002). Natural deuterium distribution in branched-chain medium-length fatty acids is nonstatistical: A site-specific study by quantitative ^2H NMR spectroscopy of the fatty acids of capsaicinoids.* Chembiochem: A European Journal of Chemical Biology, 3*(2–3), 212–218.*

the situation for the determination of $\delta^{13}C$ value, when even 1% or 2% of impurity can cause a change in the $\delta^{13}C$ value by 10s of ‰. This is principally because the impurity only affects the area under one of the peaks. As the method of calculating the distribution of heavy isotope between the peaks depends on the molar ratio (see Section 2.5.1), that is to say the extent to which each peak deviates from the statistical distribution of isotope (i.e., with zero relative fractionation) a relative change in one peak will impact on the apparent fractionation obtained for the others. This has been observed where L-methionine is contaminated with a few % of sulfoxide. The resonance position of the *S*-methyl is slightly shifted, but not enough for full resolution, causing broadening of the base of the peak and a larger area to be determined than it should be. This results in a false $\delta^{13}C_i$ value for the *S*-methyl group and distorted values for all the other positions (Fig. 2).

2.2 Comments About the NMR Spectrometer and "Absolute" vs "Relative" Values

While irm-^{2}H NMR does not show spectrometer dependencies, irm-^{13}C NMR may have an instrument response that depends on (i) the probe used, (ii) the type of experiment, and (iii) the aim of the study itself. Because of the requirement for high precision (better than 1‰), it has been shown during a

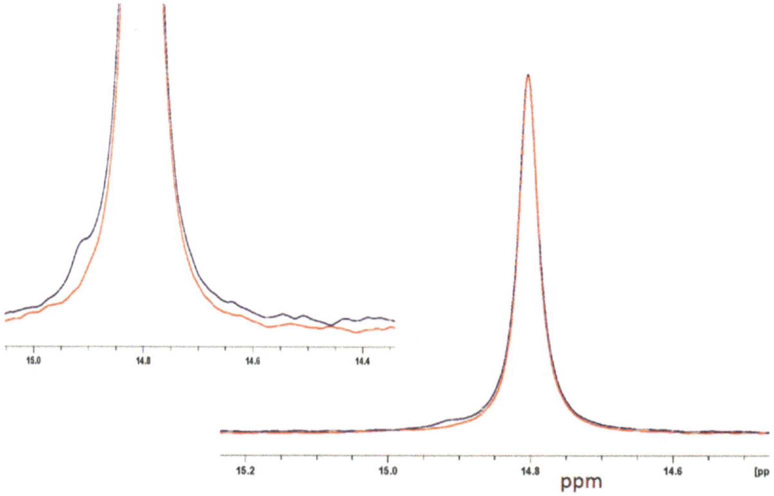

Fig. 2 Extract from the irm-^{13}C NMR spectra of two samples of L-methionine showing the peak for the *S*-methyl group of methionine without (*red line*) and with (*blue line*) some methionine-*S*-oxide present. Inset (*left*) shows a magnification on the base of the peak, illustrating the shoulder due to the oxide.

preliminary ring test that the standard deviation can reach 2‰ for the intervariability between instruments while the intravariability for each instrument was better than 1‰ when working with vanillin as the molecular probe (Chaintreau et al., 2013). This effect is increased when the range of the ^{13}C chemical shifts of the studied molecule is large (as it is for vanillin). The architecture of the probe, i.e., the respective size of the ^{13}C and ^{1}H coils, has been proposed as the explanation of the phenomenon (Bayle et al., 2014). The instrumental effect is much more pronounced when cryoprobes are used for which the baseline correction is required, a potential source of bias. In order to present true ^{13}C compositions for each isotopomer on the absolute scale of the $\delta^{13}C$ from the international reference V-PDB, corrections are applied once the response factor for the NMR spectrometer is determined. A specific instrumental qualification approach aims to determine first the repeatability within 1‰, reachable by the appropriate ^{1}H decoupling scheme (Tenailleau & Akoka, 2007) and, second, the bias from the true $\delta^{13}C_i$ values. The qualification protocol is based on the use of a bilabeled molecule containing two carbons (ethanol). In this, the areas of both doublets from the isotopologue constituted of two ^{13}C atoms should be strictly identical, independent of the level of global enrichment. Thus, any deviation from that theoretical value is indicative of a bias that can be quantified and used for the correction factor (Caytan, Remaud, Tenailleau, & Akoka, 2007). Furthermore, comparisons have been done between different techniques for position-specific isotope analysis. Consensual values for acetic acid and ethanol, for example, were found allowing both the identification of instruments giving the "true" values and the correction factors to be applied (Bayle et al., 2014).

Unlike ^{2}H, ^{13}C is a nucleus that is favorable for a series of NMR experimental techniques such as magnetization polarization transfer (known as the INEPT pulse sequence) and 2D NMR (Jézéquel, Joubert, Giraudeau, Remaud, & Akoka, 2017). These techniques are very efficient for increasing sensitivity of detection and/or improving the resolution of the peaks. It has been shown that such methodologies can reach the required precision of 1‰ while also improving the sensitivity. But, by essence, they cannot lead directly to true $\delta^{13}C$ values (Martineau, Giraudeau, Tea, & Akoka, 2011; Thibaudeau, Remaud, Silvestre, & Akoka, 2010). These methodologies are very promising for building up an isotope profile within a short analysis duration or with less material. Complete or partial profiles depending on the ^{13}C composition may be used as classification criteria, as done for any -omics study (Bussy et al., 2011; Remaud et al., 2013).

It is not always necessary to work with "absolute" true $\delta^{13}C$ values, unless comparisons between different molecules and/or different techniques have to be done. While in many cases the ^{13}C affiliation is clearly better explained by using true values, in others this level of measurement is not needed. "Relative" values may be sufficient when working on the same molecule during a transformation under investigation, for example. Hence, these more sensitive methodologies might be valuable for studies of enzymes, wherein only the change in $\delta^{13}C_i$ ($\Delta\delta^{13}C_i$) values is required, not the true value. If necessary, true values are obtained by applying appropriate correcting factors after suitable calibration.

By exploiting relative values, isotope effects or isotope enrichments have been calculated for several fractionation types (Botosoa et al., 2008; Botosoa, Silvestre, et al., 2009; Julien et al., 2015). This type of work can be achieved on any NMR spectrometer once the repeatability has been checked.

2.3 Comments About Sample Preparation for Optimal Spectral Acquisition

In the case of ^{13}C, a precision of about 1‰ is necessary for isotopic analysis because of the narrow range of specific isotopic compositions (less than 50‰ compared with 500‰ for ^{2}H). This could not be reached until an efficient and homogenous ^{1}H decoupling was set up: the optimum, so far, is found when using adiabatic pulse sequences (Tenailleau & Akoka, 2007). But, in addition to this first NMR parameter to be fixed, others have to be kept at their optimum: signal-to-noise ratio (SNR) and shape of the peaks. For a relative error on the signal area of 1‰, SNR should be of the order of 500, i.e., long duration experiments need to be conducted. In order to work with an acceptable analysis time, this means that relatively high concentrations should be used for the sample. For solid compounds, an appropriate solvent or solvent mixture has to be selected in order to dissolve a maximum mass of the molecule in the minimum mass of deuterated solvent. Criteria for this selection are the stability of the final solution (i.e., no interactions between solvent and analyte) and no overlapping of the solvent signals with those of the studied molecule. Any combination could be satisfactory. However, it is also advantageous to take advantage of the shielding or deshielding properties of the solvent, allowing a better resolution between peaks, hence a better curve fitting (see Section 2.5). Optimizing sample preparation can therefore be quite complex. Once established, the recipe should be strictly adhered to in order to ensure the repeatability of the measurements.

It should also be noted that the choice of the preparation has a direct influence on the quantitative response of the NMR through the variation of the longitudinal relaxation time T_1. Between each pulse a delay of $10 \times T_1^{max}$ should be applied to ensure that the spins have adequate time to relax and then to accumulate the signal due to the subsequent pulse without partial saturation or nOe. Intrinsically, the T_1 values of ^{13}C in a relatively small molecule (# of C < 10) are large in nonviscous media. In that case, the duration time can be excessively long. It has been shown that the use of a relaxation reagent, such as tris(2,4-pentanedionato)-chromium(III) ($Cr(Acac)_3$), makes it possible to decrease T_1 without affecting the level of repeatability (Caytan, Remaud, et al., 2007). The amount should be monitored according to the actual values of T_1 found. A compromise should be established between a very short T_1 (smaller than 500 ms) and longer ones. As a result, the preparation of the solution to put in the NMR tube should include the measurement of the T_1 values under the intended acquisition conditions, followed by an adjustment of these as required. The solvent initially chosen can potentially be satisfactory for reaching high solubility (better SNR) but catastrophic for the resulting T_1 values! High concentration in viscous solvents, such as DMSO, can lead to very efficient relaxation associated to short T_1 values ($<$500 ms). As mentioned earlier a new preparation should be envisaged. Therefore, it must be born in mind that, for a given molecule, time must be dedicated to the sample preparation for a successful irm-^{13}C NMR to be obtained.

2.4 Comments About Internal vs External Standards

In order to obtain absolute and, preferably, true values of the isotopic ratio at a given position ($^2H/^1H_i$ or $\delta^{13}C_i$), it is necessary to make reference to a known calibrated standard. As with many analytical methods, this may be internal or external.

2.4.1 Standardization for the Measurement of Positional ^2H_i

As detailed later (Section 2.5.2), the routine methodology for the determination of the positional ^2H_i (ppm) exploits the use of an internal standard. For the calculation, the molecular mass, mass added to the sample, and purity of the reference need to be known with greater precision than the target error in the spectral data. In addition, a measurement of the isotope ratio of the reference compound needs to be made by an independent means and with an accuracy greater than that of the NMR method being used.

It is advantageous that the standard carries a number of equivalent atoms of the element in consideration so that the concentration at which it is added to the sample can be kept to a minimum.

In the earlier irm-^{2}H NMR analyses, a reference calibrated by irm-MS was used, inserted as a capillary within the NMR tube (Martin et al., 1982). Subsequently, initially for the application to ethanol but then to a wider range of compounds *N,N,N′,N′*-tetramethylurea (TMU) was identified as a suitable reference that could be calibrated to the V-SMOW scale by mass spectrometry (Gonfiantini, Stichler, & Rozanski, 1995). In cases where the peaks from the analyte interfere with TMU, benzene or pyridine (Billault, Duan, Guiet, & Robins, 2005) or dioxane (Gonzalez et al., 1998), calibrated vs TMU, can be used.

2.4.2 Standardization for the Measurement of Positional $^{13}C_i$

The above protocol could, in theory, be used in the same way for the calculation of the $\delta^{13}C_i$ (‰) values obtained by irm-^{13}C NMR (see later). However, the constraints on irm-^{13}C NMR (see comments earlier) make this difficult due to the high precision required in measuring the mass of reference material present. To avoid this difficulty, a strategy has been developed in which the bulk ^{13}C ($\delta^{13}C_b$, ‰) of the analyte is determined by irm-MS, the instrument being calibrated against suitable international standards (obtained from the IAEA, Vienna, Austria or the NIST, Gaithersburg, USA). The relative nonstatistical distribution of ^{13}C between the different positions is then calculated and used, combined with the $\delta^{13}C_b$ to obtain $\delta^{13}C_i$ values (see Section 2.5.3).

This approach has two major drawbacks. First, it requires that the analyte is pure, as the measurement of $\delta^{13}C_b$ is done on the entire sample. Second, it requires access to both irm-MS and irm-NMR.

In a recent development, a method to circumvent both these difficulties by using an internal standard has been proposed. In this, the concentration of calibrated reference compound is obtained from a ^{1}H NMR spectrum measured on the same sample (Bayle et al., 2015). By this means, the sample preparation protocol is simplified as it is not necessary to know precisely the purity and/or the mass of the measure compound or of the reference material. The relative concentrations of analyte and standard are obtained from the ^{1}H spectrum. This methodology has been validated for vanillin with dimethylsulfone as an internal standard. However, it has not yet been exploited for enzyme studies.

2.5 Comments About Line-Fitting and *f/F* to Delta

Once a satisfactory spectrum is obtained, the pertinent data to calculate the ^{13}C positional values need to be extracted. This requires a high-precision determination of the areas under the peaks, preferably using a software designed specifically for this purpose.

2.5.1 Protocol: Line-Fitting

Curve fitting based on a total-line-shape analysis (deconvolution) is carried out with a mathematical model. In this procedure, line-shape parameters are optimized in terms of intensities, frequencies, line-width, line-shape (Gaussian/Lorentzian, asymmetry), phase, and baseline, by iterative fitting to a minimal residue.[a]

Imperfect shimming of the NMR spectrometer can induce distortions of the line-shape in the acquired spectrum. The contribution of the field inhomogeneity can be taken into account by introducing a Gaussian contribution and asymmetry. The percentage of Gaussian depends on the ratio between the natural line-width and the broadening induced by inhomogeneity. This parameter must therefore be independently optimized for each peak, when a high precision is required. In our experience, errors greater than 1% (10‰) can be observed for moderate (5%) maladjustment of the Gaussian percentage. This is evidently too high when a precision of 1‰ or 10‰ is required for ^{13}C or ^{2}H, respectively.

A key issue in spectral treatment is the adjustment of the phase. A default of only 4 degree (which is not detectable by the human eye) induces an error of several ‰ in the peak area. Correct adjustment of this parameter is therefore mandatory in the case of irm-^{13}C NMR, since it is not possible to phase a complete ^{13}C spectrum using zero and first-order corrections only. However, in the case of irm-^{2}H NMR, a precision of the order of 10‰ only is targeted so that the phase of a peak can be ignored as it is supposed that the processed spectra are perfectly phased.

The baseline correction has also to be performed carefully. Our experience shows that it cannot be correctly described by a polynomial function when a significant frequency range (several 10s of ppm) is considered (irm-^{13}C NMR). A local correction is therefore needed.

All line-fitting should be performed by the same operator.

[a] To our knowledge, only the TLS module of the PERCH software (PERCH™ NMR Software, http://www.perchsolutions.com) satisfies all these conditions.

2.5.2 Protocol: Spectral Data Processing for irm-^{2}H NMR

This protocol allows the calculation of the $^2H/^1H$ (ppm) values by using the surface areas under the peak obtained by irm-^{2}H NMR with the use of a calibrated internal reference material.

1. Calculate the area under the peak for the ^{2}H signal for H-atom in position i, S_i, using line-fitting (see Section 2.5.1).
2. Use the formula to calculate the $^2H/^1H_i$ ratios in the analyte:

$$\left(\frac{^2H}{^1H}\right)_i = \left(\frac{^2H}{^1H}\right)_{ref} \cdot \frac{S_i}{S_{ref}} \cdot \frac{n_{ref}}{n_i} \cdot \frac{m_{ref}}{m_S} \cdot \frac{p_{ref}}{p_S} \cdot \frac{M_S}{M_{ref}}$$

where S_i and S_{ref} are the surface areas of the signal; n_i and n_{ref} are the stoichiometric numbers of hydrogens in site i and in the reference; M_S, m_S, p_S and M_{ref}, m_{ref}, p_{ref} are, respectively, the molecular weight, mass, and purity of the sample and the reference used. In some cases, a further term, F_S, the correction factor for the content of any impurity in the sample, needs to be included.

The internal reference is frequently TMU for which the isotopic ratio, $(^2H/^1H)_{ref}$, is calibrated on the V-SMOW scale (Gonfiantini et al., 1995). Alternatively, pyridine calibrated against TMU may be used (as in the studies of long-chain fatty acids).

2.5.3 Protocol: Spectral Data Processing for irm-^{13}C NMR

This protocol allows the calculation of the $\delta^{13}C_i$ (‰) values by using the surface areas under the peak obtained by irm-^{13}C NMR combined with the $\delta^{13}C_g$ (‰) value obtained by irm-MS.

1. Calculate the area under the peak for the ^{13}C signal for C-atom in position i, S_i, using line-fitting (see Section 2.5.1).
2. Correct each S_i to compensate for the slight loss of intensity caused by satellites (^{13}C—^{13}C scalar coupling interactions) by multiplying by $(1 + n \times 0.011)$, where n is the number of carbon atoms directly attached to the C-atom position i and 1.1% (=0.011) is the average natural ^{13}C abundance.
3. Calculate S_{tot}, the sum of the areas of all ^{13}C signals from the molecule:

$$S_{tot} = \sum_n S_i$$

4. Calculate the ^{13}C mole fraction f_i, which is the area of the peak corresponding to the carbon position i divided by the sum of all the carbon sites of the molecule:

$$f_i = \frac{S_i}{\sum_n S_i}$$

5. Calculate F_i which is the statistical mole fraction for a carbon site i, in the theoretical situation where there is a homogeneous ^{13}C distribution within the molecule:

$$F_i = \frac{n_c}{N_c}$$

where n_c is the number of carbon equivalents in the molecule resonating at a given frequency and N_c is the total number of carbon atoms in the molecule (for ethanol $F_i = 1/2$ for both the methyl and the methylene positions).

6. Calculate the site-specific reduced molar fraction, which is defined as:

$$\frac{f_i}{F_i}$$

From the reduced molar fraction $\delta^{13}C_i$, the specific isotope composition of the carbon position i is obtained from the isotope composition of the whole molecule ($\delta^{13}C_g$) measured by irm-MS as follows:

A_i (%), the isotopic abundance for carbon position i, is obtained from:

$$A_i = \frac{f_i}{F_i \times A_b}$$

where A_b is the isotopic abundance of a whole molecule and is obtained from:

$$A_b = \frac{R_b}{(R_b + 1)}$$

with $R_b = (\delta^{13}C_b/1000 + 1) \times R_{V\text{-}PDB}$
where $R_{V\text{-}PDB} = 0.0112372$ (Werner & Brand, 2001).

Then:

$$R_i = \frac{A_i}{(1 - A_i)}$$

hence

$$\delta^{13}Ci(\text{‰}) = \left(\frac{R_i}{R_{V-PDB}} - 1\right) \times 1000$$

3. EXAMPLES OF METHODS

3.1 Application of irm-^{2}H NMR to the Study of Enzymes Involved in the Biosynthesis of Long-Chain Fatty Acids

3.1.1 Study of the Mechanism Involved in the Hydroxylation of Long-Chain Fatty Acids

An early example of the use of irm-^{2}H NMR to seek putative differences in the mechanism of a reaction due to the catalysis being realized by a different class of enzyme is that of the biosynthesis of ricinoleic acid [12-hydroxyoctadec-9*Z*-enoic acid]. This C18 fatty acid, which is hydroxylated at the C12 position, is a predominant component of the fatty acid profile of castor bean (*Ricinus communis*) oil. However, this compound can also be accumulated to high levels by other organisms: a case in point is the ergot fungus *Claviceps purpurea*. Different substrate/product relationships have been proposed for these two organisms. In *R. communis*, it is proposed that synthesis involves the direct hydroxyl substitution of oleate (Fig. 3), whereas in *C. purpurea* the direct addition of water to the $\Delta^{12,13}$ double bond of linoleate has been put forward (see Billault, Mantle, & Robins, 2004 for details). In neither case has a mechanism been formally established. However, it can be predicted that if the two organisms exploit the same enzymatic machinery, then the associated isotope fraction should be the same. By exploiting irm-^{2}H NMR it was shown that the distribution of ^{2}H in ricinoleic acid obtained from the two sources is very similar, indicating that the mechanism of hydroxylation is probably comparable (Billault et al., 2004). A key observation in terms of mechanistic comparisons is that the $^2H_{13}$ of ricinoleate from *C. purpurea* and from *R. communis* shows the same relationship to other sites in the molecule but not to that of linoleate, in which the $^2H_{13}$ is notably impoverished (Fig. 3). If this latter were the precursor, then this position should show equivalent impoverishment. Hence, it can be deduced that the biosynthetic precursor is oleic acid.

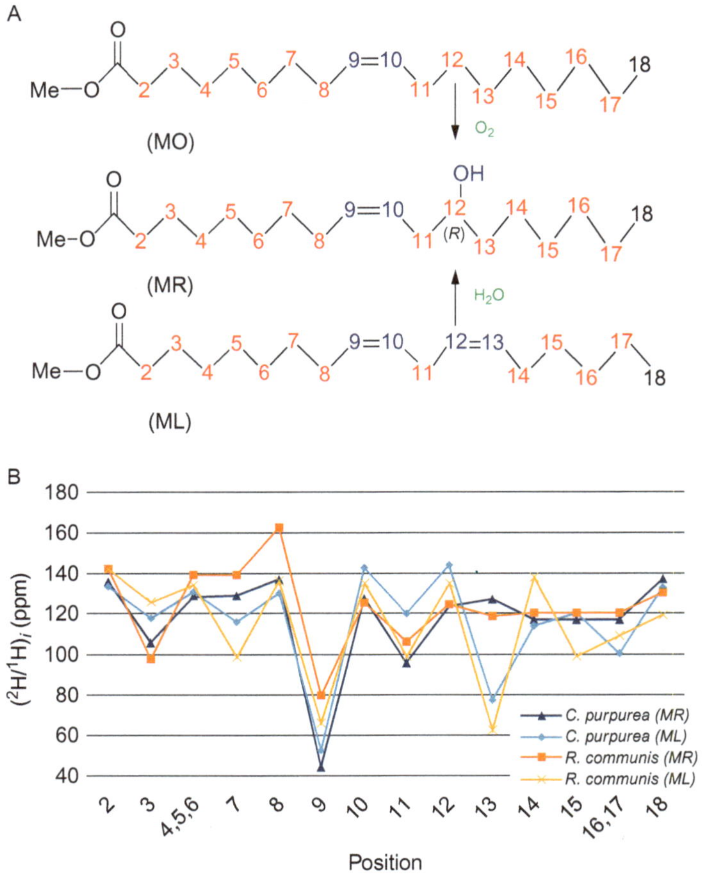

Fig. 3 The distribution of $^2H/^1H_i$ (ppm) in ricinoleic acid (methyl ester) and its proposed precursor, linoleic acid (methyl ester). (A) Structures of methyl esters of ricinoleic acid (MR), oleic acid (MO), and linoleic acid (ML) indicating their atomic numbering. (B) Comparison of the $^2H/^1H_i$ along the chain for MR and ML obtained from *R. communis* and *C. purpurea. Figure adapted from Billault, I., Mantle, P. G., & Robins, R. J. (2004). Deuterium NMR used to indicate a common mechanism for the biosynthesis of ricinoleic acid by* Ricinus communis *and* Claviceps purpurea. Journal of the American Chemical Society, 126*(10), 3250–3256.*

3.1.1.1 Protocol

1. Transesterification to obtain fatty acid methyl esters (FAMEs). Mix seed oil (10.0 g) with MeOH (150 mL) and add NaOH (3 g). Heat with stirring at reflux for 1 h. Dilute the mixture with MeOH (100 mL) and add a solution of boron trifluoride in MeOH (33 mL; 50% w/w). Heat the mixture for an additional 45 min, then cool. At ambient temperature,

add hexane (300 mL) and agitate for 30 min. Then, add a saturated NaCl aqueous solution, separate the phases, and extract the aqueous phase with $CHCl_3$. Combine the organic phases, wash with water, dry ($MgSO_4$), filter, and evaporated to get 9.2–10.3 g of mixed FAMEs.

2. Separation of methyl ricinoleate from the nonpolar FAMEs mixture by flash chromatography. Apply the FAMEs mixture to a silica gel column ($40 \times 5.0\ cm^2$) and elute this with a solvent gradient: petroleum ether (40°–70°)/Et_2O: (95/5–50/50) to obtain ricinoleic acid (8.6 g).
3. irm-^2H NMR spectroscopy: Bruker DRX 500 spectrometer operating at 76.7 MHz fitted with a ^{19}F field-frequency-locking device and a dual probe ($^1H/^2H$) 10 mm in diameter.
   Internal reference material: pyridine with the $^2H/^1H$ isotopic ratio calibrated relative to TMU on the V-SMOW scale.
   Sample preparation: methyl ricinoleate, 0.90 g; pyridine, 0.19 g; $CHCl_3$, 0.87 g; CCl_4, 2.4 g; and C_6F_6, 60 μL.
   Acquisition conditions: temperature, 310 K; flip angle, 90 degree; sampling period (AQ), 5.5 s; observed spectral range (SW), 1200 Hz; repetition time (T_R) fivefold greater than the longest relaxation time (T_1), 5.7 s; number of scans (NS), 12,000. An exponential filtering (LB) of 1 Hz is applied and FIDs are zero-filled to 32 k before Fourier transformation.
4. Spectral processing: protocol 2.5.2.

3.1.1.2 Notes

1. In this example, a greater amount of data was obtained by chemically degrading the methyl ricinoleate to yield 9,9′-bis(methoxy)nonanoic acid methyl ester and (*R*)-1,1′-bis(methoxy)-3-acetylnonane. This allowed the measurement of the $^2H_9$ and $^2H_{10}$ positions separately. This refinement of the protocol is specific to the application and, if required, details are described in Billault et al. (2004).

3.1.2 Study of the Mechanism Involved in Epoxidation of Long-Chain Fatty Acids

In a number of species, such as *Vernonia galamensis* and *Euphorbia lagascae*, the seed oil contains a long-chain fatty acids with an epoxide group: vernolic acid [12(*S*),13(*R*)-12,13-epoxyoctadec-9*Z*-enoic acid]. As with the case of ricinoleic acid, two possible mechanisms for the introduction of the epoxide have been put forward: either hydroxylation of oleic acid, followed by oxidation, or the addition of the oxygen to linoleic acid (Fig. 4). In a parallel

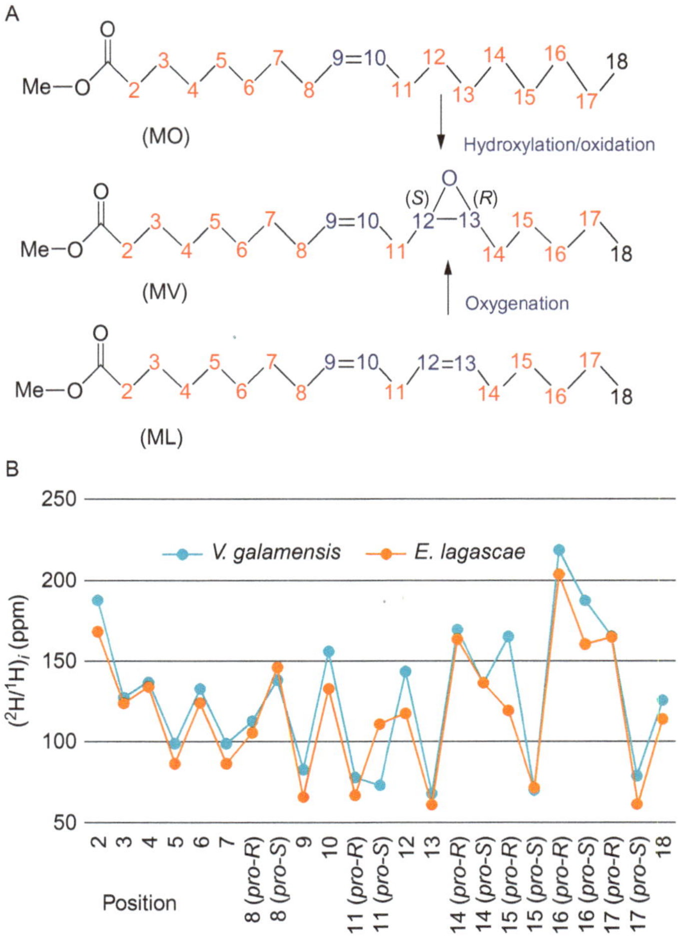

Fig. 4 Analysis of the $^2H/^1H_i$ (ppm) distribution in vernolic acid methyl ester and its proposed precursors. (A) Structures of methyl esters of oleic acid (MO), vernolic acid (MV), and linoleic acid (ML) showing their atomic numbering. (B) Comparison of the $^2H/^1H_i$ along the chain for MV isolated from *V. galamensis* and *E. lagascae*. *Figure adapted from Billault, I., Ledru, A., Ouetrani, M., Serhan, Z., Lesot, P., & Robins, R. J. (2012). Probing substrate-product relationships by natural abundance deuterium 2D NMR spectroscopy in liquid-crystalline solvents: the case of the epoxidation of linoleate to vernoleate by two different plant enzymes.* Analytical and Bioanalytical Chemistry, 402, *2985–2998.*

study to that described for ricinoleic acid, a comparison of the position-specific ^2H_i values in methyl linoleate and methyl vernoleate isolated from *V. galamensis* oil (Billault et al., 2005) indicated that it was more probable that linoleic acid is the immediate precursor of vernolic acid (Fig. 4). However, within the context of the study of the relationship between isotope fractionation and enzyme mechanism is that the two plants *V. galamensis* and *E. lagascae* have been shown to carry out this Δ^{12}-epoxidation exploiting a different enzyme: a diiron dioxygenase in *V. galamensis* and a CYP450 epoxidase in *E. lagascae*. The working hypothesis, then, is that, although both reactions involve coordinated iron in the reaction center, the differences in environment might lead to different ^2H_i fractionation patterns related to isotopic fractionation at the carbon positions involved in the reaction. The vernolic acid isolated from each of the two plant species was subjected to analysis by irm-^{2}H 2D NMR spectroscopy in a liquid-crystalline solvent (Billault et al., 2012). This methodology makes possible the resolution of the *pro-R* and *pro-S* hydrogen positions in the "substrate" and "product" molecules, linoleic and vernolic acids, respectively (Remaud, Giraudeau, Lesot, & Akoka, 2016). The observed profiles of ^2H_i in the vernolic acid from the two sources differ minimally (Fig. 4) suggesting that overall there is no major difference in the reaction mechanism. From this, it was inferred that the role of Fe as electron captor/supplier was more dominant in the mechanism than was the environment of the enzyme active site.

3.1.2.1 Protocol

1. Transesterification to obtain FAMEs. A mixture of the FAMEs is obtained from seed oils by transesterification (Billault et al., 2005). To a solution of *Vernonia* oil (25 g) in hexane (100 mL) add a fresh solution of sodium methoxide (12 mL, 25% w/w) in MeOH. Keep the mixture at ambient temperature for a further 5 h then treat with an aqueous saturated NaCl solution. Recover the organic phase and wash with water. Extract the aqueous phase with hexane (2 × 250 mL) then $CHCl_3$ (1 × 50 mL). Combine the organic phases, dry ($MgSO_4$), filter, then concentrate to give 23.2 g of mixed FAMEs. Check the purity of the mixture by GC and ^{1}H NMR.
2. Separation of methyl vernoleate from the nonpolar FAMEs mixture by flash chromatography. Mixed FAMEs (19.8 g) are applied to a column (30 cm × 8 cm) of silica gel 60 (750 mL). The column is eluted with a solvent gradient as follows: 1.0 L petroleum ether/Et_2O (100:1); 1.0 L petroleum ether/Et_2O (100:2); 1.0 L petroleum ether/Et_2O (100:3); 1.0 L petroleum ether/Et_2O (100:4); 2.0 L petroleum ether/Et_2O (100:10). Fractions of 60 mL volume are collected and the presence of FAMEs checked by GC and TLC. Fractions containing methyl vernoleate are pooled and solvent removed to give 14.8 g of methyl vernoleate.

3. irm-^{2}H NMR spectroscopy: ^{2}H-{^{1}H} NMR spectra are recorded on a Bruker Avance II 600 MHz NMR spectrometer equipped with a 5 mm selective ^{2}H cryogenic probe (92.1 MHz) operating without a ^{19}F lock device.

   Acquisition conditions: Decoupling, WALTZ-16 sequence; temperature, 310 K for the liquid solvents (a) and 300 K for the oriented media (b).

Sample preparation

- **(a)** Liquid solvents: Methyl linoleate (150 mg) or methyl vernoleate (150 mg), pyridine (0.07 g), and solvent $CHCl_3/CCl_4$ (1:4.5; 2.1 g) are mixed then filtered into a 5 mm NMR tube.
- **(b)** Liquid-crystalline solvents: Methyl linoleate (100 ± 1 mg) is dissolved in chiral mesophase poly-γ-benzyl-L-glutamate (DP: 767, 130 ± 1 mg) and dry pyridine (350 ± 1 mg). Methyl vernoleate (100 ± 1 mg) is dissolved in achiral mesophase poly-γ-benzyl-DL-glutamate (50 ± 1 mg) of L-enantiomer (DP: 767) and 50 ± 1 mg of D-enantiomer (DP: 914) and dry $CHCl_3$ (650 ± 1 mg). The sample is centrifuged and rehomogenized several times to remove matter concentration gradients. Then the tube is fire-sealed to prevent solvent evaporation.

Acquisition conditions

- **(a)** Liquid solvents: temperature, 310 K; T_R, flip angle, 90 degree; sampling period (AQ), 3.5 s; T_R fivefold greater than the longest relaxation time (T_1), 5.7 s; NS, 3200. An exponential filtering (LB) of 1.5 Hz is applied.
- **(b)** Liquid-crystalline solvents: sequence, Q-COSY Fz NMR sequence; 96 free induction decays for each t_1 increment; a 2D matrix of 1530 (t_2) × 512 (t_1) data points; spectral width, 1700 Hz in both dimensions; recycling time for each scan, 1 s; total acquisition time, ca. 15 h. An exponential filtering (LB) of 1.5–2 Hz is applied in both spectral dimensions. The 2D experiments are zero-filled to 2048 (t_1) × 2048 (t_2) data points prior to 2D Fourier transformation, then maps are symmetrized and tilted.

Three replicates are run for each sample.

4. Spectral processing: Use protocol 2.5.2 to obtain the $(^{2}H/^{1}H)_i$ ratios. The $(^{2}H/^{1}H)_i$ at all the CH_2 and some resolved *pro-R*/*pro-S* positions are calculated using:

$$\left(\frac{^{2}\mathrm{H}}{^{1}\mathrm{H}}\right)_{i}^{\mathrm{aniso}} = \left[\frac{(\%\mathrm{area})}{100} \times n \times \left(\frac{^{2}\mathrm{H}}{^{1}\mathrm{H}}\right)_{i}^{\mathrm{iso}}\right]$$

where n is the number of ^{2}H positions with a coincident chemical shift contributing to the ^{2}H NMR signal recorded in liquid medium and the value of $(^{2}\mathrm{H}/^{1}\mathrm{H})_{i}^{\mathrm{iso}}$ is determined from NMR experiments in liquid medium. By combining the curve-fitted data from both sets of NMR spectra—the area percentage of quadrupolar doublets detected at the same chemical shift on the 2D map of the Q-COSY Fz 2D NMR experiment and the $(^{2}\mathrm{H}/^{1}\mathrm{H})_{i}$ ratios determined by isotropic NMR—the individual $(^{2}\mathrm{H}/^{1}\mathrm{H})_{i}$ ratios are obtained. For further details see Lesot, Aroulanda, and Billault (2004) and Lesot, Baillif, and Billault (2008).

3.1.2.2 Notes

1. This is a sophisticated NMR methodology and should be best carried out in association with a group in which it is regularly practiced.

3.2 Interconversion of Glucose and Fructose: The Exploitation of irm-^{13}C NMR to Explain Differences in the ^{13}C Positional Distribution in These Two Sugars

Glucose and fructose are interconverted as well as being combined to make sucrose. When the $^{13}C_{i}$ isotope profiles of these sugars were compared, it was clear that isotopic fractionation was taking place at the enzyme level. Of particular interest was that the fructosyl moiety of sucrose had ^{13}C-isotope ratios at C1 and C2 inverse to those in the glucosyl moiety (Fig. 5A) (Gilbert, Silvestre, Robins, Remaud, & Tcherkez, 2012). In order better to understand this fractionation at the enzymatic level, the fractionation introduced by the enzymes involved in the interconversion of glucose and fructose—glucose isomerase and invertase—was investigated (Gilbert, Robins, Remaud, & Tcherkez, 2012). It was revealed that position-specific kinetic and thermodynamic isotope effects are associated with glucose ↔ fructose interconversion and that this causes an overall enrichment in ^{13}C at the C1 position in glucose and at the C2 in fructose (Fig. 5). These findings on the enzyme system in vitro were in good agreement with the isotope fractionation observed in sucrose (Gilbert et al., 2012).

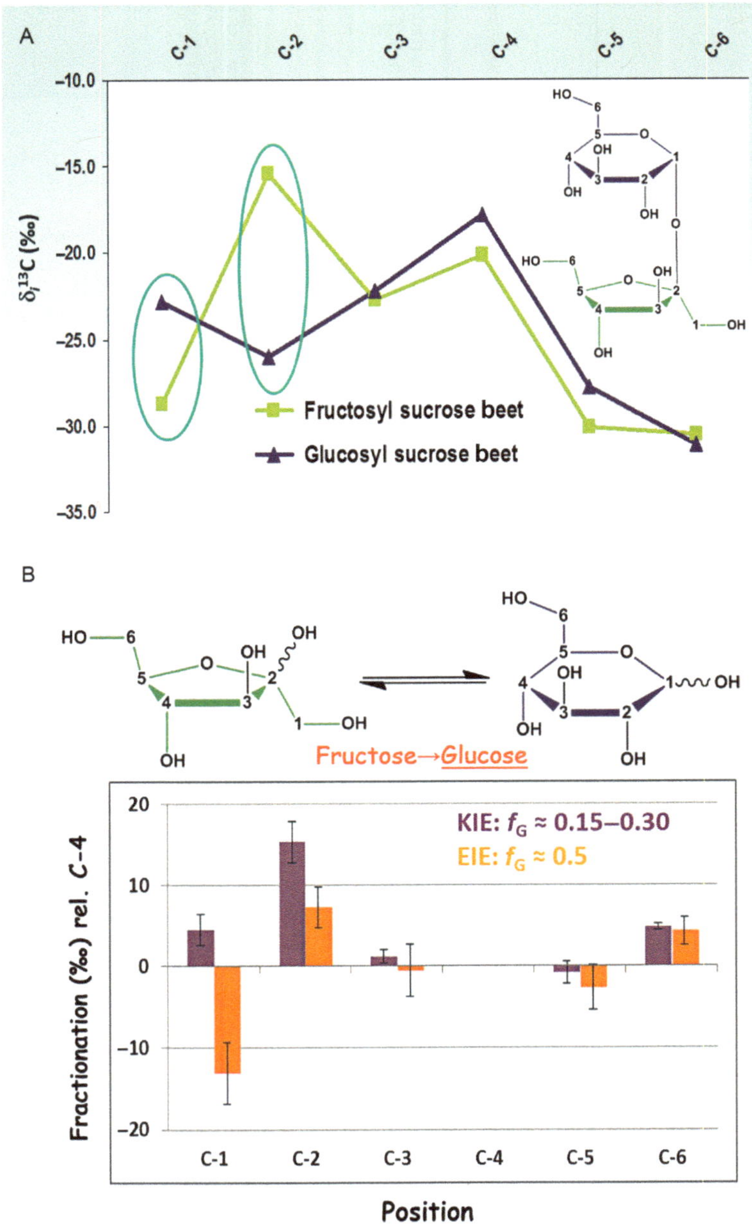

Fig. 5 Position-specific $^{12}C/^{13}C$ isotope fractionation associated with the interconversion of glucose and fructose. (A) The $\delta^{13}C_i$ (‰) values for the glucosyl and fructosyl moieties of sucrose from sugar beet. (B) The glucose isomerase reaction assayed in vitro in the direction from fructose to glucose. The kinetic isotope effect (KIE) is measured early in the reaction before the back reaction is significant. The equilibrium isotope effect (EIE) is measured when equilibrium has been reached. Values are expressed as $\Delta\delta^{13}C_i$ (‰) relative to the C4 position, for which it is assumed that no fractionation occurred.

(Continued)

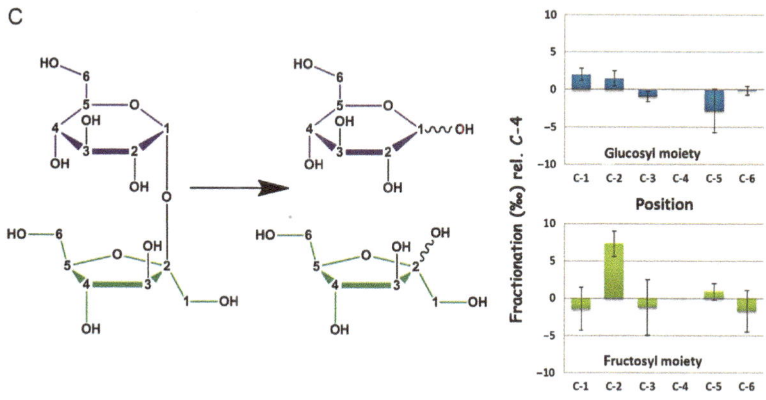

Fig. 5—Cont'd (C) The invertase reaction assayed in vitro in the direction sucrose to glucose + fructose. Values are expressed as $\Delta\delta^{13}C_i$ (‰) relative to the C4 position, for which it is assumed that no fractionation occurred. For invertase, there is no thermodynamic measurement (EIE) since the reaction is not reversible. Positive values indicate fractionation against ^{13}C, negative values indicate fractionation in favor of ^{13}C. *Figures adapted from Gilbert, A., Robins, R. J., Remaud, G. S., & Tcherkez, G. (2012). Intramolecular ^{13}C-pattern in hexoses from autotrophic and heterotrophic C_3 plant tissues.* Proceedings of the National Academy of Sciences of the United States of America, 109, *18204–18209.*

3.2.1 Protocol

The study of ^{13}C fractionation in glucose and fructose in metabolism requires that the sugar is derivatized prior to spectral acquisition. Gilbert, Silvestre, Robins, and Remaud (2009) and Gilbert, Silvestre, Robins, Tcherkez, and Remaud (2011) describe the protocols to use for glucose and fructose, respectively. Further details are given in Remaud et al. (2016). The following protocol is specific to the analysis of the isotope fractionation associated with the interconversions.

1. Isomerase assay
   - **(a)** Prepare the reaction medium: distilled water (500 mL); $MgSO_4 \cdot 5H_2O$ (1.5 g); adjust acidity to pH 7.5 (with $NaHCO_3$).
   - **(b)** To 100 mL of the medium added either glucose (17 g) or fructose (17 g); stir at 60°C for 30 min to allow isomerization. Add immobilized glucose isomerase (400 mg, 350 U/mg).
   - **(c)** Follow the reaction by HPLC (NH_2 column; refractometric detector; elution with CH_3CN/H_2O (80:20); flow rate 1 mL/min; ambient temperature). Typically, chemical equilibrium is reached in ~24 h (ratio of fructose:glucose, 1:1).
   - **(d)** Stop the reaction by filtration (0.22 μm filter). Evaporated the filtrate to yield a syrup.
   - **(e)** Perform derivatization of glucose and fructose on the syrup (see later).

2. Invertase assay
   (a) To an aqueous solution of invertase (1 mL of a 5 mg/mL (300 U/mg)) add sucrose solution (100 mL of 0.43 *M*) at room temperature.
   (b) Quench the reaction at different times (between 50 and 120 min) by adding NaOH (0.5 mL of 1 m*M*). Follow sucrose, glucose, and fructose concentrations by HPLC (see earlier).
   (c) Add to the quenched invertase assay activated charcoal (15 g) and stir (15 min at room temperature), then place at 4°C for 1 h.
   (d) To recover sucrose, deposit the mixture obtained on a charcoal celite column (40 g charcoal; 40 g celite, conditioned with distilled water) equipped with a vacuum flask (flow rate of ca. 10 mL/min). Elute with pure water, ethanol/water (1:99), ethanol/water (5:95), and then pure ethanol until all the sucrose had been recovered.
   (e) Evaporate the fractions, dry overnight over P_2O_5 in a desiccator before derivatization.
   (f) Perform derivatization (see later).
3. Carbohydrate derivatization and NMR analysis. The carbohydrates (glucose, fructose, and sucrose) were converted to glucose and fructose acetonide derivatives and were analyzed by quantitative ^{13}C NMR spectrometry as described in Gilbert et al. (2009, 2011) for glucose or fructose, respectively.

3.2.2 Notes

1. For the determination of equilibrium isotope effects, follow the reaction for 3–10 days with either fructose or glucose as substrates.
2. For the determination of kinetic isotope effects, run the reaction with fructose as substrate and stop at various times over 1–2 h.

3.3 Differentiating Between Different Pathways of Glucose Fermentation by Isotopic Fractionation Measured by irm-^{13}C NMR

In the anaerobic fermentation of glucose to ethanol, a number of different pathways are exploited. Depending on the organism, the bioconversion may occur via the Embden–Meyerhof–Parnas, the Entner–Doudoroff, or the phosphoketolase (PK) pathway. Examples of organisms that exploit these fermentative pathways are *Saccharomyces cerevisiae*, *Zymomonas mobilis*, and *Leuconostoc mesenteroides*, respectively. As can be seen in Fig. 6, the relationship between the individual six carbon position atoms of glucose

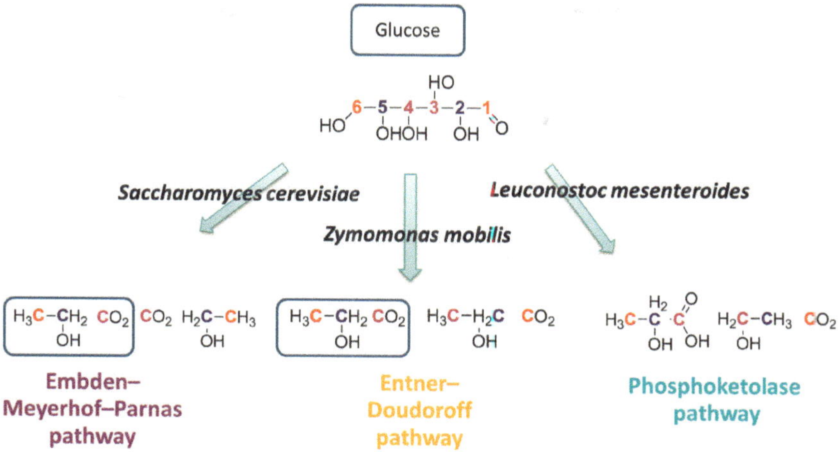

Fig. 6 The affiliation of the carbon atoms in ethanol produced from glucose by different fermentation strategies. *Figure adapted from Bayle, K., Silvestre, V., Akoka, S., Remaud, G. S., & Robins, R. J. (2015). Non-statisitcal ^{13}C distribution during carbon transfer from glucose to ethanol during fermentation is determined by the pathway exploited.* Journal of Biological Chemistry, 290*(7), 4118–4128. doi: 10.1074/jbc.M114.621441.*

and ethanol is different depending on the pathway exploited. Thus, when ethanol is recovered from a given fermentation, the $^{13}C_{CH2}$ and $^{13}C_{CH3}$ values should reflect (a) the $^{13}C_i$ values of the starting substrate, glucose, and (b) differences in the overall pathway, which will be determined by differences in the isotope effects of the enzymes involved. In this example, the technique of irm-^{13}C NMR is applied not to the isotope effect of an individual enzyme, but to probe the overall isotopic fractionation that is associated with a whole pathway (Bayle, Silvestre, Akoka, Remaud, & Robins, 2015). By measuring by irm-^{13}C NMR the $^{13}C_{CH2}$ and $^{13}C_{CH3}$ values of ethanol extracted from each fermentation, the isotopic redistribution associated with each pathway is obtained. A unique isotopic signature for each pathway is found, which can be related to weak apparent KIEs (^{app}KIEs) in the reaction pathway. The fermentation of glucose by the EMP pathway did not indicate any significant pathway-related fractionation. Although there is a relative enrichment in the CH_2 group and a relative impoverishment in the CH_3, this can largely be related back to the $^{13}C_i$ values of the substrate. Nonetheless, a small normal ^{13}C ^{app}KIE is found for the CH_3 (1.0014 ± 0.0002), which will contribute to impoverishment in the methyl group. When glucose is catabolized by the ED pathway, a normal ^{13}C ^{app}KIE is again observed for the CH_3.·This is stronger (1.0035 ± 0.0002) and can be tentatively assigned to 2-keto-3-deoxy-6-phosphogluconate aldolase (EC 4.1.2.14), the key enzyme in this pathway. A rather contrasting situation is seen for the PK pathway with isotope fractionation with normal ^{13}C $^{app}KIEs$ occurring at both the CH_2 (1.0058 ± 0.0007) and CH_3 (1.0057 ± 0.0011) positions of ethanol. In this case, the probable source of this is the key enzyme, PK (EC 4.1.2.9), which cleaves the bond between the C3 and C4 positions of glucose and could give rise to a primary KIE.

3.3.1 Protocol

1. Microbial culture conditions. All fermentations are carried out using the same batch of D-glucose under standardized conditions. Fermentations are best made using a stirred bioreactor in which automatic control and environmental monitoring are available.
   - **(a)** For each organism, two or three fermentations are carried out anaerobically at 30°C in appropriate medium with gentle mixing (100 rpm) and 300 g/L D-glucose.
   - **(b)** Samples of an appropriate volume (1000 to 250 mL) are taken to give between 10% and 75% utilization of D-glucose. Supernatant is recovered by centrifugation (4250 × *g*; 10 min; 4°C) and kept at −20°C.
2. Quantification of ethanol
   - **(a)** An aliquot of medium is withdrawn and diluted to give an ethanol concentration in the range 1–10 mg/mL.
   - **(b)** To 1 mL of medium in a 2 mL "Safelock" Eppendorf tube is added ethyl acetate (0.5 mL) and the closed tube vigorously agitated (3 min). Following phase separation by centrifugation (13,500 × *g*, 1 min), the ethyl acetate phase recovered and placed in a GC autosampler vial.
   - **(c)** Ethanol is quantified by GC: column, HP-1 or PTA-5 (30 m × 0.32 mm, film thickness 0.5 μm); carrier gas, He; gas flow, constant flow 1.2 mL/min; injector temperature, 250°C; gradient, 70°C for 4 min, 20°C/min to 200°C, 200°C for 2 min; detection, FID at 250°C; calibration with known concentrations of ethanol (0.67–8.34 mg/mL) treated as described in (2b).
3. Extraction of ethanol. Ethanol is extracted from the fermentation medium by distillation using a Cadiot column equipped with a Teflon turning band. Care must be taken to ensure that a minimum of 90% of the ethanol present is recovered so as to avoid isotopic fractionation.

4. irm-^{13}C NMR spectroscopy of ethanol
   (a) ^{13}C NMR spectra are recorded at 100.6 MHz with a Bruker 400 NMR spectrometer fitted with a 5 mm ^{1}H/^{13}C dual^{+} probe.
   (b) Sample preparation. Prepare a solution (0.1 *M*) of relaxation agent Cr(Acac)$_3$ by dissolving 34.9 mg Cr(Acac)$_3$ in 1 mL DMSO-d_6 in a 4 mL vial. To extracted ethanol (600 μL) add DMSO-d_6 (100 μL) and Cr(Acac)$_3$ solution. Mix, leave the solution for 2–4 h, then filter to remove undissolved relaxation agent. Transfer to a 5 mm NMR tube.
   (c) Acquisition conditions: probe temperature, 303 K; offsets for both ^{13}C and ^{1}H, middle of the frequency range; decoupling, inverse-gated decoupling using adiabatic full-passage pulses with cosine square amplitude modulation ($\nu_2^{max} = 17.6$ kHz) and offset independent adiabaticity with optimized frequency sweep (see Tenailleau & Akoka, 2007 for details); repetition delay between each 90° pulse, $10 \times T_1^{max}$ of ethanol.
   (d) Record five independent spectra for each sample.
5. Spectral processing. See 2.5.3 earlier.
6. irm-^{13}C MS of ethanol. The bulk values of $\delta^{13}C_b$ (‰) is determined by encapsulation and analysis using a Finnigan MAT Delta-S isotope ratio mass spectrometer coupled to an elemental analyzer (NA2100 Fisons Instruments).
   (a) Ethanol (1–1.5 mg) is sealed in a tin capsule. Care is taken to avoid evaporation. Alternatively, ca. 0.5 mL is sealed into an autosampler vial (depending on the input system available). Sample is introduced into the elemental analyzer.
   (b) The δ^{13}C is determined by reference to calibrated international reference materials (IAEA-N1 or IAEA-N2, IAEA, Vienna).

3.3.2 Notes

1. In (2c), the column is heated to 200°C to clean it between injections. Ethanol elutes prior to the peak of solvent (ethyl acetate) at 70°C.

3.4 irm-^{13}C NMR to Probe the Mechanism of the Biotransformation of Ferulic Acid to Vanillin

Vanillin is a very commonly used flavoring compound and is provided from a range of sources. The best and most costly of these is the natural source, fruit pods of *Vanilla planifolia*, but it also can be produced by semi-synthetic means from natural lignin or ferulic acid, or purely synthetically from guaiacol. Each of these origins shows a characteristic isotope pattern in the $\delta^{13}C_i$ values, which can be exploited to distinguish between different origins (Chaintreau et al., 2013; Tenailleau, Lancelin, Robins, & Akoka, 2004). Vanillin can be readily produced by biotransformation of natural ferulic acid by a number of microbial systems, offering the opportunity to study the isotope fractionation in vivo associated with this bioconversion. For this, cultures of *Streptomyces setonii* can be used, and the affiliations between the positional $\delta^{13}C_i$ values for vanillin and ferulic acid can be measured (Botosoa, Blumenstein, et al., 2009). These data in turn can provide insights into the mechanism, which more recently has partially been studied theoretically (Ma, Li, Wei, Liu, & Liu, 2014). Interestingly, although a hydrolyase enzyme carrying out this bioconversion has been well described for *Pseudomonas fluorescens* (Gasson et al., 1998; Narbad & Gasson, 1998) and its crystal structure determined (Bennett et al., 2008), the equivalent activity in *V. planifolia* has proved elusive. An earlier report of a "vanillin synthase" (Gallage et al., 2014) has recently been questioned (Yang et al., 2017).

By following the biotransformation of ferulic acid to vanillin (Botosoa, Blumenstein, et al., 2009), it could be shown that nearly all the carbon positions undergo isotope fractionation and have a $\Delta\delta^{13}C_i$ that evolves during the biotransformation (Fig. 7A). It is particularly important to note that even those carbon positions distal to the cleavage of the 2=3 bond show significant $\Delta\delta^{13}C_i$ values, which can be used to obtain information to support the putative reaction mechanism. A possible mechanism was advanced for the *P. fluorescens* hydroxycinnamoyl–CoA hydratase/lyase based on the X-ray structure (Bennett et al., 2008) in which a crucial role for one or two tyrosine residues in abstracting the H from the 4′-hydroxyl of the phenolic ring is proposed. We showed by theoretical calculation that this leads to a delocalization of the π electrons, generates a quinone structure, and decreases electron density in the 2=3 bond (Fig. 7B) (Botosoa et al., 2007; R.J. Robins, R.A. Kwiecien, & P. Paneth, unpublished data). In a recent theoretical study, it has been suggested that both tyrosine residues are involved, and a mechanism has been advanced for bond cleavage (Ma et al., 2014).

3.4.1 Protocol

1. Microbial culture conditions. All fermentations are carried out using the same batch of *S. setonii* strain ATCC39116 (obtained from the American Type Culture Collection). Culture is stored at −80°C.
   (a) Grow a culture of *S. setonii* in sterile *Streptomyces* biotransformation broth: 250 mL (pH 7.2) in a 1 L polycarbonate baffled shake flask;

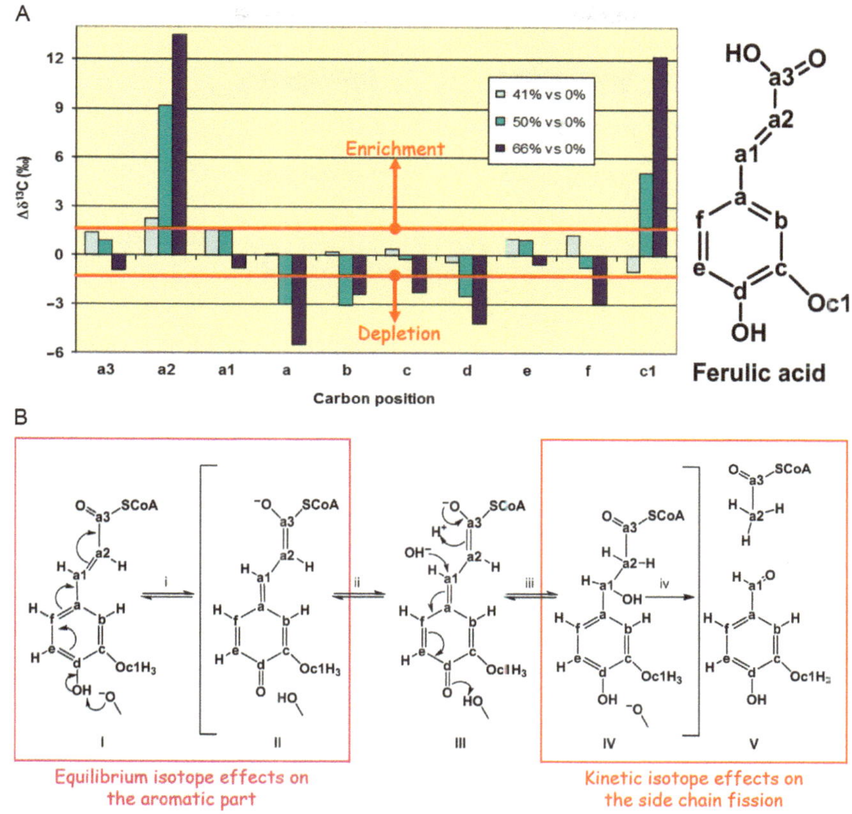

Fig. 7 Biotransformation of ferulic acid to vanillin by cultures of *Streptomyces setonii*. (A) Variation in $^{13}C_i$ of residual ferulic acid. Values are expressed as the isotope shift $\Delta\delta^{13}C$ at each carbon position of residual ferulic acid after a defined degree of advancement of degradation (41%, 50%, and 66%). Note that the $\Delta\delta^{13}C$ is for the substrate, so enrichment means that the reaction is selecting against ^{13}C in that position. (B) The proposed mechanism for ferulic acid hydrolysis to vanillin, indicating the probable origins of the different $\Delta\delta^{13}C_i$ values due to different steps of the reaction. *Figures adapted from Botosoa, E. P., Blumenstein, C., McKenzie, D. A., Silvestre, V., Remaud, G. S., Kwiecien, R. A., & Robins, R. J. (2009). Quantitative isotopic ^{13}C nuclear magnetic resonance at natural abundance to probe enzyme reaction mechanisms via site-specific isotope fractionation: The case of the chain-shortening reaction for the bioconversion of ferulic acid to vanillin. Analytical Biochemistry, 393(2), 182–188.*

temperature, 37°C; agitation, (130–150 rpm) until glucose is consumed (ca. 18 h).

(b) Add filter-sterilized solution of ferulic acid (2.5 g, freshly dissolved in water at ca. 100 g/L, pH 7.2–7.4 (NaOH)).

(c) Follow biotransformation by HPLC (see later). Samples are centrifuged (12,000 × *g*, 1 min) and the clarified supernatant diluted 20–100 times in HPLC solvent.

   (d) At appropriate time points, harvest the culture, flush with N_2 gas, and stored at −20°C.

2. Compounds present in the biotransformation medium are quantified by HPLC: column, Supelcosil LC-18 column (20 × 4.6 mm, 5 μm); solvent, 15/85 acetonitrile/m*M* trifluoroacetic acid; flow rate, 1 mL/min; temperature, 30°C; detection, 290 nm. On a new column, elution times (±0.2 min) were vanillyl alcohol, 5.3 min; vanillic acid, 8.9 min; vanillin, 14.1 min; ferulic acid, 20.7 min.

3. Extraction of substrates and products

   (a) Rapidly thaw frozen culture (45°C), acidify (pH 3–5, conc. HCl), centrifuge (4550 × *g*, 10 min), and retain the supernatant (ca. 250 mL).

   (b) To this add diethyl ether (1 × 250 mL, then 1 × 100 mL) and extract by vigorous shaking. Repeat with a further vigorous extraction with ethyl acetate (2 × 150 mL). Combine the organic phases, dry (anhydr. $MgSO_4$), and taken to dryness.

4. Isolation of Ferulic Acid and Vanillin

   (a) Separation uses silica gel 60 flash chromatography. Prepare a column (ca. 430 × 30 mm) with 80 g of silica gel in ca. 220 mL cyclohexane (height ca. 210 mm after low-pressure gas packing).

   (b) Apply the residue of extraction (max. 3 g) in a minimal volume of acetone. Elute the column under low pressure with acetone/cyclohexane (6:94 v/v) and collect fractions (200 mL). Identify fractions containing vanillin (confirm by TLC on Si gel, 50:50 chloroform/ethyl acetate). Elute the column with acetone/cyclohexane (15:85 v/v) to recover vanillic acid and vanillyl alcohol, then with a gradient of 15%–85% acetone in cyclohexane to recover ferulic acid. At the end of elution, purge the column with pure methanol.

   (c) Pool all fractions containing vanillin. Take to dryness under slight vacuum at 40°C (to avoid sublimation of the vanillin) then dry for several days over P_2O_5.

   (d) Take the dry residue into ca. 25 mL hot water/ethanol (95:5 v/v), sufficient to recrystallize between 0.25 and 1.0 g. Cool slowly (16 h at 4°C), collect the precipitate, wash, and dry. Recrystallize to recover pure vanillin from hot water/ethanol (95:5 v/v).

   (e) Pool all fractions containing ferulic acid. Take to dryness under slight vacuum at 40°C then dry for several days over P_2O_5. Check the purity by ^{1}H and ^{13}C NMR. If traces of vanillyl alcohol remain, carry out step (f).

(f) Take ferulic acid residue into 150 mL distilled water and adjust the pH to 9–10 (1 *M* NaOH). Extract vigorously with ethyl acetate (3–5 × 100 mL) until no vanillyl alcohol is detected in the organic phase (spot tests). Then, adjust the pH to 2–3 (1 *M* HCl) and extract vigorously with ethyl acetate (3–5 × 100 mL) until no ferulic acid is detected in the organic phase (spot tests).

(g) Check the purity by ^{1}H and ^{13}C NMR. If traces of vanillyl alcohol still remain, repeat steps (d), (e), and, if necessary, (f).

5. irm-^{13}C NMR spectroscopy of vanillin and ferulic acid. Samples for ^{13}C NMR are prepared in deuterated solvent filtered into a 5 mm i.d. NMR tube. Tris(2,4-pentanedionato)chromium(III) ($Cr(Acac)_3$) is added as relaxation agent.
   (a) Sample preparation conditions:
      (i) Ferulic acid (250 mg) in 200 μL DMSO-d_6 plus 400 μL methanol plus 100 μL of a solution of 0.1 *M* ($Cr(Acac)_3$) in methanol (for preparation, see earlier).
      (ii) Vanillin (250 mg) in 400 μL acetone-d_6 plus 100 μL of a solution of 0.1 *M* ($Cr(Acac)_3$) in methanol (for preparation, see earlier).
   (b) Acquisition conditions: Quantitative ^{13}C NMR spectra are acquired as described previously (Botosoa et al., 2008; Caytan, Botosoa, et al., 2007). Five spectra are recorded for each sample. Following Fourier transformation, spectra are treated with a line broadening of 2 Hz.
6. Spectral processing. See 2.5.3 earlier.
7. irm-^{13}C MS of vanillin and ferulic acid. The bulk values of $\delta^{13}C_b$ (‰) is determined by encapsulation and analysis using a Finnigan MAT Delta-S isotope ratio mass spectrometer coupled to an elemental analyzer (NA2100 Fisons Instruments).
   (a) Analyte (1–1.5 mg) is sealed in a tin capsule and introduced into the elemental analyzer.
   (b) The $\delta^{13}C$ is determined by reference to calibrated international reference materials (IAEA-N1 or IAEA-N2, IAEA, Vienna).

4. SUMMARY AND CONCLUSIONS

The methodology outlined earlier must be considered as only giving examples for a small number of compounds. Each situation requires that the method is adapted to the individual analytes. A special care is required to ensure purity, at least to the level of no peak overlap, and to establish the correct NMR conditions for acquisition. These will vary in function of the nature of the analyte, the solvent, the quantity, and the spectrometer used. The presented examples for $^2H/^1H$ ratio analysis make reference to a calibrated internal standard. As indicated earlier, this is not straightforward for the analysis of $^{13}C/^{12}C$ ratios. New developments (Bayle et al., 2015) should make it feasible to use an internal standard methodology, which will negate the current need for an irm-MS as well as a high-field NMR.

A key draw-back at present is the amount of analyte required and further advances in acquisition techniques are in development, which should make feasible the measurements of isotope ratios on smaller sample sizes (Jézéquel et al., 2017). Nevertheless, it will remain impractical to acquire spectra on just a few mg for the foreseeable future, without using very long acquisition times.

On the positive side, the approach using NMR avoids the need to synthesize substrates containing one or more heavy labels, as required for classical isotope dilution methods. Dilution approaches can in fact also be used at natural abundance, and for all positions in the molecule (an example of this for ^{15}N analyzed by irm-MS is given in Kosieradzka et al. 2014). Analyte of significantly differing $\delta^{13}C_i$ values can be obtained either by purchasing from various sources (Remaud et al., 2013), or by exploiting fractionation induced by column chromatography (Botosoa, Silvestre, et al., 2009). It also avoids the need to make assumptions about remote positions not undergoing fractionation, thus that can be used as an "unchanged" point of reference. The example of ferulic acid biotransformation (see Section 3.4) shows that distal atoms can undergo fractionation, particularly if they play a role in the mechanism.

REFERENCES

Bayle, K., Gilbert, A., Julien, M., Yamada, K., Silvestre, V., Robins, R. J., et al. (2014). Conditions to obtain precise and true measurements of the intramolecular 13C distribution in organic molecules by isotopic 13C nuclear magnetic resonance spectrometry. *Analytica Chimica Acta*, *846*, 1–7. http://dx.doi.org/10.1016/j.aca.2014.07.018.

Bayle, K., Grand, M., Chaintreau, A., Robins, R. J., Fieber, W., Sommer, H., et al. (2015). Internal referencing for ^{13}C position-specific isotope analysis measured by nmr spectrometry. *Analytical Chemistry*, *87*(15), 7550–7554. http://dx.doi.org/10.1021/acs.analchem.5b02094.

Bayle, K., Silvestre, V., Akoka, S., Remaud, G. S., & Robins, R. J. (2015). Non-statisitcal ^{13}C distribution during carbon transfer from glucose to ethanol during fermentation is determined by the pathway exploited. *Journal of Biological Chemistry*, *290*(7), 4118–4128. http://dx.doi.org/10.1074/jbc.M114.621441.

Bennett, J. P., Bertin, L., Moulton, B., Fairlamb, I. J., Brzozowski, A. M., Walton, N. J., et al. (2008). A ternary complex of hydroxycinnamoyl-CoA hydratase-hyase (HCHL) with acetyl-coenzyme A and vanillin gives insights into substrate specificity and mechanism. *Biochemical Journal, 414*, 281–289.

Billault, I., Duan, J.-R., Guiet, S., & Robins, R. J. (2005). Quantitative deuterium isotopic profiling at natural abundance indicates mechanistic differences for Δ^{12}-epoxidase and Δ^{12}-desaturase in Vernonia galamensis. *Journal of Biological Chemistry, 280*(18), 17645–17651.

Billault, I., Ledru, A., Ouetrani, M., Serhan, Z., Lesot, P., & Robins, R. J. (2012). Probing substrate-product relationships by natural abundance deuterium 2D NMR spectroscopy in liquid-crystalline solvents: The case of the epoxidation of linoleate to vernoleate by two different plant enzymes. *Analytical and Bioanalytical Chemistry, 402*, 2985–2998.

Billault, I., Mantle, P. G., & Robins, R. J. (2004). Deuterium NMR used to indicate a common mechanism for the biosynthesis of ricinoleic acid by *Ricinus communis* and *Claviceps purpurea*. *Journal of the American Chemical Society, 126*(10), 3250–3256.

Botosoa, E. P., Blumenstein, C., McKenzie, D. A., Silvestre, V., Remaud, G. S., Kwiecien, R. A., et al. (2009). Quantitative isotopic ^{13}C nuclear magnetic resonance at natural abundance to probe enzyme reaction mechanisms via site-specific isotope fractionation: The case of the chain-shortening reaction for the bioconversion of ferulic acid to vanillin. *Analytical Biochemistry, 393*(2), 182–188.

Botosoa, E. P., Blumenstein, C., Silvestre, V., Akoka, S., Remaud, G. S., & Robins, R. J. (2007). In *Measurement of site-specific ^{13}C isotope ratio variation in the metabolism of ferulic acid by ^{13}C NMR spectroscopy. Paper presented at the Isotopes 2007, Benicassim, Spain.*

Botosoa, E. P., Caytan, E., Silvestre, V., Robins, R. J., Akoka, S., & Remaud, G. S. (2008). Unexpected fractionation in site-specific ^{13}C isotopic distribution detected by quantitative ^{13}C NMR at natural abundance. *Journal of the American Chemical Society, 130*(2), 414–415. http://dx.doi.org/10.1021/ja0771181.

Botosoa, E. P., Silvestre, V., Robins, R. J., Rojas, J. M. M., Guillou, C., & Remaud, G. S. (2009). Evidence of ^{13}C non-covalent isotope effects obtained by quantitative ^{13}C nuclear magnetic resonance spectroscopy at natural abundance during normal phase liquid chromatography. *Journal of Chromatography A, 1216*(42), 7043–7048.

Bouchard, D., Hunkeler, D., & Höhener, P. (2008). Carbon isotope fractionation during aerobic biodegradation of n-alkanes and aromatic compounds in unsaturated sand. *Organic Geochemistry, 39*, 23–33.

Bussy, U., Thibaudeau, C., Thomas, F., Desmurs, J.-R., Jamin, E., Remaud, G. S., et al. (2011). Isotopic finger-printing of active pharmaceutical ingredients by ^{13}C NMR and polarization transfer techniques as a tool to fight against counterfeiting. *Talanta, 85*, 1909–1914.

Caytan, E., Botosoa, E. P., Silvestre, V., Robins, R. J., Akoka, S., & Remaud, G. S. (2007). Accurate quantitative ^{13}C-NMR spectroscopy: Long-term repeatability of site-specific ^{13}C isotope ratio determination. *Analytical Chemistry, 79*, 8266–8269.

Caytan, E., Remaud, G. S., Tenailleau, E., & Akoka, S. (2007). Precise and accurate quantitative ^{13}C NMR with reduced experimental time. *Talanta, 71*, 1016–1021.

Chaintreau, A., Fieber, W., Sommer, H., Gilbert, A., Yamada, K., Yoshida, N., et al. (2013). Site-specific ^{13}C content by quantitative isotopic ^{13}C nuclear magnetic resonance spectrometry: A pilot inter-laboratory study. *Analytica Chimica Acta, 788*, 108–113. http://dx.doi.org/10.1016/j.aca.2013.06.004.

Cook, P. F. (Ed.), (1991). *Enzyme mechanism from isotope effects*. Boca Raton: CRC Press.

Gallage, N. J., Hansen, E. H., Kannangara, R., Olsen, C. E., Motawia, M. S., Jørgensen, K., et al. (2014). Vanillin formation from ferulic acid in *Vanilla planifolia* is catalysed by a single enzyme. *Nature Communications, 5*, 1–14. http://dx.doi.org/10.1038/ncomms5037.

Gasson, M. J., Kitamura, Y., McLauchlan, W. R., Narbad, A., Parr, A. J., Parsons, E., et al. (1998). Metabolism of ferulic acid to vanillin: A bacterial gene of the enoyl-SCoA hydratase/isomerase superfamily encodes an enzyme for the hydration and cleavage of a hydroxycinnamic acid thioester. *Journal of Biological Chemistry, 273*(7), 4163–4170.

Gilbert, A., Robins, R. J., Remaud, G. S., & Tcherkez, G. (2012). Intramolecular ^{13}C-pattern in hexoses from autotrophic and heterotrophic C_3 plant tissues. *Proceedings of the National Academy of Sciences of the United States of America, 109*, 18204–18209.

Gilbert, A., Silvestre, V., Robins, R., & Remaud, G. S. (2009). Accurate quantitative isotopic ^{13}C NMR spectroscopy for the determination of the intramolecular distribution of ^{13}C in glucose at natural abundance. *Analytical Chemistry, 81*, 8978–8985.

Gilbert, A., Silvestre, V., Robins, R. J., Remaud, G. S., & Tcherkez, G. (2012). Biochemical and physiological determinants of intramolecular isotope patterns in sucrose from C_3, C_4 and CAM plants accessed by isotopic ^{13}C NMR spectrometry: A viewpoint. *Natural Product Reports, 29*(4), 476–486. http://dx.doi.org/10.1039/c2np00089j.

Gilbert, A., Silvestre, V., Robins, R. J., Tcherkez, G., & Remaud, G. S. (2011). A ^{13}C NMR spectrometric method for the determination of intramolecular $\delta^{13}C$ values in fructose from plant sucrose samples. *New Phytologist, 191*(2), 579–588.

Gonfiantini, R., Stichler, W., & Rozanski, K. (Eds.), (1995). *Reference and intercomparison materials for stable isotopes of light elements IAEA-Techdoc-825*. Vienna: IAEA.

Gonzalez, J., Jamin, E., Remaud, G. S., Martin, Y.-L., Martin, G. J., & Martin, M. L. (1998). Authentication of lemon juices and concentrates by a combined multi-isotope approach using SNIF-NMR and IRMS. *Journal of Agricultural and Food Chemistry, 46*, 2200–2205.

Jézéquel, T., Joubert, V., Giraudeau, P., Remaud, G. S., & Akoka, S. (2017). The new face of isotopic NMR at natural abundance. *Magnetic Resonance in Chemistry, 55*, 77–90. http://dx.doi.org/10.1002/mrc.4548.

Julien, M., Parinet, J., Nun, P., Bayle, K., Höhener, P., Robins, R. J., et al. (2015). Fractionation in position-specific isotope composition during vaporization of environmental pollutants measured with isotope ratio monitoring by ^{13}C nuclear magnetic resonance spectrometry. *Environmental Pollution, 205*(1), 299–306.

Kosieradzka, K., Le Faouder, P., Molinié, R., Gentil, E., Lebreton, J., & Robins, R. J. (2014). Tropane alkaloid metabolism by Pseudomonas AT3 cell cultures: Interchange between the nortropine and norpseudotropine catabolic pathways. *Phytochemistry Letters, 10*, lx–lxviii. http://dx.doi.org/10.1016/j.phytol.2014.04.014.

Lesot, P., Aroulanda, C., & Billault, I. (2004). Exploring the analytical potential of NMR spectroscopy in chiral anisotropic media for the study of the natural abundance deuterium distribution in organic molecules. *Analytical Chemistry, 76*, 2827–2835.

Lesot, P., Baillif, V., & Billault, I. (2008). Combined analysis of C-18 unsaturated fatty acids using natural abundance deuterium 2D NMR spectroscopy in chiral oriented solvents. *Analytical Chemistry, 80*, 2963–2972.

Ma, G., Li, Y., Wei, L., Liu, Y., & Liu, C. (2014). A density functional theory study on the catalytic mechanism of hydroxycinnamoyl-CoA hydratase-lyase. *International Journal of Quantum Chemistry, 114*(4), 249–254. http://dx.doi.org/10.1002/qua.24551.

Markai, S., Marchand, P. A., Mabon, F., Baguet, E., Billault, I., & Robins, R. J. (2002). Natural deuterium distribution in branched-chain medium-length fatty acids is nonstatistical: A site-specific study by quantitative ^{2}H NMR spectroscopy of the fatty acids of capsaicinoids. *Chembiochem: A European Journal of Chemical Biology, 3*(2–3), 212–218.

Martin, G. J. (1995). Non-statistical isotope distribution in natural compounds: Mirror of their biosynthesis and key for their origin assignment. In E. Wada, T. Yoneyama, M. Minagawa, T. Ando, & B. D. Fry (Eds.), *Stable isotopes in the biosphere* (pp. 36–58). Kyoto, Japan: Kyoto University Press.

Martin, G. J., Martin, M., Mabon, F., & Bricout, J. (1982). A new method for the identification of the origin of natural products. Quantitative ^{2}H NMR at the natural abundance level applied to the characterization of anetholes. *Journal of the American Chemical Society, 104*, 2658–2659.

Martin, G. J., Sun, X. Y., Guillou, C., & Martin, M. L. (1985). NMR determination of absolute site-specific natural isotope ratios of hydrogen in organic molecules. Analytical and mechanistic applications. *Tetrahedron*, *41*(16), 3285–3296.

Martin, G. J., Zhang, B.-L., Naulet, N., & Martin, M. L. (1986). Deuterium transfer in the bioconversion of glucose to ethanol studied by specific isotope labeling at the natural abundance level. *Journal of the American Chemical Society*, *108*, 5116–5122.

Martineau, E., Giraudeau, P., Tea, I., & Akoka, S. (2011). Fast and precise quantitative analysis of metabolic mixtures by 2D ^{1}H INADEQUATE NMR. *Journal of Pharmaceutical and Biomedical Analysis*, *54*, 252–257.

Narbad, A., & Gasson, M. J. (1998). Metabolism of ferulic acid via vanillin using a novel CoA-dependent pathway in a newly-isolated strain of *Pseudomonas fluorescens*. *Microbiology*, *144*, 1397–1405.

Remaud, G. S., Bussy, U., Lees, M., Thomas, F., Desmurs, J.-R., Jamin, E., et al. (2013). NMR spectrometry isotopic fingerprinting: A tool for the manufacturer for tracking Active Pharmaceutical Ingredients from starting materials to final medicines. *European Journal of Pharmaceutical Sciences*, *48*(3), 464–473. http://dx.doi.org/10.1016/j.ejps.2012.12.009.

Remaud, G. S., Giraudeau, P., Lesot, P., & Akoka, S. (2016). Isotope ratio monitoring by NMR. Part 1 recent advances. In G. Webb (Ed.), *Modern magnetic resonance* (2nd ed., pp. 1–25). Cham, Switzerland: Springer International.

Singleton, D. A., & Thomas, A. A. (1995). High-precision simultaneous determination of multiple small kinetic isotope effects at natural abundance. *Journal of the American Chemical Society*, *117*, 9357–9358.

Tenailleau, E., & Akoka, S. (2007). Adiabatic 1H decoupling scheme for very accurate intensity measurements in 13C NMR. *Journal of Magnetic Resonance*, *185*(1), 50–58. http://dx.doi.org/10.1016/j.jmr.2006.11.007.

Tenailleau, E., Lancelin, P., Robins, R. J., & Akoka, S. (2004). Authentication of the origin of vanillin using quantitative natural abundance ^{13}C NMR. *Journal of Agricultural and Food Chemistry*, *52*(26), 7782–7787.

Thibaudeau, C., Remaud, G. S., Silvestre, V., & Akoka, S. (2010). Performance evaluation of quantitative adiabatic ^{13}C NMR pulse sequences for site-specific isotopic measurements. *Analytical Chemistry*, *82*, 5582–5590.

Werner, R. A., & Brand, W. A. (2001). Referencing strategies and techniques in stable isotope ratio analysis. *Rapid Communications in Mass Spectrometry*, *15*(7), 501–519.

Yang, H., Barros-Rios, J., Kourteva, G., Rao, X., Chen, F., Shen, H., et al. (2017). A re-evaluation of the final step of vanillin biosynthesis in the orchid *Vanilla planifolia*. *Phytochemistry*, *139*, 33–46. http://dx.doi.org/10.1016/j.phytochem.2017.04.003.

Zhang, B.-L., Yunianta, & Martin, M. (1995). Site-specific isotope fractionation in the characterization of biochemical mechanisms: The glycolytic pathway. *Journal of Biological Chemistry*, *270*, 16023–16029.

CHAPTER FOURTEEN

Applications of Isotope Ratio Mass Spectrometry in Sports Drug Testing Accounting for Isotope Fractionation in Analysis of Biological Samples

Thomas Piper[1], Mario Thevis
Center for Preventive Doping Research—Institute of Biochemistry, German Sport University Cologne, Cologne, Germany
[1]Corresponding author: e-mail address: t.piper@biochem.dshs-koeln.de

Contents

1. Introduction 404
2. Sample Preparation for Urinary Steroids 406
   2.1 Wet Chemistry 408
   2.2 High-Performance Liquid Chromatography 409
   2.3 Derivatization Process—Acetylating or Trimethylsilylation 412
3. Quality Control Measures in Sports Drug Testing 414
   3.1 Validation in IRMS 415
   3.2 Linear Mixing Models 417
   3.3 Reference Population-Based Investigations 421
4. Combination of Hydrogen and Carbon Isotope Ratios 422
   4.1 Different Sources May Have Different Isotope Ratios 422
   4.2 Enzymatic Conversions Delineated by 5α- and 5β-Reductase 424
   4.3 Testosterone Excretion Study 426
5. Conclusion 428
References 429

Abstract

The misuse of anabolic-androgenic steroids (AAS) in sports aiming at enhancing athletic performance has been a challenging matter for doping control laboratories for decades. While the presence of a xenobiotic AAS or its metabolite(s) in human urine immediately represents an antidoping rule violation, the detection of the misuse of endogenous steroids such as testosterone necessitates comparably complex procedures. Concentration thresholds and diagnostic analyte ratios computed from urinary steroid concentrations of, e.g., testosterone and epitestosterone have aided identifying suspicious doping control samples in the past. These ratios can however also be affected by confounding

Methods in Enzymology, Volume 596
ISSN 0076-6879
http://dx.doi.org/10.1016/bs.mie.2017.07.013

factors and are therefore not sufficient to prove illicit steroid administrations. Here, carbon and, in rare cases, hydrogen isotope ratio mass spectrometry (IRMS) has become an indispensable tool. Importantly, the isotopic signatures of pharmaceutical steroid preparations commonly differ slightly but significantly from those found with endogenously produced steroids. By comparing the isotope ratios of endogenous reference compounds like pregnanediol to that of testosterone and its metabolites, the unambiguous identification of the urinary steroids' origin is accomplished.

Due to the complex urinary matrix, several steps in sample preparation are inevitable as pure analyte peaks are a prerequisite for valid IRMS determinations. The sample cleanup encompasses steps such as solid phase or liquid–liquid extraction that are presumably not accompanied by isotopic fractionation processes, as well as more critical steps like enzymatic hydrolysis, high-performance liquid chromatography fractionation, and derivatization of analytes. In order to exclude any bias of the analytical results, each step of the analytical procedure is optimized and validated to exclude, or at least result in constant, isotopic fractionation. These efforts are explained in detail.

1. INTRODUCTION

According to the 2015 Anti-Doping Testing Figures published by the World Anti-Doping Agency (WADA), anabolic-androgenic steroids constitute half of all adverse analytical findings (WADA Anti-Doping Testing Figures, 2015). Within this group, testosterone (T) accounts for approx. 10% of all findings, with each result confirmed by isotope ratio mass spectrometry (IRMS). The application of IRMS is rendered necessary as T, and other so-called endogenous steroids are produced by every human and are therefore ubiquitously traceable at significant concentrations in urine.

These endogenous steroids are all de novo synthesized from cholesterol (Träger, 1977) and a simplified scheme of the human steroid biosynthetic and metabolic pathway relevant to sports drug testing is depicted in Fig. 1. The isotope ratio of the precursor cholesterol mainly depends on the diet regarding carbon and on drinking water regarding hydrogen (Ehleringer, Rundel, & Nagy, 1986). Throughout the metabolism, the isotopic composition changes only slightly, resulting in $^{13}C/^{12}C$ and H/D isotope ratios that are approx. equal for the different excreted urinary metabolites. On the right side of Fig. 1, steroids are found that will not be influenced by an application of T or T-prohormones, which can therefore serve as endogenous reference compounds (ERCs) in doping control analysis. Down on the left side, T and its main urinary metabolites androsterone (A) and etiocholanolone (ETIO) are shown, and these, among others, are commonly utilized as target compounds (TCs). As 17α-OH-progesterone represents a node in the human

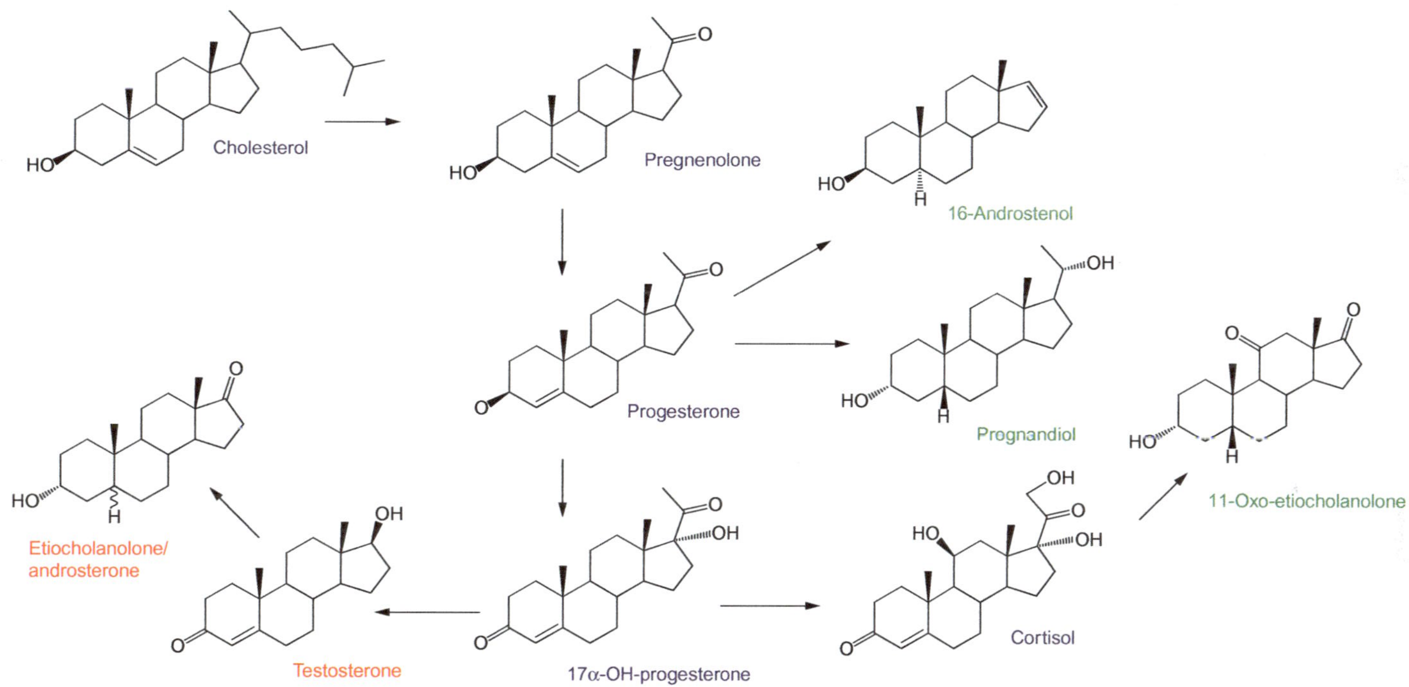

Fig. 1 Simplified overview on human steroid metabolism. Further information in the text.

steroid metabolism, the administration of any steroid from the TC side does not influence any ERC. Individual samples are considered unsuspicious as long as the isotope ratios of the ERCs are close to those of the TCs but become highly suspicious if a difference exceeding the relevant thresholds defined by WADA (WADA TD2016 IRMS, 2016) is observed. These differences between an ERC and a TC are expressed as Δ values:

$$\Delta[‰] = \delta_{ERC} - \delta_{TC} \quad (1)$$

A more in-depth description of this procedure can be found elsewhere (Piper, Emery, & Saugy, 2011).

Within this chapter, the complex sample preparation necessary to extract and purify steroids from the biological matrix urine is described. Producing chromatographic peaks composed of ideally only one compound is a prerequisite for valid IRMS determinations as all analytes are converted to CO_2 or H_2 and, consequently, all structural information is lost. Only the retention time supports identifying an analyte by comparison to the respective reference material. So, the primary objective of the sample preparation is isolation and purification of analytes with acceptable yields. The secondary objective, and at least equally important as the first one, is to avoid any alteration of the isotopic composition of the analytes during the sample preparation procedure. If this is not achievable or verifiable, the minimum acceptable requirement is to show that the isotopic fractionation that does occur is stable and reproducible. Taking these challenges into account, each step of the sample preparation will be discussed with a focus on the high-performance liquid chromatography (HPLC) cleanup, the enzymatic and chemical hydrolysis of phase-II-metabolites, and derivatization to improve the final gas chromatographic (GC) separation of the analytes.

Due to the legal importance of each result in the field of doping control analysis, different quality control measures have been implemented and will be presented here together with important steps in method validation, proving that the above-mentioned requirements are met.

Finally, the potential of isotope ratio measurements at natural abundance to investigate human steroid metabolism will be highlighted and exemplarily discussed.

2. SAMPLE PREPARATION FOR URINARY STEROIDS

The first investigations into carbon isotope ratios (CIR) of steroids in the context of doping controls were conducted in the mid-1990s and encompassed all basic sample preparation steps such as liquid–liquid extraction (LLE), solid phase extraction (SPE), and HPLC (Becchi et al., 1994). Other possibilities to extract steroids from urine like immunoaffinity chromatography (Aguilera et al., 1996; Horning et al., 1997) or anion exchange columns (Ueki & Okanao, 1999) were investigated but never implemented into routine procedures.

Several possible sample preparation procedures were published in the recent years, and they all comprise more or less the different steps schematically depicted in Fig. 2. The specific sample preparation discussed here has already been described in detail (Piper, Thomas, Thevis, & Saugy, 2012). The amount of urine necessary for sample preparation strongly depends on the urinary concentration of ERC and TC analytes and can be increased in case of extremely low concentrated steroids up to 40 mL. In Table 1, all steroids under investigation are listed together with the IUPAC names and their urinary concentration ranges (Donike, Rauth, & Wolansky, 1993).

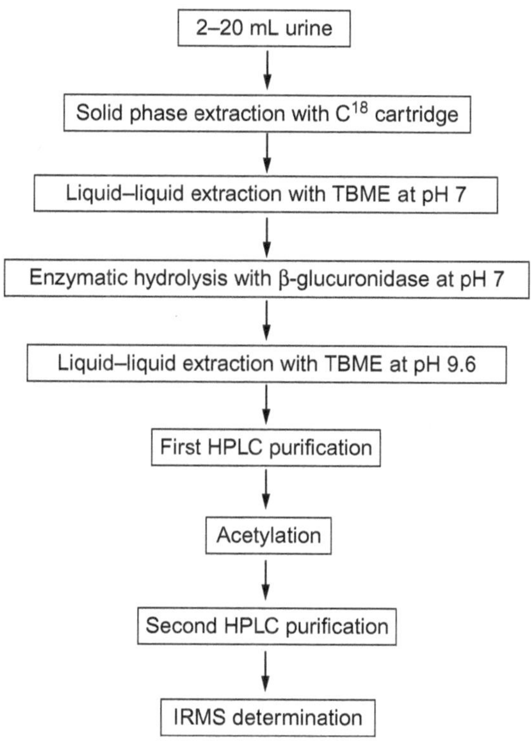

Fig. 2 The different steps in sample preparation necessary to achieve pure extracts of all steroid analytes of interest.

Table 1 Steroids Routinely Monitored for Doping Control Purposes

Steroid	Abbr.	IUPAC	Range [ng/mL]
11-Keto-etiocholanolone	11K	3α-Hydroxy-5β-androstane-11, 17-dione	50–800
Pregnanediol	PD	5β-Pregnane-3α,20α-diol	50–1000
16-Androstenol	16EN	5α-Androst-16-en-3α-ol	50–1200
Androsterone	A	3α-Hydroxy-5α-androstan-17-one	200–10,000
Etiocholanolone	ETIO	3α-Hydroxy-5β-androstan-17-one	200–10,000
Epitestosterone	E	17α-Hydroxy-androst-4-en-3-one	1–200
Testosterone	T	17β-Hydroxy-androst-4-en-3-one	1–200
5α-Androstanediol	5a	5α-Androstane-3α,17β-diol	2–200
5β-Androstanediol	5b	5β-Androstane-3α,17β-diol	10–500

The concentration range is given for the steroids excreted as glucuronides.

2.1 Wet Chemistry

The first SPE step serves mainly concentration purposes and removes salt and particulates from the urine (Fig. 2). Then the samples are reconstituted in a buffer suitable for enzymatic hydrolysis. Prior to this step, the so-called free-fraction, i.e., unconjugated steroids and less polar compounds are removed from the urinary extract by LLE with *tert*-butyl-methylether (TBME). Besides the less important additional cleaning effect of this step, its main role in the context of IRMS is to remove steroids generated from their corresponding phase-II-metabolism-derived conjugates due to sample degradation or microbial activity. Such uncontrolled cleavage processes of conjugates can be accompanied by isotopic fraction, and, therefore, unconjugated steroids are removed from the test sample before initiating the hydrolysis step (Piper, Geyer, & Schänzer, 2010).

Neither the SPE, LLE, or evaporation steps of sample preparation protocol were found to be accompanied by any measureable isotopic fraction. The first critical aspect in this context is the enzymatic hydrolysis performed with β-glucuronidase. Here, the glucuronic acid moiety is eliminated from the steroidal backbone necessitating the cleavage of a carbon–oxygen bond (Greaf, Furuya, & Nishikaze, 1977; Vestergaard, 1978). If the enzymatic reactions do not yield 100% conversion, there is the chance for isotopic fractionation (Hayes, 2001). Importantly, optimization of reaction conditions did result in yields close to 100% for steroidal compounds (Vestergaard,

1978) as the aforementioned employed sample preparation steps ensured appropriate hydrolytic conditions for all steroid glucuronides under investigation. Nevertheless, complete conversion of all analytes in each sample cannot be guaranteed. Consequently, investigations on the possible influence on CIR of incomplete hydrolysis were carried out using A-glucuronide (A-GLUC) as a model compound.

Commercially available A-GLUC (Sigma Aldrich, Darmstadt, Germany) was dissolved in phosphate buffer at pH 7 and incubated with different amounts of β-glucuronidase *(Escherichia coli)* from Roche Diagnostics GmbH (Mannheim, Germany) at different temperatures ranging from +4°C to +50°C for different time intervals ranging from 10 to 60 min. Afterward, samples were extracted with TBME, acetylated and their isotopic composition was measured. Despite significantly different yields, the observed CIR were stable over all experiments ($n=24$) with $\delta^{13}C_{VPDB}=-34.9\pm0.33$‰. For comparison, the coinjected reference standard androstanol (5α-androstan-3β-ol, RSTD) resulted in $\delta^{13}C_{VPDB}=-31.9\pm0.24$‰. Thus, the standard deviation (SD) for the liberated A was significantly larger than the ordinary measurement uncertainty. Although preliminary, these investigations strongly suggest that the error arising from the influence of the enzymatic reaction yield on resulting CIR of steroids is minor. Additionally, these results are in agreement with theory, as the carbon–oxygen bond broken during deconjugation belongs to the glucuronic acid moiety and should therefore only have little influence on the resulting CIR of the steroidal backbone.

After hydrolysis, the formerly glucuronidated steroids are extracted from the aqueous phase by LLE using TBME. The organic layer is transferred, evaporated, and the dry residue is reconstituted in methanol (MeOH) to enable the transfer to autosampler vials suitable for HPLC. The aqueous residue can be further processed concerning all steroids excreted as sulfates. As this procedure is to date solely applied for research purposes, it will not be further described here. In general, the same principles as discussed for the glucuronidated steroids apply.

2.2 High-Performance Liquid Chromatography

As the direct injection of the steroid extract from MeOH results in peak broadening under the used chromatographic conditions, samples are evaporated again and reconstituted in a mixture of acetonitrile (ACN)/water (50/50, *v*/*v*). In the literature, HPLC methods are reported that, despite minor differences, commonly consist of similar approaches utilizing a semi-preparative C^{18} column, and ACN and water as mobile phases with gradient elution. Whenever possible, pure solvents are used without addition of any acids or salts as these might interfere with the subsequent GC analysis of purified compounds.

In Fig. 3, an exemplary HPLC chromatogram of a mixture containing all relevant standards is shown, using the method described by Piper, Thomas, et al. (2012). Starting with 40% B (can), the gradient raised linearly to 60% B in 18 min, then within 1 min to 98% B, held for 11 min before reequilibration at starting conditions. Seven fractions are collected containing the steroids shown in Fig. 3. It is of utmost importance to exercise caution during this crucial step in sample cleanup due to chromatographic isotopic fractionation taking place on the analytical column. This phenomenon is well described in the literature (Braunsdorf, Hener, Przibilla, Piecha, & Mosandl, 1993; Buisson et al., 2005; Caimi & Brenna, 1997) and results in isotopic enrichment at the beginning of the peak and isotopic depletion at its end as depicted in Fig. 4. Therefore, it is necessary to collect

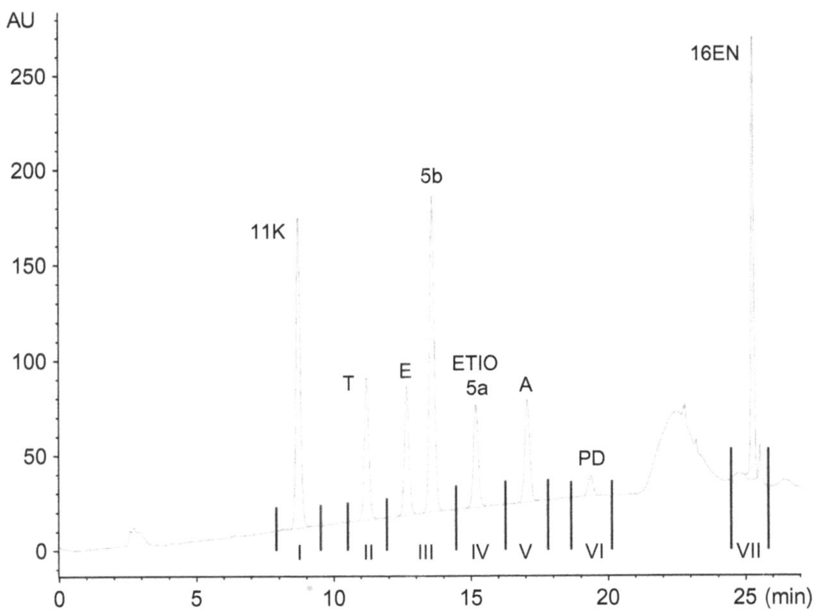

Fig. 3 UV chromatogram (195 nm) of a standard steroid mixture used to fix retention time prior to sample cleanup. Steroid concentrations range from 2 to 20 µg on column depending on their UV absorption.

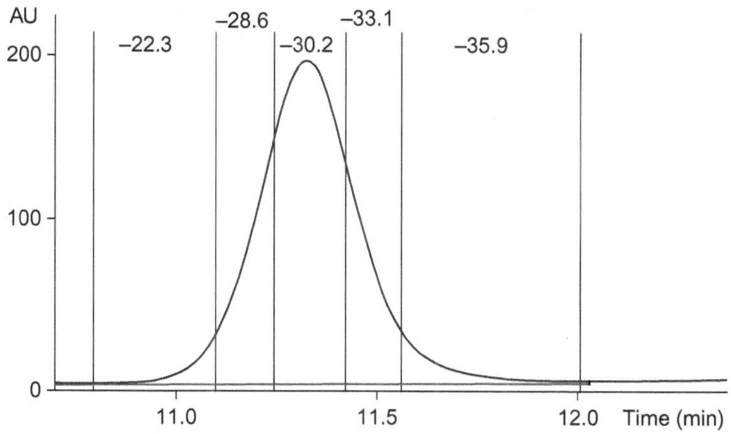

Fig. 4 Chromatographic isotopic fractionation of testosterone on a C^{18} semipreparative HPLC column. Carbon isotope ratios are given in $\delta^{13}C_{VPDB}$ [‰].

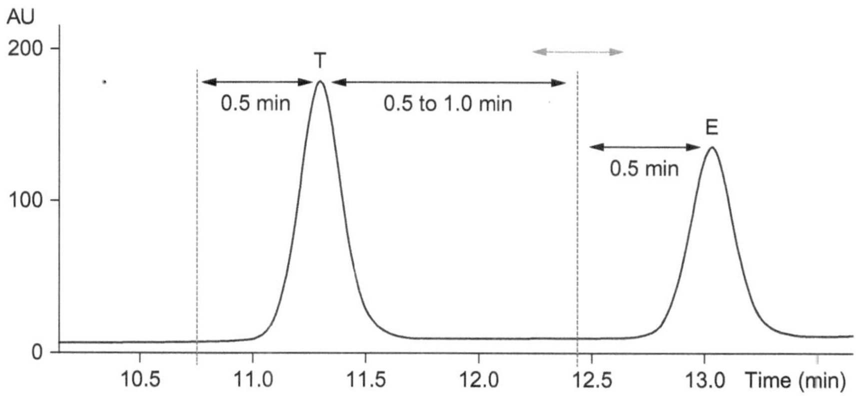

Fig. 5 UV chromatogram (195 nm) showing the separation of testosterone and epitestosterone in detail. The *red (dashed) line* represents the time points for fraction collection. Further information in the text.

the complete peak and not to miss any part as this will have a direct impact on the measured CIR and hydrogen isotope ratios (HIR) of the analyte.

Several measures are implemented into routine procedures to ensure the required completeness of the collected chromatographic peak. First, before each batch of samples, the HPLC system is thoroughly equilibrated by injection of a solvent blank before a standard mixture is analyzed twice to determine all relevant retention times. Fraction collection windows are established as depicted in Fig. 5. The collection period starts 10–20 s before the analyte elutes from the column and as it is expected that retention times under the given chromatographic conditions tend to shift backward, usually half a minute is added at the end of the peak. Of course, this rule of thumb is only applicable if the peak separation is sufficient, i.e., if retention times differ at least for 1.5 min between two adjacent peaks of interest. Otherwise it would be advisable to shift the fraction collection for the second peak closer to the start of this peak, as an overload of the column would result in delayed peak elution. If the retention times differ by less than a minute as for E and 5b shown in Fig. 3, combined collection of both peaks is the only way to ensure valid results. Both peaks can then easily be separated during a second HPLC run after acetylation of steroids.

Once collection windows are set and the sequence is started, retention time stability is monitored by retention time markers added to the urinary extract. It is also possible to use the endogenous 16EN for this purpose as it elutes late in the chromatogram (Fig. 3 at 25 min) and is visible in every sample due to its good UV response at 195 nm. Additionally the pump pressure should be recorded as sudden drops in pressure due to pump failures or air bubbles in the system are usually accompanied by retention time shifts. Depending on the HPLC setup and the fraction collector used it might also be necessary to take the time delay between the UV detector and the collector into consideration. And finally, quality control urine samples should run at the beginning and end of each sequence to enable retrospective assessment of HPLC performance by controlling their respective CIR and HIR.

After the first HPLC cleanup step, all collected fractions are concentrated to dryness and all steroids are derivatized to improve either their GC behavior or to enable additional cleanup by a second HPLC run of the now acetylated compounds taking into account the same critical variables mentioned earlier.

2.3 Derivatization Process—Acetylating or Trimethylsilylation

During derivatization, carbon and hydrogen atoms are added to the steroidal backbone contributing to the measured CIR/HIR. This operation necessitates applying an appropriate correction factor afterward to allow for comparison of underivatized, singly, and multiply derivatized compounds. During the process of derivatization, isotopic fractionation may occur; also, for the subsequent correction the employed factors have to be carefully estimated.

For steroid IRMS measurements acetylation proved to be the most suitable derivatization technique, as a minimum of carbon or hydrogen atoms is added, and the derivatives are stable under RP-HPLC conditions enabling subsequent purification (Piper, Emery, et al., 2011; Piper et al., 2008; Piper, Thevis, Flenker, & Schänzer, 2009; Piper, Thomas, et al., 2012). The correction factors applied take into account possible kinetic isotope effects (Docherty, Jones, & Evershed, 2001; Rieley, 1994). Therefore, it is necessary to conduct analyzes under comparable conditions for both underivatized and acetylated steroids. Comparing both values afterward allow for calculation of the contribution of the added acetate moiety using the following Eq. (2):

$$n_{sd}\delta^{13}C_{sd} = n_s\delta^{13}C_s + n_d\delta^{13}C_d \tag{2}$$

Here, n represents the number of carbon atoms in the molecule, and $\delta^{13}C$ stands for the CIR of the related compound. The subscript s stands for steroid, d for the acetate, and sd for the derivatized compound. A similar equation can be used for hydrogen. If it is not feasible to analyze both derivatized and underivatized steroids, the use of steroids sold as certified reference material (CRM) or secondary isotopic reference standards represents a fine working solution. One possible option is the CU/USADA 34.1 (Zhang, Tobias, & Brenna, 2009) containing ETIO, A, and PD with well-defined CIR.

The practical guidance for how to estimate the acetate correction would be as follows: Three separate aliquots of the CRM should be taken to dryness and derivatized following the established protocol. Afterward, each aliquot should be injected at least three times, and the calculated mean and SD for each aliquot should be compared to each other. If everything is working fine, no significant differences will be detected, and all measurements can be combined. Then this mean value together with the certified value is put into Eq. 2 and for each steroid the correction factor can be calculated. This apparent value for the acetate moiety will differ slightly from steroid to steroid but should fall in the same range showing a SD below ±3‰. Alternatively, the mean value can be ascertained as the correction factor for this specific acetic anhydride. As soon as a new batch or bottle of acetic anhydride will be used, the complete procedure has to be repeated as each acetic anhydride shows different and unpredictable CIR. A careful investigation on the kinetic isotope effects of acetylated steroids can be found in literature (Angelis, Kioussi, Kiousi, Brenna, & Georgakopolus, 2012). For HIR no steroid CRM is available, so here in-house reference material has to be analyzed in an underivatized form in order to enable calculation of the appropriate correction factor.

In principle the same correction method is applied for trimethylsilyl (TMS) derivatization of steroids (Guo, Luke, Lee, & Schoeller, 1993) with comparable results as for acetylation. Unfortunately, one main drawback of this derivatization technique is the formation of copper-silicon compounds during the combustion process leading to successive deactivation of the combustion tube. Additionally, very stable carbon-silicon-compounds are formed retaining carbon in the combustion tube (Shinebarger, Haisch, & Matthews, 2002). These effects necessitate a careful investigation of possible isotopic fractionation coming along with the retention of carbon, and the efficiency of combustion has to be constantly monitored during the lifetime of the combustion tube. Nevertheless, in cases where acetylation was not useful, TMS derivatization proved to be a reliable tool for CIR determinations in sports drug testing (Piper et al., 2014). For AICAR, a so-called exercise mimetic, TMS was the only working derivatization, and within this project, TMS derivatives of steroids were also successfully analyzed. The combustion tube that was employed generated robust results over a period of 6 months. The retention of carbon was assessed using slightly enriched AICAR samples analyzed on both the GC-IRMS system and on an elemental analyzer coupled to an IRMS which allowed for underivatized determinations. Comparing both results showed a statistical value for carbon retention <4%. This potential inaccuracy did not influence the CIR measurements in a significant way as the observed SD for AICAR measurements was larger than this theoretical influence.

The derivatization used strongly depends on the analytes of interest, and in each case, it is necessary to carefully investigate possible sources for isotopic fractionation to avoid, or to maintain them at a constant level, or—in the worst case—to empirically monitor such effects to enable subsequent correction of measured values.

3. QUALITY CONTROL MEASURES IN SPORTS DRUG TESTING

Due to the complicated sample preparation and the particular circumstances of CIR or HIR measurement, i.e., the complete combustion of all analytes to CO_2 or H_2 resulting in the loss of all structural information, each step has to be carefully monitored, and all methods have to be thoroughly validated prior to the implementation of the assay into routine doping controls. In recent years, WADA addressed this issue in a more and more detailed manner resulting in strict requirements that shall be met during validation (WADA TD2016 IRMS, 2016). Herein, we will focus on the basic requirements and discuss the main aspects that should be considered to ensure valid results using CIR as the role model. In principle, identical measures have to be applied during HIR method validation procedures.

3.1 Validation in IRMS

In mass spectrometry and especially in case of forensic applications, the guidelines on how to validate a method are particularly detailed and several protocols exist which may be followed. In general, requirements encompass parameters such as the limit of detection or quantification, inter- and intra-day (im)precision, accuracy, specificity, linear range, recovery, and robustness. For IRMS methods, these parameters are important, but additional relevant aspects also necessitate consideration.

A critical point is the method's repeatability, which should be determined for both short- and long-term periods. Here, the results of an initial sixfold preparation of the same blank urine followed by repeated preparation and analysis cycles over several months are recommended. This can be accomplished for instance by using the same blank urine as a quality control sample processed with each batch of specimens (Piper, Emery, Thomas, Saugy, & Thevis, 2013; Piper et al., 2008). Both results can then be evaluated, and a good repeatability is defined by statistically indistinguishable mean values of short- and long-term CIR measurements. The SDs are expected to increase over time but should remain comparable, i.e., up to ±0.4‰ and ±0.8‰ for short- and long-term determinations, respectively. The observed increase can be interpreted as a sum of different factors. While the underlying measurement uncertainty of the IRMS should be comparable and below ±0.3‰ over time, slight drifts of the mass spectrometer, minor changes in HPLC retention times, or the aging process of the combustion tube and the GC column can contribute to greater error as reflected in increased SD values.

Within routine doping control analysis, these factors have to be monitored and evaluated for each sequence. This is accomplished by processing a positive and a negative quality control urine (QC) with each batch under investigation. The CIR of each QC is evaluated by Shewhart charts to detect outliers or drifts over time (Westgard et al., 1981). In Fig. 6, a TC of the positive QC is shown as an example. The CIR of each sample preparation is plotted against the consecutive number of each batch. With 1 or 2 batches prepared each week, the time range shown encompasses approx. 1 year.

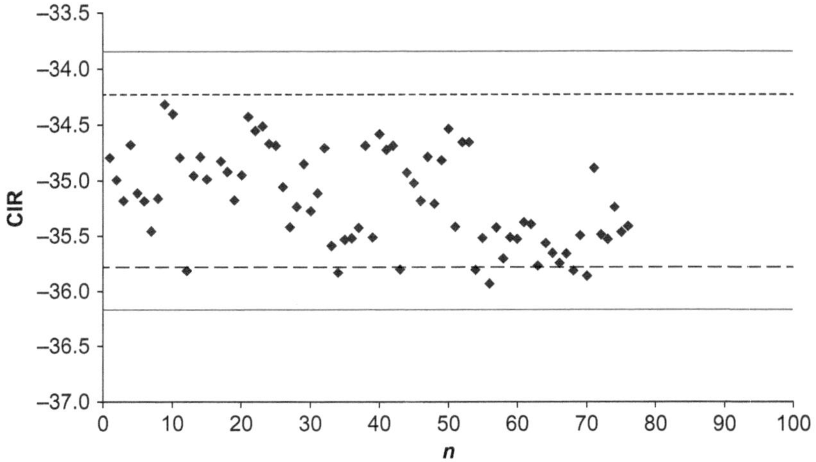

Fig. 6 Shewhart control chart for 5a extracted from the positive QC with each batch of samples. The *broken lines* represent the twofold SD (warning limit), the *bold red lines* the threefold SD (action limit). Carbon isotope ratios are given in $\delta^{13}C_{VPDB}$ [‰].

The mean value for the diacetylated 5a was found at -35.0 ± 0.39‰ ($n = 20$ initial determinations). By definition, 95% of all measurements should fall within the twofold SD. Values above the warning limit should be carefully inspected but do not necessarily trigger any action. Values outside the action limit, i.e., which fall above or below the threefold SD, necessitate at least a reinjection of the sample, and if the value remains suspicious, the complete batch of sample will have to be reprepared or the instrument needs further maintenance. For a detailed explanation on how to use these charts, please refer to the published protocols (Westgard et al., 1981). Using these charts within IRMS measurements readily allows for the detection of long-term drifts. As shown in Fig. 6, the CIR values have a slight trend toward more depleted values. As the other monitored steroids did not show a similar trend at the same time, a general drift of the IRMS system was excluded. The 5a extracted from the negative QC also showed some slightly more depleted values. So the HPLC cleanup was carefully checked, and fraction collection times were adopted to collect broader fractions, which solved the issue thereby demonstrating the great utility of these quality control measures.

The limit of detection equals the limit of quantification in IRMS as it defines the minimum amount of substance required on column to achieve sufficient signal height in the MS falling within the linear range of the instrument. Nonlinearity may have different origins arising from the ion source, the combustion process, and/or chromatographic issues. All these influencing factors must be considered during method validation and should be checked frequently under routine operation conditions. The nonlinearity of the ion source can be minimized for CIR determinations by appropriate source tuning and can be corrected with regards to HIR by accurate determinations of the so-called H_3-factor (Sessions, 2006; Sessions, Burgoyne, & Hayes, 2001). Together with stability tests, the ion source linearity can be controlled with each sequence. Nonlinearities arising from the combustion and chromatographic system should be carefully investigated during method validation by injecting each steroid of interest at different concentration levels to fix the amount on column necessary for a valid analysis. Even if the linearity proves to be acceptable, IRMS measurements will start to suffer from larger SDs below a certain steroid concentration due to statistical effects and increasing contributions from the background, mainly due to column bleed (Piper, Emery, et al., 2013; Piper, Thomas, et al., 2012). Interestingly, different steroids may show different limits of detection on the same instrument or will be affected by nonlinearities in different ways. This behavior is shown in Fig. 7. While 5a demonstrates a perfectly linear behavior with just some more scattering of the lower concentrated values, PD seems to be affected and here further investigations seem advisable before performing routine injection of PD falling below 2V.

Accuracy is difficult to ascertain as no appropriate reference material is available, but linear mixing models (LMMs) proved to be a very good substitute here as discussed in the next section. Specificity can be addressed by reference population (RP) studies (vide infra). For recovery it might not be necessary to perform detailed investigations as for IRMS measurements the absence of isotopic fractionation during the sample preparation is of higher priority than analyte recovery. For the sake of completeness it should be mentioned that due to the extensive HPLC cleanup, recoveries are usually found to be <60%.

3.2 Linear Mixing Models

The standard addition technique employed in IRMS as LMMs is based on so-called Keeling Plots (Keeling, 1958; Pataki et al., 2003). Exemplarily, the LMM will be discussed again only focusing on CIR measurements

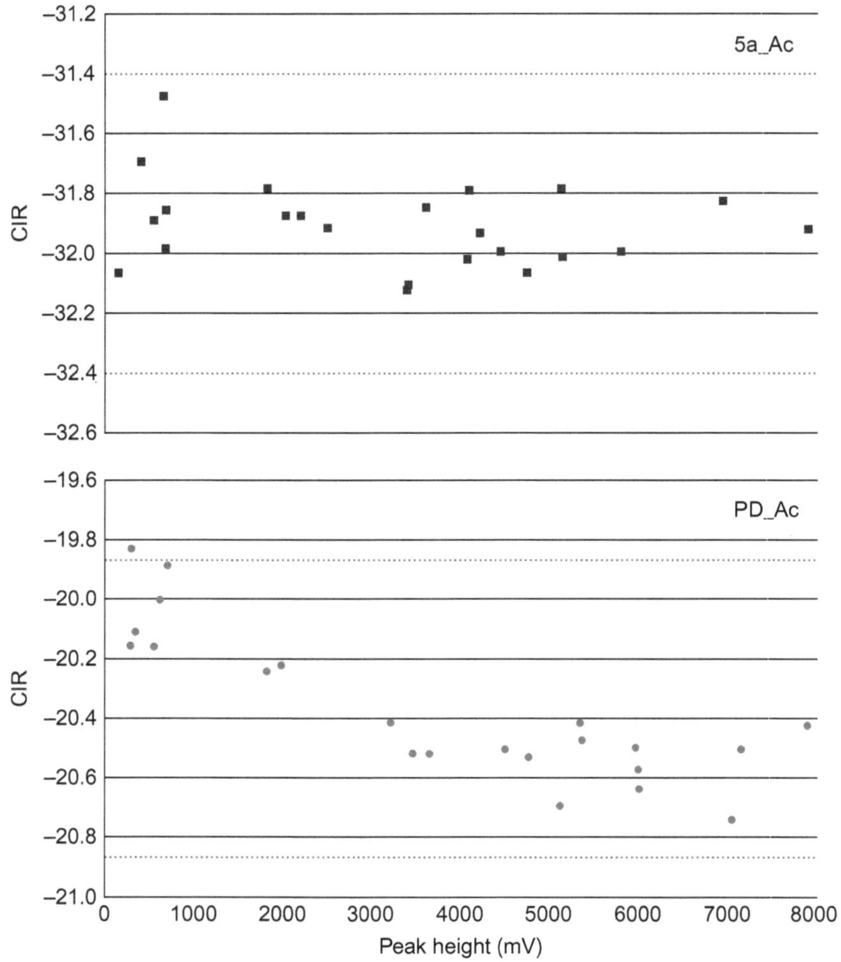

Fig. 7 Linearity test for two different steroids, performed on the same instrument during the same alternating sequence of injections. The *dashed orange lines* represent the mean value ±0.5‰. Carbon isotope ratios are given in $\delta^{13}C_{VPDB}$ [‰].

but are equally applicable to HIR determinations (Piper, Thevis, et al., 2009). Derived from a two pool mixing model, the following equations can be applied:

$$c_m = c_e + c_a \quad (3)$$

$$c_m \delta^{13}C_m = c_e \delta^{13}C_e + c_a \delta^{13}C_a \quad (4)$$

with c_e representing the endogenous concentration of the steroid under investigation, c_a the variable amount added, and c_m the resulting concentration of the mixture and $\delta^{13}C$ the corresponding CIR value. Combining and rearranging both equations results in:

$$\delta^{13}C_m = \left(\delta^{13}C_e - \delta^{13}C_a\right)\frac{c_e}{c_m} + \delta^{13}C_a \qquad (5)$$

Comparing this to a basic linear equation ($y = ax + b$) shows that if the measured CIR of the mixture is plotted against the concentration ratio of c_e/c_m, a linear relationship will become apparent with the differences in CIR between the added steroid and the endogenous value representing the slope of the line of best fit and the CIR of the added standard representing the intercept with the y-axis. When applying this procedure as a measure of method validation, all values are accessible and the theoretical values can be compared to the measured ones in order to estimate an overall measurement uncertainty and to check the method for any occurring bias.

In practice, this validation step will be performed in parallel for each steroid under investigation in the same urine. After all relevant urinary steroid concentrations have been determined using a validated routine method, two sets of five blank urine samples will be fortified with different increasing amounts of exogenous steroids. Equidistant concentration levels should be used up to two to three times the value of the corresponding endogenous steroid, by, for example, adding 0.4, 0.8, 1.2, 1.6, and 2 times the endogenous concentration. This procedure was described in detail in the literature (Piper et al., 2008; Piper, Riemann, et al., 2009; Piper, Thevis, et al., 2009). As an example, Fig. 8 shows results obtained for ETIO. By preparing two independent batches for the LMM, the results directly encompass interday imprecision and an overall reproducibility of the CIR data. The evaluation of such a LMM is performed visually and mathematically.

By visual inspection it is immediately possible to review whether the experiment was successful (cf. Fig. 8) or if analytical issues prevailed (in this case no linear relationship is found for the data; an example of which can be found in Piper, Riemann, et al., 2009). Another option would be a linear relationship, but the values of the spiked samples are constantly shifted to higher or lower CIR values (Piper et al., 2008). In such a case, the initially measured endogenous steroid concentration was not correctly estimated. As a standard addition technique, the LMM could also be used to determine endogenous concentrations.

The line of best fit defined by all fortified samples and the blank urine CIR values should intercept with the y-axis close to the CIR value of the steroid standard used for the experiment. If the distance is large, then

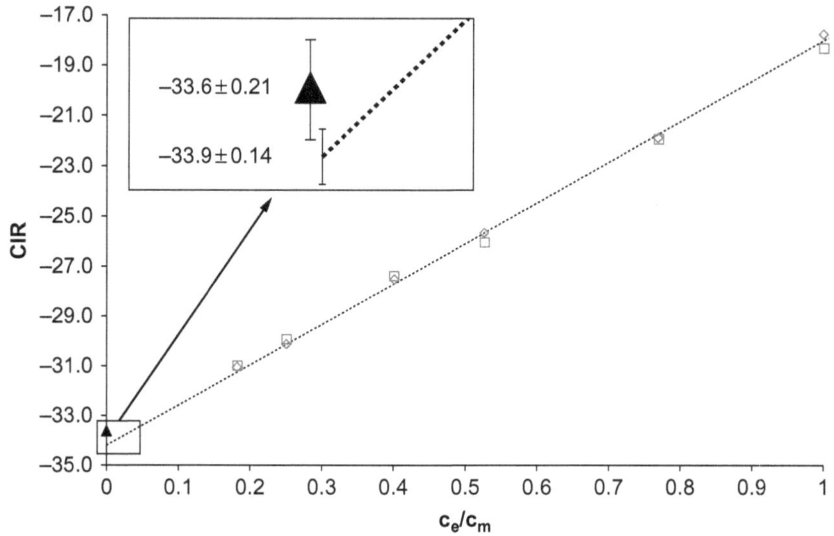

Fig. 8 Linear mixing model for ETIO. Plotted are the measured CIR ($\delta^{13}C_{VPDB}$ [‰]) of each spiked and prepared urine sample against the ratio of the endogenous concentration (c_e) divided by the concentration of the mixture (c_m). The *open squares* represent the first sample preparation, the *open diamonds* the second sample preparation, and the *black triangle* the CIR of the spiked standard. The magnification shows the intercept of the line of best fit with the *y*-axis compared to the standard. Further information in the text.

significant isotopic fractionation is likely to have occurred during sample preparation. In order to estimate if the bias is significant or not, the Gaussian least square fit has to be calculated. The average residuals determined for the slope of the line of best fit give a characteristic value for the scatter of the data. The resulting line of best fit represents a mean that can vary within certain limits, resulting in slightly different values for the slope and, accordingly, for the intercept with the *y*-axis. Therefore, this value is also a mean with an ascertainable SD that can be calculated as indicated in the enlargement of Fig. 8. As the value of the steroid standard used to fortify the samples is given as a mean with SD, both values can be tested for similarity. As a rule of thumb, if both values show an overlap when adding the twofold SD to both, they are most probably not significantly different with $P < 0.05$.

Currently, LMM seems to be the most suitable way to test the developed methods for any isotopic fractionation during sample preparation. Even the otherwise complex monitoring of the enzymatic cleavage is covered as isotopic fractionation here would result in a constant offset of the endogenous values and, therefore, in a bias in the LMM. Good results obtained during the LMM experiment does not, however, prove the complete absence of any isotopic fractionation, but it does corroborate a constant effect below the expected measurement uncertainty.

3.3 Reference Population-Based Investigations

All quality control measures described earlier enable one to estimate the contribution from sample preparation and IRMS determinations on the results. What is missing are the specificity and the endogenous factors that may contribute to the isotopic ratios of endogenous steroids. As shown in Fig. 1, all urinary steroids originate from the same precursor cholesterol. During metabolism, different enzymatic reactions at different sites within the body take place, and all of them may influence the isotopic composition of the products. Some of these aspects will be discussed in detail in Section 4. Regarding sports drug testing, it was not necessarily mandatory to figure out which specific factors influence the measured IRMS values but to identify overall possible influences to take them into account when evaluating an athlete's sample. Or, in other words, to ascertain what a normal or physiological isotopic ratio is and what may be due to doping.

The idea to use RP studies was borrowed from the field of clinical chemistry, where the differentiation between healthy or nonhealthy states of subjects was sought (PetitClerk & Solberg, 1987; Solberg, 1987a, 1987b; Solberg & PetitClerk, 1988). As soon as a data from a sufficiently large number of individuals are available, RP-based thresholds can be calculated. The individuals under investigation, i.e., the survey population, should closely reflect the target population to avoid inherent sampling biases. So for high-level athletes, sport students are considered a viable surrogate as for athletes' samples, although it cannot be excluded that false-negative samples may be present. The number of samples necessary has been extensively debated, and it depends on the observed distribution of the data. If a parametric approach is valid, i.e., the values are found, for example, Gaussian-distributed, ca. 50 samples are usually sufficient to describe the underlying distribution correctly. If values are not convertible and have to be assessed in a nonparametric way, at least 120 samples will be necessary. In our case both the absolute isotopic ratios (δ values) and the differences between ERCs and TCs (Δ values) were found Gaussian-distributed (Cawley, Trout, Kazlauskas, Howe, & George, 2009; de la Torre, Colamonici, Curcio, Molaioni, & Botre, 2012; Flenker, Güntner, & Schänzer, 2008; Piper, Emery, et al., 2013; Piper et al., 2008; Piper, Thomas, et al., 2012;

Sobolevsky, Prasolov, & Rodchenkov, 2010). While the δ values vary between the different investigations, all Δ values were found comparable regardless of the survey population. Further, it could be demonstrated that Δ values found in a population encompassing 60 individuals were perfectly reflected by the target population with $n > 1300$ individuals as soon as suspicious findings (most probably due to steroids administration) were excluded (Piper, Flenker, Mareck, & Schänzer, 2009). All the above cited investigations were the basis for the thresholds given by WADA today to evaluate the CIR of doping control samples (WADA, 2016). This concept is also working for HIR values, but until now too few investigations on hydrogen have been accomplished to allow for a generalization.

4. COMBINATION OF HYDROGEN AND CARBON ISOTOPE RATIOS

A closer look on the results obtained for the different RP shows that several factors influence the CIR and HIR of endogenous steroids. Regarding CIR, the main factor is the individual's diet reflected by a distinct geographical distribution of the δ values representing different amounts of C3 or C4 plants consumed (Piper, Flenker, et al., 2009). Fortunately, the Δ values were the same all over and were furthermore not significantly influenced by gender, ethnicity, and type of sport. HIR, not surprisingly, are mainly affected by the drinking water as the main hydrogen source in the body (Piper, Degenhardt, et al., 2013). So far, no differences between the genders were detectable, while for other possible factors more data will be necessary. Nonetheless, with the data at hand it will be possible to discuss some differences found between different urinary steroids which can be attributed to endogenous isotopic fractionation.

4.1 Different Sources May Have Different Isotope Ratios

Within the human body, steroids undergo both phase-I- and phase-II-metabolism. The main phase-II-conjugates synthesized are glucuronides (GLUC) and sulfates (SULF). Interestingly, a significant difference in CIR and HIR is found between those steroids excreted either as GLUC or as SULF. Under investigation were A, ETIO, and DHEA (dehydroepiandrosterone, 3β-hydroxyandrost-5-en-17-one), and the obtained results are shown in Fig. 9. The RP under investigation here has been described in literature (Piper, Emery, et al., 2013; Piper, Thomas, et al., 2012).

While the more enriched CIR values for steroids excreted as SULF have already been reported (Piper, Emery, et al., 2011; Piper, Opfermann,

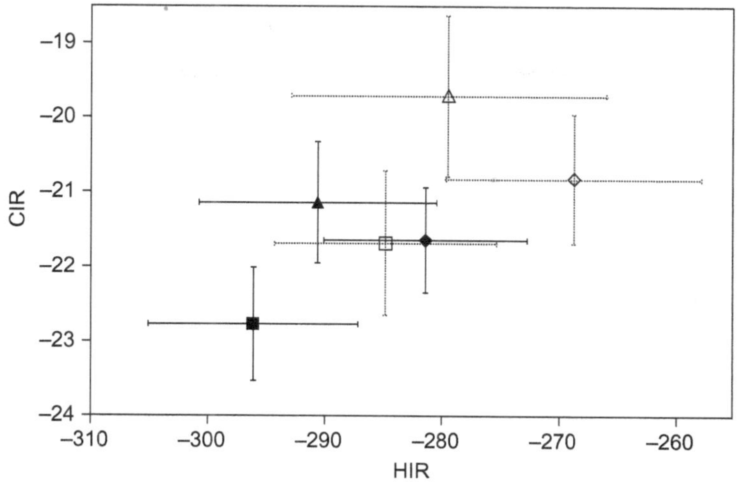

Fig. 9 Comparison of steroids excreted glucuronidated and sulfated. Mean value of HIR plotted against the mean value of CIR in a RP encompassing $n=67$ individuals. *Black symbols* represent the GLUCs (ETIO=*square*, A=*diamond*, and DHEA=*triangle*), *open symbols* the SULFs. The error bars display one single SD.

Thevis, & Schänzer, 2010), it was interesting to find a similar influence on the HIR. During phase-II-metabolism the SULF- or GLUC moiety is attached to the oxygen atom of a hydroxyl function (Burstein & Lieberman, 1958; MacKenzie, Rodbourne, & Stranks, 1992; Ramseyer, Williams, & Hirschmann, 1967). The hydrogen atom involved is eliminated and so for both, carbon and hydrogen only secondary isotope effects at vicinal atoms next to the oxygen can be expected (if any). Usually, secondary isotope effects are not very pronounced and taking into account the steroidal backbone containing at least additional 18 C-atoms and more than 30 H-atoms, this should dilute these effects strongly. More recent investigations employing ^{18}O-labeled water and ^{2}H-labeled steroids showed that the phase-II-metabolism especially regarding the SULFs may be more complicated than generally assumed (Piper, Fußhöller, & Schänzer, 2011; Piper, Schänzer, & Thevis, 2016a). As the steroids excreted as GLUC did not show similar phenomena, it is unlikely that the conjugation itself is accountable for the parallel differences detected.

Another hypothesis seems to be more likely and is supported by the first experiments conducted (Piper, Baume, Strahm, Emery, & Saugy, 2012). Here, it has been assumed that different steroid production sites in the human body may reflect different CIR values, i.e., that steroids produced in the testis may have more depleted values than steroids produced in the adrenal glands. As the proportion of produced phase-II-metabolites depends on the origin of the steroid, this can explain differences found in urine. Steroids synthesized in the adrenal gland are preferentially excreted as SULF and only partially unconjugated, while steroids derived from the testis or peripheral production are usually excreted unconjugated and subsequently are predominantly glucuronidated prior excretion. Stimulating only the production site of the testis by HCG administration resulted in more depleted CIR values of urinary steroids afterward. Unfortunately, HIR were not under investigation in that study.

The possible differences in the isotopic composition at different production sites in the body may have multiple sources. The precursor of all endogenous steroids, cholesterol, is synthesized de novo in the mitochondria directly at the different organs, and this process can potentially result in isotopic fractionation giving rise to different CIR and HIR of the starting material. The metabolism afterward from cholesterol to the different steroids encompasses several enzymatic conversions and branching points, and therefore a lot of opportunities for isotopic fractionation. A detailed description on how branching points can influence the isotope ratios of both products and educts can be found in the literature (Hayes, 2001). Unfortunately, current knowledge does not allow one to further localize the source for the described differences in phase-II-metabolites and further research will be necessary.

4.2 Enzymatic Conversions Delineated by 5α- and 5β-Reductase

After the oral administration of DHEA, one of the main urinary metabolites (ETIO) was found more depleted than the administered steroid in different excretion studies regardless of the phase-II-metabolism (Cawley, Kazlauskas, Trout, Rogerson, & George, 2005; Piper, Opfermann, et al., 2010). During the interconversion of the 5-ene-steroid DHEA toward the 5α- and 5β-reduced compounds A and ETIO, isotopic fractionation must occur resulting in depleted values for ETIO and in comparison slightly enriched values for A (Fig. 10). In this case, the cleavage of double bonds is required, and carbon atoms are directly involved in the conversion making isotopic fractionation very likely. Similar effects have been reported after the administration of nortestosterone resulting in more depleted CIR for the 5β-reduced noretiocholanolone (Flenker & Schänzer, 2001). This depletion of 5β-steroids seems to be a general trend as all individuals show a mean difference of ca. 1‰ between ETIO and A as depicted in Fig. 9 and described

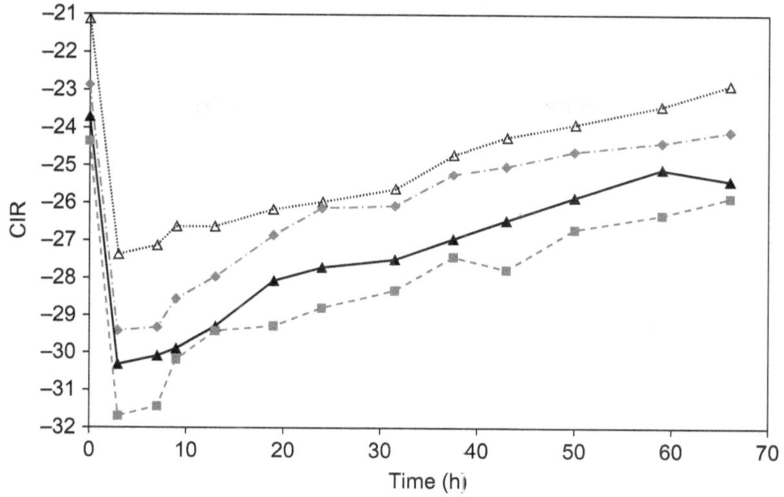

Fig. 10 CIR values (given in $\delta^{13}C_{VPDB}$ [‰]) after the oral administration of 100 mg DHEA with $\delta^{13}C_{VPDB} = -29.7 \pm 0.14$‰. *Open triangles* represent DHEA_SULF, *black triangle* DHEA_GLUC, *gray squares* ETIO_GLUC, and *gray diamonds* A_GLUC.

in the literature (Cawley et al., 2009; de la Torre et al., 2012; Flenker et al., 2008; Piper, Emery, et al., 2013; Piper et al., 2008; Piper, Opfermann, et al., 2010). Despite the fact that the 5β-reductase obviously has an influence on both isotope ratios this cannot be the only influencing factor.

Two other steroids found in urine, 5a and 5b, are presumed to be intermediates in the metabolic conversion of T or DHEA to ETIO and A; therefore they should show comparable isotopic ratios if only the reductase would be responsible for the resulting CIR and HIR values. The RP presented in Fig. 9 was also investigated regarding the isotope ratios of 5a (HIR -275 ± 12‰; CIR -22.5 ± 0.9‰) and 5b (HIR -292 ± 10‰; CIR -22.5 ± 1.0‰). While the observed change in HIR seems to be comparable, the difference in CIR disappears completely. This effect was also observed during comparable studies relying only on CIR (de la Torre et al., 2012). Obviously, other additional factors have to be considered to explain such findings. One aspect worthy of consideration is the abovementioned difference at different production sites reflected by deviating isotope ratios of the presumed precursors DHEA and T (HIR -273 ± 8‰; CIR -22.8 ± 0.9‰). DHEA is mainly excreted in the adrenal glands and therefore showing enriched values while the more depleted T derives from the testes or peripheral metabolism. As a relatively large proportion of 5a originates from T, both show closely related values. Additionally, the pool size of 5a is small resulting also in a strong influence here after small doses of T have been administered making it a sensitive marker in doping controls.

Pool sizes as well play a role regarding the different CIR values found after DHEA application for DHEA_GLUC and DHEA_SULF (Fig. 10). The size of the GLUC pool is much smaller than that of SULF, and this difference is responsible for the strong impact the exogenous DHEA has on urinary DHEA_GLUC, which perfectly reflects the CIR value of the administered DHEA while DHEA_SULF only moves to −27‰. Here, the dilution factor of the exogenous material by the endogenous analog is much higher, resulting in less depleted values in accordance with a two pool mixing model.

Besides the size of a pool, its turn-over time also plays an important role in determining the observed isotope ratios. This became apparent for epiandrosterone (3β-hydroxy-5α-androstan-17-one, EPIA) during the same excretion study (Piper, Opfermann, et al., 2010) showing a delayed response on the exogenous administration but also a prolonged influence, i.e., the CIR values remained depleted, while other steroids already returned to their basal values. Recent findings demonstrated that this prolonged influence is detectable after T administrations too (Piper, Schänzer, & Thevis, 2016b). This seems to be a great opportunity for sports drug testing to prolong the detection times of T administrations by athletes and is currently under detailed investigation.

4.3 Testosterone Excretion Study

The administration of T by athletes is a delicate topic in doping controls and presumed to go undetected in many cases due to the short detection time (usually <24 h) after administration. Besides the above-mentioned novel approach relying on EPIA, HIR were carefully investigated if they may provide an aid by showing stronger and therefore longer detectable influences on endogenous steroids. T preparations with an endogenous CIR signature sometimes available on the black market were an additional factor driving these investigations (Cawley et al., 2010).

One healthy male volunteer was administered with 100 mg of T-undecanoate, and the CIR and HIR of the T after cleaving the ester were carefully determined (Piper, Emery, et al., 2013; Piper, Thomas, et al., 2012). The complete results are shown in Fig. 11. As expected, the ERC PD is neither influenced in its CIR nor in its HIR values by the administration while T directly after the application shows a strong shift in both

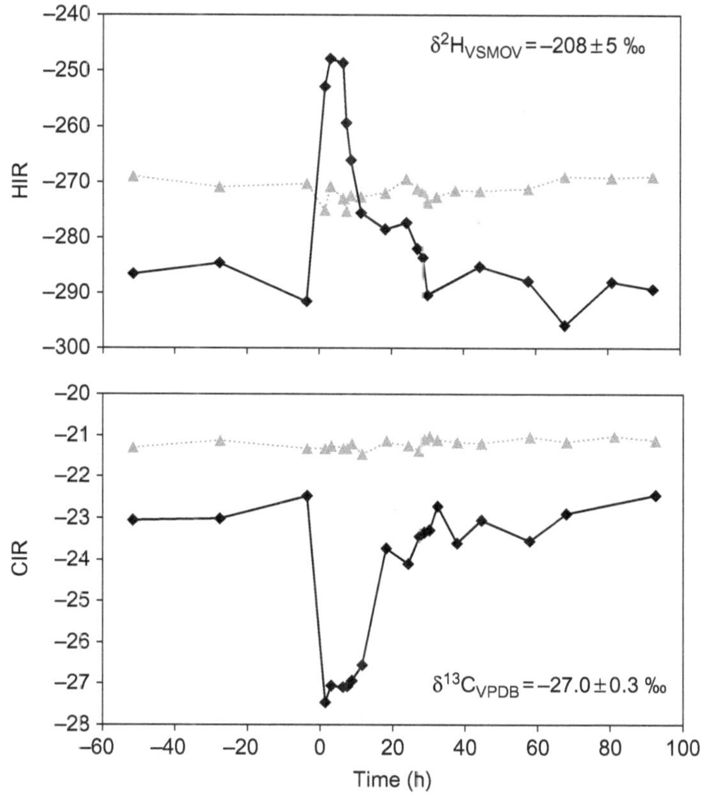

Fig. 11 Urinary HIR (*upper part*) and CIR (*lower part*) values of PD (*gray triangles*) and T (*black diamonds*) after oral application of 100 mg T-undecanoate at $t=0$. Given are the CIR and HIR values of T after cleavage of the ester.

isotope ratios. But surprisingly, the impact was different on CIR compared to HIR. For approx. 10 h after the application the urinary T perfectly reflects the CIR value of the administered steroid, suggesting that within this time period nearly no endogenous T contributes to the urinary excretion. For the HIR a similar behavior was expected but the value measured for urinary T was 40‰ less enriched than expected.

One possible explanation would be an isotopic fractionation ongoing during the cleavage of T-undecanoate to T resulting in erroneous HIR values for the liberated steroid. But as the saponification was almost complete and no difficulties have been described in literature here (Cawley et al., 2010), this hypothesis is not supported. Thus, metabolic isotopic fractionation seems to be a more likely reason for the findings. As nicely described by Hayes (2001), the isotopic composition of both the products and the starting material change over time if the conversion is affected by fractionation. Unfortunately, as there is no information available on fractionation factors associated with steroid metabolism, this hypothesis cannot yet be further substantiated. Additional investigations will be necessary to fully understand the isotopic fractionations taking place during human steroid metabolism.

Regarding the enhancement of T detection in doping control analysis, HIR were not that promising as initially expected. On the one hand is the desired stronger influence on HIR compared to CIR obviously not present, maybe due to the described isotopic fractionation. On the other hand were the endogenous HIR values found in a RP partly comparable to the published HIR values of several available T preparations (Cawley et al., 2010; Piper, Thomas, et al., 2012). This distinctive overlap of endogenous and exogenous values may prevent doping detection and results in the use of HIR only as a confirmation technique in those cases, where the CIR values were inconclusive due to the—luckily rare—misuse of T with an endogenous CIR signature.

Another option would be the combination of both CIR and HIR measurements in the same sample allowing for a longer detection of the exogenous T due to the fact that the decision limits can be narrowed as more variables are under investigation. But also here the above-mentioned new long-term metabolite EPIA seems to outperform the, additionally more complicated and time consuming, combination of both isotopic ratios.

5. CONCLUSION

IRMS applied to doping control analysis necessitates sophisticated and thoroughly validated methods to ensure high quality results for each sample. Furthermore, in-depth investigations on factors influencing endogenous CIR and HIR values were inevitable to assess the typical variation of endogenous steroids and to allow for a clear identification of those cases due to doping offenses.

As the sample preparation method encompasses several critical steps with regards to possible isotopic fractionation, the complete procedure has to be carefully monitored. The described measures taken during validation ensure that the method is fit for purpose. Especially the LMMs help to evaluate possible limitations and enable calculation of measurement uncertainties in a straight forward manner. Under routine conditions, small variations in the retention times on the HPLC may cause severe problems and need careful inspection within each sequence. Processing quality control urine samples is the only reliable measure to ensure that this critical step together with all the other extractions occurred without significant isotopic fractionation taking place. As in many other procedures applied in analytical chemistry, Shewhart charts are an invaluable aid to monitor both short- and long-term method performance.

Many possible factors influencing CIR and HIR values of endogenous steroids have been investigated mainly by determination of RP-based values. While of course the isotopic composition of food and drinking water has a strong influence on absolute δ values, within sports drug testing the more valuable Δ values are not affected. Other possible factors like gender or type of sport seem to only play an insignificant role.

Characteristic Δ values have been identified between all steroids and even between similar steroids depending on their phase-II-metabolism pointing to enzymatic isotopic fractionations ongoing during the metabolism of cholesterol to the different steroids excreted into urine. First investigations here allowed for an insight into possible mechanisms by combining RP data and individual excretion studies. Until now the data available are only scarce, but the potential of isotope ratio measurements to address questions regarding metabolism could be demonstrated. Especially a combination of both carbon and hydrogen isotopes will give a deeper insight and help to avoid misinterpretations.

REFERENCES

Aguilera, R., Becchi, M., Casabianca, H., Hatton, C. K., Catlin, D. H., Starcevic, B., et al. (1996). Improved method of detection of testosterone abuse by gas chromatography/combustion/isotope ratio mass spectrometry analysis of urinary steroids. *Journal of Mass Spectrometry*, *31*, 169–176.

Angelis, Y. S., Kioussi, M. K., Kiousi, P., Brenna, J. T., & Georgakopolus, C. G. (2012). Examination of the kinetic isotopic effect to the acetylation derivatization for the gas chromatographic-combustion-isotope ratio mass spectrometric doping control analysis of endogenous steroids. *Drug Testing and Analysis*, *4*, 923–927.

Becchi, M., Aguilera, R., Farizon, Y., Flament, M.-M., Casabianca, H., & James, P. (1994). Gas chromatography/combustion/isotope ratio mass spectrometry analysis of urinary steroids to detect misuse of testosterone in sport. *Rapid Communications in Mass Spectrometry*, *8*, 304–308.

Braunsdorf, R., Hener, U., Przibilla, G., Piecha, S., & Mosandl, A. (1993). The influence of analytical and technological procedures on the $^{13}C/^{12}C$ isotope ratio of orange oil compounds. *Zeitschrift für Lebensmittel-Untersuchung und Forschung*, *197*, 24–28.

Buisson, C., Hebestreit, M., Preiss Weigert, A., Heinrich, K., Fry, H., Flenker, U., et al. (2005). Application of stable carbon isotope analysis to the detection of 17β-estradiol administration to cattle. *Journal of Chromatography A*, *1093*, 69–80.

Burstein, S., & Lieberman, S. (1958). Kinetics and mechanism of solvolysis of steroid hydrogen sulfates. *Journal of the American Chemical Society*, *80*, 5235–5239.

Caimi, R., & Brenna, T. (1997). Quantitative evaluation of carbon isotopic fractionation during reversed-phase high-performance liquid chromatography. *Journal of Chromatography A*, *757*, 307–310.

Cawley, A., Collins, M., Kazlauskas, R., Handelsman, D. J., Heywood, R., Longworth, M., et al. (2010). Stable isotope ratio profiling of testosterone preparations. *Drug Testing and Analysis*, *2*, 557–567.

Cawley, A. T., Kazlauskas, R., Trout, G. J., Rogerson, J. H., & George, A. V. (2005). Isotopic fractionation of endogenous anabolic androgenic steroids and its relationship to doping control in sports. *Journal of Chromatographic Science*, *43*, 32–38.

Cawley, A. T., Trout, G. J., Kazlauskas, R., Howe, C. J., & George, A. V. (2009). Carbon isotope ratio ($\delta^{13}C$) values of urinary steroids for doping control in sport. *Steroids*, *74*, 379–392.

de la Torre, X., Colamonici, C., Curcio, D., Molaioni, F., & Botre, F. (2012). A comprehensive procedure based on gas chromatography-isotope ratio mass spectrometry following high performance liquid chromatography purification for the analysis of underivatized testosterone and its analogues in human urine. *Analytica Chimica Acta*, *756*, 23–29.

Docherty, G., Jones, V., & Evershed, P. (2001). Practical and theoretical considerations in the gas chromatography/combustion/isotope ratio mass spectrometry $\delta^{13}C$ analysis of small polyfunctional compounds. *Rapid Communications in Mass Spectrometry*, *15*, 730–738.

Donike, M., Rauth, S., & Wolansky, A. (1993). In M. Donike, H. Geyer, A. Gotzmann, & U. Mareck-Engelke (Eds.), *Reference ranges of urinary endogenous steroids determined by GC/MS 10th cologne workshop on dope analysis 7th to 12th June 1992* (pp. 69–86). Sport und Buch Strauß: Cologne.

Ehleringer, J. R., Rundel, P. W., & Nagy, K. A. (1986). Stable isotopes in physiological ecology and food web research. *Trends in Ecology & Evolution*, *1*, 42–45.

Flenker, U., Güntner, U., & Schänzer, W. (2008). $\delta^{13}C$-values of endogenous urinary steroids. *Steroids*, *73*, 408–416.

Flenker, U., & Schänzer, W. (2001). Kinetic isotope effects during metabolism of Δ4-steroids. In W. Schänzer, H. Geyer, A. Gotzmann, & U. Mareck-Engelke (Eds.), *Recent advances in doping analysis: Vol. 9* (pp. 179–185). Sport und Buch Strauß: Cologne.

Greaf, V., Furuya, E., & Nishikaze, O. (1977). Hydrolysis of steroid glucuronides with β-glucuronidase preparations from bovine liver, Helix pomatia, and E. coli. *Clinical Chemistry*, *23*, 532–535.

Guo, Z. K., Luke, A. H., Lee, W. P., & Schoeller, D. (1993). Compound-specific carbon isotope ratio determination of enriched cholesterol. *Analytical Chemistry*, *65*, 1954–1959.

Hayes, M. (2001). Fractionation of the isotopes of carbon and hydrogen in biosynthetic processes. *Reviews in Mineralogy and Geochemistry*, *43*, 225–277.

Horning, S., Geyer, H., Machnik, M., Schänzer, W., Hilkert, A., & Oeßelmann, J. (1997). Detection of exogenous testosterone by $^{13}C/^{12}C$ analysis. In W. Schänzer, H. Geyer, A. Gotzmann, & U. Mareck-Engelke (Eds.), *Recent advances in doping analysis: Vol. 5* (pp. 275–284). Sport und Buch Strauß: Cologne.

Keeling, C. D. (1958). The concentration and isotopic abundance of atmospheric carbon dioxide in rural areas. *Geochimica et Cosmochimica Acta*, *13*, 332–334.

MacKenzie, P. I., Rodbourne, L., & Stranks, S. (1992). Steroid UDP glucuronosyltransferases. *The Journal of Steroid Biochemistry and Molecular Biology*, *43*, 1099–1105.

Pataki, D. E., Ehlering, J. R., Flanagan, L. B., Yakir, D., Bowling, D. R., Still, C. J., et al. (2003). The application and interpretation of Keeling plots in terrestrial carbon cycle research. *Global Biogeochemical Cycles*, *17*, 1022–1037.

PetitClerk, C., & Solberg, H. E. (1987). Approved recommendation (1987) on the theory of reference values. Part 2. Selection of individuals for the production of reference values. *Clinica Chimica Acta*, *170*, S3–S12.

Piper, T., Baume, N., Strahm, E., Emery, C., & Saugy, M. (2012). Influences of β-HCG administration on carbon isotope ratios of endogenous urinary steroids. *Steroids*, *77*, 644–654.

Piper, T., Degenhardt, K., Federherr, E., Thomas, A., Thevis, M., & Saugy, M. (2013). Effect of changes in the deuterium content of drinking water on the hydrogen isotope ratio of urinary steroids in the context of sports drug testing. *Analytical and Bioanalytical Chemistry*, *405*, 2911–2921.

Piper, T., Emery, C., & Saugy, M. (2011). Recent developments in the use of isotope ratio mass spectrometry in sports drug testing. *Analytical and Bioanalytical Chemistry*, *401*, 433–447.

Piper, T., Emery, C., Thomas, A., Saugy, T., & Thevis, M. (2013). Combination of carbon isotope ratio with hydrogen isotope ratio determinations in sports drug testing. *Analytical and Bioanalytical Chemistry*, *405*, 5455–5466.

Piper, T., Flenker, U., Mareck, U., & Schänzer, W. (2009). $^{13}C/^{12}C$ ratios of endogenous urinary steroids investigated for doping control purposes. *Drug Testing and Analysis*, *1*, 65–72.

Piper, T., Fußhöller, G., & Schänzer, W. (2011). Investigation on the deconjugation processes of conjugated urinary steroids during degradation with ^{18}O labeled water. In W. Schänzer, H. Geyer, A. Gotzmann, & U. Mareck (Eds.), *Recent advances in doping analysis: Vol. 19* (pp. 296–299). Sport und Buch Strauß: Cologne.

Piper, T., Geyer, H., & Schänzer, W. (2010). Degradation of urine samples and its influence on the $^{13}C/^{12}C$ ratios of excreted steroids. *Drug Testing and Analysis*, *2*, 620–629.

Piper, T., Mareck, U., Geyer, H., Flenker, U., Thevis, M., Platen, P., et al. (2008). Determination of $^{13}C/^{12}C$ ratios of endogenous urinary steroids: Method validation, reference population and application to doping control purposes. *Rapid Communications in Mass Spectrometry*, *22*, 2161–2175.

Piper, T., Opfermann, G., Thevis, M., & Schänzer, W. (2010). Determination of $^{13}C/^{12}C$ ratios of endogenous urinary steroids excreted as sulpho conjugates. *Rapid Communications in Mass Spectrometry*, *24*, 3171–3181.

Piper, T., Riemann, P., Opfermann, G., Mareck, U., Geyer, H., Vajiala, G., et al. (2009). Determination of $^{13}C/^{12}C$ ratios of urinary epitestosterone and its main metabolites 5α- and 5β-androstane-3α, 17α-diol. *Drug Testing and Analysis*, *1*, 576–586.

Piper, T., Schänzer, W., & Thevis, M. (2016a). Revisiting the metabolism of 19-nortestosterone using isotope ratio and high resolution/high accuracy mass spectrometry. *The Journal of Steroid Biochemistry and Molecular Biology*, *162*, 80–91.

Piper, T., Schänzer, W., & Thevis, M. (2016b). Genotype-dependent metabolism of exogenous testosterone—New biomarkers result in prolonged detectability. *Drug Testing and Analysis*, *8*, 1163–1173.

Piper, T., Thevis, M., Flenker, U., & Schänzer, W. (2009). Determination of the deuterium/hydrogen ratio of endogenous urinary steroids for doping control purposes. *Rapid Communications in Mass Spectrometry*, *23*, 1917–1926.

Piper, T., Thomas, A., Baume, N., Sobolevsky, T., Saugy, M., Rodchenkov, G., et al. (2014). Determination of $^{13}C/^{12}C$ ratios of endogenous urinary 5-aminoimidazole-4-carboxamide 1β-D-ribofuranoside (AICAR). *Rapid Communications in Mass Spectrometry*, *28*, 1194–1202.

Piper, T., Thomas, A., Thevis, M., & Saugy, M. (2012). Investigations on hydrogen isotope ratios of endogenous urinary steroids: Reference-population based thresholds and proof-of-concept. *Drug Testing and Analysis*, *4*, 717–727.

Ramseyer, J., Williams, J. S., & Hirschmann, H. (1967). Carbon oxygen cleavage in the acid hydrolysis of androsterone sulfate. *Steroids*, *9*, 347–365.

Rieley, G. (1994). Derivatization of organic compounds prior to gas chromatographic-combustion-isotope ratio mass spectrometric analyses: Identification of isotope fractionation processes. *The Analyst*, *119*, 915–919.

Sessions, A. L. (2006). Isotope-ratio detection for gas chromatography. *Journal of Separation Science, 29*, 1946–1961.

Sessions, A. L., Burgoyne, T. W., & Hayes, J. M. (2001). Determination of the H_3 factor in hydrogen isotope ratio monitoring mass spectrometry. *Analytical Chemistry, 73*, 200–207.

Shinebarger, S. R., Haisch, M., & Matthews, D. E. (2002). Retention of carbon and alteration of expected 13C-tracer enrichments by silylated derivatives using continuous-flow combustion-isotope ratio mass spectrometry. *Analytical Chemistry, 74*, 6244–6251.

Sobolevsky, T., Prasolov, I. S., & Rodchenkov, G. (2010). Carbon isotope mass spectrometry in doping control. *Journal of Analytical Chemistry, 65*, 825–832.

Solberg, H. E. (1987a). Approved recommendation (1986) on the theory of reference values. Part 1. The concept of reference values. *Journal of Clinical Chemistry and Clinical Biochemistry, 25*, 337–342.

Solberg, H. E. (1987b). Approved recommendation (1987) on the theory of reference values. Part 5. Statistical treatment of collected reference values. Determination of reference limits. *Journal of Clinical Chemistry and Clinical Biochemistry, 25*, 645–656.

Solberg, H. E., & PetitClerk, C. (1988). Approved recommendation (1988) on the theory of reference values. Part 3. Preparation of individuals and collection of specimen for the production of reference values. *Clinica Chimica Acta, 177*, S1–S12.

Träger, L. (1977). *Steroidhormone*. Berlin: Springer.

Ueki, M., & Okanao, M. (1999). Analysis of exogenous dehydroepiandrosterone excretion in urine by gas chromatography/combustion/isotope ratio mass spectrometry. *Rapid Communications in Mass Spectrometry, 13*, 2237–2243.

Vestergaard, P. (1978). The hydrolysis of conjugated neutral steroids in urine. *Acta Endocrinologica. Supplementum, 217*, 96–120.

WADA. (2015). *Anti-doping testing figures*. Montreal, Canada: Anti-Doping Agency solely. www.wada-ama.org/en/resources/laboratories/anti-doping-testing-figures. Accessed 13.03.2017.

WADA Technical Document–TD2016IRMS. (2016). *Detection of synthetic forms of endogenous anabolic androgenic steroids by GC/C/IRMS*. Montreal, Canada: Anti-Doping Agency solely. www.wada-ama.org/en/resources/science-medicine/td2016-irms. Accessed 13.03.2017.

Westgard, J. O., Barry, P. L., Hunt, M. R., Groth, T., Burnett, R. W., Hainline, A., Jr., et al. (1981). A multi-rule Shewhart chart for quality control in clinical chemistry. *Clinical Chemistry, 27*, 493–501.

Zhang, Y., Tobias, H. J., & Brenna, J. T. (2009). Steroid isotopic standards for gas chromatography-combustion isotope ratio mass spectrometry (GCC-IRMS). *Steroids, 74*, 369–378.

CHAPTER FIFTEEN

Kinetic Isotope Effect Analysis of RNA 2′-*O*-Transphosphorylation

Michael E. Harris*[,1], Darrin M. York[†], Joseph A. Piccirilli[‡], Vernon E. Anderson[§]

*University of Florida, Gainesville, FL, United States
[†]Rutgers University, Piscataway, NJ, United States
[‡]The University of Chicago, Chicago, IL, United States
[§]Case Western Reserve University School of Medicine, Cleveland, OH, United States
[1]Corresponding author: e-mail address: harris@chem.ufl.edu

Contents

1. Introduction 434
   1.1 Mechanisms and Transition States of RNA 2′-*O*-Transphosphorylation Reactions 435
   1.2 Application of KIE Analyses to Investigate Transition States of RNA Strand Cleavage Reactions 438
2. Synthesis of Isotopomer RNA Substrates Enriched in ^{18}O at the Reacting Phosphoryl Oxygens 439
   2.1 [2′-^{18}O]Uridine 439
   2.2 [5′-^{18}O]Guanosine 440
   2.3 [^{18}O] Incorporation Into NPOs 441
3. Use of Electrospray Ionization Quadrupole/Time-of-Flight Mass Spectrometry to Measure ^{18}O KIEs for RNA 2′-*O*-transphosphorylation 441
   3.1 Determination of Reaction Kinetics and Purification of Substrates and Products Using HPLC 441
   3.2 Determination of ^{16}O/^{18}O Ratios in 5′-UpG-3′ by Tandem Q/TOF MS 445
   3.3 Data Fitting and Calculation of KIEs Using Internal Competition Kinetics 450
4. Future Directions 453
References 454

Abstract

The breaking of RNA strands by 2′-*O*-transphosphorylation is a ubiquitous reaction in biology, and enzymes that catalyze this reaction play key roles in RNA metabolism. The mechanisms of 2′-*O*-transphosphorylation in solution are relatively well studied, but complex and can involve different transition states depending on how the reaction is catalyzed. Because of this complexity and the lack of experimental information on transition-state structure, pinning down the chemical details of enzyme-catalyzed RNA strand cleavage has been difficult. Kinetic isotope effects (KIEs) provide information about changes in bonding as a reaction proceeds from ground state to transition state,

Methods in Enzymology, Volume 596
ISSN 0076-6879
http://dx.doi.org/10.1016/bs.mie.2017.07.017

and therefore they provide a powerful tool for revealing mechanistic detail. Application of kinetic isotope analyses to RNA 2′-*O*-transphosphorylation faces three fundamental challenges: synthesis of RNA substrate isotopomers with ^{18}O substitutions at the 2′-*O*, 5′-*O* and nonbridging phosphoryl oxygens; determination of the $^{18}O/^{16}O$ ratios in the residual unreacted substrate or product RNAs; and analyzing these data to allow calculation of the KIEs for use in evaluating different mechanistic scenarios. In this chapter, we outline methods for surmounting these challenges for solution RNA 2′-*O*-transphosphorylation reactions, and we describe their initial application to understand nonenzymatic solution reactions and reactions catalyzed by the enzyme ribonuclease A.

1. INTRODUCTION

The power of kinetic isotope effects (KIEs) combined with computation to resolve the mechanisms and characterize the transition states of enzymes is well established (Cook, 1991; Schramm, Horenstein, & Kline, 1994). However, measuring KIEs is extremely challenging due to their relatively small magnitude and the high precision that is usually required. Because of the wide range of substrates and chemical reactions catalyzed by enzymes, developing or adapting methods that allow transition-state analysis of key enzyme systems is a significant and continuing challenge. KIE analyses of enzymes that are current or potential therapeutic targets are particularly important since information on the catalytic transition states can provide guidelines for developing tight-binding competitive inhibitors (Schramm, 2015).

The breaking of RNA strands by 2′-*O*-transphosphorylation is a ubiquitous reaction in biology (Cuchillo, Nogues, & Raines, 2011; Oivanen, Kuusela, & Lonnberg, 1998), and enzymes that catalyze this reaction play key roles in RNA metabolism (Lilley, 2017; Masuda & Inouye, 2017; Saikia & Hatzoglou, 2015). Both protein ribonucleases and RNA enzymes (ribozymes) catalyze RNA strand cleavage by this mechanism making possible a comparison of biological catalysis by these two different polymers at a deep mechanistic level (Cochrane & Strobel, 2008; Lassila, Zalatan, & Herschlag, 2011; Wilcox, Ahluwalia, & Bevilacqua, 2011). The mechanisms of 2′-*O*-transphosphorylation in solution are relatively well studied, but they are complex and can involve different transition states depending on how the reaction is catalyzed (Kellerman, York, Piccirilli, & Harris, 2014; Korhonen, Koivusalo, Toivola, & Mikkola, 2013; Oivanen et al., 1998). Because of this complexity and the lack of experimental information on transition-state structure, pinning down the chemical details of enzyme-catalyzed RNA strand cleavage has been difficult. Thus, developing experimental and computational methods for measuring and interpreting KIEs for solution RNA cleavage reactions and for ribonucleases and ribozymes presents an area where success can fill key gaps in our knowledge. Accordingly, we directed significant attention toward developing the experimental and conceptual framework for transition-state analysis of RNA reactions.

This chapter focuses on the details of the mass spectrometric methods used to measure $^{16}O/^{18}O$ ratios in substrate and/or product ribonucleotides. An essential component of transition-state analysis is computational simulation and calculations of predicted KIEs, e.g., (Harris, Piccirilli, & York, 2015; Vardi-Kilshtain, Nitoker, & Major, 2015; Wong, Xu, & Xu, 2015). Together the measurement and calculation of KIEs allow modeling of transition-state structure and permit different mechanistic possibilities to be evaluated. Computational methods are important for integrating KIE and other experimental data into a comprehensive mechanistic framework. Such a combination of theory and experiment has been essential for interpreting the KIEs measured for RNA cleavage reaction. The details of the computational methods used for developing transition-state models for RNA 2′-O-transphosphorylation are beyond the scope of this chapter. Nonetheless, such efforts are integral to success and we outline the resulting framework that has been thus developed for this class of reactions (Chen et al., 2014; Chen, Piccirilli, Harris, & York, 2015; Gu et al., 2013; Huang & York, 2014; Radak, Harris, & York, 2013; Wong et al., 2012). Finally, we discuss future questions that can be addressed regarding the mechanisms and transition states of RNA 2′-O-transphosphorylation reactions by applying the general approaches described herein.

1.1 Mechanisms and Transition States of RNA 2′-O-Transphosphorylation Reactions

Attack of the 2′OH nucleophile on the adjacent phosphoryl group results in a cyclic 2′,3′-nucleotide monophosphate product and the adjacent nucleoside 5′-hydroxyl product (Fig. 1). There are large changes in the bonding and electron distribution around the 2′-O nucleophile, 5′-O leaving group, and nonbridging phosphoryl oxygens (NPOs) that may be probed by primary and secondary KIEs to gain insight into mechanism. Classic studies by Westheimer, Lönnberg, Cleland, and others established that phosphoryl transfer reactions can occur by a range of potential mechanisms and transition states (Breslow, 1993; Cleland & Hengge, 1995; Dennis & Westheimer, 1966; Kirby & Younas, 1970; Oivanen et al., 1998; Westheimer, 1968).

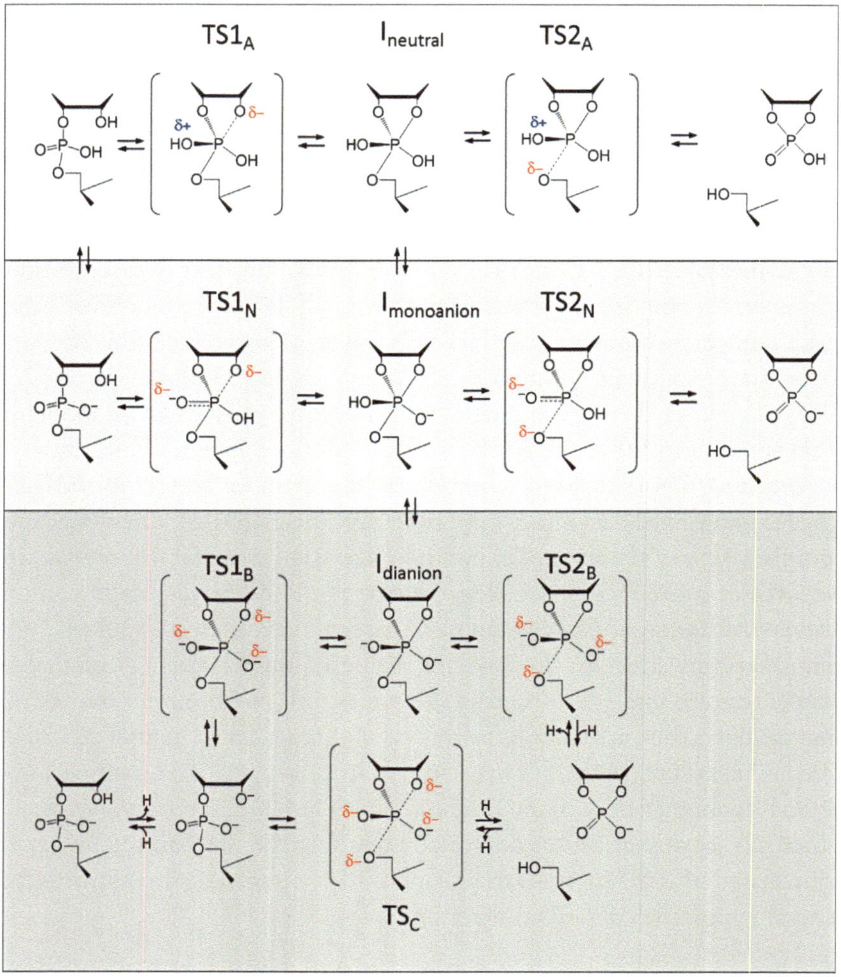

Fig. 1 Mechanisms and transition states for RNA 2′-*O*-transphosphorylation. (*Top*) Acid catalysis of neutral RNA and monoanion (*Middle*) by stepwise mechanisms via a neutral or monoanionic phosphorane. (*Bottom*) The stepwise (via a dianionic phosphorane) and concerted mechanisms for base catalysis.

Summarizing decades of research on the reactions of cyclic phosphodiesters, we know that specific acid catalysis occurs by a stepwise mechanism involving the formation of a pentacoordinate phosphorane. The structure of the phosphorane may be neutral (**I**$_{\mathbf{neutral}}$) or monoanionic (**I**$_{\mathbf{monoanion}}$) depending on pH. Accordingly there are two separate transition states, one for nucleophilic attack (**TS1**) and another for leaving group departure (**TS2**). Formation of 2′,5′-phosphodiester isomerization products, as observed for 2′-*O*-transphosphorylation reactions catalyzed by acid, is characteristic of a mechanism in which the intermediate is in at least partial equilibrium with the substrate ground state and the rate-limiting transition state is TS2.

Specific base (OH^-) catalysis is typically depicted as occurring by a concerted mechanism with a single dianionic transition state (**TSc**) (Cleland & Hengge, 1995; Gerratana, Sowa, & Cleland, 2000; Hengge, Bruzik, Tobin, Cleland, & Tsai, 2000; Oivanen et al., 1998). The KIE results obtained using the methods described herein helped to provide evidence that this is only true when the leaving group is "activated" (i.e., has a low pK_a), while those with a higher pK_a, e.g., the 5′-*O* of RNA, proceed by a stepwise mechanism through a dianionic phosphorane ($\mathbf{I}_{\text{dianion}}$) in which **TS2** is rate limiting (Chen et al., 2014; Huang & York, 2014; Lonnberg, Stromberg, & Williams, 2004). In addition to this mechanistic complexity, phosphoryl transfer reactions necessarily involve proton transfers at the nucleophile, leaving group and potentially the NPOs. These transfers can either be concerted or occur in preequilibrium steps with respect to O–P bond formation/cleavage (i.e., they may occur by general or specific acid/base mechanisms (Jencks, 1987)). This overall framework, based on KIE and physical organic chemical studies, shows that acid/base and metal ion catalysis induce large effects on the structure of the rate-limiting TS (Cassano, Anderson, & Harris, 2004a; Humphry, Forconi, Williams, & Hengge, 2002, 2004; Humphry et al., 2008; Mikkola et al., 1999; Oivanen et al., 1998; Rawlings, Cleland, & Hengge, 2003; Rishavy, Hengge, & Cleland, 2000). The characteristic plasticity of phosphodiester cleavage mechanisms suggests both that the complex active site environments of enzymes are likely to alter TS structure, and that predicting the details of the mechanism by examining ground-state structures will be problematic.

Thus, the potential for KIE analyses to help resolve catalytic mechanisms of phosphoryl transfer enzymes is very high. This potential is founded on the extensive experimental and conceptual framework derived from analyses of the KIEs of phosphodiester reactions by the Hengge and Cleland labs (Cleland, 1982; Cleland & Hengge, 1995; Feng, Tanifum, Adams, Hengge, & Williams, 2009; Grzyska, Czyryca, Purcell, & Hengge, 2003; Hengge, 2002; Hengge, Aleksandra, & Cleland, 1995; Hengge et al., 2000; Rawlings, Cleland, & Hengge, 2006; Rishavy et al., 2000). Extensive studies of nucleophile, leaving group and nonbridging oxygen KIEs demonstrate that these effects fall in the range of 0.93–1.07 (k^{16}/k^{18}) and provide important information on extents of bond formation and cleavage (Cleland & Hengge, 2006; Hengge, 2002). Additionally, the equilibrium isotope effects from the transfer of ionizable protons influence the observed KIEs. For example, the KIE for the nonbridging oxygens for hydrolysis of alkyl-nitrophenol phosphodiesters is large and inverse in the acid-catalyzed reaction due to transfer of a proton. In contrast, a secondary nonbridging oxygen KIE of near unity is observed for late, product-like transition states of base-catalyzed reactions in which there is no proton transfer to these positions (Hengge, 2002; Hengge et al., 1995). Thus, KIE analysis of model diester reactions reveals relatively large and characteristic primary and secondary effects on the phosphoryl oxygens indicating that analogous studies can resolve details of RNA 2′-*O*-transphosphorylation transition-state structures.

1.2 Application of KIE Analyses to Investigate Transition States of RNA Strand Cleavage Reactions

In order to measure KIEs by internal competition, a mixed population of RNA isotopologues containing either ^{18}O or ^{16}O at a specific phosphoryl oxygen is allowed to react, and the change in isotopic composition of the residual substrate is analyzed. For example, to determine KIE on the nucleophile the 2′-*O* position must be site-specifically enriched with ^{18}O. For determination of KIEs on the 5′-*O* leaving group or nonbridging oxygens, different RNA isotopomers are required and these positions must be individually enriched in a different population of substrates. Typically, a level of enrichment near 50% in the substrate population is targeted so that the signal-to-noise ratio for the ^{16}O and ^{18}O isotopologues is essentially identical. As the reaction proceeds, the slower reacting isotopologue becomes progressively enriched in the unreacted substrate population, and the faster reacting isotopologue becomes progressively depleted. Analysis of the kinetics of the change in isotopic composition of unreacted RNA substrate as a function of reaction progress permits the KIE to be calculated by data fitting.

These kinds of experimental data report directly on differences in activation energy that reflect differences in bonding between the ground state and the transition state. However, interpretation of KIE data in terms of transition-state structure and bonding optimally requires the use of theoretical models. The necessary goal being to provide a direct connection between observed KIEs, as well as other physical biochemistry data (primarily Brønsted coefficients) measured for these reactions, with the structure and bonding in the transition state. Establishing such relationships is essential for evaluating mechanistic alternatives. Computational simulations offer a way to estimate the effects of different contributions to observed KIEs (such as protonation, changes in hybridization, metal ion coordination, and solvation) that may be difficult or impossible to measure experimentally.

In the present example, relationships between computation and experimental data have provided a powerful tool to investigate RNA phosphoryl transfer and demonstrate the utility of building up knowledge of mechanism through the systematic study of model systems to provide insight into more complex biological systems.

2. SYNTHESIS OF ISOTOPOMER RNA SUBSTRATES ENRICHED IN ^{18}O AT THE REACTING PHOSPHORYL OXYGENS

In order to measure primary KIEs on RNA 2′-*O*-transphosphorylation reactions, it is necessary to use different RNA isotopomers as substrates that are enriched with ^{18}O at the 2′-*O* (nucleophile) and 5′-*O* (leaving group) positions. The secondary KIEs at the NPO positions are also important, and RNAs enriched at these positions can provide key mechanistic information. Fig. 2 shows the oxygen atoms that can be interrogated using ^{18}O KIEs. The synthesis of isotopically enriched nucleotide monomers was recently reviewed by Weissman, Li, York, Harris, and Piccirilli (2015), and this contribution provides a comprehensive description of synthetic strategies to target key positions in the nucleobase and ribose. For synthetic purposes, the heavy and light nucleoside monomers are chemically equivalent; therefore, conversion to phosphoramidites and subsequent incorporation into RNA oligonucleotides utilizes standard methods. Of tremendous value for future studies is extensive literature on the synthesis of isotopically enriched nucleotides stemming from early work on nucleotide metabolism and classic stereochemical studies of phosphoryl transfer and enzyme catalysis (Brody & Frey, 1981; Connolly, Eckstein, & Fuldner, 1982; Jones, Kindman, & Knowles, 1978; Webb & Trentham, 1980). The ability to make these substrates, combined with the mass spectrometric methods outlined here provides a very powerful set of tools for analyzing a wide range of reactions and enzyme systems.

2.1 [2′-^{18}O]Uridine

The synthesis of [2′-^{18}O]uridine enables synthesis of RNA substrates for isotope effect analyses of nucleophilic activation by protein and RNA enzymes that catalyze RNA 2′-*O*-transphosphorylation. Uridine containing 2′-^{18}O has been synthesized by reaction of commercially available and relatively inexpensive 2,2′-cyclouridine with an ^{18}O-enriched oxygen nucleophile that attacks the ribose C-2′ regioselectively. As reported in Dai et al. potassium [$^{18}O_2$]benzoate favors ribose attack over nucleobase attack

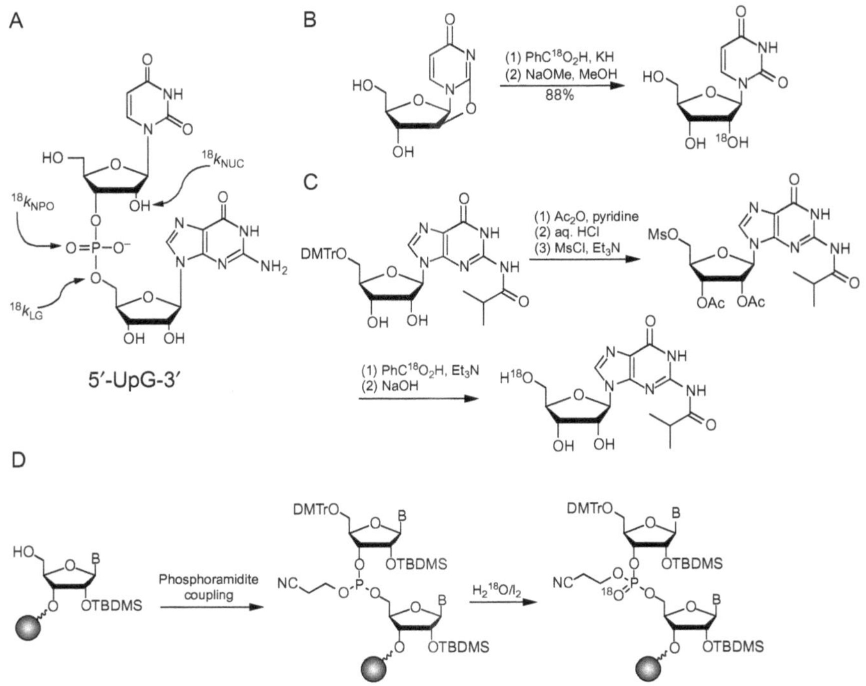

Fig. 2 (A) Structure of 5′-UpG-3′ showing the locations of the isotopic enrichment with ^{18}O required to measure KIEs for the 2′-*O* nucleophile ($^{18}k_{NUC}$), 5′-*O* leaving group ($^{18}k_{LG}$), and nonbridging oxygens ($^{18}k_{NPO}$). (B) Scheme for synthesis of 2′-[^{18}O]uridine. (C) Scheme for synthesis of 5′-[^{18}O]guanosine. (D) Scheme for synthesis of uridylyl-3′-guanosine-5′-^{18}O-phosphate. *Panels (B–D) from Weissman, B. P., Li, N. S., York, D., Harris, M., & Piccirilli, J. A. (2015). Heavy atom labeled nucleotides for measurement of kinetic isotope effects.* Biochimica et Biophysica Acta, 1854, *1737–1745.*

on 2,2′-cyclouridine by 20-fold, thus allowing for efficient synthesis of [2′-^{18}O]uridine and subsequently its phosphoramidite. This approach achieves high yield (88%) and employs a relatively inexpensive label source. The [$^{18}O_2$]benzoate is synthesized by acid hydrolysis of benzonitrile in HCl/$H_2^{18}O$. A mixture of [$^{18}O_2$]benzoic acid (1.0 equiv.) and potassium [$^{18}O_2$]benzoate (1.0 equiv.) at 140°C in DMF within 48 h gives products in a ratio of 20:1 ribo/arabino.

2.2 [5′-^{18}O]Guanosine

Synthesis of [5′-^{18}O]guanosine has been described previously in Gu et al. (2013). A similar strategy as that used for enrichment of the 2′-O position is employed to introduce ^{18}O using [$^{18}O_2$]benzoic acid to react with the

5′-*O* position. Commercially available 5′-DMT-*i*-Bu-guanosine is reacted and the 2′ and 3′ hydroxyls are protected with acetyl groups using standard methods. The 5′-DMT protecting group is subsequently removed by acid treatment. The 5′-hydroxyl is activated for nucleophilic attack as a mesylate ester, which is treated with [$^{18}O_2$]benzoic acid. The 5′-*O*-4,4-dimethoxytrityl (DMTr) protecting group is then added back resulting in a precursor compatible with phosphoramidite chemistry. As noted previously, this method should be generally applicable permitting all four 5′-^{18}O-labeled nucleosides to be prepared with appropriate protection (Weissman et al., 2015).

2.3 [^{18}O] Incorporation Into NPOs

Labeling of the nonbridging oxygen positions is accomplished by adding $H_2^{18}O$ during the oxidation step of solid-phase synthesis. Oxidation of phosphite triesters generates phosphate esters; using commercially available $H_2^{18}O$ results in incorporation of the labeled oxygen at a nonbridging position. A drawback of this approach is that it generates the diastereomeric mixture of heavy atom substitution at either the pro-S_p or the pro-R_p. Thus, the observed $^{18}k_{NPO}$ reflects the contribution from both positions. It is worth noting that an overall advantage of the use of internal competition is that the isotopically enriched monomers described here are not required to be synthesized such that the incorporation of ^{18}O is 100%. As detailed later, the optimal ratio of heavy and light isotopologue substrates in the reaction is near 50/50 in order to achieve the highest degree of precision in the calculation of KIEs from MS data. Therefore, RNA isotopomers synthesized at higher levels of enrichment are necessarily diluted with natural abundance substrates that are also synthesized by standard phosphoramidite chemistry.

3. USE OF ELECTROSPRAY IONIZATION QUADRUPOLE/TIME-OF-FLIGHT MASS SPECTROMETRY TO MEASURE ^{18}O KIES FOR RNA 2′-*O*-TRANSPHOSPHORYLATION

3.1 Determination of Reaction Kinetics and Purification of Substrates and Products Using HPLC

The base-catalyzed 2′-*O*-transphosphorylation reaction of 5′-UpG-3′ results in the formation of 2′,3′ cUMP and guanosine. Subsequent attack of hydroxide on the cyclic phosphate results in formation of a mixture of 2′-UMP and 3′-UMP. Acid catalysis occurs via a characteristic stepwise mechanism in which the phosphorane intermediate can rearrange and subsequently decompose to the 2′,5′ phosphodiester isomer of UpG. The separation of nucleotides and RNA oligonucleotides by HPLC is well developed and offers multiple alternative approaches. For the reaction of 3′,5′-UpG, the precursor dinucleotide as well as the guanosine, cUMP, and 2′,5′-UpG isomerization products can all be separated and independently quantified using this approach.

The material requirement for a KIE experiment is fundamentally determined by the sensitivity of the isotope ratio measurement and secondarily by the chemical efficiency of the delivery of the substrate or product to be analyzed to the mass spectrometer. Determination of the sensitivity and precision of the isotope ratio measurement has to be evaluated for each analyte on each instrument. In our system, initial calibration experiments indicated that 100–200 pmol of UpG are sufficient to optimally measure $^{18}O/^{16}O$ ratios for KIE measurements. Therefore, in order to be able to take multiple time points at different fractions of reaction, a total of 1–5 nmol of UpG dinucleotide are typically used in each individual reaction. For solution reactions, the kinetics are first order in substrate concentration and therefore a concentration can be chosen convenient for subsequent sample workup and HPLC injection volume. The nonenzymatic solution reactions are sufficiently slow that simple neutralization and freezing is sufficient for quenching. For example, acid catalysis of UpG transphosphorylation at pH 0 was carried out in 250–500 μL at 37°C. Aliquots were removed at specific times in order to obtain fractions of reaction ranging from 0.1 to 0.7 and neutralized with an equivalent of NaOH followed by further dilution with 100–200 μL 0.1 *M* ammonium acetate pH 8.0. Ammonium acetate is used for the subsequent HPLC separation since it can be subsequently removed by lyophilization, which effectively desalts the sample to prepare it for electrospray ionization quadrupole/time-of-flight (ESI-Q/TOF).

RP-HPLC of nucleotides and short (<15-mer) oligonucleotides is widely used (Frederiksen & Piccirilli, 2009; Gilar, 2001; McLaughlin & Bischoff, 1987) and allows quantification of reaction progress and simultaneous purification for analysis of isotopic composition by ESI-Q/TOF with near quantitative recovery. Minor losses through incomplete injection of the reaction sample are inconsequential because the extent of reaction measurement and the isotopic measurement are both ratios and independent of the recovery. A Shimadzu Prominence HPLC system was used with detection at 260 and 254 nm coupled to a C18 column (300 × 3.9 mm, 10 μm; Phenomenex, Torrence, CA), eluting isocratically at 1 mL/min using a mobile phase of 0.1 *M* ammonium acetate containing 4% acetonitrile. Peaks were identified using standards and analyses of UV absorbance and their areas were recorded and quantified using the manufacturer's software. Detection at two different wavelengths, 260 and 254 nm, identifies the position of the two mononucleoside products as these are the λ_{max} of uridine and guanosine, respectively. The difference in the A260 and A254 of the peaks eluting at 4 and 10 min is consistent with the identity of these species as cUMP and guanosine (Dawson, 1986).

The HPLC fractions containing the unreacted UpG are collected manually, taking care to collect the entire peak. In order to avoid fractionation due to interaction with the column, the collection begins before peak detection and continues until the remaining signal of the tail is well below 1% of the peak height as estimated by inspection of the signal as it is being collected. Since concentration of the sample is not difficult, there is no penalty for taking large (1–2 mL) total volumes. The lack of isotopic fractionation on this process can be demonstrated in initial experiments comparing values before and after chromatography. The recovered material is mixed, aliquoted into microfuge tubes containing <1 mL of volume, and dried by Speedvac without heating. The pellets are resuspended in 1 mL H_2O and dried down an additional four or more times to remove as much of the ammonium acetate salt as possible. The samples from each tube are progressively combined and concentrated. The recovered UpG is resuspended in water at a concentration of 20–50 μ*M* for subsequent isotopic analysis by ESI-Q/TOF. It is important to check for further significant reaction during workup, which can easily be monitored by rechromatography of the UpG recovered from drying.

For determination of KIEs an important parameter for subsequent data fitting is the fraction of reaction, $F/F0$ ($F=[P]/([P]+[S])$ and F0 is this fraction calculated for a sample taken at $t=0$). Simple integration of the chromatogram permits quantification of reaction progress. This was done both with the Shimadzu integration software as well as by exporting the data and fitting it using Origin graphing and analysis software. The advantage of the latter is that it provides much more control over the setting of baselines. Comparison of the measured values to theoretical progress curves fit to the data allows the error in measurement to be estimated. As an example Fig. 3 shows the kinetics of acid-catalyzed 2′-O-transphosphorylation and isomerization. To confirm the precision of the determinations of $F/F0$ these data are plotted vs time and fit to a standard exponential function and evaluation of the residuals tests whether the error in determination of $F/F0$ is low. The data shown are for 3′-UpG-5′ (filled squares), guanosine

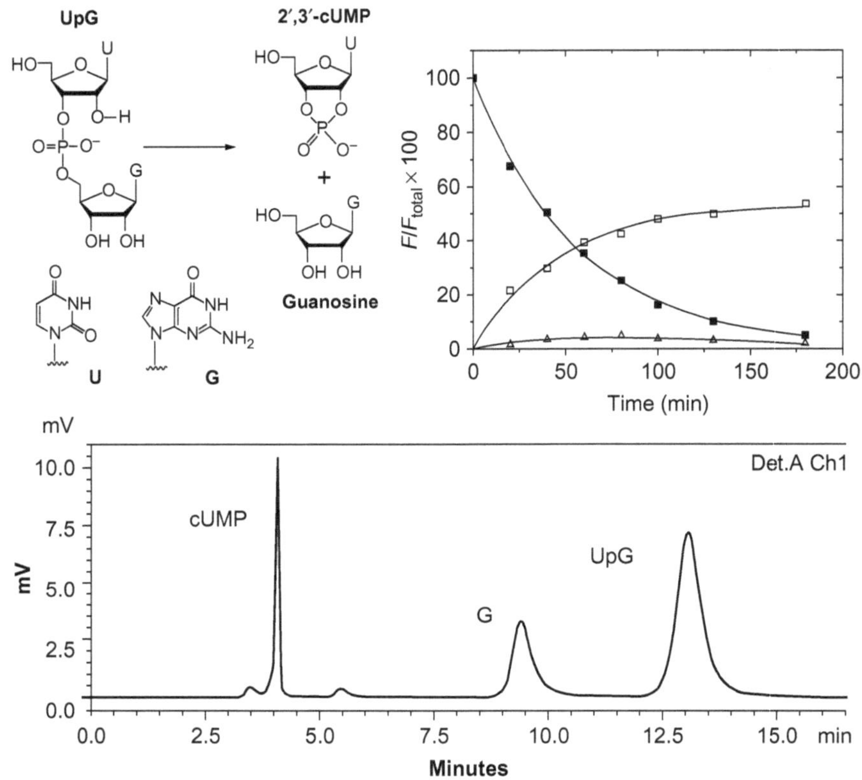

Fig. 3 (A) Scheme for 2′-*O*-transphosphorylation of 5′-UpG-3′. (B) HPLC chromatogram showing the separation of substrate (UpG) and products (cUMP and G). (C) Time course of changes in precursor and product species derived from integration of the corresponding peaks as illustrated in panel (B). The data shown are for 3′-UpG-5′ (*filled squares*), guanosine (*open squares*), and the transient formation of the isomerization product 2′-UpG-5′ is shown as *open triangles*. The cUMP product is not shown as it elutes in the void volume under these conditions and is difficult to quantify.

(open squares), and the transient formation of the isomerization product 2′-UpG-5′ is shown as open triangles. The cUMP product is not shown as it elutes in the void volume under these conditions and is difficult to quantify.

Enrichment in the slower reacting isotopologue becomes progressively greater at higher *F*/F0. Since KIEs are typically only a few percent, the change in isotope ratio is measured in the residual substrate approximate this magnitude in the *F*/F0 range of 0.2–0.9. This increase in signal to noise at high *F*/F0 is advantageous; however, the error in determination of *F*/F0 becomes increasingly greater at long reaction times as well since the signal to noise in measurement of the low concentration of residual substrate approaches zero and the value for the product(s) approaches one. A zero, unreacted time point is taken for each reaction but, the isotope ratio in the starting population is not fixed as a constant in the determination of KIEs. By fitting the data with the ratio at time = 0 as a variable rather than fixing it as a constant using the measured value, this single measurement of $^{16}O/^{18}O$ is not unduly weighted in the data fitting.

3.2 Determination of $^{16}O/^{18}O$ Ratios in 5′-UpG-3′ by Tandem Q/TOF MS

The high precision required for KIE measurements has been achieved using isotope ratio mass spectrometry (IRMS) or dual-label scintillation counting of radioactive isotopes (Hengge & Cleland, 1991). While these methods have proven highly informative in deciphering enzyme transition states, both possess inherent limitations that restrict their application. IRMS can only analyze small, nonpolar gasses and scintillation counting requires double-labeling schemes involving specific isotopic pairs limiting the molecules that can be analyzed. One alternative to these methods is whole molecule mass spectrometry using electrospray ionization (ESI) or matrix-assisted laser desorption ionization to allow larger molecules to be analyzed.

Coupled with QMS, ESI makes possible the analysis of isotope ratios for numerous biologically relevant organic molecules with molecular masses in excess of 500 Da and application has allowed KIEs to be determined for peptidyl transfer and RNA hydrolysis. However, several features limit the precision and accuracy of QMS. These features include noise (from both the ion source and detector), background, intensity fluctuations, and the inability to capture the entire ion current arising from individual isotopologues. These effects can be minimized by working with isotope ratios near unity, utilizing background subtraction, and/or injecting analyte by continuous direct infusion. Previously, we showed that the data necessary to calculate isotope effects as small as 1.005 could be obtained by analytically integrating intensity vs *m/z* data obtained by ESI-QMS to determine the isotope ratios (Cassano et al., 2007). The ability to analyze isotope ratios in nucleotides permitted the application of heavy atom isotope effect studies to RNA hydrolysis demonstrating attack by hydroxide ion and metal ion catalysis by direct coordination (Cassano, Anderson, & Harris, 2002, 2004a, 2004b).

Fundamentally, the degree of mechanistic information that can be obtained is limited by precision, which in turn is limited by the MS component of the analysis. In this regard, the use of tandem Q/TOF MS with the same ESI mode of ion generation presents several potentially transformative advantages. The TOF acquisition by design generates data in profile mode, in which the intensity over a range of *m/z* values is plotted, allowing all of the isotopomers in a cluster to be quantitatively analyzed. Also, the tandem approach has much greater signal to noise since the background of the second TOF dimension is essentially zero. Additionally, with increased sensitivity and the ability to perform long acquisition times, many individual scans can be averaged which results in a corresponding increase in precision. Also, as described in more detail later the KIE can be determined from multiple fragment ions from the same molecule which provides a check on internal accuracy. The higher sensitivity and throughput, together with analyses of multiple fragment ions allow comparison of multiple independent determinations and allow KIEs to be measured with precision approaching ± 0.001. Application of ESI-Q/TOF imposes several requirements: (1) product ions are generated that contain the isotopically labeled oxygen, (2) the isotope ratio can be measured with sufficient precision that standard errors in data fitting to determine KIE are low, and (3) the experimental error is sufficiently low that mechanistic interpretations can be made.

The first requirement is demonstrated by the TOF mass spectrum of the product ions of $m/z = 588$ (Fig. 4). In this experiment, the solution to be analyzed was injected by automated syringe pump at a flow rate of 1 μL/min and the UpG ion was identified in a single dimension of TOF MS and a mass range encompassing the entire isotopic envelope was selected. In practice, this is done empirically by varying the *m/z* range of the quadripole isolation until the isotope ratio in the product ions is constant. Fragments are generated by inert gas collision with collision energies of ca. 35 V, and the resultant molecules are separated by a second dimension of TOF MS.

Fig. 4 shows a scan of ion intensity and the resultant fragment ions analyzed by TOF MS. Fragmentation of UpG gives rise to ions at 476, 362, 305, 255, 211, 192, and 111 amu. To optimize the generation of these ions the collision energy was varied and the ion intensity quantified. The 255 amu ion was present in the unfragmented spectrum and varied in intensity and does not contain any of the enriched positions. For these reasons, this ion was considered to be a contaminant and was not analyzed further. Since the $^{16}O/^{18}O$ at the 2′-*O* of UpG is ~1.0 in this experiment, the resulting

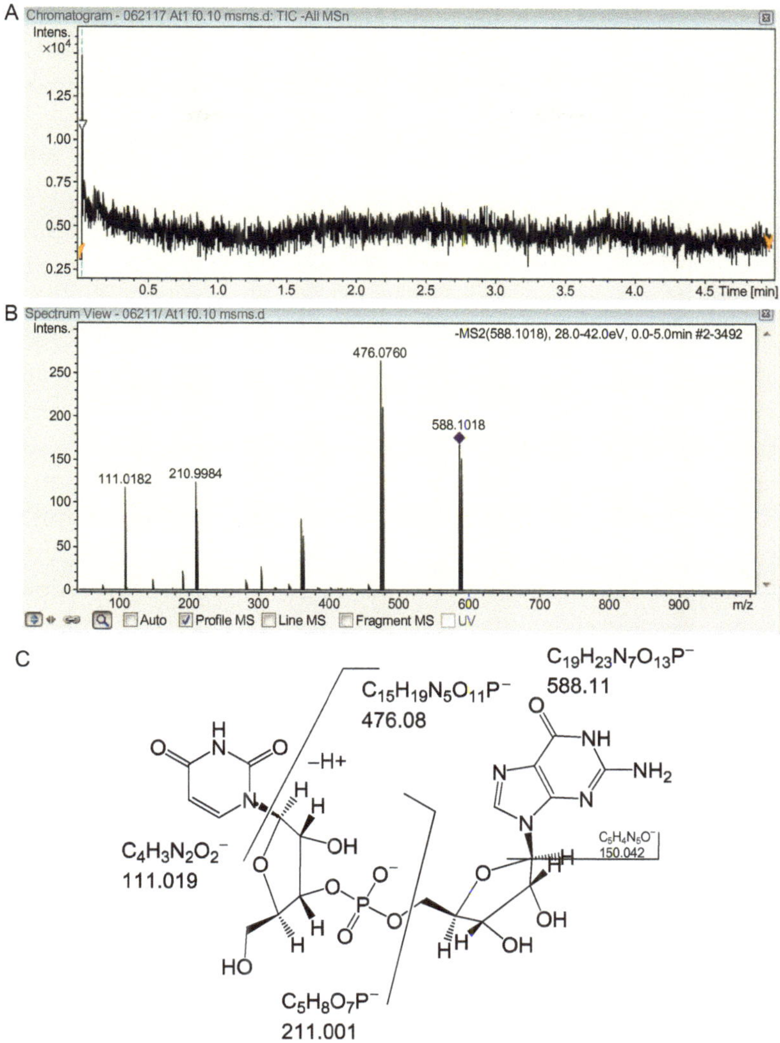

Fig. 4 (A) Total ion current time course of the direct infusion of UpG. *Shaded area* indicates the time span averaged to generate the mass spectrum. (B) Averaged negative ion tandem mass spectrum of the isotopic cluster around *m/z*. The insets demonstrate the resolution, peak shape, and lack of background signal around *m/z* 211 and 476. (C) Fragmentation products of UpG. The elemental composition and monoisotopic masses of the fragment ions are given.

product ions that contain the labeled oxygen can be identified by the pattern observed in their isotopic cluster. Only the 476, 305, 211, and 193 ions contain the enriched 2′-hydroxyl as expected from the structures of these molecules (Fig. 4). Of these fragmentation products the signal to noise of the 476 and 211 ions is clearly greater and these ions were selected for further quantitative analysis. Fortuitously, both ions also contain the 5′-*O* and NBOs and therefore are useful for determination of KIEs at all of the reacting phosphoryl oxygens. The peak shapes are largely symmetrical and the area can be integrated manually, or by using the routines included in the software (Analyst, ABI).

While all of the ions are derived from the same population in this analyzed solution, the isotope ratios determined by ESI-Q/TOF are significantly different. For example, at a collision energy of 35 the ratios for the 588, 476, and 211 ions were determined to be 1.08, 1.14, and 1.35, respectively. Clearly, the accuracy of the ratios is suspect since each are derived from the same isotopologue population. Evidence that the difference in the ratios is due to effects arising from collision is shown in Fig. 5. In this example, the observed ratio was monitored over a range of collision energies. Interestingly, the ratio varies by as much as 20% for all of the ions as the collision energy is raised. This sensitivity to collision energy suggests that subtle effects on ion flight are responsible for these differences. Multiple collisions may result in progressive displacement of the lighter ion from the ion path. For the 211 ion, at some level this displacement must result in enrichment of the secondary 211 ion. This makes sense since at higher collision energies the ^{18}O-enriched 476 ion population resulting from by multiple collisions begins to decay to the 211 ion. Therefore, the accuracy of the ratios is relatively low and as an analytic method to determine ratios of unknowns, much additional work would be necessary to establish the source of these variations.

However, for isotope effect determinations using competitive methods the absolute accuracy of the measurements is not an important factor. In the determination of the isotope effect, the same inaccuracy will be present in all of the data points analyzed. The key issue is whether changes in the ratio can be determined with sufficient precision to follow the enrichment of one or the other of the isotopologues as a function of reaction progress. To assess the precision with which $^{16}O/^{18}O$ ratios may be measured in UpG using this method, we compared the observed ratio to the ratio predicted from a set of control samples that were progressively diluted with natural abundance UpG. Fig. 5 shows a linear plot of the observed vs predicted ratio. The linearity of the data ($R^2 > 0.95$) demonstrates the ability of the method to discriminate changes in the observed ratio that is well within the precision required for KIE calculation. One factor necessary to maintain discriminate precise differences between successive samples, is to take care to minimize sample bleedover effects. Residual sample in the injector lines, sample inlet,

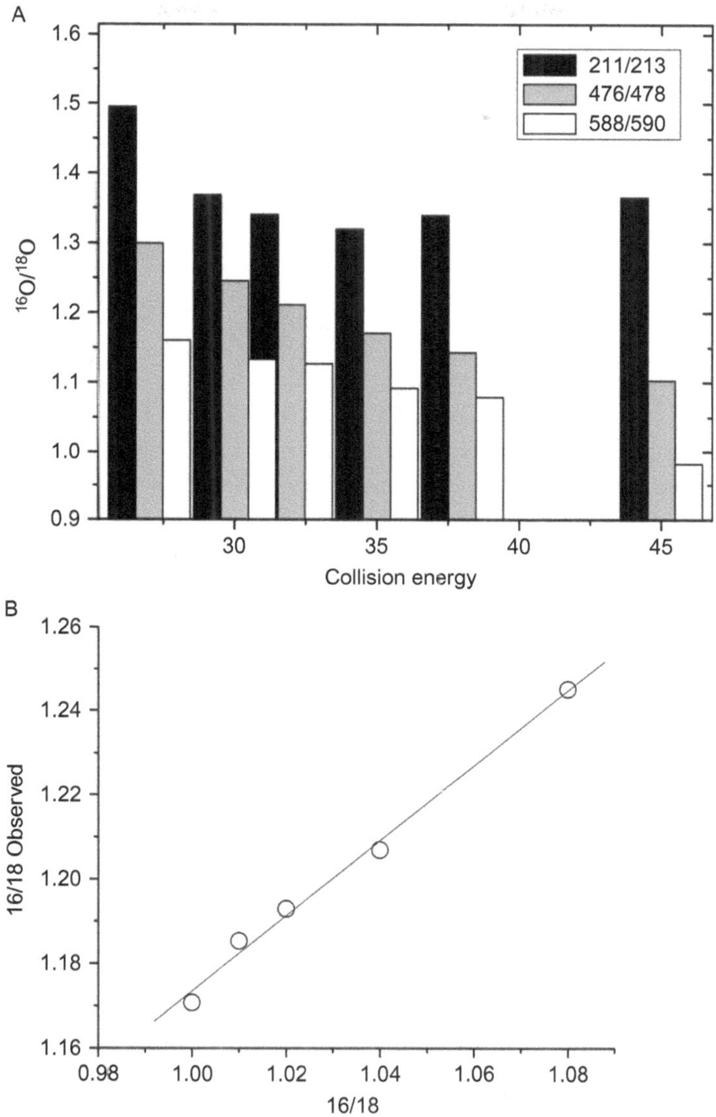

Fig. 5 (A) *Bar graph* showing the dependence on collision energy (V) of the observed $^{16}O/^{18}O$ ratio calculated from the Q/TOF MS analysis of fragments shown in Fig. 4. (B) Comparison of the intrinsic vs observed $^{16}O/^{18}O$ ratios for UpG determined from the 211 *m/z* fragment.

and ionization cell can result in significant contributions to the observed signal if these components are not sufficiently rinsed between samples. It is necessary to monitor background to insure that no bleedover will occur during acquisition of the subsequent sample.

3.3 Data Fitting and Calculation of KIEs Using Internal Competition Kinetics

To determine the isotope effect the natural log of the ratio measurements are plotted vs the fraction of reaction. Fig. 6 shows the results for determination of the KIE on the nonbridging oxygen for the acid-catalyzed 2′-*O*-transphosphorylation reaction of the UpG dinucleotide. These data are fit to the following equation:

$$\ln(R_s) = \left(1/{}^{18}k - 1\right)\ln(1-F) + \ln(R_0) \quad (1)$$

where ${}^{18}k$ is the isotope effect, F is the fraction of unreacted (residual) substrate (UpG) quantified by RP-HPLC, and R_s and R_0 are the ratios of ${}^{16}O/{}^{18}O$ for the unreacted substrate at F and its initial value, respectively. Eq. (1) is derived from internal competition kinetics and is used for fitting the precursor ratio. The derivation of this expression as well as the background and equations for fitting product ratios has recently been reviewed (Anderson, 2015).

Table 1 shows the compiled results from three independent experimental determinations of ${}^{18}k_{NBO}$. The increase in sensitivity permits an individual determination to be made using ca. 0.5 nmol of enriched nucleotide substrate, and thus multiple determinations can be made with a relatively small investment of material compared to previous whole molecule analyses and IRMS. Since the MS analysis allows the ratio to be determined in both the 211 and 476 ions each experiment yields two independent determinations of the isotope effect. Importantly, essentially the same KIE is measured whether the 211 or 476 ion is used for the analysis (see later). For each experiment, a duplicate reaction was run and thus each experiment results in four determinations of the KIE from data fitting for a total of 12 measurements of ${}^{18}k_{NBO}$. Averaging with equal weighting yields a value of 0.9944 ± 0.0074. A threshold for inclusion of individual experiments can be set if necessary at a standard error of <0.005 for determination of the KIE by fitting isotope ratio vs reaction progress data. Evaluation of the reaction kinetics can also be used to identify determinations in which the accuracy is questionable. Goodness of fit can be used as a criterion to exclude outlier data and can also be used to weight the individual KIE determinations. In this operation the individual KIEs are multiplied by the reciprocal of the standard error, averaged, and then divided by the sum of the reciprocals. If the values are weighted in this fashion and all measurements are included the ${}^{18}k_{NBO}$ is determined to be 0.9907 ± 0.0012. Given this data set the computed ${}^{18}k_{NBO}$ from using either

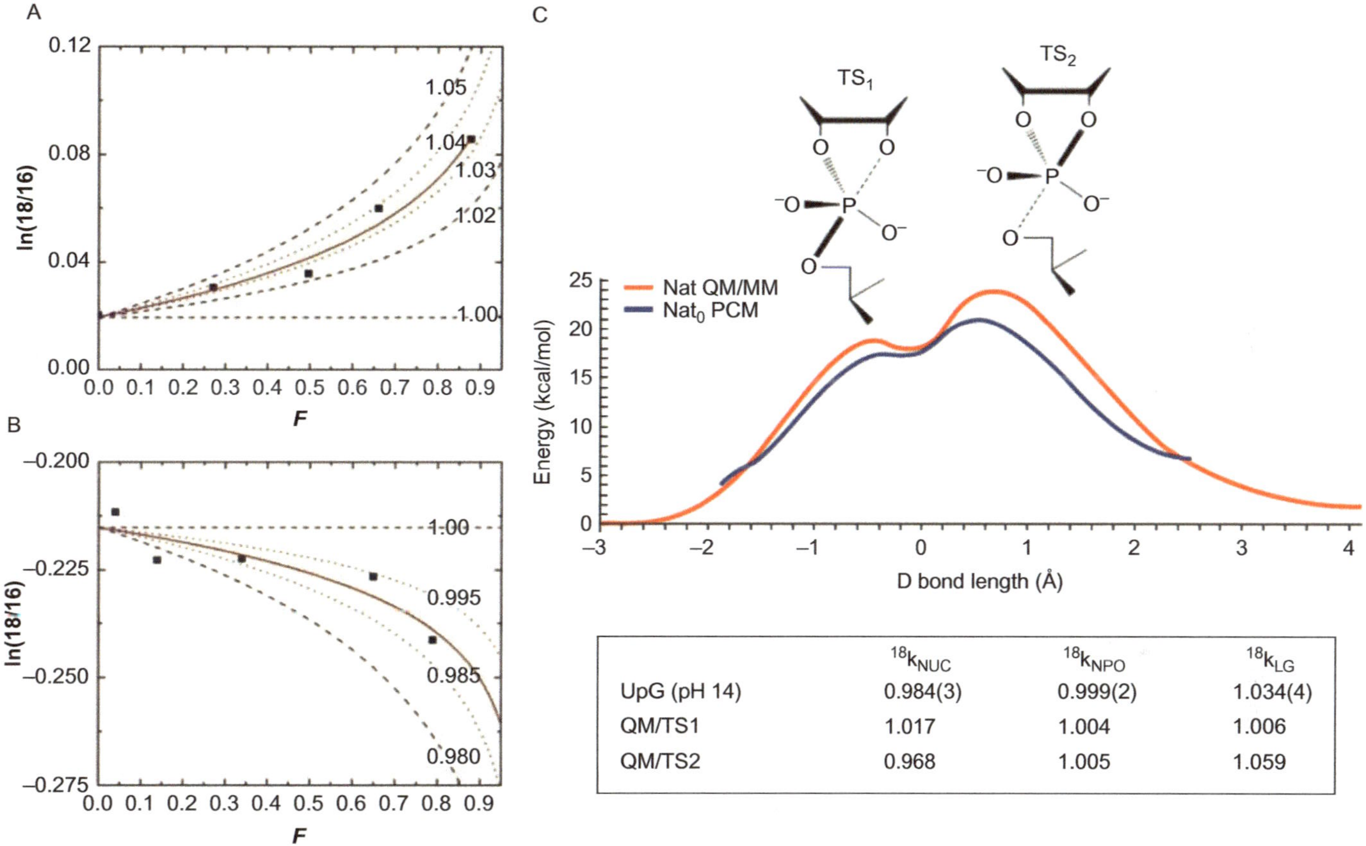

	$^{18}k_{NUC}$	$^{18}k_{NPO}$	$^{18}k_{LG}$
UpG (pH 14)	0.984(3)	0.999(2)	1.034(4)
QM/TS1	1.017	1.004	1.006
QM/TS2	0.968	1.005	1.059

Fig. 6 Determination of observed KIEs by analysis of ln($^{16}O/^{18}O$) vs *F* data. (A) Determination of $^{18}k_{LG}$ by quantification of the change in $^{16}O/^{18}O$ as a function of reaction progress (*filled symbols*). The *red line* is a fit of the data to Eq. (1) yielding a KIE of 1.032. The *blue lines* represent simulations that span the range of observable ^{18}O KIEs. (B) Determination of $^{18}k_{NUC}$. The data and fit to Eq. (1) are indicated as in panel A. The *blue lines* represent simulations of KIEs of unity to 0.980 inverse for comparison. (C) Comparison of density-functional QM/MM free energy and adiabatic PCM profiles for the native 2′-*O*-transphosphorylation reaction as a function of the difference of the bond distance (Δbond distance) between the breaking bond P-O5′ and the forming bond P-O2′. Examples of the observed and calculated KIEs for TS1 and TS2 are shown.

Table 1 Independent Determinations of $^{18}k_{NBO}$

	$18k_{NB}$	**Std Error**
Experiment-001		
Reaction-A 211 ion	0.9914	0.00189
Reaction-A 476 ion	0.991	0.00172
Reaction-B 211 ion	0.9911	0.00162
Reaction-B 476 ion	0.99121	0.00145
Experiment-002		
Reaction-A 211 ion	0.99228	0.00439
Reaction-A 476 ion	0.99136	0.0045
Reaction-B 211 ion	1.0093	0.01585
Reaction-B 476 ion	1.011	0.01914
Experiment-003		
Reaction-A 211 ion	0.9915	0.00123
Reaction-A 476 ion	0.98901	0.000205
Reaction-B 211 ion	0.9924	0.00095
Reaction-B 476 ion	0.9913	0.000239

the 211 or the 476 amu data alone can be compared. If only the 211 data is used the $^{18}k_{NBO}$ is found to be 0.991736 ± 0.000572 and if only the 476 data is used the $^{18}k_{NBO}$ is 0.990776 ± 0.000997. Importantly, use of either ion gives essentially the same value for $^{18}k_{NBO}$ within 0.001 and with comparable precision. These results indicate that the method is robust and that the inaccuracies in the observed isotope ratios do not influence the measured KIE and serve as an excellent check on internal accuracy in determination of the effect.

The cleavage of RNA chains by attack of the adjacent 2′-OH is universal in biology and has been intensively studied since the 1950s, yet these approaches allow the first characterization of RNA cleavage with primary isotope effects capable of directly informing on the extent of bond cleavage in the transition state. Comparing the results with the literature and KIE measurements of model phosphoryl transfer reactions reveals that significant differences exist in the transition states for concerted phosphoryl transfer reactions and that the high pK_a of the leaving group coupled with the enforced geometry of cleavage can significantly alter the transition state for the reaction. The KIE data indicate a concerted mechanism in which bond formation to the nucleophile and bond cleavage from the leaving group are both advanced in a "late" or product-like transition state. The nucleophile isotope effects implicate equilibrium deprotonation of the 2′-*O* prior to nucleophilic attack, and thus catalysis by hydroxide occurs by a specific base mechanism.

As reported in Wong et al., the density-functional QM/MM free energy profile and the PCM adiabatic reaction profile for the reaction of the native compound are very similar (Fig. 6) (Wong et al., 2012). Both profiles show an associative mechanism with distinct TS1 and TS2 transition states that are separated by a shallow, metastable intermediate ($I_{dianion}$). In both cases, TS2 is rate controlling. Importantly, TS2 has considerable "late" character, in which cleavage of the exocyclic P-O5′ bond is advanced. The calculated density-functional free energy barrier with PCM solvation is very similar to that estimated from the experimental rate for UpG transphosphorylation (ca. 20 kcal mol^{-1}). This result, along with the consistent values from the PCM reaction profile and from the QM/MM simulations with explicit solvent, suggests that the density-functional PCM calculations sufficiently capture the essential features of the solvation effects on the reaction profile to determine the KIE values (Chen et al., 2014; Radak et al., 2013; Wong et al., 2012). Thus, taken together the results provide an atomic-level picture of the reactions, which includes a detailed characterization of the rate-controlling transition state for RNA transphosphorylation.

4. FUTURE DIRECTIONS

An important step is application of these methods to enzymes including both protein ribonucleases and small self-cleavage ribozymes. Indeed, we have reported initial application to RNase A catalysis (Gu et al., 2013) and the results and implications for catalytic mechanism have been recently reviewed (Harris et al., 2015; Kellerman et al., 2014). The transition state analyses of RNase A represent an initial first step, but illustrate the complexity of measuring KIEs for enzymes. Unlike solution reactions, an important consideration for enzyme reactions is the determination of substrate-binding commitments, which can obscure the intrinsic magnitude of KIEs on the chemical step. A complete treatment of these factors as they apply to RNase A and other ribonucleases is beyond the scope of this chapter. However, it is apparent that the early success with RNase A bodes extremely well for application to additional systems. RNA endonucleases such as bacterial ribotoxins and ribonucleases involved in modulating gene expression are important future targets. Information about the transition states for these enzymes can potentially provide inroads into the development of transition state-based inhibitors. On a technical level, the advent of mass spectrometers that permit very high-resolution mass resolution, or atomic fine structure, raises new possibilities for isotope ratio determinations. The information gained over the course of such studies is highly likely to have broad impact by enabling the improvement of computational methods, facilitating the design of novel catalysts, and revealing the potential for development of transition state based inhibitors.

REFERENCES

Anderson, V. E. (2015). Multiple alternative substrate kinetics. *Biochimica et Biophysica Acta*, *1854*, 1729–1736.

Breslow, R. (1993). Kinetics and mechanism in RNA cleavage. *Proceedings of the National Academy of Sciences of the United States of America*, *90*, 1208–1211.

Brody, R. S., & Frey, P. A. (1981). Unambiguous determination of the stereochemistry of nucleotidyl transfer catalyzed by DNA polymerase I from Escherichia coli. *Biochemistry*, *20*, 1245–1252.

Cassano, A. G., Anderson, V. E., & Harris, M. E. (2002). Evidence for direct attack by hydroxide in phosphodiester hydrolysis. *Journal of the American Chemical Society*, *124*, 10964–10965.

Cassano, A. G., Anderson, V. E., & Harris, M. E. (2004a). Analysis of solvent nucleophile isotope effects: Evidence for concerted mechanisms and nucleophilic activation by metal coordination in nonenzymatic and ribozyme-catalyzed phosphodiester hydrolysis. *Biochemistry*, *43*, 10547–10559.

Cassano, A. G., Anderson, V. E., & Harris, M. E. (2004b). Understanding the transition states of phosphodiester bond cleavage: Insights from heavy atom isotope effects. *Biopolymers*, *73*, 110–129.

Cassano, A. G., Wang, B., Anderson, D. R., Previs, S., Harris, M. E., & Anderson, V. E. (2007). Inaccuracies in selected ion monitoring determination of isotope ratios obviated by profile acquisition: Nucleotide 18O/16O measurements. *Analytical Biochemistry*, *367*, 28–39.

Chen, H., Giese, T. J., Huang, M., Wong, K. Y., Harris, M. E., & York, D. M. (2014). Mechanistic insights into RNA transphosphorylation from kinetic isotope effects and linear free energy relationships of model reactions. *Chemistry—A European Journal*, *20*(44), 14336–14343.

Chen, H., Piccirilli, J. A., Harris, M. E., & York, D. M. (2015). Effect of Zn binding and enzyme active site on the transition state for RNA 2′-O-transphosphorylation interpreted through kinetic isotope effects. *Biochimica et Biophysica Acta*, *1854*(11), 1795–1800.

Cleland, W. W. (1982). Use of isotope effects to elucidate enzyme mechanisms. *CRC Critical Reviews in Biochemistry*, *13*, 385–428.

Cleland, W. W., & Hengge, A. C. (1995). Mechanisms of phosphoryl and acyl transfer. *FASEB Journal*, *9*, 1585–1594.

Cleland, W. W., & Hengge, A. C. (2006). Enzymatic mechanisms of phosphate and sulfate transfer. *Chemical Reviews, 106*, 3252–3278.
Cochrane, J. C., & Strobel, S. A. (2008). Catalytic strategies of self-cleaving ribozymes. *Accounts of Chemical Research, 41*, 1027–1035.
Connolly, B. A., Eckstein, F., & Fuldner, H. H. (1982). Streospecific substitution of oxygen-18 for sulfur in nucleoside phosphorothioates. *The Journal of Biological Chemistry, 257*, 3382–3384.
Cook, P. F. (1991). *Enzyme mechanism from isotope effects.* New York: CRC Press.
Cuchillo, C. M., Nogues, M. V., & Raines, R. T. (2011). Bovine pancreatic ribonuclease: Fifty years of the first enzymatic reaction mechanism. *Biochemistry, 50*, 7835–7841.
Dawson, R. M. C. (1986). *Data for biochemical research.* Oxford, UK: Clarendon Press.
Dennis, E. A., & Westheimer, F. H. (1966). The geometry of the transition state in the hydrolysis of phosphate esters. *Journal of the American Chemical Society, 88*, 3432–3433.
Feng, G., Tanifum, E. A., Adams, H., Hengge, A. C., & Williams, N. H. (2009). Mechanism and transition state structure of aryl methylphosphonate esters doubly coordinated to a dinuclear cobalt(III) center. *Journal of the American Chemical Society, 131*, 12771–12779.
Frederiksen, J. K., & Piccirilli, J. A. (2009). Separation of RNA phosphorothioate oligonucleotides by HPLC. *Methods in Enzymology, 468*, 289–309.
Gerratana, B., Sowa, G. A., & Cleland, W. W. (2000). Characterization of the transition state structures and mechanisms for the isomerization and cleavage reactions of 3′-m-nitrobenzyl phosphate. *Journal of the American Chemical Society, 122*, 12615–12621.
Gilar, M. (2001). Analysis and purification of synthetic oligonucleotides by reversed-phase high-performance liquid chromatography with photodiode array and mass spectrometry detection. *Analytical Biochemistry, 298*, 196–206.
Grzyska, P. K., Czyryca, P. G., Purcell, J., & Hengge, A. C. (2003). Transition state differences in hydrolysis reactions of alkyl versus aryl phosphate monoester monoanions. *Journal of the American Chemical Society, 125*, 13106–13111.
Gu, H., Zhang, S., Wong, K. Y., Radak, B. K., Dissanayake, T., Kellerman, D. L., et al. (2013). Experimental and computational analysis of the transition state for ribonuclease A-catalyzed RNA 2′-O-transphosphorylation. *Proceedings of the National Academy of Sciences of the United States of America, 110*, 13002–13007.
Harris, M. E., Piccirilli, J. A., & York, D. M. (2015). Integration of kinetic isotope effect analyses to elucidate ribonuclease mechanism. *Biochimica et Biophysica Acta, 1854*, 1801–1808.
Hengge, A. C. (2002). Isotope effects in the study of phosphoryl and sulfuryl transfer reactions. *Accounts of Chemical Research, 35*, 105–112.
Hengge, A. C., Aleksandra, T. E., & Cleland, W. W. (1995). Studies of transition-state structures of phosphoryl transfer reactions of phosphodiesters of p-nitrophenol. *Journal of the American Chemical Society, 117*, 5919–5926.
Hengge, A. C., Bruzik, K. S., Tobin, A. E., Cleland, W. W., & Tsai, M. D. (2000). Kinetic isotope effects and stereochemical studies on a ribonuclease model: Hydrolysis reactions of uridine 3′-nitrophenyl phosphate. *Bio-Organic Chemistry, 28*, 119–133.
Hengge, A. C., & Cleland, W. W. (1991). Phosphoryl-transfer reactions of phosphodiesters: Characterization of transition states by heavy-atom isotope effects. *Journal of the American Chemical Society, 113*, 5835–5841.
Huang, M., & York, D. M. (2014). Linear free energy relationships in RNA transesterification: Theoretical models to aid experimental interpretations. *Physical Chemistry Chemical Physics, 16*, 15846–15855.
Humphry, T., Forconi, M., Williams, N. H., & Hengge, A. C. (2002). An altered mechanism of hydrolysis for a metal-complexed phosphate diester. *Journal of the American Chemical Society, 124*, 14860–14861.

Humphry, T., Forconi, M., Williams, N. H., & Hengge, A. C. (2004). Altered mechanisms of reactions of phosphate esters bridging a dinuclear metal center. *Journal of the American Chemical Society*, *126*, 11864–11869.

Humphry, T., Iyer, S., Iranzo, O., Morrow, J. R., Richard, J. P., Paneth, P., et al. (2008). Altered transition state for the reaction of an RNA model catalyzed by a dinuclear zinc(II) catalyst. *Journal of the American Chemical Society*, *130*, 17858–17866.

Jencks, W. P. (1987). *Catalysis in chemistry and enzymology*. New York: Dover Publications.

Jones, S. R., Kindman, L. A., & Knowles, J. R. (1978). Stereochemistry of phosphoryl group transfer using a chiral [16O, 17O, 18O] stereochemical course of alkaline phosphatase. *Nature*, *275*, 564–565.

Kellerman, D. L., York, D. M., Piccirilli, J. A., & Harris, M. E. (2014). Altered (transition) states: Mechanisms of solution and enzyme catalyzed RNA 2′-O-transphosphorylation. *Current Opinion In Chemical Biology*, *21*, 96–102.

Kirby, A. J., & Younas, M. (1970). The reactivity of phosphate esters. Reactions of diesters with nucleophiles. *Journal of the Chemical Society B: Physical Organic*, *B*, 1165–1172.

Korhonen, H., Koivusalo, T., Toivola, S., & Mikkola, S. (2013). There is no universal mechanism for the cleavage of RNA model compounds in the presence of metal ion catalysts. *Organic & Biomolecular Chemistry*, *11*, 8324–8339.

Lassila, J. K., Zalatan, J. G., & Herschlag, D. (2011). Biological phosphoryl-transfer reactions: Understanding mechanism and catalysis. *Annual Review of Biochemistry*, *80*, 669–702.

Lilley, D. M. J. (2017). How RNA acts as a nuclease: Some mechanistic comparisons in the nucleolytic ribozymes. *Biochemical Society Transactions*, *45*, 683–691.

Lonnberg, H., Stromberg, R., & Williams, A. (2004). Compelling evidence for a stepwise mechanism of the alkaline cyclisation of uridine 3′-phosphate esters. *Organic & Biomolecular Chemistry*, *2*, 2165–2167.

Masuda, H., & Inouye, M. (2017). Toxins of prokaryotic toxin-antitoxin systems with sequence-specific endoribonuclease activity. *Toxins*, *9*, 1–23.

McLaughlin, L. W., & Bischoff, R. (1987). Resolution of RNA using high-performance liquid chromatography. *Journal of Chromatography*, *418*, 51–72.

Mikkola, S., Stenman, E., Nurmi, K., Yousefi-Salakdeh, E., Stromberg, R., & Lonnberg, H. (1999). The mechanism of the metal ion promoted cleavage of RNA phosphodiester bonds involves a general acid catalysis by the metal aquo ion on the departure of the leaving group. *Journal of the Chemical Society, Perkin Transactions*, *2*, 1619–1625.

Oivanen, M., Kuusela, S., & Lonnberg, H. (1998). Kinetics and mechanism for the cleavage and isomerization of phosphodiester bonds of RNA by Bronsted acids and bases. *Chemical Reviews*, *98*, 961–990.

Radak, B. K., Harris, M. E., & York, D. M. (2013). Molecular simulations of RNA 2′-O-transesterification reaction models in solution. *The Journal of Physical Chemistry. B*, *117*, 94–103.

Rawlings, J., Cleland, W. W., & Hengge, A. C. (2003). Metal ion catalyzed hydrolysis of ethyl p-nitrophenyl phosphate. *Journal of Inorganic Biochemistry*, *93*, 61–65.

Rawlings, J., Cleland, W. W., & Hengge, A. C. (2006). Metal-catalyzed phosphodiester cleavage: Secondary 18O isotope effects as an indicator of mechanism. *Journal of the American Chemical Society*, *128*, 17120–17125.

Rishavy, M. A., Hengge, A. C., & Cleland, W. W. (2000). Lanthanide catalyzed cyclization of uridine 3′-p-Nitrophenyl phosphate. *Bio-Organic Chemistry*, *28*, 283–292.

Saikia, M., & Hatzoglou, M. (2015). The many virtues of tRNA-derived stress-induced RNAs (tiRNAs): Discovering novel mechanisms of stress response and effect on human health. *The Journal of Biological Chemistry*, *290*, 29761–29768.

Schramm, V. L. (2015). Transition states and transition state analogue interactions with enzymes. *Accounts of Chemical Research*, *48*, 1032–1039.

Schramm, V. L., Horenstein, B. A., & Kline, P. C. (1994). Transition state analysis and inhibitor design for enzymatic reactions. *The Journal of Biological Chemistry*, *269*, 18259–18262.

Vardi-Kilshtain, A., Nitoker, N., & Major, D. T. (2015). Nuclear quantum effects and kinetic isotope effects in enzyme reactions. *Archives of Biochemistry and Biophysics*, *582*, 18–27.

Webb, M. R., & Trentham, D. R. (1980). Analysis of chiral inorganic [16O, 17O, 18O] thiophosphate and the stereochemistry of the 3-phosphoglycerate kinase reaction. *The Journal of Biological Chemistry*, *255*, 1775–1778.

Weissman, B. P., Li, N. S., York, D., Harris, M., & Piccirilli, J. A. (2015). Heavy atom labeled nucleotides for measurement of kinetic isotope effects. *Biochimica et Biophysica Acta*, *1854*, 1737–1745.

Westheimer, F. H. (1968). Pseudo-rotation in the hydrolysis of phosphate esters. *Accounts of Chemical Research*, *1*, 70–78.

Wilcox, J. L., Ahluwalia, A. K., & Bevilacqua, P. C. (2011). Charged nucleobases and their potential for RNA catalysis. *Accounts of Chemical Research*, *44*, 1270–1279.

Wong, K. Y., Gu, H., Zhang, S., Piccirilli, J. A., Harris, M. E., & York, D. M. (2012). Characterization of the reaction path and transition states for RNA transphosphorylation models from theory and experiment. *Angewandte Chemie (International Ed. in English)*, *51*(3), 647–651.

Wong, K. Y., Xu, Y., & Xu, L. (2015). Review of computer simulations of isotope effects on biochemical reactions: From the Bigeleisen equation to Feynman's path integral. *Biochimica et Biophysica Acta*, *1854*, 1782–1794.

CHAPTER SIXTEEN

Theory and Application of the Relationship Between Steady-State Isotope Effects on Enzyme Intermediate Concentrations and Net Rate Constants

Mark W. Ruszczycky*,[1], Hung-wen Liu*

*University of Texas at Austin, Austin, TX, United States
[1]Corresponding author: e-mail address: mwr8@utexas.edu

Contents

1. Introduction 460
2. Modeling Steady-State Isotope Effects 461
   2.1 Serial Steady-State Models 461
   2.2 Isotope Effects on the Steady-State Velocity 463
   2.3 Isotope Effects on *V/K* and *V* 465
   2.4 Isotope Effects on Net Rate Functions 466
3. An Example: DesII 472
   3.1 Background 472
   3.2 Question and Modeling 473
   3.3 Experimental Design 475
   3.4 Results 478
4. Theory 479
   4.1 Overview 479
   4.2 Basic Terminology and Notation 479
   4.3 Preliminary Definitions and Results 480
   4.4 Serial Models 482
   4.5 Steady-State Velocity and Net Rate Functions 484
   4.6 Sensitivity Indices 488
   4.7 Isomorphic Models 490
   4.8 Perturbation Effects 491
   4.9 Limits 493
   4.10 Two-Vertex Models 495
Acknowledgments 496
References 496

Methods in Enzymology, Volume 596
ISSN 0076-6879
http://dx.doi.org/10.1016/bs.mie.2017.07.022

Abstract

Steady-state kinetic isotope effects on enzyme-catalyzed reactions are often interpreted in terms of the microscopic rate constants associated with the elementary reactions of interest. Unfortunately, this approach can lead to confusion, especially when more than one elementary reaction is isotopically sensitive, because it forces one to consider the full catalytic cycle one step at a time rather than as a complete whole. Herein we argue that shifting focus from intrinsic effects to net rate constants and enzyme intermediate concentrations provides a more natural and holistic interpretation by which the effects of partial rate limitation are more easily understood. In doing so, we demonstrate how the experimental determination of isotope effects on enzyme intermediate concentrations allows a direct determination of isotope effects on net rate constants. The chapter is divided into three main sections. The first outlines the basic theory and its interpretation. The second discusses an application of the theory in the study of the radical SAM enzyme DesII. The final section then provides the complete mathematical treatment.

1. INTRODUCTION

When studying enzymatic reaction mechanisms, a common experimental approach is to introduce a perturbation that affects the values of the microscopic (or intrinsic) rate constants in a predictable manner without altering the set of elementary reactions hypothesized to make up the catalytic cycle. This is a defining feature in the measurement of steady-state kinetic isotope effects (our focus here); however, it may also be applied to the study of viscosity effects, substrate specificity, and so on. In these studies, the *perturbation effect* is observed as the ratio of the parameter of interest measured in the absence of the perturbation, say y, to that measured in its presence, say y^*, or in the reigning notation (Cook & Cleland, 2007)

$$^{*}y = y/y^{*}. \tag{1}$$

The difficulty with such an experimental design, however, is that its interpretation requires compression of an entire kinetic model for the reaction into a single number. This can lead to ambiguity as well as qualitative, vague notions of rate-limiting reaction steps (Northrop, 1981b). Furthermore, the situation is made only more confusing when an emphasis is placed on intrinsic effects rather than parameters more representative of the steady-state condition.

In this chapter, we discuss a method of conceptualizing steady-state perturbation effects using kinetic isotope effects as the example. The approach focuses not on the elementary reactions but rather the related concepts of enzyme intermediate concentrations and net rate constants. It is general for models of steady-state enzymatic reactions that may be regarded as *serial* (i.e., unbranched with a clear forward direction of reaction flux) and suggests an experimental design by which useful information may be extracted from the measurement of steady-state isotope effects despite partially rate-limiting steps. This chapter is divided into three sections. The first is an informal discussion of the theory and its interpretation. The second describes an example of how the approach has been applied in a real experimental context. The final section then provides the mathematics behind the theory that is necessary for its rigorous justification.

2. MODELING STEADY-STATE ISOTOPE EFFECTS

2.1 Serial Steady-State Models

In modeling an enzymatic reaction, we make a distinction between *enzyme species*, denoted $E_1, E_2, \ldots$, which are composed of the enzyme, and *peripheral species*, denoted $X_1, X_2, \ldots$, which are not. Peripheral species may thus be substrates, products, cofactors, and so on, between which we make no distinction. The fixed, steady-state concentrations[a] of the peripheral species will be denoted by corresponding lowercase letters (e.g., x_i would represent the concentration of X_i). Finally, the pseudo-first-order rate constant for the elementary reaction converting E_i to E_j will be denoted k_{ij}. Each pseudo-first-order rate constant is a function of the fixed peripheral species concentrations and proportional to the microscopic rate constant for the associated elementary reaction (see Definition 1 in Section 4). If there is no elementary reaction connecting E_i to E_j in the model, then the corresponding microscopic rate constant is 0, and thus so too is the pseudo-first-order rate constant for all peripheral species concentrations.

For example, an enzymatic reaction modeled by

$$E_1 \underset{-X_1}{\overset{+X_1}{\rightleftharpoons}} E_2 \underset{-X_2}{\overset{+X_2}{\rightleftharpoons}} E_3 \rightleftharpoons E_4 \rightarrow E_5 \xrightarrow{-X_3} E_1, \tag{2}$$

has five enzyme species (E_1 to E_5), three peripheral species (X_1, X_2, and X_3) and eight elementary reactions. The "irreversibility" of the $E_5 \rightarrow E_1$ elementary reaction suggests that X_3 plays no part in the kinetics being modeled. Therefore, we really only need to be concerned with X_1 and

[a] For simplicity, the concentrations of enzyme and peripheral species will be considered equivalent to their respective chemical activities.

X_2, thereby reducing the number of peripheral species *in the model* to just two. This model also has four key properties:

1. There are no interactions between the enzyme species.
2. The peripheral species interconvert only via the catalytic cycle.
3. There is one and only one connected cycle of *all* the enzyme species.[b]
4. The only other connected cycles contain just two enzyme species.

We call any steady-state kinetic model having these properties *serial*, which we can also define formally using graph theory (see Definition 4).

Serial models are *in general* unbranched and have at least one "irreversible" reaction by which a unique forward direction of reaction flux can be unambiguously identified (see Section 4.4).[c] If all the peripheral species included in a serial model have steady-state concentrations greater than 0 (however small), then we can show (see Proposition 2) that *all* of the modeled steady-state enzyme species concentrations must also be greater than 0. It follows that given any set of nonzero peripheral species concentrations, we can always assign a nonzero, steady-state velocity to the serial model according to the flux through one of the irreversible elementary reactions (see Definition 6). Moreover, it can be demonstrated (see Proposition 3) that the steady-state velocity will be *uniquely* defined in this way regardless of how many irreversible reactions there may be in the serial model.[d] This means that the velocity as well as the steady-state concentrations of each enzyme species are well-defined *functions* of the set of peripheral species concentrations.

In order to simplify the discussion of different sets of peripheral species concentrations, we will collect them together as the coordinates of a vector $\mathbf{x}$. For example, in the case of (2) we would have

$$\mathbf{x} = \begin{bmatrix} x_1 \\ x_2 \end{bmatrix}. \tag{3}$$

The condition that all of the steady-state peripheral species concentrations are greater than 0 will then be expressed as $\mathbf{x} > \mathbf{0}$. Therefore, if the velocity

[b] That is, starting at any enzyme species E_i, one can go through *all* the remaining enzyme species just once via the available elementary reactions and arrive back at E_i.

[c] This is strictly true only for serial models with more than two enzyme species. In the case that there are only two enzyme species in the model, it is necessary to impose a forward direction as described in Section 4.10.

[d] This result is not as obvious as it may seem. For example, it is easy to construct steady-state branched models that can be characterized by more than one steady-state velocity.

function is v and the concentration function for E_i is e_i, then the discussion in the preceding paragraph implies that a *net rate function* k_i' for each enzyme species E_i can be *defined* according to the ratio (see also Definition 7)

$$k_i'(\mathbf{x}) = \frac{v(\mathbf{x})}{e_i(\mathbf{x})}, \quad \mathbf{x} > \mathbf{0}. \tag{4}$$

The *evaluation* of k_i' at a given peripheral species concentration vector $\mathbf{x}$ is thus equivalent to the *net rate constant* for E_i that has been previously studied by Cleland (1975) and Noyes (1964). Furthermore, we can use the theory developed by King and Altman (1956) to derive (see Proposition 5) the familiar recursive expression,[e]

$$k_i'(\mathbf{x}) = \frac{k_{i,i+1}(\mathbf{x})k_{i+1}'(\mathbf{x})}{k_{i+1,i}(\mathbf{x}) + k_{i+1}'(\mathbf{x})}, \tag{5}$$

that was discovered by Cleland (1975) in his original work on net rate constants.

It is worth pointing out that up to this point we have not done anything particularly new. In fact, the uniqueness of the steady-state velocity as a function of the peripheral species concentrations will be familiar to anyone who has analyzed specific instances of serial models. What we have done, however, is outline a general framework for approaching serial kinetic models regardless of the number of elementary reactions and peripheral species it entails. We have also identified conditions where nonzero values of $v(\mathbf{x})$, $e_i(\mathbf{x})$, and $k_i'(\mathbf{x})$ for all E_i associated with the steady-state serial model are guaranteed to exist in a general sense, namely, whenever $\mathbf{x} > \mathbf{0}$. These results, which are justified rigorously in Section 4, are central to the validity of the following analysis of isotope effects.

2.2 Isotope Effects on the Steady-State Velocity

Suppose we have two serial models that are equivalent in all respects except for the values of the nonzero microscopic rate constants. Recognizing one of the models as an unperturbed reference, we can then designate the steady-state velocity, enzyme species concentrations, and net rate functions for the second, perturbed model with a postsuperscript "*" (e.g., v^*, e_i^*, $k_i'^*$).

[e] Here and throughout the discussion of serial models we use modular indexing, which is described in Section 4.4.

When the perturbation is understood to be induced by altering the isotopic composition of the reacting species, then it results in an isotope effect.

In Section 4.7 we refer to such models as being *isomorphic*, because there exists an invertible mapping between them that preserves structure and thus defines a pairing between the enzyme species of the first model (i.e., $E_1, E_2, \ldots$) and those of the second (i.e., $E_1^*, E_2^*, \ldots$). Consequently, given our definition (4) of a net rate function, we can express the isotope effect on the steady-state velocity at $\mathbf{x}$ according to the function ${}^*v$, where[f]

$$ {}^*v(\mathbf{x}) = \frac{v(\mathbf{x})}{v^*(\mathbf{x})} = \frac{e_i(\mathbf{x})}{e_i^*(\mathbf{x})} \cdot \frac{k_i'(\mathbf{x})}{k_i'^*(\mathbf{x})} = {}^*e_i(\mathbf{x})\,{}^*k_i'(\mathbf{x}), \tag{6} $$

whenever $\mathbf{x} > \mathbf{0}$. Note that this relationship is true for *all* pairs of corresponding enzyme species (i.e., E_i and E_i^*) described by the models. Since the total enzyme concentration is a constant and the same for both the reference and perturbed reaction systems, we can also write

$$ {}^*v(\mathbf{x}) = {}^*f_i(\mathbf{x})\,{}^*k_i'(\mathbf{x}), \tag{7} $$

where $f_i(\mathbf{x})$ is the *fractional* steady-state concentration of enzyme species E_i at $\mathbf{x}$, and ${}^*f_i(\mathbf{x})$ is the corresponding isotope effect.

Unfortunately, Eq. (7) alone does not offer much intuition regarding the underlying kinetics. Nevertheless, a more insightful equation can be obtained by rearranging Eq. (7) to obtain (see Section 4.8 for derivation)

$$ {}^*v(\mathbf{x}) = \sum_{i=1}^{n} f_i(\mathbf{x})\,{}^*k_i'(\mathbf{x}), \tag{8} $$

where n is the total number of enzyme species in the model. In other words, the isotope effect on the modeled steady-state velocity at $\mathbf{x}$ can be expressed as a weighted average of the isotope effects on each net rate constant, and the weighting terms are just the fractional concentrations of the enzyme species in the reference model.

Eqs. (7) and (8) are our first two key results. As one might expect, the latter demonstrates that if for some $\mathbf{x}$ the majority of the enzyme is piled-up in one enzyme species, then the isotope effect on the corresponding net rate constant will dominate the isotope effect on the observable steady-state velocity. The isotope effects on all other net rate constants will

[f] Note that the isomorphism extends to the domain of the steady-state velocity, enzyme species concentration, and net rate functions for both models. Therefore, we make no distinction between the vectors $\mathbf{x}$ and $\mathbf{x}^*$, which are nothing more than elements of this domain.

be obfuscated. Eq. (7), however, offers a potential solution to this problem, because it can also be rearranged to provide

$$^{*}k_i'(\mathbf{x}) = {}^{*}v(\mathbf{x})/{}^{*}f_i(\mathbf{x}). \tag{9}$$

In other words, the isotope effect on a net rate constant is obtained immediately from a measure of the isotope effects on both the steady-state velocity and the enzyme intermediate concentration of interest. Section 3 will illustrate an example where this useful property of steady-state kinetic isotope effects was used in practice to study the mechanism of a radical *S*-adenosyl-L-methionine (SAM) enzyme.

2.3 Isotope Effects on *V*/*K* and *V*

We are typically not interested in isotope effects on the steady-state velocity but rather the steady-state kinetic parameters V/K and V. However, the corresponding isotope effects $^{*}(V/K)$ and $^{*}V$ are just limiting values of $^{*}v(\mathbf{x})$ as the coordinates of $\mathbf{x}$ each approach 0^+ (i.e., 0 along the positive reals), increase without bound or approach some well-defined positive value (Northrop, 1981a). For example, the V isotope effect often refers to the limit of $^{*}v(\mathbf{x})$ as all peripheral species concentrations increase without bound. Likewise, a V/K isotope effect often refers to the limit of $^{*}v(\mathbf{x})$ as one peripheral species concentration approaches 0^+ while all the remaining concentrations increase without bound. For models with multiple peripheral species, however, these isotope effects may also be defined according to limits where only one concentration approaches 0^+ or ∞ and the remaining approach some set of finite, positive values.

It should be emphasized that these limits and hence isotope effects are not always guaranteed to exist for any serial model.[g] Therefore, some care must be taken in the construction of a serial model if it is to be useful in the study of a real enzymatic reaction. However, if a limit for $^{*}v$ does exist for all positive intrinsic isotope effects in the model, then we can show (see Propositions 11 and 12) that the corresponding limits of each fractional enzyme concentration (i.e., f_i) and isotope effect on each net rate function (i.e., $^{*}k_i'$)

[g] For example, the steady-state velocity of the serial model

$$E_1 \xrightarrow{+X_1} E_2 \xrightarrow{+X_2} E_1 + X_3$$

where X_3 may be considered a product species with steady-state concentration 0 does not converge to a unique value as x_1 and x_2 both increase without bound. Furthermore, if the microscopic rate constants for the two elementary reactions are not always equal, then the model does not necessarily even have a V isotope effect under the same limit.

either exist or their product function (i.e., $f_i \cdot {}^*k_i'$) approaches 0. As previously suggested by Northrop (1981a) this is an important result, because it means that we may interpret isotope effects on the steady-state parameters in exactly the same way that we interpret isotope effects on the steady-state velocity despite the two being different mathematical constructs.

For example, suppose we continue with the model in (2) treating X_1 and X_2 as substrates for the reaction. The isotope effect on V would then correspond to the limit of ${}^*v(\mathbf{x})$ as both x_1 and x_2 increase without bound. In this case, both $f_1(\mathbf{x})$ and $f_2(\mathbf{x})$ approach 0, and we have

$$ {}^*V = f_{3V}{}^*k'_{3V} + f_{4V}{}^*k'_{4V} + f_{5V}{}^*k'_{5V}, \tag{10} $$

where f_{iV} and ${}^*k_{iV}'$ denote the limiting values of $f_i(\mathbf{x})$ and ${}^*k_i'(\mathbf{x})$, respectively, as $(x_1,x_2) \to (\infty,\infty)$. Likewise, if we define ${}^*(V/K)_1$ as the limit of ${}^*v(\mathbf{x})$ as $(x_1,x_2) \to (0^+,\infty)$, then we expect that all the enzyme will pile-up in the E_1 species such that $f_1(\mathbf{x}) \to 1$, and indeed we find that

$$ {}^*(V/K)_1 = \lim_{\substack{x_1 \to 0^+ \\ x_2 \to \infty}} {}^*k_1'(\mathbf{x}), \tag{11} $$

where evaluation of this limit is discussed in the next section. Finally, if we define ${}^*(V/K)_2$ as the limit of ${}^*v(\mathbf{x})$ as $(x_1,x_2) \to (\infty,0^+)$, then we get

$$ {}^*(V/K)_2 = \lim_{\substack{x_1 \to \infty \\ x_2 \to 0^+}} {}^*k_2'(\mathbf{x}). \tag{12} $$

These latter two results should look familiar, since specificity constants (i.e., V/K), are generally understood (Cook & Cleland, 2007) to be equal to the normalized net rate function for the enzyme species that binds with the vanishing substrate. More complicated serial models and limiting conditions can be analyzed in exactly the same way so long as the limits that define the isotope effects on the kinetic parameters exist.

2.4 Isotope Effects on Net Rate Functions

The decomposition (8) of a steady-state isotope effect into a weighted average of isotope effects on the net rate functions naturally leads to the question as to how the latter should be interpreted. It is widely recognized that a net rate function for some enzyme species E_i only depends on the elementary reactions following E_i up to and including the first downstream reaction modeled as irreversible (Cleland, 1975; Cook & Cleland, 2007), and this can be proven from the definition in (4) (see Corollary 1). Thus, each net rate function is associated with a subsequence of the serial model that can be depicted graphically using an appropriately scaled (Ray, 1983) energy diagram.[h] For example, the subsequence for E_1 that determines k_1' for the reaction in (2) corresponds to

$$E_1 \underset{-X_1}{\overset{+X_1}{\rightleftharpoons}} E_2 \underset{-X_2}{\overset{+X_2}{\rightleftharpoons}} E_3 \rightleftharpoons E_4 \rightarrow E_5,$$

and a hypothetical energy diagram for this subsequence is shown in Fig. 1A.

Along the energy diagram for each subsequence, we can identify an activation energy $\Delta G_{ij}^{\ddagger}(\mathbf{x})$ between the first enzyme species E_i in the subsequence and the j-th downstream activated complex (see Fig. 1A). This activation energy depends on the peripheral species concentration vector

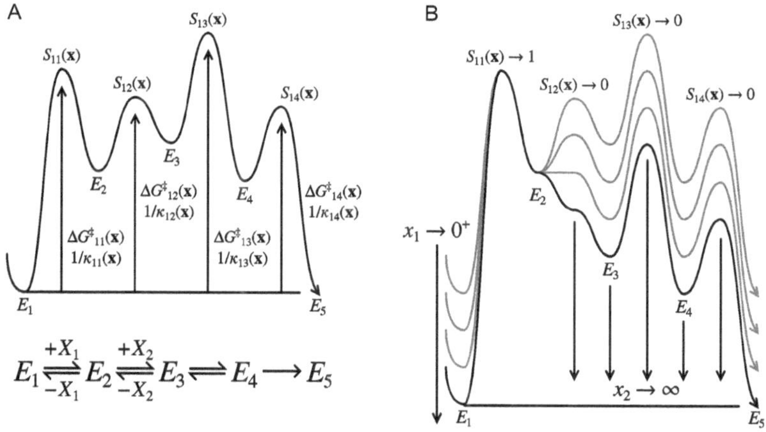

Fig. 1 Graphical interpretation of rate-limiting steps with respect to the net rate function k_1' for the example in (2). In each plot, the abscissa corresponds the subsequence of enzyme species and elementary reactions that contribute to k_1' (i.e., the reaction coordinate). The ordinate represents the scaled energy of each enzyme species and the activated complexes that connect them. (A) Hypothetical plot for the E_1 subsequence given some well-defined **x**. In this case, the $E_3 \rightarrow E_4$ elementary reaction has the largest activation energy with respect to E_1. Hence, the sensitivity index for the $E_3 \rightarrow E_4$ reaction (i.e., $S_{13}(\mathbf{x})$) is closest to 1 making it the most rate limiting with respect to $k_1'(\mathbf{x})$. (B) Graphical heuristic of how the $E_1 \rightarrow E_2$ elementary reaction alone becomes rate limiting with respect to k_1' as the concentration of X_1 decreases and X_2 becomes saturating.

[h] Within the context of transition state theory, this would correspond to a free energy diagram with the peaks and valleys corresponding to transition states and stable intermediates, respectively.

$\mathbf{x}$ and is therefore a heuristic rather than a fixed thermokinetic parameter. We can then associate what Noyes (1964) has labeled an *equilibrated first-order rate constant* $\kappa_{ij}(\mathbf{x})$ with the activation energy $\Delta G^{\ddagger}_{ij}(\mathbf{x})$ such that (Murdoch, 1981; Ray, 1983; Stein, 1981; Tian, 1992)

$$\kappa_{ij}(\mathbf{x}) = K_{ij}(\mathbf{x})k_{j,j+1}(\mathbf{x}), \tag{13}$$

where $k_{j,\ j+1}(\mathbf{x})$ is the pseudo-first-order rate constant given $\mathbf{x}$ for the $E_j \rightarrow E_{j+1}$ elementary reaction, and $K_{ij}(\mathbf{x})$ is the equilibrium constant between E_i and E_j constructed from the pseudo-first-order rate constants of the model and equal to unity if $E_i = E_j$. From the definition of a pseudo-first-order rate constant (see Definition 1), the isotope effect on κ_{ij} is a constant function for all $\mathbf{x} > \mathbf{0}$, and therefore we may simply refer to it as the constant ${}^{*}\kappa_{ij}$ independent of $\mathbf{x}$. *These isotope effects reflect progression from* E_i *to each activated complex downstream and in general do not represent the intrinsic isotope effects associated with each individual elementary reaction* (Cleland, 1982; Lewis & Schramm, 2006; Ruszczycky & Anderson, 2006; Stein, 1981; Tian, 1992).

As already pointed out by Schowen and Stein in their discussions of virtual transition states (Alvarez, Ermer, Hübner, Schellenberger, & Schowen, 1991; Schowen, 1978; Stein, 1981), the isotope effect on the net rate function k_i' is itself a weighted average of the ${}^{*}\kappa_{ij}$ isotope effects along the subsequence for E_i. More specifically, it can be shown (see Sections 4.7 and 4.8) that this isotope effect takes the form

$${}^{*}k_i'(\mathbf{x}) = \sum_{j \geq i} S_{ij}(\mathbf{x})^{*}\kappa_{ij}, \quad \mathbf{x} > \mathbf{0}, \tag{14}$$

where the sum extends over all elementary reactions in the subsequence for E_i. Each weighting term S_{ij}, which is also a function of $\mathbf{x}$, is equal to the *sensitivity index* of the i-th net rate function vs the j-th elementary reaction in the reference system (see Section 4.6 for the derivation). Sensitivity indices were defined by Ray (1983) to reflect the degree to which a steady-state parameter is affected by a change in the energy of an activated complex, all else being equal. For this reason, they may be regarded as a *quantitative* measure of rate-limitation for each elementary reaction with respect to V and V/K parameters (Murdoch, 1981; Ray, 1983). In agreement with this definition of rate-limitation, we extend this interpretation to sensitivity indices for the net rate functions.

Sensitivity indices for net rate functions can be expressed (Ray, 1983; Ruszczycky & Anderson, 2006) in terms of the familiar concepts (Cleland, 1982; Northrop, 1981a, 1981b) of catalytic commitments; however, it can also be shown (Alvarez et al., 1991; Murdoch, 1981; Ray, 1983; Stein, 1981) that they are related to the reciprocals of the κ_{ij} according to

$$S_{ij}(\mathbf{x}) = \frac{\kappa_{ij}^{-1}(\mathbf{x})}{\sum_{p \geq i} \kappa_{ip}^{-1}(\mathbf{x})}, \qquad \mathbf{x} > \mathbf{0}, \tag{15}$$

where the sum in the denominator again extends over all elementary reactions in the subsequence for E_i. This result follows immediately from Propositions 8 and 7 discussed in Sections 4.5 and 4.6. Furthermore, the activation energy $\Delta G_{ij}^{\ddagger}(\mathbf{x})$ is affinely dependent on the logarithm of $\kappa_{ij}^{-1}(\mathbf{x})$ (Noyes, 1964; Ray, 1983; Stein, 1981). Therefore, the relative contribution of each isotope effect ${}^*\kappa_{ij}$ to ${}^*k_i'(\mathbf{x})$ can readily be inferred from the difference in energy between each activated complex and the first enzyme species along the subsequence as shown in Fig. 1A (Ruszczycky & Anderson, 2006). Moreover, limits of the sensitivity indices follow the same rules as do the limits of perturbation effects on net rate functions (see Proposition 13) so that their interpretation remains the same with respect to the steady-state parameters V and V/K.

For example, returning to the model in (2), we have the following expression for ${}^*k_1'(\mathbf{x})$:

$$\begin{aligned} {}^*k_1'(\mathbf{x}) = & \frac{\kappa_{11}^{-1}(\mathbf{x})}{\sum_{j=1}^{4} \kappa_{1j}^{-1}(\mathbf{x})} {}^*\kappa_{11} + \frac{\kappa_{12}^{-1}(\mathbf{x})}{\sum_{j=1}^{4} \kappa_{1j}^{-1}(\mathbf{x})} {}^*\kappa_{12} \\ & + \frac{\kappa_{13}^{-1}(\mathbf{x})}{\sum_{j=1}^{4} \kappa_{1j}^{-1}(\mathbf{x})} {}^*\kappa_{13} + \frac{\kappa_{14}^{-1}(\mathbf{x})}{\sum_{j=1}^{4} \kappa_{1j}^{-1}(\mathbf{x})} {}^*\kappa_{14}, \end{aligned}$$

which is shown diagrammatically in Fig. 1. Furthermore, since

$$\kappa_{1j}(\mathbf{x}) \propto \begin{cases} x_1, & j = 1, \\ x_1 x_2, & j = 2, 3, 4, \end{cases}$$

we find that

$$\lim_{\substack{x_1 \to 0^+ \\ x_2 \to \infty}} S_{1j}(\mathbf{x}) = \begin{cases} 1, & j=1, \\ 0, & j=2,3,4, \end{cases}$$

and thus

$$^*(V/K)_1 = {}^*\kappa_{11},$$

which is just the kinetic isotope effect on the $E_1 \to E_2$ elementary reaction. This is what we expect, because binding of the limiting substrate (i.e., X_1) should dominate the $^*(V/K)_1$ isotope effect as the second substrate (i.e., X_2) becomes saturating and thus introduces a large forward commitment on all subsequent steps as shown in Fig. 1B.

In contrast, when we consider $k_2'(\mathbf{x})$ for the example in (2), we have the subsequence

$$E_2 \underset{-X_2}{\overset{+X_2}{\rightleftharpoons}} E_3 \rightleftharpoons E_4 \rightarrow E_5,$$

and it follows that

$$\kappa_{2j}(\mathbf{x}) \propto x_2,$$

for all values of j between 2 and 4, inclusive. Therefore, the sensitivity indices are all constant values independent of $\mathbf{x}$, because the x_2 just cancel out of the ratios in Eq. (15). Consequently, $^*k_2'(\mathbf{x})$ and by extension $^*(V/K)_2$ are fixed averages of the four different $^*\kappa_{2j}$ isotope effects, and we can write

$$^*(V/K)_2 = S_{22}{}^*\kappa_{22} + S_{23}{}^*\kappa_{23} + S_{24}{}^*\kappa_{24},$$

treating the S_{2j} as constants.

Now, suppose that the $E_3 \to E_4$ reaction alone happens to be isotopically sensitive in the model. Eq. (13) then implies that $^*\kappa_{22}$ will be unity, $^*\kappa_{23}$ will be equal to the intrinsic kinetic isotope effect on the $E_3 \to E_4$ elementary reaction, and $^*\kappa_{24}$ will be equal to the equilibrium isotope effect on this reaction as shown in Fig. 2. Therefore, if S_{22} is close to unity so that S_{23} and S_{24} are close to 0, then $^*(V/K)_2 \approx {}^*\kappa_{22} = 1$, which corresponds to a large forward commitment on the $E_3 \to E_4$ reaction (see Fig. 2A). In contrast, if S_{23} is close to unity, then $E_3 \to E_4$ is "rate limiting" for the net rate constant and $^*(V/K)_2 \approx {}^*\kappa_{23}$ (see Fig. 2B). Finally, if S_{24} is close to unity, then $^*(V/K)_2$ is dominated by the equilibrium isotope effect $^*\kappa_{24}$, which corresponds to a

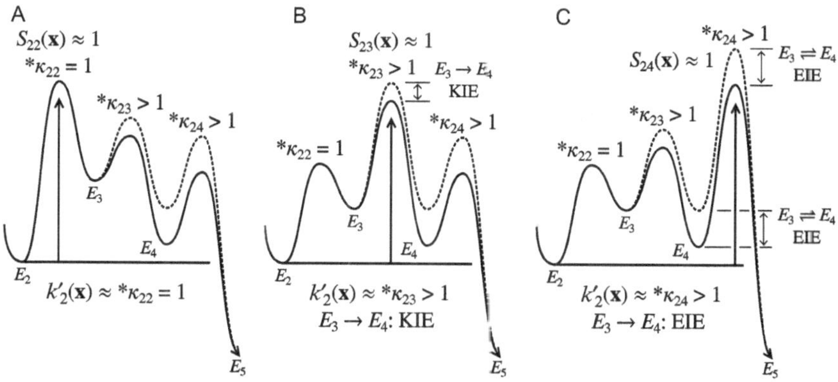

Fig. 2 Graphical interpretation of isotope effects on k'_2 for the example in (2). Plots are drawn as in Fig. 1. The *solid line* corresponds to the reference model and the *broken line* to a model where there is a normal isotope effect on the $E_3 \rightarrow E_4$ elementary reaction alone. (A) The $E_2 \rightarrow E_3$ activated complex is highest in energy vs E_2 making this reaction most rate limiting with respect to k'_2. This introduces large forward commitments on both the $E_3 \rightarrow E_4$ and $E_4 \rightarrow E_5$ reactions. (B) The $E_3 \rightarrow E_4$ reaction is rate limiting for k'_2, and the $E_3 \rightarrow E_4$ kinetic isotope effect is strongly reflected in $^*k'_2$, because there is a unit equilibrium isotope effect on $E_2 \rightleftharpoons E_3$. This places a large reverse commitment on $E_2 \rightarrow E_3$ and a large forward commitment on $E_4 \rightarrow E_5$. (C) The $E_4 \rightarrow E_5$ reaction is most rate limiting, and the isotope effect reflects the bonding changes established in the $E_3 \rightarrow E_4$ reaction, which is the $E_3 \rightleftharpoons E_4$ equilibrium isotope effect. This corresponds to large reverse commitments on the $E_2 \rightarrow E_3$ and $E_3 \rightarrow E_4$ reactions.

large reverse commitment on the $E_3 \rightarrow E_4$ reaction as shown in Fig. 2C. Additional examples using this graphical approach to interpreting isotope effects on net rate constants as shown in Fig. 2 have been previously discussed (Ruszczycky & Anderson, 2006; Tian, 1992).

The previous sections have attempted to show that an intuitive and general framework for the interpretation of steady-state isotope effects is available by shifting the focus from intrinsic effects to parameters more representative of the steady-state, namely, net rate functions and their limits. Furthermore, the fundamental equations (8) and (14) along with their limits are completely general to all serial models and do not require that only one step be regarded as isotopically sensitive. Therefore, they permit one to reason holistically about steady-state isotope effects in terms of activation energy diagrams and the distribution of enzyme species at steady-state without the need to write out the full kinetic equations or focus on individual elementary reactions.

3. AN EXAMPLE: DesII

3.1 Background

The theory outlined in the previous sections not only provides a framework for thinking about steady-state isotope effects (and perturbation effects in general) but also suggests an approach to their measurement, which is exemplified by studies on the radical SAM enzyme DesII. Radical SAM enzymes are characterized by an active site $[4Fe–4S]^{1+}$ cluster that serves to reductively homolyze SAM to produce L-methionine and a 5′-deoxyadenosyl radical (Duschene, Veneziano, Silver, & Broderick, 2009; Frey, 2014; Frey & Magnusson, 2003; Wang & Frey, 2007). The 5′-deoxyadenosyl radical can then act as a radical initiator for a broad range chemical reactions catalyzed by this enzyme superfamily (Broderick, Duffus, Duschene, & Shepard, 2014; Frey, 2014; Landgraf, McCarthy, & Booker, 2016; Mehta et al., 2015).

In the case of DesII, reductive homolysis of SAM results in abstraction of the hydrogen atom from the C3 carbon of its substrate TDP-4-amino-4,6-dideoxy-D-glucose thereby initiating a radical-mediated deamination reaction (**1**→ **2**, see Fig. 3) (Ruszczycky & Liu, 2015; Ruszczycky, Ogasawara, & Liu, 2012; Szu, Ruszczycky, Choi, & Liu, 2009). DesII can also accept TDP-D-quinovose (**3**), which has a hydroxyl rather than an amino group at C4, as a substrate with relatively little change in its kinetic profile (Ruszczycky, Choi, & Liu, 2010; Ruszczycky & Liu, 2015; Szu et al., 2009). However, this small structural change converts DesII from a lyase into a radical-mediated dehydrogenase whereby the substrate radical reduces the $[4Fe–4S]^{2+}$ cluster back to the $[4Fe–4S]^{1+}$ state with each turnover

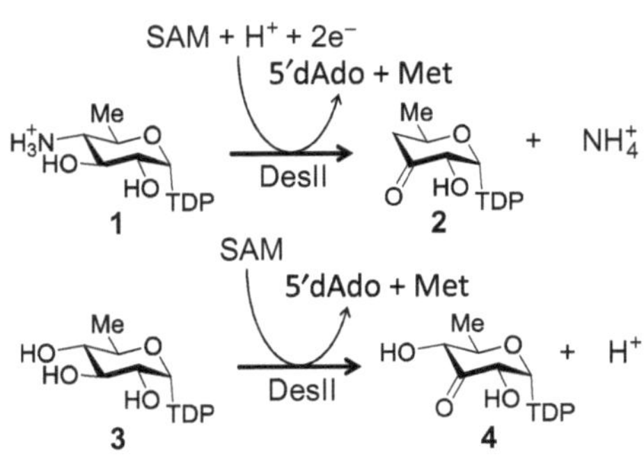

Fig. 3 Reactions catalyzed by DesII.

yielding the product **4** (see Fig. 3) (Ruszczycky et al., 2010). DesII has thus attracted attention as a model system for understanding how the fate of a radical intermediate is controlled in an enzyme active site (Ko, Ruszczycky, Choi, & Liu, 2015).

3.2 Question and Modeling

An initial electron paramagnetic resonance (EPR) investigation of the dehydrogenation steady-state intermediates indicated that the substrate radical is an α-hydroxyalkyl radical (**5**, see Fig. 4) (Ruszczycky, Choi, Mansoorabadi, & Liu, 2011). This raised the question as to whether deprotonation of the α-hydroxyalkyl radical precedes (model A), follows (model B) or occurs concerted (model C) with electron transfer to the $[4Fe–4S]^{2+}$ cluster. These three models, which are shown in Fig. 4, also take into account the observed pH-dependence of the DesII-catalyzed dehydrogenation reaction that implicated titratable forms of the enzyme with decreasing activity as the pH is lowered (Ruszczycky, Choi, & Liu, 2013).

While these models are clearly not serial due to the acid–base equilibria, they can be converted into serial form by treating the acid–base equilibria as rapid and applying Cha's method (Cha, 1968). Furthermore, since the steady-state velocity exists and is nonzero in the limit as the concentrations of both SAM and **3** increase without bound, so too does the limit corresponding to the V isotope effect. Therefore, the discussion in Section 2.3 implies that we can consider these models under saturating conditions more simply as

$$ESH \underset{k_{RQ}^{app}}{\overset{k_{RG}^{app}}{\rightleftharpoons}} ES^{\bullet}H \underset{k_{HR}}{\overset{k_{HT}^{app}}{\rightleftharpoons}} ES^{\bullet -} \xrightarrow{k_{ET}} EP \xrightarrow{k_D} ESH, \tag{A}$$

$$ESH \underset{k_{RQ}^{app}}{\overset{k_{RG}^{app}}{\rightleftharpoons}} ES^{\bullet}H \xrightarrow{k_{ET}} EPH^{+} \xrightarrow{k_{HT}} EP \xrightarrow{k_D} ESH, \tag{B}$$

$$ESH \underset{k_{RQ}^{app}}{\overset{k_{RG}^{app}}{\rightleftharpoons}} ES^{\bullet}H \xrightarrow{k_{HE}^{app}} EP \xrightarrow{k_D} ESH, \tag{C}$$

Here, the rate constants for radical generation (k_{RG}^{app}), quenching (k_{RQ}^{app}), and hydron transfer (k_{HT}^{app} and k_{HE}^{app}) elementary reactions are apparent due to the acid–base equilibria such that,

$$k^{app}(\mathbf{x}) = \frac{K}{K+h} k(\mathbf{x}), \tag{16}$$

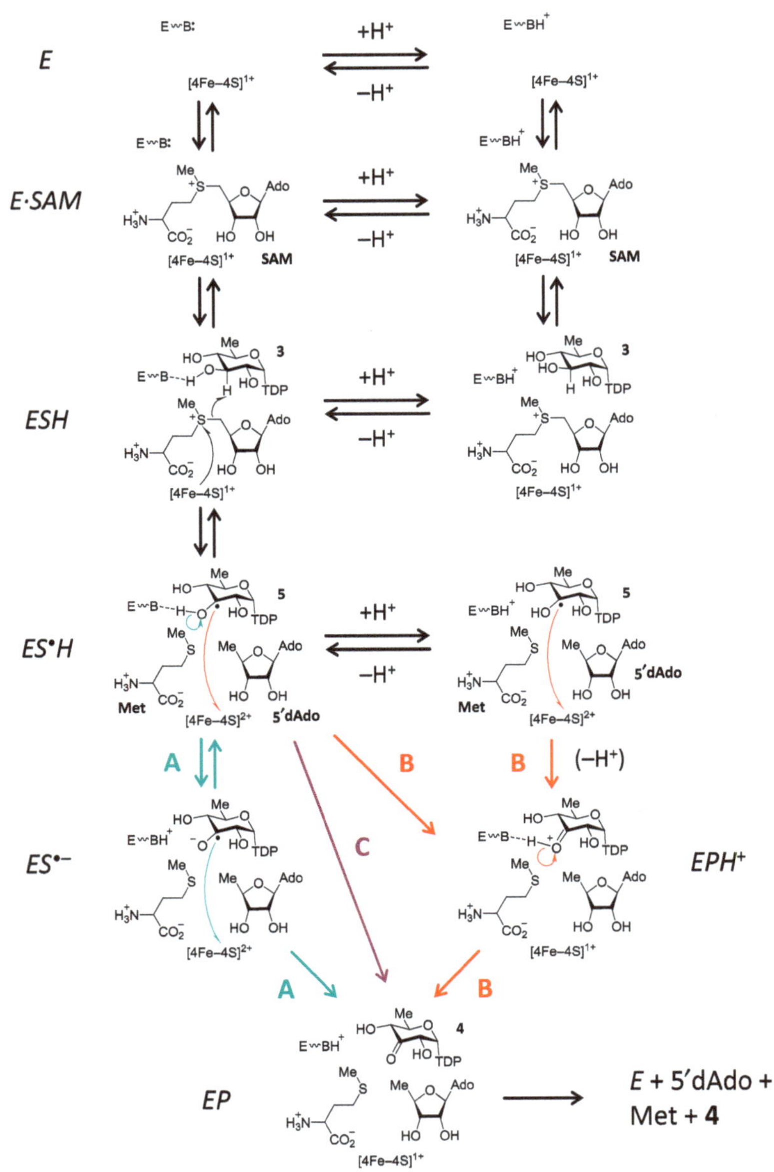

Fig. 4 Models of catalysis for the DesII-catalyzed dehydrogenation of TDP-D-quinovose (**4**). L-Methionine and 5′-deoxyadenosine are abbreviated Met and 5′dAdo, respectively.

where K is the acid dissociation constant, and h is the fixed concentration of H_3O^+. The radical generation step (i.e., $ESH \rightleftharpoons ES^{\bullet}H$) is considered pH-dependent to account for the possibility that the active site needs to be in the correct ionization state to properly orient SAM and **3** for reductive homolysis and H-atom abstraction.

3.3 Experimental Design

The three models can be distinguished experimentally if the solvent deuterium isotope effect on the net rate constant for $ES^{\bullet}H$ (i.e., the limit of ${}^{D_2O}k'_{ES\bullet H}$ as **3** and SAM become saturating) can be measured as shown in Fig. 5. If proton abstraction precedes electron transfer (i.e., model A), then this isotope effect is given by the limiting weighted average (see Eq. (14))

$$ {}^{D_2O}k'_{ES\bullet H} = S_{HT}\,{}^{D_2O}\kappa_{HT} + S_{ET}\,{}^{D_2O}\kappa_{ET}, \tag{17} $$

where S_{HT} and S_{ET} are the sensitivity indices for the hydron transfer and electron transfer steps, respectively, for $k'_{ES\bullet H}$ as the enzyme becomes saturated. After correction of the apparent rate constants for differences due to the acid–base equilibria (see below), the isotope effect ${}^{D_2O}\kappa_{HT}$ reflects bonding changes between $ES^{\bullet}H$ and the activated complex for hydron transfer and thus represents a primary deuterium kinetic isotope effect. In contrast, the isotope effect ${}^{D_2O}\kappa_{ET}$ reflects bonding changes between the subsequent activated complex for electron transfer and $ES^{\bullet}H$ and thus largely represents the equilibrium isotope effect on the $ES^{\bullet}H \rightleftharpoons ES^{\bullet -}$ reaction (see Eq. (13)). In other words, ${}^{D_2O}k'_{ES\bullet H}$ should be large and normal if $S_{HT} \approx 1$.

In model B, where electron transfer occurs before proton transfer, the isotope effect on the net rate constant for $ES^{\bullet}H$ is given by

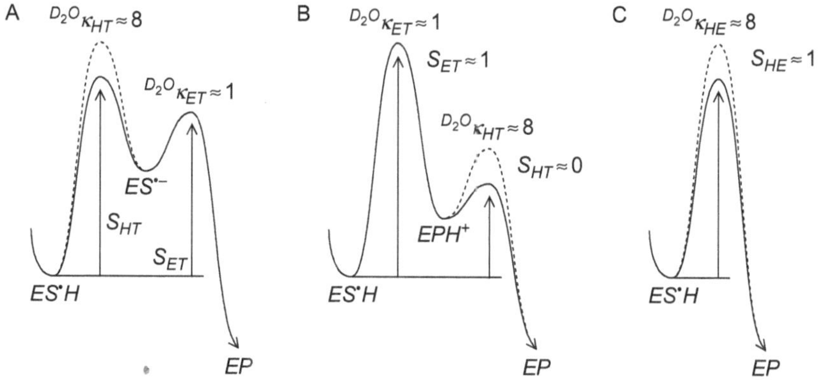

Fig. 5 Energy diagrams of the $ES^{\bullet}H$ subsequence for each model of the DesII reaction. The *solid curve* denotes the reaction in H_2O and the *broken curve* in D_2O. The energy levels of the $ES^{\bullet}H$ intermediate in H_2O and D_2O are aligned to make it easier to see the origin of the ${}^{D_2O}\kappa$ isotope effects. The deprotonation reaction is modeled with a near unit equilibrium isotope effect and a kinetic isotope of ca. 8 based on the experimental observations discussed in the text. Both the electron transfer and proton transfer steps are shown for model B for the sake of clarity; however, the electron transfer is treated as essentially irreversible. This is reflected in the values of the sensitivity indices.

$$^{D_2O}k'_{ES^\bullet H} = {}^{D_2O}\kappa_{ET}, \tag{18}$$

where the sensitivity index is 1, because electron transfer is the only step contributing to $k'_{ES^\bullet H}$ in this model. In this case, $^{D_2O}\kappa_{ET}$ should be essentially unity, because there is neither a solvent hydron in flight in the corresponding activated complex nor is there a preceding reaction to introduce an equilibrium isotope effect with respect to $ES^\bullet H$. Hence, model B always predicts $^{D_2O}k'_{ES^\bullet H} \approx 1$. In contrast, a concerted process (i.e., model C) predicts

$$^{D_2O}k'_{ES^\bullet H} = {}^{D_2O}\kappa_{HE}. \tag{19}$$

Here, $^{D_2O}\kappa_{HE}$, after correcting for acid–base equilibria (see below), reflects bonding changes between $ES^\bullet H$ and the activated complex for concerted electron–hydron transfer. Thus, the concerted model always predicts a normal primary deuterium kinetic isotope effect on $k'_{ES^\bullet H}$.

The solvent deuterium isotope effect ^{D_2O}V on the DesII-catalyzed dehydrogenation of **3** was measured by direct comparison and found to be 2.7 ± 0.4 at the suboptimal pH 8.0 HEPES buffer ratio,[i] where the α-hydroxyalkyl substrate radical could be observed by EPR (Ruszczycky et al., 2013). This isotope effect seemed to suggest that there was indeed a hydron transfer reaction taking place during the catalytic cycle. However, the full expression for ^{D_2O}V is given by

$$^{D_2O}V = f_{ESH}{}^{D_2O}k'_{ESH} + f_{ES^\bullet H}{}^{D_2O}k'_{ES^\bullet H} + f_{EPH^+}{}^{D_2O}k'_{EPH^+} + f_{EP}{}^{D_2O}k'_{EP}, \tag{20}$$

for model B with similar expressions for models A and C as per the limit of Eq. (8). While proton transfer from the protonated ketone in EPH^+ should be fast and thus f_{EPH^+} low, there is no guarantee that other steps (e.g., $EP \rightarrow ESH$) would not involve transfer of solvent exchangeable hydrons. In other words, the modest, normal ^{D_2O}V isotope effect alone did not convincingly rule out model B.

What one really wants to do is to increase the value of $f_{ES^\bullet H}$ closer to unity so that ^{D_2O}V is more representative of $^{D_2O}k'_{ES^\bullet H}$. In the case of DesII,

[i] In other words, the ratio of ionized forms of HEPES in H_2O at pH 8.0. As discussed by Schowen and Schowen (1982), this helps to ensure experimentally that solvent deuterium isotope effects are measured at *equivalent* values of the pH and pD as opposed to simply setting the pH *equal* to the pD, which can lead to mismatches in enzyme activity. The approach is not perfect, however, and additional corrections may be necessary as discussed in the text.

it was possible to *reduce* the value of $f_{ES \bullet H}$ by running the reaction at the more optimal pH of 10 where the substrate radical could no longer be observed by EPR (Ruszczycky et al., 2013). Consequently, the value of ${}^{D_2O}V$ decreased to 1.8 ± 0.2 at this pH, thereby further supporting models A and C. This approach is identical to that first proposed by Cook and Cleland (1981a, 1981b), whereby the degree to which an elementary reaction is rate limiting can be modulated by examining the isotope effect at suboptimal pH. However, if the isotope effect on the steady-state concentration of $ES^{\bullet}H$ under saturating conditions can be determined (i.e., ${}^{D_2O}f_{ES \bullet H}$), then Eq. (9) implies that ${}^{D_2O}k'_{ES \bullet H}$ can be calculated from the ratio

$$ {}^{D_2O}k'_{ES \bullet H} = \frac{{}^{D_2O}V}{{}^{D_2O}f_{ES \bullet H}}. \tag{21} $$

Note that this expression is independent of the exact form of the serial model. This was the approach taken with DesII, where ${}^{D_2O}f_{ES \bullet H}$ was measured by integrating the EPR signal intensities for the $ES^{\bullet}H$ enzyme species in both H_2O and D_2O buffers and then taking their ratio.

In order to do this experiment with reasonable precision and minimal bias, a number of controls need to be in place in addition to proper replication. First, the enzyme must have the same specific activity in each sample preparation so that the concentration ratio reflects the isotope effect and not differences in total active enzyme. In the case of DesII, comparable H_2O and D_2O EPR samples were prepared from the same batch of enzyme at the same time. Furthermore, the same buffer exchange protocol was applied to the preparation of both the D_2O and H_2O reaction samples, and small aliquots of each enzyme preparation were extracted to determine the specific activity following buffer exchange. It is also important to make sure that the enzyme intermediate is trapped under the same conditions for both reaction systems. For example, if the reaction in H_2O buffer is significantly faster than in D_2O buffer, then the intermediate concentration differences could represent different levels of enzyme saturation and not the isotope effect. This was controlled for by both measuring the ratio at different freeze-quench times and also examining control quenches by HPLC to determine the extent of reaction in each system. The DesII enzyme also had a number of additional peculiarities requiring attention such as substrate inhibition and ensuring proper integration of the EPR signals, and these aspects are carefully addressed in the original work (Ruszczycky et al., 2013).

3.4 Results

The final value of ${}^{D_2O}f_{ES^\bullet H}$ was thus 0.22 ± 0.03 measured at the pH 8.0 HEPES buffer ratio indicating approximately 4–5 times more enzyme in the $ES^\bullet H$ state in D_2O vs H_2O (Ruszczycky et al., 2013). Consequently, ${}^{D_2O}k'^{app}_{ES^\bullet H}$ was found to be 12 ± 3 prior to correction for acid–base equilibria. In the case of model B, k_{ET} and thus the net rate constant do not depend on the protonation state of the enzyme, such that[j]

$$ {}^{D_2O}k'_{ES^\bullet H} = {}^{D_2O}k'^{app}_{ES^\bullet H} = 12 \pm 3. \tag{22} $$

This is inconsistent with the model B prediction that this isotope effect be near unity. In contrast, the net rate constant for models A and C does depend on the protonation state of $ES^\bullet H$ via the apparent rate constants k^{app}_{HT} and k^{app}_{HE}, respectively. In both cases, the dependence manifests as

$$ {}^{D_2O}k'^{app}_{ES^\bullet H} = {}^{D_2O}k'_{ES^\bullet H}\left(\frac{{}^{D_2O}K/(h/d) + K/h}{1 + K/h}\right), \tag{23} $$

where d is the concentration of D_3O^+ at the pH 8.0 HEPES buffer ratio. In the original work, the correction factor on the right-hand side was estimated to be approximately 1.6 ± 0.5, such that the corrected isotope effect on ${}^{D_2O}k'_{ES^\bullet H}$ was 8 ± 3, which is consistent with either of these models.

Collectively, these results ruled out model B indicating that proton transfer occurs either prior to or concerted with electron transfer during the DesII-catalyzed dehydrogenation of TDP-D-quinovose. From a mechanistic perspective, the result implied that deprotonation of the α-hydroxyalkyl radical may be necessary to increase its oxidation potential to at least match the reduction potential of the active site $[4Fe–4S]^{2+}$ cluster (Daley & Holm, 2003; Duschene et al., 2009; Hayon & Simic, 1974; Hinckley & Frey, 2006; Rao & Hayon, 1974; Rao & Holm, 2004). This would then facilitate what is presumably an outer-sphere electron transfer to yield the dehydrogenated product (**4**) and regenerate the active $[4Fe–4S]^{1+}$ form of the enzyme.

When investigating the steady-state of an enzymatic reaction, the arguably most useful and experimentally accessible kinetic parameters are not the

[j] In the original report, this number was reported as 13 ± 4. The discrepancy is not meaningful but is real. It arises due to an approximation that was used at the time to simplify some of the arithmetic when comparing the result between the competing models. The details of that approximation are provided in the original work.

hypothesized microscopic rate constants but rather the net rate functions and enzyme intermediate concentrations. Hence, the study and manipulation of net rate functions and enzyme intermediate concentrations in their own right represent an important feature for developing and testing steady-state kinetic models of real enzymatic reactions. This thesis has been previously recognized in the works of Tittmann et al. (Schütz, Golbik, König, Hübner, & Tittmann, 2005; Tittmann et al., 2003; Tittmann, Neef, Golbik, Hübner, & Kern, 2005; Tittmann, Vyazmensky, Hübner, Barak, & Chipman, 2005), who have studied intermediate concentrations directly, as well as those of Bollinger et al. (Hoffart, Barr, Guyer, Bollinger, & Krebs, 2006; Peck et al., 2017; Price, Barr, Glass, Krebs, & Bollinger, 2003; Tamanaha et al., 2016; Xing et al., 2006), who have also used isotope effects to manipulate the relative populations of intermediates observable by EPR. This conceptualization extends naturally to the general study of perturbation effects themselves providing well-defined, quantitative concepts of partial rate limitation that can be easily represented with activation energy diagrams.

4. THEORY

4.1 Overview

This section provides a rigorous justification of the theory discussed informally in Section 2. As such, we take a completely mathematical approach and use the language of graph theory and vector calculus with no reliance on chemical intuition, despite some suggestive naming conventions. The section begins by establishing careful definitions of the mathematical constructs and then deriving their properties. A model that meets these definitions for any perturbation effect (whether it be an isotope effect, viscosity effect, etc.) will then have the derived properties. We first provide proofs that can be applied to steady-state serial models with more than two enzyme species and then show how this can be extended to models with only two enzyme species.

4.2 Basic Terminology and Notation

We use standard notation, with bold symbols representing real vectors and matrices or their functions. One exception is the use of the prime to denote a net rate function rather than single-variable differentiation. The sets $\mathbb{R}$, $\mathbb{R}^n$, and $\mathbb{R}^{n\times n}$ correspond to the sets of all real numbers, all vectors of n real numbers and all $n \times n$ matrices of real numbers, respectively. Standard notation is used to define functions and build sets. We use the basic definition of a digraph (i.e., a directed graph) that does not include multigraphs so that directed edges can only be represented once and there are no self-loops. Finally, if $\mathbf{x}$ is an element of $\mathbb{R}^n$, then we will continue with the notation $\mathbf{x} > \mathbf{0}$ from Section 2 to indicate that *all* n coordinates of the vector $\mathbf{x}$ are greater than 0.

4.3 Preliminary Definitions and Results

Definition 1 (Rate Function).
A *rate function* k parameterized in terms of a set of m nonnegative integer constants $B = \{b_i\}_{i=1}^{m}$ and a nonnegative real number β is a function

$$k: \mathbb{R}^m \rightarrow \mathbb{R},$$
$$k(\mathbf{x};\, \beta, B) = \beta \prod_{i=1}^{m} x_i^{b_i}, \tag{24}$$

where x_i is the i-th coordinate of the real column vector $\mathbf{x}$ in $\mathbb{R}^m$.

In the context of an elementary enzymatic reaction, a rate function is nothing more than a generalized pseudo-first-order rate constant that accounts for all possible peripheral species concentrations. The parameter β may then be regarded as the associated microscopic rate constant and the integer constant b_i as the molecularity of the elementary reaction with respect to peripheral species X_i. Note that if k is independent of x_i, then $b_i = 0$, and x_i^0 is defined to be 1 for all $x_i \in \mathbb{R}$ such that k is a continuous function.

Definition 2 (Steady-State Model).
A *steady-state model* is an ordered triple (D_n, A_n^m, e_0) where $n > 1$ and
D_n is a digraph with n vertices labeled $E_1, E_2, \ldots, E_n$,
A_n^m is a set of $n(n-1)$ rate functions $\{k_{ij}\}_{i \neq j}^{n,n}$ defined on $\mathbb{R}^m$,
e_0 is a positive real number,
such that for each rate function $k_{ij}(\,\cdot\,; \beta_{ij}, B_{ij})$ in A_n^m, $\beta_{ij} > 0$ if the ordered pair of vertices (E_i, E_j) exists in D_n, and $\beta_{ij} = 0$ if (E_i, E_j) does not exist in D_n.

Note that the ordered pairs (E_i, E_j) in a steady-state model represent the directed edges in the graphical representation of D_n. Therefore, the steady-state kinetics of an enzymatic reaction can be represented using a steady-state model by assigning each enzyme species to a vertex and defining $\mathbf{x}$ as a vector of m peripheral species concentrations. Note that this definition of a steady-state model establishes the first two conditions discussed in Section 2.1.

The next definition follows the analysis of King and Altman (1956), which is built upon that by Matsen and Franklin (1950).

Definition 3 (King–Altman Function).
The *King–Altman function* **A** is defined for a steady-state model (D_n, A_n^m, e_0) as

$$\mathbf{A}:\mathbb{R}^m \rightarrow \mathbb{R}^{n\times n},$$
$$[\mathbf{A}(\mathbf{x})]_{ij} = \begin{cases} k_{ji}(\mathbf{x}), & i \neq j, \\ -\sum_{p\neq i}^{n} k_{ip}(\mathbf{x}), & i = j, \end{cases} \tag{25}$$

where $[\mathbf{A}(\mathbf{x})]_{ij}$ is the element in the j-column of the i-th row of $\mathbf{A}(\mathbf{x})$, and the k_{ij} are the elements of A_n^m.

Given a steady-state model (D_n, A_n^m, e_0), if the rank of $\mathbf{A}(\mathbf{x})$ is $n-1$ and $\mathbf{x} > \mathbf{0}$, then there is a unique n-dimensional column vector $\mathbf{e}(\mathbf{x})$ that solves the linear system of equations given by

$$\mathbf{A}(\mathbf{x})\mathbf{e}(\mathbf{x}) = \mathbf{0}, \tag{26a}$$
$$\mathbf{1}^{\mathsf{T}}\mathbf{e}(\mathbf{x}) = e_0, \tag{26b}$$

where **1** is a column vector of n ones, **0** is a column vector of n zeros, and ^{T} denotes vector/matrix transposition. We can therefore treat **e** as an n-dimensional vector-valued function of **x**. Furthermore, if we denote the i-th coordinate of $\mathbf{e}(\mathbf{x})$ as $e_i(\mathbf{x})$, then e_i is a real-valued function of **x**.

King and Altman (1956) have provided a detailed theoretical analysis of the solutions to Eq. (26). In particular, they have shown that if a solution $\mathbf{e}(\mathbf{x})$ exists, then it is given according to Cramer's rule as

$$e_i(\mathbf{x}) = \frac{e_0 \det \widetilde{\mathbf{A}}_i(\mathbf{x})}{\sum_{j=1}^{n} \det \widetilde{\mathbf{A}}_j(\mathbf{x})}, \quad 1 \leq i \leq n, \tag{27}$$

where $\widetilde{\mathbf{A}}_i(\mathbf{x})$ is $\mathbf{A}(\mathbf{x})$ with the i-th row replaced by ones, the i-th column replaced by zeros, and $[\widetilde{\mathbf{A}}_i(\mathbf{x})]_{ii} = 1$. The notation $\det \widetilde{\mathbf{A}}_i(\mathbf{x})$ refers to the determinant of $\widetilde{\mathbf{A}}_i(\mathbf{x})$. A defining achievement of King and Altman's analysis is the theorem they proved for determining $\det \widetilde{\mathbf{A}}_i(\mathbf{x})$ from direct inspection of the digraph D_n. We will refer to this as the *King–Altman theorem* and state it here without proof. Note that this theorem along with Eq. (27) also establishes a one-to-one correspondence ($E_i \Leftrightarrow e_i$) between the vertices of D_n and the elements of **e**.

Theorem 1 (King–Altman).
Let (D_n, A_n^m, e_0) be a steady-state model and P_i be a subgraph of D_n with n vertices and $n - 1$ edges such that vertex E_i is reachable from all other vertices. The determinant of $\widetilde{\mathbf{A}}_i(\mathbf{x})$ is then given by

$$\det \widetilde{\mathbf{A}}_i(\mathbf{x}) = (-1)^{n-1} \sum \alpha(\mathbf{x}, P_i), \tag{28}$$

where the sum extends over all subgraphs P_i and $\alpha(\mathbf{x}, P_i)$ is the product of the rate functions associated with the edges of P_i evaluated at $\mathbf{x}$. If there is no such subgraph P_i, then the determinant is 0.

4.4 Serial Models

Definition 4 (Serial Model).
A steady-state model (D_n, A_n^m, e_0) is called *serial* if D_n has one and only one n-cycle C_n and no p-cycle where $2 < p < n$.

We can label the vertices of a serial model such that the unique n-cycle is represented by any circular shift of the sequence of n edges,

$$C_n = (E_1, E_2)(E_2, E_3)\ldots(E_{n-1}, E_n)(E_n, E_1). \tag{29}$$

Since how we choose to index the vertices of a graph is arbitrary, we will always use an indexing such that C_n can be written according to Eq. (29). This will allow us to use modular arithmetic when writing indices given a serial kinetic model, namely, index i in k_i represents the least positive residue of i modulo n. For example, if $n = 5$, then $E_2 = E_2$, $E_0 = E_5$, $E_{-2} = E_3$, and $E_7 = E_2$. Two other properties of a serial model are worth emphasizing. First, there are no "branches" as this would otherwise introduce a cycle with more than 2 vertices. Second, if $n > 2$, then C_n defines an *orientation* for the model, which corresponds to the *forward direction* when applied to an enzymatic reaction.

Definition 5 (Irreversible Edge).
Given a directed graph D, if the ordered pair (E_i, E_j) exists in D and (E_j, E_i) does not exist in D, then (E_i, E_j) is called an *irreversible edge* of D.

Proposition 1. *If a steady-state model (D_n, A_n^m, e_0) is serial and $n > 2$, then there is at least one irreversible edge in D_n.*

Proof. Since the model is serial, there exists a sequence of edges forming the n-cycle C_n as in Eq. (29). If there are no irreversible edges in D_n, then for every edge (E_i, E_j) in D_n the reverse edge (E_j, E_i) is also in D_n. The existence of C_n (see Eq. (29)) then implies that all n edges in the set

$$\{(E_2,E_1),(E_3,E_2),\ldots,(E_n,E_{n-1}),(E_1,E_n)\}, \tag{30}$$

exist in D_n, and thus D_n also contains the n-cycle given by the sequence

$$C_n' = (E_1,E_n)(E_n,E_{n-1})\ldots(E_3,E_2)(E_2,E_1). \tag{31}$$

However, if $n > 2$, then there is no permutation such that the sequences in Eqs. (29) and (31) become equal, hence $C_n' \neq C_n$ and there must be two different n-cycles in D_n. This contradicts the hypothesis that the model is serial. Hence, there must be at least one irreversible edge in D_n. □

Proposition 2. *Given a serial model* (D_n, A_n^m, e_0), *if* $\mathbf{x} > 0$, *then* $e_i(\mathbf{x})$ *and* $k_{i,\ i+1}(\mathbf{x})$ *are both positive for all vertices* E_i *in* D_n.

Proof. The definition of a rate function implies that $k_{ij}(\mathbf{x}) > 0$ whenever $\mathbf{x} > 0$ and (E_i, E_j) exists in D_n. Thus, $k_{i,\ i+1}(\mathbf{x}) > 0$. From the King–Altman theorem, the determinants $\det \widetilde{\mathbf{A}}_i(\mathbf{x})$ are all composed of sums of products of rate functions, and for every vertex there is a path of $n - 1$ edges corresponding to a subsequence of C_n by which the vertex is reachable from all others. Therefore, for all E_i, $\det \widetilde{\mathbf{A}}_i(\mathbf{x})$ cannot be 0 when $\mathbf{x} > \mathbf{0}$. Furthermore, the King–Altman theorem asserts that all of these determinants have the same sign; hence, Eq. (27) implies $e_i(\mathbf{x}) > 0$. □

Proposition 3. *Given a serial model* (D_n, A_n^m, e_0), *if* (E_i, E_{i+1}) *and* (E_j, E_{j+1}) *are both irreversible edges in* D_n *and* $\mathbf{x} > 0$, *then*

$$e_i(\mathbf{x})k_{i,i+1}(\mathbf{x}) = e_j(\mathbf{x})k_{j,j+1}(\mathbf{x}). \tag{32}$$

Proof. Recognizing our use of modular indexing, the King–Altman theorem implies

$$\det \widetilde{\mathbf{A}}_i(\mathbf{x}) = (-1)^{n-1} k_{i+1,i+2}(\mathbf{x})\cdots k_{i-2,i-1}(\mathbf{x})k_{i-1,i}(\mathbf{x}), \tag{33a}$$

$$\det \widetilde{\mathbf{A}}_j(\mathbf{x}) = (-1)^{n-1} k_{j+1,j+2}(\mathbf{x})\cdots k_{j-2,j-1}(\mathbf{x})k_{j-1,j}(\mathbf{x}). \tag{33b}$$

Since every factor of the form $k_{p,\ p+1}(\mathbf{x})$ is found in each product except $k_{i,\ i+1}(\mathbf{x})$ in the first and $k_{j,\ j+1}(\mathbf{x})$ in the second, we have

$$\det \widetilde{\mathbf{A}}_i(\mathbf{x})k_{i,i+1}(\mathbf{x}) = \det \widetilde{\mathbf{A}}_j(\mathbf{x})k_{j,j+1}(\mathbf{x}), \tag{34}$$

and it follows immediately from Eq. (27) that

$$e_i(\mathbf{x})k_{i,i+1}(\mathbf{x}) = e_j(\mathbf{x})k_{j,j+1}(\mathbf{x}). \blacksquare \tag{35}$$

4.5 Steady-State Velocity and Net Rate Functions

Propositions 1–3 are of particular importance for two reasons. First, they provide conditions under which $e_i(\mathbf{x})\neq 0$ so that $1/e_i(\mathbf{x})$ exists. Second, they provide a mechanism for unambiguously defining a property of a serial model that we may equate with the steady-state velocity of an enzymatic reaction.

Definition 6 (Velocity).

If (D_n, A_n^m, e_0) is a serial model with $n > 2$, and (E_i, E_{i+1}) is any irreversible edge in D_n, then the *velocity* v of the model is the function

$$\begin{aligned} v &: \{\mathbf{x}\in\mathbb{R}^m : \mathbf{x}>\mathbf{0}\}\rightarrow\mathbb{R}, \\ v(\mathbf{x}) &= e_i(\mathbf{x})k_{i,i+1}(\mathbf{x}). \end{aligned} \tag{36}$$

Definition 7 (Net Rate Function).

If (D_n, A_n^m, e_0) is a serial model with $n > 2$ vertices, then the *net rate function* for the i-th vertex is defined as the function

$$\begin{aligned} k_i' &: \{\mathbf{x}\in\mathbb{R}^m : \mathbf{x}>\mathbf{0}\}\rightarrow\mathbb{R}, \\ k_i'(\mathbf{x}) &= v(\mathbf{x})/e_i(\mathbf{x}). \end{aligned} \tag{37}$$

Proposition 4. *If (D_n, A_n^m, e_0) is a serial model with $n > 2$ and (E_i, E_{i+1}) is an irreversible edge in D_n, then $k_i'(\mathbf{x}) = k_{i,\ i+1}(\mathbf{x})$ whenever $\mathbf{x} > 0$.*

Proof. Since $\mathbf{x} > 0$ we have $e_i(\mathbf{x}) > 0$ by Proposition 2. Therefore, $k_i'(\mathbf{x})$ exists and according to Proposition 3 is given by

$$k_i'(\mathbf{x}) = \frac{v(\mathbf{x})}{e_i(\mathbf{x})} = \frac{e_i(\mathbf{x})k_{i,i+1}(\mathbf{x})}{e_i(\mathbf{x})} = k_{i,i+1}(\mathbf{x}).\blacksquare \tag{38}$$

We are now in a position to prove the recursive expression for net rate constants first identified by Cleland (1975).

Proposition 5 (Cleland). *Given a serial model with $n > 2$ vertices, if $\mathbf{x} > 0$, then the net rate function at $\mathbf{x}$ for the i-th vertex E_i is given by*

$$k_i'(\boldsymbol{x}) = \frac{k_{i,i+1}(\boldsymbol{x})k_{i+1}'(\boldsymbol{x})}{k_{i+1,i}(\boldsymbol{x}) + k_{i+1}'(\boldsymbol{x})}. \tag{39}$$

Proof. Let n be any integer greater than 2. Proposition 1 asserts that there is at least one irreversible edge in the digraph D_n of the serial model. We permute the indices of the vertices such that (E_n, E_1) is an irreversible edge. Therefore, D_n must contain the edges (E_1, E_2), (E_n, E_1), and possibly (E_2, E_1) but not (E_1, E_n). No other edges in D_n include E_1. Thus,

$$[\mathbf{A}(\mathbf{x})]_{1i} = \begin{cases} -k_{12}(\mathbf{x}), & i=1, \\ k_{i1}(\mathbf{x}), & i=2,n, \\ 0, & \text{otherwise}, \end{cases} \tag{40}$$

and it follows from Eq. (26) that

$$k_{21}(\mathbf{x})e_2(\mathbf{x}) + k_{n1}(\mathbf{x})e_n(\mathbf{x}) - k_{12}(\mathbf{x})e_1(\mathbf{x}) = 0. \tag{41}$$

Since $\mathbf{x} > \mathbf{0}$, Proposition 2 implies that the net rate functions for E_1, E_2, and E_n are all positive at $\mathbf{x}$, and we may write

$$k_{21}(\mathbf{x})\frac{\nu(\mathbf{x})}{k_2'(\mathbf{x})} + k_{n1}(\mathbf{x})\frac{\nu(\mathbf{x})}{k_n'(\mathbf{x})} - k_{12}(\mathbf{x})\frac{\nu(\mathbf{x})}{k_1'(\mathbf{x})} = 0. \tag{42}$$

Proposition 4 implies that $k_n'(\mathbf{x}) = k_{n1}(\mathbf{x})$, so Eq. (42) rearranges to

$$k_1'(\mathbf{x}) = \frac{k_{12}(\mathbf{x})k_2'(\mathbf{x})}{k_{21}(\mathbf{x}) + k_2'(\mathbf{x})}, \tag{43}$$

since $\nu(\mathbf{x}) > 0$. This proves the result for $i = 1$.

If $1 < i < n$, then D_n must contain the edges (E_{i-1}, E_i), (E_i, E_{i+1}), and possibly the reverse of either. No other edges in D_n include E_i so that

$$[\mathbf{A}(\mathbf{x})]_{ij} = \begin{cases} -(k_{i,i-1}(\mathbf{x}) + k_{i,i+1}(\mathbf{x})), & j=i, \\ k_{ji}(\mathbf{x}), & j=i\pm 1, \\ 0, & \text{otherwise}. \end{cases} \tag{44}$$

We now assume the inductive hypothesis that Eq. (39) holds for some vertex E_i where $1 \leq i < n-1$ and consider the vertex E_{i+1}. Since $1 < i+1 < n$, we may use (44) to write Eq. (26) for $e_{i+1}(\mathbf{x})$ as

$$k_{i,i+1}(\mathbf{x})e_i(\mathbf{x}) - (k_{i+1,i}(\mathbf{x}) + k_{i+1,i+2}(\mathbf{x}))e_{i+1}(\mathbf{x}) + k_{i+2,i+1}(\mathbf{x})e_{i+2}(\mathbf{x}) = 0. \tag{45}$$

Since $\mathbf{x} > \mathbf{0}$, Proposition 2 implies that the net rate functions for E_i, E_{i+1}, and E_{i+2} evaluated at $\mathbf{x}$ are all positive, and we may write

$$\begin{aligned} k_{i,i+1}(\mathbf{x})\frac{\nu(\mathbf{x})}{k_i'(\mathbf{x})} - (k_{i+1,i}(\mathbf{x}) + k_{i+1,i+2}(\mathbf{x}))\frac{\nu(\mathbf{x})}{k_{i+1}'(\mathbf{x})} \\ + k_{i+2,i+1}(\mathbf{x})\frac{\nu(\mathbf{x})}{k_{i+2}'(\mathbf{x})} = 0. \end{aligned} \tag{46}$$

However, the inductive hypothesis implies that

$$k_{i,i+1}(\mathbf{x})\frac{v(\mathbf{x})}{k_i'(\mathbf{x})}=\frac{k_{i+1,i}(\mathbf{x})+k_{i+1}'(\mathbf{x})}{k_{i+1}'(\mathbf{x})}v(\mathbf{x}). \tag{47}$$

Combining Eqs. (46) and (47) and rearranging we get

$$k_{i+1}'(\mathbf{x})=\frac{k_{i+1,i+2}(\mathbf{x})k_{i+2}'(\mathbf{x})}{k_{i+2,i+1}(\mathbf{x})+k_{i+2}'(\mathbf{x})}. \tag{48}$$

Therefore, it follows inductively that Eq. (39) holds for all $i < n$. All that is left is to show that it holds for $i = n$. Now, since (E_n, E_1) is an irreversible edge, the definition of a model implies that $k_{1n}(\mathbf{x}) = 0$ for all $\mathbf{x}$. Therefore, by Proposition 4 we have

$$k_n'(\mathbf{x})=k_{n1}(\mathbf{x})=\frac{k_{n1}(\mathbf{x})k_1'(\mathbf{x})}{0+k_1'(\mathbf{x})}=\frac{k_{n1}(\mathbf{x})k_1'(\mathbf{x})}{k_{1n}(\mathbf{x})+k_1'(\mathbf{x})}. \tag{49}$$

Thus Eq. (39) holds for all $i \leq n$. Furthermore, since our choice of $n > 2$ was arbitrary, the result holds for all serial models with $n > 2$. □

Definition 8 (Equilibrium Function).
Given a serial model with n vertices, the subgraph defined by the sequence of edges along C_n given by

$$(E_i,E_{i+1})(E_{i+1},E_{i+2})\ldots(E_{j-2},E_{j-1})(E_{j-1},E_j), \tag{50}$$

where $1 \leq i < j \leq n$ has an associated *equilibrium function* given by

$$K_{ji}:\{\mathbf{x}\in\mathbb{R}^m : k_{p,p+1}(\mathbf{x})\neq 0 \text{ for all } i\leq p<j\}\rightarrow\mathbb{R},$$
$$K_{ji}(\mathbf{x})=\prod_{p=i}^{j-1}\frac{k_{p+1,p}(\mathbf{x})}{k_{p,p+1}(\mathbf{x})}, \tag{51}$$

and $K_{ii}(\mathbf{x}) = 1$ for all $\mathbf{x}$.

Proposition 6. *Given a serial model* (D_n, A_n^m, e_0) *with* $n > 2$, *if* $\mathbf{x} > 0$, *then* $K_{ji}(\mathbf{x}) \geq 0$ *for any two vertices* E_i *and* E_j *where* $1 \leq i \leq j \leq n$. *Furthermore, if* $i < j$ *and the sequence*

$$(E_i,E_{i+1})(E_{i+1},E_{i+2})\ldots(E_{j-1},E_j), \tag{52}$$

does not contain any irreversible edges in D_n, *then* $K_{ji}(\mathbf{x}) > 0$.

Proof. Index the vertices such that (E_n, E_1) is irreversible. By Proposition 2,

$$0 < v(\mathbf{x}) = e_n(\mathbf{x})k_{n1}(\mathbf{x}) = \frac{e_0 \det \widetilde{\mathbf{A}}_n(\mathbf{x})k_{n1}(\mathbf{x})}{\sum_{i=1}^{n} \det\widetilde{\mathbf{A}}_i(\mathbf{x})}, \tag{53}$$

and therefore since $\mathbf{x} > \mathbf{0}$, we have by the King–Altman theorem

$$\begin{aligned} 0 &< \det \widetilde{\mathbf{A}}_n(\mathbf{x})k_{n1}(\mathbf{x}), \quad n \text{ is odd,} \\ 0 &> \det \widetilde{\mathbf{A}}_n(\mathbf{x})k_{n1}(\mathbf{x}), \quad n \text{ is even,} \end{aligned}$$

which implies

$$0 < k_{12}(\mathbf{x})k_{23}(\mathbf{x})\cdots k_{n-1,n}(\mathbf{x})k_{n1}(\mathbf{x}),$$

so $k_{p,p+1}(\mathbf{x}) > 0$ for all $1 \leq p \leq n$, and thus $K_{ji}(\mathbf{x})$ exists. The hypothesis that $\mathbf{x} > \mathbf{0}$ also implies $k_{p+1,p}(\mathbf{x}) \geq 0$ for all p so that $K_{ji}(\mathbf{x}) \geq 0$. If the sequence in (52) does not contain any irreversible edges, then $k_{p+1,p}(\mathbf{x}) > 0$ for all $i \leq p < j$ and $K_{ji}(\mathbf{x}) > 0$. □

The next proposition provides a nonrecursive expression for net rate functions. It was this form that Noyes (1964) discussed in his analysis of reaction kinetics under a uniform flux (i.e., steady-state) approximation. Our proof follows from the recursive expression provided by Proposition 5.

Proposition 7. *Let (D_n, A_n^m, e_0) be a serial model with $n > 2$ and indexed such that (E_n, E_1) is an irreversible edge. If $\mathbf{x} > 0$, then*

$$\frac{1}{k_i'(\mathbf{x})} = \sum_{j=i}^{n} \frac{K_{ji}(\mathbf{x})}{k_{j,j+1}(\mathbf{x})}. \tag{54}$$

Proof. By hypothesis, Proposition 2 implies that $k_i'(\mathbf{x}) > 0$ for all i, so the reciprocals $1/k_i'(\mathbf{x})$ exist for all i. Suppose first that $i = n$, then Proposition 4 asserts that

$$\frac{1}{k_n'(\mathbf{x})} = \frac{1}{k_{n1}(\mathbf{x})} = \frac{K_{nn}(\mathbf{x})}{k_{n,n+1}(\mathbf{x})} = \sum_{j=n}^{n} \frac{K_{jn}(\mathbf{x})}{k_{j,j+1}(\mathbf{x})}, \tag{55}$$

such that Eq. (54) holds for $i = n$. Now, assume the inductive hypothesis that expression (54) holds for some enzyme species E_i where $1 < i \leq n$ and consider the species E_{i-1}. Since $k_{i-1}'(\mathbf{x}) > 0$ we may write (see Proposition 5)

$$\frac{1}{k_{i-1}'(\mathbf{x})} = \frac{k_{i,i-1}(\mathbf{x})}{k_{i-1,i}(\mathbf{x})} \cdot \frac{1}{k_i'(\mathbf{x})} + \frac{1}{k_{i-1,i}(\mathbf{x})}. \tag{56}$$

Applying the inductive hypothesis, we get

$$\frac{1}{k'_{i-1}(\mathbf{x})} = \frac{k_{i,i-1}(\mathbf{x})}{k_{i-1,i}(\mathbf{x})} \sum_{j=i}^{n} \frac{K_{ji}(\mathbf{x})}{k_{j,j+1}(\mathbf{x})} + \frac{1}{k_{i-1,i}(\mathbf{x})}. \tag{57}$$

However, according to the definition of the equilibrium function,

$$\frac{k_{i,i-1}(\mathbf{x})}{k_{i-1,i}(\mathbf{x})} K_{ji}(\mathbf{x}) = \frac{k_{i,i-1}(\mathbf{x})}{k_{i-1,i}(\mathbf{x})} \prod_{p=i}^{j-1} \frac{k_{p+1,p}(\mathbf{x})}{k_{p,p+1}(\mathbf{x})} = \prod_{p=i-1}^{j-1} \frac{k_{p+1,p}(\mathbf{x})}{k_{p,p+1}(\mathbf{x})} = K_{j,i-1}(\mathbf{x}). \tag{58}$$

Substitution of this expression into Eq. (57), we get

$$\frac{1}{k'_{i-1}(\mathbf{x})} = \sum_{j=i}^{n} \frac{K_{j,i-1}(\mathbf{x})}{k_{j,j+1}(\mathbf{x})} + \frac{1}{k_{i-1,i}(\mathbf{x})} = \sum_{j=i-1}^{n} \frac{K_{j,i-1}(\mathbf{x})}{k_{j,j+1}(\mathbf{x})}. \tag{59}$$

Thereby proving the inductive case. Since our choice of $n > 2$ was arbitrary, the result is true for all serial models where $n > 2$. □

Corollary 1. *If $i \le q < n$ and (E_q, E_{q+1}) is an irreversible edge, then*

$$\frac{1}{k'_i(x)} = \sum_{j=i}^{q} \frac{K_{ji}(x)}{k_{j,j+1}(x)}. \tag{60}$$

Proof. By definition, $k_{q+1,q}(\mathbf{x}) = 0$ for all $\mathbf{x}$, because (E_{q+1}, E_q) does not exist in D_n. Therefore, $K_{ji}(\mathbf{x}) = 0$ for all $j > q$. □

4.6 Sensitivity Indices

In the previous development, we have treated net rate functions as functions of $\mathbf{x}$. However, they depend on $\mathbf{x}$ via the rate functions alone. Therefore, we can treat values of the rate functions as independent, parametric variables in their own right when analyzing the net rate functions. Following the example of Ray (1983), we can then define the sensitivity index S_{ij} of the net rate function k_i' with respect to the rate function $k_{j,\,j+1}$ where $i \le j$ as

$$S_{ij}(\mathbf{x}) = \frac{1/k_{j,j+1}(\mathbf{x})}{1/k'_i(\mathbf{x})} \cdot \left(\frac{\partial(1/k'_i)}{\partial(1/k_{j,j+1}(\mathbf{x}))} \right)(\mathbf{x}), \quad \mathbf{x} > \mathbf{0}, \tag{61}$$

where the partial derivative holds $K_{j+1,\,j}(\mathbf{x})$ as well as all evaluations of the rate functions other than $k_{j,\,j+1}(\mathbf{x})$ and $k_{j+1,\,j}(\mathbf{x})$ as constant values.

Proposition 8. *Let (D_n, A_n^m, e_0) be a serial model with $n > 2$. If $\mathbf{x} > 0$, then the sensitivity index of the i-th net rate function with respect to rate function $k_{j,\,j+1}$ where $i \le j$ is given by*

$$S_{ij}(\boldsymbol{x}) = \frac{k_i'(\boldsymbol{x})K_{ji}(\boldsymbol{x})}{k_{j,j+1}(\boldsymbol{x})}. \tag{62}$$

Proof. In order to evaluate the derivative in Eq. (61), we replace the parametrically defined variable $k_{j+1,j}(\mathbf{x})$ in Eq. (54) with

$$k_{j+1,j}(\mathbf{x}) = k_{j,j+1}(\mathbf{x})K_{j+1,j}(\mathbf{x}), \tag{63}$$

noting that $K_{j+1,j}(\mathbf{x})$ exists by Preposition 6 and is to be treated as a constant when taking the partial derivative, to get

$$\frac{1}{k_i'(\mathbf{x})} = \sum_{p=i}^{j-1} \frac{K_{pi}(\mathbf{x})}{k_{p,p+1}(\mathbf{x})} + \frac{K_{ji}(\mathbf{x})}{k_{j,j+1}(\mathbf{x})} + \sum_{p=j+1}^{n} \frac{K_{pi}(\mathbf{x})}{k_{p,p+1}(\mathbf{x})}, \tag{64}$$

where the equilibrium functions depend on $k_{j+1,j}(\mathbf{x})$ and $k_{j,j+1}(\mathbf{x})$ only via the constant ratio $K_{j+1,j}(\mathbf{x})$. Therefore, the first and third terms on the right-hand side of (64) are 0 when evaluating the derivative, and it follows that

$$\left(\frac{\partial(1/k_i')}{\partial(1/k_{j,j+1}(\mathbf{x}))}\right)(\mathbf{x}) = K_{ji}(\mathbf{x}). \tag{65}$$

Substituting this result into the definition Eq. (61), we get for all $\mathbf{x} > 0$,

$$S_{ij}(\mathbf{x}) = \frac{k_i'(\mathbf{x})K_{ji}(\mathbf{x})}{k_{j,j+1}(\mathbf{x})}. \blacksquare \tag{66}$$

Corollary 2. *If* $\mathbf{x} > 0$, *then* $0 \leq S_{ij}(\mathbf{x}) \leq 1$ *for all* $j \geq i$, *and*

$$\sum_{j=i}^{n} S_{ij}(\boldsymbol{x}) = 1. \tag{67}$$

Proof. The result that $S_{ij}(\mathbf{x}) \geq 0$ follows immediately from the result that $k_i'(\mathbf{x})$, $K_{ji}(\mathbf{x})$ and $k_{j,j+1}(\mathbf{x})$ are all nonnegative. If we consider the sum of $S_{ij}(\mathbf{x})$ over all $j \geq i$, then

$$\sum_{j=i}^{n} S_{ij}(\mathbf{x}) = \left(\sum_{j=i}^{n} \frac{K_{ji}(\mathbf{x})}{k_{j,j+1}(\mathbf{x})}\right) k_i'(\mathbf{x}). \tag{68}$$

However, using the result from Preposition 7, we find that

$$\sum_{j=i}^{n} S_{ij}(\mathbf{x}) = \left(\sum_{j=i}^{n} \frac{K_{ji}(\mathbf{x})}{k_{j,j+1}(\mathbf{x})}\right) \Big/ \left(\sum_{j=i}^{n} \frac{K_{ji}(\mathbf{x})}{k_{j,j+1}(\mathbf{x})}\right) = 1. \quad \square \tag{69}$$

Corollary 3. *If* $i \le q < n$ *and* (E_q, E_{q+1}) *is an irreversible edge in* D_n, *then* $S_{ij} = 0$ *for all* $j > q$, *hence* $\sum_{j=i}^{q} S_{ij} = 1$.

Proof. The proof follows from Corollary 2 and is essentially identical to that for Corollary 1. □

4.7 Isomorphic Models

Definition 9 (Isomorphic Models).

Suppose $M = (D_n, A_n^m, e_0)$ and $M^* = (D_n^*, A_n^{*m}, e_0^*)$ are steady-state models where $e_0 = e_0^*$. We say that M and M^* are *isomorphic* if there exists a set of $n(n-1)$ positive real numbers $\{{}^*k_{ij}\}_{i \ne j}^{n,n}$ and an isomorphism g from D_n onto D_n^* such that whenever (E_i, E_j) is an edge in D_n and $g(E_i) = E_p^*$ and $g(E_j) = E_q^*$, we have $k_{ij}(\mathbf{x}) = {}^*k_{ij} k_{pq}^*(\mathbf{x})$ for all $\mathbf{x} \in \mathbb{R}^m$.

Note that if two steady-state models are isomorphic, there may be more than one isomorphism between the two digraphs where the condition on A_n^m and A_n^{*m} is fulfilled. When working with isomorphic steady-state models, we will always assume an indexing that follows one of these isomorphisms g so that $g(E_i) = E_i^*$ for all indices i. Furthermore, it is straightforward to show that if models 1 and 2 are isomorphic, and models 2 and 3 are isomorphic, then models 1 and 3 are isomorphic. In other words, the relation is transitive.

Proposition 9. *Suppose* (D_n, A_n^m, e_0) *and* (D_n^*, A_n^{*m}, e_0) *are isomorphic serial models. If* (E_i, E_{i+1}) *is an irreversible edge in* D_n, *then*

$$\frac{k_i'}{k_i'^*} = {}^*k_{i,i+1}, \tag{70}$$

and if there are no irreversible edges in the sequence

$$(E_i, E_{i+1})(E_{i+1}, E_{i+2}) \ldots (E_{q-1}, E_q), \tag{71}$$

with (E_q, E_{q+1}) *being an irreversible edge in* D_n, *then*

$$\frac{k_i'}{k_i'^*} = S_{ii}{}^*k_{i,i+1} + \sum_{j=i+1}^{q} \left(S_{ij}{}^*k_{j,j+1} \prod_{p=i}^{j-1} \frac{{}^*k_{p,p+1}}{{}^*k_{p+1,p}} \right), \tag{72}$$

where the ratio function $k_i'/k_i'^*$ *is defined for all* $\mathbf{x} > 0$.

Proof. Let $\mathbf{x}$ be any vector in $\mathbb{R}^m$ such that $\mathbf{x} > 0$. Proposition 8 implies

$$S_{ii}(\mathbf{x}) = \frac{k_i'(\mathbf{x})}{k_{i,i+1}(\mathbf{x})} = \frac{k_i'(\mathbf{x})}{{}^*k_{i,i+1} k_{i,i+1}^*(\mathbf{x})} = \frac{k_i'(\mathbf{x})}{k_i'^*(\mathbf{x})} \cdot \frac{S_{ii}^*(\mathbf{x})}{{}^*k_{i,i+1}}. \tag{73}$$

However, if (E_i, E_{i+1}) is irreversible, Corollary 3 implies

$$S_{ii}(\mathbf{x}) = S_{ii}^*(\mathbf{x}) = 1,$$

for all $\mathbf{x} > \mathbf{0}$ and therefore

$$\frac{k_i'(\mathbf{x})}{k_i' * (\mathbf{x})} \cdot \frac{1}{*k_{i,i+1}} = 1, \tag{74}$$

which rearranges immediately to Eq. (70). If, however, (E_i, E_{i+1}) is not irreversible, then for any $i < j \leq q$ in sequence (71), Proposition 8 asserts

$$S_{ij}(\mathbf{x}) = \frac{k_i'(\mathbf{x})K_{ji}(\mathbf{x})}{k_{j,j+1}(\mathbf{x})} = \frac{k_i'(\mathbf{x})}{k_{j,j+1}(\mathbf{x})}\prod_{p=i}^{j-1}\frac{k_{p+1,p}(\mathbf{x})}{k_{p,p+1}(\mathbf{x})}. \tag{75}$$

Therefore, it follows that

$$S_{ij}(\mathbf{x}) = k_i'(\mathbf{x})\left(\frac{1}{k_{j,j+1}^*(\mathbf{x})}\prod_{r=i}^{j-1}\frac{k_{r+1,r}^*(\mathbf{x})}{k_{r,r+1}^*(\mathbf{x})}\right)\left(\frac{1}{*k_{j,j+1}}\prod_{p=i}^{j-1}\frac{*k_{p+1,p}}{*k_{p,p+1}}\right). \tag{76}$$

Applying the results of Proposition 8 for (D_n^*, A_n^{*m}, e_0), we find that

$$S_{ij}(\mathbf{x}) = \frac{k_i'(\mathbf{x})}{k_i'^*(\mathbf{x})}S_{ij}^*(\mathbf{x})\left(\frac{1}{*k_{j,j+1}}\prod_{p=i}^{j-1}\frac{*k_{p+1,p}}{*k_{p,p+1}}\right), \tag{77}$$

which we can rearrange to get

$$\frac{k_i'(\mathbf{x})}{k_i'^*(\mathbf{x})}S_{ij}^*(\mathbf{x}) = S_{ij}(\mathbf{x})\left(*k_{j,j-1}\prod_{p=i}^{j-1}\frac{*k_{p,p+1}}{*k_{p+1,p}}\right), \tag{78}$$

where $i < j \leq q$. Summing (78) over all j for $i < j \leq q$ and adding Eq. (73), we have

$$\frac{k_i'(\mathbf{x})}{k_i'^*(\mathbf{x})}\sum_{j=i}^{q}S_{ij}^*(\mathbf{x}) = S_{ii}(\mathbf{x})^*k_{i,i+1} + \sum_{j=i+1}^{q}\left(S_{ij}(\mathbf{x})^*k_{j,j+1}\prod_{p=i}^{j-1}\frac{*k_{p,p+1}}{*k_{p+1,p}}\right), \tag{79}$$

and use of Corollary 3 yields Eq. (72). □

4.8 Perturbation Effects

Proposition 9 is our key result, and while the notation is useful for ensuring that all of the ratios exist (i.e., there is no division by zero), it is rather ungainly. Having established this result, however, we can simplify the notation considerably to match that in Section 2.4 by defining the constants ${}^*\kappa_{ij}$ (where $j \geq i$) such that

$$ {}^*\kappa_{ij} = \begin{cases} {}^*k_{i,i+1}, & j = i, \\ {}^*k_{j,j+1} \prod_{p=i}^{j-1} \dfrac{{}^*k_{p,p+1}}{{}^*k_{p+1,p}}, & j > i. \end{cases} \tag{80} $$

This will allow us to write both Eqs. (70) and (72) together using the more compact expression

$$ \frac{k_i'}{k_i'^*} = \sum_{j \geq i} S_{ij} {}^*\kappa_{ij}, \tag{81} $$

where the sum is understood to extend from i to q for the first irreversible edge (E_q, E_{q+1}). It is also worth pointing out for the sake of clarity that we have defined the ${}^*\kappa_{ij}$ in Eq. (80) differently than we did in Section 2.4, though the results are equivalent. In Section 2.4, they represented constant-valued functions defined informally based on chemical intuition. Here they are constants defined rigorously with respect to two isomorphic serial models.

The definition of a net rate function also implies that whenever $\mathbf{x} > \mathbf{0}$, we can write for any pair of isomorphic serial models

$$ \frac{v(\mathbf{x})}{v^*(\mathbf{x})} = \frac{k_i'(\mathbf{x})}{k_i'^*(\mathbf{x})} \cdot \frac{e_i(\mathbf{x})}{e_i^*(\mathbf{x})}, \tag{82} $$

for any vertex E_i, since Proposition 2 asserts that $v(\mathbf{x})$ and $v^*(\mathbf{x})$ are both positive. We can rearrange this expression to also get

$$ \frac{v(\mathbf{x})}{v^*(\mathbf{x})} e_i^*(\mathbf{x}) = \frac{k_i'(\mathbf{x})}{k_i'^*(\mathbf{x})} e_i(\mathbf{x}), \tag{83} $$

and summing over all the vertices, we have (see Eq. (26b))

$$ \sum_{i=1}^{n} \frac{k_i'(\mathbf{x})}{k_i'^*(\mathbf{x})} e_i(\mathbf{x}) = \sum_{i=1}^{n} \frac{v(\mathbf{x})}{v^*(\mathbf{x})} e_i^*(\mathbf{x}) = \frac{v(\mathbf{x})}{v^*(\mathbf{x})} \sum_{i=1}^{n} e_i^*(\mathbf{x}) = \frac{v(\mathbf{x})}{v^*(\mathbf{x})} e_0, \tag{84} $$

which rearranges immediately to provide

$$ \frac{v(\mathbf{x})}{v^*(\mathbf{x})} = \sum_{i=1}^{n} \frac{e_i(\mathbf{x})}{e_0} \cdot \frac{k_i'(\mathbf{x})}{k_i'^*(\mathbf{x})}. \tag{85} $$

This defines a function on all $\mathbf{x} > \mathbf{0}$ that we can write

$$\begin{aligned} &{}^*\nu: \{\mathbf{x} \in \mathbb{R}^m : \mathbf{x} > \mathbf{0}\} \rightarrow \mathbb{R}, \\ &{}^*\nu(\mathbf{x}) = \nu(\mathbf{x})/\nu^*(\mathbf{x}). \end{aligned} \tag{86}$$

This function is thus defined for any pair of isomorphic serial models.

The expression for ${}^*\nu(\mathbf{x})$ can be simplified if we introduce the notation

$$f_i(\mathbf{x}) = e_i(\mathbf{x})/e_0, \tag{87}$$

so that for all $\mathbf{x} > \mathbf{0}$, we have

$${}^*\nu(\mathbf{x}) = \sum_{i=1}^{n} f_i(\mathbf{x}) k_i'(\mathbf{x})/k_i'^*(\mathbf{x}) = \sum_{i=1}^{n} \sum_{j \geq i} f_i(\mathbf{x}) S_{ij}(\mathbf{x})^* \kappa_{ij}. \tag{88}$$

Proposition 2 implies $0 < f_i(\mathbf{x}) < 1$ for all $\mathbf{x} > \mathbf{0}$, and from Eq. (26b),

$$\sum_{i=1}^{n} f_i(\mathbf{x}) = 1. \tag{89}$$

Combining this result with Corollary 3, it follows that

$$\sum_{i=1}^{n} \sum_{j \geq i} f_i(\mathbf{x}) S_{ij}(\mathbf{x}) = 1. \tag{90}$$

Hence, ${}^*\nu(\mathbf{x})$ is a weighted average of the constant values ${}^*\kappa_{ij}$.

4.9 Limits

We now want to consider limits of the function ${}^*\nu$ as discussed in Section 2.3. Since the entire vector $\mathbf{x}$ is to be considered in the limit, we will express limits using the compact notation

$$\mathbf{x} \rightarrow \mathbf{y} = [y_1, y_2, \ldots, y_m]^{\mathsf{T}}, \tag{91}$$

where each y_i represents 0^+, a positive real number or ∞. As discussed in Section 2.3, it is not always guaranteed that such a limit exists for ${}^*\nu$ given any pair of isomorphic serial models. Nevertheless, results useful for the interpretation of the limits of ${}^*\nu$ are given by the following propositions.

Proposition 10. *Let ν_1 be the velocity function for a serial model $(D_n, A_n^m, e_0)_1$ and let ν_2 be the velocity function for an isomorphic model $(D_n, A_n^m, e_0)_2$. If*

$$\lim_{x \rightarrow y} \nu_1(x)/\nu^*(x) \tag{92}$$

exists for all models (D_n^*, A_n^{*m}, e_0) *isomorphic to* $(D_n, A_n^m, e_0)_1$, *then*

$$\lim_{x \to y} v_2(x)/v^*(x) \tag{93}$$

exists for all models (D_n^*, A_n^{*m}, e_0) *isomorphic to* $(D_n, A_n^m, e_0)_2$.

Proof. Since $(D_n, A_n^m, e_0)_2$ is isomorphic to $(D_n, A_n^m, e_0)_1$, the limit

$$\lim_{\mathbf{x} \to \mathbf{y}} v_1(\mathbf{x})/v_2(\mathbf{x}), \tag{94}$$

exists by hypothesis. Furthermore, Proposition 9 and Eqs. (88) and (90) imply that the ratio $v_1(\mathbf{x})/v_2(\mathbf{x})$ is bounded below on $\mathbf{x} > \mathbf{0}$ by a positive value such that the limit must also be positive. Therefore, the limit

$$\lim_{\mathbf{x} \to \mathbf{y}} v_2(\mathbf{x})/v_1(\mathbf{x}) \tag{95}$$

also exists. If (D_n^*, A_n^{*m}, e_0) is any model isomorphic to $(D_n, A_n^m, e_0)_2$, then it is also isomorphic to $(D_n, A_n^m, e_0)_1$, and the existence of the limit in Eq. (93) follows from

$$\lim_{\mathbf{x} \to \mathbf{y}} \frac{v_2(\mathbf{x})}{v^*(\mathbf{x})} = \lim_{\mathbf{x} \to \mathbf{y}} \frac{v_2(\mathbf{x})v_1(\mathbf{x})}{v_1(\mathbf{x})v^*(\mathbf{x})} = \left(\lim_{\mathbf{x} \to \mathbf{y}} \frac{v_2(\mathbf{x})}{v_1(\mathbf{x})}\right)\left(\lim_{\mathbf{x} \to \mathbf{y}} \frac{v_1(\mathbf{x})}{v^*(\mathbf{x})}\right). \blacksquare \tag{96}$$

Proposition 11. *Let* (D_n, A_n^m, e_0) *be a serial model with* $n > 2$. *If the limit* $\lim_{\mathbf{x} \to \mathbf{y}} {}^*v(\mathbf{x})$ *exists for some condition* $\mathbf{y}$ *and all isomorphic models, then the limit* $\lim_{\mathbf{x} \to \mathbf{y}} f_i(\mathbf{x})$ *exists for all* i.

Proof. Let (D_n^*, A_n^{*m}, e_0) be a serial model isomorphic to (D_n, A_n^m, e_0) such that all elements of $\{{}^*k_{up}\}$ are unity except ${}^*k_{i,i+1} = {}^*k_{i,i-1} = 2$, and thus

$${}^*\kappa_{up} = \begin{cases} 1, & u \neq i, \\ 2, & u = i. \end{cases} \tag{97}$$

Eqs. (88) and (90) then imply that for all $\mathbf{x} > \mathbf{0}$,

$$\begin{aligned} {}^*v(\mathbf{x}) &= 2f_i(\mathbf{x}) + \sum_{u \neq i} \sum_{p \geq u} f_u(\mathbf{x}) S_{up}(\mathbf{x}), \\ &= 2f_i(\mathbf{x}) + (1 - f_i(\mathbf{x})), \\ &= f_i(\mathbf{x}) + 1. \end{aligned} \tag{98}$$

Hence, the existence of $\lim_{\mathbf{x} \to \mathbf{y}} f_i(\mathbf{x})$ follows from

$$\lim_{\mathbf{x} \to \mathbf{y}} f_i(\mathbf{x}) = \lim_{\mathbf{x} \to \mathbf{y}} ({}^*v(\mathbf{x}) - 1). \blacksquare \tag{99}$$

Proposition 12. *Let (D_n, A_n^m, e_0) be a serial model with $n > 2$. If the limit $\lim_{\mathbf{x}\to\mathbf{y}} {}^*v(\mathbf{x})$ exists for some condition $\mathbf{y}$ and all isomorphic models, then for any isomorphic serial model and any i, $\lim_{\mathbf{x}\to\mathbf{y}} k_i'(\mathbf{x})/k_i'^*(\mathbf{x})$ exists or*

$$\lim_{x\to y} f_i(x)k_i'(x)/k_i'^*(x) = 0. \tag{100}$$

Proof. From Eq. (81) and Corollary 2, we have for any isomorphic model

$$f_i(\mathbf{x})\min\{{}^*\kappa_{ij}\} \leq f_i(\mathbf{x})k_i'(\mathbf{x})/k_i'^*(\mathbf{x}) \leq f_i(\mathbf{x})\max\{{}^*\kappa_{ij}\}. \tag{101}$$

If $\lim_{\mathbf{x}\to\mathbf{y}} f_i(\mathbf{x}) = 0$, then Eq. (100) follows from the Squeeze Theorem. Furthermore, Eq. (83) and Propositions 10 and 11 imply the existence of

$$\lim_{\mathbf{x}\to\mathbf{y}} f_i(\mathbf{x})k_i'(\mathbf{x})/k_i'^*(\mathbf{x}) = \lim_{\mathbf{x}\to\mathbf{y}} f_i^*(\mathbf{x})^*v(\mathbf{x}). \tag{102}$$

Therefore, if $\lim_{\mathbf{x}\to\mathbf{y}} f_i(\mathbf{x}) \neq 0$, then $\lim_{\mathbf{x}\to\mathbf{y}} k_i'(\mathbf{x})/k_i'^*(\mathbf{x})$ exists according to

$$\lim_{\mathbf{x}\to\mathbf{y}} \frac{k_i'(\mathbf{x})}{k_i'^*(\mathbf{x})} = \lim_{\mathbf{x}\to\mathbf{y}} \frac{f_i(\mathbf{x})k_i'(\mathbf{x})}{f_i(\mathbf{x})k_i'^*(\mathbf{x})} = \frac{\lim_{\mathbf{x}\to\mathbf{y}} f_i(\mathbf{x})k_i'(\mathbf{x})/k_i'^*(\mathbf{x})}{\lim_{\mathbf{x}\to\mathbf{y}} f_i(\mathbf{x})}. \blacksquare \tag{103}$$

Proposition 13. *Let (D_n, A_n^m, e_0) be a serial model with $n > 2$ and suppose the limit $\lim_{\mathbf{x}\to\mathbf{y}} {}^*v(\mathbf{x})$ exists for some condition $\mathbf{y}$ and all isomorphic models. If $\lim_{\mathbf{x}\to\mathbf{y}} f_i(\mathbf{x}) \neq 0$, then $\lim_{\mathbf{x}\to\mathbf{y}} S_{ij}(\mathbf{x})$ exists for all $j \geq i$.*

Proof. Choose an isomorphic model such that all elements of $\{{}^*k_{up}\}$ are unity except for ${}^*k_{j,j+1} = {}^*k_{j+1,j} = 2$. It then follows from Eq. (81) that

$$k_i'(\mathbf{x})/k_i'^*(\mathbf{x}) = 2S_{ij}(\mathbf{x}) + (1 - S_{ij}(\mathbf{x})) = S_{ij}(\mathbf{x}) + 1. \tag{104}$$

Therefore, the existence of $\lim_{\mathbf{x}\to\mathbf{y}} S_{ij}(\mathbf{x})$ is implied by Proposition 12, because

$$\lim_{\mathbf{x}\to\mathbf{y}} S_{ij}(\mathbf{x}) = \lim_{\mathbf{x}\to\mathbf{y}} \left(k_i'(\mathbf{x})/k_i'^*(\mathbf{x}) - 1\right). \blacksquare \tag{105}$$

4.10 Two-Vertex Models

The development in Sections 4.5–4.9 only considered serial mechanisms where $n > 2$, because directed graphs with only two vertices are unoriented. Nevertheless, it is easy to show that the preceding results are also applicable to two-vertex serial models of the form

$$E_1 \underset{k_{21}(\mathbf{x})}{\overset{k_{12}(\mathbf{x})}{\rightleftharpoons}} E_2. \tag{106}$$

To do so, a forward direction to the model must be imposed by splitting the $E_2 \rightarrow E_1$ reaction into reverse and forward components governed by rate functions k_r and k_ν, respectively, where $k_\nu \neq 0$ such that

$$k_{21}(\mathbf{x}) = k_r(\mathbf{x}) + k_\nu(\mathbf{x}). \tag{107}$$

Next, we construct a three-vertex model of the form

$$\tilde{E}_1 \underset{2k_r(\mathbf{x})}{\overset{k_{12}(\mathbf{x})}{\rightleftharpoons}} \tilde{E}_2 \xrightarrow{2k_\nu(\mathbf{x})} \tilde{E}_3 \xrightarrow{2k_\nu(\mathbf{x})} \tilde{E}_1. \tag{108}$$

It is straightforward to show that the solutions to Eq. (26) for the models in (106) and (108) are related according to

$$e_1(\mathbf{x}) = \tilde{e}1(\mathbf{x}), \tag{109a}$$

$$e_2(\mathbf{x}) = \tilde{e}_2(\mathbf{x}) + \tilde{e}_3(\mathbf{x}), \tag{109b}$$

and that $\tilde{\nu}(\mathbf{x}) = \nu(\mathbf{x}) = k_\nu(\mathbf{x})e_2(\mathbf{x})$. It follows then that the net rate function for E_1 is equal to that for $\tilde{E}_1$, and we have

$$\begin{aligned} {}^*\nu(\mathbf{x}) &= \tilde{f}_1(\mathbf{x})\frac{k_1'(\mathbf{x})}{k_1'^*(\mathbf{x})} + \tilde{f}_2(\mathbf{x})\frac{k_\nu'(\mathbf{x})}{k_\nu'^*(\mathbf{x})} + \tilde{f}_3(\mathbf{x})\frac{k_\nu'(\mathbf{x})}{k_\nu'^*(\mathbf{x})}, \\ &= f_1(\mathbf{x})\frac{k_1'(\mathbf{x})}{k_1'^*(\mathbf{x})} + f_2(\mathbf{x})\frac{k_\nu'(\mathbf{x})}{k_\nu'^*(\mathbf{x})}, \end{aligned} \tag{110}$$

since the factor of two cancels in the ratio functions.

ACKNOWLEDGMENTS

This work was supported by a grants from the National Institutes of Health (R01 GM035906) and the Welch Foundation (F-1511).

REFERENCES

Alvarez, F. J., Ermer, J., Hübner, G., Schellenberger, A., & Schowen, R. L. (1991). Catalytic power of pyruvate decarboxylase. Rate-limiting events and microscopic rate constants from primary carbon and secondary hydrogen isotope effects. *Journal of the American Chemical Society*, *113*, 8402–8409.

Broderick, J. B., Duffus, B. R., Duschene, K. S., & Shepard, E. M. (2014). Radical *S*-adenosylmethionine enzymes. *Chemical Reviews*, *114*, 4229–4317.

Cha, S. (1968). A simple method for derivation of rate equations for enzyme-catalyzed reactions under the rapid equilibrium assumption or combined assumptions of equilibrium and steady state. *The Journal of Biological Chemistry*, *243*(4), 820–825.

Cleland, W. W. (1975). Partition analysis and the concept of net rate constants as tools in enzyme kinetics. *Biochemistry*, *14*(14), 3220–3224.

Cleland, W. W. (1982). Use of isotope effects to elucidate enzyme mechanisms. *Critical Reviews in Biochemistry*, *13*, 385–429.

Cook, P. F., & Cleland, W. W. (1981a). pH variation of isotope effects in enzyme-catalyzed reactions. 1. Isotope- and pH-dependent steps the same. *Biochemistry, 20*, 1797–1805.

Cook, P. F., & Cleland, W. W. (1981b). pH variation of isotope effects in enzyme-catalyzed reactions. 2. Isotope-dependent step not pH dependent. Kinetic mechanism of alcohol dehydrogenase. *Biochemistry, 20*, 1805–1816.

Cook, P. F., & Cleland, W. W. (2007). *Enzyme kinetics and mechanism*. New York: Garland Science.

Daley, C. J. A., & Holm, R. H. (2003). Reactions of site-differentiated $[Fe_4S_4]^{2+, 1+}$ clusters with sulfonium cations: Reactivity analogues of biotin synthase and other members of the *S*-adenosylmethionine enzyme family. *Journal of Inorganic Biochemistry, 97*, 287–298.

Duschene, K. S., Veneziano, S. E., Silver, S. C., & Broderick, J. B. (2009). Control of radical chemistry in the AdoMet radical enzymes. *Current Opinion in Chemical Biology, 13*, 74–83.

Frey, P. A. (2014). Travels with carbon-centered radicals. 5′-Deoxyadenosine and 5′-deoxyadenosine-5′-yl in radical enzymology. *Accounts of Chemical Research, 47*, 540–549.

Frey, P. A., & Magnusson, O. T. (2003). *S*-Adenosylmethionine: A wolf in sheep's clothing, or a rich man's adenosylcobalamin? *Chemical Reviews, 103*, 2129–2148.

Hayon, E., & Simic, M. (1974). Acid-base properties of free radicals in solution. *Accounts of Chemical Research*, 7(4), 114–121.

Hinckley, G. T., & Frey, P. A. (2006). Cofactor dependence of reduction potentials for $[4Fe\text{-}4S]^{2+/1+}$ in lysine-2,3-aminomutase. *Biochemistry, 45*, 3219–3225.

Hoffart, L. M., Barr, E. W., Guyer, R. B., Bollinger, J. M., Jr., & Krebs, C. (2006). Direct spectroscopic detection of a C–H-cleaving high-spin Fe(IV) complex in a prolyl-4-hydroxylase. *Proceedings of the National Academy of Sciences of the United States of America, 103*, 14738–14743.

King, E. L., & Altman, C. (1956). A schematic method of deriving the rate laws for enzyme-catalyzed reactions. *The Journal of Physical Chemistry, 60*, 1375–1378.

Ko, Y., Ruszczycky, M. W., Choi, S.-H., & Liu, H.-w. (2015). Mechanistic studies of the radical *S*-adenosylmethionine enzyme DesII with TDP-D-fucose. *Angewandte Chemie International Edition, 54*, 860–863.

Landgraf, B. J., McCarthy, E. L., & Booker, S. J. (2016). Radical *S*-adenosylmethionine enzymes in human health and disease. *Annual Review of Biochemistry, 85*, 485–514.

Lewis, B. E., & Schramm, V. L. (2006). Enzymatic binding isotope effects and the interaction of glucose with hexokinase. In A. Kohen & H.-H. Limbach (Eds.), *Isotope effects in chemistry and biology* (pp. 1019–1053). Boca Raton: CRC Taylor & Francis.

Matsen, F. A., & Franklin, J. L. (1950). A general theory of coupled sets of first order reactions. *Journal of the American Chemical Society, 72*, 3337–3341.

Mehta, A. P., Abdelwahed, S. H., Mahanta, N., Fedoseyenko, D., Philmus B., Cooper, L. E., … Begley, T. P. (2015). Radical *S*-adenosylmethionine (SAM) enzymes in cofactor biosynthesis: A treasure trove of complex organic radical rearrangement reactions. *The Journal of Biological Chemistry, 290*, 3980–3986.

Murdoch, J. R. (1981). What is the rate-limiting step of a multistep reaction? *Journal of Chemical Education, 58*, 32–36.

Northrop, D. B. (1981a). The expression of isotope effects on enzyme-catalyzed reactions. *Annual Review of Biochemistry, 50*, 103–131.

Northrop, D. B. (1981b). Minimal kinetic mechanism and general equation for deuterium isotope effects on enzymatic reactions: Uncertainty in detecting a rate-limiting step. *Biochemistry, 20*, 4056–4061.

Noyes, R. M. (1964). Kinetic treatment of consecutive processes. *Progress in Reaction Kinetics, 2*, 337–362.

Peck, S. C., Wang, C., Dassama, L. M. K., Zhang, B., Guo, Y., Rajakovich, L. J., ... van der Donk, W. A. (2017). O–H activation by an unexpected ferryl intermediate during catalysis by 2-hydroxyethylphosphonate dioxygenase. *Journal of the American Chemical Society, 139*, 2045–2052.

Price, J. C., Barr, E. W., Glass, T. E., & Krebs, C.Bollinger, Jr., J. M.. (2003). Evidence for hydrogen abstraction from C1 of taurine by the high-spin Fe(IV) intermediate detected during oxygen activation by taurine:α-ketoglutarate dioxygenase (TauD). *Journal of the American Chemical Society, 125*, 13008–13009.

Rao, P. S., & Hayon, E. (1974). Redox potentials of free radicals. I. Simple organic radicals. *Journal of the American Chemical Society, 96*(5), 1287–1294.

Rao, P. V., & Holm, R. H. (2004). Synthetic analogues of the active sites of iron-sulfur proteins. *Chemical Reviews, 104*, 527–559.

Ray, W. J. (1983). Rate-limiting step: A quantitative definition. Application to steady-state enzymatic reactions. *Biochemistry, 22*, 4625–4637.

Ruszczycky, M. W., & Anderson, V. E. (2006). Interpretation of *V/K* isotope effects for enzymatic reactions exhibiting multiple isotopically sensitive steps. *Journal of Theoretical Biology, 243*, 328–342.

Ruszczycky, M. W., Choi, S.-h., & Liu, H.-w. (2010). Stoichiometry of the redox neutral deamination and oxidative dehydrogenation reactions catalyzed by the radical SAM enzyme DesII. *Journal of the American Chemical Society, 132*, 2359–2369.

Ruszczycky, M. W., Choi, S.-h., & Liu, H.-w. (2013). EPR-kinetic isotope effect study of the mechanism of radical-mediated dehydrogenation of an alcohol by the radical SAM enzyme DesII. *Proceedings of the National Academy of Sciences of the United States of America, 110*, 2088–2093.

Ruszczycky, M. W., Choi, S.-h., Mansoorabadi, S. O., & Liu, H.-w. (2011). Mechanistic studies of the radical *S*-adenosyl-L-methionine enzyme DesII: EPR characterization of a radical intermediate generated during its catalyzed dehydrogenation of TDP-D-quinovose. *Journal of the American Chemical Society, 133*, 7292–7295.

Ruszczycky, M. W., & Liu, H.-w. (2015). Mechanistic enzymology of the radical SAM enzyme DesII. *Israel Journal of Chemistry, 55*, 315–324.

Ruszczycky, M. W., Ogasawara, Y., & Liu, H.-w. (2012). Radical SAM enzymes in the biosynthesis of sugar-containing natural products. *Biochimica et Biophysica Acta, 1824*, 1231–1244.

Schowen, K. B., & Schowen, R. L. (1982). Solvent isotope effects on enzyme systems. *Methods in Enzymology, 87C*, 551–606.

Schowen, R. L. (1978). Catalytic power and transition-state stabilization. In R. D. Gandour & R. L. Schowen (Eds.), *Transition states of biochemical processes* (pp. 77–114). New York: Plenum Press.

Schütz, A., Golbik, R., König, S., Hübner, G., & Tittmann, K. (2005). Intermediates and transition states in thiamin diphosphate-dependent decarboxylases. A kinetic and NMR study on wild-type indolepyruvate decarboxylase and variants using indolepyruvate, benzoylformate, and pyruvate substrates. *Biochemistry, 44*, 6164–6179.

Stein, R. L. (1981). Analysis of kinetic isotope effects on complex reactions utilizing the concept of the virtual transition state. *The Journal of Organic Chemistry, 46*, 3328–3330.

Szu, P.-H., Ruszczycky, M. W., Choi, S.-h., & Liu, H.-w. (2009). Characterization and mechanistic studies of DesII: A radical *S*-adenosyl-L-methionine enzyme involved in the biosynthesis of TDP-D-desosamine. *Journal of the American Chemical Society, 131*, 14030–14042.

Tamanaha, E., Zhang, B., Guo, Y., Chang, W.-c., Barr, E. W., Xing, G., ... Krebs, C. (2016). Spectroscopic evidence for the two C–H-cleaving intermediates of *Aspergillus nidulans* isopenicillin *N* synthase. *Journal of the American Chemical Society, 138*, 8862–8874.

Tian, G. (1992). Effective rate constants and general isotope effect equations for steady state enzymatic reactions with multiple isotope-sensitive steps. *Bioorganic Chemistry, 20*, 95–106.

Tittmann, K., Golbik, R., Uhlemann, K., Khailova, L., Schneider, G., Patel, M., … Hübner, G. (2003). NMR Analysis of covalent intermediates in thiamin diphosphate enzymes. *Biochemistry, 42*, 7885–7891.

Tittmann, K., Neef, H., Golbik, R., Hübner, G., & Kern, D. (2005). Kinetic control of thiamin diphosphate activation in enzymes studied by proton-nitrogen correlated NMR spectroscopy. *Biochemistry, 44*, 8697–8700.

Tittmann, K., Vyazmensky, M., Hübner, G., Barak, Z., & Chipman, D. M. (2005). The carboligation reaction of acetohydroxyacid synthase II: Steady-state intermediate distributions in wild type and mutants by NMR. *Proceedings of the National Academy of Sciences of the United States of America, 102*, 553–558.

Wang, S. C., & Frey, P. A. (2007). Binding energy in the one-electron reductive cleavage of *S*-adenosylmethionine in lysine 2,3-aminomutase, a radical SAM enzyme. *Biochemistry, 46*, 12889–12895.

Xing, G., Diao, Y., Hoffart, L. M., Barr, E. W., Prabhu, K. S., Arner, R. J., … Bollinger, J. M., Jr. (2006). Evidence for C–H cleavage by an iron-superoxide complex in the glycol cleavage reaction catalyzed by *myo*-inositol oxygenase. *Proceedings of the National Academy of Sciences of the United States of America, 103*, 6130–6135.

CHAPTER SEVENTEEN

Determining Carbon Kinetic Isotope Effects Using Headspace Analysis of Evolved CO_2

Scott O.C. Mundle*, Barbara Sherwood Lollar†, Ronald Kluger†,[1]
*Great Lakes Institute for Environmental Research, University of Windsor, Windsor, ON, Canada
†University of Toronto, Toronto, ON, Canada
[1]Corresponding author: e-mail address: r.kluger@utoronto.ca

Contents

1. Basic Experimental Procedures 504
2. Continuous Pressure-Monitored Headspace Analysis 510
3. Control Reactions 514
4. Improving Data Quality 515
   4.1 Calculating ε Values That Account for the Magnification of Error at Later Stages of the Reaction 517
   4.2 Assessment of the Impact of Experimental Error on Calculated ε Values 518
5. Conclusions 519
Acknowledgments 519
References 519

Abstract

Isotope ratio mass spectrometry (IRMS) provides accurate measurements of relative abundance of isotopes of heavy atoms for reactions that are subject to kinetic isotope effects (KIEs). The recent development of compound-specific isotope analysis (CSIA) allows the use of multiple time points that provide data for a rate plot as well as isotope ratios. Utilizing CSIA in enzymology presents opportunities for obtaining heavy atom KIEs in diverse areas.

Gas chromatography coupled to on-line combustion isotope ratio mass spectrometry (GC-C-IRMS) and continuous flow compound-specific isotope analysis (CSIA) offer convenient and accurate methods for obtaining carbon kinetic isotope effects (KIEs) in enzymology. This modern approach should be widely applicable to diverse enzyme reactions in which carbon–carbon bonds cleave in kinetically significant steps. Isotope ratio mass spectrometry (IRMS) measures the relative abundance of stable isotopes with great accuracy and sensitivity (Brand, 1996). The theoretical framework of KIEs accounts precisely for the specific effects of isotopic substitutions on rates of chemical

Methods in Enzymology, Volume 596
ISSN 0076-6879
http://dx.doi.org/10.1016/bs.mie.2017.07.024

and biochemical reactions (Bigeleisen, 1949a). Applying those fundamental principles to data obtained by modern methods provides insights on reaction processes that would otherwise remain inaccessible (Bigeleisen, 1949a; Urey, 1947). The key feature of IRMS is that it measures isotope ratios as a function of time in a way that is inherently highly precise without any need for adding tracers or isotopic enrichment. It is also especially suited to working with non-radioactive isotopes of heavier nuclei that are at the core of biologically important compounds ($^{13}C/^{12}C$, $^{15}N/^{14}N$, $^{18}O/^{16}O$).

O'Leary's articles in this series describe how to apply the conventional methods of determination of heavy atom KIEs based on the Nier/Bigeleisen approach to IRMS (O'Leary, 1980). O'Leary, a pioneering expert in the field, introduces the concepts and the equations necessary for data analysis. This is based on his extensive experience in applications of IRMS to carbon KIEs in enzymology (O'Leary, 1969, 1976, 1980, 1984, 1988, 1989; O'Leary & Brummund, 1974; O'Leary & Piazza, 1981; O'Leary & Westheimer, 1968; Paneth & O'Leary, 1985; Roeske & O'Leary, 1984). The work that followed those publications exemplifies the many critical applications of those methods to mechanistic analysis. The sophistication of multiisotopic probes utilized by O'Leary, Cleland, and their coworkers extended the range of possibilities with confidence based on acquisition of high-quality IRMS data (O'Leary, Yamada, & Yapp, 1981; Parmentier, Weiss, O'Leary, Schachman, & Cleland, 1992; Weiss et al., 1991).

The ability to use data from samples analyzed for their content of stable isotopes of carbon at natural abundance is the result of the efforts of Bigeleisen and coworkers that readily extended to analysis of heavy atom isotope effects in enzymic reactions (Bigeleisen, 1949a, 1949b; Bigeleisen & Allen, 1951; Bigeleisen & Friedman, 1949). Applying the approach for carbon KIEs utilizes data on the composition CO_2 for only two points along the reaction coordinate. These are determined with sufficiently high accuracy to provide a confident result for differentiation of the rates of isotopic materials (Bigeleisen & Allen, 1951; Bigeleisen & Friedman, 1949). Most of the enzymes studied by this approach are decarboxylases, where reaction solutions for both states are prepared. The resulting CO_2 is collected and then transferred into the dual-inlet of an IRMS (Fig. 1) (O'Leary, 1980). The continued use of this classical method is the result of the confidence provided by results of the highest possible sensitivity and precision. The method continues in its original form as it deals readily with evolved CO_2 by conventional chemical techniques that produce reliable, precise data (Maccoll, 1975).

Pioneering work successfully focused on the collection and measurement of CO_2 from chemical (Bigeleisen & Friedman, 1949; Gelles &

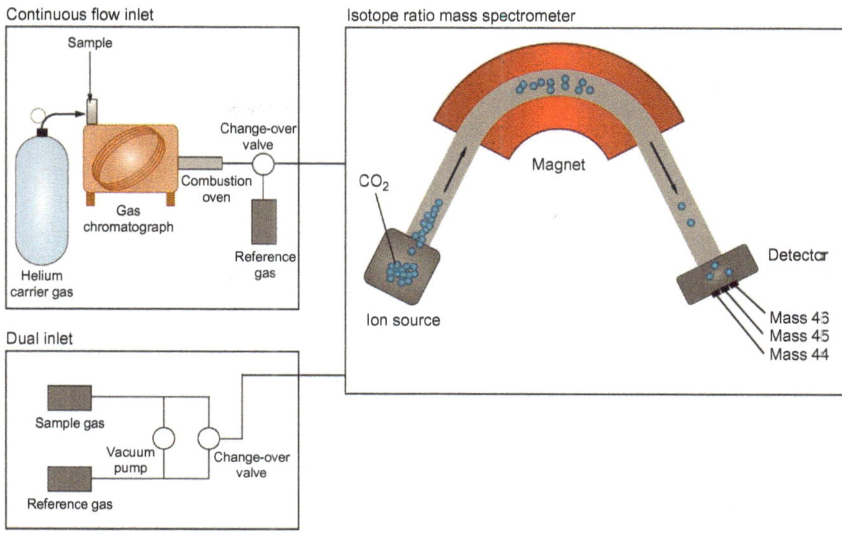

Fig. 1 Simplified schematic of an isotope ratio mass spectrometer with a dual-inlet and continuous flow inlet denoted.

Reed, 1955) and enzymatic (O'Leary, 1969; Seltzer, Hamilton, & Westheimer, 1959) decarboxylation reactions. The resulting isotopic ratios can generally be evaluated using the Rayleigh equation for measuring isotopic distributions ($R/R_0 = f^{\varepsilon/1000}$) (Bigeleisen, 1949a). Based on this equation, isotope fractionation is expressed as enrichment factors (ε), fractionation factors (α, where $\varepsilon = 1000(\alpha^{-1})$), or carbon KIEs (CKIE $= \alpha^{-1}$) (Melander & Saunders, 1980). For decarboxylation, these parameters are typically calculated by measuring the $^{13}C/^{12}C$ isotope ratio of CO_2 at two time points from individual sacrificial samples that are quenched in acid to terminate the reaction (Bigeleisen & Allen, 1951). The process requires a large amount of substrate as the CO_2 that is formed must also serve as the carrier gas (Goodman & Brenna, 1992). In an important contrast, CSIA technology utilizes gas chromatography with a helium carrier gas (Fig. 1). In this process, CO_2 for isotopic analysis is present at lower levels than where it also fulfills the carrier gas function. The use of an added carrier gas provides an easier and more sensitive operation for isolation and measurement of the isotopic ratio (Sessions, 2006). This dramatically reduces the quantity of substrate necessary for analysis and allows measurement of single analytes from mixtures that are separable via gas chromatography. Although this improvement increases the versatility of the spectrometer, it also decreases the precision of the measured isotope ratios by an order of magnitude (Brand, 1996). The loss of precision results from the inability to control the pressure of the sample relative to the standard gas, which increases the error in a typical CKIE from ±0.05‰ to ±0.5‰ (Hunkeler, Meckenstock, Sherwood Lollar, Schmidt, & Wilsont, 2008). The trade-off for the improved precision of the dual-inlet system is the need for greater amounts of the CO_2-producing reactant (Sherwood Lollar, Hirschorn, Chartrand, & Lacrampe-Couloume, 2007).

CSIA is now in widespread use and the overall precision of the isotopic measurements is impressive. The purpose of this chapter is to highlight the improved accessibility of these types of experiments and analyses. This should lead to a wider range of researchers applying these methods by bridging the gap between the classical methods and the technological advances that have led to dramatic improvements in the ease, efficiency, and resources needed to study diverse enzymatic reactions.

1. BASIC EXPERIMENTAL PROCEDURES

The experimental approach in the classical method for chemical and enzymatic decarboxylation reactions can be modified and applied to CSIA. As outlined by O'Leary (1980), in the classical method, isotope effects can be measured using a two-point approach, where the decarboxylation of a substrate for one point proceeds to approximately 10% completion (the "low conversion" sample), and the other point is taken from the CO_2 at completion (the "complete conversion" sample). The benefit of using CSIA with this approach is that the CO_2 from each case does not need to be purified on a vacuum line. Rather, it can be separated from other volatile components in the headspace in the gas chromatograph. The isotopic composition can then be analyzed using CSIA (Mundle, Lacrampe-Couloume, Sherwood Lollar, & Kluger, 2010; Vandersteen, Mundle, & Kluger, 2012; Vandersteen, Mundle, Lacrampe-Couloume, Sherwood Lollar, & Kluger, 2013).

In the classical approach, the scale of the isotope effect experiment is limited by the amount of starting material required for purification on a standard vacuum line and introduction into the dual-inlet of the isotope ratio mass spectrometer. Larger samples reduce errors from the different stages of the analytical workup. Contamination from atmospheric carbon dioxide, where final concentrations for analysis of roughly 100 μmol per sample are a minimum reliable threshold. This amount of CO_2 scaled up for the low-conversion sample point (~10%) of substrate requires roughly 1.0 mmol of CO_2-producing starting material per experiment (O'Leary, 1980).

Using CSIA, the scale of the isotope effect experiment is limited by the concentration of CO_2 in the headspace that can be reliably extracted from the reaction vessel and injected into the GC-C-IRMS and the linear range of the instrument (Mundle, Vandersteen, Lacrampe-Couloume, Kluger, & Sherwood Lollar, 2013). Unlike the classical method, if other volatile components are present in the headspace of the reaction vessel, they can be separated and purified in the gas chromatograph prior to being measured in the IRMS. If other volatile chemical species are components of the reaction, an added benefit of this approach is that the isotope values for each component can often be measured in a single GC-C-IRMS run. A wide range of syringe volumes and the possibility of using a low split flow or splitless injection on the GC can provide methods that measure very low concentrations of CO_2. However, we have found that using 1-mL injection volumes with low split flow rates provides the best balance between practicality and reliability for routine measurements using this approach (Mundle et al., 2013). The experimental detection limit using CSIA will vary slightly for different instrument configurations; however, it can be anticipated to be orders of magnitude lower than the classical method. The smaller amount of substrate required for each experiment allows a greater number of replicates to be sampled over the course of the experiment. Although the analytical precision for the classical dual-inlet approach is an order of magnitude better than CSIA, the access to many more samples improves confidence in the results and provides a reliable evaluation of total error (experimental and analytical error) for an experimental design. Ultimately, the total error is typically greater than the reliable detection limit for both dual-inlet and CSIA; therefore, there is often no added benefit to the overall precision of an experiment using the dual-inlet approach.

The reaction vessels used in the classical method are conventional round-bottom flasks with greased fittings and stopcocks (O'Leary, 1980). For the method in this chapter, we adapted the equipment used for environmental applications of CSIA (Chan et al., 2012; Elsner, Zwank, Hunkeler, & Schwarzenbach, 2005; Hunkeler et al., 2008; Liang et al., 2014; Mundle et al., 2012; Sherwood Lollar et al., 2010). These utilize readily accessible and inexpensive glass serum bottles with butyl rubber gas-tight stoppers for sampling gases and volatile organic compounds (Eby, Gibson, & Yi, 2015; Hunkeler et al., 2008; Sherwood Lollar, Westgate, Ward, Slater, & Lacrampe-Couloume, 2002; Slater, Dempster, Sherwood Lollar, & Ahad, 1999; Vandersteen et al., 2013). Butyl rubber is impermeable to gases, making it appropriate for critical gas sampling applications. The clean and dry borosilicate glass bottles are inert for the purposes of the conditions used in enzymology. They also do not induce isotopic exchange for a broad range of sampled gases (Nelson, 2000). As a result, sealed borosilicate glass vessels are the choice for distributing isotopic reference samples in instrumental calibration (Coleman, Risatti, & Schoell, 1981; Nelson, 2000).

The butyl blue rubber stoppers in vessels used for decarboxylation reactions prevent the influx of atmospheric gases and the efflux of CO_2 for the duration of the experiment. Hungate, Smith, and Clarke (1966) reported tests on butyl rubber stoppers for their impermeability to atmospheric oxygen (influx) and for the retention of hydrogen and CO_2 gases (outflux) from glass sample vessels. The rubber stoppers were evaluated in tests of oxygen composition from gas chromatography and the retention pressure in the sample vessels that correspond to the concentration of hydrogen and carbon dioxide. In every experiment, oxygen concentrations were below detection limits in butyl rubber-stoppered glass vessels (<0.025 μL/mL). The impermeability of butyl stoppers to hydrogen and carbon dioxide was demonstrated by the negligible overall pressure decreases with butyl stoppers compared to significant pressure drops observed with black rubber stoppers (~10 days). Longer-term stability (20 months) was investigated by Fulghum and Worthington (1977) in a study that assessed the use of butyl blue rubber stoppers and glass serum bottles for the storage of anaerobic culture media. Their study demonstrated that the combination of butyl blue rubber stoppers and glass serum bottles prevented the influx of low molecular weight gases with a success rate of 80%–100% after 12 months. Butyl blue rubber stoppers should be pretreated by boiling in 0.1 mol/L sodium hydroxide solution for an hour as per Oremland and Desmarais (1983).

Isotope fractionation in samples with low concentrations of CO_2 has been reported during storage in glass bottles sealed with butyl rubber stoppers (Nelson, 2000). These effects were observed to have a greater impact on the $\delta^{18}O$ values compared to $\delta^{13}C$ for CO_2. Applications involving lower concentrations of CO_2 in decarboxylation reactions should be evaluated with caution, especially where oxygen isotope measurements are integrated into the mechanistic interpretation. We recommended that analysis should take place within a short time frame (<20 days) in order to obtain reliable isotope data for samples with low concentrations of CO_2.

The use of serum bottles and butyl rubber stoppers with CSIA provides a means to reduce the potential leaking components in the experimental system. An additional benefit of the permanent seal and stability of the headspace gases in the serum bottles allows the experimental component of the decarboxylation reaction to be completed at a location that is remote from the site for isotopic analysis. This provides greater access to these techniques as reactions can be completed and sent to a facility with an appropriate mass spectrometer for analysis. It is important to evaluate the effects of containers and to establish reliable controls as per Ward et al. (2004).

An overview of the procedure is outlined in a flow chart (Fig. 2). CO_2 dissolved prior to reactions must be removed from each component. Upon assembly of these components in the reaction vessel, the headspace must be

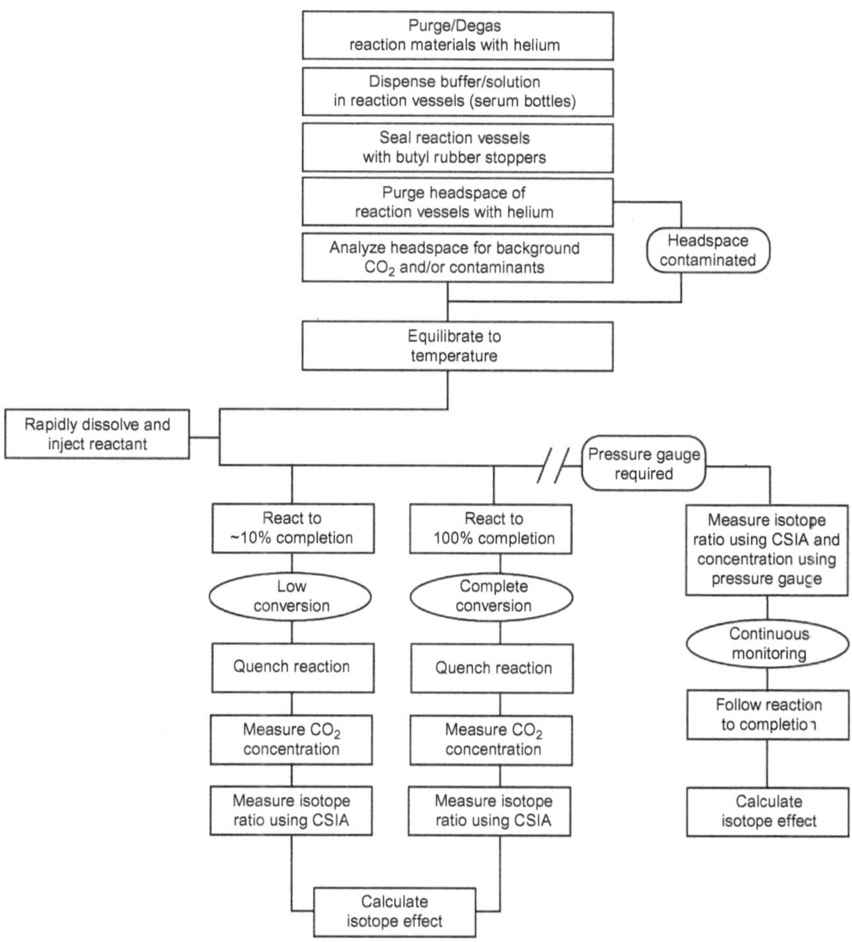

Fig. 2 The processes for determining CKIEs using CSIA analysis of evolved CO_2 in the headspace of a dedicated reaction vessel.

purged to remove CO_2. In the classical approach, this is achieved using several freeze–thaw–evacuate cycles (O'Leary, 1980). For enzyme-catalyzed reactions, instead of evacuating the reaction vessel, ultra high purity nitrogen sparges solutions for 30 min or more using a fritted disk. With CSIA, utilizing helium as a carrier gas for the analysis of CO_2, the most reliable approach sparges the solution components and the headspace of the reaction vessels with helium. Purging times vary based on the specific nature of the components used in the experiment and the headspace volume. These should be evaluated for each experimental system. If the pH of the solution is below ~5, helium purging is sufficient to remove all traces of CO_2 (Mundle, Rathgeber, Lacrampe-Couloume, Sherwood Lollar, & Kluger, 2009). At higher pH, it is necessary to increase purging times to several hours. Although a fritted disk is a good approach to remove CO_2 from larger volumes, syringe needles are more convenient for use in smaller reaction vessels.

Prepare a serum bottle with a degassed pH-buffered reaction solution that must contain all the media required for the decarboxylation reaction prior to sealing the vessel with a butyl rubber stopper. The buffer must maintain pH against the acidity generated from the production of carbonic acid from the evolved CO_2. Volumes of 25–50 mL with equal amounts of headspace in each reaction vessel (50- to 100-mL serum bottles with butyl rubber gas-tight stoppers) are convenient; however, the reaction can be scaled up or down as required. Once sealed, the headspace of the prepared reaction vessel must be purged with helium (an inlet needle and exhaust needle inserted through the septum). The syringe needles are quickly removed; the headspace can be analyzed for the absence of CO_2 or any other volatile species.

After the reaction vessel is prepared and the headspace is purged with helium, the vessel can be placed in a temperature-controlled environment (such as a water bath) to achieve thermal equilibrium. Once the reaction vessels reach a stable temperature, the substrate or enzyme can be added to each using a syringe needle through the septum. Unstable chemical species, cofactors, and enzymes are added to initiate the reaction using a syringe. The substrate may be dissolved in an inert solution or in the same buffer as the reaction. The sample is then injected into the reaction vessel. The substrate solution can be injected into low-conversion samples at a minimum concentration to achieve reliable analysis using CSIA. Alternatively, it can be placed into high-conversion samples at a lower concentration, so that the final yields of CO_2 in the two samples are comparable. The two samples should be treated identically, except for longer reaction times that are needed for the complete conversion sample.

It is critical for the contents of the reaction vessel to be rapidly and thoroughly mixed without delay. The amount of time the reaction will proceed to achieve ~10% conversion of starting material to CO_2 can be approximated using several methods. For materials that are only available in small quantities, independent techniques such as UV–Vis spectroscopy can be used to evaluate the reaction kinetics to determine the amount of time needed to complete the low-conversion sample point. If the amount of material available is not an issue, a separate experiment can be completed using gas chromatography to measure CO_2 concentration changes over time for a reaction. The amount of substrate or enzyme added to the low-conversion sample should be adjusted to achieve completion within a few hours. The exact extent of reaction is later confirmed by measuring the concentration of CO_2. The complete conversion sample should use either excess catalyst or enzyme to ensure that all the substrate is completely converted to CO_2 during the decarboxylation. As outlined by O'Leary, it is possible to correct for incomplete decarboxylation; however, the error in isotope ratio can be large. At a predetermined time, the low-conversion samples are quenched. This can be completed by rapidly freezing the sample, injecting acid, base, or any other chemical agent that will stop the decarboxylation process. Since CSIA does not purify CO_2 on a vacuum line, a good practice is to try to quench the reaction at a pH value between 2 and 4 to encourage the release of dissolved CO_2, or hydrated CO_2 (carbon dioxide in the form of carbonic acid) into the headspace. It is critical that the decarboxylation ceases immediately upon quenching and that all possible reaction process, such as acid-catalyzed decarboxylation, do not provide an undesired source of CO_2. The complete-conversion samples are exposed to identical conditions with the only difference being longer reaction times.

Once the reaction is quenched, the concentration of CO_2 is measured in each reaction vessel. The concentration can be measured using gas chromatography. The concentration of CO_2 can be compared with the values observed from the complete-conversion samples. If gas chromatography is not possible, a less precise approach can be developed by generating a calibration curve for CO_2 and using the mass intensity from the peak generated during isotope analysis. The less precise approach is not advisable when investigating small KIEs, where the larger error of this approach would diminish the value of the results. Reversible reactions can complicate measuring the isotope effects using this approach. Conditions should be controlled to minimize the reverse reaction in the low-conversion samples.

An added benefit of CSIA is the ability to measure many replicate analyses from the same sample and to complete more experiments with less material. This allows a more robust assessment of the reproducibility of the measured effect between experiments. Small differences in the reaction progress will be observed for the low-conversion samples, which upon measuring the concentration differences associated with each sample vessel should result in a comparable KIE. This approach can be extended in separate experiments to evaluate a larger range of isotope effect values from 1% reaction to 80% reaction (limitations beyond 80% are discussed later). The isotope ratio for each sample will change markedly as the percent reaction changes but the calculated isotope effect should be the same for all samples. If all conditions are constant with high quality materials, the isotope ratio in the complete conversion sample should be constant and ultimately independent of the conditions under which the reaction was conducted (isotopic mass balance is achieved).

2. CONTINUOUS PRESSURE-MONITORED HEADSPACE ANALYSIS

The ability to directly measure CO_2 from the headspace of a reaction vessel provides a unique opportunity to monitor the change in isotope values over the course of the entire reaction (Mundle et al., 2013). CSIA and headspace analysis can be used to monitor decarboxylation reactions that lead to a net change in pressure in closed systems. We developed an approach to determine the reaction progress by monitoring the change in headspace pressure directly using a digital pressure gauge attached to a 60-mL serum bottle modified with a glass sidearm (Fig. 3). The reaction progress follows the change in headspace pressure using a digital pressure gauge (Fig. 4). The analytical error associated with pressure gauges is often around 0.25% of the reported pressure. The observed first-order rate constant for decarboxylation reactions can be calculated from the increase in pressure due to evolution of CO_2 as a function of time using the integrated first-order rate expression. The acidity of a solution combined with vigorous stirring (DeTar, 1956) can ensure that the CO_2 produced in the aqueous phase partitions quickly into the headspace. Although the acidity of the solutions used in this study favors partitioning of the CO_2 to the headspace, we have

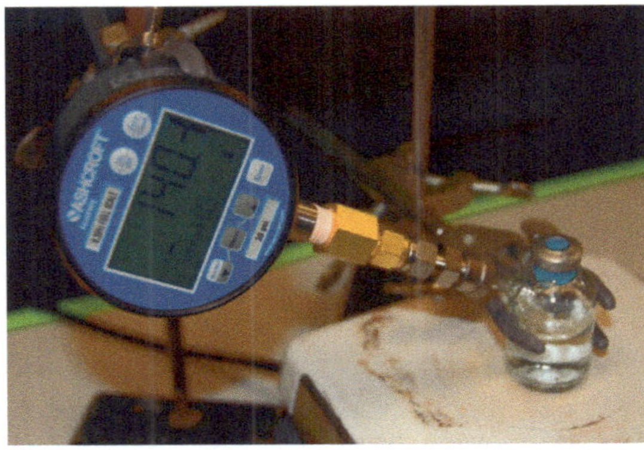

Fig. 3 Reaction vessel used to determine the enrichment factors for decarboxylation using pressure-monitored headspace analysis (recirculating water bath removed for purpose of the photograph).

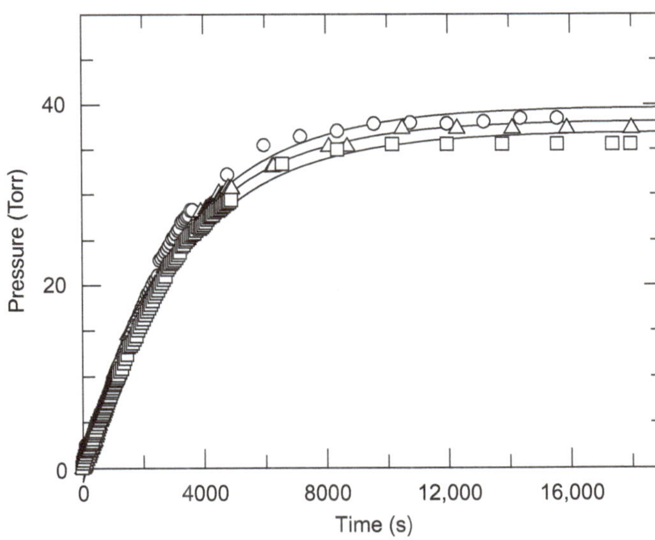

Fig. 4 Example of an experiment that monitors the evolution of CO_2 from a decarboxylation reaction by the increase in pressure in the headspace of the reaction vessel shown in Fig. 3. Pressure (Torr) vs time (s) for the evolution of CO_2 in the decarboxylation of pyrrole-2-carboxylic acid in 26% (w/w) perchloric acid at 25°C ($k_{obs} = 3.2 \times 10^{-4}\ s^{-1}$ (*open circles*), $k_{obs} = 3.3 \times 10^{-4}\ s^{-1}$ (*open triangles*), $k_{obs} = 3.1 \times 10^{-4}\ s^{-1}$ (*open squares*)). Different symbols denote individual experimental runs. The *error bars* are smaller than the plotted symbols. *Data from Mundle, S. O. C., Vandersteen, A. A., Lacrampe-Couloume, G., Kluger, R., Sherwood Lollar, B. (2013). Pressure-monitored headspace analysis combined with compound-specific isotope analysis to measure isotope fractionation in gas-producing reactions.* Rapid Communications in Mass Spectrometry, *27(15), 1778–1784.*

applied this approach in less acidic solutions to show that the retention of CO_2 in solution (via reaction with water to form bicarbonate) is not significant up to pH 4–5. Hence, this approach should be feasible in solutions up to that acidity (Mundle et al., 2009).

The $^{13}C/^{12}C$ ratio is determined from headspace sampling using a pressure-lock analytical syringe with a side port taper needle. The volume of headspace sampled (50–500 μL) will vary based on the CO_2 concentration throughout the experiment. The carbon isotope value will change over the course of the experiment according to the magnitude of the KIE associated with the reaction (Fig. 5). As each headspace sample is removed for CSIA, an equal volume of helium can be added to compensate and maintain evolved pressure. Over the course of an experiment, the composition of the headspace is depleted in the target substrate. This depletion can be approximated by accounting for the dilution of the headspace caused by the inert gas ($Area_{CORR} = ((V_{HS} + V_{DIL}) \times Area_{GC})/V_{HS}$). V_{HS} is the total volume of

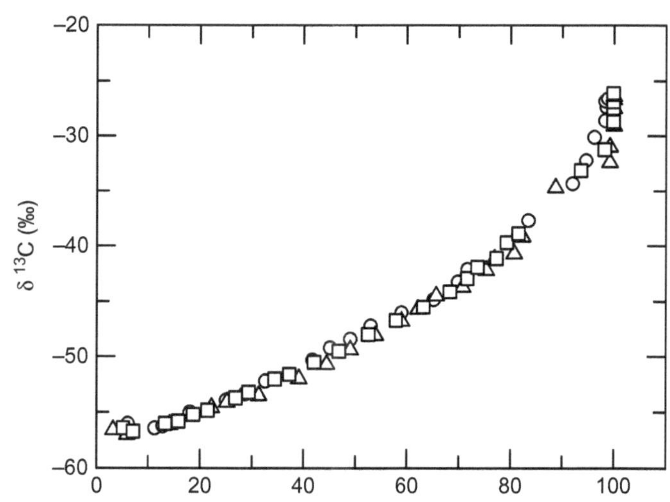

Fig. 5 Example of an experiment that monitors the change in $\delta^{13}C$ (‰) of CO_2 in the headspace of the reaction vessel shown in Fig. 3 from a decarboxylation reaction. $\delta^{13}C$ (‰) vs the reaction progress (%) for the decarboxylation of pyrrole-2-carboxylic acid in 26% (w/w) perchloric acid at 25°C. Different symbols denote individual experimental runs. The *error bars* are smaller than the plotted symbols. *Data from Mundle, S. O. C., Vandersteen, A. A., Lacrampe-Couloume, G., Kluger, R., Sherwood Lollar, B. (2013). Pressure-monitored headspace analysis combined with compound-specific isotope analysis to measure isotope fractionation in gas-producing reactions.* Rapid Communications in Mass Spectrometry, *27(15), 1778–1784.*

headspace in the closed reaction vessel, V_{DIL} is the volume of inert gas used to maintain consistent headspace pressure, $Area_{GC}$ is the peak area obtained directly from gas chromatography, and $Area_{CORR}$ is the peak area corrected for dilution of the headspace with an inert gas. This approach requires that the pressure remain constant throughout the experiment. Reactions that produce or consume volatile compounds or gas will lead to a net change in the pressure within the headspace. Therefore, as the pressure changes during the experiment, the concentration of substrate within a volume of gas sampled will vary, which is a source of potential error. The reaction should be monitored throughout the experiment with a final series of measurements taken after the reaction is complete. The reaction progress (f) is conveniently measured directly from the pressure gauge readings, which can be stored in modern computer-controlled systems.

Measuring the reaction progress directly from the pressure change in the headspace eliminates uncertainty in the CO_2 measurement and reaction progress parameter for a Rayleigh model calculation (see details below). It also dramatically reduces the amount of substrate needed to produce reliable ε values. For example, we have successfully achieved accurate ε values using 15 mg of pyrrole-2-carboxylic acid and have successfully monitored reactions with as little as 6 mg of this substrate (Mundle et al., 2010). The traditional approach with sacrificial samples typically requires a minimum of 150 mg of substrate for a two-point determination for a single ε value calculation (O'Leary, 1980). The CSIA approach also permits simultaneous determination of the rate constant directly from the pressure change.

The results are fit to the Rayleigh equation ($R/R_0 = f^{(\varepsilon/1000)} = f^{(\alpha-1)}$) that describes a system with low natural abundance of the heavy isotope $(1+R)/(1+R_0) \approx 1$, where R and R_0 are the $^{13}C/^{12}C$ ratios of CO_2 at time t and time zero, respectively, f is the fraction of CO_2 remaining, and ε (denoted in parts per thousand) is the enrichment factor equal to $1000(\alpha - 1)$, where $\alpha = (\text{CKIE})^{-1}$ (Elsner et al., 2005; Mariotti et al., 1981). Bigeleisen (Bothner-By & Bigeleisen, 1951) reported a derivation of this equation $((N_x/N_{x0}) = [1 - (1-f)(k_H/k_L)]/f)$, for determining the isotope fractionation for the accumulated CO_2 produced in decarboxylation reactions. k_H and k_L are the observed first-order rate coefficients for reaction of the heavy and light isotopes, respectively. In this case, R and R_0 are replaced with N_x and N_{x0} ($\delta^{13}C$ values are converted into abundance ratios using $\delta^{13}C = (N_{sample} - N_{standard})/N_{standard}$) (Mariotti et al., 1981). A plot of log $[1 + ((N_{x0} - N_x)/N_{x0})(f/1 - f)]$ vs $-\log(1-f)$ generates a straight line with a slope of $\varepsilon/1000 = (\alpha - 1)$, where $\alpha = \text{CKIE}^{-1}$. The $\delta^{13}C$ values can be converted into abundance ratios (N) where $\delta^{13}C = (N_{sample} - N_{standard})/N_{standard}$ (Mariotti et al., 1981). In both cases, for the equation for accumulated CO_2 product should be fit without forcing the line through the origin, as suggested by Scott, Lu, Cavanaugh, and Liu (2004) (Fig. 6).

3. CONTROL REACTIONS

Isotope fractionations can occur during sample preparation and the product can be contaminated by external sources of CO_2. Since CSIA separates the injected sample on a gas chromatograph, issues with product contamination by other materials that interfere with mass spectrometry are not an issue. A proper preparative control evaluates each component independently as well as combinations of each component that will be assessed. CO_2 should not be present in any of the control reactions. A subset of samples

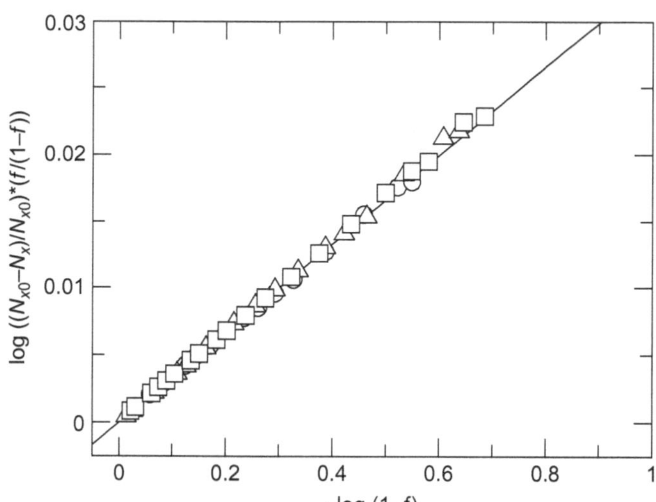

Fig. 6 Example of a KIE calculated using Bigeleisen's derivation of the Rayleigh equation. Fits of from pressure-monitored headspace analysis and CSIA data for decarboxylation of pyrrole-2-carboxylic acid in 26% (w/w) perchloric acid at 25°C. In this case, the linear fit has been completed with data points beyond 80% reaction progress (20% starting material remaining) removed (discussed later). Individual experimental replicates denoted by different symbols. Linear regression analysis fit to combined data up to 80% reaction progress ($\varepsilon = -33.9‰ \pm 0.4‰$, $n = 51$). The *error bars* are smaller than the plotted symbols. *Data from Mundle, S. O. C., Vandersteen, A. A., Lacrampe-Couloume, G., Kluger, R., Sherwood Lollar, B. (2013). Pressure-monitored headspace analysis combined with compound-specific isotope analysis to measure isotope fractionation in gas-producing reactions.* Rapid Communications in Mass Spectrometry, *27(15), 1778–1784.*

should be used to test the quenching method used in the experiment. If possible, the isotope value should be carried out to characterize the stability of the materials quenched. If the stability is not sufficient for the time interval needed for isotope analysis, the quenching method must be improved. In some cases, this could involve an additional temperature change or freezing process to lengthen the window of stability required for analysis. Additional controls and caution should be used to make sure that these manipulations do not fractionate the isotope values of CO_2 as many chemical and physical processes can introduce their own isotope effects. The most effective way to avoid these issues is to ensure that any manipulation results in 100% recovery of the CO_2 produced in the reaction. For continuous monitoring from a single sample vessel the entire reaction sequence should be repeated without the substrate, and carbon dioxide should not be detected in the headspace in the control. For longer times, leaks in the reaction vessel can lead to significant errors in the concentration of CO_2 measured by the pressure.

Avoiding potential sources of error in isotope effect experiments requires controls and experimental designs that ensure the statistical independence of individual experiments and each series of experiments. For example, several batches of substrate and/or buffer solution should be prepared independently and divided it into sets of experiments, where 3–5 low-conversion experiments and 2–3 complete conversion experiments are used for a given batch of materials. Subsequent experiments must use independently prepared materials. This approach will help avoid the interdependence in the data points within individual experimental replicates. Although increasing the number of data points used to calculate an ε value for an individual replicate will generally lead to a smaller 95% confidence interval, it does not provide any insight into the total variability of the experiment. The total uncertainty for an experiment can only be reasonably estimated from an ε value. This is calculated by combining all the data from multiple independent replicates with overlapping confidence intervals into a single regression analysis.

4. IMPROVING DATA QUALITY

The largest source of experimental error is not typically from the measured $^{13}C/^{12}C$ ratio but from uncertainty in the reaction progress (f). In this case, the larger component of the error arises from the x-axis of the Rayleigh plot ($R/R_0 = f^{\varepsilon/1000}$), which is used to calculate ε values and KIEs (Bigeleisen & Allen, 1951). The reaction progress can be determined for slower reactions by sampling the headspace sequentially for CSIA and for compositional analysis using gas chromatography, where the typical GC error is <5% of the peak area (Slater et al., 1999). Faster reactions, where a change in the reaction progress can occur between sequential collection of samples for CSIA and GC, can be measured by the change in pressure of the headspace during a decarboxylation reaction.

The precision and accuracy of an ε value can be improved by reducing the experimental error in measuring the reaction progress (f) for the accumulation of CO_2 monitored over the course of a reaction. Measuring ε values based on the accumulation of CO_2 as a product of the reaction can be complicated. Since CO_2 produced from decarboxylation is initially present in the solution phase, it must be rapidly transferred into the headspace for analysis. Any delay in the movement of CO_2 from the aqueous phase to the headspace will lead to mixing between instantaneously formed CO_2 with later stage CO_2 that will suppress the magnitude of the observed ε value. In addition, as the system equilibrates over the course of the reaction, a pH-dependent distribution of CO_2 between the headspace and the solution phase may affect the concentration of CO_2 in the headspace. In such a case, the reaction progress would need to be calculated from the total concentration of CO_2 in both the headspace and aqueous phases, which will necessarily increase the error associated with the ε value. The magnitude of the potential error associated with any concentration measurement or corrections to those measurements will have the greatest impact when large volume samples are taken from low volume headspaces. Since pressure-monitored headspace analysis determines the reaction progress (f) directly from the evolution of gas into the solution and then to the headspace, it completely eliminates the need to account for substrate removal and headspace dilution. However, removal of larger volumes of headspace at early stages of the reaction can impact the pressure reading and must be accounted for in the calculation.

The benefit of being able to analyze larger numbers of samples and collect more data points for CSIA-based KIE experiments on decarboxylation reactions opens new doors for improved statistical analysis of these results. However, we have found that the magnitude of the statistical confidence intervals does not typically account for the experimental error in both the [CO_2] and $^{13}C/^{12}C$ inputs of the Rayleigh calculation (Mundle et al., 2013). The statistical approximation of the confidence interval is often based solely on the data point (Hunkeler et al., 2008). This can lead to the appearance of artificially small confidence intervals that become problematic when small differences in reported ε values are used to elucidate different processes or reaction mechanisms.

4.1 Calculating ε Values That Account for the Magnification of Error at Later Stages of the Reaction

Bigeleisen and Allen (1951) reported that the error in ε values associated with determining reaction progress is magnified at later stages of the reaction. They estimated that 0.5% error in the reaction progress in the region above 80% reaction completion (or <20% starting material remaining) will produce an error of 4%–7% or larger in the ε value. This is consistent with the scatter observed at >80% reaction completion in Fig. 7. The reaction progress cannot be reliably determined in the region above 80% conversion. Based on Fig. 7, data points from beyond 75% reaction have a greater impact on the calculation of the ε values; however, the ε value including all data

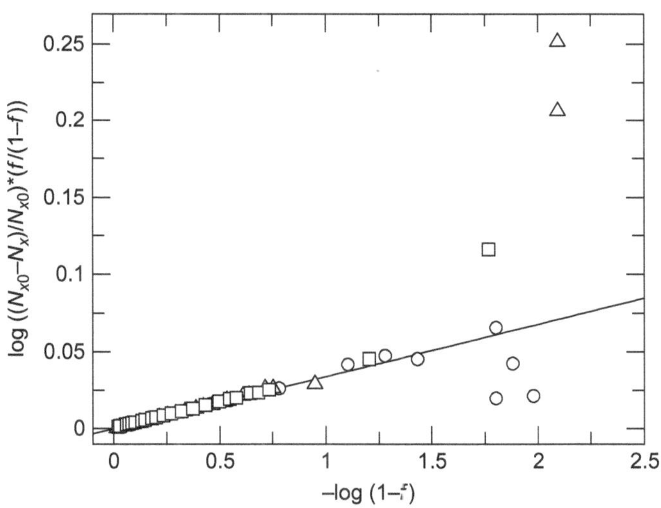

Fig. 7 Example of a KIE calculated using Bigeleisen's derivation of the Rayleigh equation. Fits of from pressure-monitored headspace analysis and CSIA data for decarboxylation of pyrrole-2-carboxylic acid in 26% (w/w) perchloric acid at 25°C. This plot shows the complete data set from Fig. 6. In this case, the linear fit is poor for data points beyond 80% reaction progress (20% starting material remaining), a phenomenon described by Bigeleisen and Allen for late stages of a reaction. Individual experimental replicates denoted by different symbols. The *error bars* are smaller than the plotted symbols. *Data from Mundle, S. O. C., Vandersteen, A. A., Lacrampe-Couloume, G., Kluger, R., Sherwood Lollar, B. (2013). Pressure-monitored headspace analysis combined with compound-specific isotope analysis to measure isotope fractionation in gas-producing reactions.* Rapid Communications in Mass Spectrometry, *27(15), 1778–1784.*

points up to 80% reaction completion still falls within the calculated 95% confidence intervals for every interval except 30% and 60%, where there is only a 0.1‰ disparity between the confidence intervals. As the reaction approaches completion, the increment of change in the concentration of substrate approaches a minimum. This stage of the reaction is more sensitive to error because it has the largest change in $\delta^{13}C$ values with the smallest change in concentration. Since an experimental error of only ±0.5% will lead to major shifts in the calculated ε value, most analytical techniques will not give better than ±1% to 5% uncertainty in concentration measurements. Data collected beyond 80% reaction (<20% starting material remaining) should not be used for calculating ε values without specific justification.

4.2 Assessment of the Impact of Experimental Error on Calculated ε Values

We previously suggested an approach that can be used to evaluate the "robustness" of a calculated ε value by comparing statistical analyses for shorter intervals of the reaction (Mundle et al., 2013). This approach demonstrated that the slope of the regression analysis used to calculate ε values is highly sensitive to variability in the final data points used in the calculation. Large variations in ε values that arise from data points at later stages of the reaction are a direct indication that the error determined for reaction progress is having a significant impact on the calculated ε value. This phenomenon is not necessarily reflected in the correlation coefficient (r^2) and can produce inaccurate KIE values that appear statistically valid. In these cases, recalculating the ε values on shorter reaction progress intervals across the entire reaction will provide a means to evaluate the additional variation in the calculated ε values that arise from the greater uncertainty in data from later stages of the reaction. This procedure is important when attempting to interpret different CO_2-producing reactions based on different ε values (Liang et al., 2011; Mundle et al., 2012) and the role of masking effects (Chan et al., 2012; Kluger & Mundle, 2010; Sherwood Lollar et al., 2010). This is especially important in cases with few data points, or in samples with more scatter at later stages of the reaction (Bloom, Aravena, Hunkeler, Edwards, & Frape, 2000; Slater, Sherwood Lollar, Sleep, & Edwards, 2001). In cases where the range of ε values calculated on shorter reaction progress intervals is larger than the 95% confidence interval that is based on all of the data points, the total error associated with this ε value is likely to be greater than is predicted by the statistical analysis.

5. CONCLUSIONS

While methods that use cryogenic separation of CO_2 on a vacuum line provide ε values for decarboxylation reactions (Breider & Hunkeler, 2011; O'Leary, 1980), the classical methodrequires extensive sample manipulations and provides only two data points. In this article we have explained how as an alternative approach, CSIA, provides a significant improvement through decreased sample volume requirements and increased sample throughput. The use of pressure-monitored headspace analysis and CSIA is an accurate, robust approach to determine isotope fractionation for decarboxylation reactions and can be extended to other applications. This modern approach allows collection of data at many points during the course of a reaction. The use of a distinct carrier gas and coupled GC greatly reduces the amount of product needed for each analysis. The use of heavy atom isotope effects has provided important insights into the fundamental questions of enzymology. The approach we have outlined should lead to new applications that add to the value of information from isotope effects in the area.

ACKNOWLEDGMENTS

We thank NSERC Canada for support through Discovery Grants to B.S.L. and R.K. and an NSERC Strategic Grant to B.S.L. We thank Georges Lacrampe-Couloume, Adelle Vandersteen, and Steven Rathgeber for their input and efforts in the development of these methods.

REFERENCES

Bigeleisen, J. (1949a). The validity of the use of tracers to follow chemical reactions. *Science*, *110*, 14–16.

Bigeleisen, J. (1949b). The relative reaction velocities of isotopic molecules. *The Journal of Chemical Physics*, *17*(8), 675–678.

Bigeleisen, J., & Allen, T. L. (1951). Fractionation of the carbon isotopes in decarboxylation reactions. IV. The relative rates of decomposition of 1-C12 and 1-C13 trichloroacetate ions. *The Journal of Chemical Physics*, *19*, 760–764.

Bigeleisen, J., & Friedman, L. (1949). C-13 isotope effects in the decarboxylation of malonic acid. *The Journal of Chemical Physics*, *17*(10), 998–999.

Bloom, Y., Aravena, R., Hunkeler, D., Edwards, E , & Frape, S. K. (2000). Carbon isotope fractionation during microbial dechlorination of trichloroethene, cis'1,2-dichloroethene, and vinyl chloride: Implications for assessment of natural attenuation. *Environmental Science & Technology*, *34*(13), 2768–2772.

Bothner-By, A. A., & Bigeleisen, J. (1951). Fractionation of the carbon isotopes in decarboxylation reactions. III. The relative rates of decomposition of carboxyl-C12 and -C13 mesitoic acids. *The Journal of Chemical Physics*, *19*, 755–759.

Brand, W. A. (1996). High precision isotope ratio monitoring techniques in mass spectrometry. *Journal of Mass Spectrometry*, *31*(3), 225–235.

Breider, F., & Hunkeler, D. (2011). Position-specific carbon isotope analysis of trichloroacetic acid by gas chromatography/isotope ratio mass spectrometry. *Rapid Communications in Mass Spectrometry*, *25*(24), 3659–3665.

Chan, C. C. H., Mundle, S. O. C., Eckert, T., Liang, X., Tang, S., Lacrampe-Couloume, G., et al. (2012). Large carbon isotope fractionation during biodegradation of chloroform by Dehalobacter cultures. *Environmental Science & Technology*, *46*(18), 10154–10160.

Coleman, D. D., Risatti, J. B., & Schoell, M. (1981). Fractionation of carbon and hydrogen isotopes by methane-oxidizing bacteria. *Geochimica et Cosmochimica Acta*, *45*(7), 1033–1037.

DeTar, D. F. (1956). The precise manometric determination of rate data. *Journal of the American Chemical Society*, *78*(16), 3911–3915.

Eby, P., Gibson, J. J., & Yi, Y. (2015). Suitability of selected free-gas and dissolved-gas sampling containers for carbon isotopic analysis. *Rapid Communications in Mass Spectrometry*, *29*(13), 1215–1226.

Elsner, M., Zwank, L., Hunkeler, D., & Schwarzenbach, R. P. (2005). A new concept linking observable stable isotope fractionation to transformation pathways of organic pollutants. *Environmental Science & Technology*, *39*(18), 6896–6916.

Fulghum, R. S., & Worthington, J. M. (1977). Butyl rubber stoppers increase shelf-life of prereduced, anaerobically sterilized media. *Applied and Environmental Microbiology*, *33*(5), 1220–1221.

Gelles, E., & Reed, R. I. (1955). Carbon-13 isotope effects in the decarboxylation of some oxaloacetates. *Nature*, *176*(4496), 1262.

Goodman, K. J., & Brenna, J. T. (1992). High-sensitivity tracer detection using high-precision gas-chromatography combustion isotope ratio mass-spectrometry and highly enriched [U-^{13}C]-labeled precursors. *Analytical Chemistry*, *64*(10), 1088–1095.

Hungate, R. E., Smith, W., & Clarke, R. T. J. (1966). Suitability of butyl rubber stoppers for closing anaerobic roll culture tubes. *Journal of Bacteriology*, *91*(2), 908–909.

Hunkeler, D., Meckenstock, R. U., Sherwood Lollar, B., Schmidt, T. C., & Wilson, J. T. (2008). *A guide for assessing biodegradation and source identification of organic groundwater contaminants using compound specific isotope analysis (CSIA)*. Ada, Oklahoma: United States Environmental Protection Agency. p 68 p.

Kluger, R., & Mundle, S. O. C. (2010). The role of pre-association in Bronsted acid-catalyzed decarboxylation and related processes. *Advances in Physical Organic Chemistry*, *44*, 357–375.

Liang, X. M., Howlett, M. R., Nelson, J. L., Grant, G., Dworatzek, S., Lacrampe-Couloume, G., et al. (2011). Pathway-dependent isotope fractionation during aerobic and anaerobic degradation of monochlorobenzene and 1,2,4-trichlorobenzene. *Environmental Science & Technology*, *45*(19), 8321–8327.

Liang, X., Mundle, S. O. C., Nelson, J. L., Passeport, E., Chan, C. C. H., Lacrampe-Couloume, G., et al. (2014). Distinct carbon isotope fractionation during anaerobic degradation of dichlorobenzene isomers. *Environmental Science & Technology*, *48*(9), 4844–4851.

Maccoll, A. (1975). Heavy-atom kinetic isotope effects. *Annual Reports on the Progress of Chemistry, Section A: Physical and Inorganic Chemistry*, *71*, 77–101.

Mariotti, A., Germon, J. C., Hubert, P., Kaiser, P., Letolle, R., Tardieux, A., et al. (1981). Experimental-determination of nitrogen kinetic isotope fractionation: Some principles; illustration for the denitrification and nitrification processes. *Plant and Soil*, *62*(3), 413–430.

Melander, L. C. S., & Saunders, W. H. (1980). *Reaction rates of isotopic molecules*. New York, NY: Wiley. p xiv, 331 p.

Mundle, S. O. C., Johnson, T., Lacrampe-Couloume, G., Perez-de-Mora, A., Duhamel, M., Edwards, E. A., et al. (2012). Monitoring biodegradation of ethene and bioremediation of chlorinated ethenes at a contaminated site using compound-specific isotope analysis (CSIA). *Environmental Science & Technology*, *46*(3), 1731–1738.

Mundle, S. O. C., Lacrampe-Couloume, G., Sherwood Lollar, B., & Kluger, R. (2010). Hydrolytic decarboxylation of carboxylic acids and the formation of protonated carbonic acid. *Journal of the American Chemical Society*, *132*(7), 2430–2436.

Mundle, S. O. C., Rathgeber, S., Lacrampe-Couloume, G., Sherwood Lollar, B., & Kluger, R. (2009). Internal return of carbon dioxide in decarboxylation: Catalysis of separation and C-12/C-13 kinetic isotope effects. *Journal of the American Chemical Society*, *131*(33), 11638–11639.

Mundle, S. O. C., Vandersteen, A. A., Lacrampe-Couloume, G., Kluger, R., & Sherwood Lollar, B. (2013). Pressure-monitored headspace analysis combined with compound-specific isotope analysis to measure isotope fractionation in gas-producing reactions. *Rapid Communications in Mass Spectrometry*, *27*(15), 1778–1784.

Nelson, S. T. (2000). Sample vial influences on the accuracy and precision of carbon and oxygen isotope ratio analysis in continuous flow mass spectrometric applications. *Rapid Communications in Mass Spectrometry*, *14*(4), 293–297.

O'Leary, M. H. (1969). Carboxyl carbon isotope effect on enzymatic decarboxylation of glutamic acid. *Journal of the American Chemical Society*, *91*(24), 6886–6887.

O'Leary, M. H. (1976). Carbon isotope effect on enzymatic decarboxylation of pyruvic-acid. *Biochemical and Biophysical Research Communications*, *73*(3), 614–618.

O'Leary, M. H. (1980). Determination of heavy-atom isotope effects on enzyme-catalyzed reactions. *Methods in Enzymology*, *64*, 83–104. Enzyme Kinet. Mech., Part B.

O'Leary, M. H. (1984). Measurement of the isotope fractionation associated with diffusion of carbon-dioxide in aqueous-solution. *Journal of Physical Chemistry*, *88*(4), 823–825.

O'Leary, M. H. (1988). Transition-state structures in enzyme-catalyzed decarboxylations. *Accounts of Chemical Research*, *21*(12), 450–455.

O'Leary, M. H. (1989). Multiple isotope effects on enzyme-catalyzed reactions. *Annual Review of Biochemistry*, *58*, 377–401.

O'Leary, M. H., & Brummund, W. (1974). pH jump studies of glutamate decarboxylase—Evidence for a pH-dependent conformation change. *Journal of Biological Chemistry*, *249*(12), 3737–3745.

O'Leary, M. H., & Piazza, G. J. (1981). Medium effects in enzyme-catalyzed decarboxylations. *Biochemistry*, *20*(10), 2743–2748.

O'Leary, M. H., & Westheimer, F. H. (1968). Acetoacetate decarboxylase. Selective acetylation of enzyme. *Biochemistry*, *7*(3), 913–919.

O'Leary, M. H., Yamada, H., & Yapp, C. J. (1981). Multiple isotope effect probes of glutamate-decarboxylase. *Biochemistry*, *20*(6), 1476–1481.

Oremland, R. S., & Desmarais, D. J. (1983). Distribution, abundance and carbon isotopic composition of gaseous hydrocarbons in Big-Soda Lake, Nevada—An alkaline, meromictic lake. *Geochimica et Cosmochimica Acta*, *47*(12), 2107–2114.

Paneth, P., & O'Leary, M. H. (1985). Carbon isotope effect on dehydration of bicarbonate ion catalyzed by carbonic-anhydrase. *Biochemistry*, *24*(19), 5143–5147.

Parmentier, L. E., Weiss, P. M., O'Leary, M. H., Schachman, H. K., & Cleland, W. W. (1992). C-13 and N-15 isotope effects as a probe of the chemical mechanism of Escherichia coli aspartate-transcarbamylase. *Biochemistry*, *31*(28), 6577–6584.

Roeske, C. A., & O'Leary, M. H. (1984). Carbon isotope effects on the enzyme-catalyzed carboxylation of ribulose bisphosphate. *Biochemistry*, *23*(25), 6275–6284.

Scott, K. M., Lu, X., Cavanaugh, C. M., & Liu, J. S. (2004). Optimal methods for estimating kinetic isotope effects from different forms of the Rayleigh distillation equation. *Geochimica et Cosmochimica Acta*, *68*(3), 433–442.

Seltzer, S., Hamilton, G. A., & Westheimer, F. H. (1959). Isotope effects in the enzymatic decarboxylation of oxalacetic acid. *Journal of the American Chemical Society*, *81*(15), 4018–4024.

Sessions, A. L. (2006). Isotope-ratio detection for gas chromatography. *Journal of Separation Science*, *29*(12), 1946–1961.

Sherwood Lollar, B., Hirschorn, S. K., Chartrand, M. M. G., & Lacrampe-Couloume, G. (2007). An approach for assessing total instrumental uncertainty in compound-specific carbon isotope analysis: Implications for environmental remediation studies. *Analytical Chemistry*, *79*(9), 3469–3475.

Sherwood Lollar, B., Hirschorn, S., Mundle, S. O. C., Grostern, A., Edwards, E. A., & Lacrampe-Couloume, G. (2010). Insights into enzyme kinetics of chloroethane biodegradation using compound specific stable isotopes. *Environmental Science & Technology*, *44*(19), 7498–7503.

Sherwood Lollar, B., Westgate, T. D., Ward, J. A., Slater, G. F., & Lacrampe-Couloume, G. (2002). Abiogenic formation of alkanes in the Earth's crust as a minor source for global hydrocarbon reservoirs. *Nature*, *416*(6880), 522–524.

Slater, G. F., Dempster, H. S., Sherwood Lollar, B., & Ahad, J. (1999). Headspace analysis: A new application for isotopic characterization of dissolved organic contaminants. *Environmental Science & Technology*, *33*(1), 190–194.

Slater, G. F., Sherwood Lollar, B., Sleep, B. E., & Edwards, E. A. (2001). Variability in carbon isotopic fractionation during biodegradation of chlorinated ethenes: Implications for field applications. *Environmental Science & Technology*, *35*(5), 901–907.

Urey, H. C. (1947). The thermodynamic properties of isotopic substances. *Journal of the Chemical Society*, 562–581.

Vandersteen, A. A., Mundle, S. O. C., & Kluger, R. (2012). Protonated carbonic acid and reactive intermediates in the acidic decarboxylation of indolecarboxylic acids. *Journal of Organic Chemistry*, 77(15), 6505–6509.

Vandersteen, A. A., Mundle, S. O. C., Lacrampe-Couloume, G., Sherwood Lollar, B., & Kluger, R. (2013). Carbon kinetic isotope effects reveal variations in reactivity of intermediates in the formation of protonated carbonic acid. *Journal of Organic Chemistry*, *78*(23), 12176–12181.

Ward, J. A., Slater, G. F., Moser, D. P., Lin, L. H., Lacrampe-Couloume, G., Bonin, A. S., et al. (2004). Microbial hydrocarbon gases in the Witwatersrand Basin, South Africa: Implications for the deep biosphere. *Geochimica et Cosmochimica Acta*, *68*(15), 3239–3250.

Weiss, P. M., Gavva, S. R., Harris, B. G., Urbauer, J. L., Cleland, W. W., & Cook, P. F. (1991). Multiple isotope effects with alternative dinucleotide substrates as a probe of the malic enzyme reaction. *Biochemistry*, *30*(23), 5755–5763.

CHAPTER EIGHTEEN

Kinetic Isotope Effect Studies to Elucidate the Reaction Mechanism of RNA-Modifying Enzymes

Govardhan R. Veerareddygari, Eugene G. Mueller[1]
University of Louisville, Louisville, KY, United States
[1]Corresponding author: e-mail address: eugene.mueller@louisville.edu

Contents

1. Introduction 524
   1.1 Pseudouridine 524
   1.2 Pseudouridine Synthases 524
2. KIE Studies on TruB and RluA 526
3. Equipment 526
4. Chemicals 527
5. Protocol 528
   5.1 Synthesis of UTP 528
   5.2 In Vitro Transcription 537
   5.3 ^{2}KIE Analysis 542
6. Concluding Thoughts 544
Acknowledgments 545
References 545
Further Reading 546

Abstract

The synthesis of specifically deuterated uridine, its incorporation into an RNA oligonucleotide substrate, and the use of the labeled substrate to determine the deuterium kinetic isotope effect for the reaction catalyzed by the pseudouridine synthases (enzymes that isomerize uridine to pseudouridine in RNA) are described. Both enzymes—TruB and RluA—display a primary kinetic isotope effect, which indicates the formation of a glycal intermediate in the ribose ring during turnover. Although the details of the protocols are specific to these two enzymes, the general methodology is readily adaptable to the synthesis and incorporation of other labeled nucleosides into any RNA molecule by in vitro transcription.

Methods in Enzymology, Volume 596
ISSN 0076-6879
http://dx.doi.org/10.1016/bs.mie.2017.07.018

1. INTRODUCTION

1.1 Pseudouridine

Pseudouridine (Ψ) is the C-glycoside isomer of uridine (U) and the most common modified nucleoside in cellular RNAs (Grosjean & Benne, 1998). Its existence in tRNA extracts was reported as early as in 1957 (Davis & Allen, 1957) and Ψ was subsequently found in all noncoding RNAs, where it was thought to primarily impart structural rigidity to the backbone of the RNAs (Johnson & Söll, 1970). This structural rigidity can primarily be attributed to the chemically distinct nature of Ψ when compared to U—the extra imino proton that can form an additional hydrogen bond with the backbone of the RNA through a coordinated water molecule (Charette & Gray, 2000). Furthermore, the next-generation sequencing techniques have now facilitated the detection of Ψ in both yeast and human mRNA (Carlile et al., 2014; Schwartz et al., 2014), where pseudouridylation is shown to be heavily regulated in response to the environmental stress conditions (Carlile et al., 2014; Schwartz et al., 2014). This shows that Ψ in RNA not only holds a structural significance but also probably helps regulate gene expression in higher organisms.

1.2 Pseudouridine Synthases

Pseudouridine synthases catalyze the isomerization of U to Ψ in RNA. These enzymes use both sequence and structural information from their RNA substrates to achieve this deceptively simple transformation (Mueller & Ferré-D'Amaré, 2009). Ψ synthases are divided into six different families that do not share significant global sequence similarity, and they are thought to be a result of divergent evolution (Mueller & Ferré-D'Amaré, 2009). Structural studies show that all Ψ synthases share a core β-fold and a conserved catalytic domain (Mueller & Ferré-D'Amaré, 2009). An active site aspartic acid (Asp) is conserved among all Ψ synthases, and it is absolutely essential for the catalytic activity of these enzymes (Mueller & Ferré-D'Amaré, 2009). All bacterial Ψ synthases are stand-alone, but some eukaryotic Ψ synthases such as Cbf5p and its orthologs require a functional guide RNA component for catalysis (Grosjean & Benne, 1998). Additionally, the stand-alone bacterial pseudouridine synthase TruB that is a focus of the described work has also been recently shown to act as a tRNA chaperone (Keffer-Wilkes, Veerareddygari, & Kothe, 2016).

Historically, mutations associated with Ψ synthases have been implicated in several human diseases. For example, mutations in human gene dyskerin are associated with the X-linked form of dyskeratosis congenita, a rare autosomal recessive disorder (Heiss et al., 1998; Luzzatto & Karadimitris, 1998; Mitchell, Wood, & Collins, 1999). Mutations in human pseudouridine synthases 1 (PUS1) and 3 (PUS3) are associated with mitochondrial myopathy, lactic acidosis, and sideroblastic anemia (Bykhovskaya, Casas, Mengesha, Inbal, & Fischel-Ghodsian, 2004).

The mechanism followed by the pseudouridine synthases has remained elusive. Most of the early mechanistic work used substrate analogs containing 5-fluorouridine in place of U and suggested one of two mechanistic possibilities: the "Michael mechanism" that involves the addition of the catalytic Asp to C6 of the pyrimidine ring of the isomerized uridine; or the "acylal mechanism" that involves the nucleophilic attack of the catalytic "Asp" on C1′ of the ribose ring of uridine (Gu, Liu, & Santi, 1999). However, more recent mechanistic work on bacterial pseudouridine synthases TruB and RluA using RNA containing 5-fluorouridine provided strong evidence of a new mechanistic possibility, the "glycal mechanism" (Fig. 1) (Miracco & Mueller, 2011). This new mechanism was posited based

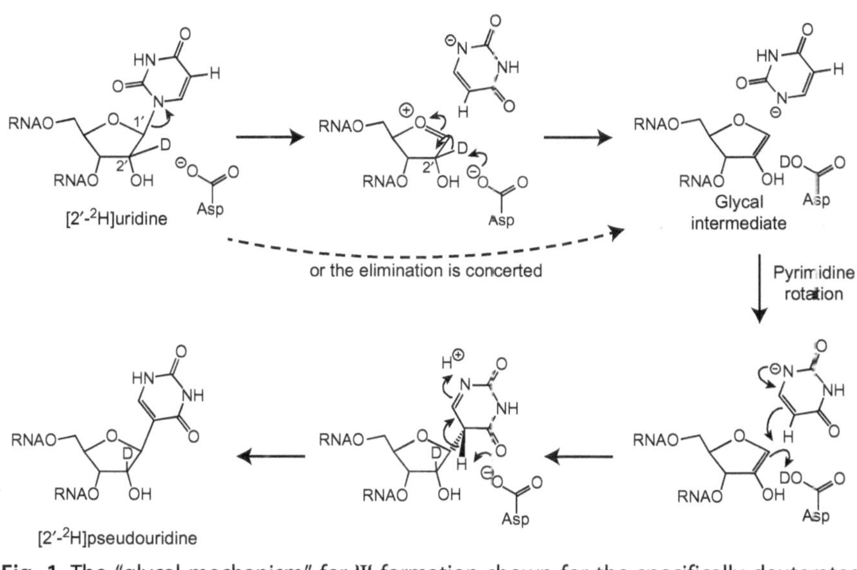

Fig. 1 The "glycal mechanism" for Ψ formation shown for the specifically deuterated substrate. The catalytic Asp initiates the reaction by deprotonating C2′, which leads to a glycal intermediate. After rearrangement to the C-glycoside, C2′ is reprotonated by the catalytic Asp, which is the only acid/base group positioned next to C2′. The sequestration of the protonated Asp from exchange with solvent hydron has been experimentally determined (Veerareddygari, Singh, & Mueller, 2016).

on the observed epimerization at C2′ of the ribose ring of 5-fluorouridine in RNA, which can only be achieved through deprotonation of C2′ and then reprotonation of the resulting glycal intermediate from the opposite face (Miracco & Mueller, 2011). The methods described in this chapter were utilized to investigate the occurrence of the glycal mechanism by seeking the predicted kinetic isotope effect (KIE) using RNA substrates containing 2′-deuterated uridine (Veerareddygari et al., 2016).

2. KIE STUDIES ON TruB AND RluA

If the glycal mechanism (Fig. 1) were to operate for pseudouridine synthases, it requires the proton on C2′ of uridine to be removed and then put back by the catalytic Asp. If either of these steps is partially rate limiting, one expects a primary KIE with a labeled substrate (one containing [2′-^{2}H]U in the isomerized position). Such labeled nucleoside analog ([2′-^{2}H]U) was commercially unavailable, so it was prepared from [2′-^{2}H]-ribose by enzymatic synthesis (Fig. 3) (Scott, Tolbert, & Williamson, 2000; Tolbert & Williamson, 1997). By means of in vitro transcription, [2′-^{2}H]UTP was incorporated into TSL[a] (GCUGUGUUCGAUCCACAG; the isomerized U is underlined) and ASL[a] (GCGGAUUGAAAUCCGC), which are the substrates for *E. coli* pseudouridine synthases TruB and RluA, respectively (Veerareddygari et al., 2016). A direct comparison of V_{max} and V_{max}/K_m (Cleland, 1987) of the initial reaction rates for labeled vs unlabeled substrates for TruB and RluA revealed a primary KIE, thereby showing that the glycal mechanism operates in both cases (Veerareddygari et al., 2016). Because all Ψ synthases share a conserved catalytic domain, they likely follow a similar mechanistic pathway. This chapter presents the methodologies adapted for such studies.

3. EQUIPMENT

1. UV/visible spectrophotometer
2. Rotary evaporator
3. Lyophilizer
4. Standard SDS-PAGE and urea-PAGE gel apparatus

[a] T-arm stem loop (TSL) from yeast tRNA^{Phe} serves as a substrate for *Escherichia coli* Ψ synthase TruB (Gu et al., 1999), and anticodon stem loop (ASL) from *E. coli* tRNA^{Phe} is likewise a good substrate for bacterial Ψ synthase RluA. (Hamilton et al., 2006). In both cases, the sequences were tweaked from those purchased earlier in order to facilitate in vitro transcription (Veerareddygari et al., 2016).

5. Gel imager
6. TLC plate impregnated with a fluorophore
7. Ultrafiltration units (10,000 NMWCO)
8. Bond Elut C8 solid-phase extraction cartridges (5 g, 20 mL) (Agilent)
9. Elutrap electroelution system (Whatman)
10. EMD Millipore Steriflip™ Vacuum Filter Units (Fisher Scientific)
11. Corning Costar™ Spin-X™ centrifuge tube filters (Fisher Scientific)

4. CHEMICALS

1. BLR(DE3) pLysS competent cells (Novagen)
2. Ni-NTA superflow resin (Qiagen)
3. D-Ribose (Sigma-Aldrich)
4. D-[2′-^{2}H]ribose (Omicron Biochemicals, South Bend, IN)
5. Ribose-5-phosphate barium salt (Sigma-Aldrich)
6. Dithiothreitol (DTT)
7. Tris base
8. Sodium phosphate dibasic
9. Triethylammonium acetate
10. Ethylenediaminetetraacetic acid (EDTA)
11. Phosphoenolpyruvate (PEP) (Chem-Impex International, Wood Dale, IL)
12. Pyruvate kinase/lactate dehydrogenase enzyme cocktail (PK/LDH) (Sigma-Aldrich)
13. Darco G-60 activated charcoal
14. Dowex-50 (H^{+} form) (Sigma-Aldrich)
15. Thermostable inorganic pyrophosphatase (New England Biolabs)
16. Pyruvate kinase II (PK) (Sigma-Aldrich)
17. Yeast myokinase (MK) (CalBiochem)
18. Ribonucleotide triphosphates (rNTPs) (Promega)
19. T7 RNA polymerase (Epicentre Biotechnologies, Madison, WI)
20. Stop RNase Inhibitor RX (5 PRIME, Inc. now part of ThermoFisher)
21. TURBO™ DNase (Fisher Scientific)
22. Calf intestinal alkaline phosphatase (Promega)
23. β-Nicotinamide adenine dinucleotide, reduced disodium salt hydrate (NADH) (Sigma-Aldrich)
24. Adenosine 5′-triphosphate disodium salt hydrate (ATP) (Sigma-Aldrich)

5. PROTOCOL

5.1 Synthesis of UTP

5.1.1 Required Enzymes

5.1.1.1 Overexpression

1.1. Separately transform BLR(DE3) pLysS with the overexpression constructs for His_6-tagged *E. coli* ribokinase (Veerareddygari et al., 2016), phosphoribosyl pyrophosphate synthetase (PRPPS) (Van Lanen et al., 2003), and uracil phosphoribosyl transferase (UPRT) (Van Lanen et al., 2003). Grow colonies on LB solid media supplemented with the appropriate antibiotics.

1.2. Transfer cells from a single colony to liquid LB media with appropriate antibiotics and grow with agitation. At OD_{600} 0.4–0.6, induce overexpression by the addition of IPTG (1 m*M* final concentration). Grow the cells for 3 h postinduction to allow for the overexpression of proteins, and confirm overexpression of the target proteins by SDS-PAGE (10% gel).

1.3. Harvest the cells by centrifugation (6000 × *g* for 30 min). Quick-freeze the cell pellet for storage at −80°C or proceed directly to enzyme purification.

1.4. Suspend the cell pellet in 50 m*M* sodium phosphate buffer (5 mL), pH 8.0, containing NaCl (300 m*M*) and imidazole (10 m*M*). Perform all the subsequent steps at 4°C.

1.5. Induce cell lysis by sonication (6 × 15 s pulses with 45 s on ice in between), and pellet the cell debris by centrifugation (18,000 × *g*, 30 min).

1.6. Add a slurry of Ni-NTA superflow resin (3 mL of the resuspended resin provided by the manufacturer) and nutate the mixture for 60 min at 4°C.

1.7. Centrifuge the resin (6000 × *g*, 30 min), and resuspend the pelleted resin in the same volume (10 mL) of buffer used to prepare the cell suspension; load the slurry into a column, and let it settle, then allow the buffer to drain until it is just at the level of the settled resin.

1.8. Wash the packed column (3 × 5 mL) with 50 m*M* sodium phosphate buffer, pH 8.0, containing sodium chloride (300 m*M*) and imidazole (20 m*M*). Collect all the washes. Elute the bound enzyme with (2 × 5 mL each) 50 m*M* sodium phosphate buffer, pH 8.0, containing sodium (300 m*M*) and imidazole (250 m*M*).

1.9. Check the wash and elution fractions by SDS-PAGE to locate the enzyme.

1.10. Store the purified enzyme fractions in 50 m*M* sodium phosphate buffer, pH 7.5, containing sodium chloride (150–300 m*M*) and glycerol (50%, v/v) at −20°C.

5.1.1.2 Assay of Ribokinase Activity

This spectrophotometric coupled assay is based on the consumption of NADH (Fig. 2A).

2.1. Prepare a 1 mL reaction mix in 50 m*M* Tris•HCl buffer, pH 7.5 with D-ribose (5 m*M*), potassium chloride (100 m*M*), magnesium chloride (10 m*M*), NADH (0.2 m*M*), ATP (3 m*M*), PEP (1 m*M*), and PK/LDH mix (2 U) (Table 1).

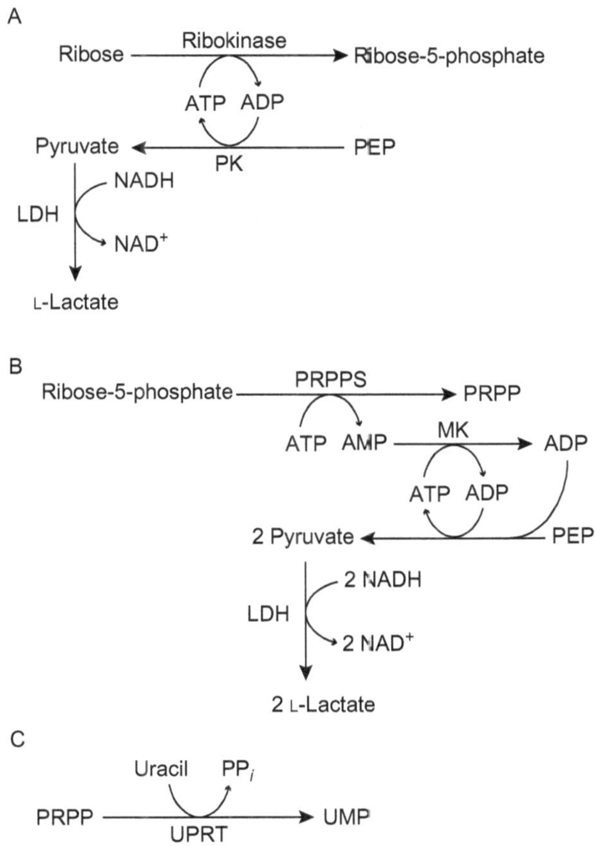

Fig. 2 Activity assays for the overexpressed enzymes. (A) Ribokinase. (B) PRPPS. (C) UPRT. *LDH*, lactate dehydrogenase; *MK*, myokinase; *PK*, pyruvate kinase.

Table 1 Ribokinase Activity Assay

Component	Final Concentration	Stock	Amount
Tris•HCl, pH 7.5	50 m*M*	500 m*M*	100 μL
D-Ribose	5 m*M*	100 m*M*	50 μL
Potassium chloride	100 m*M*	2 *M*	50 μL
Magnesium chloride	10 m*M*	200 m*M*	50 μL
NADH	0.2 m*M*	4 m*M*	50 μL
ATP	3 m*M*	60 m*M*	50 μL
PEP	1 m*M*	20 m*M*	50 μL
PK/LDH mix	2 U	600 U mL^{-1}	3.3 μL
Ribokinase	1 μg	—	≤1 μL

Add water to 1 mL.

2.2. Place the cuvette containing the reaction mixture into the spectrophotometer held at 25°C. Once a steady baseline is observed at 340 nm, add ribokinase (<1 μL, 1 μg) and quickly mix the contents of the cuvette by inversion.

2.3. Monitor the change in absorbance at 340 nm with time. Use this rate of change along with ε_{340} of NADH (6220 M^{-1} cm^{-1}) to calculate ribokinase activity using the following equation

$$U = \frac{V}{(K^{*}\varepsilon^{*}l)} * \left(\frac{\Delta A}{t}\right) \tag{1}$$

where, V is the total reaction volume in liters, ε is the extinction coefficient of NADH at 340 nm (6220 M^{-1} cm^{-1}), l is the path length in cm, ΔA is the change in absorbance at 340 nm, t is the assay time in minutes, and $K = 1$ (reflecting the stoichiometry of NADH consumption to R5P production).

Tip: Solutions of NADH must always be prepared fresh, right before the assay.

5.1.1.3 R5P•Na

The sodium salt of R5P (R5P•Na) is expensive, so it is more economical to perform anion exchange on the less expensive barium salt of R5P (R5P•Ba) using Dowex-50 (Na^{+} form).

3.1. Convert Dowex-50 (H^+ form) to the Na^+ form by the careful addition of solid sodium hydroxide to a stirred slurry of Dowex-50 (H^+ form) in a very generous volume of water until pH is ~14; the resin will undergo a noticeable change in hue. Let the resin settle, decant the supernatant, add water to form a slurry, and load the resin into a column. Wash exhaustively with water until the eluate is pH ~7. Let the column run dry, then unpack the resin, and air-dry it for storage.

3.2. Dissolve R5P•Ba (100 m*M*, 1 mL) in water.

3.3. Prepare a column (2 × 1 cm) of Dowex-50 (Na^+ form) in a disposable syringe (5 mL) plugged with cotton wool. Load the R5P•Ba solution onto the Dowex column and collect the eluate. Wash the column with additional water and collect the eluate in 1 mL fractions. R5P•Na is usually in the first fraction.

3.4. Use a small aliquot of each eluted fraction to locate R5P•Na and determine its concentration. Dilute an aliquot (10 μL) of the load eluate and collected fractions with alkaline phosphatase digestion buffer (1 ×, as provided by the manufacturer) to a final volume of 100 μL. Add alkaline phosphatase and incubate the mixture at 37°C for 1 h. Withdraw a small aliquot from this reaction mixture and use it to calculate the concentration of the inorganic phosphate using the Ames assay (Ames, 1966) and thus the amount of R5P•Na present in each fraction. After concentration (if necessary), aliquots can be quick-frozen and stored at −80°C.

Tip: All the apparatus must be scrupulously clean and devoid of other organic and inorganic phosphates.

5.1.1.4 Assay of PRPPS Activity

This spectrophotometric coupled assay is also based on the consumption of NADH (Fig. 2B).

3.5. Prepare the assay mixture (1 mL) in 100 m*M* sodium phosphate buffer, pH 7.6 with R5P•Na (5 m*M*), magnesium chloride (7 m*M*), NADH (0.2 m*M*), ATP (3 m*M*), PEP (1 m*M*), PK/LDH mix (2 U), myokinase (4 U) (Table 2).

3.6. Monitor the absorbance at 340 nm until a steady baseline is observed.

3.7. Add PRPPS (1 μg), mix the contents of the cuvette by inversion, and monitor the change in A_{340} as a function of time. Use the rate of change and ε_{340} of NADH (6220 M^{-1} cm^{-1}) to calculate the activity of PRPPS using Eq. (1), where $K = 2$.

Table 2 PRPPS Activity Assay

Component	Final Concentration	Stock	Amount
Sodium phosphate buffer, pH 7.6	50 m*M*	500 m*M*	100 μL
R5P•Na	5 m*M*	100 m*M*[a]	50 μL
Magnesium chloride	10 m*M*	200 m*M*	50 μL
NADH	0.2 m*M*	4 m*M*	100 μL
ATP	3 m*M*	60 m*M*	50 μL
PEP	1 m*M*	20 m*M*	50 μL
Pyruvate kinase	2 U	600 U mL^{-1}	3.3 μL
Lactate dehydrogenase	2 U	600 U mL^{-1}	3.3 μL
Myokinase	4 U	750 U mL^{-1}	5.3 μL
PRPPS	1 μg	—	≤1 μL

[a]Adjust the volume of R5P•Na per the stock concentration.
Add water to 1 mL.

5.1.1.5 PRPP Synthesis

PRPP is highly unstable at room temperature and must be freshly synthesized before assaying UPRTase.

4.1. Dissolve R5P•Na (100 μmol), PEP (200 μmol), magnesium chloride (16 μmol), ATP (6 μmol), and EDTA (6 μmol) in 100 m*M* sodium phosphate, pH 7.5 (1 mL). Add PRPPS (1 U), pyruvate kinase (1 U), and myokinase (1 U); incubate the reaction mixture 3 h at 37°C.

4.2. Filter the reaction mixture through activated charcoal to remove ATP. Use PRPP within a day of synthesis to assay UPRT activity.

Tip: If the PRPP displays significant absorbance at 271 nm, residual ATP is present and must be removed by repeating the charcoal filtration step.

5.1.1.6 Assay of UPRT Activity

The spectrophotometric assay (Fig. 2C) is based on monitoring the conversion of uracil to UMP using the change in extinction coefficient at 271 nm, $\Delta\varepsilon_{271} = 2763\,M^{-1}\,\text{cm}^{-1}$ (Arthur, Alvarado, & Dayie, 2011; Veerareddygari et al., 2016).

4.3. Prepare the assay mixture (1 mL) in 50 m*M* Tris•HCl buffer, pH 7.5, containing magnesium chloride (5 m*M*), PRPP (1.5 m*M*), and uracil (0.1 m*M*) (Table 3).

Table 3 UPRT Activity Assay

Component	Final Concentration	Stock	Amount
Tris•HCl, pH 7.5	50 m*M*	500 m*M*	100 μL
Magnesium chloride	5 m*M*	200 m*M*	25 μL
PRPP	5 m*M*	100 m*M*	50 μL[a]
Uracil	0.1 m*M*	10 m*M*	10 μL

[a]Adjust the volume of PRPP per the stock concentration.
Add water to 1 mL.

4.4. After a steady baseline is observed at 271 nm, add UPRT (<1 μL, 1 μg), mix the contents by inversion, and monitor the change in A_{271}.

4.5. Use $\Delta\varepsilon_{271} = 2763\,M^{-1}\,\text{cm}^{-1}$ for uracil vs UMP (greater ε_{271}) to calculate the activity of UPRT using Eq. (1), where $K = 1$ (Arthur et al., 2011; Veerareddygari et al., 2016).

5.1.2 *Enzymatic Synthesis of [2′-^{2}H]UTP*

5.1.2.1 Synthetic Reaction Mixture

5.1. Adapt the scale of synthesis (400 μmol) as required.

5.2. In a 50-mL tube, dissolve D-[2′-^{2}H]ribose and uracil in 50 mL of 50 m*M* potassium phosphate buffer, pH 7.5, containing magnesium chloride (10 m*M*), ATP (0.5 m*M*), PEP (40 m*M*), DTT (20 m*M*), kanamycin (0.1 m*M*), and ampicillin (5 m*M*) (Table 4).

5.3. Initiate the synthesis by adding ribokinase (50 U), PRPPS (2 U), UPRT (2 U), nucleoside monophosphate kinase (2 U), myokinase (37.5 U), and pyruvate kinase (75 U). Incubate the reaction mixture at 37°C.

5.4. Monitor the reaction progress by ^{31}P NMR and a coupled assay for pyruvate formation based on LDH as described later (Scott et al., 2000; Tolbert & Williamson, 1997). Stop the reaction when no further increase in the amount of [2′-^{2}H]UTP or pyruvate is detected (Fig. 3).

5.1.2.2 Pyruvate Assay

To a 1 mL reaction mixture containing NADH (0.2 m*M*), LDH (2 U) in 50 m*M* Tris, pH 7.6, add a small aliquot (5–10 μL) of the reaction mixture that contains pyruvate. Mix the contents well and monitor the change in A_{340}. Use Eq. (1) to calculate the amount of NADH consumed and thereby

Table 4 Enzymatic Synthesis of [2′-^{2}H]UTP

Component	Final Concentration	Stock	Amount
Sodium phosphate buffer, pH 7.6	50 m*M*	500 m*M*	50 mL
D-[2′-^{2}H]ribose	—	400 μμmol	60.7 mg
Uracil	—	400 μmol	44.8 mg
DTT	20 m*M*	1 *M*	1 mL
Magnesium chloride	10 m*M*	—	101.7 mg
ATP	0.5 m*M*	—	22 mg
PEP	40 m*M*	—	1.1 g
Kanamycin	0.1 m*M*	50 m*M*	100 μL
Ampicillin	5 m*M*	100 m*M*	2.5 mL
Ribokinase	≅50 U	1100 U mL^{-1}[a]	50 μL
PRPPS	≅2 U	14.6 U mL^{-1}[a]	137 μL
UPRTase	≅2 U	47 U mL^{-1}[a]	43 μL
Myokinase	37.5 U	750 U mL^{-1}	50 μL
Pyruvate kinase	75 U	5 U μL^{-1}	17.5 μL
Nucleoside monophosphate kinase	1 U	5 U mL^{-1}	200 μL

[a]Adjust the volume of these enzymes based on their stock concentration.
Add water to 50 mL.

the concentration of pyruvate. Four equivalents of pyruvate are formed for every equivalent of [2′-^{2}H]UTP produced.

Tip: To avoid buildup of ADP, the activity of myokinase needs to be kept to half of that of pyruvate kinase.

Tip: As the reaction progresses, the mixture turns cloudy because of the precipitation of inorganic pyrophosphate released from PRPP.

Tip: Check the pH of the reaction mixture prior to adding the enzymes to initiate the synthesis.

5.1.3 Purification of [2′-^{2}H]UTP

6.1. Filter the reaction mixture through an Amicon DIAFLO YM-10 membrane (10,000 MWCO) to remove enzymes.

6.2. Add activated charcoal (2 g) to the filtrate (~48 mL) and nutate the resulting mixture for 15 min at 4°C. Pellet the charcoal by

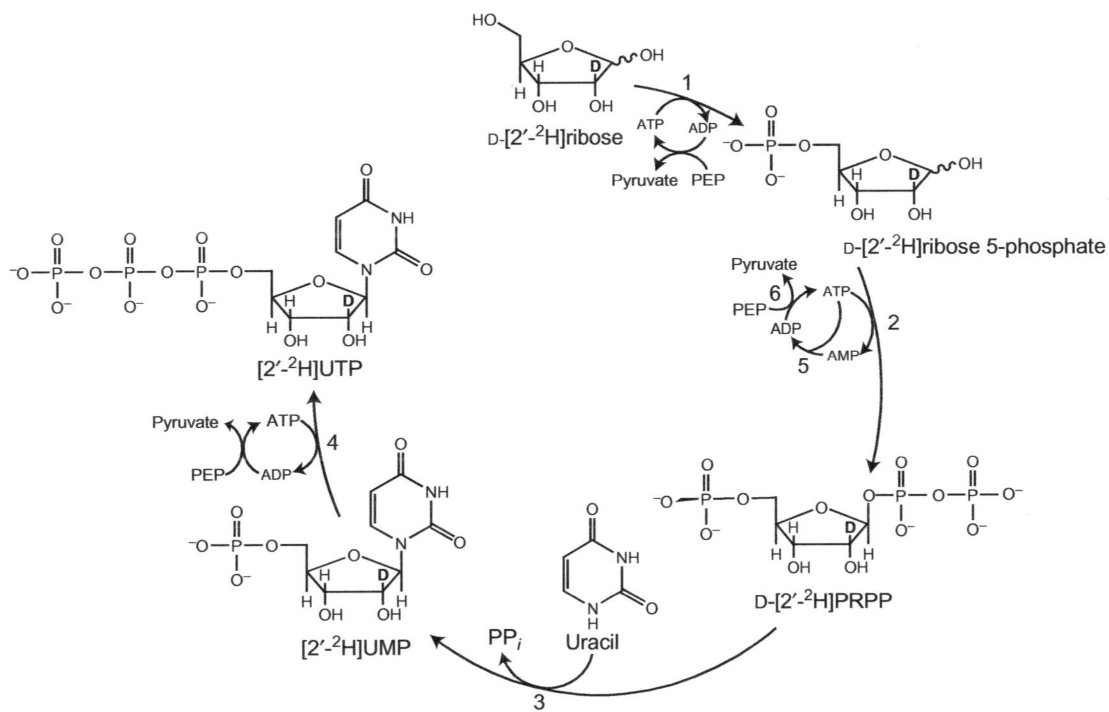

Fig. 3 Enzymatic synthesis of [2′-^{2}H]UTP. ATP is generated in situ at the expense of PEP by pyruvate kinase and myokinase. 1, ribokinase; 2, PRPPS; 3, UPRT; 4, nucleoside monophosphate kinase; 5, myokinase; 6, pyruvate kinase.

centrifugation (8000 × *g*, 15 min) and measure A_{260} of the supernatant. Continue the nutation until $A_{260} \asymp 0$ indicates complete adsorption of [2′-^{2}H]UTP and ATP.

6.3. Pellet the charcoal by centrifugation (8000 × *g*, 15 min) and decant the supernatant.

6.4. Wash the charcoal once with ice-cold water (20 mL) and then with aqueous ethanol (20 mL; 2.5%, v/v) containing ammonium hydroxide (0.5%, v/v). Collect the washes and check A_{260} to ensure that [2′-^{2}H]UTP is still bound to the charcoal.

6.5. Elute [2′-^{2}H]UTP and other "sticky" reaction components (primarily PEP and ATP) from the charcoal by washing twice with aqueous ethanol (30 mL; 50%, v/v) containing ammonium hydroxide (0.5%, v/v) (Stambaugh & Wilson, 1960).

6.6. Combine the ethanolic elutions (60 mL) and filter them through a Steriflip® vacuum filter unit to remove residual charcoal.

6.7. Adjust the pH to 6.5 using acetic acid (1 *M*) and then remove the solvent in vacuo.

6.8. [2′-^{2}H]UTP is obtained as a yellowish white powder with typical yields of 60%–70%. Check the purity of [2′-^{2}H]UTP by ^{31}P NMR.

6.9. If residual PEP and ATP are found, proceed with the remaining purification steps.

6.10. Dissolve the yellowish white powder ([2′-2H]UTP) in 5 m*M* triethylammonium acetate buffer (1 mL), pH 7.0.

6.11. Preequilibrate a Bond Elut C8 SPE cartridge with the same buffer and load the sample.

6.12. Use a step gradient (5 mL for each step) of aqueous acetonitrile (0%–50%, v/v in 10% steps). [2′-^{2}H]UTP elutes typically between 0% and 2.5% acetonitrile.

6.13. Monitor the A_{260} to locate fractions bearing [2′-^{2}H]UTP. Pool them and use a rotary evaporator to reduce the volume in vacuo.

6.14. Remove the residual triethylammonium acetate by repeated addition and removal of water and isopropanol in vacuo while maintaining pH ~8.0 by the addition of aqueous sodium hydroxide (1 *M*) as necessary.

6.15. Confirm the purity of [2′-^{2}H]UTP (recovered as a white powder) by ^{31}P and ^{1}H NMR (Veerareddygari et al., 2016).

Tip: One could repeat step 6.4 if necessary to wash residual pyruvate and PEP off the charcoal.

Tip: When removing solvent in vacuo (steps 6.7 and 6.14), make sure that the pH of the solution never exceeds 8.0 as higher pH may result in UTP hydrolysis.

Tip: Solutions of TEAA buffer (step 6.14) often "bump" in the rotary evaporator. Care must be taken to avoid excessive bumping as it results in considerable inconvenience and potentially a significant loss of product. Apply the vacuum gradually, stopping when vigorous bubbling is observed and continuing to higher vacuum only when bubbling ceases. The added isopropanol speeds the removal of water, and isopropanol is preferable to ethanol for this purpose because the azeotrope with isopropanol has a higher mole fraction of water.

Tip: While removing ammonium acetate in vacuo (step 6.7), reduce the volume of the solution until the smell of acetic acid has greatly faded.

Final yields typically range from 40% to 50% relative to D-ribose (i.e., 150–200 μmol of [2′-^{2}H]UTP from 400 μmol D-ribose). For in vitro transcription of an 18-mer RNA such as the TSL, 30 μmol of labeled NTP is generally required to obtain 2–3 nmol of the purified in vitro transcript.

5.2 In Vitro Transcription

5.2.1 Overview

Chemical synthesis is a rapid and reliable route to obtain short oligonucleotides with the desired sequence, but it requires nucleotides in a chemically suitable form (usually a phosphoramidite). However, chemical synthesis might not be the best route to obtain isotopically labeled oligonucleotides because the conversion of a nucleoside to its phosphoramidite may prove challenging on the required small scale or impractical for biochemists who lack synthetic equipment or expertise. In these cases, an alternate and more efficient way to obtain short RNA oligonucleotides is in vitro transcription.

Standard in vitro transcription is based on the recognition of a double-stranded promoter sequence within a DNA template by T7 RNA polymerase and subsequent transcription of the RNA encoded by the template. When transcribing full-length tRNA, the 3′-terminal CCA sequence is accomplished by run-off transcription with T7 RNA polymerase using a template generated by digestion with a restriction enzyme (typically *Bst*NI) that leaves a complementary 3′-GGT-5′ end. Furthermore, unlike SP6 RNA polymerase (Krieg & Melton, 1987), T7 RNA polymerase is highly useful in generating small RNAs because it can utilize a single-stranded

DNA template once it initiates at a double-stranded promoter (Milligan, Groebe, Witherell, & Uhlenbeck, 1987; Milligan & Uhlenbeck, 1989). The transcription template can thus be formed by annealing a longer oligonucleotide with a 3′-end complementary to a shorter oligonucleotide that contains the T7 promoter; the rest of the longer oligonucleotide is the template for the transcribed RNA.

5.2.2 Template Considerations

Care must be taken in designing such partially double-stranded template sequence to generate short RNA transcripts. T7 RNA polymerase requires the first transcribed base to be a purine, preferably a G (Milligan et al., 1987; Milligan & Uhlenbeck, 1989). If the first base of the RNA is C or U, the transcription start site is often altered and results in erroneous transcription (Milligan et al., 1987; Milligan & Uhlenbeck, 1989). Furthermore, it is necessary to avoid three or more consecutive Gs in the early part of the RNA-coding sequence because T7 RNA polymerase tends to "slip" back by one nucleotide when it encounters successive Gs during the early stages of transcription, which often results in a "ladder" of transcripts (Kuzmine, Gottlieb, & Martin, 2001; Macdonald, Zhou, & McAllister, 1993). Initiation is the slowest step of transcription and directly affected by the concentration of DNA template, NTPs, and T7 RNA polymerase. In the transcription of oligonucleotides shorter than 20-mers, a higher concentration of DNA template (0.5–1 μ*M*) and T7 RNA polymerase (2500 U mL^{-1}) yields better results because of an increased number of initiation events. As T7 RNA polymerase tends to produce abortive transcripts that are 3–12 nucleotides in length before switching into its hardier processive mode, there is a chance for the depletion of the NTPs, especially GTP as it is often the first nucleotide of the transcript (Kuzmine et al., 2001; Macdonald et al., 1993).

5.2.3 Protocol

7.1. Prepare the double-stranded template by annealing both strands (Fig. 4) together (1 m*M* each in a volume of 100 μL) for 5 min at 65°C followed by cooling on ice for 5 min.

7.2. Prepare the transcription mixture (15 mL), by dissolving (in order) magnesium chloride, spermidine, Triton X-100, rNTPs, GMP, DTT, Prime RNase Inhibitor, and inorganic pyrophosphatase in 40 m*M* Tris•HCl buffer, pH 8.1. Add the prepared double-stranded template to this reaction mixture (Table 5).

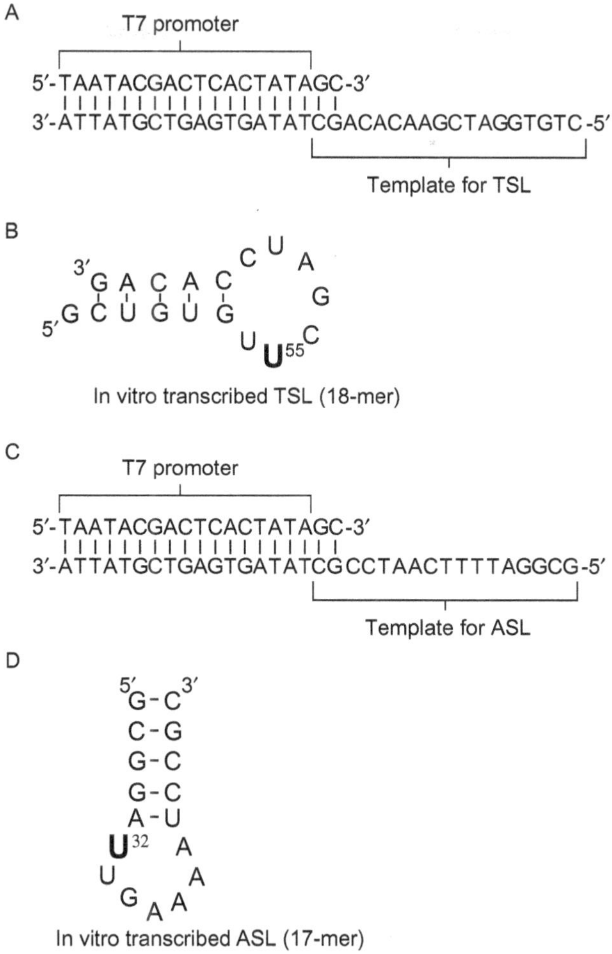

Fig. 4 In vitro transcription of TSL and ASL. (A) Annealed template for the transcription of TSL with a double-stranded promoter region and single-stranded template region. (B) In vitro transcribed TSL. (C) Annealed template for the transcription of ASL w th a double-stranded promoter region and single-stranded template region. (D) In vitro transcribed TSL.

7.3. Add T7 RNA polymerase to initiate the in vitro transcription and incubate the mixture at 37°C.

7.4. Monitor the reaction progress by urea-PAGE. Remove an aliquot (3 μL) of the reaction mixture, dilute it with water (10 μL final volume) and then an equal volume of 2 × RNA loading dye (10 μL) that contains bromophenol blue and xylene cyanol as tracking dyes. Heat the sample for 5 min at 94°C, cool on ice for 5 min, and load it onto a

Table 5 In Vitro Transcription

Component	Final Concentration	Stock	Amount
Tris•HCl buffer, pH 8.1	40 m*M*	1 *M*	600 μL
Magnesium chloride	25 m*M*	1 *M*	375 μL
ATP	2 m*M*	100 m*M*	300 μL
CTP	2 m*M*	100 m*M*	300 μL
GTP	2 m*M*	100 m*M*	300 μL
[2′-^{2}H]UTP	2 m*M*	48.8 m*M*	650 μL
GMP	8 m*M*	100 m*M*	1.2 mL
Spermidine	1 m*M*	100 m*M*	150 μL
DTT	5 m*M*	500 m*M*	150 μL
Triton X-100	0.01%	0.50%	300 μL
Prime RNase Inhibitor	1 U mL^{-1}	1 U μL^{-1}	15 μL
Inorganic pyrophosphatase	5 U mL^{-1}	1 U μL^{-1}	75 μL
Double-stranded template	1 m*M*	1 μ*M*	15 μL
T7 RNA polymerase	25,000 U	2500 U μL^{-1}	10 μL

Add water to 15 mL.

urea-PAGE gel (15%, 7 *M* urea, 10 × 10 cm). Run the gel for 90 min at 150 V, which should result in bromophenol blue migrating two thirds the length of the gel; transcripts between 15 and 30 nucleotides in length will lie between the bromophenol blue and xylene cyanol dyes.

7.5. Visualize the transcription products by staining with SYBR® Gold nucleic acid stain.

7.6. When the gel analysis reveals no further accumulation of TSL (~9 h), digest the template strand by adding TURBO™ DNase (5 μL, 10 U) and incubation for 15 min at 37°C.

7.7. Extract the RNA from the transcription mix by mixing with equal volume of phenol/chloroform (1:1, v/v). Vortex briefly (15 s) followed by centrifugation (10,000 × *g*, 5 min) to speed phase separation. Remove the aqueous (upper) layer.

7.8. To the aqueous layer, add 1/10 volume of 3 *M* sodium acetate buffer, pH 5.2, and mix well. Add ethanol (3 × volume) and incubate the solution for at least 1 h at −80°C to precipitate the RNA. Pellet the RNA by centrifugation (8000 × *g*, 30 min). Wash the pelleted RNA with aqueous ethanol (70%, v/v), briefly air-dry it, and then dissolve the pellet in 10 m*M* Tris•HCl buffer, pH 7.5, containing EDTA (1 m*M*). Determine the concentration of the transcript by measuring A_{260} and using a calculated extinction coefficient for the transcript; online calculators are easily found by a web search; in this case, the calculator at http://dharmacon.gelifesciences.com/custom-rna returned $\varepsilon = 173{,}000\,M^{-1}\,\text{cm}^{-1}$ for TSL and $\varepsilon = 169{,}300\,M^{-1}\,\text{cm}^{-1}$ for ASL.

7.9. If necessary, further purify the transcript by gel electrophoresis (15% PAGE, 7 *M* urea; 15 cm × 15 cm; generally a comb resulting in a small well for size markers and a large single well for sample is used). Mix the crude transcript (1 mL) with equal volume of 2× RNA dye (1 mL), heat for 5 min at 94°C, cool it on ice for 5 min, and load it onto the gel. Run the gel at 165 V for 6 h, remove the gel, and use "UV shadowing" to locate the RNA band. Place the gel on plastic wrap atop a TLC plate impregnated with fluorophore, and shine a handheld UV lamp on the gel; the RNA absorbs the UV light and thus makes a shadow on the TLC plate.

7.10. Using a fresh scalpel, excise the RNA band and dice it into fine pieces (~1 mm squares). Load the gel pieces into an Elutrap electrophoresis unit (GE Healthcare; Pittsburgh, PA) and electroelute the RNA from the diced gel slice by running at 200 V for 3 h. Alternatively, use the crush and soak method (Chen & Ruffner, 1996).

7.11. Precipitate the electroeluted RNA as described in step 7.8, and redissolve the pellet in RNase-free water. Divide the purified RNA transcript into aliquots for quick-freezing and store at −80°C.

Tip: Maintain an RNase-free environment while working with RNA. Use fresh disposable gloves at every step.

Tip: Use of excess GMP in the reaction is not necessary but is recommended to avoid depletion of GTP stocks in the reaction because of abortive transcripts that are often associated with T7 transcription initiation and to generate transcripts nearly homogeneous in 5′-monophosphate ends.

Tip: It may not be necessary to remove the template DNA by digestion with DNase (step 7.6), and one can directly proceed to step 7.7 if neither of the template DNA strands is similar in size to the transcribed RNA. Comigration of either template DNA strand with RNA on the gel obviously interferes with the purification of RNA transcript.

Tip: Gel purification of the transcript is necessary to remove impurities that absorb at 260 nm (such as short abortive transcripts and free NTPs) because a precise determination of RNA concentration is necessary to measure the KIE (Cleland, 1987).

Tip: Check for the purity of the transcripts by reverse-phase HPLC and/or mass spectroscopy (Veerareddygari et al., 2016). The presence of impurities that are competitive inhibitors or alternate substrates for the enzyme would compromise the accuracy of the determined KIE (Cleland, 1987).

5.2.4 Sequence Verification and Purity Check

8.1. Dilute an aliquot of the transcript (10 μL) into RNase-free water (40 μL) to a final concentration of 15–20 μ*M*. Add RNase T1 (2 μL, 200 U) and incubate the mixture for 1 h at 37°C. Add alkaline phosphatase (5 μL, 5 U) to the digestion mixture and incubate it for an additional 3 h at 37°C. Use mass spectrometry (generally MALDI-MS) to verify the sequence and composition of the transcript (Veerareddygari et al., 2016).

8.2. Dilute an aliquot of the transcript (10 μL) into S1 nuclease buffer (1 ×, 90 μL) to a final concentration of 5–10 μ*M*. Add S1 nuclease (1 μL, 100 U) and incubate the mixture for 1 h at 37°C. Heat the mixture for 5 min at 94°C, cool it on ice on ice for 5 min, and add another aliquot of S1 nuclease (1 μL, 100 U) and alkaline phosphatase (3 μL, 3 U). Incubate the mixture for 3 h at 37°C. Analyze the composition of resulting nucleosides by reverse-phase HPLC (Veerareddygari et al., 2016).

Tip: If necessary, digest the transcript with alkaline phosphatase to remove heterogeneity of the phosphate group at the 5′-end by the complete dephosphorylation of mono-, di-, and triphosphates.

5.3 ^{2}KIE Analysis

5.3.1 Overview

To determine a KIE, one can either use a direct comparison method or an internal competition method (Cleland, 1987). The direct comparison method is a straightforward comparison of the rates obtained for unlabeled vs labeled substrates by incubating the enzyme with varying substrate concentrations and fitting the data to the Briggs–Haldane equation. Simple division of the V_{max} or V_{max}/K_m values for unlabeled substrate by those for labeled substrate yields the KIE for each kinetic parameter (Veerareddygari et al., 2016). In the internal competition method, a ratio of unlabeled to labeled substrate is used to start the reaction and the KIE is determined by monitoring the mass ratio (unlabeled vs labeled) of either the substrate or the product (Cleland, 1987). The direct comparison method is often sufficient when probing for a primary deuterium KIE (KIE $\geq$ 1.4) because this method is accurate for measuring KIE greater than 1.05 (Cleland, 1987). For a secondary deuterium (KIE $\leq$ 1.4) or heavy element KIE, one must use the internal competition method that is more sensitive and can measure isotope effects smaller than 1.05 (Cleland, 1987).

5.3.2 Direct Comparison Method

The described method was conducted with unlabeled TSL and ASL prepared as described for the labeled oligonucleotide substrates rather than relying on published values of k_{cat} and K_m with the purchased, chemically synthesized substrates used in earlier studies. Unlabeled substrates will generally be prepared in advance of the more precious labeled ones in order to familiarize the experimenter with the procedures.

9.1. Overexpress and purify the enzymes TruB and RluA using the previously published protocols (Hamilton et al., 2006; Ramamurthy, Swann, Paulson, Spedaliere, & Mueller, 1999).

9.2. For TruB, prepare the assay mixture (400–2600 μL) by dissolving either unlabeled or labeled TSL (1–10 μ*M*) in standard reaction buffer, which is a 50 m*M* HEPES buffer, pH 7.5, containing ammonium chloride (100 m*M*), EDTA (1 m*M*), and DTT (5 m*M*) (Ramamurthy et al., 1999). Preincubate the reaction mixture for 3 min at 37°C (Ramamurthy et al., 1999).

9.3. Initiate the reaction by adding a small volume (<2 μL) of TruB (to 10 n*M*). Withdraw aliquots (75–650 μL) at various time intervals and quench the reaction by adding the aliquots directly into preheated (100°C) microcentrifuge tubes containing 10 × S1 nuclease digestion buffer (1/10 aliquot volume).

9.4. Leave the tubes for 5 min in a boiling water bath and then cool them on ice for at least 5 min.

9.5. Add S1 nuclease (0.5 μL, 50 U) to each quenched aliquot and incubate at 37°C for 1 h. Heat the samples for 5 min at 94°C, cool them on ice for 5 min, and add fresh S1 nuclease (1 μL, 100 U) and alkaline phosphatase (3 μL, 3 U). Incubate the samples for 3 h at 37°C.

9.6. Filter the digestion mixtures by passage through a Corning Costar™ spin filter (0.22 μm). Collect the flow-through and analyze it by reverse-phase HPLC (Veerareddygari et al., 2016).

9.7. The procedure was the same for RluA except that the substrate was ASL and sodium chloride replaced by ammonium chloride in the reaction buffer (Hamilton et al., 2006).

9.8. Plot the measured initial rates vs the substrate concentration. Determine the kinetic parameters by fitting the data to the Briggs–Haldane equation using nonlinear regression to find V_{max} and K_m. In KaleidaGraph, the equation to be used is

$$y = (m1 \times x)/(m2 + x) \tag{2}$$

where y is the rate, x is [S], $m1$ is V_{max}, and $m2$ is K_m. One may alternatively fit the data to a recast form of the Briggs–Haldane equation such as the Lineweaver–Burk or Eadie–Hofstee equations.

9.9. Compare the kinetic parameters obtained with labeled vs unlabeled substrates to obtain ^{2}KIE.

5.3.3 Results

For TruB with in vitro transcribed unlabeled TSL, $k_{cat} = 0.34\,s^{-1}$ and $K_m = 1.3\ \mu M$. Comparison to the values for the deuterated substrate ($k_{cat} = 0.12\,s^{-1}$; $K_m = 1.7\ \mu M$) revealed KIEs of 2.5 for V_{max} and 3.6 for V_{max}/K_m. Similarly, the determined kinetic parameters for RluA with in vitro transcribed ASL (unlabeled: $k_{cat} = 1.4\,s^{-1}$, $K_m = 1.3\ \mu M$; labeled: $k_{cat} = 0.78\,s^{-1}$, $K_m = 1.6\ \mu M$) returned KIEs of 1.8 for V_{max} and 2.2 for V_{max}/K_m. For both enzymes, then, the results clearly indicate the deprotonation (or reprotonation since the removed hydron is occluded from exchange with solvent) of C2′ is a partially rate-determining step in the reaction and thus demonstrate that the pseudouridine synthases operate through a glycal intermediate, either by the mechanism offered in Fig. 1 or some near variant of it (Veerareddygari et al., 2016).

6. CONCLUDING THOUGHTS

Although the procedures described here are for short RNA substrates (17-mers and 18-mers), they should prove readily adaptable to longer RNA substrates. In fact, very little of the methodology was novel but developed by others for production of RNA to suit their purposes. It is offered as an example of how labeled nucleosides can be incorporated to conduct isotope effect studies with oligonucleotide substrates. When designing such experiments, however, one must bear in mind that in vitro transcription results in the universal substitution of the unlabeled nucleoside with a labeled one, so interpretation may be clouded if multiple sites are reactive rather than only a single site as in the case of the pseudouridine synthases TruB and RluA. The determination of the KIEs for these two enzymes clearly demonstrates the power of this methodology to elucidate the mechanisms followed by enzymes that work on RNA substrates.

ACKNOWLEDGMENTS

This work was supported by NIH grants GM059636 and GM107820 (to E.G.M.), the Commonwealth of Kentucky Research Challenge Trust Fund (Bucks for Brains), and the Charles L. Bloch, M.D. Professorship.

REFERENCES

Ames, B. N. (1966). Assay of inorganic phosphate, total phosphate and phosphatases. *Methods in Enzymology*, *8*, 115–118.

Arthur, P. K., Alvarado, L. J., & Dayie, T. K. (2011). Expression, purification and analysis of the activity of enzymes from the pentose phosphate pathway. *Protein Expression and Purification*, *76*(2), 229–237.

Bykhovskaya, Y., Casas, K., Mengesha, E., Inbal, A., & Fischel-Ghodsian, N. (2004). Missense mutation in pseudouridine synthase 1 (PUS1) causes mitochondrial myopathy and sideroblastic anemia (MLASA). *The American Journal of Human Genetics*, *74*(6), 1303–1308.

Carlile, T. M., Rojas-Duran, M. F., Zinshteyn, B., Shin, H., Bartoli, K. M., & Gilbert, W. V. (2014). Pseudouridine profiling reveals regulated mRNA pseudouridylation in yeast and human cells. *Nature*, *515*, 143–146.

Charette, M., & Gray, M. W. (2000). Pseudouridine in RNA: What, where, how, and why. *IUBMB Life*, *49*(5), 341–351.

Chen, Z., & Ruffner, D. (1996). Modified crush-and-soak method for recovering oligodeoxynucleotides from polyacrylamide gel. *BioTechniques*, *21*(5), 820.

Cleland, W. W. (1987). The use of isotope effects in the detailed analysis of catalytic mechanisms of enzymes. *Bioorganic Chemistry*, *15*(3), 283–302.

Davis, F. F., & Allen, F. W. (1957). Ribonucleic acids from yeast which contain a fifth nucleotide. *The Journal of Biological Chemistry*, *227*(2), 907–915.

Grosjean, H., & Benne, R. (1998). *Modification and editing of RNA 1998*. Washington, DC: ASM Press.

Gu, X., Liu, Y., & Santi, D. V. (1999). The mechanism of pseudouridine synthase I as deduced from its interaction with 5-fluorouracil-tRNA. *Proceedings of the National Academy of Sciences of the United States of America*, *96*(25), 14270–14275.

Hamilton, C. S., Greco, T. M., Vizthum, C. A., Ginter, J. M., Johnston, M. V., & Mueller, E. G. (2006). Mechanistic investigations of the pseudouridine synthase RluA using RNA containing 5-fluorouridine. *Biochemistry*, *45*(39), 12029–12038. http://dx.doi.org/10.1021/bi061293x.

Heiss, N. S., Knight, S. W., Vulliamy, T. J., Klauck, S. M., Wiemann, S., & Mason, P. J. (1998). X-linked dyskeratosis congenita is caused by mutations in a highly conserved gene. *Nature Genetics*, *19*(1), 32–38.

Johnson, L., & Söll, D. (1970). In vitro biosynthesis of pseudouridine at the polynucleotide level by an enzyme extract from Escherichia coli. *Proceedings of the National Academy of Sciences of the United States of America*, *67*(2), 943–950.

Keffer-Wilkes, L. C., Veerareddygari, G. R., & Kothe, U. (2016). RNA modification enzyme TruB is a tRNA chaperone. *Proceedings of the National Academy of Sciences of the United States of America*, *113*(50), 14306–14311.

Krieg, P. A., & Melton, D. (1987). In vitro RNA synthesis with SP6 RNA polymerase. *Methods in Enzymology, 155*, 397–415.

Kuzmine, I., Gottlieb, P. A., & Martin, C. T. (2001). Structure in nascent RNA leads to termination of slippage transcription by T7 RNA polymerase. *Nucleic Acids Research, 29*(12), 2601–2606.

Luzzatto, L., & Karadimitris, A. (1998). Dyskeratosis and ribosomal rebellion. *Nature Genetics, 19*(1), 6–7.

Macdonald, L. E., Zhou, Y., & McAllister, W. T. (1993). Termination and slippage by bacteriophage T7 RNA polymerase. *Journal of Molecular Biology, 232*(4), 1030–1047.

Milligan, J. F., Groebe, D. R., Witherell, G. W., & Uhlenbeck, O. C. (1987). Oligoribonucleotide synthesis using T7 RNA polymerase and synthetic DNA templates. *Nucleic Acids Research, 15*(21), 8783–8798.

Milligan, J. F., & Uhlenbeck, O. C. (1989). Synthesis of small RNAs using T7 RNA polymerase. *Methods in Enzymology, 180*, 51–62.

Miracco, E. J., & Mueller, E. G. (2011). The products of 5-fluorouridine by the action of the pseudouridine synthase TruB disfavor one mechanism and suggest another. *Journal of the American Chemical Society, 133*(31), 11826–11829. http://dx.doi.org/10.1021/ja201179f.

Mitchell, J. R., Wood, E., & Collins, K. (1999). A telomerase component is defective in the human disease dyskeratosis congenita. *Nature, 402*(6761), 551–555.

Mueller, E. G., & Ferré-D'Amaré, A. R. (2009). Pseudouridine formation, the most common transglycosylation in RNA. In H. Grosjean (Ed.), *DNA and RNA modification enzymes: structure, mechanism, function and evolution* (p. 363). Austin, TX: Landes Bioscience.

Ramamurthy, V., Swann, S. L., Paulson, J. L., Spedaliere, C. J., & Mueller, E. G. (1999). Critical aspartic acid residues in pseudouridine synthases. *The Journal of Biological Chemistry, 274*(32), 22225–22230.

Schwartz, S., Bernstein, D. A., Mumbach, M. R., Jovanovic, M., Herbst, R. H., León-Ricardo, B. X., et al. (2014). Transcriptome-wide mapping reveals widespread dynamic-regulated pseudouridylation of ncRNA and mRNA. *Cell, 159*(1), 148–162.

Scott, L. G., Tolbert, T. J., & Williamson, J. R. (2000). Preparation of specifically 2 H-and 13 C-labeled ribonucleotides. *Methods in Enzymology, 317*, 18–38.

Stambaugh, R., & Wilson, D. W. (1960). The chromatography of nucleotides, nucleosides, and pyrimidines and purines on activated charcoal. *Journal of Chromatography A, 3*, 221–224.

Tolbert, T. J., & Williamson, J. R. (1997). Preparation of specifically deuterated and 13C-labeled RNA for NMR studies using enzymatic synthesis. *Journal of the American Chemical Society, 119*(50), 12100–12108.

Van Lanen, S. G., Kinzie, S. D., Matthieu, S., Link, T., Culp, J., & Iwata-Reuyl, D. (2003). tRNA modification by S-adenosylmethionine: tRNA ribosyltransferase-isomerase assay development and characterization of the recombinant enzyme. *Journal of Biological Chemistry, 278*(12), 10491–10499.

Veerareddygari, G. R., Singh, S. K., & Mueller, E. G. (2016). The pseudouridine synthases proceed through a glycal intermediate. *Journal of the American Chemical Society, 138*(25), 7852–7855. http://dx.doi.org/10.1021/jacs.6b04491.

FURTHER READING

Lowry, T., & Richardson, K. (1981). *Theory and mechanism in organic chemistry*. New York: Harper & Row.

Westheimer, F. (1961). The magnitude of the primary kinetic isotope effect for compounds of hydrogen and deuterium. *Chemical Reviews, 61*(3), 265–273.

CHAPTER NINETEEN

Measurement of Kinetic Isotope Effects by Continuously Monitoring Isotopologue Ratios Using NMR Spectroscopy

Natalia Sannikova, Andrew R. Lewis, Andrew J. Bennet[1]
Simon Fraser University, Burnaby, BC, Canada
[1]Corresponding author: e-mail address: bennet@sfu.ca

Contents

1. Introduction 548
2. Procedure 549
   2.1 Choice of NMR-Active Probe Nucleus 550
   2.2 Choosing Whether to Monitor Isotopologue Ratios in the Substrate or the Product 556
   2.3 Sample Preparation 557
   2.4 Acquisition of NMR Spectra 558
3. Data Processing 560
   3.1 1D NMR Data 560
   3.2 2D NMR Data 561
4. Data Analysis 567
5. Summary and Conclusion 568
References 568

Abstract

Nuclear magnetic spectroscopic (NMR) methods are discussed for the measurement of heavy atom (^{13}C, ^{18}O, ^{15}N) and secondary deuterium kinetic isotope effects. The discussion focuses primarily on the NMR methods that enable the measurement of quantitative spectra and not on methods to make labeled substrates. Two main techniques are considered: single-point determinations on natural abundance material and the continuous monitoring of isotopically enriched materials. The second method is described in more detail, and we include a discussion of the current state of instrumentation and computer programs for data acquisition and analysis.

Methods in Enzymology, Volume 596
ISSN 0076-6879
http://dx.doi.org/10.1016/bs.mie.2017.06.034

1. INTRODUCTION

Kinetic isotope effects (KIEs) have been used for many decades by physical organic chemists, as an integral component of their mechanistic toolbox, to probe reaction transition states (Melander, 1960). In general, for many reactions the rate-limiting step involves chemical bond making and breaking, a fact that makes analysis of mechanism using KIE data relatively straightforward (Berti & Tanaka, 2002; Kohen & Limbach, 2006; Meyer, 2012). In contrast, for enzyme-catalyzed reactions it is rarer for a chemical bond-breaking (or bond-making) step to be the only kinetically significant (rate-limiting) step during the catalytic cycle (Cleland, 2006). In addition, for enzymatic reactions two types of KIEs can be measured, those on the turnover number (V) and on the enzymatic efficiency (V/K) (Cleland, 1982). As a result, the number of researchers using KIEs to study enzymatic reactions initially lagged behind that for nonenzymatic reactions. These earlier limitations have been circumvented largely with the development of both the theory for enzyme reactions with multiple significant transition states (Ruszczycky & Anderson, 2006) and the experimental techniques for determining intrinsic KIEs on individual chemical steps during catalysis (Cleland, 2007; Cook & Cleland, 2007; Schramm, 2011).

In essence, two different approaches are used for the measurement of KIEs on enzyme-catalyzed reactions: (i) separate rate constant determinations for each isotopologue and (ii) a competitive determination of the rate ratios (Melander & Saunders, 1980). Of note, competitive techniques can be employed using either radiolabeled tracer (Cook & Cleland, 2007; Parkin, Leung, & Schramm, 1984; Sen, Stojkovic, & Kohen, 2012) or natural abundance (stable isotopes) isotopologues (Cassano, Anderson, & Harris, 2004; Lee, Bain, & Berti, 2004; Rosenberg & Kirsch, 1979; Singleton & Thomas, 1995) as well as highly enriched samples (Chan, Lewis, Gilbert, Karwaski, & Bennet, 2010; Melander & Saunders, 1980). On the other hand, for separate determinations of rate constants highly enriched isotopologues are required (Chan, Lu, & Bennet, 2011; Melander & Saunders, 1980).

Competitive techniques, in general, yield more accurate and precise KIE measurements because systematic errors, which can be caused by concentration and temperature differences when separate reactions are monitored, are minimized (Kohen & Limbach, 2006). Of note, competitive KIE experiments on enzymatic reactions only report on the KIE for the second-order rate constant (V/K), irrespective of the substrate concentration (Melander & Saunders, 1980).

The use of nuclear magnetic resonance spectroscopy (NMR) to measure deuterium KIEs in organic and biological reactions was reported twice in 1986. Specifically, Pascal et al. used NMR to measure changes in natural abundance levels of deuterium in order to evaluate several primary deuterium KIEs (Pascal, Baum, Wagner, Rodgers, & Huang, 1986), while Sawada and coworkers acquired phosphorus-31 NMR spectra to evaluate the secondary deuterium KIE for the S_N2 iodide ion demethylation (CL_3; L = H or D) of a pyridinium salt by using triphenylphosphine to trap the methyl iodide product (Sawada, Takai, Chong, Hanafusa, & Misumi, 1986). Notably, Singleton and Thomas refined the procedures of Pascal et al. markedly to yield secondary deuterium and C-13 KIEs using natural abundance starting material (Singleton & Thomas, 1995). Singleton and coworkers further refined their methods to improve sensitivity (Singleton & Szymanski, 1999). In 2010, a modified spectroscopic method was introduced by Chan et al. in which isotopologue ratio data were acquired continuously using quantitative C-13 NMR with highly enriched singly and doubly labeled starting materials (Chan et al., 2010).

The information provided herein aims to: (a) highlight the pros and cons for the use of NMR spectroscopy to measure competitive KIEs in biochemical systems, with single-point or continuous monitoring of isotopologue ratios, (b) outline practicable considerations for experimental design, and (c) describe strategies for the acquisition of quantitative NMR data for determining KIEs. Throughout this chapter, we will use IUPAC nomenclature for isotopically substituted and selectively labeled compounds (McNaught & Wilkinson, 1997). That is, an isotopically substituted compound is one in which all molecules have the specific nuclide at the assigned position (we use this definition for >98% enrichment), whereas selective labeling conventionally arises from mixing, potentially several, isotopically substituted compounds. For example, methyl α-D-[1-$^2H_{0;1}$,1-$^{13}C_{1;1}$]glucopyranoside is a mixture of doubly labeled methyl α-D-(1-^{2}H,1-^{13}C)glucopyranoside and singly substituted methyl α-D-(1-^{13}C)glucopyranoside (McNaught & Wilkinson, 1997).

2. PROCEDURE

Several points must be evaluated when deciding how to measure KIEs using continuously monitored NMR spectroscopy. These include: (a) the specific KIE(s) to be measured, for example, primary C-13, O-18, or secondary deuterium; (b) the choice of NMR-active "probe" nucleus;

(c) the sensitivity of the NMR spectrometer; (d) the rate constants for spin–lattice relaxation (T_1 values); (e) the ease of synthesis of isotopically enriched substrates; (f) the availability of labeled starting material(s); (g) the chemical shifts (and coupling constants) for the probe nucleus in the starting material and product(s); and (h) any interference from natural abundance signals in the NMR spectra. Points (a)–(d) address the ability to acquire acceptable progress curves of isotopologue ratio vs the fraction of reaction. Points (e) and (f) are relevant if complex labeling patterns are required, given that several milligrams of material are required for each triplicate KIE measurement. Points (g) and (h) are pertinent to the ease of peak integration and isotopologue ratio determinations.

2.1 Choice of NMR-Active Probe Nucleus

The main factors to consider in selecting the NMR probe nucleus are spectral dispersion, sensitivity, and ease of isotopic labeling. Shown in Table 1 are the isotopic abundances, magnetic properties, and typical chemical shift ranges for nuclei with spin = ½ (and ^{2}H, $I = 1$) that can be potentially used to measure biochemical KIEs by NMR spectroscopy. The use of the highest field instrument available will provide the best sensitivity. For each experiment, 4 × as much material would be needed when using a 300 MHz vs a 750 MHz system to obtain spectra with identical signal-to-noise ratio (SNR) using similar NMR probes.

2.1.1 ^{1}H NMR Spectral Data Acquisition

Apart from tritium, ^{1}H is the most sensitive nucleus in NMR spectroscopy; however, there are two main disadvantages inherent to KIE measurements of biological processes by proton NMR: resonances from the solvent and limited spectral dispersion. As a result, only simple organic substrates, in general, allow the use of ^{1}H NMR for the continuous monitoring of isotopologue ratios. For example, when using water as solvent the concentration of ^{1}H from the H_2O signal is 110 M, and this must be suppressed in order to achieve sufficient signal intensity (and SNRs) to detect the isotopologues of interest, which are present at significantly lower concentrations (usually 1–100 mM). In fortunate cases, signals from the proton of interest do not overlap with other signals, such as when 1-phenyl[1-$^{13}C_{0;1}$]ethene was used to measure a ^{13}C KIE for the benzylic carbon during the hydroamination of styrene with aniline (Pabis, Kaminski, Ciepielowski, Jankowski, & Paneth, 2011). Another example is the anomeric protons in monosaccharide derivatives, as these resonances generally occur separately

Table 1 Physical Properties of Potentially Useful Probe Nuclei for the Measurement of Biological KIEs by NMR Spectroscopy

Isotope	NMR Frequency[a] (MHz)	Natural Abundance[b] (%)	Relative Sensitivity	Relative Sensitivity at Natural Abundance	Spin Number	Typical Chemical Shift Range (ppm)
^{1}H	100.000	99.944	1.000	1.000	½	+10 to 0
^{2}H	15.351	0.0156	9.6×10^{-3}	1.45×10^{-6}	1	+10 to 0
^{13}C	25.144	1.108	1.59×10^{-2}	1.76×10^{-4}	½	+220 to 0
^{15}N	10.133	0.366	1.04×10^{-3}	3.85×10^{-6}	½	+620 to −420
^{19}F	94.077	100	0.834	0.834	½	+550 to −250
^{31}P	40.481	100	6.64×10^{-2}	6.64×10^{-2}	½	+1400 to −500

[a]At a field strength of ~2.35 T.
[b]De Laeter et al. (2003).

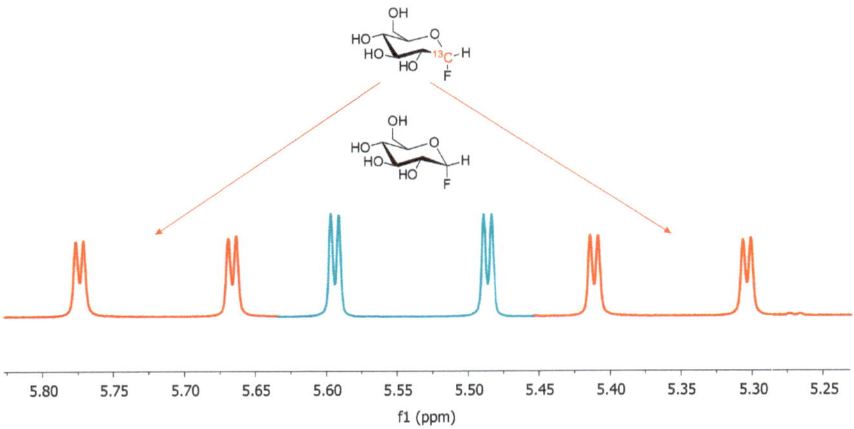

Fig. 1 Part of the ^{1}H NMR spectrum of α-D-[1-$^{13}C_{0;1}$]glucopyranosyl fluoride in D_2O showing the resonances for the anomeric protons.

from other ring C–H protons, as shown in the ^{1}H NMR resonances associated with the anomeric C–H for α-D-[1-$^{13}C_{0;1}$]glucopyranosyl fluoride in D_2O (Fig. 1).

The time required to monitor reactions by NMR spectroscopy is typically of the order of a few to tens of hours. Therefore, it is essential that magnetic field drift be eliminated during the course of reaction monitoring to ensure that the resonances in the spectrum do not move. This requirement is achieved easily with modern NMR spectrometers by "locking" the field using a ^{2}H frequency lock. In aqueous systems this means having sufficient ^2H_2O (D_2O) present; however, the presence of D_2O can influence the reaction due to solvent isotope effects, so the amount of heavy water should be kept to the minimum needed for a stable lock signal. With regular NMR probes this typically requires >5% D_2O by volume, whereas for cryogenically cooled NMR probes the higher lock channel sensitivity allows one to reduce the amount of D_2O to 2–20 μL in a 500–600 μL sample.

2.1.2 ^{13}C NMR Spectral Data Acquisition

The discontinuous method developed by Singleton and Thomas for the measurement of a full set of ^{2}H (using ^{2}H NMR) and ^{13}C KIEs, which require an explicit assumption that at least one KIE is 1.0, is useful for many reactions (Hirschi, Takeya, Hang, & Singleton, 2009; Lee et al., 2004; Singleton & Szymanski, 1999; Singleton & Thomas, 1995). However, such procedures are limited to the evaluation of KIEs for NMR-active nuclei (Singleton & Szymanski, 1999; Singleton & Thomas, 1995), which makes it unsuitable for the measurement of ^{18}O KIEs. Natural abundance measurements have grown in popularity due to increases in NMR spectrometer sensitivity (Huang et al., 2012), measurement of product-specific KIEs (Xiang & Meyer, 2014), and implementation of selective pulse sequences (Kwan, Park, Besser, Anderson, & Jacobsen, 2017).

The use of selectively labeled material for the measurement of ^{2}H, ^{13}C, and ^{18}O KIEs using a ^{13}C nucleus as the magnetic probe allows continuous monitoring of isotopic ratios in aqueous solution (Chan et al., 2010). That is, the isotopologues are distinguished by coupling to the probe nucleus (if NMR active) and/or by isotopic chemical shift perturbation (for $I=0$ nuclei such as ^{16}O and ^{18}O), as indicated in the ^{13}C NMR spectrum for 4-nitrophenyl α-D-[2-$^{13}C_{1;1}$,2-$^{18}O_{0;1}$]mannopyranoside in basic aqueous solution (NaOH = 0.1 *M*, $I=0.4$) displayed in Fig. 2.

Indeed, the use of a ^{13}C probe nucleus allows the routine measurement of ^{18}O KIEs in carbohydrate substitution reactions, for the leaving group (Chan et al., 2010, 2012; Speciale, Farren-Dai, Shidmoossavee, Williams, & Bennet, 2016), the ring oxygen, which stabilizes pyranosylium ion intermediates (Chan et al., 2010, 2012), and an intramolecular nucleophile (Speciale et al., 2016). A major limitation for the continuous monitoring protocols is the requirement for selectively labeled substrates; however, for biological systems, the availability of ^{13}C-labeled materials is improving markedly and as such the cost for each KIE measurement, in terms of labeled materials, is generally not prohibitive.

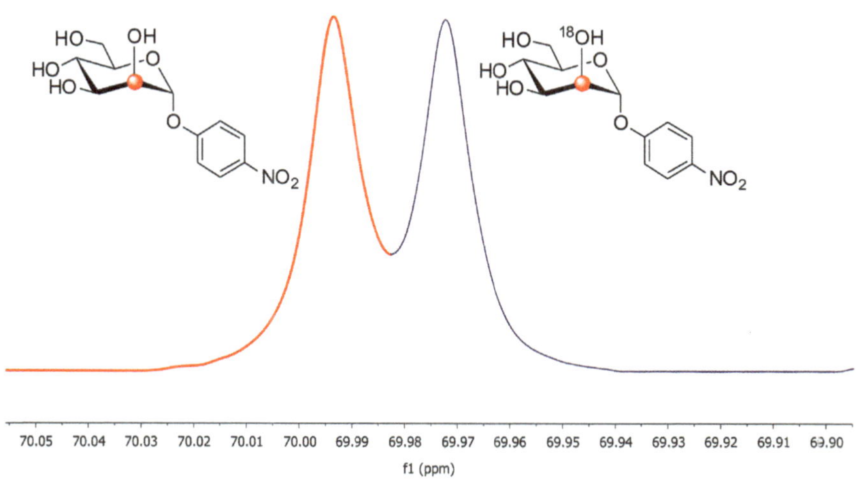

Fig. 2 Proton-decoupled ^{13}C NMR spectrum of the C-2 resonance in 4-nitrophenyl α-D-[2-$^{13}C_{1;1}$,2-$^{18}O_{0;1}$]mannopyranoside in basic aqueous solution (NaOH = 0.1 *M*, $I=0.4$).

The use of ^{13}C as a probe nucleus can be advantageous owing to a number of factors: simpler spectra due to fewer peaks and limited coupling, appreciable chemical shift differences due to the presence of nearby labeled nuclei, and elimination of problems from strong solvent signals (none in water). However, using ^{13}C NMR also has certain disadvantages: lower sensitivity in comparison to other NMR-active nuclei (^{1}H, ^{19}F, ^{31}P, see Table 1), potentially complex syntheses from available starting materials, overlapping NMR peaks leading to the need for extensive spectral deconvolution, or multiplet signals due to coupling with adjacent (also isotopically enriched) ^{13}C or other nuclei that cannot be decoupled for various reasons.

When using ^{13}C (or ^{31}P or ^{19}F) as the detected nucleus, care must be exercised to ensure that any decoupling of other nuclei (normally ^{1}H, but also potentially ^{2}H, ^{19}F, and ^{13}C) does not lead to changes in native signal intensities due to enhancement by nuclear Overhauser effects (NOEs) (Claridge, 1999). Heteronuclear enhancements can result in up to a threefold sensitivity for ^{13}C nuclei coupled to ^{1}H. NOEs occur through space (dipolar coupling) and not through bond (scalar coupling), so they can be dissimilar for the same carbon atom in isotopologues (Claridge, 1999). To minimize the NOE, any decoupling (at least for one-dimensional experiments) should be inverse gated, meaning that decoupling is only applied during the acquisition of the FID, and not during the recycle delay between successive scans as is typically done for normal ^{13}C NMR spectroscopy (i.e., power-gated decoupling).

The wider availability of higher magnetic field instruments (9.4 T and greater) and commercial cryogenically cooled probe electronics (cryoprobes) provide significant sensitivity enhancements for many nuclei (including ^{13}C) that partially mitigate these limitations (Table 2).

Another way to overcome the limitations of a ^{13}C-based approach is to use other types of NMR experiments, for example, sensitivity-enhanced two-dimensional (2D) [^{13}C, ^{1}H]-heteronuclear single-quantum coherence spectroscopy (HSQC) (Manning, Sathyamoorthy, Eletsky, Szyperski, & Murkin, 2012). The combined ^{1}H excitation and detection can potentially increase sensitivity by up to 32-fold compared to 1D ^{13}C NMR. 2D [^{13}C, ^{1}H]-HSQC requires that a proton is attached to the ^{13}C reporter nucleus. Many organic substrates meet this criterion, for either the starting material or reaction product, allowing the measurement of KIEs.

On occasions, NMR signal overlap of starting material and product can be resolved, if the product can be enzymatically modified, thus resulting in its in situ removal from the spectral region of interest (Chan et al., 2010).

Table 2 Typical Signal-to-Noise Specifications for Various 5 mm NMR Probes[a] at Different Field Strengths

Observed Nucleus	Reference Sample	9.4 T (400 MHz)	11.7 T (500 MHz)	14.1 T (600 MHz)
^{1}H	0.1% Ethylbenzene in $CDCl_3$	300 (BBO), 600 (BBI), 1050 (P-CP)	260 (TBO), 800 (TXI)	400 (BBO), 1200 (TXI), 1580 (QNP-CP), 4500 (QCI-CP), 6000 (TCI-CP)
^{19}F	0.05% Trifluorotoluene in $CDCl_3$	275 (BBO), 850 (P-CP)	200 (TBO), 450 (TXI)	4500 (QCI-CP)
^{31}P	48.5 m*M* Triphenylphosphine in acetone-d_6	150 (BBO), 300 (P-CP)	140 (TBO)	220 (BBO), 640 (QNP-CP)
^{13}C	40% Dioxane in C_6D_6	190 (BBO), 475 (P-CP)	150 (TXI), 160 (TBO)	160 (TXI), 300 (BBO), 750 (TCI-CP), 1500 (QNP-CP)
^{15}N	90% Formamide in DMSO-d_6	20 (BBO), 40 (P-CP)	30 (TBO)	50 (BBO), 110 (QNP-CP)

[a] *BBI*, broadband inverse (inverse, ^{15}N–^{31}P, ^{1}H); *BBO*, broadband observe (direct, ^{1}H, ^{15}N–^{31}P); *P-CP*, broadband observe nitrogen-cooled cryoprobe; *QCI-CP*, quad carbon inverse helium-cooled cryoprobe (^{1}H, ^{19}F, ^{13}C); *QNP-CP*, quad nucleus carbon helium-cooled cryoprobe (^{13}C or ^{31}P or ^{15}N, ^{1}H); *TBO*, triple broadband observe (^{15}N–^{31}P, 1H); *TCI*, triple carbon inverse (^{1}H, ^{13}C); *TCI-CP*, triple carbon inverse (^{1}H, ^{13}C); *TXI*, triple X inverse (^{1}H, ^{13}C, ^{15}N). *Notes*: Inverse probes have ^{1}H coil closest to the sample, direct probes have ^{1}H coil farthest from the sample, observable nuclei are indicated, SNRs are for single-scan acquisition.

2.1.3 ^{19}F NMR Spectral Data Acquisition

Fluorine-19 is another potential probe nucleus due to its high sensitivity, 100% natural abundance, and broad range of chemical shifts (Table 1). Moreover, isotopic substitution can produce significant chemical shift perturbations which allows for simple integration without multistep deconvolution. For example, the ^{19}F NMR resonances associated with α-D-[2-$^{2}H_{0;1}$,1-$^{2}H_{0;1}$] galactopyranosyl fluoride in H_2O are displayed in Fig. 3.

Introduction of the fluorine into a substrate can also result in lowering the commitment to catalysis for the chemical steps within the enzyme catalytic cycle (Williams & Withers, 2000), facilitating the measurement of intrinsic KIEs (Tanaka, Tao, Blanchard, & Hehre, 1994; Williams & Withers, 2000). Using ^{19}F as a probe can also result in simpler and cheaper synthesis of substrates compared to the preparation of ^{13}C-labeled materials.

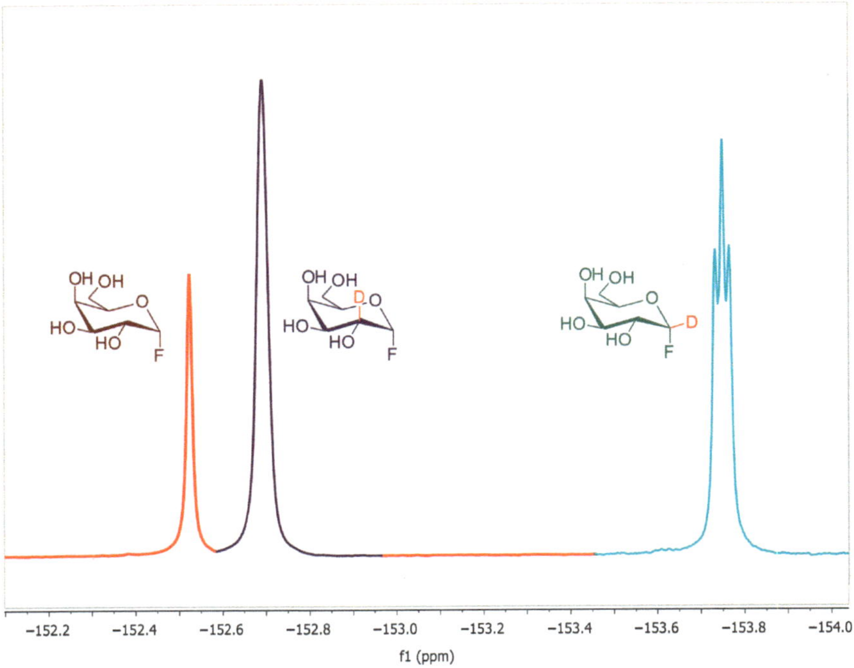

Fig. 3 Proton-decoupled ^{19}F NMR spectrum of α-D-[2-$^{2}H_{0;1}$,1-$^{2}H_{0;1}$]galactopyranosyl fluoride in H_2O.

2.2 Choosing Whether to Monitor Isotopologue Ratios in the Substrate or the Product

Competitive KIE experiments can be devised in which the change in isotopologue ratio is measured as the reaction proceeds either in the remaining starting material or in the product.

During the course of a reaction that exhibits a normal KIE, which is defined as one where the light isotopologue reacts faster ($k_{light}/k_{heavy} > 1$), the starting material becomes enriched with the heavy isotopologue, while the product becomes enriched with the lighter and more reactive isotopologue (Melander & Saunders, 1980). Thus, KIEs can be determined based on either reactant or product analysis (or both).

In general, for reactions in which two reactants form a single product the analysis of product isotopologues is more efficient as it allows the determination of KIEs with respect to both reactants in a single set of experiments (Kwan et al., 2017). Moreover, product-specific KIEs reflect isotopic fractionation that occurs at both rate- and product-determining steps in multiproduct reactions (Xiang & Meyer, 2014). On the other hand, the remaining starting material becomes increasingly enriched as the reaction progresses, which can be an advantage for the measurement of KIE values that are close to unity (Melander & Saunders, 1980).

2.3 Sample Preparation

A critical component for the successful measurement of KIE data is the choice of a suitable internal standard. An internal standard is required in order to determine the fraction of reaction and to normalize the peaks area ratios for all isotopologues. Also, the internal standard must be inert to the experimental conditions: ideally have a simple NMR spectrum that does not overlap with the resonances of the species of interest (singlet preferred), one of its corresponding NMR signal(s) should have a chemical shift close to the signals of interest, and its T_1 relaxation time should be comparable to those of the isotopologues being analyzed.

A deuterated solvent is generally used as a lock for the spectrometer frequency (i.e., D_2O is added to the aqueous buffer). However, NMR tubes with coaxial inserts that contain a deuterated solvent can be used instead. Importantly, normal manual or automated (gradient) ^{2}H-shimming techniques must be used with caution in this case because these methods will tend to optimize field homogeneity only for the ^{2}H-containing capillary. In addition, because this lock solvent is not in contact with the reaction mixture, in principle it can be any deuterated solvent and does not have to match the reaction solvent, but a matched solvent is usually preferred for the best magnetic field homogeneity. The sample volume should be enough that the liquid height in the NMR tube is at least three times the length of the r.f. observation coil in the probe being used. If only limited quantities of compounds are available, the sample volume can be reduced by using either a smaller diameter NMR tube (e.g., 3 mm) or a Shigemi tube whose susceptibility is matched to the solvent being used.

For enzymatic reactions, the amount of enzyme added should be adjusted such that the reaction goes to ~80%–90% completion in a time period that permits the acquisition of sufficient spectra (generally >20–30) with a suitable SNR, such that the time required to acquire the individual spectra is small compared to the total time of monitoring the reaction. The NMR-based measurements of KIEs described here are very versatile because the experimental parameters may be adjusted to optimize data collection depending on the (usually limited) quantities of labeled reactants available for each reaction, the sensitivity of the NMR probe used (time needed to acquire spectra with usable SNR), and the duration of the reaction (amount of enzyme added).

2.4 Acquisition of NMR Spectra

For NMR spectra to be useful for quantitative analysis, important parameters to be considered are the instrumentation setup and optimization, relaxation delay, digitization, and pulse sequence design, as well as use of appropriate postacquisition processing parameters and procedures for optimal peak area determinations (Bharti & Roy, 2012; Pauli, Jaki, & Lankin, 2005).

Shimming can provide a homogeneous magnetic field and result in symmetrical peak shapes that are very important, especially when peaks overlap and require spectral deconvolution. In some cases the best way to achieve symmetrical peak shape is to adjust the room temperature shims while directly observing the FID or real-time Fourier-transformed spectrum in single-scan mode. After shimming, the lock phase should be adjusted to ensure optimal performance and stability.

In cases where the reaction itself, spectrometer stability, or NMR laboratory environmental conditions (temperature and air pressure) cause changes in the shimming over the duration of the experiment, it may be necessary to automate shimming after the acquisition of each spectrum, or to make on-going automated adjustments to specific shims (to maintain the lock level). Such automated adjustments are easily implemented on most modern NMR spectrometers.

To obtain spectra with peak areas that accurately reflect the relative number of nuclei, the acquisition conditions must be set correctly (Pauli et al., 2005). The SNR of all peaks being integrated should be at least 10 (generally accepted minimum for limit of quantitation), but ideally much higher. There should also be at least 10 points defining each peak above half height, which is directly related to the acquisition time (dwell time and number of points acquired in the time domain), spectral width, and zero filling. To ensure that nuclei are fully relaxed between successive scans, the relaxation delay (i.e., the sum of acquisition time and interpulse delay) should be at least seven times the spin–lattice (T_1) relaxation time (>99.9% relaxation) (Pauli et al., 2005).

Approximate T_1's can be estimated using a 1D inversion-recovery pulse sequence ($T_1 \asymp 1.45 \times t_{null}$) (Claridge, 1999). T_1 relaxation times should be measured for the specific nuclei in the compounds being used to measure the KIE (reactants or products and internal standard) at the same approximate concentration, solvent, buffer solution, pH, temperature, and field strength that will be used. This operation can be done using a standard inversion-recovery ($180 - \tau - 90 - \mathrm{AQ}$) or a saturation-recovery pulse sequence (Pauli et al., 2005).

The transmitter offset for the observed nucleus should be close to the peaks of interest, but not exactly on any resonance, and ideally centered between the internal reference standard and the peaks of compounds being measured. The transmitter offset for any decoupling channels should also be set to match closely the frequencies of the directly bonded nuclei for optimal performance. Modern composite pulse decoupling methods (e.g., WALTZ or GARP) (Claridge, 1999) are preferred for their better performance and reduced radio frequency (r.f.) power (Pauli et al., 2005).

The pulse lengths for observation and decoupling channel(s) should be measured for each sample. Typically, the pulse length for a 90-degree excitation should be close to 10 μs to ensure uniform excitation over all resonances (for a 90-degree pulse of 10 μs, the excitation bandwidth is ca. $1/(4 \times 0.000010) = 25{,}000$ Hz, which is ample for the typical range of proton resonances in organic samples; 5000–7000 Hz at 500 MHz). It is also important to maintain a constant sample temperature throughout the experiment to avoid peak shifting and broadening. Sample temperature settings can be calibrated using, e.g., 99.98% deuterated methanol (Findeisen, Brand, & Berger, 2007). The use of buffers raises the ionic strength of aqueous solutions, and this means that r.f.-induced sample heating can occur, especially when multiple types of nuclei are being decoupled for several seconds during acquisition. Dummy scans may be required to reach a steady state in such cases. If the desired reaction temperature is outside the recommended range for normal sample turbines, ceramic sample spinners should be used. Sample spinning (20 Hz) can be used but is usually not recommended for systems operating at 500 MHz and higher (Pauli et al., 2005), or for 2D NMR acquisitions (Giraudeau, 2014).

Note that because an internal standard is used, the peak areas can be normalized relative to the area of the reference peak(s). Thus, the reaction can be followed from any time after the addition of the enzyme (or other reagent) because the initial spectrum defines the zero point for the fraction of reaction. This means that, in general, time can be taken to achieve stable spectrometer conditions (shimming, tuning, and sample temperature). Furthermore, the number of scans can be increased for individual spectra as the reaction proceeds to allow the acquisition of spectra having similar SNR as the concentration of starting material decreases. Alternatively, additional data points or improved SNR may be obtained by coadding successive FIDs and plotting them vs the average fraction of reaction for the summed spectrum. For the best accuracy and precision of KIEs, it is recommended that spectra are acquired until the fraction of reaction reaches ~80%.

3. DATA PROCESSING

3.1 1D NMR Data

Typically, Bruker TopSpin (Topspin User Manual, Version 3.5; Bruker BioSpin GmbH, Germany, 2016) or Mestrelab MNova (Mnova User Manual, Version 11.0; Mestrelab S.L. Spain, 2016) software is used for NMR data processing. Spectra processing parameters are set to obtain accurate integrals for precise KIE measurement, as follows:

1. Fourier transformation of FIDs is performed with two- to fourfold zero filling and appropriate apodization by applying exponential line broadening to increase the SNR. In theory, the best value for the parameter is that which approximates the peak width at half height of the peaks in Hz (Pauli et al., 2005).
2. Spectra should be properly phased and baseline corrected. This directly affects the precision and accuracy of the peak area measurements. If using automated processing, it is recommended to check each spectrum for proper phasing and manually adjust any spectra that are misphased.
3. FIDs can be summed together, for the latter stages of the reaction as the SNR decreases, which compensates for any deterioration in sensitivity.
4. All peaks should be referenced to the chemical shift of the chosen signal of the internal standard.
5. Integration regions must be set up. In order to cover 99.0% of the area of any peak, the region must spread over a range that is at least 25 times the line width of the peak in both directions (Griffiths & Irving, 1998). To achieve an accuracy of 0.1%, the regions must cover about 75 times the peak width in both directions.
6. If the peaks of interest are overlapping, deconvolution must be used. Integral regions should not overlap. For example, two singlets 1 Hz wide should be separated by at least 50 Hz (0.1 ppm at 500 MHz) to be integrated individually with an accuracy of at least 1%. Various algorithms can be used to achieve desired results. Mestrelab Mnova software has implemented deconvolution functionality, and a Mathematica (Mathematica, Version 11.1; Wolfram Research Inc., Champaign, IL, 2017) notebook has been used successfully to deconvolve $^{13}C[^{1}H]$ quantitative spectra (Chan et al., 2010; Speciale et al., 2016). An improved method for integration that removes the areas of small unrelated (impurity) peaks overlapped with the peak-of-interest area has been developed recently by Schoenberger et al. (2016). This methodology has been

implemented in the Mnova software and is called "edited sum." Ideal liquid NMR resonances have a Lorentzian shape, but in practice this is difficult to achieve (Metz, Lam, & Webb, 2000). Commonly, a mixed Lorentzian–Gaussian peak shape or the newer "generalized Lorentzian" peak shape can be used to fit peaks (Schoenberger et al., 2016). If shimming is poor and peaks are asymmetric, then reference peak deconvolution may be applied to regain symmetrical peak shapes (albeit at some expense to the SNR levels) (Morris, Barjat, & Horne, 1997).

7. The obtained peak areas are then normalized by dividing values for the peaks of interest by that obtained for the internal standard in the same spectrum. This operation can be performed automatically in Mnova.

3.2 2D NMR Data

When measuring KIEs by NMR spectroscopy, the chemical shifts of the various reactants or products are by nature often quite similar, with only small differences expected for the isotopologues. Hence, there is a high probability that at least some of the peaks will overlap to some degree in the one-dimensional NMR spectra. As the number of species that are present increases, this overlap can complicate the measurement of accurate peaks areas for the various species, even when using peak deconvolution. 2D NMR offers the possibility to simultaneously determine the relative concentrations of similar compounds in a mixture thanks to its capacity to separate overlapping resonances.

However, 2D NMR requires significantly longer acquisition times (owing to the reduced sensitivity vs 1D and the need to acquire a series of FIDs at incremented delays to create the second dimension, which can be reduced somewhat if sparse sampling methods are used), and there can be significant differences in signal response (peak volumes), which can be highly site specific (Giraudeau, 2014). The peak volumes in 2D spectra are typically not comparable between different carbon types (methyl, methine, and methylene), but in the case of KIE measurements, the reporter nuclei are all exactly the same types of carbons in all the isotopologues, so the peak volumes are directly comparable in principle as long as appropriate precautions are taken (namely, the use of inverse-gated decoupling to reduce NOE enhancements).

The use of Shigemi tubes is recommended when applying 2D NMR techniques because they can reduce the solvent volume by one-half, and thus double the concentrations of the labeled compounds if the same quantities are used. This has three direct benefits: reducing the solvent signal, maximizing the dynamic range, and providing a fourfold sensitivity gain. The latter translates to a reduction of the experiment time by a (theoretical) factor of 0.5 for an equivalent SNR, which is highly significant given the relatively long experiment times required for 2D NMR spectroscopy. Making use of ^{1}H for excitation and detection in a ^{1}H—^{13}C correlation experiment (e.g., HSQC), SNR vs direct 1D ^{13}C observation can be improved by $4^{5/2}$ (i.e., 32-fold).

In the following sections, we discuss 2D experiments in terms of ^{1}H and ^{13}C, but other nuclei can also be used (e.g., ^{19}F or ^{31}P). The best 2D NMR experiments will be those that offer high sensitivity and high resolution in both dimensions. The most efficient 2D NMR techniques are those based on correlations between heteronuclei that share a chemical bond, for example, ^{1}H–^{13}C spin pairs. Two experiments of interest are the heteronuclear single-quantum correlation (HSQC) (Bodenhausen & Ruben, 1980) and heteronuclear correlation (HETCOR) (Martin & Zektzer, 1988). To achieve the highest sensitivity and minimize spectral artifacts, one-bond ^{1}H–^{13}C scalar coupling (*J*) should be measured for the compounds being studied and the delays in both the HETCOR (¼*J*) and HSQC (¼*J* and $^{1}/_{8}J$) experiments should be set appropriately.

The optimal values of the acquisition times in the direct (F2) and indirect (F1) dimensions can be estimated based on the spin–spin (T_2) relaxation times of the appropriate nuclei. For F2 the number of acquired points in the time domain of the FID multiplied by the dwell time should be $\geq 5 \times T_2$ of the observed nucleus. In the indirect dimension the number of increments (FIDs) multiplied by the t1 increment should be $\geq 5 \times T_2$ of the nondetected nucleus. Note also that power limitations for NMR probes and amplifiers should be carefully considered when setting the FID acquisition times (number of acquisition point and spectral width) and recycle delays to avoid damaging sensitive and expensive equipment. Dummy scans (typically 16 or 32) are used to achieve a steady state, but this is only done prior to acquisition of the first FID.

Recently, it has become relatively straightforward to utilize sophisticated prediction algorithms to increase the final spectral resolution or to reduce the acquisition time (Mobli, Maciejewski, Schuyler, Stern, & Hoch, 2012; Palmer et al., 2015). Extrapolation of the FIDs using forward linear prediction can be applied to increase (usually double at least) the effective resolution in both dimensions. Interpolation techniques based on sparely sampled data (e.g., nonuniform sampling) can be used to significantly reduce (often by >75%) the number of increments acquired in a 2D NMR experiment, which may be employed to either reduce the total experiment time or to increase the resolution in the indirect dimension.

3.2.1 HSQC Experiments

Based on sensitivity arguments alone, gradient HSQC would in general be considered the superior experiment, because it detects the higher sensitivity nucleus (^{1}H rather than ^{13}C), and incorporates field gradient pulses that significantly reduce the phase cycle and minimize artifacts. Because the initial polarization in an HSQC experiment typically comes from the ^{1}H nuclei (which usually have shorter T_1 relaxation times), the interscan delays can be significantly (often >5-fold) shorter. The optimal recycle delay for HSQC is $1.27 \times T_1$ (of ^{1}H) less the FID acquisition time, but usually longer delays are used to permit probe r.f. duty cycle limits to be adhered to, and to prevent r.f. heating of the sample (Palmer et al., 2015).

However, there also disadvantages for HSQC experiments: ^{1}H detection means that there can be artifacts from incompletely suppressed ^{1}H–^{12}C signals, or signals arising from the solvent used (95% H_2O for example). Although many solvent suppression techniques have been developed, the application of these usually results in distortions of the baseline and sometimes also peak shapes. There are two (INEPT) polarization transfer steps ($^{1}H \rightarrow {}^{13}C \rightarrow {}^{1}H$) in an HSQC experiment, which means there is potential for larger variations between different isotopologues. Furthermore, to resolve the small chemical shift differences adequately (sometimes <0.05 ppm) expected in the ^{13}C spectral dimension, a high number of increments are needed which dramatically increases the experiment time. This latter requirement can be reduced somewhat by restricting the spectral width in the ^{13}C dimension (e.g., to 3–10 ppm) or by applying band-selective (shaped) ^{13}C r.f. pulses. Note that because the indirect dimension is not filtered by digital or analog filters, through judicious choice of the transmitter offset and spectral width in the indirect (F1) dimension, it may be possible to have the signal from the internal standard (usually several ppm away from the peaks of interest) fold-back into the observation window to reduce the required number of increments and consequently also the time required for the experiment.

One other problem with the signals in HSQC spectra is that they are not single, symmetrically shaped peaks, but are actually split in F2 dimension due to residual ^{1}H—^{1}H couplings which are not removed and cannot be decoupled.

3.2.2 HETCOR Experiments

HETCOR experiments have an advantage in comparison to the corresponding HSQC experiments in that the resolution achievable in the direct dimension (less sensitive ^{13}C) is greatly increased without the need for increasing the experiment time dramatically, unlike HSQC where a large number of t1 increments (and thus very long experiments) are needed to achieve high resolution in the ^{13}C dimension. In addition, carbons with no protons attached do not show up in HSQC experiments, and there is no need to suppress the magnetization of protons bonded to ^{12}C atoms, a problem that plagues all of the ^{1}H-detected sequences.

Decoupling of ^{1}H is easier than ^{13}C (as used for HQSC) mainly due to the smaller bandwidth and thus lower r.f. power requirements. This means that the number of points in the time domain (FID) can be significantly higher in HETCOR, again offering higher resolution. This also permits the use of exponential apodization in F2 rather than QSINE so less peak distortion, which effectively suppresses residual truncation artifacts. The resulting 2D spectra will have better peak shapes and minimal baseline distortions. Also, because there is usually no strong solvent signal (and none if water is the solvent), higher receiver gains can be used which result in improved dynamic range.

The major drawback with HETCOR is the significantly lower sensitivity, but as the compounds are typically ^{13}C enriched (i.e., 99-fold enhancement), this is somewhat mitigated. Cryogenically cooled NMR probes with cooled preamplifiers for the ^{13}C channel also provide another ca. 4-fold gains in sensitivity for ^{13}C detection. We recommend using a modified HETCOR with refocusing of chemical shifts (Bax & Morris, 1981) so that the final echo is balanced. This also permits ^{2}H decoupling to be applied throughout the pulse sequence (including FID acquisition), but be turned off during the recycle delay so that ^{2}H field locking can be used.

3.2.3 2D Data Processing and Fitting

For any quantitative NMR spectra, proper postacquisition data processing is critical as the values of the KIEs being measured can be very small (ca. 1.01). The choice of apodization functions is critical because the number of data points in each dimension is significantly less than for typical 1D spectra (typically at least 32 times). It is recommended that at least twofold zero filling be applied, as well as linear prediction to increase the resolution. Once the 2D Fourier transformation has been carried out, the spectrum should also be correctly phased in both dimensions, and then baseline corrections must be carried out on both the F1 and F2 dimensions. The data should be inspected by taking slices (rows or columns) from the 2D spectrum and checking that the phasing and baseline corrections are appropriately done (this operation is especially important when automated processing procedures using default methods have been used).

The choice of apodization functions will depend on the software used and the type of spectrum being processed. They should be selected so that the (truncated) FIDs are smoothed to zero intensity at the end to avoid truncation artifacts at the base of the peaks in the resulting spectrum. The resulting peak shape should also be considered, as this can simplify the next step of measuring the peak volume. Such artifacts severely interfere with the ability to accurately integrate the peak volumes.

If the peaks are all well resolved and there is no overlap, then the preferred method of determining the peak volumes is integration. This method typically involves defining rectangular boxes around the peaks of interest. In cases where there is overlap of peaks there now exists options in the Topspin and Mnova software for 3D deconvolution. Use of this option typically requires defining the peak maxima on the 2D spectrum. Fig. 4 shows an example of this approach for the HETCOR spectrum acquired on mixture of six isotopologues. The major drawback with this approach is that most commercial software implementations currently use symmetrical Lorentzian peak shapes (which are not usually a very close match to the actual peaks shapes) and unlike deconvolutions of 1D spectra, there is limited or no ability to restrain or adjust the peak-fitting parameters.

A third, practical approach for approximating peak volumes in 2D NMR spectra was recently described by Manning et al. (2012). The principle is that a series of (10–40) consecutive slices covering the width of a given peak are extracted from the 2D spectrum. These individual 1D slices are then added to create a 1D sum spectrum for each peak. The resulting 1D sum spectra for each peak in the 2D spectrum can then be integrated (or deconvoluted if there is peak overlap) and the peak area (which approximates the peak volume) can thus be determined. Fig. 5 shows an example of this approach for a group of peaks extracted from a series of columns (i.e., the F1 or ^{13}C dimension) 2D HSQC spectrum. This approach works well regardless of the peak shapes and allows individual slices to be utilized, either rows or columns from the 2D spectrum, whichever has the best peak shape in each case. Note that whatever method is used to determine peak volumes, only peaks volumes measured by the same method should be compared or used for calculations.

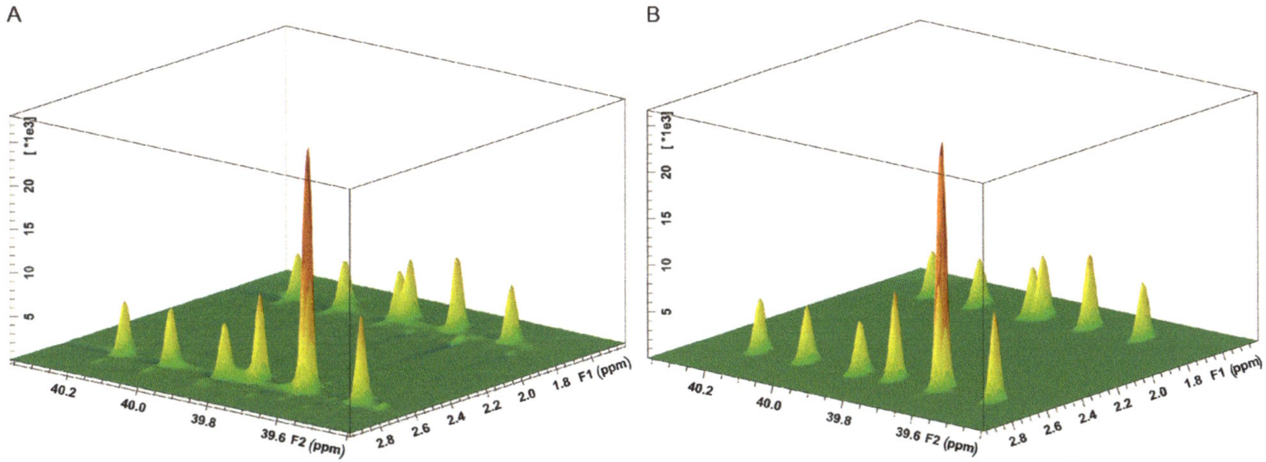

Fig. 4 2D ^{13}C–$^{1}H\{^{1}H,^{2}H\}$ HETCOR spectrum acquired for a mixture of six isotopologues with ^{13}C probe nucleus: (A) observed spectrum and (B) fit of spectrum using 3D peak deconvolution (TopSpin).

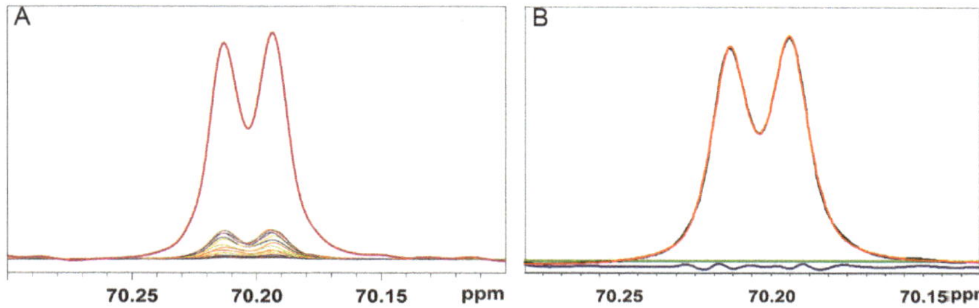

Fig. 5 (A) 20 sequential rows extracted from a peak in a 2D ^{1}H–$^{13}C\{^{2}H,^{13}C\}$ HSQC spectrum and their sum. (B) Fit of sum spectrum using peak deconvolution (Mathematica): *black* = real spectrum, *red* = sum of fitted peaks, *blue* = difference (expt–fit), and *green* = baseline.

4. DATA ANALYSIS

The two equations (Eqs. 1 and 2) that are used to fit the experimental KIE data obtained by analyzing the isotopomer ratios in the substrate and product, respectively, are shown below and discussed in detail by Melander and Saunders (1980). Where k_1 and k_2 are the rate constants for the light and heavy isotopologues, respectively. F_1 is the fraction of reaction for the light isotopologue, R is the ratio of heavy-to-light isotopologue in remaining starting material, R_0 is the ratio at $T=0$, and R_p is the ratio in the product.

$$\frac{k_1}{k_2} = \frac{\ln(1-F_1)}{\ln[(1-F_1)R/R_0]} \tag{1}$$

and

$$\frac{k_1}{k_2} = \frac{\ln(1-F_1)}{\ln\left[1-\left(F_1 R_p/R_0\right)\right]}. \tag{2}$$

These authors also give a detailed error analysis for the standard equations that are used to calculate the KIE values (Melander & Saunders, 1980). Shown in Fig. 6 are typical ^{19}F NMR spectra acquired during a reaction of α-D-[1-$^{13}C_{0;1}$]glucopyranosyl fluoride as a function of the fraction of reaction for α-D-glucopyranosyl fluoride (F_1).

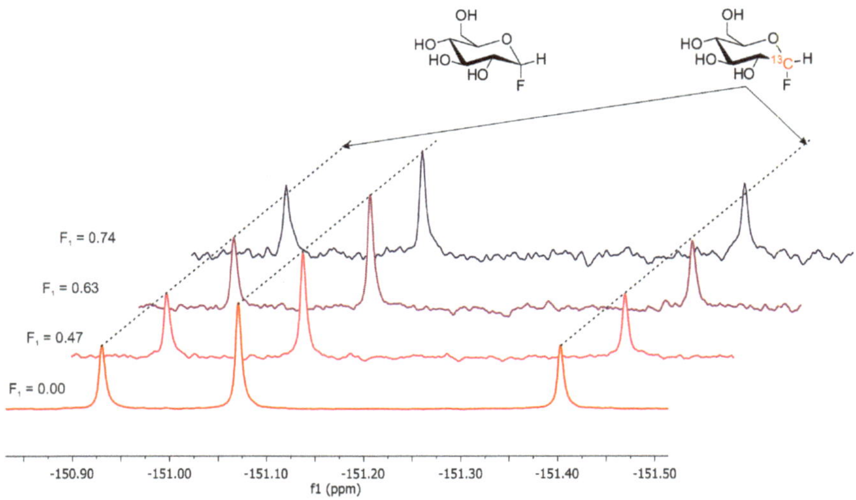

Fig. 6 Stacked plot of proton-decoupled ^{19}F NMR spectra of α-D-[1-$^{13}C_{0;1}$] glucopyranosyl fluoride as the fraction of reaction of α-D-glucopyranosyl fluoride. The *parallel dotted lines* are for visualization of the change in relative peak intensity as the reaction progresses.

5. SUMMARY AND CONCLUSION

For the measurement of competitive KIEs using NMR spectroscopy, it is imperative to decide what NMR-active nucleus will be used and whether to use isotopically enriched substrates. Knowledge of the SNR of available NMR probes will allow estimates of both the amount of material needed and the time frame for spectral acquisition. Isotopically enriched materials also allow the measurement of KIEs for NMR-inactive nuclei such as O-16 and O-18.

Recent advances in NMR spectrometer design and more importantly the advent of "cryoprobe" technology have greatly extended the number and types of experiments that can be performed without other specialized equipment.

REFERENCES

Bax, A., & Morris, G. A. (1981). An improved method for heteronuclear chemical-shift correlation by two-dimensional NMR. *Journal of Magnetic Resonance*, *42*, 501–505.

Berti, P. J., & Tanaka, K. S. E. (2002). Transition state analysis using multiple kinetic isotope effects: Mechanisms of enzymatic and non-enzymatic glycoside hydrolysis and transfer. *Advances in Physical Organic Chemistry*, *37*, 239–314.

Bharti, S. K., & Roy, R. (2012). Quantitative H-1 NMR spectroscopy. *Trends in Analytical Chemistry: TRAC*, *35*, 5–26.
Bodenhausen, G., & Ruben, D. J. (1980). Natural abundance N-15 NMR by enhanced heteronuclear spectroscopy. *Chemical Physics Letters*, *69*, 185–189.
Cassano, A. G., Anderson, V. E., & Harris, M. E. (2004). Analysis of solvent nucleophile isotope effects: Evidence for concerted mechanisms and nucleophilic activation by metal coordination in nonenzymatic and ribozyme-catalyzed phosphodiester hydrolysis. *Biochemistry*, *43*, 10547–10559.
Chan, J., Lewis, A. R., Gilbert, M., Karwaski, M. F., & Bennet, A. J. (2010). A direct NMR method for the measurement of competitive kinetic isotope effects. *Nature Chemical Biology*, *6*, 405–407.
Chan, J., Lewis, A. R., Indurugalla, D., Schur, M., Wakarchuk, W., & Bennet, A. J. (2012). Transition state analysis of *Vibrio cholerae* sialidase-catalyzed hydrolyses of natural substrate analogues. *Journal of the American Chemical Society*, *134*, 3748–3757.
Chan, J., Lu, A., & Bennet, A. J. (2011). Turnover is rate-limited by deglycosylation for *Micromonospora viridifaciens* sialidase-catalyzed hydrolyses: Conformational implications for the Michaelis complex. *Journal of the American Chemical Society*, *133*, 2989–2997.
Claridge, T. D. W. (1999). *High-resolution NMR techniques in organic chemistry* (1st ed.). Amsterdam; New York: Pergamon.
Cleland, W. W. (1982). Use of isotope effects to elucidate enzyme mechanisms. *CRC Critical Reviews in Biochemistry*, *13*, 385–428.
Cleland, W. W. (2006). Enzyme mechanisms from isotope effects. In A. Kohen & H. -H. Limbach (Eds.), *Isotope effects in chemistry and biology* (pp. 915–930). Boca Raton, FL: Taylor & Francis.
Cleland, W. W. (2007). Use of isotope effects to determine enzyme mechanisms. *Journal of Labelled Compounds & Radiopharmaceuticals*, *50*, 1006–1015.
Cook, P. F., & Cleland, W. W. (2007). *Enzyme kinetics and mechanism*. London: Garland Science.
De Laeter, J. R., Bohlke, J. K., De Bievre, P., Hidaka, H., Peiser, H. S., Rosman, K. J. R., et al. (2003). Atomic weights of the elements: Review 2000—(IUPAC technical report). *Pure and Applied Chemistry*, *75*, 683–800.
Findeisen, M., Brand, T., & Berger, S. (2007). A 1H-NMR thermometer suitable for cryoprobes. *Magnetic Resonance in Chemistry*, *45*, 175–178.
Giraudeau, P. (2014). Quantitative 2D liquid-state NMR. *Magnetic Resonance in Chemistry*, *52*, 259–272.
Griffiths, L., & Irving, A. M. (1998). Assay by nuclear magnetic resonance spectroscopy: Quantification limits. *Analyst*, *123*, 1061–1068.
Hirschi, J. S., Takeya, T., Hang, C., & Singleton, D. A. (2009). Transition-state geometry measurements from ^{13}C isotope effects. The experimental transition state for the epoxidation of alkenes with oxaziridines. *Journal of the American Chemical Society*, *131*, 2397–2403.
Huang, M., Garrett, G. E., Birlirakis, N., Bohe, L., Pratt, D. A., & Crich, D. (2012). Dissecting the mechanisms of a class of chemical glycosylation using primary ^{13}C kinetic isotope effects. *Nature Chemistry*, *4*, 663–667.
Kohen, A., & Limbach, H.-H. (2006). *Isotope effects in chemistry and biology*. Boca Raton, FL: Taylor & Francis.
Kwan, E. E., Park, Y., Besser, H. A., Anderson, T. L., & Jacobsen, E. N. (2017). Sensitive and accurate ^{13}C kinetic isotope effect measurements enabled by polarization transfer. *Journal of the American Chemical Society*, *139*, 43–46.
Lee, J. K., Bain, A. D., & Berti, P. J. (2004). Probing the transition states of four glucoside hydrolyses with C-13 kinetic isotope effects measured at natural abundance by NMR spectroscopy. *Journal of the American Chemical Society*, *126*, 3769–3776.

Manning, K. A., Sathyamoorthy, B., Eletsky, A., Szyperski, T., & Murkin, A. S. (2012). Highly precise measurement of kinetic isotope effects using ^{1}H-detected 2D [13C,1H]-HSQC NMR spectroscopy. *Journal of the American Chemical Society, 134*, 20589–20592.

Martin, G. E., & Zektzer, A. S. (1988). Long-range two-dimensional Heteronuclear chemical-shift correlation. *Magnetic Resonance in Chemistry, 26*, 631–652.

McNaught, A. D., & Wilkinson, A. (1997). *Compendium of chemical terminology: IUPAC recommendations* (2nd ed.). Oxford England; Malden, MA, USA: Blackwell Science.

Melander, L. C. S. (1960). *Isotope effects on reaction rates.* New York: Ronald Press Co.

Melander, L. C. S., & Saunders, W. H. J. (1980). *Reaction rates of isotopic molecules.* New York: Wiley.

Metz, K. R., Lam, M. M., & Webb, A. G. (2000). Reference deconvolution: A simple and effective method for resolution enhancement in nuclear magnetic-resonance spectroscopy. *Concepts in Magnetic Resonance, 12*, 21–42.

Meyer, M. P. (2012). New applications of isotope effects in the determination of organic reaction mechanisms. *Advances in Physical Organic Chemisty, 46*, 57–120.

Mobli, M., Maciejewski, M. W., Schuyler, A. D., Stern, A. S., & Hoch, J. C. (2012). Sparse sampling methods in multidimensional NMR. *Physical Chemistry Chemical Physics, 14*, 10835–10843.

Morris, G. A., Barjat, H., & Horne, T. J. (1997). Reference deconvolution methods. *Progress in Nuclear Magnetic Resonance Spectroscopy, 31*, 197–257.

Pabis, A., Kaminski, R., Ciepielowski, G., Jankowski, S., & Paneth, P. (2011). Measurements of heavy-atom isotope effects using ^{1}H NMR spectroscopy. *Journal of Organic Chemistry, 76*, 8033–8035.

Palmer, M. R., Suiter, C. L., Henry, G. E., Rovnyak, J., Hoch, J. C., Polenova, T., et al. (2015). Sensitivity of nonuniform sampling NMR. *Journal of Physical Chemistry B, 119*, 6502–6515.

Parkin, D. W., Leung, H. B., & Schramm, V. L. (1984). Synthesis of nucleotides with specific radiolabels in ribose: Primary ^{14}C and secondary ^{3}H kinetic isotope effects on acid-catalyzed glycosidic bond hydrolysis of AMP, dAMP, and inosine. *Journal of Biological Chemistry, 259*, 9411–9417.

Pascal, R. A., Baum, M. W., Wagner, C. K., Rodgers, L. R., & Huang, D. S. (1986). Measurement of deuterium kinetic isotope effects in organic and biochemical reactions by natural abundance deuterium NMR spectroscopy. *Journal of the American Chemical Society, 108*, 6477–6482.

Pauli, G. F., Jaki, B. U., & Lankin, D. C. (2005). Quantitative 1H NMR: Development and potential of a method for natural products analysis. *Journal of Natural Products, 68*, 133–149.

Rosenberg, S., & Kirsch, J. F. (1979). Measurement of heavy atom kinetic isotope effects by direct mass spectrometric analysis. *Analytical Chemistry, 51*, 1375–1379.

Ruszczycky, M. W., & Anderson, V. E. (2006). Interpretation of *V/K* isotope effects for enzymatic reactions exhibiting multiple isotopically sensitive steps. *Journal of Theoretical Biology, 243*, 328–342.

Sawada, M., Takai, Y., Chong, C., Hanafusa, T., & Misumi, S. (1986). Determination of the secondary α-deuterium kinetic isotope effect on a reverse Menschutkin reaction using phosphorus-31 nuclear magnetic resonance spectrometry. *Analytical Chemistry, 58*, 231–233.

Schoenberger, T., Menges, S., Bernstein, M. A., Perez, M., Seoane, F., Sykora, S., et al. (2016). Improving the performance of high-precision qNMR measurements by a double integration procedure in practical cases. *Analytical Chemistry, 88*, 3836–3843.

Schramm, V. L. (2011). Enzymatic transition states, transition-state analogs, dynamics, thermodynamics, and lifetimes. *Annual Review of Biochemistry, 80*, 703–732.

Sen, A., Stojkovic, V., & Kohen, A. (2012). Synthesis of radiolabeled nicotinamide cofactors from labeled pyridines: Versatile probes for enzyme kinetics. *Analytical Biochemistry*, *430*, 123–129.

Singleton, D. A., & Szymanski, M. J. (1999). Simultaneous determination of intermolecular and intramolecular ^{13}C and ^{2}H kinetic isotope effects at natural abundance. *Journal of the American Chemical Society*, *121*, 9455–9456.

Singleton, D. A., & Thomas, A. A. (1995). High-precision simultaneous determination of multiple small kinetic isotope effects at natural abundance. *Journal of the American Chemical Society*, *117*, 9357–9358.

Speciale, G., Farren-Dai, M., Shidmoossavee, F. S., Williams, S. J., & Bennet, A. J. (2016). C2-oxyanion neighboring group participation: Transition state structure for the hydroxide-promoted hydrolysis of 4-nitrophenyl α-D-mannopyranoside. *Journal of the American Chemical Society*, *138*, 14012–14019.

Tanaka, Y., Tao, W., Blanchard, J. S., & Hehre, E. J. (1994). Transition state structures for the hydrolysis of α-D-glucopyranosyl fluoride by retaining and inverting reactions of glycosidases. *Journal of Biological Chemistry*, *269*, 32306–32312.

Williams, S. J., & Withers, S. G. (2000). Glycosyl fluorides in enzymatic reactions. *Carbohydrate Research*, *327*, 27–46.

Xiang, S. H., & Meyer, M. P. (2014). A general approach to mechanism in multiproduct reactions: Product-specific intermolecular kinetic isotope effects. *Journal of the American Chemical Society*, *136*, 5832–5835.

AUTHOR INDEX

Note: Page numbers followed by "*f*" indicate figures, "*t*" indicate tables, "*np*" indicate footnotes, and "*s*" indicate schemes.

A

Abdelwahed, S.H., 472
Abe, H., 120–121
Abeysinghe, T., 69
Abolin, C., 232
Achim, C., 240
Adamczyk, P., 186–188, 299–300, 303–304, 306–307, 317
Adams, H., 437–438
Addington, T.A., 24, 46, 49–50*t*, 57*t*, 60, 62*t*, 65–66*t*, 67, 70
Adesina, A.S., 24–26, 38–39, 47–53, 49–50*t*, 67, 71–72, 74
Adkins, J.F., 313
Adrian, J.C., 344–345
Aelion, C.M., 293
Aeppli, C., 295
Aeschbacher, M., 317
Agarwal, P.K., 44–45, 62–63
Agarwal, V., 180
Agbandje-McKenna, M., 54–55
Aguilera, R., 406–407
Ahad, J., 505–506, 515–516
Ahluwalia, A.K., 434–435
Aikens, D.A., 347–348
Akoka, S., 370–398
Albe, M., 195
Albericio, F., 36
Albert, C., 114–115, 118–119*t*
Albert, K., 360–361
Alberti, B.N., 196
Albertino, A., 297
Albery, W.J., 100–101
Aleksandra, T.E., 437–438
Allain, F.H.T., 45–48, 50
Allard, A.-S., 292
Allemann, R.K., 24–39, 47, 49–50*t*, 53, 57*t*, 59, 62*t*, 63–64, 64–66*t*, 67–69, 71–74, 73*t*
Allen, F.W., 524
Allen, T.L., 502–504, 515–518
Almendros, G., 114–115
Almo, S.C., 17
Altman, C., 86, 462–463, 480–481
Alvarado, L.J., 532–533
Alvarez, F.J., 468–469
Amaya, K., 35–37
Ames, B.N., 531
Amrani, A., 313
Amyes, T.L., 164–174, 164*s*, 167*s*
Anderson, D.R., 445
Anderson, K.S., 152
Anderson, T.L., 552–553, 556–557
Anderson, V.E., 2–5, 354, 437, 445, 450, 467–471, 548
Andersson, B., 229–230
Andersson, H., 45–48
Andersson, K.K., 250–251
Angeles, R.M., 232
Angeles-Boza, A.M., 268–269
Angelis, Y.S., 413
Anisimov, V., 17, 189–191
Antheaume, I., 114–115, 116–119*t*, 126–129, 134, 136, 138–139
Antoniou, D., 24, 46–52, 49–50*t*, 57*t*, 62–64, 62*t*, 64–66*t*, 67–70, 72, 74–76
Appelman, E.H., 259
Appleby, T.C., 166–167, 173–174
Aranda, J., 203–204
Aranjuelo, I., 114–115, 116–117*t*
Araus, J.L., 114–115, 116–117*t*
Aravena, R., 293, 518
Arbogast, R.-M., 114–115, 118–119*t*
Ariza-Castolo, A., 343–344, 363–364
Armstrong, K.B., 333, 337–338
Arner, R.J., 240, 478–479
Arnold, W.A., 134, 297–298, 315–315
Aronow, L., 229
Aroulanda, C., 386–387
Arthur, P.K., 532–533
Asbury, J.B., 68–69

Atteia, O., 303–304
Attwood, D., 51–53
Attwood, P.V., 53–55
Auchus, R.J., 230–231, 230*f*
Avila-Castro, A., 363–364
Ayers, B., 38

B

Babior, B.M., 104
Baca, M., 36–37
Bachovchin, W.W., 333
Badeck, F., 120–121
Baguet, E., 371–372
Baillie, T.A., 229, 231–232
Baillif, V., 386–387
Bain, A.D., 548, 552–553
Bala, K., 192
Baldridge, K.K., 342–343
Baldwin, J., 240
Balogh, G.T., 344
Banerjee, A., 37
Banerjee, R., 44–45, 62–63, 240–286
Banga, J.R., 88–89
Barak, Z., 478–479
Barclay, A.R., 114–115, 118–119*t*
Bard, J., 240
Bar-Dagan, M., 32
Barducci, A., 203–204
Barjat, H., 560
Barkan, E., 318
Barnes, J.A., 204–205
Barnett, S.A., 169–170
Barr, E.W., 240, 478–479
Barros-Rios, J., 393–394
Barry, P.L., 415–416
Bartell, L.S., 51
Barthel, M., 114–120, 116–117*t*
Bartoli, K.M., 524
Bartolucci, S., 44–45
Bary, Y., 351–352
Baryshnikova, O.K., 335–336
Basch, H., 264–268
Bashir, S., 192
Basner, J.E., 48–52
Bates, D.J., 87
Bates, D.M., 87–88
Bauer, K., 36–37
Baum, M.W., 549
Baume, N., 414, 423–424
Bax, A., 564
Baxter, N.J., 24, 47, 49–50*t*, 52–53, 65–66*t*, 67–74, 73*t*
Bayle, K., 373–375, 377, 390–392, 397–398
Bean, K., 154
Becchi, M., 406–407
Becke, A.D., 196–197
Beckmann, B., 121–123, 127–128
Begley, T.P., 166–167, 173–174, 472
Behiry, E., 25–26
Belasco, J.G., 100–101
Bell, A.F., 354
Bell, H.M., 219–220
Bell, L.C., 225–226, 230*f*
Bell-Parikh, L.C., 225, 230*f*
Ben-Bassat, A., 36–37
Benkovic, S.J., 25–26, 38–39, 44–45, 62–63
Benne, R., 524
Bennet, A.J., 548–568
Bennett, J.P., 393–394
Bentley, H., 181
Berg, M., 141, 293, 295, 315–317
Berger, S., 559
Bergman, R.G., 355
Bernasconi, S.M., 293
Berns, D.S., 51
Bernstein, A., 184, 293–294, 313, 315
Bernstein, D.A., 524
Bernstein, M.A., 560
Berthet, J.-C., 345
Berti, P.J., 548, 552–553
Bertin, L., 393–394
Besnard, J., 128–129
Besser, H.A., 552–553, 556–557
Bevilacqua, P.C., 203–204, 434–435
Bharti, S.K., 558
Bhattacharyya, A., 220
Bianchi, E., 36
Biemann, K., 220
Bigeleisen, J., 501–504, 513–518
Bigeleisen, J.J., 67
Billault, I., 371–372, 377, 381, 383–387
Billeter, S.R., 44–45, 62–63
Bingaman, J.L., 203–204
Birck, M.R., 5, 17
Birlirakis, N., 552–553
Bischoff, R., 442–443

Björkhem, I., 228–229
Blakeley, M.P., 54–55
Blakemore, P., 363
Blanchard, J.S., 166–167, 555
Blanco-Canosa, J.B., 36
Blanquet, S., 36–37
Blessing, M., 315–316
Bligh, E., 121–123
Blomberg, M.R.A., 189–191
Bloom, Y., 518
Blum, G., 5–6
Blum, M.M., 54–55
Blumenstein, C., 393–394
Bocker, R.H., 231
Bodenhausen, G., 562
Boeing, H., 114–115
Boeker, E.A., 86
Boex-Fontvieille, E.R.A., 114–115, 116–117*t*, 134, 136, 140
Bogaard, A., 114–115, 116–117*t*
Bohe, L., 552–553
Bohlke, J.K., 551*t*
Bölcskei, A., 344
Bolhuis, P.G., 62–63
Bollag, J.-M., 196
Bollinger, J.M., 240
Bolotin, J., 134, 293–304, 305*t*, 306–308, 311–321
Bominaar, E.L., 253
Bonanno, J.B., 174
Bonin, A.S., 506–507
Bonomi, M., 203–204
Booker, S.J., 472
Boosman, A., 36–37
Bora, R.P., 44–45
Borowski, T., 189–191
Bothner-By, A.A., 513–514
Bothwell, I.R., 5–6
Botosoa, E.P., 375, 393–394, 397–398
Botre, F., 421–422, 424–426
Botti, P., 37
Bouchard, D., 370
Bouzida, D., 202
Bowden, G.T., 232
Bowling, D.R., 417–419
Boyd, D.R., 292
Boza, J.J., 130
Bradbury, J.H., 347–348
Brand, L., 228
Brand, T., 559
Brand, W., 123–124, 130
Brand, W.A., 296–297, 316, 380, 501–504
Branduardi, D., 203–204
Braunsdorf, R., 410–411
Brazeau, B.J., 241, 269–271
Brehm, L., 336–337
Breider, F., 519
Breimer, D.D., 231
Brenna, J.T., 413, 502–504
Brenna, T., 410–411
Brennwald, M.S., 312–313, 315, 318–321
Breslow, R., 435–437
Bret, C., 195
Brezovsky, J., 205–206
Briccut, J., 370–371, 377
Brittain, D.R.B., 192–194, 206–208
Brockwell, D., 51–53
Broderick, J.B., 472, 478
Brody, R.S., 439
Brooks, B.R., 186–188, 194–195
Brooks, C.L., 62–63
Brothers, E.N., 199–201
Brown, R.E., 45–48
Brueckner, A.C., 355
Bruice, T.C., 228
Brummund, W., 502
Brunckova, J., 340–341
Bruzik, K.S., 437–438
Brzozowski, A.M., 393–394
Buchmann, N., 114–120, 116–117*t*
Buckpitt, A.R., 229–230
Budayova-Spano, M., 54–55
Budisa, N., 337
Bugg, T.D.H., 292
Buisson, C., 410–411
Bunce, S.C., 347–348
Burgoyne, T.W., 416–417
Burke, K.D., 351
Burnett, R.W., 415–416
Burstein, S., 422–423
Busch, G.K., 36–37
Bush, E.D., 228
Bussy, U., 374, 398
Bykhovskaya, Y., 524–525
Byuna, K.S., 194–195

C

Cachau, R., 54–55
Caimi, R., 410–411
Calcutt, M.W., 230*f*
Caldwell, W.S., 4–5
Callender, R., 44–45
Cameron, S.A., 6, 17, 18–19*t*, 46, 49–50*t*, 57*t*, 62–63, 62*t*, 64–66*t*, 67, 74
Camin, F., 114–115, 116–117*t*
Cannio, R., 44–45
Canonica, S., 141, 315–316
Cantalapiedra-Hijar, G., 114–115, 118–119*t*
Caratzoulas, S., 44–45
Careri, G., 44–45
Carlile, T.M., 524
Carr, C.A.M., 268–269
Carrasco, M.R., 37
Carsten, K., 478–479
Casabianca, H., 406–407
Casas, K., 524–525
Cassano, A.G., 437, 445, 548
Castro, J.N., 5–6, 24, 46, 49–50*t*, 57*t*, 60, 62*t*, 65–66*t*, 67, 70
Catak, S., 186–188
Catlin, D.H., 406–407
Cavanaugh, C.M., 297–298, 513–514
Cawley, A.T., 421–422, 424–428
Caytan, E., 373–376, 397
Cha, G.S., 219–220
Cha, S., 473–474
Chadegani, F., 151–153
Chahrour, O., 45–48
Chaintreau, A., 373–374, 377, 393–394, 397–398
Chakrabarti, M., 253
Champ, M., 114–115, 118–119*t*
Chan, C.C.H., 505–506, 518
Chan, J., 548–549, 553–554, 560
Chan, K., 166–167, 167*s*, 170–174
Chanco, E., 273
Chandler, D., 62–63
Chang, E.P., 47, 49–50*t*, 64–66*t*, 67–68, 70
Chang, H.C., 196
Chang, S., 36–37
Chang, S.L., 269–271
Chang, S.Y., 36–37
Chang, T.K., 166–167
Chang, W.-c., 478–479
Charette, M., 524
Chartrand, M.M.G., 300–303, 315–316, 502–504
Chatfield, D.C., 194–195
Chattopadhyaya, J., 337–338
Cheeseman, J.R., 17, 194–197
Chen, C.-H., 51
Chen, F., 393–394
Chen, H., 435, 437, 453
Chen, H.L., 164–165
Chen, Y.P., 196
Chen, Z., 541
Cheong, P.H.-Y., 355
Chiang, Y., 166–167
Chikaraishi, Y., 128
Chipman, D.M., 478–479
Chivas, A.R., 181
Choi, P.J., 349–350
Choi, S.-H., 472–473, 476–478
Choi, S. K., 45–48
Chong, C., 549
Chovancova, E., 205–206
Chowdhury, G., 230*f*
Christianson, D.W., 54–55
Ciccotti, G., 204
Ciepielowski, G., 550–552
Cirera, J., 240
Cirpka, O.A., 295
Claridge, T.D.W., 554, 558–559
Clarke, R.T.J., 506
Clauser, K.R., 54–55
Cleland, W.W., 44, 69, 100–101, 134, 149–150, 224, 295, 350–351, 435–438, 445, 460, 462–463, 466–469, 476–477, 484, 502, 526, 542–543, 548
Cléroux, M., 114–115, 118–119*t*
Cobice, D., 45–48
Cochrane, J.C., 434–435
Cocivera, M., 333
Cohen, J.S., 347–348
Colamonici, C., 421–422, 424–426
Cole, R.B., 153
Coleman, D.D., 505–506
Coleman, J.E., 55–56
Coleman, M., 297
Coleman, M.L., 181
Collins, K., 524–525

Collins, M., 426–428
Connolly, B.A., 439
Contzen, J., 51
Cook, P.F., 44, 69, 166–167, 224, 295, 434, 460, 466–467, 476–477, 502, 548
Coon, M.J., 224–225, 229
Cooper, A., 51–53
Cooper, L.E., 472
Cooper, S., 51–53
Coplen, T.B., 296–297, 314–315
Copping, R., 345, 347
Cornic, G., 120–121
Cornish-Bowden, A., 86
Cotton, G.J., 38
Cowburn, D., 36–37
Cozzi, F., 343
Craig, J.C., 229
Cramer, C.J., 134, 297–298, 315–316
Crehuet, R., 203–204
Crespi, H.L., 51, 55–56
Crich, D., 37, 552–553
Crooks, G.P., 152
Cros, J., 114–115, 118–119*t*
Cuchillo, C.M., 434–435
Cuevas, G., 337–338
Culp, J., 528
Curcio, D., 421–422, 424–426
Czekster, C.M., 24, 47, 49–50*t*, 53, 57*t*, 59, 62*t*, 64–66*t*, 67–68, 70–72, 75–76
Czyryca, P.G., 437–438

D

Dadhwal, M., 192
Dalton, H., 249
Daly, J.W., 228
Damborsky, J., 192
Dametto, M., 68–69, 72, 75–76
Damjanovich, S., 44–45
Damsté, J.S.S., 130–131
Dappricha, S., 194–195
Darmaun, D., 114–115, 118–119*t*
Dassama, L.M.K., 240
Dauvergne, M.-T., 54–55
Dávid, A., 349
Davis, F.F., 524
Davis, S., 181
Dawson, P.E., 36–37
Dawson, R.M.C., 442–443
Dawson, W.M., 24–26, 38–39, 44–45, 47, 49–50*t*, 53, 57*t*, 59, 62–64, 62*t*, 64–66*t*, 67–69, 71–74, 73*t*
Dayie, T.K., 45–48, 532–533
Dayton, P.G., 228
De Bievre, P., 551*t*
de Groot, P.A., 120–121
de Koster, C.G., 114–115, 116–117*t*
de la Hoz, A., 360–361
de la Rosa, J.M., 114–115
de la Torre, X., 421–422, 424–426
De Laeter, J.R., 551*t*
de Rijke, E., 114–115, 116–117*t*
De Souza, D.P., 36–37
de Visser, S.P., 261–262
Dec, J., 196
Deerfield, D.W., 199–201
Dege, J.E., 244, 250–251
Degenhardt, K., 422
Dehn, R.L., 229
Deleens, E., 120–121
Dellago, C., 62–63
Dellero, Y., 140
DelMonte, A.J., 299
Dempster, H.S., 505–506, 515–516
Deng, F.-K., 37
Denisov, I.G., 240
Dennis, E.A., 435–437
Desmarais, D.J., 506
Desmurs, J.-R., 374, 398
Dessen, P., 36–37
Detar, D.F., 510–512
Devi-Kesavan, L.S., 206
Devos, J.M., 45–48
DeVoss, J.J., 228
Di Costanzo, L., 54–55
Di Donato, A., 240
di Guana, C., 36–37
Diac, Y., 240, 478–479
Dias, R.F., 142
Díaz-Fernández, M., 363–364
Diederichsen, U., 37
Dissanayake, T., 435, 440–441, 453–454
Doan, P.E., 240
Docherty, G., 124–126, 412–413
Dong, Y., 51, 356–358
Dong, Y.H., 261–262

Donike, M., 407
Dorer, C., 299–300, 306–307
Doucet, N., 55–56
Drahos, L., 344
Drimmie, R.J., 181
Driscoll, M.D., 102–104
Du, Q., 8–9
Du, X., 133–134
Duan, J.-R., 377, 383–385
Duff, A.P., 45–48, 53–55
Duffus, B.R., 472
Duggleby, R.G., 87
Duhamel, M., 505–506, 518
Duhr, A., 314–315
Dumont, E., 114–115, 116–117*t*
Dunaway-Mariano, D., 188–189
Dunietz, B.D., 264–268
Dunn, P.J.H., 114–115, 118–119*t*, 297, 317–318
Dunne, O., 45–48
Duranceau, M., 120–121
Durbin, S.M., 262
Durkin, P., 337
Duschene, K.S., 472, 478
Dutta, K., 36–37
Dworatzek, S., 518
Dybała-Defratyka, A., 44–45, 62–63, 180–210
Dyer, J., 121–123
Dyer, R.B., 44–45

E

Eaborn, C., 173–174
Ealick, S.E., 166–167, 173–174
Earhart, C.A., 250–251
Ebert, K., 184
Ebert, K.A., 294, 313, 315
Eby, P., 505–506
Eckert, D., 295
Eckert, T., 505–506, 518
Eckstein, F., 439
Edwards, E.A., 505–506, 518
Egli, L., 114–115, 118–119*t*
Ehlering, J.R., 417–419
Ehleringer, J.R., 404–406
Ehrlich, J.I., 166–167
Eiler, J.M., 294, 313
El Gamal, A.A., 180
Elango, N., 250–251
Elberson, M.A., 45–48
Eletsky, A., 554, 565
Eliel, E.L., 2
Elison, C., 218–219, 228–229, 231–232
Elliott, H.W., 218–219, 228–229, 231–232
Ellison, S.L.R., 350
Elsner, M., 114–115, 184, 188–189, 201–202, 293–295, 298–306, 313–315, 505–506, 513–514
Emenike, B.U., 346, 362
Emery, C., 406, 412–413, 415–417, 421–427
Engelke, D.R., 55–56
Ephritikhine, M., 345
Erdélyi, M., 356
Erickson, P.R., 315–316
Eriksson, G., 114–115, 118–119*t*
Eriksson, J., 186–188
Ermer, J., 468–469
Esaki, N., 181, 188–189
Escher, B.I., 292
Eurenius, K.P., 194–195
Eustaquio, A.S., 180
Evans, C.A., 337
Evans, R.M., 100–101
Evans, T.C., 35–37
Evelhoch, J.L., 333
Evershed, P., 412–413
Evershed, R.P., 114–115, 116–119*t*, 124–126
Eyer, P., 283

F

Fabian, M.A., 333–335, 337–338, 340–342, 346
Fairlamb, I.J., 393–394
Fallick, A.E., 123–124, 130
Fan, R., 273, 279–283, 285–286
Fang, Y., 186–188
Fang, Y.-R., 182
Farizon, Y., 406–407
Farquhar, G.D., 114–115, 116–117*t*, 142
Farren-Dai, M., 553, 560
Fattori, D., 36
Favreau, L.V., 220
Fay, L.-B., 114–115, 118–119*t*, 130
Fayat, G., 36–37

Federherr, E., 120–121, 131–132, 311, 314–315, 422
Fedorov, A.A., 174
Fedorov, E.V., 174
Fedoseyenko, D., 472
Feist, T., 126–127
Felshin, L.M., 196
Feng, G., 437–438
Fenner, K., 292
Fenwick, C.S., 123–124, 130
Ferchaud-Roucher, V., 114–115, 118–119*t*
Ferguson, K., 133–134
Ferrari, R.P., 196
Ferré-D'Amaré, A.R., 524
Ferreira, A.A., 294
Ferrer, S., 204–205
Fersht, A., 218, 222
Fieber, W., 373–374, 377, 393–394, 397–398
Field, M., 203–204
Field, M.J., 195
Fielding, A.J., 102, 107–108, 108*f*
Fierke, C.A., 25–26, 38–39, 55–56, 67
Findeisen, M., 559
Fischel-Ghodsian, N., 524–525
Fischer, A., 192, 204
Fisher, A.J., 5–6
Fisher, H.F., 184
Fisher, S.Z., 54–55
Fitzpatrick, P.F., 149–161
Flach, A., 358
Flament, M.-M., 406–407
Flanagan, L.B., 417–419
Flannery, B.P., 87–89
Fleig, W.E., 126–127
Flenker, U., 410–413, 415, 417–419, 421–422, 424–425
Flesch, V., 140
Floss, H.G., 264–268
Fong, J., 297
Forconi, M., 437
Forsyth, D.A., 353
Fortova, A., 205–206
Foster, A.B., 229, 232
Fowlks, W.L., 263–264
Fox, B.G., 240–242, 244, 246, 249–251, 253–256, 268
Fraczkiewicz, G., 261–262
Fraczkiewicz, R., 261–262
Francis, K., 24, 44–45, 47, 49–50*t*, 57*t*, 59, 62*t*, 64*t*, 67–68, 70–72, 75–76
Franke, D., 36–37
Franklin, J.L., 480–481
Franks, S.G., 142
Franzen, S., 188–189, 196–198
Frape, S.K., 181, 518
Fraser, R.A., 114–115, 116–119*t*
Frasquet-Darrieux, M., 114–115, 118–119*t*, 126–127, 138–139
Frederick, C.A., 250–251, 273–274
Frederiksen, J.K., 442–443
Freeman, J.P., 196
Freeman, K.H., 142
French, J.B., 68–69
Frey, P.A., 439, 472, 478
Friedien, C., 87
Friedman, L., 502–504
Friedmann, M., 337–338
Friesner, R.A., 194–197, 199–201, 264–268
Frisch, M.J., 17, 194–197
Froland, W.A., 241–242, 244, 246, 249–251, 253–256, 264–268
Frølund, B., 336–337
Frottin, F., 36–37
Fry, A., 181–182
Fry, H., 410–411
Frydenvang, K., 336–337
Fujii, H., 261–262
Fujitani, T., 181
Fukazawa, R., 45–48
Fukuda, M., 192
Fukuta, K., 52
Fuldner, H.H., 439
Fulghum, R.S., 506
Funaki, T., 231
Furuya, E., 408–409
Fußhöller, G., 422–423

G

Gaffney, P.R.J., 36–37
Gajda, L., 51, 55–56
Gallage, N.J., 393–394
Gamble, T.R., 54–55
Gamo, T., 181
Gandhi, T., 276–279
Gandolfi, A.J., 232

Ganguly, A., 203–204
Gao, J., 194–195, 206
Gao, Y., 186–188
Garcia, H., 232
García-Báez, E., 363–364
García-Meseguer, R., 24–26, 47, 49–50*t*, 53, 56, 57*t*, 62–64, 62*t*, 64–66*t*, 67, 71–72
Garcías-Morales, C., 343–344, 363–364
Garcia-Viloca, M., 204–205
Garland, W.A., 229–230
Garr, A.N., 134, 315–316
Garrett, B.C., 63–64
Garrett, G.E., 552–553
Garritt, R., 123–124, 130
Gasson, M.J., 393–394
Gates, Z.P., 30
Gauges, R., 94, 103, 105–106
Gavva, S.R., 502
Gawandi, V., 153, 159
Gaweska, H.M., 151–152
Gawlita, E., 4–5, 354
Gawlita-Kamzurska, E., 184
Gaylord, N.G., 2
Ge, X., 24, 46, 64, 64*t*, 74
Gehre, M., 124–126, 314–315
Geissler, P.L., 62–63
Gelb, M.H., 219–220
Gelles, E., 502–504
Gelman, F., 313, 315
Gentil, E., 398
Gentle, I.E., 36–37
Georgakopolus, C.G., 413
George, A.V., 421–422, 424–425
Gerfen, G.J., 253
Gerlt, J.A., 166–167, 167*s*, 169–174
Germon, J.C., 513–514
Gerratana, B., 437
Gervasio, F.L., 203–204
Gessler, A., 114–115, 116–117*t*
Geueke, B., 192, 205–206
Geyer, H., 406–408, 412–413, 415, 419, 421–422, 424–425
Ghashghaie, J., 120–121
Gherman, B.F., 264–268
Ghosh, A.K., 44–77
Ghysels, A., 186–188
Gibson, D.T., 292, 308
Gibson, J.J., 505–506
Giese, T.J., 435, 437, 453
Giglione, C., 36–37
Gilar, M., 442–443
Gilard, F., 114–115, 118–119*t*, 126–127
Gilbert, A., 142, 373–374, 387, 389–390, 393–394
Gilbert, M., 548–549, 553–554, 560
Gilbert, W.V., 524
Gilboa, H., 351–352
Gilevska, T., 314–315
Ginsbach, J.W., 240
Ginter, J.M., 526*np*, 543–544
Giovani, B., 32
Giraud, N., 114–115, 116–117*t*, 134, 136
Giraudeau, P., 374, 383–385, 389–390, 398, 559, 561
Glad, S.S., 186–188
Glasoe, P.K., 362
Glass, T.E., 478–479
Gleixner, G., 114–115
Glowacki, D.R., 24–26, 38–39, 44–45, 47, 49–50*t*, 53, 57*t*, 62–64, 64–66*t*, 67–69, 71–72
Godin, J.-P., 114–115, 118–119*t*, 311, 314–315, 322
Goenaga Infante, H., 297, 317–318
Golbik, R., 478–479
Golden, E., 53–55
Gomez, M.V., 360–361
Gonzalez, E., 223
Gonzalez, J., 377
González-Pérez, J.A., 114–115
González-Vila, F.J., 114–115
Goodman, K.J., 502–504
Goody, R.S., 86
Görlich, S., 24, 47, 49–50*t*, 52, 65–66*t*, 67–72, 89*f*
Gorsky, L.D., 224–225, 229
Gottlieb, P.A., 538
Goyal, P., 203–204
Grand, M., 377, 397–398
Grant, G., 518
Gratton, E., 44–45
Gray, M.W., 524
Gray-Owen, S.D., 344
Greaf, V., 408–409

Greco, T.M., 526*np*, 543–544
Green, M.T., 268–269
Greenfield, N.J., 53
Gregory, R., 133–134
Gregory, R.B., 44–45
Grehn, L., 351, 353–354
Grice, K., 114–115, 116–117*t*
Griffiths, L., 560
Grimsrud, E.P., 181–182
Griswold, W.R., 5–6
Groebe, D.R., 537–538
Grosjean, H., 524
Grossfield, A., 202–203
Grostern, A., 505–506, 518
Groth, T., 415–416
Groves, J.T., 229
Grün, A., 344
Grunwald, E., 333
Grzybkowska, A., 180–210
Grzyska, P.K., 259, 437–438
Gschwend, P.M., 292
Gu, H., 435, 440–441, 453–454
Gu, X., 525–526, 526*np*
Guengerich, F.P., 218–233, 227*f*, 230*f*
Guerrero-Alvarez, J.A., 343–344
Guerriero, A., 24, 47, 49–50*t*, 52–53, 65–66*t*, 67–74, 73*t*
Guiet, S., 377, 383–385
Guillou, C., 370–371, 375, 398
Güntner, U., 421–422, 424–425
Guo, H.X., 166–167
Guo, Y., 273, 279–283, 285–286, 478–479
Guo, Z.K., 414
Guroff, G., 228
Gustavsson, E., 45–48
Gutfreund, H., 55–56, 150–151, 153
Guyer, R.B., 478–479
Gyéresi, A., 336
Gyorgydeak, Z., 337–338

H

Haderlein, S.B., 184, 303–304, 315–316
Hadjiliadis, N., 349
Haegele, K.D., 229
Haertlein, M., 45–48, 54–55
Hägele, G., 335–336
Hai, L., 297, 317–318
Hainline, A., 415–416
Haino, T., 52
Haisch, M., 130, 414
Halas, S., 181–182
Halevi, E.A., 351–352
Haley, M.M., 355
Halford, B., 233
Halford, S.E., 55–56
Halicz, L., 313, 315
Halime, Z., 276–279
Hall, L.R., 226–227
Haller, J., 299
Halstead, P., 114–115, 118–119*t*
Hamilton, C.S., 526*np*, 543–544
Hamilton, G.A., 502–504
Hammes, G.G., 67
Hammes-Schiffer, S., 44–45, 48–52, 62–63, 67–68, 203–204
Han, R., 55–56
Han, W.-G., 189–191
Hanafusa, T., 549
Handelsman, D.J., 426–428
Hanff, E., 121–123, 127–128
Hang, C., 552–553
Hankard, R., 118–119*t*, 126–127
Hanna, I.H., 230*f*
Hanoian, P., 68–69
Hansen, C., 37
Hansen, E.H., 393–394
Hansen, P.E., 51
Hanson, B.L., 51, 54–55
Hanzlik, R.P., 226–228
Harada, N., 229
Hardman, S.J.O., 24, 47, 49–50*t*, 52, 65–66*t*, 67–72, 89*f*
Hari, S.P., 348–349, 362–363
Harris, B.G., 502
Harris, M.E., 434–454, 548
Hartenbach, A.E., 141, 297–298, 317
Hartridge, H., 150–151, 153
Harvey, J.N., 194–195
Hassell, A.M., 169
Hatcher, E., 48–52
Hatton, C.K., 406–407
Hattori, A., 51
Hatzoglou, M., 434–435
Hausinger, R.P., 256–259

Hay, P.J., 196–197
Hay, S., 24, 44–45, 47, 49–50*t*, 52–53, 65–66*t*, 67–74, 73*t*, 86–109, 89–90*f*, 108*f*
Hayaishi, O., 263–264
Hayden, J.A., 253
Hayes, J.M., 120–121, 124–126, 129, 416–417
Hayes, M., 408–409, 424, 427–428
Hayon, E., 478
Hazemann, I., 54–55
Hebestreit, M., 410–411
Hedwig, N., 299–300, 306–307
Heeb, N.V., 205–206
Hehre, E.J., 555
Heimbrook, D.C., 219–220
Heinrich, F., 45–48
Heinrich, K., 410–411
Heiss, N.S., 524–525
Heller, S.R., 347–348
Henckaerts, L., 114–115, 118–119*t*
Henderson Pozzi, M., 153, 159
Hendrich, M.P., 249–251, 253
Hener, U., 410–411
Hengge, A.C., 435–438, 445
Henion, J., 154
Hennig, U., 124–126
Henri, V., 86
Henry, G.E., 562–563
Henshaw, T.F., 256–259
Hentschel, P., 360–361
Heppner, D.E., 240
Herbst, R.H., 524
Hermes, J.D., 100–101
Herschlag, D., 434–435
Hess, A.S., 343
Hettmann, E., 114–115
Hewitt, M.C., 232
Heywood, R., 426–428
Hidaka, H., 551*t*
Higashi, N., 114–115, 116–117*t*
Hilkert, A., 406–407
Hilkert, A.W., 314–315
Hill, J.W., 181–182
Himo, F., 189–191
Hinck, C.S., 151–153
Hinckley, G.T., 478
Hirel, P.H., 36–37
Hirose, K., 344
Hirotsu, K., 181, 188–189
Hirschi, J.S., 552–553
Hirschmann, H., 422–423
Hirschorn, S.K., 300, 315–316, 502–506, 518
Hites, R.A., 192
Hitzfeld, K.L., 314–315
Hjelmeland, L.M., 229
Hocart, C.H., 114–115, 116–117*t*
Hoch, J.C., 562–563
Hodges, M., 142
Hodoscek, M., 194–195
Hoffart, L.M., 240, 478–479
Hoffman, B.M., 240
Hofmann, D., 124–126
Hofstetter, T.B., 134, 141, 292–304, 305*t*, 306–308, 311–321, 323
Hogberg, K., 228
Höhener, P., 293, 299–300, 303–304, 306–307, 370, 375
Holbert, M., 4*f*, 6, 15, 17, 18–19*t*
Holden, P.J., 45–48
Hollenberg, P.F., 219–220, 228
Hollender, J., 314–315
Holliger, C., 295
Holm, R.H., 478
Holtin, K., 360–361
Holz, R.C., 36–37
Honch, N.V., 114–115, 118–119*t*
Hong, C.B., 232
Hook, W.A.V., 293
Hoopes, J.T., 45–48
Hoops, S., 94, 103, 105–106
Hopfgartner, G., 154
Horenstein, B.A., 434
Horne, T.J., 560
Horning, M.G., 229
Horning, S., 114–115, 118–119*t*, 406–407
Hosoya, K., 51
Hothi, P., 44–45
Hou, W., 4*f*, 6, 15, 17, 18–19*t*
Houk, K.N., 299
Howe, C.J., 421–422, 424–425
Howlett, M.R., 518
Hrabie, J.A., 232
Hratchian, H.P., 189–191
Hsieh, J., 55–56
Huang, D.S., 549

Huang, L.Q., 152
Huang, M., 435, 437, 453, 552–553
Huang, R., 308
Huang, X., 67–68
Hubert, P., 513–514
Hübner, G., 468–469, 478–479
Huddleston, M.J., 4*f*
Huff, G.S., 268–269
Hull, W.E., 55–56
Humphry, T., 437
Hungate, R.E., 506
Hunkeler, D., 184, 293–295, 303–304, 313, 315, 370, 502–506, 513–514, 516–519
Hunt, M.R., 415–416
Huntscha, S., 314–315
Husted, S., 114–115, 116–117*t*
Hwang, C.-C., 166–167
Hynes, J.T., 44–45, 62–63

I

Ikegami, T., 44–45, 51
Imboden, D.M., 292
Imfeld, G., 114–115
Inbal, A., 524–525
Indurugalla, D., 553
Indyk, L., 184
Ingenito, R., 36
Inouye, M., 434–435
Inouyeb, H., 36–37
Iranzo, O., 437
Irving, A.M., 560
Isaakidou, V., 114–115, 118–119*t*
Islam, Z., 69–70
Iwamoto, H., 52
Iwasaki, T., 45–48
Iwata, S., 52
Iwata-Reuyl, D., 528
Iyer, S., 437

J

Jacksier, T., 313
Jacobsen, E.N., 552–553, 556–557
Jaki, B.U., 558–560
James, P., 406–407
Jamin, E., 374, 377, 398
Jankowski, S., 550–552
Janousch, M., 345, 347
Jarkovsky, J., 205–206
Jarman, M., 229, 232
Jaroszewski, J.W., 336–337
Jarret, R.M., 169
Jasniewski, A.J., 276–279
Javier Ruiz-Pernía, J., 24–26, 38–39
Jeannerat, D., 347, 361–363
Jeannottat, S., 184, 294, 313, 315
Jencks, W.P., 164–165, 437
Jennings, R.R., 86
Jensen, A.A., 336–337
Jensen, F., 186–188
Jerina, D.M., 228
Jézéquel, T., 374, 398
Jiang W., 240
Jiménez-Morillo, N.T., 114–115
Jin, B., 184, 303–304
Jin, Y., 264–268
Jitsumori, K., 181, 188–189
Jochmann, M.A., 293–294, 311, 313–316, 322
Johannissen, L.O., 24, 44–45, 47, 49–50*t*, 52–53, 65–66*t*, 67–74, 73*t*
John, H., 54–55
Johnson, C.A., 292
Johnson, C.C., 264–268
Johnson, D.W., 355
Johnson, K.A., 25–26, 38–39, 67, 86–87, 94–95, 102
Johnson, K.S., 36–37
Johnson, L., 524
Johnson, T., 505–506, 518
Johnson, W.W., 230*f*
Johnston, E.M., 240
Johnston, J.N., 343
Johnston, M.V., 526*np*, 543–544
Jollie, D.R., 241–242, 246, 249
Jones, A.R., 102, 105–107
Jones, J.P., 224
Jones, S.R., 439
Jones, V., 124–126, 412–413
Jonsson, N.B.-H., 333
Jørgensen, K., 393–394
Jorgensen, W.L., 196–197, 199–201
Jossier, M., 140
Joubert, V., 374, 398
Jovanovic, M., 524
Juaristi, E., 337–338
Juchelka, D., 114–115, 118–119*t*, 314–315

Judson, C.M., 228
Julien, M., 114–115, 118–119*t*, 126–127, 138–139, 373–375
Jung, C., 51
Jung, K., 124–126

K

Kaden, T.A., 337
Kaerner, A., 348–349, 362–363
Kaiser, P., 513–514
Kaizer, J., 268–269
Kállay, C., 349
Kalu, U., 90*f*
Kaminski, G.A., 196–197, 199–201
Kamiński, R., 180–210, 550–552
Kanaan, N., 204–205
Kanazawa, N., 181
Kanda, H., 114–115, 116–117*t*
Kania, R., 45–48
Kannangara, R., 393–394
Kapilashrami, K., 5–6
Karadimitris, A., 524–525
Karim, A., 45–48
Karri, P., 357
Karwaski, M.F., 548–549, 553–554, 560
Kasperek, G.J., 228
Katagiri, H., 263–264
Katz, J.J., 51, 55–56
Kauffmann, K.E., 240, 256
Kaufmann, R.S., 181
Kayacelebi, A.A., 121–123, 127–128
Kazlauskas, R., 421–422, 424–428
Kedderis, G.L., 219–220, 228
Keefer, L.K., 232
Keeling, C.D., 417–419
Keffer-Wilkes, L.C., 524
Keglevich, C., 344
Kellerman, D.L., 434–435, 440–441, 453–454
Kelly, R.T., 154
Kelman, Z., 45–48
Kent, S.B.H., 30, 37
Kern, D., 478–479
Khailova, L., 478–479
Kieber-Emmons, M.T., 276–279
Kiirend, E., 336, 349, 351, 353–354
Kim, D., 219–220, 317
Kimata, K., 51
Kindman, L.A., 439
King, E.L., 86, 462–463, 480–481
King, R.F.G.J., 114–115, 118–119*t*
Kinnaman, F.S., 134
Kinsland, C., 166–167
Kinzie, S.D., 528
Kiousi, P., 413
Kioussi, M.K., 413
Kipp, D.R., 5, 18–19*t*, 24, 46, 49–50*t*, 53, 57*t*, 60, 62*t*, 64–66*t*, 66, 72–74, 73*t*
Kirby, A.J., 337–338, 435–437
Kirsch, J.F., 548
Kitamura, Y., 393–394
Kitao, A., 202
Klauck, S.M., 524–525
Klibanov, A.M., 196
Kline, P.C., 434
Klinker, E.J., 268–269
Klinman, J.P., 5, 44–45, 233, 268–269
Kluger, R., 504–519
Klvana, M., 192
Knapp, M.J., 44–45
Knight, S.W., 524–525
Knight, W.B., 350–351
Knöfel, A.-K., 121–123, 127–128
Knoot, C.J., 273–274
Knowles, J.R., 100–101, 439
Ko, Y., 472–473
Koch, P., 232
Koerner, T., 182
Kohen, A., 24–25, 44–77, 49–50*t*, 57*t*, 62*t*, 64–66*t*, 204–205, 293, 548
Kohler, H.-P.E., 134, 205–206, 292–296, 299–304, 305*t*, 306–308, 311–313, 315, 318–321
Koivusalo, T., 434–435
Kökösi, J., 336
Kollman, P.A., 202
Kołodziejska-Huben, M., 186–188
Komáromia, I., 194–195
Komor, A.J., 240–286
Konermann, L., 152
König, S., 478–479
Koop, D.R., 224–225, 229
Korboukh, V., 240
Korhonen, H., 434–435
Korzekwa, K.R., 224, 228
Kosieradzka, K., 398

Kossiakoff, A.A., 54–55
Kothe, U., 524
Koudelakova, T., 205–206
Kourteva, G., 393–394
Kovaleva, E.G., 240, 273–274
Kraszni, M., 335–336, 344
Krauser, J.A., 219–220, 222, 224, 226–227, 229, 230*f*
Kraut, J., 25–26
Krebs, C., 240, 478–479
Krempf, M., 114–115, 118–119*t*
Kresge, A.J., 164–167, 169
Krieg, P.A., 537–538
Krinsky, J.L., 345, 347
Krueger, J., 69
Krummen, M., 314–315
Krzemińska, A., 46, 49–50*t*, 62–64, 64–66*t*, 66
Kubota, K., 192
Kubyshkin, V., 337
Kuchitsu, K., 51
Kujawinski, D.M., 311, 314–315, 322
Kulik, H.J., 5
Kumar, S., 202
Kumari, K., 192–194, 206–208
Kümmel, S., 314–315
Kuo, D.J., 165
Kuperman, J., 341, 353, 356
Kurihara, T., 181, 188–189
Kurt, Z., 315–316
Küster, A., 114–115, 118–119*t*
Kuusela, S., 434–437
Kuzmic, P., 87, 94, 102
Kuzmine, I., 538
Kwan, E.E., 552–553, 556–557
Kwiecień, R.A., 188–189, 393–394

L

Laatikainen, R., 335–336
Lacrampe-Couloume, G., 300–303, 315–316, 502–508, 510–513, 516–518
Lake, V., 45–48
Lal, R., 192–194, 206–208
Lalande, J., 114–115, 118–119*t*, 126–127
Lam, M.M., 560
Lancelin, P., 393–394
Landgraf, B.J., 472
Lane, R.S., 102, 107–108, 108*f*
Langan, P., 51, 54–55
Lange, L., 120–121, 131–132, 311, 314–315
Lankin, D.C., 558–560
Lantz, M., 354
LaReau, R.D., 4–5
Larsson, R., 229–230
Laskov, C., 184, 294, 313, 315
Lassila, J.K., 434–435
Laurenti, E., 196
Laursen, K.H., 114–115, 116–117*t*
Laursen, R., 218–219, 228–229, 231–232
Laux, V., 45–48
Lawson, M., 294
Layfield, J.P., 67–69
Le Faouder, P., 398
Le Guennec, A., 114–115, 118–119*t*, 126–127, 138–139
Le, M.T., 45–48
Leatherbarrow, R.J., 36–37
Lebreton, J., 398
Leckrone, K.J., 129
Lecoultre, V., 114–115, 118–119*t*
Ledru, A., 383–385
Lee, A.L., 24, 47, 49–50*t*, 57*t*, 59, 62*t* 64*t*, 67–68, 70–72, 75–76
Lee, C., 94, 103, 105–106, 196–197
Lee, D.J., 30
Lee, F.K.N., 308
Lee, J.K., 548, 552–553
Lee, S.-K., 247–249, 253–256, 268, 271–273
Lee, W.P., 414
Lees, M., 374, 398
Lehmann, M.M., 114–120, 116–117*t*
Lei, L., 230–231, 230*f*
Lemke, S., 126–127
León-Ricardo, B.X., 524
Lesot, P., 383–387, 389–390
Lessner, D.J., 308
LeTadic-Biadatti, M.-H., 264–268
Letolle, R., 513–514
Leung, H.B., 548
Lev, O., 313, 315
LeValley, S.E., 228, 231
Lewandowicz, A., 186, 188–189, 191, 206
Lewis, A.R., 548–568
Lewis, B.E., 5, 8–9, 17, 18–19*t*, 467–468
Ley, B.A., 240

Li, F., 276–279
Li, J., 189–191
Li, N.S., 439–441
Li, Y., 393–394
Liang, X., 505–506, 518
Liao, R.-Z., 189–191
Lib, P., 36–37
Liberman, J., 356
Lidén, K., 114–115, 118–119*t*
Lieberman, S., 422–423
Lienemann, P., 205–206
Lijinsky, W., 232
Lilley, D.M.J., 434–435
Limbach, H.-H., 293, 548
Lin, L.H., 506–507
Lin, M. T., 45–48
Lin, S.-S., 164*s*
Lin, Y., 253
Lincoln, D.E., 196
Lindeke, B., 229–230
Lindquist, T., 229–230
Ling, J., 240
Link, T., 528
Linscott, J.A., 5–6
Lippard, S.J., 240, 250–251, 264–268, 273–274
Lippmaa, E., 336, 349, 351, 353–354
Lipscomb, J.D., 240–286
Liu, A.T., 346, 362
Liu, C., 393–394
Liu, C.T., 67–69
Liu, D., 36–37
Liu, H.-w., 472–473, 476–478
Liu, I.W., 51
Liu, J.O., 36–37
Liu, J.S., 297–298, 513–514
Liu, K.E., 264–268
Liu, L., 36
Liu, O., 131
Liu, X., 51, 54–55, 114–115, 118–119*t*
Liu, Y., 249, 253–256, 393–394, 525–526, 526*np*
Liuni, P., 152
Lloyd, R.V., 196
Loehr, T.M., 240
Lollar, B.S., 502–519
Long, A., 181
Long, F.A., 362
Longbotham, J.E., 24, 47, 49–50*t*, 52, 65–66*t*, 67–72, 89*f*
Longworth, M., 426–428
Lonnberg, H., 434–437
Lonsdale, R., 194–195
López-Canut, V., 204–205
Loria, J.P., 55–56
Lough, A.J., 344
Lovell, C.R., 196
Lovell, T., 189–191
Loveridge, E.J., 24–26, 38–39, 44–45, 47–53, 49–50*t*, 56, 57*t*, 59, 62–64, 62*t*, 64–66*t*, 67–69, 71–74, 73*t*
Lu, A., 548
Lu, A.Y.H., 224, 229
Lu, H.G., 273
Lu, X., 297–298, 513–514
Luk, L.Y.P., 24–26, 38–39, 44–45, 47–53, 49–50*t*, 56, 57*t*, 59, 62–64, 62*t*, 64–66*t*, 67–69, 71–74, 73*t*
Luke, A.H., 414
Lumry, R., 44–45
Luo, L., 188–189
Luz, B., 318
Luzzatto, L., 524–525

M

Ma, G., 393–394
Mabon, F., 370–372, 377
Macciantelli, D., 173–174
Maccoll, A., 502
MacColl, R., 51
Macdonald, L.E., 538
Macdonald, T.L., 218–219, 227*f*, 228, 230
Mace, R., 232
Machnik, M., 406–407
Maciejewski, M.W., 562–563
MacKenzie, P.I., 422–423
Magennis, S.W., 102, 107–108, 108*f*
Maggi, F., 303–304
Magnusson, O.T., 472
Magyar, P.M., 294
Mahanta, N., 472
Mahe, A., 142
Maier, M.P., 314–315
Maiti, K.S., 45–48
Major, D.T., 435
Makris, T.M., 240, 247–249, 273–279

Malandrinos, G., 349
Malhotra, S., 192
Malinovsky, D., 297, 317–318
Mälkönen, P., 219–220
Mallot, R.J., 344
Malone, J., 45–48
Malys, N., 94
Maniscalco, S.J., 184
Manna, R.N., 180–210
Manning, K.A., 554, 565
Mansoorabadi, S.O., 473
Manta, B., 189–191
Mantle, P.G., 381, 383
Maragliano, L., 204
Marchand, P.A., 371–372
Marcus, R., 44–45, 62–63
Mareck, U., 412–413, 415, 419, 421–422, 424–425
Mariano, M., 35–36
Mariappan, S.V.S., 333, 351
Mariotti, A., 513–514
Mark, L.C., 228
Markai, S., 371–372
Marques, A.-S., 114–115, 118–119*t*
Martí, S., 24–26, 47, 49–50*t*, 53, 56, 57*t*, 62–64, 62*t*, 64–66*t*, 67, 71–72, 195, 203–206
Martin, C.T., 538
Martin, G.E., 562
Martin, G.J., 370–371, 377
Martin, M., 370–371, 377
Martin, M.L., 370–371, 377
Martin, M.V., 230*f*
Martín, Q., 114–115, 118–119*t*
Martin, Y.-L., 377
Martineau, E., 114–115, 118–119*t*, 126–127, 374
Martinez, A., 36–37
Martinez, T.J., 5
Mas-Ku, W.P., 343–344
Mason, H.S., 263–264
Mason, P.J., 524–525
Massoud, S.S., 170–172
Masterson, J.E., 47, 62–63, 64*t*, 72, 75–76
Masuda, H., 434–435
Mathew, L., 166–167
Matico, R.E., 4*f*, 6, 15, 17, 18–19*t*
Matsen, F.A., 480–481
Matsson, O., 186, 188–189, 191, 206, 222
Matsushita, S., 45–48
Matthews, D.E., 120–121, 130, 414
Matthews, M.L., 240
Matthieu, S., 528
Maurino, V.G., 114–120, 116–117*t*
Mauve, C., 114–115, 116–119*t*, 126–127, 134, 136, 140
Maxwell, D.S., 196–197, 199–201
Mayo, K.H., 269–271
Mazurkiewicz-Łazowska, H., 184
Mbughuni, M.M., 253
McAllister, W.T., 538
McCarthy, E.L., 472
McCleland, S., 51–53
McClusky, G.A., 229
McCullagh, J.S.O., 311, 314–315, 322
McCulloch, M.T., 181
McDevitt, P.J., 4*f*, 6, 15, 17, 18–19*t*
McKay Wood, B., 169–170
McKenzie, D.A., 393–394
McLauchlan, W.R., 393–394
McLaughlin, L.W., 442–443
McMahon, R.E., 229
Mcnall, M., 313
McNaught, A.D., 549
McNeill, K., 315–316
Meckenstock, R.U., 502–506, 516–517
Mehdoui, T., 345
Mehra, A.P., 472
Meier, K.K., 240, 247–249, 253–256, 273–279
Meier-Augenstein, W., 123–124, 128–130, 297
Meijer, H.A.J., 314–315
Meilleur, F., 51, 53–55
Melander, L., 180, 230–231
Melander, L.C.S., 295, 301, 303–304, 502–504, 548, 556–557, 567
Meller, J., 51
Melton, D., 537–538
Mendes, P., 88–89, 94
Menges, S., 560
Mengesha, E., 524–525
Menten, M.L., 86
Merritt, D.A., 124–126
Mersch, F.J., 123–124, 130
Merz, K.M.J., 199–201

Mesnard, F., 297
Messiha, H., 94
Metairon, S., 114–115, 118–119*t*
Metges, C.C., 114–115, 124–127
Metral, C.J., 232
Metz, K.R., 560
Meyer, A.H., 201–202
Meyer, M.P., 268–269, 548, 552–553, 556–557
Michaelis, L., 86
Michel, E., 45–48, 50
Miertus, S., 194–197
Mikkelsen, I., 336–337
Mikkola, S., 434–435, 437
Milburn, M.V., 169
Miles, Z.D., 180
Milicevic, S.D., 363
Miller, B.G., 166, 169
Miller, G.P., 230*f*
Miller, J.P., 220
Milligan, J.F., 537–538
Minasian, S.G., 345, 347
Mincer, J.S., 44–45, 48–52
Miracco, E.J., 525–526
Miska, M.E., 192
Misumi, S., 549
Mitchell, J.R., 229–230, 524–525
Mitoma, C., 228–229, 231–232
Mitra, S., 36–37
Mitschler, A., 54–55
Mitsui, K., 114–115, 116–117*t*
Miwa, G.T., 219–220, 224, 228–229
Miyajima-Nakano, Y., 45–48
Mobli, M., 562–563
Moenne-Loccoz, P., 240
Mogi, K., 264–268
Molaioni, F., 421–422, 424–426
Molero, G., 114–115, 116–117*t*
Moles, C.G., 88–89
Moliner, V., 24–26, 38–39, 44–53, 49–50*t*, 57*t*, 59, 62–64, 62*t*, 64–66*t*, 66–69, 71–74, 73*t*, 195, 204–206
Molinié, R., 297, 398
Molt, K., 311, 314–315
Monaghan, P.J., 100–101
Mondal, B., 276–279
Monincova, M., 205–206
Monterosso, G.R., 256–259
Montigon, F., 130
Montoliu, I., 114–115, 118–119*t*
Moore, B.S., 180
More O'Ferrall, R.A., 164–165, 169
Morimoto, H., 264–268
Morokuma, K., 194–195, 264–268
Morris, E.D., 196
Morris, G.A., 560, 564
Morrison, D.J., 114–115, 118–119*t*
Morrison, J.F., 87
Morrow, J.R., 437
Mosandl, A., 410–411
Moser, D.P., 506–507
Moser, M., 114–115, 118–119*t*
Motawia, M.S., 393–394
Moulin, M., 45–48, 54–55
Moulton, B., 393–394
Mueller, E.G., 524–545, 525*f*, 526*np*
Mugridge, J.S., 355
Muir, T.W., 32, 38
Mukherjee, A., 268–269
Mulholland, A.J., 194–195
Mumbach, M.R., 524
Münck, E., 240, 247–251, 253–256, 268, 273–279
Mundle, S.O.C., 504–508, 510–513, 516–518
Murdoch, J.R., 467–469
Murkin, A.S., 5, 17, 18–19*t*, 24, 45–46, 49–50*t*, 56, 57*t*, 60–61, 62*t*, 64–66*t*, 67, 70, 73*t*, 554, 565
Musaev, D.G., 264–268
Myambo, K., 36–37
Myles, D.A.A., 51, 54–55

N

Nacev, B.A., 36–37
Nadler, A., 37
Nadler, C., 37
Nagata, K., 192
Nagata, Y., 192
Nagy, K.A., 404–406
Nagy, L.D., 219–220, 230*f*
Nakamura, N., 181
Namanja-Magliano, H.A., 6, 17, 18–19*t*
Narbad, A., 393–394
Naulet, N., 370–371
Naveo, E.P., 346, 362

Neef, H., 478–479
Neilson, A.H., 292
Nelson, J.L., 505–506, 518
Nelson, S.D., 229–230, 232
Nelson, S.T., 505–506
Nesheim, J.C., 247–249, 253–256, 263–264, 268
Neto, E.V.S., 294
Neumann, A., 297–298
Newcomb, M., 264–268
Nicell, J.A., 196
Niemann, C., 86
Nijenhuis, I., 192, 314–315
Nilsson, U., 229–230
Nishikaze, O., 408–409
Nishino, S.F., 315–317
Nitoker, N., 435
Nock, S., 35–36
Nogues, M.V., 434–435
Nogués, S., 114–115, 116–117*t*
Noisier, A.F.M., 36
Noodleman, L., 189–191
Nordlund, P., 250–251, 273–274
Northcott, D., 351–352
Northrop, D.B., 2–3, 60–61, 69, 134, 220, 222, 224–226, 295, 460, 465–466, 469
Noszál, B., 335–336
Nowlin, J., 229
Noyes, R.M., 462–463, 467–469, 487
Numata, M., 181
Nun, P., 375
Núñez, S., 48–52, 62–63, 74
Nurmi, K., 437
Nussim, M., 351–352

O

O'Hara, J.P., 114–115, 118–119*t*
O'Leary, M.H., 134–136
Oakley, A.J., 192
Ochiai, N., 114–115, 116–117*t*
O'Connell, E.L., 165
O'Donoghue, A.C.M., 164*s*
Oeßelmann, J., 406–407
Offer, J., 37
Ogasawara, Y., 472–473
Ogliaro, F., 261–262
Oh, N.Y., 268–269
Ohta, B.K., 340–341, 352–353, 356
Oivanen, M., 434–437
Okai, M., 192
Okamoto, Y., 202
Okamao, M., 406–407
Okazaki, O., 219–220, 228, 230
O'Leary, M.H., 4–5, 100–101, 502–508, 513, 519
Olkhov-Mitsel, E., 152
Olsen, C.E., 393–394
Omi, R., 181, 188–189
Omura, T., 228–229
Onasch, O.F., 347–348
Opfermann, G., 419, 422–426
Orellana, A., 152
Oremland, R.S., 506
Orgován, G., 336
Orphan, V.J., 294
Ortegón-Reyna, D., 363–364
Ortigues-Marty, I., 114–115, 118–119*t*
Ortiz de Montellano, P.R., 218–219, 227*f*, 228
Ösz, K., 349
Ottesen, J.J., 32
Otyepka, M., 192
Ouetrani, M., 383–385

P

Pabis, A., 293–294, 299, 301, 306–308, 550–552
Padilla-Martínez, I., 363–364
Page, J.S., 154
Pahle, J., 94, 103, 105–106
Paiva, A.A., 152
Palfey, B.A., 184
Pallan, P.S., 230–231, 230*f*
Palmer, M.R., 562–563
Pandey, G., 192–194, 206–208
Pandey, R., 192–194, 206–208
Paneth, P., 2, 4–5, 17, 44–45, 62–63, 180–210, 293–294, 299–301, 303–304, 306–308, 317, 354, 437, 502, 550–552
Pang, J., 100–101
Paolini, M., 114–115, 116–117*t*
Papajak, E., 188–189
Parales, J.V., 308
Parales, R.E., 134, 292–296, 299–304, 305*t*, 306–308, 311, 313, 321

Parinet, J., 375
Park, Y., 552–553, 556–557
Parkin, D.W., 548
Parmentier, L.E., 502
Parmerter, S., 51, 55–56
Parr, A.J., 393–394
Parr, R.G., 196–197
Parrinello, M., 203–204
Parrish, R.M., 51
Parsons, E., 393–394
Pascal, R.A., 549
Passeport, E., 505–506
Pataki, D.E., 417–419
Patel, M., 478–479
Pati, S.G., 134, 292–323, 305*t*
Patskovsky, Y., 174
Pauli, G.F., 558–560
Paulsen, H., 337–338
Paulson, J.L., 543
Pauwels, E., 186–188
Pearson, J.A., 114–115, 118–119*t*
Peck, S.C., 478–479
Pehk, T., 336, 349, 351, 353–354
Pei, D., 36–37
Peiser, H.S., 551*t*
Pelc, A., 181–182
Pellechia, P.J., 51
Penning, H., 201–202
Peralta-Cruz, J., 363–364
Peraza-Campos, A., 363–364
Pereira, W.E., 229
Perel, J.M., 228
Perera, H., 363
Perez, M., 560
Perez-de-Mora, A., 505–506, 518
Perler, F.B., 35–37
Perrin, C.L., 51–52, 332–364
Pesch, R., 314–315
Pessi, A., 36
Peters, M., 337
Peterson, E., 263–264
Peterson, L.A., 231
PetitClerk, C., 421–422
Petterson, K.A., 349–350
Petzke, K.-J., 114–115, 124–127
Peuker, S., 45–48
Peynot, P., 36–37
Philipp, D.M., 194–195, 264–268
Phillips, R.S., 90*f*
Philmus, B., 472
Piasecki, A., 294
Piazza, G.J., 502
Piccirilli, J.A., 434–435, 439–443, 453–454
Piecha, S., 410–411
Pikus, J.D., 240
Pilkington, S.J., 249
Piper, T., 114–115, 118–119*t*, 404–429
Platen, P., 412–413, 415, 419, 421–422, 424–425
Plotnick, H., 232
Poiger, T., 192
Polenova, T., 562–563
Poulin, M.B., 2–20, 4*f*, 18–19*t*
Poulos, T.L., 261–262
Pouteau, E., 114–115, 118–119*t*
Powell, M.F., 164–167, 169
Prabhu, K.S., 240, 478–479
Prache, S., 114–115, 118–119*t*
Prasolov, I.S., 421–422
Pratt, D.A., 552–553
Preiss Weigert, A., 410–411
Press, W.H., 87–89
Preston, R.J., 45–48
Preston, T., 114–115, 118–119*t*
Previs, S., 445
Price, J.C., 240, 478–479
Priestley, N.D., 264–268
Prohaska, T., 296–297
Proshlyakov, D.A., 256–259, 261–262
Proshlyakov, Y., 256–259, 261–262
Proust-De Martin, F., 195
Przibilla, G., 410–411
Pudney, C.R., 24, 44–45, 47, 49–50*t*, 52–53, 65–66*t*, 67–74, 73*t*, 89*f*, 102, 107–108, 108*f*
Puiu, S.C., 276–279
Purcell, J., 437–438

Q

Qayyum, M.F., 240, 276–279
Qi, H., 297, 314–315
Qi, Y.-K., 36
Qin, J., 192
Qiu, S., 295
Quaytman-Machleder, S., 44–45
Que, L., 256, 273, 276–283, 285–286
Quinn, D.M., 44

R

Rabenstein, D.L., 333, 337, 348–349, 351, 362–363
Radak, B.K., 435, 440–441, 453–454
Rademann, J., 36
Radhakrishnan, R., 250–251
Radkiewicz, J.L., 62–63
Ragnarsson, U., 336, 349, 351, 353–354
Raines, R.T., 434–435
Rajagopalan, P.R., 44–45, 62–63
Rajakovich, L.J., 478–479
Ramamurthy, V., 543
Ramseyer, J., 422–423
Rao, P.S., 478
Rao, P.V., 478
Rao, X., 393–394
Rapoport, H., 218–219, 228–229, 231–232
Raposo, C., 114–115, 118–119*t*
Rathgeber, S., 507–508, 510–512
Ratti, M., 315–316
Raushel, F.M., 189–191
Rauth, S., 407
Rawlings, J., 437–438
Ray, W.J., 466–469, 488
Raymond, K.N., 355
Raz, R., 36
Rebek, J., 52
Rebelo, L.P.N., 293
Rechavi, D., 52
Recio, C., 114–115, 118–119*t*
Reddy, P.T., 45–48
Reed, R.I., 502–504
Reichardt, N., 114–115, 118–119*t*
Reichert, P., 304
Rekas, A., 45–48
Remaud, G.S., 142, 370–398
Rendic, S., 219
Renger, T., 44–45, 62–63
Renpenning, J., 314–315
Renson, J., 228
Rentergent, J., 102–107
Rentsch, D., 192
Rettie, A.E., 224
Reyes, A., 363
Richard, J.P., 164–174, 164*s*, 167*s*, 437
Richards, R., 142
Richelle, M., 114–115, 118–119*t*
Richnow, H.-H., 192, 299–300, 306–307, 314–315
Rickert, K., 44–45
Rieley, G., 123–124, 130, 412–413
Riemann, P., 419
Riggs, C.W., 232
Riggsa, P.D., 36–37
Riley, W.J., 303–304
Rinaldo, D., 264–268
Rinaldo-Matthis, A., 17
Rinehart, J.D., 345, 347
Rinne, K., 114–115, 116–117*t*
Risatti, J.B., 505–506
Rishavy, M.A., 437–438
Rittle, J., 268–269
Rivard, B.S., 240, 247–249, 273–283, 285–286
Rivero, I.A., 342, 360–361
Robb, M.A., 17, 194–197
Roberts, J.D., 333, 346, 349–350, 362
Roberts, K.M., 149–161
Robertson, R.E., 351–352
Robins, R.J., 114–115, 118–119*t*, 128–129, 142, 371–375, 377, 381, 383–385, 387, 389–394, 397–398
Robinson, J.L., 165
Robinson, M.J.T., 350
Roca, M., 24–26, 38–39, 44–45, 47, 49–50*t*, 53, 57*t*, 62–64, 64–66*t*, 67–69, 71–72, 203–204
Rodbourne, L., 422–423
Rodchenkov, G., 414, 421–422
Rodgers, L.R., 549
Roeske, C.A., 100–101, 502
Rogerson, J.H., 424–425
Rohde, J.-U., 268–269
Rojas, J.M.M., 375, 398
Rojas-Duran, M.F., 524
Rokop, S., 51, 55–56
Rolle, M., 184, 303–304
Roman, J.M., 232
Romek, K.M., 114–115, 118–119*t*, 126–127, 138–139
Romero, J., 363
Ron, A., 351–352
Roos, N., 311, 314–315
Rose, I.A., 60–61, 69, 165
Rosenberg, J.M., 202
Rosenberg, S., 548
Rosenzweig, A.C., 250–251, 273–274
Rosivatz, E., 36–37

Rosman, K.J.R., 551*t*
Rosner, M., 296–297
Ross, A.B., 114–115, 118–119*t*
Ross, D., 229–230
Rosset, R., 114–115, 118–119*t*
Roth, J.P., 268–269
Rothberg, S., 263–264
Roughton, F.J.W., 150–151, 153
Rousseau, D.L., 253
Rovnyak, J., 562–563
Roy, R., 558
Ruben, D.J., 562
Rudziński, J., 182, 186, 188–189, 191, 206
Ruffner, D., 541
Ruiz, F., 54–55
Ruiz-Pernía, J.J., 24–26, 38–39, 44–45, 47–53, 49–50*t*, 56, 57*t*, 59, 62–64, 62*t*, 64–66*t*, 67–69, 71–74, 73*t*, 186–188, 203–205
Rundel, P.W., 404–406
Rusu, A., 336
Ruszczycky, M.W., 2–3, 460–496, 548
Ruterjans, H., 54–55
Ryberg, P., 186–189, 191, 206
Ryle, M.J., 256–259

S

Sachs, D.H., 347–348
Saen-Oon, S., 44–45
Sagatys, D.S., 164–165
Sage, J.T., 262
Sahle, S., 94, 103, 105–106
Sahlin, M., 240
Said Ahmad, W., 313
Saikia, M., 434–435
Samokyszyn, V.M., 196
Sanchez-Martinez, M., 203–204
Sander, M., 317
Sanders-Loehr, J., 240
Sano, M., 120–121
Santi, D.V., 525–526, 526*np*
Sapienza, P.J., 24, 47, 49–50*t*, 57*t*, 59, 62*t*, 64*t*, 67–68, 70–72, 75–76
Sasaki, S., 120–121
Sathyamoorthy, B., 554, 565
Sato, H., 230*f*
Sato, R., 228–229
Sauer, P.E., 297
Saugy, M., 406–407, 410–414, 416–417, 421–424, 426–428
Saugy, T., 415–417, 421–422, 424–427
Saunders, M., 169
Saunders, W.H., 180, 230–231, 295, 301, 303–304, 502–504
Saunders, W.H.J., 548, 556–557, 567
Saurer, M., 114–115, 116–117*t*
Sawada, M., 549
Sawaya, M.R., 25–26
Sazinsky, M.H., 240
Scarso, A., 52
Schachman, H.K., 502
Schänzer, W., 114–115, 118–119*t*, 406–408, 412–413, 417–419, 421–426
Schechter, A.N., 347–348
Scheidt, W.R., 262
Schellenberger, A., 468–469
Schimmelmann, A., 293–294, 297, 313–315
Schindelka, J., 192
Schiphorst, A.-M., 114–115, 118–119*t*
Schlegel, H.B., 17, 189–191, 194–197
Schlichting, I., 54–55, 240
Schlüter, H.-J., 314–315
Schmidt, C., 120–121, 131–132
Schmidt, H., 240
Schmidt, T., 293–294, 311, 313–315
Schmidt, T.C., 294, 311, 313–316, 322, 502–506, 516–517
Schmitter, M.J., 36–37
Schneck, J.L., 4*f*, 6, 15, 17, 18–19*t*
Schneewind, O., 37
Schneider, G., 478–479
Schneiter, P., 114–115, 118–119*t*
Schneller, M., 283
Schoell, M., 505–506
Schoeller, D., 414
Schoenberger, T., 560
Schoenborn, B.P., 54–55
Schouten, S., 130–131
Schowen, K.B., 169, 271–273, 476*np*
Schowen, R.L., 100–101, 164–165, 169, 271–273, 468–469, 476*np*
Schramm, V.L., 2–20, 18–19*t*, 24, 44–53, 49–50*t*, 56, 57*t*, 59–64, 62*t*, 64–66*t*, 66–68, 70–76, 73*t*, 434, 467–468, 548
Schur, M., 553
Schürner, H.K.V., 188–189, 201–202

Schütz, A., 478–479
Schuyler, A.D., 562–563
Schwartz, H.M., 347–348
Schwartz, S., 524
Schwartz, S.D., 24, 44–52, 49–50*t*, 57*t*, 62–64, 62*t*, 64–66*t*, 67–70, 72, 74–76
Schwarzenbach, R.P., 141, 292–293, 295, 297–298, 317, 505–506, 513–514
Schwert, G.W., 86
Scott, K.M., 297–298, 513–514
Scott, L.G., 526, 533
Scrocco, E., 194–197
Scrutton, N.S., 24, 44–45, 47, 49–50*t*, 52, 65–66*t*, 67–72, 89*f*, 102, 105–108, 108*f*
Scuseria, G.E., 17, 194–197
Seconi, G., 173–174
Seffernick, J.L., 188–189, 201–202
Seltzer, S., 502–504
Sen, A., 548
Senevirathne, C., 5–6
Seoane, F., 560
Serhan, Z., 383–385
Sessions, A.L., 294, 313, 416–417, 502–504
Shahar, A., 181
Shaik, S., 261–262
Shanklin, J., 240, 250–251
Shanmugam, M., 240
Shao, L., 232
Sharma, P., 192–194, 206–208
Sharpless, K.B., 299
Shawar, L., 313
Shen, H., 393–394
Shepard, E.M., 472
Sheraga, H.A., 347–348
Sherrill, C.D., 51
Sherwood Lollar, B., 300–303, 315–316, 502–506, 510–512, 516–518
Shi, W., 17
Shidmoossavee, F.S., 553, 560
Shimizu, K.D., 51
Shin, H., 524
Shinebarger, S.R., 130, 414
Shinkyo, R., 230*f*
Shivapurkar, R., 347, 361–363
Short, S.A., 169
Shouakar-Stash, O., 181, 184, 294, 313, 315
Shrager, R.I., 347–348
Shu, L., 256
Shvarev, A., 363
Sicińska, D., 188–191
Siders, S., 35–36
Siegbahn, P.E.M., 189–191, 264–268
Siegel, J.S., 342–343
Siegfried, L., 337
Siegwolf, R., 114–115, 116–117*t*
Siegwolf, R.T.W., 114–120, 116–117*t*
Sieper's, H.-P., 120–121, 131–132
Sigel, H., 170–172
Silla, E., 204
Silva, R.G., 5, 18–19*t*, 24, 45–46, 49–50*t*, 53–56, 57*t*, 60–61, 62*t*, 64–66*t*, 66–67, 70, 72–74, 73*t*
Silver, S.C., 472, 478
Silverman, D.N., 54–55
Silvestre, V., 142, 373–375, 387, 389–394, 397–398
Simic, M., 478
Simon, A.E., 45–48
Simus, N., 94, 103, 105–106
Singh, P., 24, 44–45, 47, 49–50*t*, 53, 57*t*, 59, 62*t*, 64–66*t*, 67–72, 75–76
Singh, S.K., 525–526, 525*f*, 526*np*, 528, 532–533, 536, 542–544
Singleton, D.A., 295, 299, 370–371, 548–549, 552–553
Sinnott, M.L., 333, 340
Sipes, I.G., 232
Siwek, A., 181, 188–189
Sjöberg, B.M., 240
Skarpeli-Liati, M., 134, 315–316
Slater, G.F., 505–507, 515–516, 518
Sleep, B.E., 518
Sligar, S.G., 219–220, 240
Small, A.C., 114–115, 118–119*t*
Smith, D.B., 36–37
Smith, M.D., 51
Smith, R.D., 154
Smith, W., 506
Sobolevsky, T., 414, 421–422
Soininen, P., 335–336
Solberg, H.E., 421–422
Söll, D., 524
Solomon, E.I., 240
Somerville, C., 240, 250–251
Sommer, H., 373–374, 377, 393–394, 397–398

Sommer, K.R., 229
Somogyi, B., 44–45
Song, W.J., 268–269
Soons, P.A., 231
Soudackov, A.V., 48–52
Southworth, M.W., 35–37
Sowa, G.A., 437
Spahr, S., 314–315
Spain, J.C., 315–317
Speciale, G., 553, 560
Spedaliere, C.J., 543
Spehar, A., 219–220
Sprang, S.R., 133–134
Stafford, M., 229
Stambaugh, R., 536
Starcevic, B., 406–407
Statham, C.N., 229–230
Steffan, R.J., 240
Steffens, R.J., 342–343
Stein, R.L., 467–469
Steinhauser, L., 360–361
Stenman, E., 437
Stephan, J.R., 30
Stern, A.S., 562–563
Stevens, J.D., 229, 232
Stewart, R.J., 68–69
Still, C.J., 417–419
Stojkovic, V., 548
Strahm, E., 423–424
Stranks, S., 422–423
Strassner, T., 299
Strathmann, T.J., 317
Stratton, C.F., 2–20, 18–19*t*
Strobel, S.A., 434–435
Stromberg, R., 437
Stuart-Williams, H., 114–115, 116–117*t*
Stubbe, J., 240
Stubna, A., 268–269
Stucki, G., 205–206
Studts, J.M., 240
Styring, A.K., 114–115, 116–119*t*
Suarez, D., 199–201
Suarez, J., 46, 48–52, 49–50*t*, 56, 57*t*, 62–64, 62*t*, 64–66*t*, 67, 72–74, 73*t*
Sudmeier, J.L., 333
Sugita, Y., 202
Suiter, C.L., 562–563
Sullivan, H.R., 229
Sun, X.Y., 370–371
Surerus, K.K., 249–251
Sutcliffe, M.J., 44–45, 100–101
Swann, P.F., 232
Swann, S.L., 543
Swanson, P., 205–206
Swendsen, R.H., 202
Świderek, K., 2, 45–46, 49–50*t*, 62–64, 64–66*t*, 66, 68–69, 180
Swinney, D.C., 228
Sydor, J.R., 35–36
Sykes, B.D., 55–56, 335–336
Sykora, S., 560
Szakács, Z., 335–336
Szatkowski, Ł., 180–210
Szöcs, L., 336
Szu, P.-H., 472–473
Szymanski, M.J., 549, 552–553
Szyperski, T., 554, 565

T

Tagg, J., 228–229, 232
Takahara, P.M., 250–251
Takai, Y., 549
Takeya, T., 552–553
Tamanaha, E., 478–479
Tanabe, M., 228–229, 231–232
Tanaka, K.S.E., 548
Tanaka, N., 360
Tanaka, Y., 555
Tang, K., 154
Tang, S., 36, 505–506, 518
Tanifum, E.A., 437–438
Tanokura, M., 192
Tao, W., 555
Tardieux, A., 513–514
Tate, E.W., 36–37
Tauriello, C.L., 228
Taylor, E.A., 17
Taylor, J.W., 181–182
Taylor, R.R.R., 344
Tcherkez, G., 114–142, 116–117*t*, 387, 389–390
Tea, I., 114–142, 116–119*t*, 374
Teixidor, P., 114–115, 116–117*t*
Tenailleau, E., 373–376, 393–394
Tentscher, P.R., 141
Teukolsky, S.A., 87–89

Thenot, J.-P., 229
Thevis, M., 114–115, 118–119*t*, 404–429
Thibaudeau, C., 337–338, 374
Thomas, A., 195, 407, 410–417, 421–422, 424–428
Thomas, A.A., 295, 299, 370–371, 548–549, 552–553
Thomas, F., 374, 398
Thomas, P., 229, 232
Thompson, M.K., 188–189, 196–198
Thordarson, P., 344
Thuer, M., 205–206
Thuéry, P., 345
Tiainen, M., 335–336
Tian, G., 467–468, 470–471
Tienpont, B., 114–115, 116–117*t*
Tilton, R.F., 152
Tinberg, C.E., 240
Tirado-Rives, J., 196–197, 199–201
Tittmann, K., 478–479
Tobias, H.J., 413
Tobin, A.E., 437–438
Tobler, N.B., 297–298
Toivola, S., 434–435
Tolbert, T.J., 36–37, 526, 533
Tomanicek, S.J., 54–55
Tomasi, J., 194–197
Tomaszewski, J.E., 228
Tomchick, D.R., 268–269
Toney, M.D., 5–6, 24, 46, 49–50*t*, 57*t*, 60, 62*t*, 65–66*t*, 67, 70, 94
Tonge, P.J., 354
Tormos, J.R., 151–153
Toro, R., 174
Toteva, M.M., 164*s*
Tóth, G., 336
Toth, K., 167, 170–174
Toy, P.H., 264–268
Träger, L., 404–406
Trager, W.F., 224, 228–230, 232
Trentham, D.R., 439
Tresca, B.W., 355
Trevor Forsyth, V., 45–48
Tronstad, L., 186, 188–189, 191, 206
Trotta, F., 196
Trout, G.J., 421–422, 424–425
Trucks, G.W., 17, 194–197
Trudell, J.R., 229
Truhlar, D.G., 44–45, 62–64, 189–191
Truyols, V., 114–115, 116–117*t*
Tsai, M.D., 437–438
Tsang, W.-Y., 164–166, 167*s*, 168, 173
Tsikas, D., 121–123, 127–128
Tsuji, Y., 164*s*
Tuñón, I., 24–26, 38–39, 44–45, 47–53, 49–50*t*, 56, 57*t*, 59, 62–64, 62*t*, 64–66*t*, 67–69, 71–74, 73*t*, 188–189, 195, 203–206, 208–209
Turgeon, A., 134, 315–316
Turich, C., 313
Turner, A., 195–197, 199–201
Turowski, M., 51
Tuthorn, M., 114–115, 116–117*t*
Twin, H.C., 344
Tyka, R., 335–336
Tyler, P.C., 5, 18–19*t*
Tyler, S.C., 134
Tyliszczak, T., 345, 347
Tynkkynen, T., 335–336

U

Udenfriend, S., 228
Ueki, M., 406–407
Uhlemann, K., 478–479
Uhlenbeck, O.C., 537–538
Ullrich, V., 228–229
Ulrich, E.N., 192
Urbauer, J.L., 502
Urey, H. C., 501–502

V

Vaishnav, Y., 229–230
Vajiala, G., 419
Valastyán, Z., 349
Valcour, A., 335–336
Valentine, A.M., 264–268
Valentine, D.L., 134
Van Acker, M.R.M.D., 181
Van der Linde, W., 351–352
van Dongen, B.E., 130–131
Van Hook, W.A., 180
van Ingen, H., 45–48
Van Lanen, S.G., 528
Van Neck, D., 186–188
Van Speybroeck, V., 186–188
Van H.K.M., 276–279

Vanbreukelen, B., 184
Vanden-Eijnden, E., 204
Vanderslice, C.W., 219–220
Vandersteen, A.A., 504–506, 510–512, 516–518
VanStone, N., 301–303
Vardi-Kilshtain, A., 435
Varney, M., 232
Veerareddygari, G.R., 524–545, 525*f*, 526*np*
Veneziano, S.E., 472, 478
Vengosh, A., 181
Venkatasubban, K.S., 100–101
Verbeke, K., 114–115, 118–119*t*
Vestergaard, P., 408–409
Vetterling, W.T., 87–89
Viola, R.E., 51, 54–55
Vizthum, C.A., 526*np*, 543–544
Vladimirova, A., 174
Vogl, J., 296–297
Vogt, C., 299–300, 306–307
Volders, F., 120–121, 131–132
von Gunten, U., 292
Vrielink, A., 53–55
Vu, V.V., 240, 247–249, 273–279
Vulliamy, T.J., 524–525
Vyazmensky, M., 478–479

W

Wachs, T., 154
Wackett, L.P., 188–189, 201–202
Wade, D., 232
Wadt, W.R., 196–197
Wagner, C.K., 549
Wakarchuk, W., 553
Walker, S.C., 55–56
Wallar, B.J., 240, 249–251, 269–271
Walsh, C., 225–226
Walsh, J.S., 219–220, 224, 228–229
Waltho, J.P., 24, 47, 49–50*t*, 52–53, 65–66*t*, 67–74, 73*t*
Walton, N.J., 393–394
Wan, W., 4–5
Wang, B., 445
Wang, C., 230–231, 230*f*, 478–479
Wang, J., 114–115, 118–119*t*
Wang, S.C., 472
Wang, W.-H., 333
Wang, X., 100–101
Wang, Y.-T., 37
Wang, Z., 5–6, 8–9, 24, 47, 49–50*t*, 53, 57*t*, 59, 62*t*, 64–66*t*, 67–68, 70–72, 75–76, 131
Wang, Z.-P., 36
Ward, J.A., 505–507
Warnecke, G., 121–123, 127–128
Warshel, A., 44–45, 62–63, 194–195
Waterman, M.R., 230–231, 230*f*
Watt, E.D., 55–56
Watts, D.G., 87–88
Weaver, L.T., 114–115, 118–119*t*
Webb, A.G., 560
Webb, E.C., 114–115, 118–119*t*
Webb, M.R., 439
Webb, T.H., 344–345
Wegener, F., 114–120, 116–117*t*
Wehrli, B., 292
Wei, L., 393–394
Weisblat, D.A., 104
Weiss, P.M., 350–351, 502
Weissman, B.P., 439–441
Wen, W.W., 344
Werner, R.A., 114–115, 116–117*t*, 297, 312–313, 315–316, 318–321, 380
Westaway, K.C., 182, 222
Westgard, J.O., 415–416
Westgate, T.D., 505–506
Westheimer, F.H., 164, 167, 230–231, 435–437, 502–504
Westwood, J.H., 229, 232
White, R.D., 232
White, R.E., 220, 229
Whittington, D.A., 264–268
Widersten, M., 186, 188–189, 191, 206
Wiemann, S., 524–525
Wiese, F.W., 196
Wijker, R.S., 293–294, 299–300, 303–304, 306–307, 315–317, 323
Wilce, M.C.J., 192
Wilcox, C.S., 344–345
Wilcox, J.L., 434–435
Wilde, K.L., 45–48
Wilkinson, A., 549
Wilkinson, E.C., 261–262
Willach, S., 311, 314–315
Willett, K.L., 192

Williams, A., 437
Williams, I.H., 186–191, 195, 204–205
Williams, J.S., 422–423
Williams, K.B., 164–165, 164*s*
Williams, N.H., 437–438
Williams, P.G., 264–268
Williams, S.J., 553, 555, 560
Williams, T.C., 335–336
Williamson, J.R., 526, 533
Wilson, D.J., 152
Wilson, D.W., 536
Wilson, J.T., 502–506, 516–517
Winfield, C.J., 292
Wing, C., 48–52, 62–63, 74
Winston, J.M., 232
Winter, J.M., 180
Wist, J., 363
Witherell, G.W., 537–538
Withers, S.G., 555
Witkop, B., 228
Wolansky, A., 407
Wolfenden, R., 166, 169
Wolfsberg, M., 180, 293
Wong, C.-H., 36–37
Wong, F.M., 166, 167*s*, 173
Wong, K., 313
Wong, K.Y., 435, 437, 440–441, 453–454
Wong, L.C.K., 232
Wood, B.M., 166–167, 167*s*, 169–174
Wood, E., 524–525
Woodin, S.A., 196
Woodworth, R.C., 335–336
Worthington, J.M., 506
Woscholski, R., 36–37
Wu, W., 166, 167*s*, 173

X

Xia, X., 194–195
Xiang, S.H., 552–553, 556–557
Xiao, Q., 36–37
Xiao, Y.K., 181
Xing, G., 240, 478–479
Xing, Y., 114–115, 118–119*t*
Xu, L., 435
Xu, M.Q., 35–37
Xu, R., 36–38
Xu, Y., 114–115, 118–119*t*, 435

Y

Yagi, H., 228
Yakir, D., 417–419
Yamada, H., 502
Yamada, K., 373–374, 393–394
Yamakawa, N., 51
Yamashita, K., 181
Yan, L.Z., 37
Yang, C.S., 232
Yang, H., 393–394
Yang, J.-R., 353
Yang, R., 114–115, 118–119*t*
Yang, W., 114–115, 118–119*t*, 196–197
Yapp, C.J., 502
Yarnell, A.T., 232–233
Yasuda, D.M., 228–229, 231–232
Yeh, S.-R., 253
Yi, Y., 505–506
Yoder, R.A., 343
York, D.M., 434–454
Yoshida, N., 373–374, 393–394
Yoshimoto, F.K., 219–220, 230–231, 230*f*
Yoshizawa, K., 264–268
Yotsui, Y., 120–121
Younas, M., 435–437
Young, E.D., 181
Yousefi-Salakdeh, E., 437
Yu, C.-L., 308
Yu, L., 51–53
Yun, C.-H., 219–220, 228, 230, 230*f*

Z

Zakharov, L.N., 363
Zakon, Y., 313, 315
Zalatan, J.G., 434–435
Zang, Y., 261–262
Zawacki, F.J., 344–345
Zech, M., 114–115, 116–117*t*
Zektzer, A.S., 562
Zeyer, J., 293–294, 315–316, 323
Zhang, B., 478–479
Zhang, B.-L., 370–371
Zhang, C.G., 181
Zhang, F., 36–37
Zhang, H., 45–48, 131
Zhang, J., 5
Zhang, L., 37, 311, 314–315, 322

Zhang, N., 192
Zhang, S., 151–153, 203–204, 435, 440–441, 453–454
Zhang, W., 230–231
Zhang, Y., 5, 17, 413
Zhao, C., 51
Zhao, H.M., 273
Zheng, H., 268–271
Zheng, J.-S., 36
Zhou, Y., 114–115, 116–117*t*, 538
Ziller, L., 114–115, 116–117*t*
Zindel, D., 205–206
Zinovjev, K., 188–189, 203–204, 208–209
Zinshteyn, B., 524
Zoi, I., 46, 49–50*t*, 57*t*, 62–63, 62*t*, 64–66*t*, 67, 74
Zwank, L., 295, 315, 505–506, 513–514

SUBJECT INDEX

Note: Page numbers followed by "*f*" indicate figures, "*t*" indicate tables, "*b*" indicate boxes, "*np*" indicate footnotes, and "*s*" indicate schemes.

A

Acetylation, 412–414
  monosaccharides, 130–131
Acid catalysis
  neutral RNA, 436*f*
  2′-*O*-transphosphorylation, 444*f*
Acylal mechanism, 525–526
AgCl
  FAB spectrum, 183*f*
  total ion current and isotope ratio, 183*f*
A-glucuronide (A-GLUC), 409, 425*f*
Alanine racemase (AR), 46–47*t*, 60
Amino acid-specific labeling, 49–50*t*
  *Hs*PNP, 74
Ammonium acetate, 152–153, 442
Analyte, 122, 130
Anomeric effect, 337–338
Anomeric equilibrium, 340–341
Anticodon stem loop (ASL), 526*np*, 539*f*, 543–544
Anti-Doping Testing Figures, 404
Apparent isotope effects, 180, 295
Apparent kinetic isotope effects (AKIEs), 298–299
  enzymatic oxygenations
    product, 300
    substrate, 298–300
Arylamine oxygenase, 273, 275–276, 279–282
Aspartic acid (Asp), 524
Atomic properties
  CmlI oxygenase
    Mössbauer spectroscopy, 273–274
    reactive peroxo intermediate preparation, 275–276
  methane monooxygenase, 249–256
    Mössbauer spectroscopy, 249–251
Atrazine hydrolysis, 199–201, 200*s*

B

Bacillus stearothermophilus DHFR (*Bs*DHFR), 53, 63–64
BDE. *See* Bond dissociation energy (BDE)
BIEs. *See* Binding isotope effects (BIEs)
Bigeleisen's derivation, 514*f*, 517*f*
Binary complex association, 14–15
Binding isotope effects (BIEs)
  binary complex association, 14–15
  energetic considerations, 3*f*
  enzyme–substrate interactions, 2
  equilibrium measurement using RED device
    scintillation counting, 16
    spectral deconvolution, 16
    substrate equilibration time determination, 12–13
    substrate stability determination, 13–14
  equipment, 6
  ground state, 2–3
  interpretation and examples, 17–19, 18–19*t*
  kinetic isotope effects, 2–3
  materials, 6–9
  Michaelis complex, 2
  ternary complex association, 15
  transition state, 2
  ultrafiltration approach, 5
Biotransformation, ferulic acid to vanillin, 393–397
Bligh–Dyer method, 121–123
Bond dissociation energy (BDE), 240
  chiral ethane, 264–267
  P and Q formation, 271–273
  Q reactions
    deuterated methane isotopomers, rate constant for, 268–269
    quantum tunneling in, 269–271
  size selectivity detection, 269–271
  source of oxygen atom, 263–264

Born–Oppenheimer (BO) approach, 45
Born–Oppenheimer approximations, 189–191
Buffers, 7, 7–8*t*
Butyl rubber gas-tight stoppers, 505–506

C

^{13}C-AKIE, 299
CAM. *See* Chloramphenicol (CAM)
Capsicum frutescens, 371–372, 372*f*
Carbon isotope ratios (CIR), 406–407, 411*f*, 416*f*, 418*f*
5-Carboxyvanillate decarboxylase (CVDB)
  decarboxylation, 174–175*s*
  PDIE, 174
  SDIE, 174
Catalysis, 44–45
^{12}C/^{13}C isotope effect, 140–141
^{13}C/^{12}C isotope ratios, 137, 512–513, 515–517
Certified reference material (CRM), 413
CFMS. *See* Continuous-flow mass spectrometry (CFMS)
^{13}C fractionation, 389–390
Chiral ethane, 264–267
Chloramphenicol (CAM), 273
  reaction catalyzed by CmlI, 273*f*
  source of oxygen, 279–282
Chlorinated reactants, 184
  DhlA dehalogenase, 184–186
Chlorine isotopic composition
  fast-atom bombardment-isotope ratio mass spectrometer, 182–184
  inductively coupled plasma, 181
  isotope ratio mass spectrometry, 181
Chlorine kinetic isotope effects (Cl-KIEs)
  active site model, 192*f*
  apparent sotope effects, 180
  chlorine isotopic composition, 181–184
  enzyme-catalyzed dehalogenations, 180
  heavy-atom isotope effects, 180
  intrinsic isotope effects, 180
  multiply chlorinated reactants, 184–186
  QM/MM optimization calculations, 209
  theoretical predictions, 197*t*
    cluster model approach, 189–194
    full enzymatic model approaches, 194–209
  transition states, 180
Circular dichroism (CD) spectroscopy, 53, 54*f*
^{37}Cl/^{35}Cl ratios, 181
Cl-KIEs. *See* Chlorine kinetic isotope effects (Cl-KIEs)
Cluster model approach, 189–194
CmlI oxygenase
  atomic properties
    Mössbauer spectroscopy, 273–274
    reactive peroxo intermediate preparation, 275–276
  mass properties
    oxidative pathway mechanism, 279–286
    substrate-to-product conversions, 273
    vibrational spectroscopies, 276–279
  preparation, 246–247
  proposed reaction cycle, 285*f*
Competitive intermolecular experiment, 222
Competitive intramolecular experiment, 222–223
Compound-specific ^{13}C analysis
  history, 120–121
  LC-co-IRMS, 131–133
  methylation of fatty acids, 131
  monosaccharides, acetylation/methylboronation of, 130–131
  principle and precautions, 129
  sample preparation and analytical strategies, 121–124
  silylation, 130
Compound-specific isotope analysis (CSIA), 114–115, 293, 501–502
  experimental procedures
    butyl blue rubber stoppers, 506–507
    CKIEs by GC-IRMS analysis, 507*f*
    classical method, 504
    UV–vis spectroscopy, 509
  experimental workflow, 123*f*
  human and animal tissues, 118–119*t*
  instrumental coupling, 120*f*
  isotope fractionation, 295
    evaluation methods, 305*t*
    experimental procedures, 506
    multielement, 305–307
    oxygen isotope effects, 312–313

substrate *vs.* product, 304
isotope ratio mass spectrometry
instrumentation and function, 313–315
oxygen isotope, 318–321
product isotope, 321–322
substrate isotope, 315–318
KIE measurement
GC-C-IRMS data, 137–139
LC-co-IRMS data, 137–139
principle, 134–137
plant tissues, 116–117*t*
product isotope fractionation
apparent kinetic isotope effects, 300
multiple oxygenated products, 301–303
simple substrate to product, 300–301
substrate isotope fractionation
apparent kinetic isotope effects, 298–300
isotope ratios and signatures, 296–297
measurements, 295–296
quantification, 297–298
Compound-specific ^{15}N analysis, 124–126
amine functional group, *N*-pivaloylation, 126–127
nitrogen stable isotope analysis, 124–126
N,O-trifluoroacetylation, 127–128
tert-butyltrimethylsilylation, 128–129
Continuous-flow mass spectrometry (CFMS)
alternative systems, 152
basic components, 150–151
buffers used, 152–153
calibration
apparent delay volume (V_{app}), 157–158
assays, 156
buffers and reagents, 156
DBDB oxidation, 154–155, 155*f*
delay line, 153
equipments, 155
reaction time, 153
enzyme-catalyzed reaction, 149–150
KIE measurement, 158–160
restriction, 152
system design, 151–152
Continuous flow time-resolved resonance Raman spectroscopy (TR^3), 256–260, 258*f*
Continuous pressure-monitored headspace analysis
$^{13}C/^{12}C$ ratio, 512–513
decarboxylation reactions, 510–512, 511–512*f*
Rayleigh model calculation, 513
reaction vessel, 511*f*
Continuous transformation assays, 307, 310*f*
Control reactions
isotope fractionations, 514–515
Rayleigh equation, 514*f*
Conventional free energy calculation techniques, 203–204
COPASI, 94, 103
CSIA. *See* Compound-specific isotope analysis (CSIA)
Cyclic phosphodiesters, 435–437
Cytochrome P450 KIE
arene oxide formation, 228
aryl hydroxylation mechanism, 227*f*
atomic purity, 220
carbon hydroxylation *vs.* *N*-dealkylation mechanism, 227*f*
catalytic cycle, 218*f*
competitive intermolecular experiment, 222
competitive intramolecular experiment, 222–223
concept and theory, 218
deuterated substrate use, 220
drug development, 232–233
enzymatic parameters, 225–226
experiment types, 221*f*
gabriel method, 219–220
intermolecular *vs.* noncompetitive, 230*f*
interpretation, 227*f*
intrinsic KIE estimation
isotopically sensitive branching method, 224
multiple isotope effect method, 224
Northrop's method, 224
metabolic switching, 229–230
noncompetitive intermolecular experiment, 223
noncompetitive intramolecular experiment, 221–222
single-substrate concentration, 225
in vivo work, 231–232

D

DBDB. *See* N,N′-dibenzyl-1,4-diaminobutane (DBDB)
Dechlorination
β-HXH catalyzed by LinB, 205–206
n-butyl chloride, 191
postulated transformation, 206*f*
Dehalogenations, enzyme-catalyzed, 180
Dehydrochlorination, 193*s*
^{37}Cl-KIEs predicted, 209*t*
γ-HXH by LinA, 207–209
proposed reaction mechanisms, 207*s*
δ-value/isotope composition, 122
Density-functional QM/MM free energy, 453
Density functional theory (DFT), 17
Derivatization, 122
isotopic corrections, 125*b*
DESII steady-state KIE
experimental design, 475–477
modeling, 473–474
proton transfer, 478
SAM reductive homolysis, 472–473
Deuterated methane isotopomers, 268–269
Deuterium kinetic isotope effects (DKIEs). *See* Primary deuterium kinetic isotope effects (1°DKIEs)
DhlA-catalyzed dehalogenation
active site model, 192*f*
1,2-dichloroethane, 186, 191
n-butyl chloride, dechlorination of, 191
1,2-Diammoniumethylene diacetate (EDDA), 152–153
Diamond core model, 261–262
N,N′ -dibenzyl-1,4-diaminobutane (DBDB), 153
deuterium KIE, 160*f*
oxidation, 154–155*f*
reaction mass spectrum, 159*f*
1,2-Dichloroethane (DCE), 184–186, 191
Dihydrofolate reductase (DHFR), 24, 46–47*t*
Diiron oxygenases, 240, 280*f*
CmlI oxygenase
atomic properties, 273–276
mass properties, 273, 276–286
preparation, 246–247
isotopes and effects, 240
methane monooxygenase
atomic properties, 249–256
BDE shifts, dynamic isotope effects, 263–273
mass properties, 249, 256–262
NH_2-Ar to NO_2-Ar, 280*f*
oxidation state determination, 284–285
reduced enzyme preparation, 247–248
soluble form of methane monooxygenase
hydroxylase component purification, 240–244
reductase component purification, 246
regulatory component purification, 244–246
Dinuclear iron cluster, 240
Direct comparison method, KIE analysis, 543–544
DNA polymerase, KIE calculation, 102–104
Dope testing
CIR/HIR measurement
IRMS, 415–417
linear mixing models, 417–421
reference population (RP) studies, 421–422
CIR *vs*. HIR, 422
enzymatic conversions, 424–426
SULF and GLUC, 422–424
sample preparation
acetylation, 412–413
carbon isotope ratios, 406–407
HPLC, 409–412
liquid-liquid extraction, 408–409
solid phase extraction, 408
steps in, 407
trimethylsilylation, 414
steroids under investigation, 408*t*
testosterone excretion study, 426–428
DSolve function, 92–93
DynaFit, 94

E

EDDA. *See* 1,2-Diammoniumethylene diacetate (EDDA)
Edited sum, 560
Electrodeless NMR titration, 336

Electron paramagnetic resonance (EPR), 473, 476–477
Electrospray ionization (ESI), 445–446
Electrospray ionization quadrupole/ time-of-flight mass spectrometry (ESI-Q/TOF)
  isotope ratios, 448
  $^{16}O/^{18}O$ ratios in 5′-UpG-3′, 445–449
Endogenous reference compounds (ERCs), 404–407
Endogenous steroids
  CIR and HIR, 422
  metabolism, 405*f*
  precursor, 424
  trimethylsilyl derivatization, 414
Enzymatic catalysis, protein dynamics in, 44–45
Enzymatic oxygenations. *See also* CSIA; Oxygenation reactions
  AKIEs
    product, 300
    substrate, 298–300
  description, 292
  isotope fractionation
    nitroaromatic substrates, 308–311
    organic products, 311–312
  pollutant biodegradation, 292–293
Enzyme-catalyzed RNA, 434–435
Enzyme isotope effect. *See* Isotopically labeled enzymes
Enzyme kinetic isotope effect (enzyme KIE), 24
  normal, 59*f*
  theoretical, 65–66*t*
Enzyme reaction progress curves. *See* Reaction progress curves
Enzyme species, 461
Enzyme–substrate (ES) interactions, 2. *See also* Binding isotope effects (BIEs)
Epitestosterone, 408*t*
  UV chromatogram, 411*f*
Equilibrated first-order rate constant, 467–468
Equilibrium binding isotope effects (BIEs). *See* Binding isotope effects (BIEs)
Equilibrium function, 486
Equilibrium isotope effects (EIEs), 3*f*, 114–115, 134, 351
Escherichia coli dihydrofolate reductase (*Ec*DHFR), 53
  active site-labeled, 75–76
  heavy enzyme labeling, 24–26
  hybrid isotopomers
    chemical ligation, 26, 27*f*
    circular dichroism spectroscopy, 34
    C-terminal cysteine peptide production, 31–33
    expressed protein ligation, 27–28, 28*f*
    full-length formation, 33
    N-terminal thioester peptide production, 28–31
    peptide mass spectrometry, 34
    presteady-state kinetic measurements, 35
    steady-state kinetic measurements, 34–35
  segmentally labeled, 74
Ethanol
  extraction of, 392
  irm-^{13}C NMR spectroscopy, 393
  quantification, 392
Ethanolamine ammonia lyase (EAL)
  initial hydrolysis, 105–106
  model testing, 107
  partial reaction scheme, 104, 105*f*
  rate equations, 105–106
  stopped-flow transients, 106*f*
Etiocholanolone (ETIO), 404–406, 408*t*, 424–425
  linear mixing model, 420*f*
Expressed protein ligation, 27–28, 28*f*

F

FAB-IRMS. *See* Fast-atom bombardment-isotope ratio mass spectrometer (FAB-IRMS)
Fast-atom bombardment-isotope ratio mass spectrometer (FAB-IRMS), 182, 183*f*
Fatty acid methylation, 131
Fatty acid methyl esters (FAMEs), 131, 382, 385
^{57}Fe isotopes, 250

Ferulic acid, 393–397
Fluorine-19 (^{19}F), 555
NMR spectra
fluoroorotidine 5′-monophosphate, 172*f*
galactopyranosyl fluoride, 556*f*
glucopyranosyl fluoride, 568*f*
Fluoroorotidine 5′-monophosphate (FOMP)
^{19}F NMR spectrum, FUMP, 172*f*
OMPDC-catalyzed decarboxylation, 167*s*
Fmoc method, 36
Fractionation, 122
isotope, 295
evaluation methods, 305*t*
experimental procedures, 506
multielement, 305–307
oxygen isotope effects, 312–313
substrate *vs.* product, 304
product isotope
apparent kinetic isotope effects, 300
multiple oxygenated products, 301–303
simple substrate to product, 300–301
substrate isotope
apparent kinetic isotope effects, 298–300
isotope ratios and signatures, 296–297
measurements, 295–296
quantification, 297–298
Fructose interconversion, 387–390
position-specific ^{12}C/^{13}C isotope fractionation, 388–389*f*

G

Gabriel method, 219–220
Gas chromatography (GC), 115, 317, 509–510
Gas chromatography-coupled isotope ratio mass spectrometry (GC-IRMS), 501–502, 504, 507*f*
GC-coupled IRMS, 123–124, 130–131, 137–139, 504
KIE calculation, 134–136
KIE measurement, 133–141
nitrogen stable isotope analysis, 124–126
Geometric distortions, 17
Global fitting. *See* Nonlinear data fitting/regression
Glucopyranosyl fluoride
^{19}F NMR spectra, 568*f*
^{1}H NMR spectra, 552*f*
Glucose interconversion, 387–390
carbon atoms, affiliation of, 391*f*
position-specific ^{12}C/^{13}C isotope fractionation, 388–389*f*
Glucuronides (GLUC), 408–409, 422
Glutathione-*S*-transferase (GST), 36–37, 51
Glycal mechanism, 525–526, 525*f*
Glycolate oxidase (GOX), 140
GNU Octave, 94

H

Harmonic oscillator approximations, 189–191
Heavy-atom isotope effects, 180
Heavy enzyme approach, 45
Heavy enzyme labeling
cysteine peptide production, 36–37
*Ec*DHFR catalysis, 24–26
*Ec*DHFR hybrid
chemical ligation, 26, 27*f*
circular dichroism spectroscopy, 34
C-terminal cysteine peptide production, 31–33
expressed protein ligation, 27–28, 28*f*
full-length formation, 33
N-terminal thioester peptide production, 28–31
peptide mass spectrometry, 34
presteady-state kinetic measurements, 35
steady-state kinetic measurements, 34–35
native cysteine lack, 37
peptides with C-terminal α-thioester, 35–36
presteady-state kinetic measurements, 25–26
segmental ligation, 38
Heteronuclear correlation (HETCOR), 562, 564, 566*f*
Heteronuclear single-quantum coherence spectroscopy (HSQC), 554, 563, 567*f*

Hexachlorocyclohexane (HCH), 192–194
dehydrochlorination, 207*s*
$^{2}H/^{1}H_i$ distribution
in ricinoleic acid, 382*f*
in vernolic acid methyl ester, 384*f*
High-performance liquid chromatography (HPLC)
in dope testing, 409–412
substrate isotope fractionation, 295–296
High-temperature combustion (HTC), 115, 131–132, 132*f*
HIV-1 proteas, 46–47*t*, 60
^{1}H NMR titration experiment, 341–342
acidic/basic amino acid, 348–349
in DMSO-d_6, 343
1,5-oxazaspiro[5,5]undecanes, 343–344
pH indicator, 344
Human steroid metabolism, 405*f*
[2′-^{2}H]UTP
enzymatic synthesis, 534*t*, 535*f*
purification of, 534–537
Hybrid isotopomers, 25*f*, 26, 38–39
Hydrogen bonding interactions, 17, 166–167
Hydrogen isotope ratios (HIR), 410–411
steroids excreted glucuronidated *vs.* sulfated, 423*f*
urinary, 427*f*

I

IgorPro, 87–88
Inductively coupled plasma (ICP-MS) mass spectrometers, 181
Intein fusion, 35–36
Intrinsic isotope effects, 96*f*, 180
Intrinsic reaction coordinate (IRC), 189–191, 195
In vitro transcription, 540*t*
anticodon stem loop, 539*f*
double-stranded template sequence, 538
isotopic labeling, 537
protocol, 538–542
RNA oligonucleotides, 537
sequence verification and purity check, 542
T-arm stem loop, 539*f*
template considerations, 538
T7 RNA polymerase, 537–538
irm-^{13}C
MS, vanillin and ferulic acid, 397
NMR, 373–374
biotransformation of ferulic acid to vanillin, 393–397
glucose fermentation by isotopic fractionation, 390–393
vanillin and ferulic acid, 397
irm-^{2}H NMR, 379
irm-MS. *See* Isotope ratio monitoring by mass spectrometry (irm-MS)
irm-NMR. *See* Isotope ratio monitoring by nuclear magnetic resonance spectrometry (irm-NMR)
IRMS. *See* Isotope ratio mass spectrometry (IRMS)
Irreversible edge, 482
Isomorphic model, 464, 490–491
Isotope effects, 122, 136–137, 180, 370
Isotope ratio, 370
Isotope ratio mass spectrometry (IRMS), 115, 181, 445, 504
definition, 122
in dope testing, 415–417
instrumentation and function, 313–315
key feature, 501–502
^{15}N chromatogram, 127*f*
Nier/Bigeleisen approach, 502
oxygen isotope analysis, 318–321
product isotope analysis, 321–322
simplified schematic of, 503*f*
substrate isotope analysis, 315–318
Isotope ratio monitoring by mass spectrometry (irm-MS), 370, 377
Isotope ratio monitoring by nuclear magnetic resonance spectrometry (irm-NMR)
absolute *vs.* relative values, 373–375
advantage/disadvantage, 370–371
enzyme mechanism, 370
fructose interconversion, 387–390
glucose interconversion, 387–390
internal *vs.* external standards, 376
positional $^{13}C_i$ measurement, 377
positional $^{2}H_i$ measurement, 376–377
irm-^{13}C NMR
biotransformation of ferulic acid to vanillin, 393–397

Isotope ratio monitoring by nuclear magnetic resonance spectrometry (irm-NMR) (*Continued*)
glucose fermentation by isotopic fractionation, 390–393
line-fitting protocol, 378
long-chain fatty acids
epoxidation of, 383–387
hydroxylation of, 381–383
optimal spectral acquisition, 375–376
quality and suitability
line overlap, 372–373
purity, 371–372
spectral data processing
irm-^{13}C NMR, 379–381
irm-^{2}H NMR, 379
Isotopes, 114–115, 240
atomic properties, 240
bond dissociation energy, 240
mass properties, 240
Isotopically labeled enzymes
computational studies, 62–64
effects, 44
enzymatic reactions, 46–47*t*
experimental studies
chemical step, rate of, 64–69
kinetic isotope effects, 69–72
kinetic analysis, 24
labeling techniques
amino acid-specific labeling, 49–50*t*
segmental labeling, 49–50*t*
uniform labeling, 49–50*t*
Michaelis-Menten curves
dihydrofolate, 59*f*
inosine, 58*f*
physiochemical properties
acidity, 51
hydrophilicity, 51
steric effects, 52
regio-specific effects, 72–76
steady-state kinetic parameters
alanine racemase, 60
DHFRs, 59
enzyme effects, 56, 57*t*
general trends, 61–62, 62*t*
HIV-1 protease, 60
PNP, 56
structural characterization
circular dichroism spectroscopy, 53, 54*f*
NMR spectroscopy, 53, 55*f*
X-ray crystallography, 54–55
Isotopically sensitive branching method, 224
Isotopic partition function ratio (IPFR), 189–191, 204–205
Isotopologue, 122, 548, 556

K

Keeling Plots, 417–419
^{2}KIE analysis, 542–544
Kinetic isotope effects (KIEs)
Bigeleisen's derivation of Rayleigh equation, 514*f*, 517*f*
carbon, 501–502
CFMS, 149–150, 158–160
chemicals, 527
chlorine, 181
CSIA, 149–150
cytochrome P450, 218–219
data fitting and calculation, 450–453
definition, 69, 133
DESII steady-state
experimental design, 475–477
modeling, 473–474
proton transfer, 478
SAM reductive homolysis, 472–473
direct comparison method, 543–544
DNA polymerase, 102–104
equipment, 526–527
ε value calculation, 517–518
experimental procedures, 504–510
GC-C-IRMS data, 137–139
glucose isomerase reaction, 388–389*f*
in vitro transcription, 537–542, 540*t*
anticodon stem loop, 539*f*
isotopic labeling, 537
protocol, 538–542
RNA oligonucleotides, 537
sequence verification and purity check, 542
T-arm stem loop, 539*f*
template considerations, 538
T7 RNA polymerase, 537–538
In(^{16}O/^{18}O) *vs.* *F* data, 451*f*
^{2}KIE analysis, 542–544
LC-co-IRMS data, 137–139
model-free calculation, 107–108

NMR, 548–549
$^{16}O/^{18}O$ ratios, 442
phosphodiester reactions, 437–438
potential error, 139*f*
principle, 134–137
pseudouridine, 524
synthases, 524–526
rate constant of Q decay, 268*f*
reaction progress curves, 86
ribokinase activity assay, 530*t*
RNA-modifying enzymes, 524
RNA strand cleavage reactions, transition states, 438–439
substrate *vs.* enzyme, 70
temperature dependence, 70
2′-*O*-transphosphorylation, 434–439
TruB and RluA, 526
uses, 548
UTP synthesis
[2′-^{2}H]UTP purification, 534–537
overexpression of required enzymes, 528–533, 529*f*
King–Altman function, 481
King–Altman theorem, 481
KinTek Explorer, 94, 102
$^{18}k_{NBO}$, 450–452, 452*t*

L

Labeled nucleoside, 526, 544–545
Lactate dehydrogenase (LDH), 4–5, 46–47*t*
LC-C-IRMS systems, 132–133, 132*f*
LC-co-IRMS systems, 131–133, 137–139
KIE measurement, 133–141
LC-HTC-IRMS, 132–133
Limited turnover assays, 307, 309*f*
LinA enzyme, 192–194, 207*s*
γ-HXH dehydrochlorination reaction of, 207–209
LinB enzyme, 206*f*
β-HXH, dechlorination of, 205–206
Linear mixing models, 417–421
Long-chain fatty acids
epoxidation of, 383–387
hydroxylation of, 381–383

M

Magnetization polarization transfer, 374
Mass properties
CmlI oxygenase
oxidative pathway mechanism, 279–286
substrate-to-product conversions, 273
vibrational spectroscopies, 276–279
methane monooxygenase
substrate-to-product conversions, 249
vibrational spectroscopies, 256–262
Mass spectrometry. *See* Continuous-flow mass spectrometry (CFMS)
Mathematica, 87, 94, 560
MATLAB, 94
Metabolic switching, 229–230
Metabolomics, 121–123
Metal ion catalysis, 437, 445
Methane monooxygenase
atomic properties
Mössbauer spectroscopy, 249–251
stopped-flow analysis, 252–256
bond dissociation energy
chiral ethane, 264–267
P and Q formation, 271–273, 272*f*
quantum tunneling in Q reactions, 269–271
rate constant for Q reaction, deuterated methane isotopomers, 268–269
size selectivity detection, 269–271
source of oxygen atom, 263–264
mass properties
oxygen source, in product methanol, 249
substrate-to-product conversions, 249
time-resolved resonance Raman spectroscopy, 256–262
vibrational spectroscopies, 256–262
Methionine aminopeptidase (MAP), 36–37
Methylation, fatty acids, 131
Methylboronation (MBA), 130–131
3-Methylcatechol
$\delta^{13}C_{3\text{-}MC}$, 302–303
$\delta^{13}C$-value, 302–303
Michaelis complex, 2, 20, 60–61
Michaelis–Menten kinetics, 134
Michael mechanism, 525–526
Minimum free energy path (MFEP), 203–204, 208*f*
collective variables, 204
MNova, 560–561, 565

Molecular mechanics (MM) approaches, 194–195
Monosaccharides
  acetylation/methylboronation, 130–131
  silylation, 130
Mössbauer spectroscopy, 249–251, 251*f*, 273–274, 274–275*f*
  diferric, diferrous and diferric peroxo CmlI, 274–275*f*
  ^{57}Fe nuclear transition, 262, 273–274
Multicomponent NMR titration
  data analysis, 333–334
  electrodeless pH titrations, 335–337
  experimental procedure, 335
Multielement isotope fractionation analysis, 305–307
Multiple isotope effect method, 224
Multiply chlorinated reactants, 184
  DhlA dehalogenase, dechlorination of 1,2-dichloroethane, 184–186

N

N-acetyl-isopropyl esters (NAIP), 127–128
N-acetylpolyamine oxidase (PAO), 151–152
  DBDB oxidation by, 154–155, 155*f*
^{15}N-AKIE, 299
Natural isotope abundance, 114–115
NBDO. *See* Nitrobenzene dioxygenase (NBDO)
NDSolve, 95
Net rate function, 462–463, 484
Next-generation sequencing techniques, 524
NH_2-CAM, 278*f*, 282*f*, 283
Nitrobenzene
  dioxygenation, 296*f*
  εC-values, 304, 305*t*
  oxygenation, 296*f*
Nitrobenzene dioxygenase (NBDO), 295–296
  oxygen isotope effects, 312
  product oxygenation, 311
  substrate oxygenation, 308
2-Nitrotoluene
  $\delta^{13}C_{2\text{-NT},0}$, 302–303
  εC-values, 304, 305*t*
  NBDO-catalyzed oxygenation, 302, 302*f*
NMR spectrometer, 373–374, 378, 549–550
NMR titration. *See also* pH-free multicomponent NMR titration
  acidity constants, 350
  advantages, 359–360
  intramolecular comparisons, 347–349
  of $(CH_2D)_3NH^+$ and $(CH_2D)_3NH^+$ mixture, 358*f*
  secondary isotope effects, 350
  single-point, 346–347
  two dimensional, 362–363
$^{14}N/^{15}N$ isotope effect, 128–129, 134
Nonbridging phosphoryl oxygens (NPOs), 435–437
  [^{18}O] incorporation, 441
Noncompetitive intermolecular experiment, 223
Noncompetitive intramolecular experiment, 221–222
Nonexchangeable isotopes, 24, 45
Nonlinear data fitting/regression
  advantages, 89–91
  commercial fitting softwares, 87–88
  vs. local, 88–91, 90*t*
  optimization algorithms, 88–89
$N^{16}O^{18}O$-CAM, 282–283
Normal KIE, 556
Northrop's method, 224
N,O-trifluoroacetylation, 127–128
N-oxygenase CmlI, 240
N-pivaloyl-*O*-isopropyl esters, 126
NSD2, 6, 9*f*, 18–19*t*
Nuclear magnetic resonance spectroscopy (NMR), KIE measurement
  competitive techniques, 548
  continuous method, 549–550
  data analysis, 567
  data processing
    and fitting, 2D, 564–566
    HETCOR, 562, 564, 566*f*
    HSQC, 562–563, 567*f*
    one-dimensional, 560–561
    two-dimensional, 561–566
  isotopologue ratio, 548
  probe nucleus
    ^{13}C NMR Spectrum, 552–554
    fluorine-19, 555
    ^{1}H NMR spectrum, 550–552

properties, 551*t*
sample preparation, 557
spectrum acquisition, 558–559
Nuclear Overhauser effects (NOEs), 554, 561
Nuclear resonance vibrational spectroscopy (NRVS), 262
Numerically integrated rate equations
enzyme behavior, 94
Michaelis–Menten curve, 98, 99*f*
NDSolve, 95
software packages, 94

O

[5′-^{18}O]guanosine, 440–441
^{18}O isotopes, 249, 263–264
1D-PMF, 206
ONIOM model, 194–195
On-the-fly string method, 204
$^{16}O/^{18}O$ ratios, 435
collision energy, 449*f*
intrinsic *vs.* observed, 449*f*
kinetic isotope effects, 442
signal-to-noise ratio, 438
5′-UpG-3′ by tandem Q/TOF MS, 445–449
Origin, 87–88
OriginPro, 87–88
Orotidine 5′-monophosphate (OMP), 166
^{1}H NMR spectrum, UMP, 171*f*
OMPDC-catalyzed decarboxylation, 167*s*
Orotidine 5′-monophosphate decarboxylase (OMPDC), 166
[2′-^{18}O]uridine, 439–440
Overexpressed enzymes, 528–533
activity assays for, 529*f*
Oxygenase enzymes, 240, 263–264
Oxygenation reactions
isotope fractionation
continuous transformation assays, 307, 310*f*
cosubstrate *vs.* substrate isotope analysis, 312
enzyme assays, substrate, 308–311
limited turnover assays, 307, 309*f*
whole cell assays, product, 311–312
nitrobenzene, 296*f*
Oxygen isotope analysis
enzyme assays, 312–313
by GC/IRMS, 318–321

P

PAO. *See N*-acetylpolyamine oxidase (PAO)
PDIE. *See* Product deuterium isotope effect (PDIE)
Pentaerythritol tetranitrate reductase (PETNR), 46–47*t*, 52
N-Pentafluoropropyl anhydride (PFPA), 124–128
Peripheral species, 461
Perturbation effect, 460, 491–493
pET-31b plasmid, 31
pH-free multicomponent NMR titration
advantages, 359–360
binding constants measurement
binary complexes, 344–345
diacids H_2A, 349–350
intermolecular *vs.* intramolecular, 347
intramolecular comparisons, 347–349
single-point, 346–347
toluene-d_8, 345
chemical shifts, 332–333
data analysis, 333–334
electrodeless pH titrations, 335–337
experimental procedure, 335
^{1}H and ^{13}C, 341–344
heavy-atom isotope effects, 350–351
practical considerations, 360–364
protonation/deprotonation, 360
reverse anomeric effect, 337–341
secondary deuterium isotope effects
on acidity, 351–354
on amine basicities, 353*t*, 356
on binding, 355
of carboxylic acidities, 356–357
enthalpy and entropy, 359*t*
formic and acetic acid, 358
pyridine and 2,6-lutidine-2,6-$(CD_3)_2$, 357
trimethylamine, 357–358
two dimensional, 362–363
Phosphoribosyl pyrophosphate synthetase (PRPPS), 528, 529*f*
activity assay, 532*t*
synthesis, 529*f*, 532
PicoGreen, 102–103

P450 KIE. *See* Cytochrome P450 KIE
Plant peroxidases, 196
Potential of mean force (PMF), 202–204
Pressure-monitored headspace analysis, 517*f*
Primary deuterium kinetic isotope effects (1°DKIEs)
  5-carboxyvanillate decarboxylase, 174
  experimental procedure
    enzyme-catalyzed decarboxylation, 169–173
    SDIE *vs.* PDIE, 168–169
  OMPDC-catalyzed decarboxylation, 166–167
  PDIE interpretation, 173–174
  proton transfer, 164–165
  pyruvate kinase, 165
  vinyl ether hydrolysis, 164–165
Primary isotope effects, 350
  hydron transfer, 165*s*, 167*s*
Product deuterium isotope effect (PDIE), 165
Product isotope analysis
  apparent kinetic isotope effects, 300
  by LC/IRMS, 321–322
  multiple oxygenated products, 301–303
  simple substrate to product, 300–301
  whole cell assays, 311–312
Protein–ligand system, 17
Protein-mediated atomic distortions, 17
Protein ribonuclease, 434–435, 453–454
Proton inventory-style experiments, 100–101, 101*f*
Pseudo-first-order rate constant, 467–468
Pseudouridine, 524
  synthases, 524–526
Pseudouridine synthases 1 (PUS1) and 3 (PUS3), 524–525
pTWIN-1 expression vector, 28, 29*f*
Purine nucleoside phosphorylase (PNP), 45, 46–47*t*
Pyruvate assay, 533–534
Pyruvate kinase, 1°DKIEs, 165

Q

Q reactions
  deuterated methane isotopomers, rate constant for, 268–269
  quantum tunneling in, 269–271
Quantitative reaction, 122

R

Radical clock chemistry, 264–268
Radical SAM enzyme, 472
Radiolabeled substrates, 8–9
Rapid equilibrium dialysis (RED), 6
  dialysis insert at equilibrium, 9*f*
Rapid freeze-quench (RFQ) methods, 250–251
  freezing device, 254*f*
Rate equations, 91–92
Rate function, 480
Rayleigh equation, 514*f*, 517*f*
Rayleigh model calculation, 513
Reactant state (RS), 189–191
Reaction progress curves
  deuterium KIE, 87
  enzyme kinetics, 86
  initial velocity approach, 86
  integrated rate equations, 91–94
  KIE calculation
    DNA polymerase, 102–104
    ethanolamine ammonia lyase, 104–107
    model-free, 107–108
  model selection and error analysis, 98–102
  nonlinear data fitting/regression, 87–91
  numerically integrated rate equations, 94–98
  rate and integrated rate laws, 86*t*
RED. *See* Rapid equilibrium dialysis (RED)
Resonance Raman spectroscopy (rRaman), 256–261, 261*f*, 276–279, 277–278*f*
Reverse anomeric effect, 337–341
Ribokinase, 529*f*
  activity assay, 530*t*
Ribonuclease, 434–435, 453–454
Ribozymes, 434–435, 453–454
Rieske nonheme iron oxygenases, 293–295
RluA, 525–526
  kinetic isotope effects, 526
RNA oligonucleotides, 537
RNA 2′-*O*-transphosphorylation
  electrospray ionization quadrupole/time-of-flight mass spectrometry
    HPLC, 441–445

$^{16}O/^{18}O$ ratios in 5′-UpG-3′, 445–449
kinetic isotope effects, 371–373, 439
mechanisms and transition states, 435–438, 436*f*
[5′-^{18}O]guanosine, 440–441
[^{18}O]incorporation into NPOs, 441
[2′-^{18}O]uridine, 439–440
RNase A catalysis, 453–454
RP-HPLC, 442–443
R5P • Na, 530–531

S

S-adenosyl- L -methionine (SAM), 6, 9*f*
substrate stability, HPLC method, 14*t*
Sage, 94
SDIE. *See* Substrate deuterium isotope effect (SDIE)
Secondary deuterium isotope effects
on acidity, 351–354
on amine basicities, 353*t*, 356
on binding, 355
of carboxylic acidities, 356–357
enthalpy and entropy, 359*t*
formic and acetic acid, 358
pyridine and 2,6-lutidine-2,6-$(CD_3)_2$, 357
trimethylamine, 357–358
Segmental labeling, 49–50*t*
*Ec*DHFR, 74
Self-cleavage ribozymes, 453–454
Sensitivity index, 488–490
Serial steady-state models, 461–463
definition, 482
irreversible edge, 482
SigmaPlot, 87–88
Signal-to-noise ratio (SNR), 375
for DBDB, 154–155
NMR probes, 555*t*
Silylation, 130
Single-point NMR titration, 346–347
sMMO. *See* Soluble form of methane monooxygenase (sMMO)
sMMO hydroxylase component (sMMOH)
assay methods, 241
buffers and reagents, 241
equipment, 240
^{57}Fe, 250, 251*f*
nitrobenzene assay procedure, 241
protons role, 271–273
purification procedure, 242
single turnover reactions, 252–256
sMMO reductase component (MMOR), 246
sMMO regulatory component (MMOB)
buffers and reagents, 244
equipment, 244
Solid phase microextraction (SPME), 307, 315–316
Soluble form of methane monooxygenase (sMMO), 240
ethanol products, stereochemical distribution of, 265*f*
hydroxylase component purification, 240–244
isotopic substitution, 268
proposed catalytic cycle, 255*f*
reductase component purification, 246
regulatory component purification, 244–246
Solvent deuterium kinetic isotope effects (SDKIEs), 164
inverse, 164–165
Spectral data processing protocol
irm-^{13}C NMR, 379–381
irm-^{2}H NMR, 379
Spectrophotometric coupled assay, 531–533
SPME. *See* Solid phase microextraction (SPME)
Sports drug testing. *See* Dope testing
Standard Mean Ocean Chloride (SMOC), 181–182
Steady-state KIE, enzymatic reactions
DESII
experimental design, 475–477
modeling, 473–474
proton transfer, 478
SAM reductive homolysis, 472–473
function limits, 493–495
isomorphic model, 490–491
on kinetic parameters, 465–466
on net rate functions
enzyme species, 466–467
kinetic parameter, 469–470
rate-limiting steps, 467–468, 471*f*
sensitivity indices, 469

Steady-state KIE, enzymatic reactions (*Continued*)
  perturbation effect, 460, 491–493
  sensitivity index, 488–490
  serial model
    enzyme species, 461
    irreversible reaction, 462
    net rate function, 462–463
    peripheral species, 461
    properties, 462
  on steady-state velocity, 463–465
  two-vertex serial model, 495–496
Steady-state model, 480
Steady-state velocity
  definition, 484
  KIE, 463–465
Stopped-flow spectrophotometry, 250–251
Streptomyces setonii, 393–394, 395*f*
Substrate deuterium isotope effect (SDIE), 168–169, 174
Substrate isotope analysis
  apparent kinetic isotope effects, 298–300
  enzyme assays, 308–311
  by GC/IRMS, 315–318
  isotope ratios and signatures, 296–297
  measurements, 295–296
  quantification, 297–298
Sulfates (SULF), 422

T

Tandem Q/TOF MS, 445–449
T-arm stem loop (TSL), 526*np*, 539*f*, 543–544
Ternary complex association, 15
tert-butyl-methylether (TBME), 408–409
tert-butyltrimethylsilylation (*t*-BDMS), 128–130
Testosterone (T), 404, 408*t*
  chromatographic isotopic fractionation, 411*f*
  excretion study, 426–428
Thermotoga maritima DHFR (*Tm*DHFR), 53
Time course datasets. *See* Reaction progress curves
Time-resolved resonance Raman spectroscopy
  equipment, 257
  procedure, 259
  reagents and buffers, 257
  resonance Raman data collection and analysis, 260–261, 261*f*
  schematic diagram, 258*f*
  Stokes-shifted, scattered radiation, 256–259
  TR^3 instrument setup, 259–260
TOF MS, 446–448
TopSpin, 560–561, 565, 566*f*
Total-line-shape analysis, 378
Transition path sampling (TPS), 62–63
Transition states (TSs), 2–3, 180, 186–191
  structure protocol, 194–209
2′-*O*-Transphosphorylation, 434–435
  RNA metabolism, 434–435
2,4,6-Trichlorophenol (TCP), 196–197, 197*t*
N-Trifluoroacetyl anhydride (TFA), 124–128
Trifluoroacetylation, 130
Trimethylsilyl (TMS) derivatives, 130
Trimethylsilyl (TMS) derivatization, 414
tRNA chaperone, 524
T7 RNA polymerase, 537–538
TruB, 525–526
  kinetic isotope effects, 526
2D NMR, 374, 561–566
2D-PMF, 203, 206
Two-vertex serial model, 495–496

U

Uniform labeling, 49–50*t*
5′-UpG-3′
  structure of, 440*f*
  tandem Q/TOF MS, 445–449
  2′-*O*-transphosphorylation, 444*f*
UpG substrate, 444*f*, 447*f*
Uracil phosphoribosyl transferase (UPRT), 528, 529*f*
  activity assay, 533*t*
UTP synthesis
  [2′-^{2}H]UTP purification, 534–537
  overexpression of required enzymes, 528–533, 529*f*

V

Vanillin, 393–397
Velocity, 484
Vibrational spectroscopies, 262
  resonance Raman spectroscopy, 276–279, 277–278*f*
  time-resolved resonance Raman spectroscopy, 256–262, 258*f*, 261*f*
Vienna Pee Dee Belemnite (VPDB), 296–297
Vinyl ether, hydrolysis mechanism, 164–165
V isotope effect, 465
V/*K* isotope effect, 465

W

Weighted histogram analysis method (WHAM), 202–203
Wild-type TrzN–Atz
  atrazine hydrolysis, 200*s*
  potential energy profiles, 201*f*
World Anti-Doping Agency (WADA), 404, 414–415

Kolibri 360 GmbH
Ferdinand-Jühlke-Straße 7
99095 Erfurt, Deutschland
produktsicherheit@kolibri360.de